CONSTRUCTION CONTRACTS

Law and Practice

CONSTRUCTION CONTRACTS

Law and Practice

SECOND EDITION

RICHARD WILMOT-SMITH QC
39 Essex Street Chambers;
Recorder;
One of Her Majesty's Counsel;
Bencher Middle Temple

OXFORD
UNIVERSITY PRESS

OXFORD
UNIVERSITY PRESS

Great Clarendon Street, Oxford OX2 6DP

Oxford University Press is a department of the University of Oxford.
It furthers the University's objective of excellence in research, scholarship,
and education by publishing worldwide in

Oxford New York

Auckland Cape Town Dar es Salaam Hong Kong Karachi
Kuala Lumpur Madrid Melbourne Mexico City Nairobi
New Delhi Shanghai Taipei Toronto

With offices in

Argentina Austria Brazil Chile Czech Republic France Greece
Guatemala Hungary Italy Japan Poland Portugal Singapore
South Korea Switzerland Thailand Turkey Ukraine Vietnam

Oxford is a registered trade mark of Oxford University Press
in the UK and in certain other countries

Published in the United States
by Oxford University Press Inc., New York

First published 2006
Second edition published 2010

British Library Cataloguing in Publication Data
Data available

Library of Congress Cataloging in Publication Data

Wilmot-Smith, Richard.
Construction contracts: law and practice / Richard Wilmot-Smith. — 2nd ed.
p. cm.
Includes index.
ISBN 978–0–19–957959–4
1. Construction contracts—Wales. 2. Construction contracts—England. I. Title.
KD1615.W55 2010
343.42'07869—dc22 2009053052

Typeset by Glyph International, Bangalore, India
Printed in Great Britain
on acid-free paper by
CPI Antony Rowe, Chippenham, Wiltshire

ISBN 978-0-19-957959-4

1 3 5 7 9 10 8 6 4 2

FOREWORD

In a foreword to the first edition, written almost exactly four years ago, I stated that this book was a welcome addition to construction law literature; that it would take its place amongst the standard textbooks on the subject; and that it would go into many future editions. I stand by all of those views and welcome the opportunity to write a foreword to the second edition.

Construction law is a fast moving subject and generates a steady stream of significant judicial decisions. There is no sign of that stream abating. Indeed, as from this month, the number of High Court judges assigned to the Technology and Construction Court ('TCC') is increased to four. The regular updating of any textbook in this field is to be welcomed.

This edition of 'Construction Contracts: Law and Practice' contains a new chapter on the important subject of partnering. The later chapters of the book have been re-organized. All chapters have been updated, with many new sections added in.

During my time as a TCC judge I found this book a most useful source of reference. I am sure that the second edition will prove of even greater value and commend it to both practitioners and judges in the construction law field.

Rupert Jackson
23 October 2009

PREFACE TO THE SECOND EDITION

A second edition of any textbook should update and improve upon the first and it is the objective of this volume to do both. The aims are, as before, to provide an exposition of the modern law in a form which is accessible and relevant.

As you can see from the acknowledgement page I have been helped by no less than twenty people in that task of updating and improvement. Many hours have been spent by all of them and it would have been impossible to produce this edition without their help and support. To single a few people out would be invidious, but equally doing full justice to all involved would produce a litany. I owe an enormous debt to all of the contributors.

The timing of the passing of the Local Democracy, Economic Development and Construction Act 2009 has not been easy for us, given that only small adjustments to the text of the book could be made after October 2009 and the Bill had not passed all its stages in Parliament by then. Whilst some of the essentials could be covered in Chapters 1 and 23 and its provisions are contained in Appendix 3, whether the parts relevant to the construction industry are actually brought into force by the government of the day remains to be seen.

The task of updating the first edition by way of the website was a bit ambitious and I had no proper system in place to do it.

I do now and the second edition's website <http://www.oup.com/lawonline/wilmotsmith> is live, should be up to date and now contains the revisions to the Housing Grants, Construction and Regeneration Act 1996 by the Local Democracy, Economic Development and Construction Act 2009 in their final form. The site will also give notification of the date that the amendments come into force, if indeed they do, together with a short commentary on the provisions themselves.

I must thank Tony Blackler for reading the whole of the manuscript. He has corrected many errors. Also John Rushton who took a great deal of his time to correct four large chapters. Those that remain merely demonstrate that I can defeat even the most patient critic.

Finally many thanks to Roxanne Selby, Fiona Sinclair, and Lucy Page of OUP.

I have aimed to set out the law as at 5 October 2009.

ACKNOWLEDGEMENTS

This book could not have been written without the considerable help of those whose names appear below:

Caroline Allen, BA (Oxon), Barrister.

Jonathan Bellamy, MA (Oxon), FCIArb, Chartered Arbitrator, Barrister.

Tony Blackler, BA, LLB (Cantab), MCIArb.

Jess Connors, BA (Oxon), Barrister.

Charlie Cory-Wright, QC, BA (Oxon).

Andrew Deakin, BA (Cantab), MPhil, Barrister.

John Denis-Smith, MA (Oxon), MSc, FCIArb, Barrister.

Karim Ghaly, BA (Oxon), Barrister.

Karen Gough, LLB, FCIArb, Dip IC Arb, Chartered Arbitrator, Barrister, Past President of the Chartered Institute of Arbitrators.

Kate Grange, MA (Cantab), Barrister.

Martin Hirst, BSc, MSc, LLM Solicitor.

Adrian Hughes, QC, MA (Oxon).

Simon Nurney, LLM, MSc, Solicitor.

Adam Robb, BA (Oxon), BCL. Barrister.

John Rushton, BA (Cantab), FCIArb, the ICC United Kingdom Consultant on Arbitration and Alternative Dispute Resolution.

Marion Smith, LLB, LLM, Barrister.

Douglas Wass, LLB (Hons), MSc, Solicitor.

Ellen Wiles, BA (Oxon), LLM, Barrister.

Sean Wilken, BA (Oxon), Barrister.

Claudia Wilmot-Smith, BA (Cantab), Barrister.

CONTENTS—SUMMARY

VI PROCUREMENT, COMPETITION IN THE EUROPEAN COMMUNITY, AND PARTNERING

VII DISPUTES

CONTENTS

TABLE OF CASES

Treasure v Dawes [2009] EWHC 1932 (TCC); (2009) CILL 2781 . . . . 21.33, 23.67
Trendtex Trading Corp v Credit Suisse [1982] AC 679; [1981] 3 WLR 766; [1981] 3 All ER 520; [1981] Com LR 262; 125 SJ 761, HL 4.08
Trident Turboprop (Dublin) Ltd v First Flight Couriers Ltd [2009] EWCA Civ 290; [2009] Bus LR 1189; [2009] 1 Lloyd's Rep 702; [2009] 1 CLC 490 . . . . 7.38, 8.13
Trollope & Colls Ltd (t/a Nuclear Civil Constructions) v Atomic Power Constructions Ltd [1963] 1 WLR 333; [1962] 3 All ER 1035; 107 SJ 254, QBD. 1.43, 23.412
Trollope & Colls Ltd v North West Metropolitan Regional Hospital Board [1973] 1 WLR 601; [1973] 2 All ER 260; 9 BLR 60; 117 SJ 355, HL. 1.119, 1.120
TTI Team Telecom International Ltd v Hutchison 3G UK Ltd [2003] EWHC 762 (TCC); [2003] 1 All ER (Comm) 914. . . . . 9.31
Tufano v Vincenti [2006] EWHC 1496 . . . . 22.136
Turcan, Re (1889) LR 40 Ch D 5, CA. 2.14
Turkiye IS Bankasi AS v Bank of China [1998] 1 Lloyd's Rep 250; [1998] CLC 182; The Times, December 17, 1997, CA (Civ Div); affirming [1996] 2 Lloyd's Rep 611; [1996] 5 Bank LR 241; The Times, March 8, 1996, QBD (Comm) 9.31
Turriff Construction Ltd v Regalia Knitting Mills Ltd 9 BLR 20; (1971) 222 EG 169, QBD 1.18, 1.20, 2.09
Tychy, The. See MSC Mediterranean Shipping Co SA v Polish Ocean Lines (The Tychy) (No.2)
Tyco Fire & Integrated Solutions (UK) Ltd (formerly Wormald Ansul (UK) Ltd) v Rolls Royce Motor Cars Ltd (formerly Hireus Ltd) [2008] EWCA Civ 286; [2008] 2 All ER (Comm) 584; [2008] 1 CLC 625; [2008] BLR 285; 118 Con LR 25; [2008] Lloyd's Rep IR 617; [2008] 2 EGLR 113; [2008] 14 EG 100 (CS); (2008) 105(15) LSG 25; (2008) 152(14) SJLB 31 . . . . 7.12

UBAF Ltd v European American Banking Corp (The Pacific Colocotronis) [1984] QB 713; [1984] 2 WLR 508; [1984] 2 All ER 226; [1984] 1 Lloyd's Rep 258; (1984) 81 LSG 429; (1984) 128 SJ 243, CA (Civ Div) 15.31
UCB Corporate Services Ltd v Thomason [2005] EWCA Civ 225; [2005] 1 All ER (Comm) 601, CA (Civ Div) 8.11
Udal v Dutton [2007] EWHC 2862 (TCC); [2008] 1 P & CR DG13 . . . . 16.10
Ultramares Corp v Touche, Niven & Co. (1931) 255 NY 170. . . . . 6.10
Unilever Plc v Procter & Gamble Co [2000] 1 WLR 2436; [2001] 1 All ER 783; [2000] FSR 344; (2000) 23(1) IPD 23001; (1999) 96(44) LSG 40; (1999) 143 SJLB 268; The Times, November 4, 1999; Independent, November 5, 1999, CA (Civ Div). 20.24
United Brands Co v Commission of the European Communities (Banana Market) (27/76) [1978] ECR 207; [1978] 1 CMLR 429, ECJ 18.52
United International Pictures v Cine Bes Filmcilik ve Yapimcilik AS; Sub Nom: Cine Bes Filmcheck ve Yapimcilik AS v United International Pictures [2003] EWCA Civ 1669; [2004] 1 CLC 401; (2003) 147 SJLB 1396. . . . . 11.24
United Scientific Holdings Ltd v Burnley BC; Cheapside Land Development Co Ltd v Messels Service Co [1978] AC 904; [1977] 2 WLR 806; [1977] 2 All ER 62; 75 LGR 407; (1977) 33 P & CR 220; (1977) 243 EG 43; 121 SJ 223, HL 13.16
United States v Behan, 110 US 338 (1884) 12.52
Universal Steam Navigation Co Ltd v James McKelvie & Co; Sub Nom: Ariadne Steamship Co Ltd v James McKelvie & Co [1923] AC 492; (1923) 15 Ll L Rep 99, HL. 1.90
Universale-Bau AG v Entsorgungsbetriebe Simmering GmbH (C-470/99) [2002] ECR I-11617, ECJ 18.30
University of Glasgow v William Whitfield; Sub Nom: University Court of the University of Glasgow v William Whitfield and John Laing (Construction) Ltd; University of

TABLES OF PRIMARY LEGISLATION

ALPHABETICAL

CHRONOLOGICAL

1999 Contracts (Rights Against Third Parties) Act 4.38, 4.39, 4.40
s 1 . . . . 4.40
(2) 4.41
s 8 . . . . 4.40
2001 Anti-Terrorism, Crime and Security Act 18.77, 18.78
s 109 . . . . 18.77
2002 Enterprise Act 18.40, 18.61, 18.65, 18.67
ss 228–290 . . . . 18.67
2006 Companies Act . . . .
s 860 . . . . 17.06, 17.12
s 874 . . . . 17.06
s 878 . . . . 17.06
s 889 . . . . 17.06
s 1295 . . . . 17.06
2009 Local Democracy, Economic Development and Construction Act 1.101, 1.108, 1.110, 1.112, 1.117, 23.03, 23.48, 23.63, 23.71, 23.79, 23.405, 23.456
s 143(1) 1.101

TABLES OF SECONDARY LEGISLATION

ALPHABETICAL

CHRONOLOGICAL

Part I

THE CONTRACT

1

CONTRACTS: FORMATION AND INTERPRETATION

Introduction

1.01 In this book we shall be examining the law relating to construction contracts and the creation of obligations arising therefrom which the courts will enforce. Some obligations are enforced and some are not. The law has developed a series of rules which guide the courts into making decisions about whether obligations shall be enforced. Certain types of contract are not enforced as a matter of policy, such as betting contracts. These are outside the scope of this book.

1.02 The courts will enforce most contracts in circumstances where the parties had an intention to create legal relations, there has been an offer which is accepted, and the obligations are supported by consideration.[1]

[1] Consideration is not necessary, however, when the parties have contracted under seal.

1.03 **However, parties to a contract must have capacity to contract.** Adults usually have the capacity to contract, save for those with certain mental health problems. Those who do not have the mental capacity to conduct their own affairs will generally speaking (subject to exceptions) not have the capacity to contract. Minors do not have capacity to contract. Corporations will usually have capacity. Companies can contract through their agents. An unincorporated association has no legal personality (in contradistinction to a company or Limited Liability Partnership) and therefore members of the association will not be liable under a contract unless they expressly ratify the contract. However, members of the committee who contract will be liable. Government departments can contract and can sue or be sued pursuant to their contracts.[2] An enemy alien may not enforce a contract, but an alien may.

1.04 The minimum requirement for an enforceable construction contract to be entered into is an agreement between two or more parties whereby one party offers to execute work for the other for a price.

Invitations to Tender or Treat

1.05 Prior to the formation of a construction contract, employers often invite tenders for the execution of work. Invitations to tender are not normally offers which are capable of acceptance, but are invitations to treat.

1.06 Provided the invitation to tender is expressed to be exactly that, no legal consequences flow from the submission of a tender. The tender is received and the employer then has the option either to accept or to reject it. For clarification, invitations to tender usually state expressly that the putative employer does not bind itself to accept the lowest tender.

1.07 The tender process is different to a fixed bidding sale where a party invites fixed bids and undertakes to accept the lowest offer. In such circumstances the lowest price submitted in conformity with the invitation will constitute an acceptance of an offer.[3]

1.08 Modern tenders are expensive for contractors to compile and require a great deal of management and professional time. Whilst the employer may not be bound to accept the lowest tender, there is likely to be a contractual obligation upon the employer inviting tenders which are expensive to prepare, to examine and consider any valid tender. Bingham LJ gave an authoritative explanation of the process in *Blackpool and Fylde Aero Club Ltd v Blackpool Borough Council*:[4]

> A tendering procedure of this kind is, in many respects, heavily weighted in favour of the invitor. He can invite tenders from as many or as few parties as he chooses. He need not tell any of them who else, or how many others, he has invited. The invitee may often, although not here, be put to considerable labour and expense in preparing a tender, ordinarily without recompense if he is unsuccessful. The invitation to tender may itself, in a complex case,

[2] See the Crown Proceedings Act 1947.

[3] See *Harvela Ltd v Royal Trust Co* [1986] AC 207; and *South Hetton Coal Co v Haswell, Shotton and Easington Coal and Coke Co Ltd* [1898] 1 Ch 465 (CA).

[4] [1990] 1 WLR 1195 at 1202.

> although again not here, involve time and expense to prepare, but the invitor does not commit himself to proceed with the project, whatever it is; he need not accept the highest tender; he need not accept any tender; he need not give reasons to justify his acceptance or rejection of any tender received. The risk to which the tenderer is exposed does not end with the risk that his tender may not be the highest or, as the case may be, lowest. But where, as here, tenders are solicited from selected parties all of them known to the invitor, and where a local authority's invitation prescribes a clear, orderly and familiar procedure—draft contract conditions available for inspection and plainly not open to negotiation, a prescribed common form of tender, the supply of envelopes designed to preserve the absolute anonymity of tenderers and clearly to identify the tender in question, and an absolute deadline—the invitee is in my judgment protected at least to this extent: if he submits a conforming tender before the deadline he is entitled, not as a matter of mere expectation but of contractual right, to be sure that his tender will after the deadline be opened and considered in conjunction with all other conforming tenders or at least that his tender will be considered if others are. Had the club, before tendering, inquired of the council whether it could rely on any timely and conforming tender being considered along with others, I feel quite sure that the answer would have been 'of course'. The law would, I think, be defective if it did not give effect to that.

The consequences of a breach of contract of this nature may be difficult to establish. A party **1.09**
has an obligation to consider a tender, provided it complies with the published rules of submission.[5] If it fails to do so, it will be open to a claim for damages. It may be that no loss can be established.[6] However, if the party losing out is able to establish that its tender would have been accepted had it been considered, it could recover damages on that basis.[7] As for the scope of the obligation to consider a tender: a party has an obligation to act fairly and in good faith.[8] However, the meaning of fairness in this context is somewhat intangible.

---

[5] In *JB Leadbitter & Co Ltd v Devon County Council* [2009] EWHC 930 (Ch), Richards J held that the local authority was entitled to refuse to consider the incomplete, submitted tender of a construction company to participate in a framework agreement from which public bodies could procure construction contracts, as the rules for submission clearly stated that only complete tenders would qualify for consideration.

[6] See, for example, *Fairclough Building Ltd v Port Talbot Borough Council* (1992) 33 Con LR 24. The Court of Appeal upheld the Council's entitlement to remove Fairclough from the list of tenderers to avoid the conflict of interest arising from the wife of a Fairclough director being on the tender appraisal team, holding that they had a duty *honestly to consider the tenders of those whom they had placed on the short-list, unless there were reasonable grounds for not doing so.*

[7] This is just like a contractor which loses the opportunity to tender because of a breach of the Public Works Contracts Regulations 1991. See, for example, *Harmon CFEM Facades (UK) Ltd v Corporate Officer of the House of Commons* (1999) 67 Con LR 1.

[8] In the Privy Council case *Pratt Contractors Ltd v Transit New Zealand* (2003) 100 Con LR 29, it was accepted that there was an obligation to act fairly and in good faith in the consideration of tenders. However, Lord Hoffman commented that the duty of fairness was considered in Commonwealth authorities to be a 'rather indefinable term' [para 45]. On the facts of the case, he stated:

> It is . . . necessary to identify exactly what standard of conduct was required of the TET [tender evaluation team] in making its assessment. In their Lordships' opinion, the duty of good faith and fair dealing as applied to that particular function required that the evaluation ought to express the views honestly held by the members of the TET. The duty to act fairly meant that all the tenderers had to be treated equally. One tenderer could not be given a higher mark than another if their attributes were the same. But Transit was not obliged to give tenderers the same mark if it honestly thought that their attributes were different. Nor did the duty of fairness mean that Transit were obliged to appoint people who came to the task without any views about the tenderers, whether favourable or adverse. It would have been impossible to have a TET competent to perform its function unless it consisted of people with enough experience to have already formed opinions about the merits and demerits of roading contractors. The obligation of good faith and fair dealing also did not mean that the TET had to act judicially.

It has been held to include treating the tenderers equally, and using competent decision makers.[9] In the public law sphere, where a council puts a contract out to tender and complies with the statutory tendering procedure, it is entitled to apply its own commercial judgment as to what is the most effective and economic tender, even if that results in some unfairness to the applicant.[10]

1.10 If a tender is sought ostensibly for the purpose of letting a contract but in fact turns out to have been sought for another purpose, then the tenderers will be able to recover their tendering costs.[11] Whether this arises by way of an implied contract or by way of restitution is not always easy to see from the cases. The principle is that the party which has invited the tender has obtained a benefit by reason of the receipt of the tender and has thus received value for which it should pay.[12] The quid pro quo which normally attends the receipt of tenders is that the tenderers have a realistic chance of getting the contract. The matter can be analysed as a matter of contract as follows: (a) when the invitor asks for tenders, there is an implied representation that the tender when received will be considered; (b) the sending of the tender by the invitee is an acceptance of the invitor's offer to consider the tender. The invitor has therefore contracted to consider the offer. Should it not do so, for example because it never expected the tenderers to get the work, then it will be in breach of contract.

1.11 As a matter of quasi contract or restitution, the analysis is simpler. The invitor of tenders gets the benefit of the tender and must make restitution.

1.12 There is a lack of modern cases dealing with this situation. However, the increasingly heavy expense that tenderers incur makes this a potential area for significant claims in the future by disappointed contractors who consider that their tenders are used for ulterior purposes.

1.13 Nevertheless, it remains the case that the circumstances discussed above are exceptions to the general rule that work done pursuant to an invitation to tender is at the risk of the tenderer. There is normally no legitimate expectation that the party compiling the tender will receive anything. The preparation of the tender is a normal function of commercial life undertaken in the expectation that, of several tenders submitted, only one will succeed.

> It did not have to accord [each tenderer] a hearing or enter into debate with him about the rights and wrongs of [any particular] contract. It would no doubt have been bad faith for a member of the TET to take steps to avoid receiving information because he strongly suspected that it might show that his opinion on some point was wrong. But that is all. [para 47].

See also *Fairclough Building Ltd v Port Talbot Borough Council* (1992) 33 Con LR 2, discussed above.

9 Ibid.

10 *R v Bridgend County Borough Council (Respondent), ex parte Alison Jones (t/a Shamrock Coaches)* (QBD) (Crown Office List) (Kay J) 1 October 1999.

11 *William Lacey (Hounslow) v Davis* [1957] 1 WLR 932; *Craven-Ellis v Canons Ltd* [1936] 2 KB 403.

12 See *Countrywide Communications Ltd v ICL Pathway Ltd and International Computers Ltd* (2000) CLC 324; and see also *Sabemo Pty Ltd v North Sydney Municipal Corp* [1977] 2 NSWLR 880; but cf *Jennings and Chapman Ltd v Woodman Matthews & Co* [1952] TLR 409; and *Brewer Street Investments Ltd v Barclays Woollen & Co Ltd* [1964] 1 QB 428.

There is no commercial expectation on the part of the invitor to pay the several tenderers' costs and neither is there any expectation by the invitees that they will be paid either.[13]

Where there is an issue as to whether the invitation to tender is in fact an offer to accept the lowest submitted bid, this will be a question of construction where all the relevant surrounding circumstances are taken into account. Invitations which do give such undertakings of acceptance are rare. The inclusion of words which state expressly that the employer is not bound to accept the lowest tender are sufficient to exclude any contractual obligation to do so. **1.14**

This topic overlaps with the concepts set out in the *Walford v Miles* and *Courtney & Fairbairn* cases discussed below. **1.15**

Subject to Contract

Work is often asked to be done 'subject to contract'.[14] Sometimes, work may be asked to be done subject to contract and in anticipation of a contract. In such cases it becomes a question of construction of the 'agreement' as to whether any obligation to pay for the work done arises. **1.16**

However, the term 'subject to contract' ought not to be confused with agreements which contemplate that a more formal contract will be agreed later, but where the parties wish to bind themselves in the meantime. The ICE Conditions, for example, contemplate a formal deed being executed, but that does not mean that a contract has not already been made. For recent discussions on this topic see *Harvey Shopfitters Ltd v ADI Ltd*[15] *and Bryen & Langley Ltd v Martin Boston.*[16] It is a question of construction as to whether the parties intend to be bound by their agreement pending the execution of a later and formal agreement, or whether they do not wish to be bound until the formal agreement is executed. In the two cases cited above, the Court of Appeal decided that the parties were bound by the deal they had made even though they contemplated the execution of a formal agreement at a later date. It is very common for parties to exchange letters where all of the terms essential to a contract have been agreed and then agree further that they will execute a formal document. It will take clear words to show that in those common circumstances the execution of the formal document is required before the parties are in fact contractually bound. Obviously if the parties make it clear that the execution of a formal agreement is a condition precedent to being bound, then different considerations apply. See the further discussion under 'Intention to Create Legal Relations', paragraph 1.62 et seq below. **1.17**

[13] *Regalian Properties plc v London Docklands Development Corp* [1995] 1 WLR 212. Cf the decision of Judge Bowsher QC in *Marston Construction Co Ltd v Kigass Ltd* (1989) 15 Con LR 116 criticized by Rattee J in *Regalian* and probably wrongly decided.

[14] This dealt with below under the topic 'Intention to Create Legal Relations', para 1.62 et seq.

[15] [2003] EWCA Civ 1757, [2004] 2 All ER 982.

[16] [2005] EWCA Civ 973. See also *Rossiter v Miller* (1878) 3 App Cas 1124 at 1151 *per* Lord Blackburn; and *Von Hatzfeldt-Wildenburg v Alexander* [1912] 1 Ch 284 at 288–9.

Letters of Intent

1.18 A letter of intent is a document which is sent to a contractor whereby the employer states that it intends to enter into a contract for the works and, pending such contract being entered into, asks for work to be done. It is intended to give some measure of security to the party commencing work, pending the conclusion of the contract. Letters of intent normally seek an agreement that a limited amount of work will be done for either a fixed price or up to an amount which is capped, or occasionally for a reasonable sum for the work done (a quantum meruit).[17] They may be appropriate in circumstances where either the price is agreed, or the contract terms are very likely to be agreed, and there are good reasons to start work in advance of the finalization of all the contractual documents: *Cunningham v Collett* [2006] EWHC 1771 (TCC), (2007) 113 Con LR 143.

1.19 The effect of a letter of intent depends upon the objective meaning of the words used. It may have no binding effect. This may well be its objective. For example, when considering the effect of the letter of intent sent to the contractor in *British Steel v Cleveland Bridge* [1984] 1 All ER 504, Robert Goff J said: 'In these circumstances, if the buyer asks the seller to commence work "pending" the parties entering into a formal contract, it is difficult to infer from the buyer acting on that request that he is assuming any responsibility for his performance, except such responsibility as will rest on him under the terms of the contract which both parties confidently anticipate they will shortly enter into . . .'.[18]

1.20 On the other hand, a letter of intent can create an ancillary contract for preliminary works which is enforceable in the ordinary way. In *Turriff Construction v Regalia* (1971) 9 BLR 20, the letter of intent included the words 'subject to agreement on an acceptable contract', but it was held that this referred only to the full contract, and it amounted to a binding contract in respect of the preliminary works. Alternatively, a letter of intent can be held to constitute acceptance of an offer leading to a binding contract. This was held to be the case in *Wilson Smithett v Bangladesh Sugar* [1986] 1 Lloyd's Rep 378, despite the requirement for the submission of a security deposit/performance bond.

Another alternative is that the letter of intent can constitute a contractual offer to the effect that if the recipient undertakes the proposed action, he will be remunerated reasonably where he acts to the benefit of the sender pursuant to the letter (a quantum meruit). The legal basis for recovery of a reasonable sum is the law of unjust enrichment, rather than by a contractual promise of payment: *BSC v Cleveland Bridge* [1984] 1 All ER 504, at page 511.

---

[17] Examples of such letters of intent producing binding obligations to pay either a price or a quantum meruit include *Turriff Construction v Regalia* (1971) 9 BLR 20; and *British Steel Corp v Cleveland Bridge and Engineering Co Ltd* [1984] 1 All ER 504, (1983) 24 BLR 94.

[18] At 510j–511a.

Offer and Acceptance

No contract can be legally enforceable without an offer and an acceptance. Usually these elements will be documented in writing and the parties will sign or seal their agreement in a document which will record what they have agreed.[19] **1.21**

A written contract has additional consequences, namely the right of adjudication[20] as an enforceable dispute resolution procedure. It also means that the terms of the contract will not normally require a court to decide what words the parties used to express their agreement, although the court may be required to decide what the chosen wording actually means. **1.22**

Contracts are not always formed by the signing of a written document. The agreement can be formed entirely orally (or verbally as it is sometimes described), or it can be formed partly orally and partly in writing. For example, an offer can be made by email and accepted by telephone. An offer can also be accepted by conduct. This is common in the field of construction. A main contractor may send an order form to a sub-contractor (the offer) and that may be accepted by the sub-contractor commencing work. The commencement of the work will usually be taken as an acceptance if there has been no intermediate communication. Sometimes the conduct of the parties will lead the court to infer a contract was either entered into or continued by conduct. The objective actions of the parties will be determinative in deciding whether a contract was entered into in such a way.[21] **1.23**

The essential characteristic is the offer and the acceptance, whatever modes are used. An offer which is replied to by a counter offer (often called a 'cross offer'), the law states is rejected. **1.24**

Once an offer is rejected it is incapable of being accepted. Once the recipient of an offer (the offeree) has replied with a counter offer, the recipient cannot then accept the previous offer, unless the original offeror expressly allows it. But that allowance must normally be by way of a fresh offer which is capable of being accepted. **1.25**

Whilst the concept of offer and acceptance might appear simple, in practice in many cases it is not at all. It can often be difficult to establish whether or not there has been an offer and an acceptance in circumstances where the passing of numerous offers and counter offers mounts up before at some point a consensus is reached. For example, the passing of **1.26**

---

[19] A contract can be deemed a written contract for the purposes of s 107 of the Housing Grants, Construction and Regeneration Act 1996 provided it is evidenced in writing, for example by the exchange of several written communications; there is no requirement for the entire agreement to be contained in one document: *Hatmet Ltd v Herbert (t/a LMS Lift Consultants)* [2005] EWHC 3529 (TCC).

[20] Provided for by s 107 of the Housing Grants, Construction and Regeneration Act 1996.

[21] For example, in *Baird Textiles Holdings Ltd v Marks & Spencer plc* [2002] All ER (Comm) 737 it was held that you could infer a common intention to be bound by a contract which has legal effect in order to establish an agreement based on conduct. This case was applied by Ramsey J in *Skanska Rashleigh Weatherfoil v Somerfield Stores Ltd* [2006] EWHC 947, (TCC) where he held in para 92: 'The important factor in this case is that the parties continued to conduct themselves as before in circumstances where they had a pre-existing agreement'. The appeal was allowed against part of his judgment [2006] EWCA Civ 1732, but not that part.

correspondence between putative contracting parties may be a succession of letters or emails chipping away at disagreements until at some point an agreement is reached on enough of the individual items contained within one or other party's terms and conditions sufficient to permit a court to hold that there is a consensus.

1.27 The structural scheme of offer, acceptance, and counter offer described above is classic law.[22] However, it has been called into question by the courts in recognition of the fact that the conduct of the parties is not always susceptible to clear analysis in that classic way.

1.28 Lord Denning MR in *Butler Machine Tool v Ex-Cell-O Corp*,[23] a case dealing with a 'battle of the forms', said:

> In many of these cases our traditional analysis of offer, counter-offer, rejection, acceptance and so forth is out of date. This was observed by Lord Wilberforce in *New Zealand Shipping Co Ltd v A M Satterthwaite & Co Ltd* [1975] AC 154, at 167. The better way is to look at all the documents passing between the parties—and glean from them, or from the conduct of the parties, whether they have reached agreement on all material points—even though there may be differences between the forms and conditions printed on the back of them. As Lord Cairns said in *Brogden v Metropolitan Railway Co* (1877) 2 App Cas 666, at 672:
>
> > ... there may be a *consensus* between the parties far short of a complete mode of expressing it, and that *consensus* may be discovered from letters or from other documents of an imperfect and incomplete description; ...
>
> Applying this guide, it will be found that in most cases when there is a 'battle of forms', there is a contract as soon as the last of the forms is sent and received without objection being taken to it.[24]

1.29 The *Ex-Cell-O* case is often used to deal with battle of the forms arguments, and is illustrative of the point that the traditional analysis outlined above, whilst a correct starting point, is not always determinative.

1.30 This is particularly the case when the work has been completed, and where the parties have exchanged correspondence throughout the period whilst works were being carried out. It should be noted that formalities are not necessary for the parties to have entered into a contract.[25]

1.31 The commercial objective of some contractors and sub-contractors may be to avoid entering into a contract so that, at the completion of the words, a quantum meruit can be claimed. In those circumstances the classic offer and acceptance mechanism may not be apt.[26]

---

[22] *Per* Steyn LJ in *G Percy Trentham Ltd v Archital Luxfer Ltd* [1993] 1 Lloyd's Rep 25 at 27.

[23] [1979] 1 WLR 401 at 404.

[24] For a more recent TCC case in which the *Ex-Cell-O* principles were applied in relation to the terms and conditions of a sub-contract, see *AE Yates Trenchless Solutions Ltd v Black & Veatch Ltd* [2008] EWHC 3183 (TCC).

[25] For example, Coulson J in *Fitzpatrick Contractors Ltd v Tyco Fire & Integrated Solutions (UK) Ltd* [2008] EWHC 1301 (TCC) had no hesitation on the facts of that case in holding that a contract had been entered into despite there being no signed document because it was clear that the parties had intended to create legal relations.

[26] See *Brogden v Metropolitan Railway* (1877) 2 App Cas 666; New *Zealand Shipping Co v Satterthwaite* [1975] AC 154 at 167; and *Gibson v Manchester City Council* [1979] 1 WLR 294.

In those circumstances the courts will look hard to see whether in fact there is a contract, by reference to any oral and written exchanges and the performance of the obligations as to construction and payment. It can be difficult to contend against the existence of a contract when performance has taken place. In *G Percy Trentham Ltd v Archital Luxfer Ltd*,[27] Steyn LJ said: **1.32**

> The fact that the contract was performed on both sides will often make it unrealistic to argue that there was no intention to enter into legal relations. It will often make it difficult to submit that the contract is void for vagueness or uncertainty. Specifically, the fact that the transaction is executed makes it easier to imply a term resolving any uncertainty, or, alternatively, it may make it possible to treat a matter not finalised in negotiations as inessential.

Difficulties will arise however if, for example, there is no agreement on a term such as the time for completion, or the rate of liquidated damages, or a cap on liability for damages for defective work, in circumstances where it is plain that these terms are important. If the court concludes that an essential term important to both parties has not been agreed, it will not make the bargain for the parties. A good example of this is *British Steel Corp v Cleveland Bridge and Engineering Co Ltd*,[28] where a failure to agree the amount of damages payable in the event of delay was held by Robert Goff J to relate to an essential term of the putative contract and, since the term was not agreed, he concluded that no contract was entered into. Without a contract, there could be no breach of contract claim, therefore defective work or late completion of the work would have (and in this case had) no remedy. **1.33**

Battles of forms, whereby each party seeks to have its terms and conditions apply over those of the other, are common in the field of building contracts. Care needs to be taken to see whether the 'final shot' wins simply because it is not objected to, or whether a more sophisticated analysis is needed in the circumstances. **1.34**

In some circumstances where there has been a long chain of correspondence, the parties may both be uncertain as to if or when an agreement was reached, and/or whether the agreement pertained as a result of the content of their continuing correspondence. **1.35**

In that event, the courts will analyse the correspondence and decide that at the point where agreement is reached a curtain comes down on the correspondence, a contract is made, and all that takes place subsequently is irrelevant.[29] Subsequent acts of parties do not assist the court in deciding how a contract is to be interpreted, and nor do they affect a contract once it has been agreed. What parties think and do subsequent to the formation of a written contract is irrelevant, unless a fresh contract is entered into.[30] That does not apply if the contract has been entered into orally or partly orally and partly in writing. In that event subsequent conduct might be examined to help determine the parties' original intentions. In *Brian Maggs v Guy Marsh* [2006] EWCA Civ 1058, Marsh employed Maggs to carry out work at his shop. The bill was disputed and came to trial where a major issue was the extent **1.36**

[27] [1993] 1 Lloyd's Rep 25 at 27.

[28] [1984] 1 All ER 504, (1983) 24 BLR 504.

[29] *Harmony Shipping Co SA v Saudi Europe Line Ltd 'The Good Helmsman'* [1981] 1 Lloyd's Rep 377 *per* Ackner LJ.

[30] *Whitworth Street Estates v James Miller* [1969] 1 WLR 377.

to which it was permissible for the judge to receive evidence of conduct subsequent to the formation of the contract. The court held, citing Lewison on *The Interpretation of Contracts*, 3rd edn,[31] that the principles laid down in *Whitworth Street Estates v James Miller* [1969] 1 WLR 377 that subsequent conduct is inadmissible did not apply to an oral contract or a contract partly oral and partly in writing. The rationale is that the subsequent words or conduct may assist in ascertaining the terms agreed to but not written down. This doctrine is consistent with the principles now applicable to the remedy of Rectification. In *Chartbrook Ltd v Persimmon Homes Ltd and others* [2009] UKHL 38, the House of Lords considered the requirements of the remedy, namely a prior consensus between the parties. In paragraph 65 of his judgment Lord Hoffmann considered the evidence which could be adduced to establish that prior consensus as follows:

> A party may have had a clear understanding of what was agreed without necessarily being able to remember the precise conversation or action which gave rise to that belief. Evidence of subsequent conduct may also have some evidential value. On the other hand, where the prior consensus is expressed entirely in writing, (as in *George Cohen Sons & Co Ltd v Docks and Inland Waterways Executive* (1950) 84 Ll L Rep 97) such evidence is likely to carry very little weight. But I do not think that it is inadmissible.

1.37 The exchange of correspondence leading to an agreement should not be analysed too strictly. The law does not require every matter to be agreed upon before a contract is formed. The law only requires that the essential and important terms be agreed.

1.38 What the essential and important terms are is a matter for decision in each case. There is no hard and fast rule because the applicable test is what was important to the parties, as opposed to what would have been important to an objective and cool bystander. Nevertheless, the important terms are usually:

(a) the definition of the work to be done;
(b) the price to be paid for that work;
(c) the time in which the work is to be done (often referred to as the contract period, as it shall be in this book); and
(d) the amount of damages which are to be paid in the event of delay.

1.39 It has been suggested that the price to be paid for the work is invariably an essential element which must be agreed upon for there to be a contract.[32] But this is too simple a position. If the price is not agreed and it is not an essential term, then the absence of agreement as to price will not prevent an agreement coming into existence.[33] In that event the parties will

---

[31] Particularly at pp 89 and 91.

[32] *Courtney & Fairbairn v Tolaini Brothers (Hotels) Ltd* [1975] 1 WLR 297; *Foley v Classique Coaches Ltd* [1934] 2 KB 1 at 13 *per* Maugham LJ; *May and Butcher Ltd v R* [1934] 2 KB 17 at 20 *per* Lord Buckmaster and 21 *per* Viscount Dunedin.

[33] *Furmans Electrical Contractors v Elecref Ltd* [2009] EWCA Civ 170; and *British Steel Corp v Cleveland Bridge and Engineering Co Ltd* [1984] 1 All ER 504. Here the plaintiffs had performed work pursuant to a request by the defendants, in anticipation of an agreement as to price. The work done was not referable to any contractual terms as to payment or performance. However, in the circumstances, the defendants were obliged to pay a reasonable sum for the work by quantum meruit. See also Rix LJ in *Mamidoil-Jetoil Greek Petroleum Co SA v Okta Crude Oil Refinery AD* [2001] EWCA Civ 406, [2001] 2 All ER (Comm) 193, [2001] 2 Lloyd's Rep 76 and sub-paragraph (viii) of the propositions cited by him and which are set out in paragraph 1.45 below.

have made an agreement and all the court has to do is ascertain what is fair and reasonable. For example, in *Furmans Electrical Contractors v Elecref Ltd* [2009] EWCA Civ 170 the Court of Appeal held that an electrical sub-contractor could have entered into a contract even though no price had been agreed on the basis that the price was to be a reasonable price. In that event recovery would not be for a quantum meruit (and therefore a restitutionary remedy) but for sums due under a contract. A bare agreement to negotiate a price is not enforceable.[34] But an express agreement to negotiate a price in good faith may well be.[35] If there is an agreement that a fair price will be paid, then the courts can enforce the agreement and quantify the amount.[36] If the agreement is a long-term one and the matter of price is left for agreement in the future, then the courts will seek to enforce the agreement, if they can. They cannot do so if the contract contains no objective criteria by reference to which the matter still to be agreed can be ascertained.[37] The contract has to contain sufficient indications of objective criteria to enable that which has still to be agreed or calculated to be arrived at by the parties—or if they cannot agree, by a court or arbitrator. If there are those objective indications, then the contract will be enforceable.[38] The presence of an arbitration clause may not provide the necessary objective indications, and it may not be possible to imply a term that the arbitrator could arrive at what the parties were unable to agree. On the other hand the courts could look upon an arbitration clause as providing the appropriate objective machinery to allow what appears to be an incomplete bargain to have contractual certainty by the operation of that arbitration clause.[39] This would particularly be the case if the arbitration clause provided machinery for completion of the bargain.[40] The courts will not make bargains for the parties. Thus if the parties agree that

---

See also *Monavon Construction Ltd v Davenport* [2006] EWHC 1094 (TCC) in which HH Judge Thornton QC held that there was no enforceable agreement containing a cost limit because the precise scope of work to which such a limit would relate was not clearly defined or definable.

[34] *Walford v Miles* [1992] AC 128; see also *P&O Property Holdings Ltd v Norwich Union Life Insurance Society* (1994) 68 P&CR 261. See also the discussion below on the intention to create legal relations and *Jackson & Ors v Thakrar & Ors* [2007] EWHC 271 (TCC).

[35] See the obiter discussion in the Court of Appeal in *Petromec Inc v Petroleo Brasileiro SA Petrobas* [2005] EWCA Civ 891 at [115]–[121] particularly at [121] referring to Lord Steyn's Sultan Azlan Shah lecture on 24 October 1996 reprinted in 113 LQR 433, (concerning the reasonable expectations of honest men) saying that *Walford v Miles* did not bind the Court of Appeal to hold that an express obligation to negotiate in good faith was completely without legal substance. But see *Little v Courage Ltd*, The Times, 4 January 1995; *Phillips Petroleum Co UK Ltd v Enron Europe Ltd* (1997) CLC 329 cited by Jackson J in *Multiplex Constructions (UK) Ltd v Cleveland Bridge UK Ltd and another* [2006] EWHC 1341 (TCC).

[36] *The Didymi* [1988] 2 Lloyd's Rep 108 particularly at 115–6 and 119; and *Mamidoil-Jetoil Greek Petroleum Co SA v Okta Crude Refinery AD* [2001] 2 Lloyd's Rep 76 at 91.

[37] *Malozzi v Carapelli* [1976] 1 Lloyd's Rep 407. See also *Walford v Miles* [1992] AC 128 where an open-ended agreement by a seller to deal with only one potential purchaser lacked certainty and was therefore unenforceable.

[38] *The Didymi*, n 36 above, and *Sudbrook Trading Estate Ltd v Eggelton* [1983] 1 AC 444.

[39] For the two sides to this coin See *Beer v Bowden* [1981] 1 WLR 522 at 526 *per* Goff LJ (to the effect that such a clause does not complete the bargain) and Rix LJ in *Mamidoil-Jetoil Greek Petroleum Co SA v Okta Crude Oil Refinery AD* [2001] EWCA Civ 406, [2001] 2 All ER (Comm) 193, [2001] 2 Lloyd's Rep 76 and proposition (x) cited in para 1.45 below (to the effect that the clause might do so).

[40] See for example *Queensland Electricity Generating Board v New Hope Collieries Pty Ltd* [1988] 1 Lloyd's L Rep 205, in which Sir Robin Cooke said:

> At the present day, in cases where the parties have agreed on an arbitration or valuation clause in wide enough terms, the Courts accord full weight to their manifest intention to create continuing

they shall agree a price, then the courts will not usurp their contractual position and make an agreement for them.[41] There is a distinction between an agreement to pay a reasonable price (enforceable) and an agreement that the price is to be agreed (probably unenforceable).[42]

1.40 The following situations could potentially provide sufficient certainty for a court to decide whether the parties have reached a sufficient degree of agreement to give rise to a contract. Where the price is ascertainable by reference to a formula subsequent to the execution of the work, then a contract can be entered into with no price agreed. If the exact price is not important to the parties, then a contract can be formed without a price being agreed. If the contract price is to be ascertained by a third party, be it an expert or an arbitrator, then the contract is enforceable.

1.41 Difficulties frequently arise when the parties do not actually agree, but agree that they will enter an agreement. Such an agreement is not enforceable.[43]

1.42 The reason that the four elements discussed above (namely: (a) definition of work to be done; (b) price of the work; (c) time in which the work is to be done; and (d) the damages to be paid in the event the works are not completed to time) are normally essential ingredients within the body of the offer and acceptance, is that they are usually essential to the parties' commercial objective.

1.43 But if lesser terms are not agreed, then a contract can still be formed. It is not the case that all terms must be agreed, but that all important terms must. Contracting parties can, however, reach an enforceable agreement on some matters, leaving the rest to be

---

> legal relations. Arguments invoking alleged uncertainty, or alleged inadequacy in the machinery available to the Courts for making contractual rights effective, exert minimal attraction . . .

41 *Willis Management (Isle of Man) Ltd v Cable and Wireless plc* [2005] EWCA Civ 806. See also *Little v Courage Ltd* (1994) 70 P&CR 469 at 476 *per* Millet LJ:

> An undertaking to use one's best endeavours to obtain planning permission or an export licence is sufficiently certain and is capable of being enforced. An undertaking to use one's best endeavours to agree, however, is no different from an undertaking to agree, to try to agree, or to negotiate with a view to reaching agreement; all are equally uncertain and incapable of giving rise to an enforceable legal obligation.

See also *Phillips Petroleum Co Ltd v Enron Europe Ltd* [1997] CLC 329.

42 See Rix LJ in *Willis Management (Isle of Man) Ltd v Cable and Wireless plc* [2005] EWCA Civ 806 at para. 33:

> It is like the distinction between agreeing, whether expressly or by implication, that the price of goods to be sold is to be a fair or reasonable price and on the other hand agreeing that the price is to be a price 'to be agreed'. In the first case the price will be set by the court (if none is previously agreed by the parties), but in the second case the parties have said that they, and not the court, will agree the price: see *May and Butcher Ltd v R* [1934] 2 KB 17, at 21, 22.

43 See *Courtney*, n 32 above; *Albion Sugar Co Ltd v Williams Tankers Ltd* [1977] 2 Lloyd's Rep 457; *Scandinavian Trading Tanker Co AB v Flota Petrolera Ecuatoriana* [1981] 2 Lloyd's Rep 425; *Voest Alpine Intertrading GmbH v Chevron International Oil Co Ltd* [1987] 2 Lloyd's Rep 547; *Walford v Miles* [1992] 2 AC 128; *Willis Management (Isle of Man) Ltd v Cable and Wireless plc* [2005] EWCA Civ 806. But see the decision of Peter Pain J in *Donwin Productions Ltd v EMI Films Ltd, The Times*, 9 March 1984, in which he held that an oral contract which contained a term that the parties would negotiate in good faith a fuller agreement with further terms which would be reduced to writing was a proper contractual agreement in law. This might be doubted however having regard to Millett LJ in *Little v Courage Ltd* (1994) 70 P&CR 469 at 476 but, as the recent cases illustrate, may be explained on the basis that the further terms were not sufficiently important to mean that no agreement had been reached.

determined by further agreement or the process of litigation, for example in *Tomlin v Standard Telephones & Cables Ltd* [1969] 1 WLR 1378.[44] The courts will enforce a bargain when the essential terms are agreed even if some unimportant ones are not and the parties intended to be bound. However, they will not make a bargain for parties by filling in gaps where the parties have failed to agree terms which are essential to the working of their agreement.[45]

Lloyd LJ in *Pagnan SpA v Feed Products Ltd*[46] summarized the position as follows: **1.44**

> As to the law, the principles to be derived from the authorities, some of which I have already mentioned, can be summarised as follows:
>
> (1) In order to determine whether a contract has been concluded in the course of correspondence, one must first look to the correspondence as a whole (see *Hussey v Horne-Payne* (1878–1879) 4 App Cas 311).
>
> (2) Even if the parties have reached agreement on all the terms of the proposed contract, nevertheless they may intend that the contract shall not become binding until some further condition has been fulfilled. That is the ordinary 'subject to contract' case.
>
> (3) Alternatively, they may intend that the contract shall not become binding until some further term or terms have been agreed; see *Love and Stewart v Instone*,[47] where the parties failed to agree the intended strike clause, and *Hussey v Horne-Payne*,[48] where Lord Selbourne said at p 323:
>
>> . . . the observation has often been made, that a contract established by letters may sometimes bind parties who, when they wrote those letters, did not imagine that they were finally settling the terms of the agreement by which they were to be bound; and it appears to me that no such contract ought to be held to be established, even by letters which would otherwise be sufficient for the purpose, if it is clear, upon the facts, that there were other conditions of the intended contract, beyond and besides those expressed in the letters, which were still in a state of negotiation only, *and without the settlement of which the parties had no idea of concluding any agreement.* [Lloyd LJ's emphasis].
>
> (4) Conversely, the parties may intend to be bound forthwith even though there are further terms still to be agreed or some further formality to be fulfilled (see *Love and Stewart v Instone* per Lord Loreburn at p 476).
>
> (5) If the parties fail to reach agreement on such further terms, the existing contract is not invalidated unless the failure to reach agreement on such further terms renders the contract as a whole unworkable or void for uncertainty.
>
> (6) It is sometimes said that the parties must agree on the essential terms and that it is only matters of detail which can be left over. This may be misleading, since the word 'essential' in that context is ambiguous. If by 'essential' one means a term without which the contract cannot be enforced then the statement is true: the law cannot enforce an incomplete contract. If by 'essential' one means a term which the parties have agreed to be essential for the formation of a binding contract, then the statement is tautologous. If by 'essential' one

[44] See, for example, *Western Broadcasting Services v Edward Seaga* [2007] UKPC 19 (29 March 2007).

[45] *Trollope & Colls Ltd, Holland and Hannen and Cubitts Ltd v Atomic Power Constructions Ltd* [1963] 1 WLR 333 (Megaw J); and *Mmecen SA v Inter Ro-Ro SA and Gulf Ro-Ro Services SA, 'The Samah and Lina V'* [1981] 1 Lloyd's Rep 40 (Parker J); and *Western Broadcasting*, ibid.

[46] [1987] 2 Lloyd's Rep 601 at 619. Applied in *Haden Young Ltd v Laing O'Rourke Midlands Ltd* [2008] All ER (D) 49 (Jun).

[47] (1917) 33 TLR 475.

[48] (1878–1879) 4 App Cas 311.

means only a term which the Court regards as important as opposed to a term which the Court regards as less important or a matter of detail, the statement is untrue. It is for the parties to decide whether they wish to be bound and, if so, by what terms, whether important or unimportant. It is the parties who are, in the memorable phrase coined by the Judge, 'the masters of their contractual fate'. Of course the more important the term is the less likely it is that the parties will have left it for future decision. But there is no legal obstacle which stands in the way of the parties agreeing to be bound now while deferring important matters to be agreed later . . .

1.45 The two principal cases cited above, *Pagnan SpA* and *G Percy Trentham Ltd,* have had very little citation in other cases outside the field of law relating to construction contracts. One additional authority gives helpful guidance on this topic. It is the judgment of Rix LJ in *Mamidoil-Jetoil Greek Petroleum Co SA v Okta Crude Oil Refinery AD* [2001] EWCA Civ 406, [2001] 2 All ER (Comm) 193, [2001] 2 Lloyd's Rep 76 in which he set out the following propositions:

(i) Each case must be decided on its own facts and on the construction of its own agreement. Subject to that,

(ii) Where no contract exists, the use of an expression such as 'to be agreed' in relation to an essential term is likely to prevent any contract coming into existence, on the ground of uncertainty. This may be summed up by the principle that 'you cannot agree to agree'.

(iii) Similarly, where no contract exists, the absence of agreement on essential terms of the agreement may prevent any contract coming into existence, again on the ground of uncertainty.

(iv) However, particularly in commercial dealings between parties who are familiar with the trade in question, and particularly where the parties have acted in the belief that they had a binding contract, the courts are willing to imply terms, where that is possible, to enable the contract to be carried out.

(v) Where a contract has once come into existence, even the expression 'to be agreed' in relation to future executory obligations is not necessarily fatal to its continued existence.

(vi) Particularly in the case of contracts for future performance over a period, where the parties may desire or need to leave matters to be adjusted in the working out of their contract, the courts will assist the parties to do so, so as to preserve rather than destroy bargains, on the basis that what can be made certain is itself certain. *Certum est quod certum reddi potest.*

(vii) This is particularly the case where one party has either already had the advantage of some performance which reflects the parties' agreement on a long term relationship, or has had to make an investment premised on that agreement.

(viii) For these purposes, an express stipulation for a reasonable or fair measure or price will be a sufficient criterion for the courts to act on. But even in the absence of express language, the courts are prepared to imply an obligation in terms of what is reasonable.

(ix) Such implications are reflected but not exhausted by the statutory provision for the implication of a reasonable price now to be found in section 8(2) of the Sale of Goods Act 1979 (and, in the case of services, in section 15(1) of the Supply of Goods and Services Act 1982).

(x) The presence of an arbitration clause may assist the courts to hold a contract to be sufficiently certain or to be capable of being rendered so, presumably as indicating a commercial and contractual mechanism, which can be operated with the assistance of experts in the field, by which the parties, in the absence of agreement, may resolve their dispute.

These propositions were applied by Ramsey J in *Bell Scaffolding (Aust) Pty Ltd v Rekon Ltd* and *Alba Hire & Sales Ltd* [2006] EWHC 2656 (TCC).

It is in the sphere of construction contracts, where arguments as to whether there is a contract or not (a 'contract v no contract' argument), that the problems are felt most acutely.[49] Often the questions of fact which must be applied to the above principles by the judge, will be looked at through the viewpoint of whether there can be divined from them objectively speaking a desire to be contractually bound. If one party appears to the judge to be astute to ensure that there is no agreement in the classical offer and acceptance sense in order that it can avoid damages for breach of contract or obtain a quantum meruit, the judge is likely to find that there was a contractual agreement on the basis that the reasonable expectations of honest men would, on the objective evidence, require it. Alternatively, the judge may find that there is no contract, but may value the quantum meruit at the rates which the contract provided for.

Consideration

In English law it is still thought to be the case that there must be 'consideration' within the contract before the law will enforce a bargain.[50] **1.46**

Normally this presents no difficulty. In a normal building contract consideration might be described in the following way: *in consideration of A building the structure, then B will pay to A the contract price.* **1.47**

Consideration is the thing which is given in exchange for the contractual promise. Therefore, in the above example, the consideration for the construction of the structure is the payment of the contract price and the consideration for the payment of the price is the construction of the structure. **1.48**

Absent consideration, the law will regard a 'contract' as a worthless promise which is unenforceable at law. **1.49**

Thus, taking the above example, if A had already built the structure on B's land with no promise by A to pay, then there will be no consideration in a subsequent promise by A to pay the contract price. This situation has been characterized by the legal maxim *past consideration is no consideration.* In that situation, because the structure has already been built before a promise to pay has been made, there is no mutual exchange of obligations and **1.50**

---

49 First instance decisions which seek to apply these principles do not do more than just that; there has been no real development of them. An example of how these principles are applied is in *J Murphy & Sons Ltd v ABB Daimler Benz Transportation (Signal) Ltd* (HHJ Hicks QC), 2 December 1998 [1999] CILL 1461.

50 It was not always so as a rule of substantive law. Lord Mansfield in *Pillans v Van Mierop* (1765) 3 Burr 1663 at 1668, 1669 said: 'I take it that the ancient notion about want of consideration was for the sake of evidence only; for when it is reduced to writing, as in covenants, specialties, bonds etc., there was no objection to want of consideration.' This extract from the judgment is quoted by Oliver Wendell Holmes Jnr, *The Common Law*, Lecture VII (1881). In the same lecture Holmes shows from the Yearbooks that the doctrine was one of proof and evidence only.

A's subsequent promise to pay the contract price will be unenforceable for want of consideration.

1.51 Consideration must be real. It cannot be illusory. A promise to repay a debt of £100 with a cheque for £90 will not therefore discharge the debt, even if both parties agree that should be so. That is because the pre-existing contractual obligation to repay a debt of £100 is clearly of greater value than the payment of £90. So the bargain: *in consideration of A paying B £90, B agrees to forgive A's debt of £100 in full*, will not be an enforceable contract.[51]

1.52 However, the courts do not look too closely into the adequacy of consideration. In the above example, where money alone is the commodity traded, it is easy to see that the consideration is not consideration at all. But if the bargain is expressed: *in consideration of A paying B £90 and a peppercorn, B agrees to forgive A's debt of £100 in full*, then it would be enforceable because the court will not inquire into the exact worth of the peppercorn.

1.53 This is obviously artificial. Everyone knows that a peppercorn is virtually worthless. But the distinction remains part of English law.

1.54 In construction contracts, such bargains as those identified above will attend the completion of the works where the builder is owed money by the employer and the employer wishes to reduce the amount payable. If the builder claims £100 and the employer disputes the claim but agrees to pay £90 in settlement of the claim, then there will be an enforceable bargain. The claim will have been bought off with the promise of payment. But if the builder's claim is agreed in the sum of £100 and then the employer agrees to pay £90 in settlement, then the bargain is unenforceable for want of consideration.

1.55 Upon such distinctions many cases turn. But at the end of the works, when tempers on both sides are frayed, it will often be the case that the contractor wishes a swift payment and the employer wishes to minimize its payment, because of real or perceived inadequacies of the performance of the builder. Such distinctions become important in that event.

1.56 An analogous situation is when the builder is not performing its obligations in a way which gives the employer confidence that the works will be properly done, and the employer agrees to pay to the builder a sum greater than the contract price in order to persuade it to complete its work properly. In such a situation, which is not uncommon, a strict view might be that the promise to pay a sum in addition to the contract price was unsupported by consideration and was therefore unenforceable. That would be, so the argument would go, because the employer was not getting anything in exchange for its promise beyond what it was already entitled to.

1.57 In that situation the courts have held that the promise to pay the further sum was enforceable because it was in fact supported by consideration. The leading case of *Williams v Roffey Bros & Nichols (Contractors) Ltd*,[52] is authority for the proposition that the promise to execute a pre-existing contractual obligation can be good consideration. The judgment manifests a tension between the court's wish to enforce a bargain and its wish not to reward

[51] See *Foakes v Beer* (1884) 9 App Cas 605.
[52] [1991] 1 QB 1.

a builder by awarding more money than the original bargain allowed.[53] Nevertheless, the court set out the applicable principles as follows:

> (i) if A has entered into a contract with B to do work for, or to supply goods or services to, B in return for payment by B; and (ii) at some stage before A has completely performed his obligation under the contract B has reason to doubt whether A will, or will be able to, complete his side of the bargain; and (iii) B thereupon promises A an additional payment in return for A's promise to perform his contractual obligations on time; and (iv) as a result of giving his promise, B obtains in practice a benefit, or obviates a disbenefit; and (v) B's promise is not given as a result of economic duress or fraud on the part of A; then (vi) the benefit to B is capable of being consideration for B's promise, so that the promise will be legally binding. [15G–16A]
>
> Consideration there must still be but, in my judgment, the courts nowadays should be more ready to find its existence so as to reflect the intention of the parties to the contract where the bargaining powers are not unequal and where the finding of consideration reflect (sic) the true intention of the parties.[18H]

The current position is that the courts will rarely find that a commercial bargain is unenforceable because of a lack of consideration. Such situations where the courts will do so usually involve unfair commercial pressure; a want of consideration can be invoked when seeking to set aside a bargain obtained in unfair circumstances. However, the law relating to economic duress would seem to be more than sufficient to deal with any such injustice. **1.58**

The practical result of this is that it would be a brave litigant now who would seek to defend a case on the basis that the bargain lacked consideration. The doctrine is ripe for reconsideration by the House of Lords. It is more likely, however, that it will not be revisited since very few cases will turn on the question of whether or not there has been consideration. **1.59**

Economic Duress

This topic is a close relation of the doctrine of consideration and so is examined here. An agreement which is achieved under economic duress is voidable, that is avoidable by the party who was under the duress. However, economic duress is a hard matter to prove. It involves the coercion of will such that there is no true consent. It is not merely a situation where a party has or feels he has no choice.[54] Economic duress is often a matter which has been considered by those in the construction field, particularly when contractors or subcontractors are starved of cash and feel they have no choice but to bend the knee in the particular situation in which they are in. But a weak bargaining position does not create economic duress. The requirement is of coercion of will, which will usually mean some kind of moral turpitude on the part of the party doing the coercing. The two areas where it can be a factor are first where the employer seeks to get the builder to take a lesser sum than **1.60**

[53] See *Ward v Byham* [1956] 1 WLR 496; *Williams v Williams* [1957] 1 WLR 148; *New Zealand Shipping Co Ltd v Satterthwaite & Co Ltd* [1975] AC 154; *Pao On v Lau Yiu Long* [1980] AC 614. But for a contrary view see *Stilk v Myrick* (1809) 2 Camp 317; and *North Ocean Shipping Co Ltd v Hyundai Construction Co Ltd* [1979] QB 705.

[54] See *Pao On v Lau Yiu Long* [1980] AC 614 particularly at 635, the advice of Lord Scarman in the Privy Council being the classic statement in this field.

he is truly entitled to just so that he gets paid; secondly where the contractor or sub-contractor states that it will not proceed unless an enhanced rate is paid.

1.61 Whilst the concept of economic duress has received academic criticism,[55] it is not something which can be disregarded as not being part of the law. Cases of economic duress mostly involve trade union 'blacking' and other practices which were prevalent in the 1970s. Such cases savour a little of the distaste for the methods of the trade unions.[56] In the construction sphere, there are relatively few. The most famous one, *D&C Builders v Rees*,[57] involved a threat not to pay, and a lesser sum than that which was owed was accepted in an agreement which was set aside. That case, as many others no doubt, could have been solely decided on the basis of a lack of consideration.[58] Only one TCC case in recent times has upheld economic duress.[59] But, generally speaking, claims for economic duress fail unless some kind of moral turpitude can be shown.[60] In cases involving withdrawal of credit or payment of a smaller sum than that claimed, for example, much may depend on the genuineness or good faith of the party applying the duress.[61]

Intention to Create Legal Relations

1.62 Apart from offer, acceptance, and consideration, the final ingredient for a contract to be entered into which is enforceable at law is that the parties must have an intention to create legal relations. In *Jackson & Ors v Thakrar & Ors* [2007] EWHC 271 (TCC), Coulson J stated in this regard 'It is trite law that, in order for there to be an enforceable contract or a binding compromise, the parties must have intended to create legal relations: see, by way of example, *Baird Textile Holdings v Marks and Spencer* [2002] 1 All ER (Comm) 737 . . .'

1.63 Without an intention to create legal relations, there is no binding contract. A lack of intention can be inferred from a lack of specificity in the work to be done as explained above.[62] The intention is presumed in nearly every commercial situation and, save for the one situation looked at in the next paragraph, it would be an unusual case where parties had made an agreement concerning the demolition, alteration, or creation of a structure and they did not intend that agreement to have legal force.

1.64 The common situation where parties do not intend to create legal relations is where an agreement is expressly stated to be 'subject to contract'. This is taken from conveyancing

---

55 As for example by PS Atiyah in his article, 'Economic Duress and the Overborne Will' (1982) 98 LQR 197. See also D Tiplady, 'Concepts of Duress' (1983) 99 LQR 188 and PS Atiyah, 'Duress and the Overborne Will Again' (1983) 99 LQR 353.

56 A more up to date example is *Dimskal Shipping Co SA v International Transport Workers Federation (The Evia Luck)* [1992] 2 AC 152.

57 [1966] 2 QB 617.

58 *Foakes v Beer* (1884) 9 App Cas 605.

59 *Carillion Construction Ltd v Felix (UK) Ltd* [2001] BLR 1.

60 *DSND Subsea Ltd v Petroleum Geoservices ASA* [2000] BLR 530; *Dorimex SRL v Visage Imports* (CA, 18 May 1999); *Huyton SA v Peter Cremner GmbH* [1999] 1 Lloyd's Rep 620; *Amsalem (t/a MRE Building Contractors) v Raivid* [2008] EWHC 3028 (TCC).

61 *CTN Cash and Carry Ltd v Gallaher Ltd* [1994] 4 All ER 714.

62 An example of this would be *University of Plymouth v European Language Centre Ltd* [2009] EWCA Civ 784.

practice where the courts have been astute to ensure that bargains which are made 'subject to contract' or 'subject to a formal agreement' are not enforceable. Since people are not made legally bound to contractual bargains without an intention of being so bound, the express statement that the agreement is not legally enforceable without a further step being taken ensures that the bargain is not so enforceable.

The two principal times when it is expressly stated that a contract is expressed to be 'subject to contract' are during negotiations prior to the execution of a contract and during negotiations for settlement of a contractor's final account or other claims. In those events one or other of the parties will wish to ensure that it is not legally bound until a defining event, usually signature upon a contractual document, has happened.[63] One of the defining cases where this doctrine is set out is *Chillingworth v Esche*[64] where Warrington LJ said: **1.65**

> It has been undoubted ever since the decision of Sir George Jessel in *Winn v Bull* (1877) 7 Ch D 29 that the words 'subject to the preparation and approval of a formal contract' in a document prevented the document from being held to be a final agreement of which specific performance could be enforced, and it has been the practice of estate agents to insert these words to prevent parties being imposed upon. In many cases it is important to avoid the disastrous results of entering into open contracts, and I think it would be most mischievous to throw any doubt on the effect and meaning of such expressions. I do not overlook what was said by Lord Sterndale in *Rossdale v Denny* [1921] 1 Ch 57, at 66 in this court:
>
>> I am far from saying that there may not be an unconditional offer and acceptance of a binding contract although the letters may contain the words 'subject to a formal contract,' but certainly those words do point in the direction of the offer or acceptance being conditional. I do not think it can be put higher than that; I think he is well founded in saying that the general trend of the decisions has been, where those words occurred, to hold that the offer or acceptance was conditional.
>
> But it seems to me that too much importance has been attributed to those expressions of Lord Sterndale, and I think what he meant to say was that the words in question indicate in themselves no binding bargain, and are merely conditional, but that there might be other circumstances which would induce the court not to give them that meaning in a particular case.

In the same case at 114 Sargant LJ said: **1.66**

> To my mind the words 'subject to contract' or 'subject to formal contract' have by this time acquired a definite ascertained legal meaning—not quite so definite a meaning perhaps as such expressions as f.o.b. or c.i.f. in mercantile transactions, but approaching that degree of definiteness. The phrase is a perfectly familiar one in the mouths of estate agents and other persons accustomed to deal with land; and I can quite understand a solicitor saying to a client about to negotiate for the sale of his land: 'Be sure that to protect yourself you introduce into

---

[63] See *Smallman v Smallman* [1972] Fam 25; *Brown v Gould* [1972] Ch 53; *Tiverton Estates Ltd v Wearwell Ltd* [1975] Ch 146; *Munton v GLC* [1976] 1 WLR 649; *Secretary of State for Transport v George William Christos* (CA, 25 July 2003) *per* Mummery LJ at [37]–[40]. But cf *Alpenstow Ltd v Regalian plc* [1985] 1 WLR 721 in which it was held that the contract was nevertheless binding, but that case may be said to be the exception which proves the rule. Cf also *Richards (Michael) Properties Ltd v St Saviour's Parish, Southwark* [1975] 3 All ER 416 in which the words 'subject to contract' were added by mistake and were therefore surplusage. See also *Lockett v Norman-Wright* [1925] Ch 56; and *Raingold v Bromley* [1931] 2 Ch 307.

[64] [1924] 1 Ch 97 at 110–11.

> any preliminary contract you may think of making the words "subject to contract".' I do not say that the phrase makes the contract containing it necessarily and whatever the context a conditional contract. But they are words appropriate for introducing a condition, and it would require a very strong and exceptional case for this clear prima facie meaning to be displaced.

1.67 The other defining case is *Tiverton Estates Ltd v Wearwell*[65] where Lord Denning MR said '. . . for over a hundred years, the courts have held that the effect of the words "subject to contract" is that the matter remains in negotiation until a formal contract is executed . . .'.

1.68 The heading on a contractor's negotiating correspondence that the negotiations are subject to contract will, therefore, usually ensure that it is not contractually bound to its opposite number by accident. A situation where there has been an offer and an acceptance for work to be done for consideration will not bring about a binding agreement enforceable by the courts where the correspondence is expressed to be 'subject to contract'. The words have a special meaning in that the documents which are headed 'subject to contract' cannot themselves bespeak a contract. They mean that a further agreement which does not have those words within it must be entered into.

1.69 The cases where the term 'subject to contract' is contained in an exchange of correspondence is to be contrasted with apparent agreements where the parties have stated that the agreement shall be set out in a formal agreement. Unless the execution of a formal agreement is expressed to be a condition precedent to the creation of a contract, the mere fact that the parties expect to or agree to enter into a formal agreement (perhaps under seal) will not prevent the contract which they have entered into from being valid and enforceable.

1.70 Once there has been an offer and an acceptance together with an intention to create legal relations, then there are no real policy reasons why the bargain should not be enforceable by the courts, whether there is consideration given or not.

What Happens When Work is Done and There is No Contract

1.71 Every request for work done (be it for work and materials or professional work by an architect or an engineer) will normally have implied within it a promise to pay for the work. So if an architect is asked to draw up plans and no agreement is made but the architect does so, then the request carries within it the implied promise to pay for the work, and the law will therefore require payment to be made if the work is done.

1.72 This is a very common situation. If a builder is asked to do work in an emergency and no agreement is made, for example, then the builder will be entitled to payment. With professionals, like architects, engineers, or surveyors, the legal solution is the same. However, the situation is normally less likely to be prompted by an emergency and more likely to be brought about by inattention to the formalities of an agreement or indeed their essentials (particularly price). A surveyor may be asked to carry out work with no price mentioned in situations where the employer is not focused on the essentials of an agreement because the

[65] [1975] Ch 146 at 159H.

project is still in its infancy. In that situation the professional is entitled to be paid for the work done.

1.73 Given that there is no contract price agreed, the law implies or infers that there is an obligation to pay a reasonable price for the work. This is frequently referred to (even now that Latin is unfashionable) as quantum meruit. It is a remedy in quasi contract and is part of the law of restitution and is dealt with in Chapter 3. However that remedy is only in the event that no contract has been entered into. If, however, a contract is found to have been entered into notwithstanding that no price has been agreed, then the recovery is not for a quantum meruit but is for sums due under a contract. Frequently when parties fall out when small works are done, an employer will seek to allege that it had overpaid previously submitted invoices and accounts under the contract which has the obligation to pay a reasonable price. In this situation it will usually be difficult to re-open invoices already paid.[66]

The Interpretation of Contracts: Rules as to How They Are Interpreted and What Background It Is Permissible to Look at in Interpreting a Contract

1.74 There are no magic rules of construction applying to building contracts. Rules of contractual interpretation apply to all contracts and there are some clear rules and statements applicable to contractual interpretation which assist in understanding how the courts will interpret a contract which is before them. There are several maxims which are applicable; some must be read with others and there is the whole question of context. In this section we will consider first literalism and reasonableness together, then context, and finally more particular rules of, or aids to, construction and interpretation.

Literalism and reasonableness

1.75 The first and basic maxim is that the courts will give effect to the objective meaning of the words which the parties have chosen to use. This is because the courts are not trying to make a bargain for the parties, but instead are seeking to understand what the parties meant by reference to the words which they chose.[67]

1.76 However that basic rule is tempered by common sense. Lord Diplock said in *Antaios Compania Naviera SA v Salen Rederierna AB*:[68] 'if detailed semantic and syntactical analysis of words in a commercial contract is going to lead to a conclusion that flouts business commonsense, it must be made to yield to business commonsense.' Therefore the words used in their objective sense must be interpreted or construed. The words of Lord Diplock are an example of construction. If the express words of the contract will lead to an unreasonable

[66] See for example *Furmans Electrical Contractors v Elecref Ltd* [2009] EWCA Civ 170.

[67] *Wickman Machine Tool Sales Ltd v L Schuler AG* [1974] AC 235. See, for example in a case applying *Wickman* in an insurance context, Lord Denning MR in *Attica Sea Carriers Corp v Ferrostaal Poseidon Bulk Reederei GmbH* [1976] 1 Lloyd's L Rep 250 at 253: 'The parties can, by clear words, provide that complete performance of a particular stipulation can be a condition precedent. But, in the absence of clear words the court should look to see which of the rival interpretations gives the more reasonable result.'

[68] [1985] AC 191 at 201.

result, this is unacceptable. The courts have adopted the rule that they will not interpret the contract so as to produce a result which is unreasonable, unless it is clear that is what the parties actually want. That doctrine is succinctly contained in the speech of Lord Reid in *Wickman Machine Tool Sales Ltd v L Schuler AG*:[69]

> The fact that a particular construction leads to a very unreasonable result must be a relevant consideration. The more unreasonable the result the more unlikely it is that the parties can have intended it, and if they do intend it the more necessary it is that they shall make that intention abundantly clear.

1.77 It is fair to say that the above rules overarch all contractual interpretation. On the one hand you have the express words used and, on the other, you may have an absurd result if there is too much literalism.[70] There is a balance to be struck. Lord Mustill in *Charter Reinsurance Co Ltd v Fagan*[71] referred to the above statement by Lord Reid and said:

> This practical rule of thumb (if I may so describe it without disrespect) must however have its limits. There comes a point at which the court should remind itself that the task is to discover what the parties meant from what they have said, and that to force upon the words a meaning which they cannot fairly bear is to substitute for the bargain actually made one which the court believes could better have been made. This is an illegitimate role for a court. Particularly in the field of commerce, where the parties need to know what they must do and what they can insist on not doing, it is essential for them to be confident that they can rely on the court to enforce their contract according to its terms.

1.78 Subject to the assorted aids to construction which are dealt with below, reasonableness (and its twin, absurdity) must be considered within context. There have been many instances where the literal meaning of words has been modified or even rejected because of context, if a literal approach to interpretation would lead to an unreasonable result. In *Wickman Machine Tool Sales Ltd v L Schuler AG*,[72] it was held that the words 'it shall be a condition of this agreement' did not mean that it was a fundamental term of the agreement which is what lawyers would understand by the term 'a condition'. The context here was that S Co, a German company, entered into a contract with W Co, an English company, giving W Co the sole rights to sell S Co's panel presses in England. A contractual clause provided that 'it shall be a condition of this agreement' that W Co's representatives should visit six named firms each week to solicit orders. W Co's representatives failed on a few occasions to do so. S Co claimed to be entitled to repudiate the agreement, on the basis that a single failure was a breach of condition, giving them an absolute right to treat the contract as at an end. The House of Lords held that such a breach did not entitle S Co to repudiate, since such a construction of the clause was so unreasonable that the parties could not have intended it. Conversely in *Antaios Compania Naviera SA v Salen Rederierna AB*,[73] it was held that the words 'on any breach of this charterparty' did not mean any breach, but any repudiatory breach. The context of that case was that the owners of a vessel purported to withdraw it

[69] [1974] AC 235 at 251.
[70] *Charter Reinsurance Co Ltd v Fagan* [1997] AC 313.
[71] Ibid at 388.
[72] [1974] AC 235.
[73] [1985] AC 191.

13 days after discovering that the charterers had mistakenly issued incorrect bills of lading. The House of Lords held that, on a true construction of the relevant clause, it only applied to repudiatory breaches and therefore did not apply to this breach which was non-repudiatory. Therefore it can be seen that, whilst the touchstone of contractual interpretation might be the words themselves, context and reasonableness may lead to results which a literal reading of the contract would not permit.

The best discussion of literalism and its folly comes from Lord Steyn in *Sirius International Insurance Co v FAI General Insurance Ltd*:[74] **1.79**

> There has been a shift from literal methods of interpretation towards a more commercial approach. In *Antaios Compania Naviera SA v Salen Rederierna AB* [1985] AC 191, Lord Diplock, in an opinion concurred in by his fellow Law Lords, observed (at 201): 'if detailed semantic and syntactical analysis of a word in a commercial contract is going to lead to a conclusion that flouts business common sense, it must be made to yield to business common sense.' In *Mannai Investment Co Ltd v Eagle Star Life Assurance Co Ltd* [1997] AC 749, at 771, I explained the rationale of this approach as follows:
>
> > In determining the meaning of the language of a commercial contract . . . the law . . . generally favours a commercially sensible construction. The reason for this approach is that a commercial construction is more likely to give effect to the intention of the parties. Words are therefore interpreted in the way in which a reasonable commercial person would construe them. And the standard of the reasonable commercial person is hostile to technical interpretations and undue emphasis on niceties of language.
>
> The tendency should therefore, generally speaking, be against literalism. What is literalism? It will depend on the context. But an example is given in *The Works of William Paley*, 1838 edn, Vol III, 60. The moral philosophy of Paley influenced thinking on contract in the 19th century. The example is as follows: the tyrant Temures promised the garrison of Sebastia that no blood would be shed if they surrendered to him. They surrendered. He shed no blood. He buried them all alive. This is literalism. If possible it should be resisted in the interpretative process.

Context and 'factual matrix': what is admissible and what is inadmissible evidence of context

Context has always been troubling for the courts. A court must decide what evidence it can look at outside the four corners of the contract documents in question in order to ascertain the meaning of its contents. On the one hand, it must be remembered that a court is not seeking to rewrite the contract which the parties have made; it is seeking to interpret that contract. But, on the other hand, as a general principle, the meaning of words cannot be ascertained if those words are divorced from their context. Lawyers, who might describe themselves as rigorous and be described as austere, frequently try to exclude as much extraneous matter as possible. It might be thought that this would lead to greater certainty. But it often does not, given the infinite variety of ways in which people can express themselves. **1.80**

---

[74] [2004] UKHL 54 at [19].

1.81 There have been many distillations of what is admissible extrinsic evidence. One of the most helpful is in *Arbuthnott v Fagan*[75] where Sir Thomas Bingham MR said:

> Courts will never construe words in a vacuum. To a greater or lesser extent, depending on the subject matter, they will wish to be informed of what may variously be described as the context, the background, the factual matrix or the mischief. To seek to construe any instrument in ignorance or disregard of the circumstances which gave rise to it or the situation in which it is expected to take effect is in my view pedantic, sterile and productive of error. But that is not to say that an initial judgment of what an instrument was or should reasonably have been intended to achieve should be permitted to override the clear language of the instrument, since what an author says is usually the surest guide to what he means. To my mind construction is a composite exercise, neither uncompromisingly literal nor unswervingly purposive: the instrument must speak for itself, but it must do so in situ and not be transported to the laboratory for microscopic analysis.[76]

1.82 Admissible surrounding circumstances must at least be known, or capable of being known, to both parties at the time when the contract is made.[77] The admissible evidence is in order to obtain the objective or aim or genesis of the transaction.[78] The admissible evidence does not include the parties' subjective intentions. These may change from time to time. Nor does it include the negotiations between the parties, because negotiations are fluid matters of give and take where also the parties' positions may change from time to time. The courts are looking for the common intention of the parties divined by what they expressed and not what each party may have thought.[79] Lord Wilberforce in *Prenn v Simmonds*[80] put it this way:

> In my opinion, then, evidence of negotiations, or of the parties' intentions, . . . ought not to be received, and evidence should be restricted to evidence of the factual background known to the parties at or before the date of the contract, including evidence of the 'genesis' and objectively the 'aim' of the transaction.

However in *Chartbrook Ltd v Persimmon Homes Ltd and others* [2009] UKHL 38 it was pointed out that the remedy of Rectification was available if the written contract did not properly reflect the parties' consensus. Evidence of negotiations is admissible in an action for Rectification.

1.83 The ambit of admissible evidence has been explained by Lord Hoffmann in *Investors Compensation Scheme Ltd v West Bromwich Building Society*[81] where he set out five principles of construction which have been regarded as stating the modern law relating to the interpretation of contracts:

(1) Interpretation is the ascertainment of the meaning which the document would convey to a reasonable person having all the background knowledge which would reasonably

[75] [1996] LRLR 135 (CA, 30 July 1993).

[76] See also *Charrington & Co Ltd v Wooder* [1914] AC 71 at 82 *per* Lord Dunedin.

[77] *Reardon Smith Line Ltd v Yngvar Hansen-Tangen (trading as HE Hansen-Tangen)* [1976] 1 WLR 989.

[78] See *Chartbrook Ltd and another v Persimmon Homes Ltd and another* [2009] UKHL 38 affirming *Prenn v Simmonds* [1971] 1 WLR 1381.

[79] Staughton LJ put it succinctly in *Scottish Power v Britoil*, The Times, 2 December 1977 '. . . the court is looking for the *common* intention of the parties, and not what intention each had *in pectore*'.

[80] [1971] 1 WLR 1381 at 1385.

[81] [1998] 1 WLR 896.

have been available to the parties in the situation in which they were at the time of the contract.

(2) The background was famously referred to by Lord Wilberforce as the 'matrix of fact', but this phrase is, if anything, an understated description of what the background may include. Subject to the requirement that it should have been reasonably available to the parties and to the exception to be mentioned next, it includes absolutely anything which would have affected the way in which the language of the document would have been understood by a reasonable man.

(3) The law excludes from the admissible background the previous negotiations of the parties and their declarations of subjective intent. They are admissible only in an action for Rectification.[82] The law makes this distinction for reasons of practical policy and, in this respect only, legal interpretation differs from the way we would interpret utterances in ordinary life. The boundaries of this exception are in some respects unclear. But this is not the occasion on which to explore them.

(4) The meaning which a document (or any other utterance) would convey to a reasonable man is not the same thing as the meaning of its words. The meaning of words is a matter of dictionaries and grammars; the meaning of the document is what the parties using those words against the relevant background would reasonably have been understood to mean. The background may not merely enable the reasonable man to choose between the possible meanings of words which are ambiguous but even (as occasionally happens in ordinary life) to conclude that the parties must, for whatever reason, have used the wrong words or syntax: see *Mannai Investments Co Ltd v Eagle Star Life Assurance Co Ltd* [1997] AC 749.

(5) The 'rule' that words should be given their 'natural and ordinary meaning' reflects the common sense proposition that we do not easily accept that people have made linguistic mistakes, particularly in formal documents. On the other hand, if one would nevertheless conclude from the background that something must have gone wrong with the language, the law does not require judges to attribute to the parties an intention which they plainly could not have had.

This formulation has recently been reaffirmed as definitive.[83] In the recent case of *Chartbrook* **1.84**
Ltd and another v Persimmon Homes Ltd and another [2009] UKHL 38, the House of Lords was invited by the defendant, a developer, to take into account the parties' pre-contractual negotiations for the purpose of interpreting the contract. The claimant company was the owner of development land and the defendants were developers. They entered into a contract whereby the defendant agreed to obtain planning permission for the claimant's land and then to enter into possession, construct a mixed residential and commercial development, and sell the properties on long leases. The claimant agreed to grant the leases at the direction of the defendant, who would receive the proceeds for their own account and pay the claimant an agreed price for the land. Planning permission was granted and the development was built. The dispute arose about a term of the contract which provided for

[82] See *Chartbrook Ltd and another v Persimmon Homes Ltd and another* [2009] UKHL 38.

[83] See, for example, *Mannai Investment Co Ltd v Eagle Star Assurance Co Ltd* [1997] AC 749 at 779 *per* Lord Hoffmann.

an 'additional residential payment', which was a term defined in the contract. The defendant calculated the sum due under that heading as £897,051 whereas the claimant claimed to be entitled to £4,484,862. The claimant brought proceedings against the defendant for the unpaid balance of the additional residential payment which it claimed to be owed. The defendant sought to persuade their Lordships to take account of pre-contractual negotiations. In the alternative, it contended that, in light of the context and background, it would be clear to a reasonable person there must have been a mistake in the language of the contract. Their Lordships allowed the defendant's appeal on the latter ground, refusing to depart from the long-standing rule that excluded evidence of pre-contractual negotiations. It was emphasized that the courts would not easily accept that people had made linguistic mistakes, particularly in formal documents. However, in some cases, the context and background drove a court to conclude that something must have gone wrong with the language. In such a case the law did not require a court to attribute to the parties an intention which a reasonable person would not have understood them to have had. In this case, to interpret the definition of 'additional residential payment' in accordance with ordinary rules of syntax made no commercial sense. It was clear that something had gone wrong with the language and the defendant's interpretation of the contract was that which a reasonable person would have understood the parties to have meant.

1.85 The law as it stands today, therefore, is that subject to those two exceptions (negotiations and subjective intentions), commercial contracts, and with them construction contracts, are construed in the light of all the background which could reasonably have been expected to be available to the parties in order to ascertain what would objectively have been understood to be their intention.[84] There is a limitation upon this, namely, that before evidence is sought to be admitted, care must be taken to consider exactly why the evidence is said to assist and to consider whether its relevance is sufficiently cogent to the determination of the joint intention of the parties to justify having regard to it.[85]

Individual Rules of Construction and Interpretation

1.86 The law has accumulated many maxims of legal interpretation and these are examined below. However, each must be looked at carefully against the over-arching rules of construction set out above that: (a) the words the parties use is the primary guide to intention, but (b) those words will be looked at in context, and (c) interpreted in line with the intention of avoiding absurdity. But it is to be noted that it has been held that the JCT Practice Notes are no aid to the construction of a JCT Minor Works standard form.[86] Therefore it can be inferred that the contractual context does not include such documents which are issued by the publishers of standard forms.

---

[84] *Mannai Ltd v Eagle Star Assurance Co Ltd* [1997] AC 749 at 779.

[85] See Lord Phillips MR in *The Tychy* [2001] 2 Lloyd's Rep 403 at [29]. See also Staughton LJ in *Scottish Power v Britoil, The Times*, 2 December 1977, in which he decried the citation of masses of pre-contract documentation.

[86] *TFW Printers Ltd v Interserve Project Services Ltd* [2006] EWCA Civ 875.

The contract will be read as a whole This is almost axiomatic. 1.87

Errors and mistakes of meaning and syntax will be corrected or accommodated to make the agreement in line with the intention of the parties.[87] 1.88

An uncommercial bargain will not be taken to be the intention of the parties without very clear language to indicate that it should be.[88] 1.89

Print over type, writing over printing Most construction contracts use standard terms and have other special or bespoke terms supplementing them, particularly Bills of Quantities, Specifications, and Drawings. In those circumstances the bespoke terms will normally prevail over the standard terms.[89] The exception is where the printed terms clearly state otherwise, as for example they did in clause 12 of the 1963 Edition of the Joint Contracts Tribunal (JCT) Standard Form of Building Contract.[90] The reason is fairly prosaic, namely that the courts will give effect to the words which the parties have chosen specially for their contract. 1.90

Inconsistencies and repugnancy It has been stated that where there is inconsistency which cannot be reconciled, the earlier clause will prevail over the later one. This is incorrect. The principle is taken from Lord Wrenbury in the Privy Council[91] where he said: 'If in a deed an earlier clause is followed by a later clause which destroys altogether the obligation created by the earlier clause, the later clause is to be rejected as repugnant and the earlier clause prevails.' 1.91

This statement might be thought of as too wide given that most contracts today are made with the aid of computers which means that the order of the stated obligations is often cut and pasted. The principle, however, may be a good one if all it means is that the parties will not be taken to have destroyed their obligations through the addition of later wording. But that is an aspect of the doctrine of repugnancy, which is that the courts will not give effect to a term of the contract if it is repugnant to the primary obligation as set out in the contract documents. The better view today is that parties should construe the contract as a whole and deal with inconsistencies on their individual merits without reference to the place of the inconsistent clauses in the contract documents. 1.92

Recitals may be resorted to in cases of ambiguity Most construction contracts contain recitals and articles of agreement which are printed ahead of the detailed provisions of the contract conditions and other documents. Usually these articles of agreement or recitals do no more than state the primary essential terms, namely, that the builder has agreed to build, who the contract professional is, and what the contract sum is. Recitals usually state the 1.93

[87] *Investors Compensation Scheme Ltd v West Bromwich Building Society* [1998] 1 WLR 896 at 914; and *Mannai Ltd v Eagle Star Assurance Co Ltd* [1997] AC 749 at 775.

[88] *Beaufort Developments v Gilbert Ash* [1999] AC 266 at 282.

[89] *Homburg Houtimport BV v Agrosin Private Ltd* [2004] 1 AC 715. See also *Glynn v Margetson & Co* [1893] AC 351; and *Universal Steam Navigation Co Ltd v James McKelvie & Co* [1923] AC 492.

[90] *English Industrial Estates Corp v George Wimpey & Co Ltd* (1972) 7 BLR 22, the particular provision stating 'nothing in the Contract Bills shall override or modify that which is contained in these Conditions'.

[91] *Forbes v Git* [1922] AC 256 at 259.

objective or aim of the transaction. It has been said in an ancient case that the recitals cannot be resorted to if the express words of the contract are clear or there is no doubt about what they mean.[92] Such a statement does not stand analysis with the modern law on the admission of extrinsic evidence outlined above, in which it has been held that the extrinsic evidence is admissible in all cases and not simply in cases of ambiguity.[93]

1.94 It must follow that it would be illogical if the recitals to a contract were not looked at.

1.95 **Latin Part 1: the ejusdem generis 'rule'** (Latin for 'of the same kind'). This rule of construction applies to resolve the problem of giving meaning to groups of words where one of the words is ambiguous or inherently unclear. It provides that, in a contract where a list of words is set out referring to items of a particular class and they are followed by general words, then the general words are treated as referring to matters of the same class. The rule springs from a 'presumption against surplusage', namely, if the general words have an unrestricted meaning, then the enumerated items are surplusage.[94] In fact the 'rule' has no place in commercial contracts because there is no reason to suppose that commercial parties will not allow for surplusage in their contracts.[95] The arrival of the computer has made the 'rule' even less compelling. It is a perfectly permissible aid to the construction of wills and legislation (primary and secondary) because such intentions can be expected to come from the drafters of wills and statutes, but not for commercial contracts. In relation to the JCT Standard Form, the rule was not applied to the then clause 23(h), the extension of time clause giving an extension to artists and tradesmen engaged by the employer, for that reason.[96] The rule can be seen to have been applied in old cases but there is no case where it has been applied in a commercial context. Indeed when the rule is sought to be applied in a commercial context, it is usually ignored or distinguished, see for example *Effort Shipping Co Ltd v Linden Management SA*.[97]

1.96 **Latin Part 2: contra proferentem 'rule'** This is a rule which means in English 'against the party putting forward the document'. If there is ambiguity or doubt, then the party which drafted any particular clause or which put forward any clause will have the ambiguity resolved against it. The rule was applied, for example by Edmund Davies LJ in *Billyack v Leyland Construction Co Ltd*,[98] who said: 'If there be any ambiguity in the building contract drawn up by the defendants in the present case, it must, of course, be interpreted

92 See Brett LJ in *Leggott v Barrett* (1880) 15 Ch D 306 at 311.

93 *Arbuthnott v Fagan* (CA, 30 July 1993) where Steyn LJ said:

> I readily accept [Counsel's] submission that the starting point of the process of interpretation must be the language of the contract. But [Counsel] went further and said that, if the meaning of the words is clear, as he submitted it is, the purpose of the contractual provisions cannot be allowed to influence the court's interpretation. That involves approaching the process of interpretation in the fashion of a black-letter man. The argument assumes that interpretation is a purely linguistic or semantic process until an ambiguity is revealed. That is wrong.

94 See Devlin J in *Chandris v Isbrandtsen-Moller Co Inc* [1951] 1 KB 240 at 245.

95 Ibid.

96 *Henry Boot Construction Ltd v Central Lancashire New Town Development Corp* (1980) 15 BLR 1.

97 [1998] AC 605.

98 [1968] 1 WLR 471 at 477.

contra proferentem.'[99] This rule is to be contrasted with the rules of construction relating to exemption clauses and extensions of time and liquidated damages which are dealt with separately. Given the appellate endorsement of Edmund Davies LJ, it might be thought that the rule has some force. But the modern approach to contractual construction is against the application of the rule, which was more appropriate to be used when there was a much more restricted recourse to the surrounding circumstances and extrinsic evidence in that context.[100] The 'rule' also gives problems to the court in deciding who the 'proferens' (the party putting forward the document) is.

In the construction context in particular, where parties use standard form contracts that have been produced by bodies fully representative of the interests of prospective parties, the 'rule' has become largely defunct.[101] For building works, the JCT standard form contracts are used as principal or main contracts and sub-contracts. For civil engineering projects the ICE Conditions of Contract are commonly used or adapted for use by major employers. The Government generally uses its own sets of general conditions for works contracts, which encompass all aspects of building works, in conjunction with its own standard forms for inviting and accepting tenders and for recording non-standard terms. As a result, there is no real place for the 'rule' in modern construction law. The modern approach to contractual construction is set out in *Investors Compensation Scheme Ltd v West Bromwich Building Society*,[102] namely, that its use will only be in special circumstances. It has been applied by the Court of Appeal in *Lexi Holdings plc v Stainforth* [2006] EWCA Civ 988 but that decision discussed the rare nature of the application of the 'rule'. There have also been other applications of the doctrine. But although it is more often prayed in aid and rejected it has found favour occasionally.[103] **1.97**

Implied Terms and Impermissible Terms

Terms are implied into contracts by operation of law. The law operates in two ways, first by statute and, secondly, by the application of common law principles. Just as terms are implied into contracts by operation of law, so too are some terms outlawed. Outlawed terms are dealt with in this section, albeit that there is overlap with the non-enforcement of certain terms. (Those are dealt with separately, although reference is made to them.) **1.98**

---

99 See also the rule being applied in *Cook v Shoesmith* [1951] 1 KB 752; and *Skillion plc v Celltech Industrial Research Ltd* [1992] EGLR 123 at 125.

100 See for example *C&J Clark International Ltd v Regina Estates Ltd* [2003] EWHC 1622 (Ch) *per* Neuberger J at [40]–[43].

101 See *Tersons Ltd v Stevenage Development Corpn* [1963] 2 Lloyd's Rep 333 at 368, CA *per* Pearson LJ.

102 [1998] 1 WLR 896.

103 More recently Coulson J in *William Hare Ltd v Shepherd Construction Ltd* [2009] EWHC 1603 (TCC) was faced with the suggestion of its application but found the contract clear enough not to need to use it. Akenhead J applied it in *CJP Builders Ltd v William Verry Ltd* [2008] EWHC 2025 (TCC) (15 August 2008). As did Coulson J in *YJL London Ltd v Roswin Estates* [2009] EWHC 3174 (TCC). The Court of Appeal has applied it in an insurance context, in *Pratt v Aigaion Insurance Co SA* [2008] EWCA Civ 1314; and also in a commercial context in *Whitecap Leisure Ltd v John H Rundle Ltd* [2008] EWCA Civ 429; and *Elektrim SA v Vivendi Holdings 1 Corp* [2008] EWCA Civ 1178. Aside from a holiday case and those cited here, the rule has not been applied by the Court of Appeal, the language of the contract being sufficient.

1.99 There is a separate category of implied terms, namely, one implied by a course of dealing. Terms will be dealt with in the following order: first, terms implied by course of dealing; second, terms implied by statute; third, terms implied by the common law; and finally, outlawed terms.

A course of dealing

1.100 Terms can be implied into a contract by a course of dealing. This is obtained by way of an objective examination of contractual intention from the actions of the parties. If over a period the parties have always dealt on certain terms and conditions, then when they make a further bargain the terms and conditions upon which they have previously contracted are likely to be implied.[104] This is not invariable, however, because the objective contractual intention might be otherwise. If, for example, the parties expressly agree that the former terms and conditions will not apply to the transaction under consideration, then the course of dealing will be of no effect. A course of dealing normally establishes the implication of terms and conditions when there is a long history of contracting on those terms and conditions. The omission of express reference is taken to be an error or omission which they will 'of course' agree to rectify.[105] Most of the cases on a course of dealing are concerned with exemption clauses. A course of dealing is likely to be established with the use of delivery notes always referring to particular terms and conditions, for example with crane hire or the supply of materials to site.[106]

Terms implied by statute

1.101 There are two major pieces of legislation which cause terms to be implied into construction contracts and contracts between employers and professionals. The first is the Supply of Goods and Services Act 1982 and the second is the Housing Grants, Construction and

---

[104] *McCutcheon v David Macbrayne Ltd* [1964] 1 WLR 125; and applied in *Hardwick Game Farm v SAPPA* [1969] AC 31 at 90 where Lord Morris said:

> . . . the learned judge, after considering the case of *McCutcheon v David Macbrayne Ltd* [1964] 1 WLR 125, HL, held that the conditions in the contract note were not incorporated into the contracts of sale. In agreement with all the members of the Court of Appeal I consider that they were. Over the course of a long period prior to the three oral contracts which are now in question SAPPA knew that when Grimsdale sold they did so on the terms that they had continuously made known to SAPPA. In these circumstances it is reasonable to hold that when SAPPA placed an order to buy they did so on the basis and with the knowledge that an acceptance of the order by Grimsdale and their agreement to sell would be on the terms and conditions set out on their contract notes to the extent to which they were applicable.

In that case see also Sellers LJ [1966] 1 WLR 287 at 308, and Diplock LJ (at 339).

[105] In *British Crane Hire v Ipswich Plant Hire* [1975] QB 303 at 311 Lord Denning MR said:

> From that evidence it is clear that both parties knew quite well that conditions were habitually imposed by the supplier of these machines: and both parties knew the substance of those conditions. In particular that if the crane sank in soft ground it was the hirer's job to recover it: and that there was an indemnity clause. In these circumstances, I think the conditions on the form should be regarded as incorporated into the contract. I would not put it so much on the course of dealing, but rather on the common understanding which is to be derived from the conduct of the parties, namely, that the hiring was to be on the terms of the plaintiffs' usual conditions.

[106] Ibid.

Regeneration Act 1996.[107] Those in the 1982 Act are more traditional implied terms relating to performance. The 1996 Act is more wide ranging and seeks to deal with long-standing issues. Those who adhere to a freedom of contract approach might find this difficult. Others think it is a helpful attempt to deal with long-standing vices in the construction industry and the courts' processes.

Supply of Goods and Services Act 1982

Sections 13 to 16 provide for four implied terms. The Act applies to contracts for the supply of a service and this catches contracts for professional services and building contracts. **1.102**

Section 13 provides that there is an implied term that the work will be carried out with reasonable skill and care. **1.103**

Section 14 provides that there is an implied term that the service will be carried out within a reasonable time. **1.104**

Section 15 provides that, where a price is not agreed, there is an implied term that there will be a reasonable charge for the service. **1.105**

Section 16 provides that the above implied terms yield to any express term to the contrary and any course of dealing between the parties. The section also provides that the terms do not affect any rule of law which imposes stricter duties than those set out in ss 13 to 15. **1.106**

These implied terms, yielding as they do to the express terms of the contract, are uncontroversial and accord with anyone's reasonable and honest expectation of what is intended. Few could argue with much conviction, for example, that reasonable skill and care would not be expected of anyone who provides services. The contrary would be uncommercial. **1.107**

Housing Grants, Construction and Regeneration Act 1996: Part II

The relevant sections and the statutory scheme are set out in full in Appendices 1–3[108] to this book. The 1996 Act modifies considerably several traditional rights and obligations. It does not apply to operations outside the United Kingdom. It provides for rights in a way which is not traditional, in that it permits the parties, by their contract, to make express provision for the matters dealt with by the Act, but provides that if they do not do so or do not do so adequately, then the provisions of the Act will apply irrespective of the provisions of the contract or the intentions of the parties. **1.108**

The 1996 Act applies to all construction contracts, as defined by the Act, entered into after 1 May 1998. It gives certain rights which are statutory in nature. Therefore, strictly speaking, the provisions of the statute are not implied terms but are obligations which are **1.109**

---

[107] Note that this Act is subject to prospective amendment by the Local Democracy, Economic Development and Construction Act 2009. The 1996 Act, as intended to be amended is set out in App 2 to this book; the original text of the Act is set out in App 1. The new provisions are to come into force on a date to be prescribed by the Secretary of State. See s 143(1) of the Local Democracy, Economic Development and Construction Act 2009. It is possible that the new provisions will not be brought into force. If they are, then the fact will appear on this book's website.

[108] See n 107 above relating to prospective amendment by the Local Democracy, Economic Development and Construction Act 2009.

impressed upon the parties by statute. However, where the parties make an agreement which makes sufficient provision for the statutory requirement, then the statutory requirement need not apply and the terms of the contract will. At all events, rights are given to the contractor, which cannot be abrogated by agreement. These are set out below.

1.110 **The right to adjudication (section 108).** This is dealt with in detail in Chapter 23, so no more is said about it here. Suffice it to say that, with the exception of a contract with a residential occupier of a building and particular construction contracts (usually with a PFI element to them),[109] in all construction contracts and all contracts for professional services (such as architectural work, design work, engineering work, surveying work, interior and exterior decoration, and landscaping work), which are made or evidenced in writing as defined by the 1996 Act,[110] there is a right to adjudication. The requirement that the contract be in writing is subject to prospective statutory repeal on a date to be fixed by the Secretary of State.[111]

1.111 **The right to stage payments (section 109).** The old law used to be that the contractor had to complete its obligations before being paid the contract sum. The terminology used was that of the 'entire contract'.[112] This was a harsh doctrine which resulted in the doctrine of 'substantial completion', whereby the courts would permit a claim for the contract price if substantial rather than absolute completion had been reached.[113] The entire contract doctrine was, of course, totally uncommercial. All standard forms now provide for interim or stage payments to the contractor as the work progresses. Whilst there have been occasions where the courts have had to consider a single lump sum payment in exchange for completion, these are very rare. Apart from the obvious reason, namely, that contractors are not finance houses able to finance the construction of expensive structures, there is also the problem of confidence, namely the confidence of the builder that he will actually get paid when the works are completed given the possible financial situation of the employer. The 1996 Act enshrines a right to stage payments in all contracts, unless the contract provides that the work shall take less than forty-five days or it is agreed that the work is estimated to take less than forty-five days. The section permits the parties a right to agree the quantity and interval of stage payments in the absence of which the Scheme (particularly paragraphs 2 to 4 thereof) applies.[114]

---

[109] As provided by s 106(1)(b) of the 1996 Act and, if ever brought into force by the Secretary of State pursuant to the Local Democracy, Economic Development and Construction Act 2009, by s 106A of the 1996 Act.

[110] s 107 of the 1996 Act defines a contract made in writing as including: (a) an agreement in writing; (b) an agreement made by an exchange of correspondence in writing; (c) an agreement evidenced in writing, including being recorded in writing by a third party with the authority of the parties to the agreement; (d) an agreement which is made by reference to terms which are in writing; and (e) the situation where an agreement in writing is alleged in adjudication proceedings and not denied by the other party.

[111] See the prospective repeal of the requirement in the Local Democracy, Economic Development and Construction Act 2009.

[112] *Cutter v Powell* (1795) 6 Term Rep 320.

[113] *H Dakin & Co Ltd v Lee* [1916] 1 KB 566; and *Hoenig v Isaacs* [1952] 2 All ER 176. See particularly Denning LJ at 180–1: 'When a contract provides for a specific sum to be paid on completion of specified work, the courts lean against a construction of the contract which would deprive the contractor of any payment at all simply because there are some defects or omissions . . .'

[114] The Scheme is set out in App 2 to this book.

The right to have an adequate mechanism for determining what payments become due and when, and a final date for payment (section 110(1)).[115] If the contract does not make provision for this, then the Scheme[116] will apply. The section permits the parties to agree how long the period is between the date a sum becomes due and the final date for payment. 1.112

The right to receive payment unless a proper notice has been served (section 111).[117] Payment may not be withheld after the final date payment becomes due unless an effective notice of intention to withhold payment, in proper form and in time, has been made (such notice may be given in accordance with s 110(2) of the 1996 Act or otherwise in compliance with s 111).[118] The notice to withhold must be given in time. Such time can be agreed or, failing agreement, is to be that provided for by the Scheme, namely that it must be no later than seven days before the final date for payment of that sum.[119] The notice must be in a form which specifies the amount to be withheld and the grounds for withholding payment. If more than one ground is relied upon, then each ground must be specified. In the absence of a proper notice, there is no defence to a claim for payment. A set off is a withholding of payment.[120] Section 111 does not remove the obligation of the contractor to prove that the sum claimed is in fact due.[121] But if a certificate is issued in accordance with the contract showing a sum due, then that is sufficient.[122] The purpose of the legislation is to ensure quick payment on an interim basis and thus employers cannot go behind the certificate in resisting payment when a notice has not been served. Nevertheless, if they serve a notice in accordance with s 111 then the prohibition against non-payment by the section does not apply. The mischief behind this provision is the habit employers of contractors and sub-contractors had of asserting a right to set off against sums due long after the sum became due. In order to avoid the draconian provisions prohibiting non-payment, it is relatively easy for the employer to give notice. A well-planned assertion of set off can still succeed. But when this provision is taken with the entitlement to adjudication, it has considerable force. Originally, prior to this legislation, the asserted set off, at the start of any argument between employer and contractor, would often change as the original arguments raised would be seen to be unmeritorious. The shifting sands of the employer's resistance to payment would not attract any real penalty as the litigation process went on its way. But this provision makes it less easy to alter the reasons for non-payment to suit perceived circumstances. If the grounds are not sustainable or are abandoned, then notice will be ineffective. As previously noted the Local Democracy, Economic Development and 1.113

[115] Note the prospective amendment of this section in the Local Democracy, Economic Development and Construction Act 2009. These provisions include the outlawing of pay-when-certified clauses. See the prospective s 110(1A) of the 1996 Act.

[116] See paras 2 to 8 of the Scheme.

[117] Note the prospective amendment of this section (to be on a date prescribed by the Secretary of State) in the Local Democracy, Economic Development and Construction Act 2009. The contractual requirements as to notices are potentially subject to considerable amendment by the Local Democracy, Economic Development and Construction Act 2009 if the provisions are brought into force. See ss 110A and 110B as well as the new s 111.

[118] *SL Timber Systems Ltd v Carillion Construction Ltd* [2001] BLR 517.

[119] See para 10 of the Scheme.

[120] *VHE Construction v RBSTB Trust Co* [2000] BLR 187.

[121] *SL Timber Systems Ltd v Carillion Construction Ltd* [2001] BLR 517.

[122] *Rupert Morgan Building Services v David and Harriet Jervis* [2004] BLR 19, [2004] 1 WLR 1867.

Construction Act 2009 provides for this section to be amended on a day to be appointed by the Secretary of State.

1.114 In the recent case of *Reinwood Ltd v L Brown & Sons Ltd* [2008] UKHL 12, the House of Lords held an employer was, in accordance with a withholding notice served pursuant to provisions in a JCT construction contract, entitled to deduct from a payment certificate a sum representing damages for non-completion, even if the contract had subsequently been granted an extension of time. The extension did not operate retrospectively to invalidate the withholding notice. The appellant contractor (B) had been engaged by R pursuant to a JCT standard form contract. The contract specified a completion date and provided for extensions of time, damages for non-completion, and determination by the contractor on specified defaults. In December 2005, B applied for an extension of time. The project's architect issued a certificate of non-completion and an interim certificate for payment. Pursuant to the contract, the final date for payment of the sum due under that certificate was 25 January 2006. On 17 January, R served a withholding notice stating its intention to withhold liquidated and ascertained damages from B and, on 20 January, R paid B the sum stipulated in the certificate, less damages. Three days later the architect, pursuant to B's earlier application, retrospectively granted an extension of time until 10 January. B argued that the effect of the extension of time was to reduce the amount that R was entitled to withhold, but R made no further payment. B sought to determine the contract on the basis of R's failure to pay the full sum due under the interim certificate by the final date for payment. Their Lordships held that the extension of time did not have the effect of denying R the right to rely on the non-completion certificate. There was no doubt that if the extension had been granted before 11 January, R would not have been entitled to deduct the damages resulting from the non-completion because that certificate would have been cancelled as a result of the extension by the time of the issue of the interim certificate. On the face of it, the clause providing for damages for non-completion was clear. Provided certain preconditions were satisfied, the employer was entitled to give notice specifying the amount it proposed to withhold. Although not spelt out, it had to be the case that both parties should be entitled to proceed on the basis that payment could properly be made in accordance with that notice. Any other conclusion would fly in the face of commercial common sense. The notices required by the contract were those specified in the Housing Grants, Construction and Regeneration Act 1996, ss 110 and 111, and part of the purpose of those sections was to enable parties to a construction contract to know where they stood. B's submissions were inconsistent with that purpose. Whilst the effect of the extension of time was to cancel the non-completion certificate, such a cancellation was not retrospective in its effect.

1.115 If there is a dispute as to non-payment, then the best remedy is an adjudication followed by enforcement proceedings. An adjudicator ought to be able to get a decision out quickly on the question. If there is an arbitration clause, then an action in the courts for the sum withheld is vulnerable to an application for a stay pursuant to s 9(4) of the Arbitration Act 1996.[123] There is no objection in principle to court action where summary judgment can

[123] *Collins (Contractors) Ltd v Baltic Quay Management (1994) Ltd* [2004] EWCA Civ 1757.

be given as it was in *VHE Construction v RBSTB Trust Co.*[124] But where there is an arbitration clause, as many construction contracts have, court proceedings will be met by an application for a stay which will succeed and much time will be wasted. Arbitration today can be a swift process, and is recognized as having that potential by the courts.[125]

Section 111 does not permit the giving of a withholding notice in respect of sums awarded by way of an adjudicator's decision.[126] The adjudicator's decision is directly enforceable by summary judgment and is not subject to a stay to arbitration. **1.116**

The right to suspend work if there is non-payment (section 112).[127] Prior to this legislation there was no right to suspend work in the event of non-payment.[128] This is because you cannot approbate and reprobate a contract. If you have promised to perform the contract, you cannot yourself refuse to perform if the other party has not done so. You may accept a repudiation, but unless you do so there is nothing for it but to perform the contract. Section 112 provides that if a sum is due under a construction contract and there is no effective notice to withhold payment then the contractor has the right to suspend performance of its obligations. At least seven days' notice must be given of the intention to suspend and the right to suspend ceases when payment in full is made. The suspension period and delay caused thereby (directly or indirectly) is to be disregarded in computing any contractual time limit including the time to complete. Therefore the time when the works are suspended will effectively be the subject of an extension of time. Equally if there is a knock-on delay (the suspension causing a contractor to lose its 'slot' in a sub-contractor's workshop for example), that too will be the subject of an extension of time. There has, so far, been hardly any litigation concerning the operation of this section.[129] This is surprising because the drafting is somewhat opaque. The sense of the section is clear and that might be the explanation. However, in the context of a difficult economic climate, the acts of withholding payments and suspending performance may well become more common, and litigation on the matter may increase. **1.117**

Terms Implied at Common Law

As has already been seen the courts do not make contracts for the parties: they interpret them. Parties' freedom of contract can be interfered with by statute. Terms can also be implied into contracts by common law. It is necessary to see by what touchstone the courts **1.118**

---

[124] [2000] BLR 187.

[125] See the remarks of Saville LJ in *Hayter v Nelson and Home Insurance Co* [1990] 2 Lloyd's Rep 265; and Clarke LJ in *Collins (Contractors) Ltd v Baltic Quay Management (1994) Ltd* [2004] EWCA Civ 1757 at [46].

[126] *Construction Centre Group Ltd v Highland Council* [2002] BLR 476.

[127] Note that this is subject to the possibility of a small prospective amendment by the Local Democracy, Economic Development and Construction Act 2009.

[128] *Canterbury Pipelines v Christchurch Drainage* (1979) 16 BLR 76 (decision of the Court of Appeal in New Zealand with a particularly impressive judgment by Lord Cooke).

[129] An exception is *Levolux At Ltd v Ferson Contractors Ltd* [2002] BLR 341, in which HHJ Wilcox held that a sub-contractor had been entitled to suspend works under s 112(1) as the respondent contractor had failed to pay monies owing under the contract and had issued a defective withholding payment notice which was non-compliant with s 111.

will imply terms, and then examine some familiar terms in common form which have been traditionally implied into construction contracts. However, care should be taken with some of those terms. Many 'traditional terms' may not be implied into particular construction contracts simply because they do not pass the fundamental test which brings about the implication of terms. Although in mercantile contracts terms are implied by usage or common understanding, these are not relevant here. Therefore the question of the common law implying terms will be approached in two stages. First, an examination of the basic criteria for implication of terms, and secondly, an examination of common or 'traditional' terms.

Criteria for the implication of a term into a contract: the basic criterion

1.119 The law in relation to implied terms has been recently re-stated by Lord Hoffmann in the Privy Council in *Attorney General of Belize & Ors v Belize Telecom Ltd & Anor* [2009] UKPC 11. The opinion of Lord Hoffmann can only be described as seminal.[130] Whether a term is to be implied is a matter of interpretation or construction of the contract and is to be considered as such as a single test rather than, as was thought of before, a series of tests. Overturning or re-interpreting much of the conventional wisdom in relation to the implication of terms Lord Hoffmann expressed himself in the following terms:

> The question of implication arises when the instrument does not expressly provide for what is to happen when some event occurs. The most usual inference in such a case is that nothing is to happen. If the parties had intended something to happen, the instrument would have said so. Otherwise, the express provisions of the instrument are to continue to operate undisturbed. If the event has caused loss to one or other of the parties, the loss lies where it falls.
>
> In some cases, however, the reasonable addressee would understand the instrument to mean something else. He would consider that the only meaning consistent with the other provisions of the instrument, read against the relevant background, is that something is to happen. The event in question is to affect the rights of the parties. The instrument may not have expressly said so, but this is what it must mean. In such a case, it is said that the court implies a term as to what will happen if the event in question occurs. But the implication of the term is not an addition to the instrument. It only spells out what the instrument means.
>
> The proposition that the implication of a term is an exercise in the construction of the instrument as a whole is not only a matter of logic (since a court has no power to alter what the instrument means) but also well supported by authority. In *Trollope & Colls Ltd v North West*

---

[130] Sir Anthony Clarke MR said in *Mediterranean Salvage and Towage Ltd v Seamar Trading and Commerce Inc* [2009] EWCA Civ 531 at paras 8 and 9 of his judgment:

> The correct approach to the question when to imply a term into a contract or other instrument, including therefore a charterparty, has recently been considered by Lord Hoffmann, giving the judgment of the Judicial Committee of the Privy Council, which also comprised Lord Rodger, Baroness Hale, Lord Carswell and Lord Brown, in *Attorney General of Belize v Belize Telecom Ltd* [2009] UKPC 11. I predict that his analysis will soon be as much referred to as his approach to the construction of contracts in *Investors Compensation Scheme v West Bromwich Building Society* [1998] 1 WLR 896 at 912–3. His analysis in the *Belize* case is extensive: see [16] to [27].
>
> It repays detailed study but for present purposes it is I think sufficient to say that the implication of a term is an exercise in the construction of the contract as a whole: see *Trollope & Colls Ltd v North West Metropolitan Hospital Board* [1973] 1 WLR 601, at 609 *per* Lord Pearson, with whom Lord Guest and Lord Diplock agreed and *Equitable Life Assurance Society v Hyman* [2002] 1 AC 405, at 459.

Metropolitan Regional Hospital Board [1973] 1 WLR 601, at 609 Lord Pearson, with whom Lord Guest and Lord Diplock agreed, said:

> [T]he court does not make a contract for the parties. The court will not even improve the contract which the parties have made for themselves, however desirable the improvement might be. The court's function is to interpret and apply the contract which the parties have made for themselves. If the express terms are perfectly clear and free from ambiguity, there is no choice to be made between different possible meanings: the clear terms must be applied even if the court thinks some other terms would have been more suitable. An unexpressed term can be implied if and only if the court finds that the parties must have intended that term to form part of their contract: it is not enough for the court to find that such a term would have been adopted by the parties as reasonable men if it had been suggested to them: it must have been a term that went without saying, a term necessary to give business efficacy to the contract, a term which, though tacit, formed part of the contract which the parties made for themselves.

More recently, in *Equitable Life Assurance Society v Hyman* [2002] 1 AC 408, at 459, Lord Steyn said 'If a term is to be implied, it could only be a term implied from the language of [the instrument] read in its commercial setting.'

It follows that in every case in which it is said that some provision ought to be implied in an instrument, the question for the court is whether such a provision would spell out in express words what the instrument, read against the relevant background, would reasonably be understood to mean. It will be noticed from Lord Pearson's speech that this question can be reformulated in various ways which a court may find helpful in providing an answer—the implied term must 'go without saying', it must be 'necessary to give business efficacy to the contract' and so on—but these are not in the Board's opinion to be treated as different or additional tests. There is only one question: is that what the instrument, read as a whole against the relevant background, would reasonably be understood to mean?

There are dangers in treating these alternative formulations of the question as if they had a life of their own. Take, for example, the question of whether the implied term is 'necessary to give business efficacy' to the contract. That formulation serves to underline two important points. The first, conveyed by the use of the word 'business', is that in considering what the instrument would have meant to a reasonable person who had knowledge of the relevant background, one assumes the notional reader will take into account the practical consequences of deciding that it means one thing or the other. In the case of an instrument such as a commercial contract, he will consider whether a different construction would frustrate the apparent business purpose of the parties. That was the basis upon which *Equitable Life Assurance Society v Hyman* [2002] 1 AC 408 was decided. The second, conveyed by the use of the word 'necessary', is that it is not enough for a court to consider that the implied term expresses what it would have been reasonable for the parties to agree to. It must be satisfied that it is what the contract actually means.

The danger lies, however, in detaching the phrase 'necessary to give business efficacy' from the basic process of construction of the instrument. It is frequently the case that a contract may work perfectly well in the sense that both parties can perform their express obligations, but the consequences would contradict what a reasonable person would understand the contract to mean. Lord Steyn made this point in the *Equitable Life* case (at p 459) when he said that in that case an implication was necessary 'to give effect to the reasonable expectations of the parties.'

The same point had been made many years earlier by Bowen LJ in his well known formulation in *The Moorcock* (1889) 14 PD 64, 68:

> In business transactions such as this, what the law desires to effect by the implication is to give such business efficacy to the transaction as must have been intended at all events by both parties who are business men.

Likewise, the requirement that the implied term must 'go without saying' is no more than another way of saying that, although the instrument does not expressly say so, that is what a reasonable person would understand it to mean. Any attempt to make more of this requirement runs the risk of diverting attention from the objectivity which informs the whole process of construction into speculation about what the actual parties to the contract or authors (or supposed authors) of the instrument would have thought about the proposed implication. The imaginary conversation with an officious bystander in *Shirlaw v Southern Foundries (1926) Ltd* [1939] 2 KB 206, at 227 is celebrated throughout the common law world. Like the phrase 'necessary to give business efficacy', it vividly emphasizes the need for the court to be satisfied that the proposed implication spells out what the contact would reasonably be understood to mean. But it carries the danger of barren argument over how the actual parties would have reacted to the proposed amendment. That, in the Board's opinion, is irrelevant. Likewise, it is not necessary that the need for the implied term should be obvious in the sense of being immediately apparent, even upon a superficial consideration of the terms of the contract and the relevant background. The need for an implied term not infrequently arises when the draftsman of a complicated instrument has omitted to make express provision for some event because he has not fully thought through the contingencies which might arise, even though it is obvious after a careful consideration of the express terms and the background that only one answer would be consistent with the rest of the instrument. In such circumstances, the fact that the actual parties might have said to the officious bystander 'Could you please explain that again?' does not matter.

In *BP Refinery (Westernport) Pty Ltd v Shire of Hastings* (1977) 180 CLR 266, at 282–283 Lord Simon of Glaisdale, giving the advice of the majority of the Board, said that it was 'not . . . necessary to review exhaustively the authorities on the implication of a term in a contract' but that the following conditions ('which may overlap') must be satisfied:

> (1) it must be reasonable and equitable; (2) it must be necessary to give business efficacy to the contract, so that no term will be implied if the contract is effective without it; (3) it must be so obvious that 'it goes without saying' (4) it must be capable of clear expression; (5) it must not contradict any express term of the contract.

The Board considers that this list is best regarded, not as a series of independent tests which must each be surmounted, but rather as a collection of different ways in which judges have tried to express the central idea that the proposed implied term must spell out what the contract actually means, or in which they have explained why they did not think that it did so. The Board has already discussed the significance of 'necessary to give business efficacy' and 'goes without saying'. As for the other formulations, the fact that the proposed implied term would be inequitable or unreasonable, or contradict what the parties have expressly said, or is incapable of clear expression, are all good reasons for saying that a reasonable man would not have understood that to be what the instrument meant.

1.120 The cases referred to by Lord Hoffmann included *Liverpool City Council v Irwin*;[131] *Trollope & Colls v North West Metropolitan Regional Hospital Board*;[132] and *BP Refinery (Westernport) v Shire of Hastings*.[133] In the first edition of this book it was submitted that each said the

[131] [1977] AC 239.

[132] [1973] 1 WLR 601.

[133] [1978] 52 ALJR 20, (1990) 106 LQR 179 at 182. Approved by the Court of Appeal in *Countryside Residential (North Thames) Ltd v Christina Tugwell* (2000) 34 EG 87; and *Rosehaugh Stanhope v Redpath Dorman Long* (1990) 50 BLR 75 at 87.

same thing in different ways and that the fundamental touchstone was not whether the term was reasonable, but whether it was necessary.[134] This must now be regarded as either incorrect or too rigid a formulation. Necessity may be a touchstone, but it is one which arises from a process of interpretation and understanding what the parties meant. That does not mean that the courts will no longer look at necessity as being decisive in most cases.[135] But it will do so as a matter of interpretation having regard to the limits of how far the courts will go in interposing matters which the parties themselves have not specifically addressed. As Lord Bingham MR said in in *Phelps Electronique Grand Public SA v British Sky Broadcasting Ltd* [1995] EMLR 472:

> The courts' usual role in contractual interpretation is, by resolving ambiguities or reconciling apparent inconsistencies, to attribute the true meaning to the language in which the parties themselves have expressed their contract. The implication of contract terms involves a different and altogether more ambitious undertaking: the interpolation of terms to deal with matters for which, ex hypothesi, the parties themselves have made no provision. It is because the implication of terms is potentially so intrusive that the law imposes strict constraints on the exercise of this extraordinary power.

Criteria for the implication of a term into a contract: traditional terms examined

There are several implied terms which might be described as 'traditional' in that they are frequently said to be implied into construction contracts. However, care must be taken to see whether in fact a term is implied as a matter of construction in each case. That a term may be implied into one contract does not mean that it will be implied into every contract. Each contract must be looked at having regard to its express terms. The express terms of the contract may mean that a term which would traditionally be implied is not in fact implied in the particular case. The more basic or 'incomplete' the agreement, the more likely that a term will need to be implied as the court will be filling a gap to either make the contract work or to accord with the intention of the parties. When there are detailed contract conditions, then the room for implied terms recedes. **1.121**

'*Mackay v Dick* terms' are terms which involve the employer agreeing that it will do all that is necessary on its part to permit the completion of the contract works. As Lord Blackburn said in his judgment:[136] **1.122**

> I think I may safely say, as a general rule, that where in a written contract it appears that both parties have agreed that something shall be done, which cannot effectually be done unless both concur in doing it, the construction of the contract is that each agrees to do all that is necessary to be done on his part for the carrying out of that thing, though there may be no express words to that effect. What is the part of each must depend on circumstances.

[134] See also *The Moorcock* (1889) 14 PD 64.

[135] As the Court of Appeal did in *Mediterranean Salvage and Towage Ltd v Seamar Trading and Commerce Inc* [2009] EWCA Civ 531 but applying Lord Hoffmann's Opinion in *Belize.*

[136] *Mackay v Dick* (1881) 6 App Cas 251 at p 263.

Lord Watson said:[137]

> The respondents were only entitled to receive payment of the price of the machine on the condition that it should be tried at a proper working pace provided by the appellant, and that on trial it should excavate a certain amount of clay or other soft substance within a given time. They have been thwarted in the attempt to fulfil that condition by the neglect or refusal of the appellant to furnish the means of applying the stipulated test; and their failure being due to his fault, I am of opinion that, as in a question with him, they must be taken to have fulfilled the condition. The passage cited by Lord Shand from *Bell's Principles of the Laws of Scotland*, 8th ed. (1884), paragraph 50 to the effect that, 'If the debtor bound under a certain condition have impeded or prevented the event, it is held as accomplished. If the creditor had done all that he can to fulfil a condition which is incumbent on himself, it is held sufficient implement,' expresses a doctrine, borrowed from the civil law, which has long been recognised in the law of Scotland, and I think it ought to be applied to the present case.

1.123 It should be noted that this case is a Scottish case in which the language of Lord Watson is not that of an implied term but seems to relate to civil law concepts in Scots law, and the language of Lord Blackburn is of construction rather than implied term. Nevertheless, the case has since been looked at as an implied term case, usually in relation to charterparty cases and not in relation to construction contracts. A better view is that the case relates to fictional non-fulfilment of a condition subsequent.[138]

1.124 One of the '*Mackay v Dick* terms' often referred to in the textbooks is possession of the site. However, that is not always an implied term. Many standard forms require possession of the site to be given expressly. Indeed, some standard forms and bespoke forms expressly state that the contractor shall not have exclusive possession of the site. It will depend on the circumstances as to what degree of possession is in fact required for the contract works to be carried out. However, in the absence of any specific mention of it in the contract, reasonable possession of the site will have to be implied, otherwise there is no sensible way that the contractor can execute its obligations. But that analysis should be one of necessity and not non-hindrance. A non-hindrance term is circular and unnecessary, as pointed out by Devlin J in *Mona Oil Equipment and Supply Co Ltd v Rhodesia Railways Ltd*:[139]

> The second term alleged is that the defendants would do nothing to prevent or obstruct the performance of the condition of payment in the Messrs. Turnbull, Gibson & Co. clause. This is a common form of allegation, but I venture to think that in the light of the decision in *Luxor (Eastbourne) Ltd v Cooper* [1941] AC 108 its use could now with advantage be restricted to the exceptional case. It is generally regarded as originating from the dictum of Willes J, in *Inchbald v Neilgherry Coffee, Tea & Cinchona Plantation Co, Ltd* (1864) 17 CBNS 733. It was applied in *Trollope (George) & Sons v Martyn Bros* [1934] 2 KB 436 but with the qualification that the prevention must be 'without just cause or excuse.' In *Luxor (Eastbourne) Ltd v Cooper* [1941] 1 All ER 33 the right qualification was said by Lord Wright to be that the prevention must be wrongful. Lord Wright explains what this means in the following passage—and then Devlin J cited a passage from Lord Wright's judgment that I have cited and went on—'This

[137] Ibid, p 271.

[138] See *New Zealand Shipping Co Ltd v Société des Ateliers et Chantiers de France* [1919] AC 1; and *Thompson v Asda* [1988] Ch 241. See also *Luxor (Eastbourne) Ltd v Cooper* [1941] AC 108; and *Mona Oil Equipment and Supply Co Ltd v Rhodesia Railways Ltd* [1949] 2 All ER 1014.

[139] [1949] 2 All ER 1014 at 1016–18.

> shows that, except possibly in the rare cases where the wrongful act alleged is independent of the contract, the allegation of prevention is only circumlocution. Where the wrongful act is a breach of the contract, it can stand alone. There is no advantage in alleging an implied term not to break another term. Indeed, it is absurd, since it is implicit in this form of allegation that the breach of the second term is objectionable only if it prevents the performance of a condition . . .' In truth, the proposed term, like all other implied terms, must be judged by the test whether or not it is necessary for the business efficacy of the contract. The fact that an act, if not prohibited by the contract, is one which would result in a party being robbed of the benefits which otherwise the contract would give him is certainly an important matter to be considered in relation to the business efficacy of the contract, but it is not necessarily the most important, and it is certainly not the only matter. There are many decided cases in which it has not prevailed . . . Since *Luxor (Eastbourne) Ltd v Cooper* it is clear that the court cannot by means of an implied term prohibit arbitrary behaviour or acts done 'without just cause or excuse,' for the reason that such matters are not capable of sufficiently precise definition. For the same reason the court cannot by such means exact a higher degree of co-operation than that which can be defined by reference to the necessities of the contract.

That analysis is impeccable.[140] Therefore, again, the touchstone is necessity as a matter of construction. This and other *Mackay v Dick* terms can be analysed in that light. **1.125**

Sometimes the term is expressed more widely than possession, as a 'duty of necessary cooperation.' A particular act or omission is then examined to see whether it is a breach of that term or not.[141] For example, an obligation sometimes expressed is that the employer will not interfere with the certifier and must ensure that its architect or certifier applies the contract properly.[142] One might ask whether this is necessary today when there is the right to adjudication or arbitration, where the adjudicator or arbitrator can review and revise certificates.[143] Equally it can simply be a breach of the obligation to provide a contract professional who carries out the proper functions of the contract as assigned to the employer and its agent under the contract. The contract professional must act honestly and reasonably. As a matter of interpretation, the parties cannot intend that the contract professional (architect, engineer, or surveyor) need not act honestly and reasonably.[144]Another aspect of cooperation is the giving of instructions to the contractor so that it can execute the work on time and so as not to hinder its performance.[145] Again express terms of standard forms usually deal with this aspect of the matter. **1.126**

It is also common to plead an implied term that the employer will not hinder the contractor in the execution of its work. This is too broad a plea. Hindrance takes many forms and is **1.127**

[140] Applied by Cooke J in *Imamovic v Cinergy Global Trading Ltd* [2006] EWHC 323 (Comm) para 175.

[141] For discussion and application of these terms, see *Hamlyn v Wood* [1891] 2 QB 488; *Holland Hannen & Cubitts v Welsh Health Technical Services Organisation* (1983) 18 BLR 80 particularly at 117; and *London Borough of Merton v Stanley Hugh Leach* (1985) 32 BLR 51 particularly at 81.

[142] See *Croudace Ltd v London Borough of Lambeth* (1986) 33 BLR 20; and see also *Panamena v Leyland* [1947] AC 428; *Perini v Commonwealth of Australia* (1969) 12 BLR 82; and *Minster Trust v Trapps Tractor* [1954] 1 WLR 963 at 975 *per* Devlin J.

[143] See *Lubenham v South Pembrokeshire District Council* (1986) 33 BLR 39.

[144] See, for example, a similar term being implied by Gloster J in *Socimer International Bank Ltd (in liquidation) v Standard Bank London* [2006] EWHC 718 (Comm).

[145] See *Glenlion Construction Co Ltd v Guiness Trust* (1987) 39 BLR 89; *J Fee Ltd v Express Lift Co* (1993) 34 Con LR 147; and *Neodox v Swinton and Pendlebury Borough Council* (1958) 5 BLR 38 (Diplock J).

seldom anything other than specific. For example, lack of proper possession of the site is an act of hindrance. But it need not be a breach of an implied term, since, as discussed, express terms will deal with site possession. Lack of information is similarly usually dealt with by way of express terms.

1.128 It is sometimes pleaded as an implied term that information will be given so that the contractor can proceed 'expeditiously, economically and in accordance with the contract'.[146] Again, this is too wide a term. There is nothing which requires a contractor to proceed expeditiously and economically albeit that it is axiomatic that the contractor ought to proceed in accordance with the contract.[147] There is no implied term that a contractor will proceed with diligence.[148] It has been said that a term will be necessarily implied that neither party will prevent the other from performing the contract.[149] The conclusive test for *MacKay v Dick* terms, in summary, is that they must pass the test of necessity (as a matter of construction). In the words of Stocker LJ, in *Rosehaugh Stanhope v Redpath Dorman Long* dealing with a specific contract where express terms covered things:[150]

> It is very well known that the courts readily imply a *MacKay v Dick* term, but I am not for my part persuaded that it would be appropriate to do so here. A term will only be implied where it is necessary to give business efficacy to the contract, so that no term will be implied if the contract is effective without it . . . It seems to me that on a reasonably commercial construction of clause 21(1), and particularly sub-clause (1)(b), any lack of co-operation by the plaintiffs would entitle the defendants to reimbursement under that clause. If that is so, then there is no necessity to imply a term for breach of which damages would lie.

1.129 The older cases which do not apply the rigour of this position ought to be treated with caution.

1.130 Whilst it has been seen above that necessity must be shown before a non-hindrance term can be implied, in one case a non-hindrance term has been implied into a licence agreement where the licensor's parent company asked the supplier of the licencee not to supply essential equipment to it.[151] It must be said that the case, decided in a hurry by the Court of Appeal ten days after the issue of the writ, has all the hallmarks of haste and enthusiasm that Lord Denning MR could bring to cases. The case has not been applied in relation to the implied term point, although it is useful on such matters as extra territorial jurisdiction and interference with contractual rights. It cannot be regarded as authoritative in relation to an implied term of non-hindrance.

1.131 Fitness of work and materials terms are truly traditional and, some may argue, do not fall into the necessity category. If that view is correct (a debate which may be better suited to take place amongst academics) then it is the only exception to the necessity rule. It has been

---

146 A similarly formulated term was contended for in *Martin Grant and Co Ltd v Sir Lindsay Parkinson and Co Ltd* (1984) 29 BLR 31 and rejected on the basis that the express terms of the contract did not leave any room for such a term.

147 See for example *Glenlion*, n 142 above.

148 *Greater London Council v Cleveland Bridge and Engineering Co Ltd* (1986) 34 BLR 50, 8 Con LR 30 (CA); and *Multiplex Constructions (UK) Ltd v Cleveland Bridge UK Ltd and another* [2006] EWHC 1341 (TCC).

149 See the obiter dicta of Lord Asquith in *William Cory and Sons Ltd v London Corp* [1951] 2 KB 476 at 484.

150 (1990) 50 BLR 75 at 87.

151 *Acrow Ltd v Rex Chainbelt Inc* [1971] 1 WLR 1676.

established by authority for hundreds of years that there is normally an implied term in construction contracts that materials will be of good quality and reasonably fit for their intended purpose. Once stated, however, it is necessary to know that this is not an invariable rule and may not in fact be applicable, depending upon the circumstances.

The basic term has been described by du Parcq J, sitting in the Divisional Court, in *GH Myers and Co v Brent Cross Service Co*[152] as follows: **1.132**

> I think that the true view is that a person contracting to do work and supply materials warrants that the materials which he uses will be of good quality and reasonably fit for the purpose for which he is using them, unless the circumstances of the contract are such as to exclude any such warranty.

That statement has been approved by the House of Lords[153] and applied ever since and can be described as settled law. **1.133**

The sting in the tail of the warranty relates to whether the circumstances are such as to exclude any warranty. Those circumstances have, to date, been twofold. **1.134**

The first exclusion relates to skill and judgment. Implicit in the implied term is reliance upon the skill and judgment of the party supplying the work and materials. If there is not reliance upon the skill and judgment of the builder, then the warranty as described is not given.[154] This has been best described by Bayley J in *Duncan v Blundell*[155] (a case about the supply and fitting of a stove) when he said: 'Where a person is employed in a work of skill, the employer buys both his labour and his judgment; he ought not to undertake the work if it cannot succeed, and he should know whether it will or not. Of course it is otherwise if the party employing him chooses to supersede the workman's judgment by using his own.' It follows that if the selection of materials is by the employer, then the builder will not be liable if the materials turn out not to be fit for their purpose because their skill and judgment has not been relied upon in that respect. Of course that does not mean that if the supplied materials suffered from a latent defect the builder would not be liable. The latent defect would be caught by the warranty as to merchantability.[156] Those words of Bayley J are also applicable to the question of the warranty of the builder that it can execute the work, in a traditional type of contract. **1.135**

The second exclusion of the warranty relates to other special circumstances in which the employer requires the builder to divest itself of effective recourse against suppliers or subcontractors for breach of warranty.[157] The rationale behind this exclusion is that normally an employer is in contractual relations with its builder which is in turn in contractual relations with its supplier, which in turn is in contractual relations with its sub-supplier and so on. This means that if the builder supplies materials which are defective, then it has **1.136**

---

[152] [1934] 1 KB 46 at 55.

[153] *Young & Marten v McManus Childs* [1969] 1 AC 454.

[154] See *Cammell Laird & Co Ltd v Manganese Bronze and Brass* [1934] AC 402; and *Rotherham MBC v Frank Haslam Milan* (1996) 78 BLR 1.

[155] (1820) 3 Stark 6.

[156] See *Young & Marten v McManus Childs* [1969] 1 AC 454; and *Rotherham MBC v Frank Haslam Milan* (1996) 78 BLR 1.

[157] *Gloucestershire County Council v Richardson* [1969] 1 AC 480 is a case where a nominated supplier's conditions substantially limited its liability to the contractor.

recourse against its supplier and so on down the chain. There is no direct remedy for the employer against the supplier. The chain of responsibility therefore provides remedy, each member of the chain bearing the risk of the insolvency of the link below it. If, however, the employer deprives the builder of its remedy against the supplier, then it cannot have been the intention of the parties that the employer would have a remedy against the builder but that the builder should have no remedy against its supplier. The Court of Exchequer Chamber in *Readhead v Midland Railway Co*[158] said: 'Warranties implied by law are for the most part founded on the presumed intention of the parties, and ought certainly to be founded on reason, and with a just regard to the interests of the party who is supposed to give the warranty, as well as of the party to whom it is supposed to be given'.

1.137 Therefore, applying the law as stated then, the parties cannot be said to have contracted on terms which would have exposed the builder to the risk that its supplier would not perform and that it would have no recourse. The better view is that it would be unreasonable to require a term of quality and fitness to be implied when the builder has no recourse, because such a term would not be equitable and reasonable. Reliance on skill and judgment is often a matter of degree. Sometimes there is a large amount of reliance on skill and judgment, in which case fitness for purpose of the completed works is to be implied.[159] Sometimes there is little or none, as for example, when the contractor is asked to build to a detailed design specified by the employer's professional.[160] There are many cases in the books where the courts have had to look on one or other side of the line, but each case must be looked at on its own facts in order to see how far the warranties go so far as fitness for purpose is concerned. Obviously if there is reliance upon the contractor's skill and judgment, then fitness for purpose is more likely to be implied. If, however, the design is that of the employer or its architect, then the responsibility for the defective design does not lie with the contractor because the contractor's skill and judgment has not been relied upon.

1.138 The concept (in the first edition described as a heresy) of 'duty to warn' implied terms grew up in several first instance decisions in the 1980s and early 1990s.[161] The concept was that it was an implied term of a building contract that the builder would act with the reasonable skill and care of an ordinary competent builder, and allied with that that the builder would warn the employer's professional or the employer of any defects in design of which it was aware or which it believed to exist. These implied terms have no support from appellate English authority and do not pass the necessity test and it is submitted that they cannot be held to truly exist. Not only is the necessity test not complied with, but also there is clearly no reliance on the skill and judgment of the contractor in construction cases let in a traditional form where the employer's professional designs and the builder builds. There can be no warrant for the imposition of an implied term that the builder will obtrude into

[158] (1869) LR 4 QB 379 at 392.

[159] See, for example, *Greaves & Co v Baynham Meikle* [1975] 1 WLR 1095; *Test Valley BC v Greater London Council* (1979) 13 BLR 63; *IBA v EMI and BICC* (1980) 14 BLR 1; and *Basildon v Lesser* [1985] 1 All ER 20.

[160] *Lynch v Thorne* [1956] 1 WLR 303.

[161] *Equitable Debenture Life Assurance Society v William Moss* (1980) 2 Const LR 1; and *The Victoria University of Manchester v Hugh Wilson and Lewis Womersley* (1985) 2 Con LR 43; and *Edward Lindenburg v Joe Canning* (1992) 62 BLR 147.

the sphere of design when the contractual structure excludes the builder from that role. Coulson J in *J Murphy & Sons v Johnson Precast Ltd* [2008] EWHC 3024 (TCC) did hold that such a duty existed. That case related to the supply of pipe and his reasoning was shortly stated as follows:

> In my judgment, there is no reason why, as a matter of law, and as a matter of common sense, a supplier in the position of JP would not owe a similar duty to the contractor who was buying his GRP pipe.

It is respectfully submitted that the decision is wrong. Suppliers' obligations to contractors can be, and are, dealt with by the Sale of Goods Act 1979. However, it has been held by the Court of Appeal in *CGA Brown v Carr and another* [2006] EWCA Civ 785 that an allegation of a failure to carry out work with reasonable skill and care comprehended an allegation that an instruction given to a contractor in drawings and plans was inadequate. However that case's decision was not just based upon the failure to spot a defect in the drawings but also the making of a defective joint. The case is, however, the nearest there is to appellate authority on the issue of the duty to warn.

Impermissible Terms

Penalties and exemption clauses

The law prohibits penalty clauses and unfair exemption clauses. These are dealt with in separate chapters. **1.139**

'Pay when paid' clauses

Section 113 of the Housing Grants, Construction and Regeneration Act 1996 prohibits 'conditional payment provisions' unless the condition is of payment by a third party who is insolvent. These are termed 'pay when paid' clauses and normally state that the contractor is not obliged to pay the sub-contractor unless and until it has received payment by the employer in respect of the sub-contractor's work. The terms used to be in the form of a proviso following the payment clause. The larger national contractors in the 1980s and 1990s had 'pay when paid' clauses as standard terms and they were nearly always accepted by sub-contractors in a competitive situation. The clauses were a source of dispute and were frequently a last resort by a main contractor in the event that it was resisting payment to the sub-contractor. Ironically, whilst the clauses attracted some academic commentary in the construction law journals and whilst they caused a great deal of anger in the industry, they never were tested by a court's decision except in one case.[162] The reason for this may have been an unwillingness of the larger national contractors to have these clauses the subject of judicial decision. American cases suggested that they were not in fact enforceable in that they would be construed as not having the effect of preventing payment of the sub-contractor. But the main reason for the lack of case law may have been the difficulty, on the part of main contractors, in showing that the non-payment by the **1.140**

[162] *Durabella Ltd v J Jarvis and Sons* (2001) 83 Con LR 145.

employer was not its own fault or the fault of other sub-contractors. Therefore if the main contractor pleaded that it had not been paid by the employer as a reason for non-payment, the sub-contractor was often able to show that the employer's non-payment was brought about by the default of the main contractor. In those circumstances the 'pay when paid' provision would not bite because there was not actually non-payment, but rather an appropriation of sums due by the employer against a claim for damages against the contractor. Additionally there was always the risk that the court would hold that the 'pay when paid' clause was ineffective because the contractor failed to obtain its remedies against the employer. All these arguments were present in *Durabella Ltd v J Jarvis and Sons*.[163]

1.141 Given the prevalence of these clauses, and their outlawing by s 113 of the 1996 Act, it must be doubtful whether there will be any further decisions on their effectiveness. They now only bite in the case of insolvency: if a main contractor has a 'pay when paid' clause and the employer becomes insolvent, then the clause may be effective.

Entire agreement clauses

1.142 In the 1970s and 1980s, in contracts involving mergers and acquisitions or the sale of companies and businesses, there grew up the habit of inserting into them contract terms known as 'entire agreement' clauses. A typical entire agreement clause would say:

> This agreement contains the entire agreement between the parties in relation to the subject matter thereof and supersedes all prior agreements and arrangements in connection with the works, and the parties acknowledge that they have not entered into this agreement in reliance on warranties, representations or undertakings howsoever and to whomsoever made.

1.143 There are many variants of such clauses. Their commercial purpose in construction contracts is never obvious. They clearly do have a commercial purpose in contracts for the sale of companies or their businesses. They are principally an attempt to exclude any subsequent claims in misrepresentation or any allegations that prior statements or representations became collateral contracts or are to be read into the agreement. There is first instance authority to the effect that an entire agreement clause means that the parties bind themselves to the position that the full contract terms are to be found in the written agreement they have made and nowhere else.[164] However it may be the case that a prior oral agreement, if proved, can find its way into the written agreement, even if there is an entire agreement clause.[165] These alleged oral agreements may be difficult as a matter of evidence to prove, but it is possible that they may still have an effect even in the case of an entire agreement clause. It is suggested that such clauses are not effective to exclude liability for misrepresentation in the event of a fraudulent misrepresentation.[166] But query the effectiveness of such a clause in the event of an innocent misrepresentation. Here the courts do not speak with one voice. One view is that no representation has been made; the other is that the putative evidential bar provided by the clause is no more than an exclusion clause in substance and

[163] Ibid.

[164] *The Inntrepreneur Pub Co v East Crown Ltd* (2000) 41 EG 209, *The Times*, 5 September 2000. See also *Ravennavi Spa v New Century Shipping Co Ltd* [2006] EWHC 733 (Comm).

[165] *Reeds v Norwich Union Insurance Ltd* [2005] EWCA Civ 343.

[166] *Peart Stevenson Associates Ltd v Brian Holland* [2008] EWHC 1868 (QB).

so must pass a reasonableness test. The question must be considered open, though it is submitted that commercial certainty, in the case of parties of equal or almost equal bargaining power, would dictate that the evidential bar is to be effective either way.[167]

Entire agreement clauses do not, however, affect the existence or otherwise of implied terms. Some clauses try and exclude implied terms. However, if the implied term is necessary to make the contract work, then it seems a little difficult to conclude that the parties thought that their contract should not in fact work at all. **1.144**

Rectification

The courts will rectify a written agreement if it can be shown that the written contract does not set out what the parties in fact agreed. This has been re-stated by the House of Lords in *Chartbrook Ltd and another v Persimmon Homes Ltd and another* [2009] UKHL 38. The signed and formal document may be rectified to conform to the true agreement between the parties.[168] It has been stated that convincing proof must be tendered before rectification is granted.[169] But all that really means is that the courts will take as a starting point the view that the parties in fact did write the agreement as they intended, and the burden of proof is upon the party alleging rectification. Necessarily such proof must be to a high standard because the proof will have the effect of unseating the words used by the parties in their written deed. Rectification can be granted if there is common mistake[170] or unilateral mistake.[171] However, where there has been unilateral mistake the test is severe because the courts recognize that there must be a limit beyond which they will not go to assist parties, who are involved in arms length commercial transactions, from their misapprehensions. Effectively this means that there must be some element of unconscionability in the conduct of the other party before the courts will intervene to rectify a commercial agreement.[172] The best example of this is when one of the parties knows about the other's misapprehension and takes advantage of it.[173] **1.145**

The essential ingredient of rectification is that both parties failed to get the agreement to reflect what they intended it to mean.[174] If one of the parties thought it meant one thing and the other party thought it meant another, this does not bring about a right to have the agreement rectified, subject, of course, to unconscionability. For discussion of unilateral mistake and rectification, see Chapter 8. **1.146**

[167] See *Cremdean Properties Ltd v Nash* (1977) 241 EG 837; *EA Grimstead & Son Ltd v McGarrigan* (1998–9) Info TLR 384 (CA); *Watford Electronics Ltd v Sanderson CFL Ltd* [2001] EWCA Civ 317, [2001] 1 All ER (Comm) 696; and *SAM Business Systems Ltd v Hedley and Co* [2003] 1 All ER (Comm) 465.

[168] *American Airlines Inc v Hope* [1974] 2 Lloyd's Rep 301 at 307; and *Jocelyne v Nissen* [1970] 2 QB 86.

[169] *Ernest Scragg and Sons v Perseverence Banking Ltd* [1973] 2 Lloyd's Rep 101, see particularly 103–4; *Thomas Bates and Son Ltd v Wyndhams Ltd* [1981] 1 WLR 505, see particularly 514; *Oceanic Village Ltd v Shirayama Shokusan Co Ltd* [1999] EGCS 83.

[170] *Jocelyne v Nissen* [1970] 2 QB 86.

[171] *Thomas Bates and Son Ltd v Wyndhams Ltd* [1981] 1 WLR 505.

[172] *Oceanic Village Ltd v Shirayama Shokusan Co Ltd* [1999] EGCS 83; and cf *A Roberts & Co Ltd v Leicestershire County Council* [1961] 1 Ch 555.

[173] *Riverlate Properties v Paul* [1975] Ch 133.

[174] For the modern approach to the remedy of Rectification see *Chartbrook Ltd and another v Persimmon Homes Ltd and another* [2009] UKHL 38.

1.147 The standard of proof in rectification cases gets higher the more detailed and complex the commercial agreement is, and where there have been prolonged negotiations over the terms.[175]

1.148 Most arbitration clauses are sufficiently widely drawn so as to allow for the arbitrator to rectify the contract,[176] but each must be looked at to ensure that it is widely drawn enough to allow for this. The question of whether a dispute falls within the arbitrator's jurisdiction turns upon the construction of the relevant clause, and such construction is an objective exercise in contractual interpretation.[177] Arbitration clauses which give the arbitrator power to decide 'disputes arising out of the agreement' or 'in relation to the agreement' will normally give the arbitrator power to rectify the contract.[178] But arbitration clauses which merely give power to decide disputes under the agreement do not.[179] Arbitrators can, under s 30 of the Arbitration Act 1996, rule on their jurisdiction. Therefore the arbitrator may be able to rule that he or she has jurisdiction to rectify the agreement.

1.149 The position is not the same with regard to adjudicators. They do not have power to rule upon their own jurisdiction in the same way that arbitrators do. It may be doubted whether adjudicators will have jurisdiction to order Rectification.[180] But it is submitted that, as a matter of statutory construction they will do so. The intention of the 1996 Act is that all construction contracts which are in writing or evidenced in writing have the right to adjudication. Therefore there ought, in most cases, to be the power to appoint an adjudicator pursuant to an agreement which is susceptible to rectification or indeed to an agreement which is subject to rectification. To hold otherwise would be to deprive certain construction contracts, namely, agreements which are rectified or susceptible to rectification, of the benefit of the 1996 Act.

Estoppel and Waiver[181]

1.150 Estoppel and waiver are closely allied doctrines which govern the situation where the parties' obligations have altered—but there is no consideration—or one party wants, usually for commercial reasons, to resile from a previously agreed state of affairs. Although the two doctrines are frequently viewed as one, they are not.[182]

---

175 *James Hay Pension Trustees Ltd v Kean Hird* [2005] EWHC 1093, Lawrence Collins J.

176 *Ashville Investments v Elmer Construction Ltd* [1989] QB 488; cf *Crane v Hegeman-Harris* [1939] 4 All ER 68. See also *El Nasharty v J Sainsbury plc* [2003] EWHC 2195 (Comm).

177 See *Ashville Investments v Elmer Construction Ltd* [1989] QB 488 at 506; and *X Ltd v Y Ltd* [2005] EWHC 769 (QB) (TCC) at [36].

178 *Faghirzadeh v Rudolph Wolff (SA) (Pty) Ltd* [1977] 1 Lloyd's Rep 630 (a case which had to decide whether a varied agreement came within an arbitration clause); and *A and B v C and D* [1982] 1 Lloyd's Rep 166 (a case dealing with whether a third agreement came within the arbitration clause of the previous two) both decided that the arbitration clauses so worded permitted what amounted to different agreements being arbitrable under the earlier arbitration agreement. This would be consistent with a claim for rectification because ipso facto a rectified agreement is not the original agreement. This was decided authoritatively in *Ashville Investments v Elmer Construction Ltd* [1989] QB 488; and see also *El Nasharty v J Sainsbury plc* [2003] EWHC 2195 (Comm) at [17].

179 *Crane v Hegeman-Harris* [1939] 4 All ER 68; and *Fillite (Runcorn) Ltd v Aqua-Lift* (1989) 45 BLR 27.

180 *Project Consultancy Group v Trustees of the Gray Trust* [1999] All ER (D) 842.

181 A detailed consideration of these doctrines lies outside the scope of this work. Reference should be had to Wilken & Villiers, *The Law of Waiver, Variation and Estoppel*, 2nd edn, OUP, 2002.

182 For a detailed explanation of the differences between the two, see Wilken at Ch 1.

Estoppel involves the concept of someone's mouth being stopped up. The term is taken from the Norman word denoting stopping up. Estoppel is a varied concept. A person can be estopped from denying that they made a particular promise, from denying that a particular set of facts existed or from denying that they had entered into a contract.[183] Denning LJ (who was the midwife of the revival of the concept of estoppel) said in *Charles Rickards Ltd v Oppenhaim*:[184] **1.151**

> If the defendant, as he did, led the plaintiffs to believe that he would not insist on the stipulation as to time, and that, if they carried out the work, he would accept it, and they did it, he could not afterwards set up the stipulation as to the time against them. Whether it be called waiver or forbearance on his part, or an agreed variation or substituted performance, does not matter. It is a kind of estoppel. By his conduct he evinced an intention to affect their legal relations. He made, in effect, a promise not to insist on his strict legal rights. That promise was intended to be acted on, and was in fact acted on. He cannot afterwards go back on it. I think not only that that follows from *Panoutsos v Raymond Hadley Corp of New York* [1917] 2 KB 473, a decision of this court, but that it was also anticipated in *Bruner v Moore* [1904] 1 Ch 305. It is a particular application of the principle which I endeavoured to state in *Central London Property Trust Ltd v High Trees House Ltd* [1947] KB 13.

In *Amalgamated Investment & Property Co Ltd (in liquidation) v Texas Commerce International Bank Ltd*[185] Lord Denning MR said: **1.152**

> When the parties to a transaction proceed on the basis of an underlying assumption—either of fact or of law—whether due to misrepresentation or mistake makes no difference—on which they have conducted the dealings between them—neither of them will be allowed to go back on that assumption when it would be unfair or unjust to allow him to do so. If one of them does seek to go back on it, the courts will give the other such remedy as the equity of the case demands.

Thus, classically for an estoppel to arise, a party (the representor) must make a representation (usually of fact) to the other party (the representee), on which the representee then relies to its detriment. Such an estoppel has been held to arise in relation to practical completion. It was held that parties could not go back on an assumption that practical completion had been achieved because all parties had acted upon a common understanding that practical completion had in fact been achieved.[186]

Waiver by contrast does not require detriment. A waiver will arise where the representor has full knowledge of the facts and makes the unequivocal representation that it will not rely on a particular contractual right or provision. The doctrine was described by Lord Blackburn as follows: **1.153**

> . . . where a party in his own mind has thought that he would choose one of two remedies, even though he has written it down in a memorandum or has indicated it in some other way, that alone will not bind him; but so soon as he has not only determined to follow one of his remedies but has communicated it to the other side in such a way as to lead the opposite party to believe that he has made that choice, he has completed his election and can go no further;

---

[183] See *Mitsui Babcock v John Brown Engineering* (1997) 51 Con LR 129 (QBD).
[184] [1950] 1 KB 617 at 623.
[185] [1982] QB 84 at 122.
[186] *Menolly Investments 3 Sarl v Cerep Sarl (Rev 1)* [2009] EWHC 516 (Ch).

> and whether he intended it or not, if he has done an unequivocal act—I mean an act which would be justifiable if he had elected one way and would not be justifiable if he had elected another way—the fact of his having done that unequivocal act to the knowledge of the persons concerned is an election.[187]

1.154 Waiver or estoppel are often pleaded in construction cases as a matter of last resort. The reason they fail is that the evidence does not support the plea rather than that the dicta are wrong.[188] The pleas usually involve departures from strict legal rights, as for example the dispensing with a condition precedent (such as notice) or barring a party from denying that they agreed to a particular method of performance.

1.155 In relation to waiver of defects, certification or inspection will not create the necessary waiver to disentitle the employer from claiming damages for a number of reasons.[189] First, as set out above, waiver requires knowledge—if the certificate or inspection does not locate the defects there can be no waiver. Second, the certificate or inspection is usually a trigger for payment to the contractor: it is not usually a representation that there are no defects.

1.156 In forfeiture clauses there are often provisions which permit the employer to determine the contractor's employment if it has defaulted in some way. Sometimes the entitlement follows a repetition of a previous default which is subject to a notice. If, however, the contractor does default and the employer does not determine but allows the work to continue, then the employer will be taken to have waived its right to determine in respect of that default.[190] The position will be different, however, where the contractor defaults again. In those circumstances, the employer's ability to rely on subsequent defaults will be entirely unfettered by any waiver. In charterparty cases it is open to argument that there is an implied term that when there is a default the innocent party must exercise its rights to determine within a reasonable time, failing which it loses its rights.[191] However, the criteria as to the business efficacy of an implied term are not the same with charterparties as they are with construction contracts. It must be doubted that there is any need for an implied term to that effect, and it is submitted that absent a waiver or estoppel the right to determine will not be abated once accrued, and if it is abated, it will not be so by an implied term.

---

[187] *Scarf v Jardine* (1882) 7 App Cas 345 at 360–1 *per* Lord Blackburn.

[188] See Wilken and Villiers, *The Law of Waiver Variation and Estoppel*, 2nd edn, OUP, 2002, Ch 19, p 411 'Estoppel'.

[189] See *Clayton v Woodman* [1962] 1 WLR 585; see also *AMF International v Magnet Bowling* [1968] 1 WLR 1025; and *East Ham Corp v Bernard Sunley* [1966] AC 406 at 444. See also Wilken and Villiers, op cit, Ch 19.

[190] See *CMA CGM SA v Beteiligungs-Kommanditgesellschaft MS 'Northern Pioneer' Schiffahrtgesellschaft mbH & Co* [2003] 1 WLR 1015, [2003] 1 Lloyd's Rep 212 at para 55, a charterparty case, but to which the principles in construction cases are equally applicable.

[191] See *KKKK v Bantham Shipping* [1939] 2 KB 554, (1939) 63 Ll L Rep 175 at 193; and *CMA CGM SA*, above, at para 53.

2

STANDARD FORMS OF CONTRACT AND BESPOKE CONTRACTS

Introduction

For many years in the United Kingdom most substantial building contracts were let under standard forms of contract, the RIBA JCT forms, the standard forms issued by the Institution of Civil Engineers, and the government contract known as GC Works 1. **2.01**

2.02 Each form was traditional in style in that the employer's agent, be they architect or engineer, was responsible for the design of the works and the administration of the contract terms, particularly certifying payment as the works progressed, certifying completion, and certifying the date the works ought to have been completed.

2.03 The forms are often described as 'traditional' because of the distinction between the role of the architect or engineer, who was responsible for the design of the works and the role of the contractor whose job was to build what was designed. This was embodied in the distinction between the so-called professional, employed by the employer, and the artisan who was to execute the works.

2.04 During the second half of the 20th century this distinction became less and less apt. Sub-contractors of a specialist nature carried out design and their design and the responsibility for it had to be fitted into the contractual structure. Furthermore, contractors had professionals on their staff that were often as good as or better than the professional engineers and surveyors who were employed by the firm of architects or engineers which was engaged by the employer. Unsurprisingly, contractors were better at contract management than many professionals who were supposed to administer the contracts.

2.05 There is therefore a proliferation of styles and types of contract, aside from the traditional type of contract which is structured around a professional, such as architect or engineer, who provides the design for the contractor to build to. The principal types are management contracts and design and build contracts. But there are many variants which come about through bespoke or 'home made' contracts and sub-contracts which are written by the larger developers and contractors.

2.06 In this chapter we look at the elements which make up the traditional type of contract and then the principal characteristics of the management contract and the design and build contract.

The Elements of the 'Traditional Contract'

2.07 The basic structure of the contract is that the employer provides the design of the works to the contractor which undertakes by its contract to build them.

The basic warranties

2.08 This leads to two basic warranties which are either expressly or impliedly given by each of the parties. The contractor's warranty is that it can build the works as designed. The employer's warranty is that it will provide to the contractor the information which it needs in order to construct the works.

2.09 The contractor's warranty is not usually expressed, but it is implied into any construction contract of a traditional type. The implication of ability comes from the clause contained in all traditional contracts that the builder will carry out and complete the works shown on the contract drawings and described in any specification or bills of quantities. If the contractor is undertaking to do the work, it is necessarily warranting that it can do so. There is normally no excuse of impossibility, save for frustration. The exception to this lies with engineering contracts; the ICE conditions provide that the work is to be executed 'Save insofar as it is physically and legally impossible . . .'. The scope of this clause in permitting impossibility to be used as a reason for the work not being executed has been considered in the courts at first instance.[1] The reason for the clause is obvious when it is understood that the risk of adverse ground conditions which are unforeseeable, or not reasonably foreseeable, lies with the employer in such contracts.

The employer's warranty is often expressed by a clause which provides that, as and when from time to time as necessary, the employer will, through its agent (architect, engineer, or other professional and in this chapter referred to as the contract professional), give to the contractor such further levels, details, drawings, or other information which it shall need to construct the works. Such a term would not be necessary if all the design of the works was completed at the outset, but this is rarely the case in practice. The late provision of information by employers is often a rich source of contention in litigation, arbitration, or adjudication. **2.10**

The balance of provisions set out in this section are terms which will normally appear in one guise or another in a traditional form of contract. **2.11**

Sub-letting and assignment

Some forms will prohibit the sub-letting of any of the work. This is a less practicable provision than formerly. Today most work is done by sub-contractors and the tendering contractor's skill lies in its coordination of sub-contractors. However the clauses which survive will allow sub-contracting (often called sub-letting) provided that the contract professional approves the sub-contractor concerned and provided approval is not unreasonably withheld. **2.12**

The obligations of the contractor are vicariously performed by sub-contractors. However the contractor could, unless there is a prohibition, assign to another the benefit of the contract. That will usually be the assignment of the right to receive the contract sum and periodic payments under the sub-contract. If a contractor has so assigned its rights, there are obvious dangers of insolvency and inability to actually complete the works, which is why such assignment is often prohibited. **2.13**

When there is such a prohibition on assignment, and a contractor has nevertheless assigned the benefit of the contract, the assignee cannot enforce its rights to receive payment against the employer.[2] However a contractor can make an agreement with a third party to account to him for the fruits he receives from the other contracting party.[3] **2.14**

[1] *Turriff Ltd v Welsh National Water* (1979) 1994 Con LYB 122 and noted briefly (1985) 32 BLR 114 at 117; *Yorkshire Water Authority v Sir Alfred McAlpine & Son (Northern) Ltd* (1985) 32 BLR 114.

[2] *Linden Gardens Trust Ltd v Lenesta Sludge Disposals Ltd* [1994] 1 AC 85.

[3] See *In re Turcan* (1889) 40 Ch D 5; and *Don King Productions Inc v Warren* [2000] Ch 291 at 322 *per* Lightman J and 335 *per* Morritt LJ.

Prescription of the standard of work and quality and specification of materials

2.15 Although the specification or bills of quantity and drawings will provide for the identification of the materials and workmanship which is to be provided, there is usually an express term that the work shall be carried out in accordance with the contract documents.

2.16 Allied with this is a requirement that the contract professional may check that this is so. This can be done by normal inspection by a clerk of works or resident engineer (see below). But there can also be a provision that the contract professional may order the work to be opened up for checking. This might involve the uncovering of already executed works (eg taking down part of a wall to check that the cavities are free of debris) but it might also include the taking of samples (eg concrete cores). The clauses will provide that if the work inspected or tested is not in accordance with the contract, then the testing is at the contractor's cost. But if, on the other hand, the work is contract compliant, then the cost of opening up and testing is to be at the employer's expense.[4]

Setting out

2.17 In traditional forms the contractor is responsible for setting out its work and is responsible for errors in so doing, but the employer is obliged to give to the contractor the information which it needs in order to set out the works.

Obedience to instructions

2.18 There will be a provision that the contractor will comply with instructions given to it by the contract professional. The clause is usually limited in scope, namely that the instructions given must be pursuant to an express power given to the contract professional by the terms and conditions of the contract.

2.19 Such instructions are normally required to be given or confirmed in writing. This is designed to avoid confusion and ensure that each party knows exactly what the other is supposed to do, or at least to ensure that if there is ambiguity it is plain on the face of one or more documents.

Compliance with statute

2.20 The contractor will be expressly required to comply with all statutes and statutory instruments. In construction contracts the most important ones are the Building Regulations and the CDM Regulations.

2.21 So far as compliance with Building Regulations is concerned, there is a potential danger for a contractor insofar as the employer's design may provide for a structure which is non-compliant. The two most significant departures in practice by builders are Regulation C (that the building shall be waterproof) and Regulation D (that the foundations shall carry the structure safely).

---

[4] The early editions of the JCT 1980 Standard Form provided an upwards-only mechanism for the adjustment of the contract sum in the event of instructions being issued pursuant to this clause. It was a surprising state of affairs and was, eventually, remedied. Nobody was able to explain why it was remotely equitable that the contractor should be paid for the opening up of defective work and its removal.

Standard forms have developed as a result of this. Were it otherwise the contractor would be in danger of being held liable to the employer in the event of the design being defective in that the foundations were inadequate or the walls were not a sufficient barrier to moisture. The first development was the inclusion of a provision that in the event of a discrepancy between the design and the Building Regulations, then the contractor should give the contract professional notice and then the contract professional must issue instructions to the contractor remedying the divergence thus ensuring that the structure as built was compliant. This provision usually provided that the instruction so given would be a variation (as to which see below). **2.22**

The second development was to provide for the event that the contractor did not spot the divergence. Not spotting would not be surprising in the case of inadequate foundations or even inadequate acoustic attenuation. So far as foundations are concerned, contractors which built in the 1970s found themselves sued for breach of this type of clause even though they did not realize, and had no reasonable way of realizing, that the foundations were taken down to an inadequate depth. There is no reported case of a contractor being held strictly liable for a breach of this clause. The matter was argued in several cases, but there was never a concluded judgment holding the contractor so liable in a case where the contractor complied with the contract in all other respects save for the obligation to comply with statute and the Building Regulations. The reason for this is not hard to find, namely that to hold the contractor so liable would be to defeat the structure of the contract which placed design responsibility upon the employer. The forensic answer to this was that the contractor had expressly warranted such compliance, and thus the argument got off the ground. The true view is, it is submitted, that if the contractor does comply in all other respects with the contract, it will not be liable to the employer for a breach of the Building Regulations or other statute brought about by such compliance. The juridical reason for this would be that such a result defied business common sense and the clear intent of the contract. The legal route whereby this could be achieved would be interpreting or construing the contractual provision of compliance with the statute as to include the additional words 'save insofar as compliance is contrary to the express provisions of the contract'. The way in which the clauses have provided for the event, that the contractor did not spot the divergence, is by way of a proviso that the contractor shall not be liable for a breach of the clause in the event that it has otherwise complied with the express terms of the contract, and particularly the drawings and the specification or bills of quantity, if applicable. **2.23**

Persons on site

There is provision for the employer being permitted to have on site a clerk of works or resident engineer or engineer's representative. The clerk of works will have no power to issue instructions (or if he does issue them they are of no effect unless confirmed in writing by the contract professional) but he will have the right to inspect the work as they are being carried out and thus enable the employer to spot defective work before it is covered up. In addition to this is an express requirement that the contract professional should have access to the site and all off-site workshops, etc. Implicit in a building contract is an implied licence to the contractor to have exclusive possession of the site. Without such an express term, there might be difficulties in the contract professional being allowed on site by right. **2.24**

Furthermore, access to the steel fabrication workshops and quarries of the contractor or its sub-contractor is necessary so that the contract professional is satisfied that the work and materials being brought to site are contract compliant.

2.25 There is usually a requirement that the contractor has on site a competent person in charge to run the site and through whom instructions from the contract professional can be given. This person is variously described as a 'foreman in charge' or 'site agent'. There is no need for such a provision to be expressed in the contract if the purpose of the clause is to ensure that the contractor has someone on site who is competent, because the contractor could not execute the works in accordance with the contract if there was not such a person. However it is a useful clause which permits oversight by the contract professional of the contractor's personnel.

2.26 Allied with the requirement that there be a site agent, or similar such person on site, is a provision that the contract professional may require the contractor to remove from the site any person. This will apply from incompetent workmen up to the site agent himself. Some forms provide express criteria which permit an instruction requiring removal (safety of the site and competence in the ICE conditions); some do not. Implied, if not expressed, is the requirement that such instruction must be reasonable and with good cause.

2.27 Standard forms and bespoke contracts will usually provide that the employer may have its own workmen on site. This is a necessary clause should the employer need, or potentially need, its own work done on site aside from that provided for in the contract. Without it there would be the same difficulties which would obtain in relation to the professionals on site because of the implied licence which gives the contractor exclusive possession of the site.

Sufficiency of tender

2.28 Given that the contract sum which is payable is set out in the contract, this type of clause may be unnecessary in standard 'lump sum contracts'. However standard forms do state that the contract sum is the unalterable consideration for the completion of the works. Nevertheless, the contract conditions will provide for its alteration and so the effect of such a clause is to confine recovery to those conditions. Furthermore, if the contract is a 'measure and value' contract,[5] there is no contract sum agreed at the outset: all that is agreed are the rates at which the work will be done, the quantities of work not being agreed.

2.29 In engineering contracts there will be a provision that the contractor has inspected the site and satisfied itself as to the nature of the work which has to be done and will be satisfied that its prices are sufficient for the execution of the work. This provision will often have to be read alongside the provisions for additional payment when the contractor incurs unexpected adverse physical conditions in the ground.

[5] The ICE conditions, now in its 7th edition, being the paradigm measure and value contract.

Measurement: the standard method of measurement: contract sum adjustment

In contracts where there are bills of quantities, there is usually a provision which provides how the contract bills are prepared (almost invariably the Standard Method of Measurement) and the effect of departures from how they should be prepared. **2.30**

This has come in for criticism as an avenue for 'loophole engineering' given the provision in such clauses that departures from the Standard Method of Measurement, described as any 'error in description or in quantity or omission from the contract bills' shall be corrected and deemed to be a variation. **2.31**

The effect of such a clause is to provide for a concealed price increase if the work is not sufficiently described in the bills of quantity. In one case the contract bills did not provide for the temporary works required for the creation of working space for the contractor and the standard method of measurement required that there be such provision. The end result was that the contractor was paid additionally for the working space notwithstanding that there was no provision for it in the bills of quantity. The omission was deemed to be an error and corrected as a variation.[6] The effect of this is that bespoke contracts seek to have a provision which excludes such deemed additions. This is easier to do when the contract does not have bills of quantity and even more so when the contract is a design and build contract. The criticism of such deemed variations has force, but must be set against the fact that these contracts are commercial contracts like any other and contractors are all able to put in competitive tenders with the same opportunity to spot omissions in the bills as the other tenderers. Whilst it can be said that the contractor ought to deliver up the structure for the price regardless of the difficulties it faces in doing so and the costs it incurs, there is no reason why what is not priced for and what ought ordinarily should be priced for should not be paid for. Were it otherwise, the employer obtains a quantity of work beyond what it has strictly speaking agreed to pay for. **2.32**

The right to vary or change the works

The employer, through the contract professional, will have the right to vary the work. This might be described as a right to issue an instruction requiring a variation or it might use the nomenclature 'Change Order'. The right to issue such instructions is time limited and will not, without express provision, persist beyond practical completion of the works.[7] **2.33**

The definition of variation is slightly different from form to form. The definition 'Alteration or modification of the design, quality or quantity of the works, including the addition, omission or substitution of any work' or similar is standard. No or virtually no construction contract runs its course without variations. These may be changes of mind by the employer in reaction to things found on site in the case of work done to existing structures. There is **2.34**

---

[6] See *Farr v Ministry of Transport* (1977) 5 BLR 94.

[7] *TFW Printers Ltd v Interserve Project Services Ltd* [2006] EWCA Civ 875 particularly paras 31–4. This was a case under the JCT Minor Works form, but the reasoning will apply to all other standard forms of building contract. Ex hypothesi the works are completed at practical completion and the contractor's obligations thereafter are to make good defects only.

no doubt that variations and their effect on the progress of the work are the most contentious aspects of construction contracts.

2.35 The power to order variations does not amount to a power to exclude work from the contract in order that others may do the work. Neither is the power so wide that it permits the alteration of the character of the work to be done beyond that which is recognizable under the contract as let. Variations may not change the character of the work to be done so that the contractor can say: 'this is not what I undertook to do'. If the variation or variations do have that effect, then the contractor may refuse to execute the work since it is outside the scope of what was taken on. If the contractor does execute work which is outside the scope of the variations clause, then it may recover a quantum meruit instead of a valuation of the work in accordance with the contractual machinery.

2.36 Variations must be valued and paid for. Variation clauses provide for a mechanism for their valuation. Under standard forms this may mean that a schedule of rates which was included within the contract documents shall be used as a guide. Usually the provision is for the rates in the contract for similar work or work of a similar character executed under similar conditions to be applicable. If there is no such work described anywhere, then the provision is for a 'fair' valuation to be made.

2.37 A 'fair' valuation or a valuation based upon contract rates for similar work may not give the contractor sufficient recompense. The work done may be delayed and disrupted, or other work may be delayed and disrupted by the variation. Thus many forms provide for 'direct loss and/or expense' to be recoverable in addition to the fair valuation or the valuation at contract rates.

2.38 As can be seen in Chapter 14 dealing with delay and disruption, this is a contentious area and fraught with practical difficulty. In many cases contractors have been slow to make claims for recovery of loss and expense, and when they have done so their claims have been above and beyond what the contract professional reasonably expected. The resulting litigation and arbitration has been expensive and, many have felt, avoidable.

2.39 This has resulted in two developments in bespoke contracts. The first has been the creation of conditions of contract which state that claims must be made within a set period, failing which the claims are deemed to have been waived. These involve stringent notice provisions which state that notice of claims must be given within a set period, failing which the right to recovery is lost. These have created their own practical difficulties when contractors which have failed to give notice realize their difficulties and seek to recover their losses anyway.[8]

2.40 The other effect of late claims on contract drafting is the emergence of requirements that variations be valued in advance of their execution and that the contractor should agree the value of the variation prior to the work starting. Again this is often found in bespoke contracts.

[8] See, for example, *Cleveland Offshore Ltd v Britoil*, unreported, Staughton J.

The effect of variations on the time for completion must be catered for and suitable provisions are found in the standard forms.[9] **2.41**

Insurance

All standard forms of contract do, and all bespoke contracts should, make provision for insurance. Insurance relates to the works, injury to people as a result of the works being done, and damage to property, aside from the structures being built, as a result of the carrying out of the building works. **2.42**

Allied with insurance is the question of risk. **2.43**

Insurance of the works, and the risk in the works against the perils provided for in the contract as those to be insured against, is a matter of contractual choice. The only real question is 'who is to insure' and therefore who is to pay the premium. The employer, realistically, pays the premium anyway, either by way of a direct payment to the insurance company or by way of an addition to the contract price since the premium would fall to be considered as part of the contractor's costs in the execution of the work. **2.44**

The JCT Standard Form gives the choice as being: (a) under clause 22A in which the contractor assumes the risk of the perils to be insured against; (b) under clause 22B in which the employer assumes the risk and must insure; and (c) under clause 22C in which again the employer is responsible for insurance when the work to be done is to existing structures because the work involves the alteration and extension of those existing structures. **2.45**

Standard forms provide that, if these insured events actually happen, the contractor is entitled to an extension of time for the completion of the works and payment for the re-execution of the damaged works, including loss and expense arising therefrom. **2.46**

This might be thought to be equitable because the question is not 'who is to blame' but 'who is to insure'. Nevertheless there is scope for difficulty when the insured event has been brought about by the negligence of the contractor or its sub-contractors. There are several cases decided in the courts where this problem has had to be grappled with. Whilst there have been cases where the contractor responsible for fire or flood, through negligence, has been held liable notwithstanding the fact that risk and the obligation to insure lay with the employer, the preponderance of authority is to the effect that the risk of the insured event lies as provided for contractually. **2.47**

In contrast to the risk and insurance provisions relating to the works, those related to injury to people or other property are nearly always with the contractor. At the time the works are practically completed, the contractor usually leaves site only to return to remedy notified defects. Each contract will provide for when the insurance obligations transfer from the contractor to the employer. Usually this will be at practical completion as the Court of Appeal found was the case under the JCT Minor Works Form.[10] **2.48**

---

[9] See the reference to extension of time provisions in contracts below and the discussion of extension of time claims and recovery of liquidated and ascertained damages in Ch 11.

[10] *TFW Printers Ltd v Interserve Project Services Ltd* [2006] EWCA Civ 875.

Possession and time (and extensions of time) for completion

2.49 Provision is made for the time when the works are to begin and the time they are to be completed. Upon the commencement of the works, the contractor will have possession of the site and will be obliged to complete the works within the contract period, as adjusted by the extension of time clause.[11]

2.50 It is easier for the parties if the events which give rise to an extension of time are all set out in a single clause. The JCT Standard Form does so. The ICE conditions do not, and it is therefore necessary to look carefully around the form to see what events do give rise to an extension of time.

2.51 Along with the time for completion comes the provision which dictates the amount of damages payable in the event of delay to completion. These are often liquidated damages and are also often set at a level which is far below the likely loss to be suffered by the employer in the event of delay. This is particularly the case in process engineering contracts.[12] The limitation of liability relates to the contract price. Often the losses to be suffered in the event of delay are far beyond the amount of profit which the contractor expects to receive and sometimes far beyond the contract price. The limitation of liability, brought about by a low level of liquidated damages, is a necessary commercial price to pay for a contract sum which does not have a large contingency item within it to provide for the payment of a huge amount of damages for delay.

Certificates of completion and making good defects

2.52 Most standard and bespoke forms provide for the contract professional to certify when the works are practically completed, that is completed aside from defects of a snagging nature which do not prevent the beneficial occupation of the building. The certificate of completion provides the point at which the employer can occupy the structure and thereby stops liquidated and ascertained damages from running (if applicable). The certificate of making good defects brings about the point when any remaining retention is released. Up to that moment the contractor's obligation in respect of defective work, is to remedy the defect. It is not to pay damages and the employer will not ordinarily be able to recover consequential damages in respect of defective work prior to the certificate of making good defects. The contractor may be liable to the employer for the cost to it of remedying the defective work.[13] After that certificate has been issued, the contractor will have no right to execute work at all and as a result its obligation moves from the primary one (to execute the work in accordance with the contract) to the secondary one (to pay damages for breach of contract).

---

[11] For the position in relation to time for completion, extensions of time, and loss and expense, see Chs 11 and 14.

[12] See, for example, MF/1 Model Form of General Conditions of Contract recommended by the Institution of Mechanical Engineers, cl 44.

[13] See *Pearce & High Ltd v Baxter* [1999] BLR 101; and *Michael A Johnston v WH Brown Construction (Dundee) Ltd* [2000] BLR 243. These decisions are inconsistent. The first is a decision of the Court of Appeal in England which gave the employer the right to damages in the limited sense outlined, whereas in the second case the Scottish Court of Session (Inner House) did not.

It is sometimes argued that a contractor is not in breach of contract when it executes defective work because it is only in 'temporary disconformity' with the contract and its obligation is to hand over the completed building only and cannot be liable in damages or at all for executing defective work which it has a right and an obligation to put right.[14] This has never been accepted as the correct analysis by any court.[15] **2.53**

Programmes and method statements

There is provision in engineering forms for the contractor to provide a programme of works (and revised programmes from time to time) and a method statement. These have little contractual effect in that the provision of either document does not affect the time for completion of the works. Programmes have had a mixed press so far as contractors are concerned. They can be used to measure progress and thus be perceived as helpful. Sometimes, however, contractors feel that the programme is a stick to beat them with and thus resist their production. So far as method statements are concerned, there is a fine line to be drawn between seeking the information which will satisfy the contract professional that the contractor is setting about the work correctly and telling the contractor how it is to carry out the work. In the normal construction contract the 'what' is for the employer and the 'how' is for the contractor. **2.54**

Provisional Sums

Contracts frequently have within them elements called 'Provisional Sums'. Those are sums which are no more than estimates. The contract professional will issue an instruction to expend that sum (normally for a specialist item to be carried out by a specialist sub-contractor or for materials to be supplied by a specialist supplier). When the final account comes to be drawn up, the provisional sum is deducted from the contract sum and the true value of the work is added. It has received a judicial definition by May LJ[16] as follows: **2.55**

> The term 'provisional sum' is generally well understood in the construction industry. It is used in pricing construction contracts to refer either to work which is truly provisional, in the sense that it may or may not be carried out at all, or to work whose content is undefined, so that the parties decide not to try to price it accurately when they enter into their contract. A provisional sum is usually included as a round figure guess. It is included mathematically in the original contract price but the parties do not expect the initial round figure to be paid without adjustment. The contract usually provides expressly how it is to be dealt with. A common clause in substance provides for the provisional sum to be omitted and an appropriate valuation of the work actually carried out to be substituted for it. In this general sense, the term 'provisional sum' is close to a term of art but its precise meaning and effect depends on the terms of the individual contract.

[14] See *per* Lord Diplock in *Hosier & Dickinson v Kaye* [1971] 1 WLR 146.

[15] *Lintest Builders Ltd v Roberts* (1980) 13 BLR 38 (CA); and *Surrey Heath BC v Lovell Construction Ltd* (1988) 42 BLR 25 and (1990) 48 BLR 108, CA.

[16] *Midland Expressway Ltd v Carillion Construction Ltd and another* [2006] EWCA Civ 936.

Loss and expense

2.56 The JCT Standard Form provides for the payment of loss and expense in the event of certain specified defaults of the employer or the employer's agents or contractors (late possession, variations, late instructions).[17]

Determination or forfeiture

2.57 Most forms provide, with varying degrees of sophistication, for the right of either the employer or the contractor to determine the employment of the contractor.[18]

2.58 This is in contra distinction to the right to accept a repudiation of either the employer or the contractor.

2.59 The clauses provide for the sum to be paid to either side in the event of the exercise of the clause. These are known as clauses which provide for determination for cause because it involves the reaction by the innocent party to the default of the other.

2.60 In bespoke contracts there are also provisions for determination by the employer without cause at all. The clause will provide that the employer can stop the work and determine the contract whether or not the contractor is in default.

2.61 Each type of determination provision, whether it is for cause on either side or without cause, will have provision for the sums which are payable to either side in the event of the exercise of the determination provision. The clauses which provide for determination for cause will usually follow a formula which replicates or nearly replicates the position which would obtain if the party determining had instead accepted a repudiation. The clauses which permit determination without cause are often less sweeping and may provide that the contractor is entitled to be paid for work done plus its demobilization costs but not loss of profit on work not carried out. It is important to note the damages or consequential provisions of these clauses because they provide the sole remedy for either side. If, for example, the determination provision does not provide for damages for lost profit (whether the determination is for cause or not), then the contractor will not be entitled to lost profit and the determination clause will provide a complete code governing recovery. This may make a difference when it comes to deciding whether a repudiation ought to be accepted or whether the express terms of the contract should be relied upon.

Fossils

2.62 If fossils, 'antiquities or other objects of interest' are found on site, then there is usually provision in the standard forms that they become the property of the employer and must be preserved.

War

2.63 Nearly all standard forms have a war clause, that is a clause which makes provision for the event of war damage to the works and even to materials which are intended to go to site.

[17] See loss and expense in Ch 14.
[18] See determination and forfeiture in Ch 13.

These clauses have had no case law governing their interpretation because war, as understood by most of the clauses, has not been part of the construction scene since 1945. **2.64**

The ICE conditions define war as involving 'general mobilization of the armed forces for a period of 28 days'. **2.65**

The 'war on terror' does not produce 'war damage' as defined or implied in standard forms. Whilst the civilian population has been on the alert for random acts of violence against it since 1969, these acts of violence do not create war damage. **2.66**

Given the increased ability of terrorists to inflict significant damage, it is suggested that there will have to be increasingly sophisticated provision for the event of damage to people or property in the event of a terrorist attack on either the works or those engaged upon them, whether on or off site. **2.67**

Fluctuations or increased costs

Who is to bear the risk of price inflation for labour or materials? If the employer is to do so, then the tender need not have within it a contingency sum for inflation. **2.68**

In the 1970s there was high inflation, threatening the liquidity of several companies which did not have contracts providing for the payment of additional sums should the price of labour or materials increase. There are extensive clauses, particularly in the JCT Standard Form, which apply formulae to wage and materials costs set out in the tender. These clauses have not been a source of litigation. However their operation did lead to an argument as to whether the contractor had an obligation to proceed regularly and diligently in a case where the contractor was able to obtain the benefit of the formula rules governing fluctuations by delaying the commencement of a key site operation and was thereby entitled to take the benefit of the clause which provides for payment of increased costs.[19] The clauses are complicated in draftsmanship but simple in effect, namely that the contract sum is based upon rates and prices for labour and materials applicable at a base date and increases in those rates and prices will result in an increase in the contract sum to take into account the increased costs. **2.69**

Nominated Sub-contractors and Nominated Suppliers

Until the July 2005 edition, the JCT Standard Form made provision for nominated sub-contractors. The contract professional has the authority to issue instructions to the contractor requiring that it enters into sub-contracts for work to be done by sub-contractors nominated or chosen by the employer. The trades which involve nominated sub-contracts are defined in the specification or bills of quantity. This enables the employer to choose a specialist contractor or supplier of proprietary equipment. Windows, cladding, curtain walling, lifts, and many other types of specialist work are done by nominated sub-contractors. **2.70**

[19] See *Cleveland Bridge and Engineering v Greater London Council* (1986) 34 BLR 50; 8 Con LR 30.

2.71 The main contract provides for the terms upon which the sub-contractor is to execute the works and these terms appear in the JCT Standard Form of Sub-Contract for use when the sub-contractor is nominated.

2.72 The particular terms included in the sub-contracts include:

(a) A requirement on the sub-contractor to execute the sub-contract works and discharge the contractor's obligations to the employer in relation to that work.
(b) An indemnity given to the contractor in relation to any liabilities that the contractor may have to the employer arising out of any breach of sub-contract by the nominated sub-contractor.
(c) An indemnity to the contractor in relation to claims in respect of negligence or other defaults or other omissions of the sub-contractor.
(d) A provision whereby the sub-contract works will be completed in such a period so as to allow the main contractor to complete its work within the main contract period.
(e) A requirement that the sub-contractor shall pay liquidated damages in respect of its delay to completion of the sub-contract works.
(f) Payment shall be made to the sub-contractor by the contractor within a short period (fourteen days) of the issue of the architect's certificate in respect of the work.
(g) Payment is to be subject to a retention and such retention is to be the same percentage as that provided for under the main contract.
(h) The contract professional or his representatives shall have a right of access to the sub-contractor's workshops.
(i) The nominated sub-contractor's employment determines at the same time as the main contractor's employment determines.

2.73 The nominated sub-contractor will be nominated because it will have proprietary design. Because the traditional form of contract requires the employer to bear the design risk, the employer will enter into a direct warranty with the nominated sub-contractor which warrants the design of the sub-contract works. This warranty is two-fold. First that the design shall be a fit and proper one. Secondly that the sub-contractor will give its design information to the employer timeously so as not to delay the progress of the works.

2.74 The main contract will provide for an extension of time for completion of the work if the sub-contractor is in delay. No loss and expense is payable, however. The risk of those delay costs lies with the main contractor. However the provision of information to the main contractor is ultimately the employer's responsibility, so if the sub-contractor is late with design information, then the contractor will be able to recover loss and expense arising therefrom because the provision of design information rests with the employer.

2.75 If the nominated sub-contractor's employment is determined, then the employer has an obligation to re-nominate a fresh sub-contractor.[20]

2.76 Delay by a nominated sub-contractor gives rise to an extension of time for the benefit of the main contractor. However delay in re-nomination also gives rise to an extension of time, as

[20] *Bickerton & Sons v Northwestern Metropolitan Hospital Board* [1970] 1 WLR 607.

a late instruction. However the risk of delay caused by the normal delay between the expulsion of a sub-contractor and getting a fresh one on site lies with the contractor.[21]

The provision for nominated suppliers (of materials and not work) are similar but simpler reflecting the fact that they supply materials only. **2.77**

Payment and Certificates

It is a rare contract which makes no express provision for interim payment. Contracts which do will be for reasons of tax avoidance.[22] **2.78**

Payment is normally provided to be made on certificates.[23] The contractor will have an obligation to submit a periodic account, usually monthly, and the employer's contract professional will then be obliged to value the work on an interim basis for the purposes of making a monthly payment to the contractor. **2.79**

The monthly certificates and payment are not designed to settle the final sum due to the contractor, but merely an amount which is appropriate to pay on account. **2.80**

There is a twin track obligation to pay on the part of the employer. First there is the obligation to pay the contract price when the works are completed. That obligation is the consideration for the handing over of the completed structure. Secondly there is an obligation to make a payment on account periodically, such payment being related to the value of work done and materials brought to site or appropriated to the works off site. These two obligations make for separate causes of action in the hands of the contractor. It is thus inappropriate for an employer to plead that a claim for work done early on in a long-running contract (for foundations for example) is statute barred if six years elapse from the time when the work was done. The employer has an obligation to pay for the completed works and therefore it will have a separate obligation to pay for the foundations work at the time the final certification was made or ought to have been made.[24] **2.81**

These clauses will provide for retention to be withheld until defects are remedied following practical completion. Retention will be released half at practical completion and half upon the making good of defects, as certified by the contract professional. The clause will also set out separately the types of item which will add to the contract sum (variations for example) and the types of item which will cause there to be subtraction from the contract sum (defective work, for example, or the cost to the employer of remedying defective work when the contractor has been instructed but has failed to do so). **2.82**

[21] See *Percy Bilton Ltd v Greater London Council* [1982] 1 WLR 794; and *James Longley & Co Ltd v Council of the Borough of Reigate and Banstead* (1982) 22 BLR 31.

[22] See, for example, *The Wardens and Commonality and Mystery of the Mercers v New Hampshire Insurance* [1992] 2 Lloyd's Rep 365, but note the provisions of the 1996 Act, s 109 of which requires all contracts to provide for stage payments, failing which the Scheme shall apply.

[23] See discussion on certificates in Ch 10.

[24] See *Henry Boot Construction v Alstom* [2005] EWCA Civ 814, [2005] 3 All ER 932 *per* Dyson LJ. See also the discussion regarding limitation of action in Ch 15.

2.83 The JCT Standard Form makes provision for a Final Certificate, which has the effect of certifying what the final contract sum is and exactly what is finally due and payable thereunder. It also has provisions within it designed to ensure that it has finality unless adjudication or arbitration are commenced within a specified period. The provisions for a final certificate have also provided for a discharge of the contractor in respect of defective work unless adjudication or arbitration are commenced within a set period. The old forms (particularly the 1963 edition) provided for a complete discharge as a matter of evidence in relation to defective work.[25] They were amended in order to allow for a more limited finality, but the Court of Appeal[26] has since held that such a clause, even though apparently limited in ambit, did operate as a discharge relating to the quality of the work. The final certificate clause has since been amended further, but it is potentially troublesome.

Arbitration, Adjudication, Expert Determination, and Litigation: Dispute Resolution

2.84 All standard forms now provide for adjudication. This is either express or required by statute. This is discussed in detail in Chapter 23.

2.85 Standard forms provide for arbitration which is designed to provide a final and binding decision in relation to any dispute. Until 1999 ICE conditions provided for an intermediate form of dispute resolution, namely a reference to the engineer under the old clause 66 of the seventh edition of the ICE conditions. Whereas the adjudicator and the arbitrator are bound by the requirements of natural justice, the engineer in making his decision under clause 66 is not.

2.86 Dispute resolution clauses in bespoke contracts are now becoming more and more sophisticated with the different types of dispute resolution available. Some provide for intermediate stages which allow for settlement discussions. These usually require that senior executives meet with a view to resolving the dispute. These executives are usually provided to be at one remove from the immediate management of the project and therefore ought to provide some distance from day-to-day involvement. They may also have a more detached view of where the dispute sits commercially. These provisions have some limited success but often the executives concerned, unsurprisingly, stick to backing up their side. A similar form of intermediate dispute resolution, which is particularly popular in international contracts, is the dispute resolution board. The members of the board are normally named in the dispute resolution clause and they are called upon to give intermediate advice (and sometimes temporarily binding rulings) as the works progress. These can be expensive as professionals may have to go long distances at short notice. They have the

[25] See *Hosier & Dickinson v Kaye* [1971] 1 WLR 146.

[26] See *Crown Estate Commissioners v John Mowlem and Co plc* (1995) 70 BLR 1. See also *National Coal Board v WM Neill & Son (St Helens) Ltd* [1985] QB 300; *Guinness plc v CMD Property Development Ltd* (1995) 46 Con LR 48; *Beaufort Developments (NI) Ltd v Gilbert Ash (NI) Ltd* [1998] 2 All ER 778, and the excellent articles by IN Duncan Wallace QC (see (1995) 11 Const LJ 184 and (1997) 13 Const LJ 15).

advantage of providing the parties with the perspective which an independent view of experts can offer.

More and more common are expert determination clauses. These sometimes provide that all disputes are to be decided by an expert, who is not governed by the Arbitration Act 1996. Though that does not mean that the courts cannot intervene in the right case and, for example, ask the expert to give further reasons for his or her decision.[27] Nevertheless the theory is that the parties get speed and certainty. The courts are reluctant to interfere with the experts' decisions.[28] But these clauses are often uncertain in scope. If part of an overall dispute is provided for by way of expert determination, another allied part might have to be decided by way of arbitration. Sophisticated clauses provide that certain experts will decide certain disputes with lists of potential experts for each category of dispute. The disadvantage of expert determination is its finality. If an expert makes a wrong decision within his or her scope of instruction, then the decision is final and binding even though plainly wrong. If, on the other hand, the expert does not decide the question referred to him or her, or the question referred is outside the ambit of what the expert determination clause provides for, then the decision will be a nullity.[29] **2.87**

Arbitration is still the preferred method of dispute resolution. It has the disadvantage of potential delay, since arbitrations can take a long time. There has been floated recently the concept of fast-track arbitration, in which the arbitrator must go from reference to award in a short time (between three and six months is usually cited). This has not yet got off the ground in standard forms but it may be an acceptable compromise in the construction industry between the speed of adjudication which has no finality and the relative slowness of arbitration, which does. **2.88**

Care must be taken to ensure that the time limits provided for the service of a notice of arbitration (or rather a notice to concur in the appointment of an arbitrator) are complied with. An engineer's decision under clause 66 of the ICE conditions is final and binding unless the arbitration is commenced within three months of the decision. Similarly the effect of the final certificate under the JCT Standard Form is that it is binding on the parties unless arbitration is commenced within the time provided for by the clause. **2.89**

---

27 *Halifax Life Ltd v Equitable Life Assurance Society* [2007] EWHC 503 (Comm).

28 *Jones v Sherwood Computer Services Plc* [1992] 1 WLR 277.

29 *Veba Oil Supply & Trading GmbH v Petrotrade Inc* [2001] EWCA Civ 1832, [2002] 1 Lloyd's Rep 295, Simon Brown LJ. See Cresswell J's helpful statement in *Halifax Life Ltd v Equitable Life Assurance Society* [2007] EWHC 503 (Comm), at paragraph 42:

> (1) Putting on one side cases of fraud, collusion or partiality . . . the principal ground on which a party to an expert determination may succeed in a challenge to the determination is that the expert has materially departed from his instructions, so that the determination is not a determination made in accordance with the terms of the contract: *per* Dillon LJ in *Jones v Sherwood Services Limited plc* [1992] 1 WLR 277 at 287.
>
> (2) In the absence of terms of the contract which provide otherwise (such as 'save for manifest error'), an expert determination cannot be challenged on the ground that the decision was mistaken, so long as the expert has answered the right question and has not otherwise materially departed from his instructions: *per* Knox J in *Nikko Hotels v MEPC* [1991] 2 EGLR 103 at 108B. . .

Notices

2.90 Provision is made, either generally or in specific clauses that notices must be given in writing and in a specific way. These must be differentiated from provisions which deem service of any notice in a prescribed period (for example three working days after postage). Normally a provision that a notice must be sent by registered post or recorded delivery does not mean that a notice which is given in some other way is invalid.[30]

2.91 It is surprising how often notice provisions, particularly with forfeiture or determination clauses, are not followed properly and notices are held to have been given out of time when a little attention to what the express words of the contract require would ensure that this did not happen. If a notice is required, particularly in respect of a determination or forfeiture, employers and contractors should set out a programme in advance which tells them what to do and when.

Prime Cost Contracts

2.92 These follow in most particulars the traditional form of contract with the proviso that the main contractor is to be paid the 'prime cost', that is the cost to it of executing the work. This means that the valuation machinery of the contract will be substantially different in that it will be geared to ascertaining the contractor's costs. There will, therefore, be no need for schedules of rates or bills of quantity or similar valuation procedures and techniques.

2.93 These contracts are designed for when the employer wants a quick start on site and the design is not worked out and also for when the full extent of the work to be done is not known because much of the work to be done is hidden from view (as with an extensive refurbishment). The obvious disadvantage of this, from an employer's point of view, is that there is less incentive than a normal fixed price contract to carry out the work efficiently. This will remain the case notwithstanding that the contract, if in the JCT Standard Form or similar, will have a clause which requires the main contractor to work efficiently and not waste man hours or materials. However inefficiency is a very difficult thing to prove and such a clause will not give an employer total peace of mind on that front. Nevertheless, if the commercial priority is to start on site, this might be the best way to go. An alternative for an employer is to ensure that the design is worked up quickly after a certain stage of the work and to let work up to that stage on a prime cost contract and thereafter on a traditional form of contract. This may not be practical.

[30] See *Yates Building Co Ltd v RJ Pulleyn (York) Ltd* (1975) 237 EG 183; and *Goodwin & Sons v Fawcett* (1965) 175 EG 27. But see also *Central Provident Fund Board v Ho Bock Kee* (1981) 17 BLR 21, a decision of the Court of Appeal in Singapore. The import of the decisions is that the purpose of the mode of delivery is to see that the giver of the notice can prove that the notice was given. If the notice has been given, albeit by a different route to that prescribed, then the courts will not usually disregard it. However, if the giver of the notice cannot provide that proof, then that is its own misfortune by not using the prescribed mode of service.

Design and Build Main Contracts

Many of the provisions contained within traditional contracts also appear in standard forms of design and build contracts. The structure is changed slightly because the contract requires the contractor to build to its own design to fulfil the employer's requirements. **2.94**

Thus variations can be instructed by the employer, but normally these will be made by way of a change to the employer's requirements. **2.95**

In a straightforward design and build contract the contractor will be responsible for the structure when completed being fit for its purpose.[31] This would mean that the many, from the employer's point of view, tiresome disputes involving the contractor blaming poor design and the contract professional blaming poor workmanship in defective structure cases will become irrelevant. **2.96**

However in the JCT Standard Form, with contractor's design, this is not quite the case. Although the contractor is responsible for design, its responsibility does not extend to the provision of a fitness for purpose warranty. Instead it provides that the contractor warrants that it has exercised reasonable skill and care in its design as if it were a contract professional. There is also provision for insurance against such negligence. This means that in the case of poor design the employer must prove negligence just as it would normally have to do against a contract professional. This decreases significantly the attraction of a design and build contract which uses such a form because of the need to prove negligence against the builder. In practice, however, there has not been a decided case in which a contractor under a design and build contract has been found to have put forward a design which was defective but nevertheless was not negligent in the provision of that design and thus escaped liability. This is probably because in practice most defective design will be negligent design, there being very few cases (and none reported) of a defect in design leading to problems with the suitability of a structure for the purpose for which it was constructed but for which nevertheless the designer was not liable. It is also because once the designer and builder are all in the same action and are the same party, there is little incentive for the contractor to 'blame' design, because it is responsible to the putative claimant for that design. **2.97**

Management Contracts

In the last twenty-five years, there has been a move towards management contracting in which the 'main contractor' or 'management contractor' does little actual construction itself and instead sub-contracts the work to the different trades, now called 'packages'. The skill of the management contractor is to coordinate the different trades and many think that management is a skill in itself which can be the subject of a substantive construction contract. **2.98**

[31] See *Greaves & Co Ltd v Beynham Meikle* [1975] 1 WLR 1095; and *IBA v EMI and BICC* (1980) 14 BLR 1.

2.99 The schemes of the contracts vary. Often they will be closely akin to the traditional standard form of contract. They will usually have a fee payable to the management contractor and it is through the fee that the management contractor obtains its profit. In that event the management contractor will be paid the 'prime cost' of the work, in other words the sub-contract cost to the contractor. Sometimes, instead of a fee, the management contractor will put together a contract price which makes the contract more akin to a standard main contract.

2.100 What each of these contracts will provide is a limitation of liability of the management contractor for the damage which the employer will suffer because of a breach of contract by any of the trade contractors. That limitation is most often expressed as being that there can be recovery from the management contractor only of such damages as the management contractor is able to recover from the sub-contractor concerned. Some contracts are described as management contracts but in fact have the same scheme in form and substance as the traditional main contract, albeit that the prime costs of the sub-contractors are payable to the contractor rather than a main contract price.[32]

Forms of Sub-Contract

2.101 These have similar provisions which reflect the need for the sub-contract to follow the main contract in ensuring that the primary and secondary obligations of the main contractor are discharged.

The provision for the principal obligation of the sub-contractor

2.102 This provides that the sub-contractor is obliged to carry out the sub-contract works in accordance with the sub-contract documents. There is often an additional requirement which states that, to the extent that the contract professional has a discretion as to the acceptance or rejection of the quality and standard of the work, then the sub-contractor is bound by the opinion of the contract professional. This will reflect the obligations of the main contractor who will also be bound by such opinions.

Back to back and indemnity

2.103 It is often expressed by non lawyers that the sub-contractors' obligations are 'back to back' with those of the main contractor. The concept is self-explanatory but worth re-stating. The objective of the sub-contract is that, so far as possible, the sub-contractor must discharge the obligations of the main contractor insofar as they relate to its works. These obligations will be both the primary one (the execution of the work) and the secondary one (the obligation to pay damages).

2.104 Thus the form of words to the effect that the sub-contractor will 'observe, perform and comply with all the provisions of the main contract on the part of the contractor to be observed insofar as they relate to the sub-contract works' appear in most standard and bespoke forms of sub-contract. Given that the objective of the sub-contract is for the

[32] See, for example, *John Mowlem and Co plc v Eagle Star Insurance Co Ltd* (1995) 44 Con LR 134.

vicarious performance of the sub-contractor's element of the main contract work, this is unexceptional.

However, additionally the forms provide for an indemnity to be given to the main contractor insofar as the sub-contractor does not 'observe, perform and comply with all the provisions of the main contract on the part of the contractor to be observed insofar as they relate to the sub-contract works'. There is not normally any advantage given to the main contractor by indemnity clauses such as these save for the purposes of limitation of action. The main contractor will not have to sue the sub-contractor prior to its discovery of what damages are payable by it to the employer because the sub-contractor can be sued on the indemnity within either six or twelve years from the date that the liability of the main contractor to the employer is ascertained.[33] **2.105**

It will be comparatively rare when a main contractor will wish to delay suing its sub-contractor if it is sued or arbitrated against by the employer. The legal and commercial imperatives are all one way, namely the disposal of any disputes involving all relevant parties in one forum so that there are not inconsistent findings of fact between two different tribunals. **2.106**

Insurance

The insurance provisions will reflect the main contract. Risks relating to personal injury or damage to property (aside from the works themselves) are normally to be insured by the sub-contractor. **2.107**

Other damage to the works or materials (usually now referred to as 'all risks' insurance) will be at the risk of the main contractor in relation to the insurable risks. Provisions may be made for the risk of loss of materials which are off site (often to remain at the sub-contractor's risk) as opposed to on site (to be at the employer's risk). **2.108**

The effect of these clauses upon the responsibility of the sub-contractor for loss to the works caused by the insured events is, as remarked upon in relation to the main contract's insurance provisions, problematical. There will be no contract between the sub-contractor and the employer so the question may arise as to the liability of the sub-contractor in tort to the employer when the sub-contractor carelessly causes loss and damage to the employer's works. Therefore, theoretically the employer can sue the sub-contractor in tort for such loss. The suit is often theoretical because the employer will have either its own insurance or the insurance of the main contractor to fall back upon. Nevertheless, the courts have had to adjudicate upon such claims; in this regard reference is made to the cases set out above in relation to the main contractor's obligation to insure. It is submitted that when the contractual structure places the risk of loss through an insurable event on one party, be it the employer or the main contractor, then that risk ought not to be transferred to the sub-contractor merely because the sub-contractor's carelessness has been responsible for the insured event (usually flood or fire). A set of provisions which deals with such risks ought not to be set aside merely because the court thinks that the sub-contractor is 'to blame'. The contractual structure could, equally, ensure that the risk lay with the sub-contractor either **2.109**

[33] See Ch 15, Limitation of Action.

absolutely or in relation to damage caused by its own negligence. If it does so, then the sub-contractor can insure against the risk.

2.110 However the state of the law on this topic is unclear with findings by the courts which are not easy to reconcile, albeit that they (and the inconsistencies) are expressed to relate to the particular terms of the contract under consideration.

Variations

2.111 Just as the main contract will have a provision for variations, so will the sub-contract, together with a provision for the valuation of the variations (often in a separate clause). The concept of 'back to back' will ensure that the valuation rules in the sub-contract will follow those of the main contract. In order to ensure consistency of valuation under both the main and the sub-contract, there is usually provision for joinder of the sub-contractor in a main contract arbitration where the disputes relate to the same subject matter. Therefore, although the sub-contract prices will often differ from the main contract prices with a domestic sub-contract, the ability for there to be joinder allows for a 'back to back' valuation. When the sub-contractor is nominated there is no such price differential and the main contractor's profit will be a fixed percentage rather than a turn on individual rates and prices. In that event, the valuation of main and sub-contract prices for the direct cost of the work will follow one to the other.

Right of access to the site of the contract professional

2.112 Standard and bespoke forms make provision to allow access to work (on or off site) of the contract professional. This will reflect the obligation of the main contractor to allow access of the contract professional to its sub-contractor's workshops and site work.

Progress and completion: extensions of time and damages for delay

2.113 There will be a sub-contract period and attendant obligations of the sub-contractor to complete the works within that time together with an obligation to pay damages to the main contractor in the event of delay. Unlike those of the main contract, these are often not liquidated damages but damages at large. The losses which the main contractor will suffer will include:

(a) the main contractor's on and off-site overheads;
(b) disruption costs of the main contractor and its sub-contractors;
(c) any delay damages payable to the employer; and
(d) other loss and expense suffered by the main contractor.

Thus the delay damages payable by the sub-contractor are likely to be pro rata greater than those payable by the main contractor to the employer in the event of delay.

2.114 The sub-contractor's obligations to complete will often be in sections as well as related to an overall sub-contract period.

2.115 There is a proviso in nominated sub-contract forms, which prohibits the main contractor from recovering damages for delay unless the contract professional has certified that the sub-contractor has not completed to time and ought reasonably to have completed by the sub-contract completion date or any contractually provided for later date for completion.

Loss and expense

Like the main contract, there is normally provision in sub-contracts for the recovery of loss and expense by the sub-contractor. The events giving rise to loss and expense are the same as those which obtain under the main contract. **2.116**

Those events are usually the responsibility of the employer (late instructions, variations, etc) and therefore there are two important provisions in relation to nominated sub-contractors. The first is that there will be a provision whereby the main contractor must obtain for the sub-contractor the benefits of the main contract. That provision will be matched by an obligation of the sub-contractor to do what is required to enable the main contractor to obtain that benefit. Thus the sub-contractor can recover its loss and expense as part of the main contractor's claims. The second is the name-borrowing provision. That enables the sub-contractor to take the name of the main contractor in order to recover loss and expense against the employer. This will take place when the main contractor has decided that it will not pursue the employer either for commercial reasons or because it has not actually suffered loss and expense. That will be because its own areas of operation are not disrupted, whereas those of the sub-contractor are. **2.117**

Defects

Similar provisions as appear in the main contracts as applicable to defects apply to most sub-contracts. The essential difference is that the defects liability period of the main contract does not (or may not) have the same significance in relation to the sub-contract. Thus the responsibility for remedying defects will usually be pursuant to a notice in writing given by either the contract professional or the main contractor. **2.118**

Set off and 'pay when paid'

The most emotive and longest running difficulties relate to the entitlement of the main contractor to deduct monies from the account of the sub-contractor. The detailed provisions of the 1996 Act in relation to set off are dealt with elsewhere in this book. This chapter is concerned only with contractual provisions. During the 1970s and 1980s several provisions were introduced into the industry sub-contracts which were designed to limit the right of the main contractor to set off against sums due to the sub-contractor. **2.119**

In relation to delay there was the provision that there could be no such set off unless the contract professional issued a certificate to the effect that the sub-contractor was in delay (known originally and colloquially as the 'clause 8(a) certificate' long after it ceased to be the appropriate appellation). **2.120**

Generally in relation to set off there were (and are) three additional provisions in standard forms. First, that there could be no set off unless a notice was given setting out the grounds of set off and such notice should be given within a set period before the sums became due. This is designed to stop set off which is last minute in character. **2.121**

The second provision was for an adjudicator to be appointed to decide on an interim basis as to whether the set off should be allowed. The adjudication provisions were used but surprisingly rarely, in contrast to the position today when statutory adjudication is routine. **2.122**

2.123 The third provision was for the appointment of a 'trustee stakeholder' to hold disputed sums pending resolution of the underlying dispute which brought about the set off. The adjudicator would be given power to order that such disputed sums be paid to the stakeholder.

2.124 The business purpose of these provisions is to ensure that there are at least some obstacles in the way of main contractors failing to pay sub-contractors. This set of provisions was put into the standard forms on the insistence of the sub-contractor's side of the industry.

2.125 These set off restrictions were not contained in bespoke forms issued on behalf of main contractors, which usually added further provisions in relation to payment. The most notorious were 'pay when paid' clauses.

2.126 Sub-contracts of the bespoke variety, which in other ways followed standard forms of sub-contract, until 1997 would often have obnoxious payment provisions within them. These were more designed to ensure that the financial position of the main contractor was both protected and enhanced than to ensure that what was reasonable for the completion of the works was included. Thus main contractors would often make more money managing their cash flow than actually executing the works. The most infamous of these provisions was the 'pay when paid' clause. Such clauses did not often find their way into standard forms. These gave the main contractor a powerful weapon against the sub-contractor because, construed literally, they transferred the risk of the non-payment of the main contractor to the sub-contractor. These have been substantially outlawed by the 1996 Act.

Sub-letting and assignment

2.127 There is often a prohibition against sub-letting and assignment. Standard forms (and particularly nominated sub-contract forms) will provide that assignment and sub-letting cannot take place without consent and the consent will have to come from both the main contractor and the contract professional.

Forfeiture or determination

2.128 There is usually a provision for a determination 'for cause' in the hands of the main contractor. If the sub-contractor suspends the work, does not proceed regularly and diligently with the work, does not remove defective work, or commits an act of bankruptcy or insolvency, then the main contractor will have the right to determine the sub-contract. Those clauses will have a provision giving the main contractor the recovery of all its losses suffered as a result of the determination, reflecting closely the damages which would be payable in the event of a repudiation by the sub-contractor (including loss of profit on work not carried out being often described as 'loss and expense incurred by the main contractor by reason of the determination of the employment of the sub-contractor'). These determinations for cause need not be because of a breach of the sub-contract. A failure to proceed regularly and diligently is not a breach of sub-contract, unless there is an express term to that effect. Also an act of bankruptcy or the appointment of a receiver is not a breach of sub-contract.

2.129 Some forms make no provision at all for a sub-contractor's determination for cause. Others will do so and make the same provision for the accounting process as between main contractor and sub-contractor as with determination by the main contractor for cause. But it must be noted that the accounting and payment provisions for a determination in accordance with express terms of a sub-contract (whether for cause or otherwise) are the complete remedy for the determination.

Most forms will make provision for determination in the event of the main contract being determined. This is provided for as being automatic. If there is no provision for such determination, then the sub-contract will be held to have been repudiated by the main contractor with the attendant right to damages. That is why many clauses which provide for such automatic determination do not provide for the payment to the sub-contractor of the loss and damage it will have suffered by reason of the determination. **2.130**

Fluctuations

Like main contracts, there will be elaborate provision for the payment of increased costs as a result of inflation, usually by reference to indices and formula rules. For the same reasons which obtain in relation to the main contracts there has been virtually no litigation concerning how these formulae are to be applied. **2.131**

Payment and direct payment

Aside from the provisions which relate to set off, pay when paid, and the right to recover loss and expense and the benefit of the main contract, the basic obligation to pay the sub-contractor will be provided for. Usually for nominated sub-contractors this will be pursuant to certificates issued by the contract professional where the contract professional will have valued the sub-contract work. Where there is such a method of valuation of payment to be carried out by the contract professional, there will be a provision permitting the sub-contractor to use the name of the main contractor in the event that there is a dispute by the sub-contractor of the main contractor's valuation. Under domestic sub-contracts, standard or bespoke, the provision usually is that the main contractor is responsible for the valuation of the sub-contractor's work. **2.132**

Elaborate provision is now made in nominated sub-contract forms for the employer to have the right to pay the sub-contractor direct in the event that the main contractor has not done so. This is usually by way of a procedure whereby the sub-contractor informs the contract professional that it has not been paid and there is no good reason for such non-payment, and following the service of notices the employer will have the right, at the time of the next payment to the contractor, to deduct the sums unpaid from the main contractor's certified payment and pay that sum direct to the sub-contractor. **2.133**

The procedure is complicated and only works if the employer has signed a direct warranty to the sub-contractor that it will operate the direct payment provisions of the sub-contract. **2.134**

The direct payment provisions are usually expressed not to be applicable in the event of insolvency. The difficulty with direct payment provisions is that, should they be used in an insolvency situation where it is not known that there is in fact an insolvency situation, the employer is in danger of paying for the same work twice: once to the sub-contractor and once to the liquidator of the main contractor. **2.135**

Notices

2.136 As with the main contract, provision is made for notices to be given in writing and more important notices (like forfeiture or determination) to be given by registered post or recorded delivery.

Arbitration, adjudication, expert determination, and litigation: dispute resolution

2.137 Dispute resolution clauses are as complex with standard forms of sub-contract as they are with main contracts. See the discussion above in relation to main contracts. Additionally nominated sub-contract forms and some domestic sub-contract forms make provision for 'joinder'[34] in which the arbitration involving a dispute between main contractor and sub-contractor can be joined with an arbitration between main contractor and employer. This, in theory, permits consistency in findings between the three or more elements of the chain: employer, contractor, and sub-contractor (and sometimes sub-sub-contractor).

2.138 The clauses often cause practical difficulties, particularly in relation to confidentiality and the use of evidence. It is one thing to have the same arbitrator hearing main and sub-contract disputes. It is another to have all the matters tried at the same time. Arbitrations are private agreements involving the private resolution of a commercial dispute such as to bind the parties legally. If a dispute arises between main contractor and sub-contractor in relation to the same or similar subject matter as a subsisting arbitration between employer and contractor, there remains for consideration the status of evidence which is heard in the first arbitration (between main contractor and employer) and particularly whether it is receivable in the second arbitration (between main contractor and sub-contractor). Rules 2.6–2.8 of the Construction Industry Model Arbitration Rules (incorporated into JCT Standard Forms of main and sub-contract) seek to deal with these problems.

[34] As do the JCT Construction Industry Model Arbitration Rules particularly r 3.

3

QUANTUM MERUIT

The Meaning of Quantum Meruit

The English translation of quantum meruit is 'what it is worth'.[1] That is not the same as 'what it has cost the contractor' or 'what the surveyor normally charges' for the work. It is frequently the case that a contractor will seek recovery of a quantum meruit on the basis of the cost to it plus profit of the labour and materials supplied. It may be that is the appropriate measure of payment for the contractor in most cases, but there are situations where it will not be so. In most situations quantum meruit is a restitutionary claim. There are four essential ingredients to such a claim:[2] **3.01**

1. The defendant must have gained a benefit;
2. The benefit must have been obtained at the claimant's expense;
3. It must be legally unjust for the defendant to retain the benefit; and
4. There must be no defence available to extinguish or reduce the defendant's liability to make restitution.

A quantum meruit may be payable in the following broad categories: **3.02**

(a) When there has been an express or implied agreement to pay a reasonable price;
(b) When there has been no agreement but a claim may be made in quasi-contract;
(c) When there is an agreement but the work falls outside its scope; or
(d) Where there has been a void or unenforceable contract.

---

[1] For a discussion of this concept see *Costain Civil Engineering Ltd v Zanen Dredging and Contracting Co Ltd* (1996) 85 BLR 77.

[2] As Lightman J said in *Rowe v Vale of White Horse* [2003] 1 Lloyd's Rep 418 para 11. Cited with approval by the Court of Appeal in *Chief Constable of the Greater Manchester Police v Wigan Athletic AFC Ltd* [2008] EWCA Civ 1449, para 38.

When There Has Been an Express or Implied Agreement to Pay a Reasonable Price

3.03 An express agreement is easy to recognize. An implied agreement may arise from all the circumstances. But it may also arise by statute, for example under s 8 of the Sale of Goods Act 1979 and s 15 of the Supply of Goods and Services Act 1982. Both sections provide that a reasonable charge or a reasonable price is a question of fact. There is normally no difficulty of law about the assessment. The court has to set a reasonable price. The difficulty arises in practice when the court has to make an assessment. Unlike the next category (where there has been no agreement), there is no guidance from the failed contract negotiations which might point to a method of assessment because the contract expressly provides for reasonable remuneration.

3.04 For a contract professional, a quantum meruit as assessed is not subject to definitive judicial guidance. There is an obscure judgment by Birkett J in *Brewin v Chamberlain*[3] but that is almost meaningless save for the proposition that a good architect of standing will be able to receive a higher quantum meruit than a less good one of no standing, and rejecting the contentions of each side that scale fees or time-based charges would be applicable but that both factors would be an aid to ascertaining the proper amount payable. Such an approach is too whimsical and old-fashioned for today's world where time charges are the norm for all professionals and if the matter came up for consideration today, the court would turn to look at the professional's normal charge-out rate for the purposes of assessment.[4]

3.05 For a contractor reasonable cost plus reasonable profit is the best guide to what is reasonable.[5] However it must be remembered that the definition of quantum meruit is what the work is worth, not what the work cost to do.[6] So it would be possible as a cross check for the court to look at comparable tender sums and prices.In *Serck Controls Ltd v Drake & Scull Engineering Ltd* (2000) 73 Con LR 100, Judge Hicks QC in discussing quantum meruit said:

> The site conditions and other circumstances in which the work was carried out, including the conduct of the other party, are relevant to the assessment of reasonable remuneration. The conduct of the party carrying out the work may be relevant. If the value is being assessed on a 'costs plus' basis then deduction should be made for time spent in repairing or repeating defective work or for inefficient working. If the value is being assessed by reference to quantities, such matters are irrelevant to the basic valuation. A deduction should be made on either basis for defects remaining at completion because the work handed over at completion is thereby worth less.

---

[3] To be found in the almost unobtainable *Rimmer's Law Relating to the Architect*, 2nd edn.

[4] See also the implied obligation to pay reasonable fees in litigation in *Kellar v Carib West Ltd* [2004] UKPC 30.

[5] In *Sanjay Lachhani v Destination Canada (UK) Ltd* (1997) 13 Const LJ 279, Mr Recorder Colin Reese QC said: 'In my opinion, as a general rule, a 'fair value' of work undertaken by X for and at the request of Y on Y's property (real or personal) ought to recognize an entitlement to a reasonable or normal profit margin over and above the costs actually and properly incurred in carrying out the work in question', a dictum approved by HH Judge Humphrey LLoyd QC in *ERDC Group Ltd v Brunel University* [2006] EWHC 687 (TCC).

[6] See Judge Wilcox in *Costain Civil Engineering v Zanen Dredging* (1996) 85 BLR 77 at 94.

Generally there are no hard and fast rules for the assessment of a quantum meruit. All the factors have to be considered. In cases where there is no contract the assessment of a quantum meruit is usually based on reasonable and necessarily incurred actual cost including reasonable on and off site overheads plus an appropriate addition for profit. The actual costs can be checked by the use of standard rates and prices from publications such as Spon. In any case expert evidence as to market rates may be needed.[7] **3.06**

Whether the courts are determining a reasonable sum payable under a contract or by way of quantum meruit for services by way of restitution, it is often relevant to have regard to what the parties contemplated should be paid for such services.[8] Thus in *Way v Latilla*[9] the basis of assessment included a '*reasonable participation*' as that was the sort of basis which had been contemplated by the parties.[10] However the weight to be attached to the evidence of the negotiations between the parties depends on the particular facts. In *Benedetti v Sawiris*[11] Patten J drew a distinction between cases involving an ineffective agreement whether due to a failure to attain the necessary certainty or a formal defect where great weight was often attached to such evidence and those, such as *Benedetti* where the contract or agreement was abandoned as the basis of the parties' dealings as the parties ceased to regard the original contract price as an appropriate value for the services. In such circumstances there was no justification for the court imposing it as a measure of value unless it could be justified by other evidence as the market value of the services at the time when they were performed. **3.07**

The extent to which the anticipated contract price acts as a cap on recovery is controversial and there is little English authority.[12] **3.08**

If a letter of intent which fixes rates and prices expires, and the work continues, then the question arises as to how the work is to be valued. It seems that unless it would be unfair the work will be valued, as a quantum meruit, at the same rates and prices as the letter of intent.[13] **3.09**

---

7 *ERDC Group Ltd v Brunel University* [2006] EWHC 687 (TCC) at para 42, HHJ Humphrey LLoyd QC.

8 See for example *Benedetti v Sawiris* [2009] EWHC 1330 (Ch); *ENE Kos v Petroleo Brasileiro SA* [2009] EWHC 1843 (Comm) para 46; and Goff and Jones, *The Law of Restitution*, 7th edn, 2007, para 1-036.

9 [1937] 3 All ER 759.

10 See also *Vedatech Corp v Crystal Decisions (UK) Ltd* [2002] EWHC 818 (Ch) and *Benedetti v Sawiris* [2009] EWHC 1330 (Ch), para 547.

11 [2009] EWHC 1330 (Ch).

12 See Goff and Jones, *The Law of Restitution*, 7th edn, 2007, paras 1-022, 1-036, 20-020 to 20-023, and 23-002. Cases suggesting that recovery in excess of the contract price is permissible include *Lodder v Slowey* [1904] AC 442 (PC); *Boomer v Muir* 24 P 2d 570 (1933); *Rover International Ltd v Cannon Film Sales (No 3)* [1989] 1 WLR 912; and *Renard Constructions (ME) Pty Ltd v Minister for Public Works* (1992) 26 NSWLR 234. Cases suggesting that such recovery is not permissible include *Inchbald v Western Neilgherry Coffee, Tea and Cinchona Plantation Co Ltd* (1864) 17 CBNS 733; *Noyes v Pugin* 27 P 548 (1891); *Wuchter v Fitzgerald* 163 P 819 (1917); *Pavey and Matthew Property Ltd v Paul* (1987) 162 CLR 221 at p 257; and *Taylor v Motability Finance Ltd* [2004] EWHC 2619 (Comm).

13 See *ERDC Group Ltd v Brunel University* [2006] EWHC 687 (TCC).

Where There Has Been No Agreement

3.10 Sometimes work is done in anticipation of a contract. In that event it has been held that there are four factors to be taken into account in deciding whether to impose an obligation to pay for the work:[14]

(1) the nature of the services—are the services of a kind which would normally be given free of charge?

(2) the terms of the request for services—to what extent (if any) did the claimant take the risk that its services would be unrecompensed? Were the parties simply negotiating, expressly or impliedly 'subject to contract', or was some kind of assurance or indication given to the claimant that the other side would not withdraw, or would only withdraw in certain circumstances?

(3) the nature of benefit to the defendant—is the benefit real? The courts are more inclined to impose an obligation to pay for a real benefit, otherwise the abortive negotiations will leave the defendant with a windfall and the claimant out of pocket.

(4) the circumstances in which the anticipated contract did not materialize—in particular can it be said to involve 'fault' on the part of the defendant, or (potentially of more relevance) to be outside the scope of the risk undertaken by the claimant at the outset? This may often be decisive.[15]

3.11 Generally speaking when someone seeks to enter into a contract with another but fails to do so, that person cannot normally recover the costs that he has incurred in: (a) estimating; (b) deciding on a price; (c) bidding for the contract; and (d) showing the other party his capabilities or skills even though, if there was a contract, he would be paid for them.[16]

3.12 The most frequent situation, where this type of dispute as to the value of the work done occurs, is where a contractor and an employer have failed to reach an agreement but the contractor has executed a substantial quantity of work. This is common because it is often to a contractor's advantage not to actually reach an agreement with the employer. So long as the contractor is on site executing work, it will have the commercial advantage of being on site and therefore expensive to dislodge should agreement not be reached. That leads to the temptation of contractors so arranging the pre-contractual correspondence that no actual agreement is reached because one important item is still at large.

3.13 The commercial advantage to the contractor is the following: (a) it is in a good bargaining position when it comes to the agreement of the contract price the longer it is on site, because it becomes more and more expensive for the employer to replace the contractor; and (b) if there is no contract, then there are no contractual provisions as to quality of work

---

[14] *Countrywide Communications Ltd v ICL Pathway Ltd* [2000] CLC 324.

[15] See also *MSM Consulting Ltd v The United Republic of Tanzania* [2009] EWHC 121 (QB), paras 170–1.

[16] *MSM Consulting Ltd v The United Republic of Tanzania* [2009] EWHC 121 (QB) ibid.

and time for completion to breach and therefore the risk of being sued for breach of contract is abated.[17]

The lack of agreement on price is the most common situation where no agree-ment is actually made. The next most common important term which is unagreed is the contract period or the rate of liquidated damages. **3.14**

There have been many cases before the courts where work has been completed, a building built for example, and one or other of the parties (usually the contractor) suing for a quantum meruit and asserting that the work was not done pursuant to a contract. These cases frequently involve a great deal of correspondence passing between the employer and the contractor arguing about the terms of the contract which are acceptable to one or other of the parties. Sometimes both parties will be referring to 'the contract', but at the end of the construction period or when a dispute arises one of the parties will be saying that in fact there is no contract. **3.15**

The contractor then argues that, since there is no contract, and therefore there is no contract price, it is entitled to a quantum meruit which is quantified on the basis of the costs incurred by the contractor, as opposed to the contract rates of the kind being discussed between employer and contractor. **3.16**

However, the courts will have to look in each case to see whether there is a contract. A dialogue which falls short of a contract actually being entered into, but where the parties say to one another that they have entered into a contract, does not mean that a contract has in fact been entered into. **3.17**

The courts are in this situation tempted to find a contract simply because the construction of correspondence, which is designed actually not to result in an agreement, is unattractive and not in accordance with the legitimate expectations of honest commercial men. The courts' reaction is often simply to find that a contract has been entered into. An alternative analysis in this situation is for the courts to find that there is an estoppel preventing the party concerned from denying the existence of a contract. This has rarely been done, if at all.[18] The problem with such a course is that it involves the court in making a bargain for the parties as opposed to interpreting and enforcing the bargain which the parties have in fact made. It also suffers from the conceptual difficulty that if the parties have reached the point that they are estopped from saying that there is no contract, then they will have in fact made the bargain. **3.18**

Given the importance of certainty in commercial law, it is submitted that the temptations of finding a contract where, strictly speaking, none exists should be resisted. This phenomenon is now well known and commercial parties can arrange their affairs so that they are not suddenly in the position of resisting an argument that they are not in contractual relations. The courts have in many cases held that they will not make a bargain for the parties. It is submitted that straining to make a bargain where the essentials of a contract are not in **3.19**

---

[17] For a good example of this in operation, see *British Steel Corp v Cleveland Bridge and Engineering Co Ltd* [1984] 1 All ER 504 and (1981) 24 BLR 94.

[18] See the analysis of Ramsey J in *Haden Young Ltd v Laing O'Rourke Midlands Ltd* [2008] EWHC 1016 (TCC), paras 163–94.

place may be an understandable temptation, but it amounts to a court-made bargain and ought not to be made. The better solution in order to do justice in any particular case, is for the court to hold that the quantum meruit should be calculated on the basis of the kind of contract price which the parties were discussing.[19] Essentially this may amount to a finding of what price the parties would have agreed had a contract been entered into. That is easier to do when the contract price is agreed and other terms are not. It is less easy to do when it is the contract price which is not agreed. But it may be made easier to do when the parties are not very far apart on what the contract price should be. There is no court guidance as to a quantum meruit calculation because essentially it involves a reasonable sum which must have regard to many factors unique to the situation in each case. In one case the court awarded 'reasonable rates' because that was felt better than cost plus in the circumstances of that case.[20] But that case, as all of them will be, was decided on its own facts.

When There Is an Agreement But the Work Falls Outside Its Scope

3.20 This is a theoretical eventuality because most construction contracts have a power to order varied work. Therefore additional work will fall within the scope of the variations clause and fall to be valued in accordance with the valuation machinery of the contract. However, it is said that work may fall outside the scope of the works if either: (a) what is ordered to be done is outside the scope of the type of work contracted for; or (b) the varied work is of the same type as that contracted for but the volume of additions instructed, changes the nature of the contract such that what is done bears no relation at all to the original work. It is submitted that only the first is recognizable in law. The second is based upon a misreading of the authorities, or rather a wrongly decided case.

3.21 The first category is easy to recognize, for example if there is a contract to build a house and an instruction is issued to build a power station, or even a block of flats. In those circumstances the contractor can either refuse to carry out the instruction or carry it out under protest. There is an additional remedy, namely to seek an adjudicator's decision that the instruction is outside what the contract contemplates. The best example of such a case recently is *Industrial Service (Telford) Ltd v Charles Ransford & Son Ltd*[21] in which there was a simple contract (with no variation clause) to clean and dispose of contaminated waste in tanks. On the true construction of the contract only about 22,000 litres of waste would be disposed of. In fact over 200,000 litres of liquid (not all of it waste) was disposed of. It was held by the Court of Appeal (paragraph [11] of the judgment) that only a limited amount of

---

[19] See Lord Atkin in *Way v Latilla* [1937] 3 All ER 759 at 764; Goff and Jones, The Law of Restitution (2007) 7th edn, para 1-036 as to the Court being able to take into account 'the bargainings between the parties not with a view to completing the bargain for them, but as the evidence of the value which each of them puts upon the services'. See also *Hall & Tawse (South) Ltd v Ivory Gate Ltd* (1997) 62 Con LR 117 at para 64; *AC Controls v British Broadcasting Corp* [2002] EWHC 3122 (TCC); *Stephen Donald Architects Ltd v King* [2003] EWHC 1867 (TCC); *ERDC Group Ltd v Brunel University* [2006] EWHC 687 (TCC); *Claymore Services Ltd v Nautilus Properties Ltd* [2007] EWHC 805 (TCC); *ENE Kos v Petroleo Brasileiro SA (Petrobas)* [2009] EWHC 1843 (Comm).

[20] *Laserbore Ltd v Morrison Biggs Wall Ltd* [1993] CILL 896.

[21] [2005] EWCA Civ 662; [2005] All ER (D) 311 (Apr).

work had been agreed to be done in the original lump sum price and on the true construction of the contract, work beyond that limit would be paid for and a quantum meruit was the appropriate measure of payment. This right arises out of the true construction of the contract and subsequent requests for work. The contract provides for a payment in accordance with its terms. The additional work falls outside the scope of the contract and a quantum meruit is payable pursuant to the express or implied request that the work be done.

The second category is based upon reliance upon the case of *Parkinson (Sir Lindsay) & Co Ltd v Commissioners of His Majesty's Works and Public Buildings*,[22] which was a decision of the Court of Appeal demonstrably wrongly decided and which cannot stand scrutiny with later authority. Because textbooks take the case at face value, it is necessary to take a little space making this good. In that case the employer, the Ministry of Works, contracted with Parkinson for the construction of a munitions factory under a standard form of contract which permitted the ordering of variations and their valuation in the ordinary way. The war was approaching and the Ministry wanted work accelerated. So they entered into a fresh agreement with Parkinson which provided for Parkinson to accelerate and work uneconomically so as to complete the work. The contract provided that the estimated value of that work was £5m and it was further agreed that there would be payment of profit on top of cost in a range of between 3 and 6% of the £5m, namely £150,000 to £300,000. A great deal of extra work was ordered at a cost of £6.68m and the Ministry said that Parkinson was entitled only to the profit of £150,000 to 300,000 as agreed. Parkinson claimed a quantum meruit on the profit element. The Court of Appeal held that Parkinson was entitled to that extra profit. Singleton LJ said at 673: **3.22**

> I find myself unable to agree with the submission of [Counsel] that, under the contract as varied by the deed of variation, the contractors would have been bound to continue making alterations and additions, if ordered, for years and years, without any extra payment by way of profit. That would have led to manifest absurdity and injustice, as Mathew J said in *Bush v Whitehaven Trustees*.[23] There must be a limit. I think that Lewis J drew the right inference and that it was an inference which he was entitled to draw, namely, that at the date of the deed of variation both parties contemplated work which, with the introduction of exceptional methods, would cost approximately £5 million, and that it was on that basis that the maximum profit was agreed at the figure of £300,000. The additional work called for and executed later was something outside the contemplation of the parties in August, 1937. The contractors executed it at the request of the Commissioners, and they are entitled to be paid a reasonable profit or remuneration in respect of it.

The objection the Court of Appeal had was to the contention that a contractor would agree to take on an infinite amount of work for a fixed and unvaried profit. The case can be explained on the facts on the basis of *Industrial Service (Telford) Ltd v Charles Ransford & Son Ltd*[24] that the additional work did not fall within any power to vary the work. However, the contract did contain within it the power to vary the work. It is hard to see how, in principle, that power can apply to some variations and not others unless there truly is a principle that additional work to a volume beyond that which is contemplated is to be rewarded on **3.23**

[22] [1948] 2 KB 632.
[23] (1888) 52 JP 392 (Div Ct); *Hudson on Building Contracts*, 4th edn, Vol 2, at p 122 (Div Ct and CA).
[24] [2005] EWCA Civ 662.

a quantum meruit. However, there is no such principle. As can be seen from the extract from Singleton LJ's judgment, the Court of Appeal relied upon *Bush v Whitehaven Trustees*;[25] it was described by Cohen LJ as 'the principle in *Bush v Whitehaven Trustees*'. The so-called principle in that case[26] has been well and truly destroyed by the House of Lords in *Davis Contractors Ltd v Fareham UDC*[27] in which the Law Lords had a field day in stating that it was inapplicable and not a citable case. See the following extracts from the speeches:

3.24 Viscount Simmonds at 714:

> The case of *Bush v Whitehaven Port and Town Trustees*, which is reported only in Hudson on Building Contracts, if it can be said to embody any principle, illustrates an early stage in the development of the doctrine of frustration which has since been the subject of many decisions in this House. And it was, I think, to this issue that the contentions of the parties and the further findings of the arbitrator were directed.

3.25 Lord Reid at 719:

> This form of award gives rise to considerable difficulty. But I have some sympathy with the arbitrator: he may have found as much difficulty as I have in discovering 'the principle in the case of *Bush v Whitehaven Port and Town Trustees*'

and at 724:

> *Bush v Whitehaven Port and Town Trustees* appears to me to be a very special case, and it must be read in light of the development of the law in later cases. I agree with your Lordships' comments on it and I can get little assistance from it for the decision of the present case. I agree that this appeal should be dismissed.

3.26 Lord Radcliffe at 731:

> The arbitrator had put upon him the duty of making further findings of fact relevant to the application of the principle in the case of *Bush v Whitehaven Port and Town Trustees*. This may have been hard upon him, for it implies that that decision can be ascribed to one ascertainable principle. In my opinion, it cannot . . . My Lords, I think that *Bush v Whitehaven Port and Town Trustees* may be worth remembering as an instance of what can happen to a case during its passage through successive courts, but I do not think it worth recording as an exposition of any principle of law. In that regard the editors of the Law Reports, who ignored it, showed a sounder judgment than Mr. Hudson, who enshrined it.

3.27 Lord Somervell at 734:

> But for the decision in *Bush v Whitehaven Port and Town Trustees* I doubt if the issue of frustration would ever have been raised. I will give my reasons as shortly as I can for thinking that

---

25 52 JP 392 (Div Ct); *Hudson on Building Contracts*, 4th edn, Vol 2, at p 122 (Div Ct and CA).

26 Summarized by Lewis J in *Parkinson* as

> . . . that you are to look and see what were the circumstances under which the contract was made in construing the contract, and you have to see whether what has happened was an event which was in the contemplation of the parties when they contemplated it—that, I think, must be a slip and should be 'when they made it'—and it goes on: and if it was not you are entitled to construe that particular clause of the contract (having looked at the whole of the contract also, of course) to see whether or not it can clearly be said that the interpretation put upon it by one mind of the parties when they entered into the contract.

27 [1956] AC 696.

> that case should not hereafter be citable as a decision . . . It appears that Lord Esher had some doubt whether the answer to the fifth question should be taken as a binding finding. Findings by juries on mixed questions of law and fact are not precedents. That is, no doubt, why the case was not reported in any law report.

Indeed *Bush v Whitehaven Port and Town Trustees* has not been cited in any case in the law reports since. It must follow that there is no principle to be obtained from *Parkinson* either to like effect. It is noteworthy also that in *Parkinson* Singleton LJ said that but for the variation of the contract terms and had the parties remained on the original standard terms, then the contractor's case would have gone nowhere.[28] Therefore there is no principle that a large or excessive number of variations for work of the same kind contemplated under the contract will permit the recovery of a quantum meruit. If the contract is based upon a standard form which calls for variations to be measured and valued, there is no basis for such a recovery. If, as in *Industrial Service (Telford) Ltd v Charles Ransford & Son Ltd*[29] there is a simple contract with no variations clause, then additional work will be recoverable as a quantum meruit as if it was work instructed with no agreed price and was in the categories of recovery set out in paragraph 3.02(a) and (b) above. **3.28**

The value of the quantum meruit may be cost plus a reasonable profit. But it may also be the value to the receiver of the benefit of the construction work. The remedy is restitutionary. It ought not to be less than the cost plus profit and ought to adopt the principles applicable to a void contract identified in paragraph 3.31 below. **3.29**

Where There Has Been a Void or Repudiated Contract

In Chapter 12, Damages, we deal with *Lodder v Slowey*[30] and the likelihood that it is not authority for the proposition that a contractor can recover a quantum meruit in the event of a repudiation of the contract but instead can recover the contract value of work done plus the loss of profit on work not carried out. The juridical basis of a quantum meruit is that upon the acceptance of a repudiation the contract 'goes' and the parties are therefore potentially in the position as if there was no contract.[31] **3.30**

In the case of a void contract, this does not fall within the category of work done pursuant to a request, and a cost plus may be an appropriate quasi contractual remedy. The court is entitled to balance all the factors (the cost to the contractor and the benefit to the recipient of the work) and arrive at a figure which takes all these factors into account. The court is not bound to any particular approach save what will produce a broad justice.[32] One question **3.31**

[28] At 671: 'If one had to consider this matter on the original contract alone, the contractors' difficulties would be almost insurmountable. The Commissioners were entitled to call upon them to carry out additional works or alterations and payment for any such works would have been based on the prices in the bills of quantities, which included profit or remuneration for the contractors.'

[29] [2005] EWCA Civ 662.

[30] [1904] AC 442.

[31] See *Taylor v Motability Finance Ltd* [2004] EWHC 2619 (Comm), paras 24–5.

[32] As the arbitrator did in *Costain Civil Engineering Ltd v Zanen Dredging and Contracting Co Ltd* (1996) 85 BLR 77 and upheld by the judge. See that case and the cases cited therein. See also for a good summary of these principles *Countrywide Communications Ltd v ICL Pathway Ltd* [2000] CLC 324.

remains live and that is whether a quantum meruit can be reduced by being able to show inadequacies in the mode and manner of performance. In other words whether what would amount to cross claims, in a case where there was a contract in being, can be relied upon to reduce the quantum meruit. This has been discussed but there is no clear answer to the point on the authorities. It is quite common for a defendant to make allegations of defective or delayed performance. The question is how, if at all, such allegations enter into or affect the assessment of remuneration. Does the claimant's alleged sub-standard performance affect the amount of remuneration it is entitled to? Does the claimant's alleged sub-standard performance give rise to a set-off or counterclaim? What duties does the claimant owe the defendant? Absent a contract, does the party providing the services owe a duty of care to the party receiving the services?

3.32 The only time these issues have been to the Court of Appeal the issue was left open. Slade LJ said in *Crown House Engineering Ltd v Amec Projects Ltd*:[33]

> I am not convinced that either [Goff & Jones], or any of the other reported cases cited to us, affords a clear answer to the crucial question of law: On the assessment of a claim for services rendered based on a quantum meruit, may it in some circumstances (and, if so, what circumstances) be open to the defendant to assert that the value of such services falls to be reduced because of their tardy performance, or because the unsatisfactory manner of their performance has exposed him to extra expense or claims by third parties? In my judgment, this question of law is a difficult one, the answer to which is uncertain and may depend on the facts of particular cases.

Bingham LJ said:[34]

> Whether, and to what extent, Amec could rely on the matters previously pleaded as contractual cross-claims to diminish the reasonable remuneration recoverable by Crown became a matter of genuine controversy. Crown has argued, and the judge below accepted, that these matters are wholly irrelevant to the assessment of what reasonable remuneration Crown should recover. It may very well be that they are right and will ultimately be held to be so. But the answer does not seem to me to be clear and obvious.

3.33 There are four first instance decisions which assist.

3.34 In *Sanjay Lachhani v Destination Canada (UK) Ltd*[35] The Recorder said:

> If the legitimate interests of a building contractor, such as the plaintiffs, are also to be taken into account then it seems to me that, in principle, as a minimum, all costs *reasonably* and *necessarily* incurred in *properly* carrying out the works ought to be reimbursed . . . [Recorder's emphasis]
>
> If, instead of ascertaining a 'market value', it is thought appropriate to calculate a 'fair value' from the costs actually incurred by the person or organization which carried out the works, then the three words highlighted would provide the key to the necessary effective control over the level of reimbursable costs if the legitimate interests of the owner are also to be respected. Having established such an objective base figure, a margin of some sort will fall to be calculated and added thereto. If the building contractor works inefficiently and/or if the building contractor leaves defective work then, quite obviously, the actual costs incurred by

[33] (1989) 48 BLR 32, see para 54.

[34] See para 57.

[35] (1997) 13 Const LJ 279.

> the building contractor must be appropriately adjusted and/or abated to ensure that the owner will not be required to pay more than the goods and services provided are truly (objectively) worth . . .

In *Serck Controls Ltd v Drake & Scull Engineering Ltd*[36] Judge John Hicks QC felt that distinctions needed to be drawn between the basic valuation and cross-claims. In relation to the basic valuation he said: **3.35**

> a. If the value is assessed on a 'cost plus' basis, eg from time sheets and hourly rates for labour, deductions should be made for time spent in repairing or repeating defective work, and for inefficient working or excessive tea-breaks and the like.
> b. If the value is assessed by reference to quantities the claimant stands to gain nothing from such activities or inactivities and,
>   i. if attributable to the claimant or his sub-contractors, they are irrelevant to the basic valuation;
>   ii. if not attributable to the claimant's fault the general answer must be that the site conditions and other circumstances in which the work was carried out, including the conduct of the defendant, are relevant;
> c. defects remaining at completion give rise to a deduction whatever the mode of valuation, simply because the work as handed over is worth less.

On Judge Hick's analysis cross-claims, even if expressed in terms of a 'reduction' or 'diminution' of the valuation, were really matters where the defendant was seeking compensation for loss or expense suffered or liabilities incurred by reason of the claimant's conduct. These claims must depend upon breach of some duty by the claimant. In Serck there was clearly no duty to adhere to any particular contractual programme, for there was no contract and thus the claim that Serck was under a duty to comply with the main contractor's programme failed. However, Judge Hicks QC was not persuaded that at the other extreme a firm working on a quantum meruit basis on a complex construction site can wholly ignore the desirability of co-operation with others at work there. In his view there must at least be a duty: **3.36**

- not unreasonably to interfere with the carrying out of other works;
- to be aware of the progress of other trades; and
- so far as consistent with the firm's own legitimate commercial interests, to co-operate in efficient working practices.

There might also be more specific duties, for example in relation to compliance with health and safety legislation and the directions of persons having powers under it.

In *ERDC Group Ltd v Brunel University* [2006] EWHC 687 (TCC), HH Judge Humphrey LLoyd QC held that Brunel could not reduce ERDC's quantum meruit by something like a set-off or cross-claim equal to the costs of putting the work right, except perhaps where as a result of what ERDC did or did not do, there was no benefit or value. Even in such circumstances there could be no negative result. ERDC would not have to pay Brunel or forego what it would otherwise have received. However, since the benefit has to be assessed overall, if, for example, work which was otherwise up to standard could not be used because other work was not done or was not up to standard then the value must reflect that result. **3.37**

[36] (2000) 73 Con LR 100.

The judge did not accept ERDC's case that it was entitled to an allowance for snagging. In his view, rectifying defects was necessary to confer full value or benefit. The risk of having to do so was included in ERDC's rates and prices which on the facts of this case were the basis of the assessment of the quantum meruit. There were, broadly, three types of snagging: that carried out by a contractor during the works with a view to minimizing apparent defects; that carried out formally or semi-formally, usually in conjunction with whoever was supervising the work in order to attain practical completion (this category may merge with the first category); and the identification of defects after practical completion. All were covered by the contractor's rates and prices (unless of course the contract requires the contractor to put right other people's mistakes in which case that work is additional). He concluded that even if the quantum meruit was not being assessed by reference to rates and prices care would need to be taken to exclude any snagging that was beyond the norm. Furthermore, the Judge dismissed the concept of mitigation as relevant to the valuation of a restitutionary claim. The fact that Brunel had not allowed ERDC or its sub-contractors an opportunity of putting things right did not affect the net benefit to Brunel. He said: 'The net benefit to Brunel cannot be affected by whether ERDC was or was not given the chance of putting the work right or an investigation as to whether it was willing to do so.'[37]

3.38 In *Haden Young Ltd v Laing O'Rourke Midlands Ltd*[38] there was no contract between the parties and Ramsey J refused to find that Haden Young ('HY') owed Laing a duty of care to carry out the design, supply, and installation of the works in accordance with the sub-contract documents or with reasonable skill and care so that Laing could recover extra expense or claims by third parties if there was tardy or unsatisfactory performance of the M&E works. In his view, where the parties had attempted and failed to reach agreement as to an essential term, duties of care in negligence at common law could not fill the gap by imposing an obligation. To do so the court would have to establish a basis in terms of scope and extent for the price, period, and standard of performance. Duties of care in negligence could not take on the burden of setting these standards of performance when the parties and the law of contract could not do so.[39]

Thus, it is suggested that the cases so far show:

a. the court may well take account of defective work when assessing the sum that should be paid to the claimant. How the courts will do it will depend on the facts of the particular case but at its broadest the defendant should not have to pay for defective work or for any work done by the claimant in repairing it. However this exercise will not result in the claimant having to pay the defendant a sum: a negative value;
b. the court is resistant to any form of cross-claim or set off. One of the main reasons is that it is difficult to establish the basis on which to assess the key elements of liability, such as the duties owed.

[37] See para 128.
[38] [2008] EWHC 1016 (TCC).
[39] Haden Young submitted that the only duty of care it owed in carrying out the works was a duty of care not to cause damage to persons or to other property.

4

ASSIGNMENT AND SUB-LETTING AND THE CONTRACTS (RIGHTS AGAINST THIRD PARTIES) ACT 1999

Assignment and What Cannot be Assigned

There can be no assignment of the burden of a contract without the consent of the other contracting party. So by way of example in a construction law context: a builder cannot assign to another builder its obligation to build[1] and an employer cannot assign to another party its obligations under a contract to pay.[2] **4.01**

It follows that if a party to a contract desires to relieve itself of its burden under that contract there must be a novation. In other words, where A has entered into a contract with B and B wishes to pass its burden under that contract to C, the original contract between A and B must be discharged by agreement and a fresh contract made between A and C. It should be noted that this operation is not an assignment proper and the new contracting party may find itself unable to sue in respect of pre-novation breaches.[3] **4.02**

Novations are common when either a builder or an employer wishes to transfer its obligations to an associated company. Such operations can be carried out with full explanation and with a fresh agreement being entered into by the new employer/contractor. However it **4.03**

[1] See *Tolhurst v Associated Portland Cement* [1902] 2 KB 660.

[2] See *Nokes v Doncaster Amalgamated Collieries Ltd* [1940] AC 1014.

[3] *Blyth and Blyth Ltd v Carillion Construction Ltd* [2001] CILL 1789.

often happens that an employer or contractor will simply make an announcement that its obligations are to be taken over by an associated company. Such announcements are usually accompanied by a statement of the benefits of this change to the other contracting party. However care is advised in these situations. The original contracting party should ensure that such a transfer is in fact beneficial and, if not, the original contracting party must make it clear that it does not consent to the proposed novation. If there is no objection to the transfer then the prudent course is for the contracting party to insist upon a new agreement being entered into.

4.04 However, while it is not possible to assign the burden of a contract, the benefit of a contract can be assigned in principle (subject to a number of prohibitions).

Where assignment is prohibited

4.05 Most standard forms contain an express prohibition against an assignment. A purported assignment of a building contract in cases where assignment is prohibited does not constitute a breach of contract as it is an act without legal effect[4] and the assignment remains ineffective unless a later consent is given.[5] However, it must be noted that in certain circumstances assignment pursuant to statute can override contractual prohibitions.[6] Matters stand differently where leases are concerned. A condition against alienation of an estate in land is repugnant to the right of disposal and therefore void. However, such an assignment will nevertheless place the assignor in breach of covenant and liable to forfeiture.[7] It is therefore common for leases to prohibit assignment without consent with the stipulation that such consent is not to be unreasonably withheld. This formulation is also effective in cases of contractual assignment. In such cases, if an assignment is made without consent, it will be invalid. It is no answer that no reasonable objection could have been made to that assignment had consent been sought.[8] It is well settled that if consent is not sought there can be no unreasonable refusal.[9]

---

[4] *Hendry v Chartsearch Ltd* [1998] CLC 1382 at 1384.

[5] See *Helstan Securities Ltd v Hertfordshire County Council* [1978] 3 All ER 262; and *Linden Gardens v Lenesta Sludge Disposals Ltd* [1994] 1 AC 85. In the latter case Lord Browne-Wilkinson said at 107 'A party to a building contract, as I have sought to explain, can have a genuine commercial interest in seeking to ensure that he is in contractual relations only with the person whom he has selected as the other party to the contract . . . In the circumstances, I can see no policy reason why a contractual prohibition on assignment of contractual rights should be held contrary to public policy.'

[6] See *Stansell Ltd v Co-operative Group (CWS) Ltd* [2006] EWCA Civ 538. The principle in *Stansell* has recently been qualified in *London & South Eastern Railway Ltd v British Transport Police Authority etc* [2009] EWHC 460 (Admin).

[7] *Hendry v Chartsearch Ltd* [1998] CLC 1382 at 1383–4 *per* Millett LJ. However it is possible for the lessor to take a covenant against assignment and to reserve a power of re-entry for breach of covenant. An assignment in breach of covenant is effective to vest the legal estate in the assignee, but the assignee takes a defeasable interest only which is liable to forfeiture for breach of contract.

[8] *Hendry v Chartsearch Ltd* [1998] CLC 1382 *per* Millett LJ 'Consent is not withheld if it is not asked for; and if it is not withheld it cannot be said to be unreasonably withheld.'

[9] *Barrow v Isaacs* [1891] 1 QB 417; and *Eastern Telegraph v Dent* [1899] 1 QB 835.

Contracts of a truly personal nature

4.06 Assignment is also prohibited in contracts which are personal in nature or where the personality of one or more of the contracting parties is important.[10] The classic case of *Nokes v Doncaster Amalgamated Collieries Ltd*[11] involved a working man purportedly being given a new master by an order made under s 154 of the Companies Act 1929. This order provided for a transfer (without further act or deed) of all the property, rights, liabilities, and duties from one company to another. It was held that this transfer did not operate to assign the contract of service.

4.07 As it is unusual for a construction contract to be dependent on personalities this rule is unlikely to be of significance in a construction law context.[12] More difficult questions arise when professionals or celebrities are involved. In these cases it is necessary to examine the nature of the contract with care to establish whether it is of a sufficiently personal nature to rule out assignment. It is the fact that the arrangement depends upon personalities which might bring about the objection, not that the obligations are personal.[13]

Assignment savouring of maintenance and champerty

4.08 Maintenance arises where a person supports litigation in which he or she has no legitimate concern without just cause or excuse.[14] Champerty is an aggravated form of maintenance[15] and occurs where a person supports the litigation *and* agrees to share in its proceeds. An assignment which savours of 'maintenance and champerty' will not be effective. However the law relating to maintenance has become less forbidding over time. The law as it stands can be summarized as follows:[16]

(a) Maintenance can be justified if the maintainer has a genuine commercial interest in the result of the litigation.

(b) There is no difference between the interest required to justify maintenance of an action, and the interest required to justify the taking of a share of the proceeds, or the interest required to support an out and out assignment.

---

[10] *Robson & Sharpe v Drummond* [1828–34] 4 All ER Rep 158; *Kemp v Baerselman* [1906] 2 KB 604; *Cooper v Micklefield Coal & Lime Co Ltd* (1912) 107 LT 457; and *Nokes v Doncaster Amalgamated Collieries Ltd* [1940] AC 1014.

[11] [1940] AC 1014.

[12] *Thirty (Charlotte) Ltd and Bison Ltd v Croker Ltd* (1990) 24 Con LR 46.

[13] s 136 of the Law of Property Act 1925 replaced the equivalent provision in the Supreme Court of Judicature Act 1873 and made assignable in law any debt or other legal chose in action. These had previously only been assignable in equity as the legal right was considered to be 'personal' to the creditor and therefore not capable of legal assignment. Equity did not treat the fact that the law categorized the right as personal as a relevant objection to the enforceability of the assignment in equity by the assignee. The effect of the Act of 1873 was to require the legal recognition (subject to the stated formalities) of the assignment and enabled the assignee to sue on the assigned debt and obtain legal remedies in respect of it.

[14] *Hill v Archbold* [1968] 1 QB 686; *Trendtex Trading Corp v Credit Suisse* [1980] 1 QB 629, at 663.

[15] *Giles v Thompson* [1994] 1 AC 142, 161.

[16] *Trendtex Trading Corp v Credit Suisse* [1982] AC 679; *Brownton Ltd v Edward Moore Inbucon Ltd* [1985] 3 All ER 499 particularly at 506–8. See also *Ellis v Torrington* [1920] 1 KB 399 at 412–14.

(c) Assignment of bare rights to litigate remains prohibited. A bare right to litigate is a cause of action (whether in tort or contract) in which the assignee has no genuine commercial interest in the outcome.

(d) In judging whether the assignee does have a genuine commercial interest, the transaction must be looked at as a whole.

(e) If the assignee does have a genuine commercial interest in enforcing the cause of action, it is not fatal that the assignee may make a profit out of its assignment.

4.09 The most frequent forms of assignment are: (a) assignments by liquidators of construction companies of the rights under construction contracts to former directors; and (b) assignments of the benefit of construction contracts and contracts with professionals as well as warranties. It often happens that a former director will indeed have a great interest in the contract. However, where the benefit to a purported assignee of a right to litigate is disproportionate to his commercial interest to an 'absurd' degree, the assignment will fail.[17] The Court of Appeal struck down as champertous an assignment of a cause of action to a former director so that he could obtain legal aid to pursue the claim.[18] However, the House of Lords subsequently held that this point was wrongly decided.[19] In short, whilst pleas of champerty may be made in these circumstances by disgruntled employers, establishing that this is the case will be difficult in light of recent authority.

Statutory and Equitable Assignment

Statutory assignment

4.10 Assignment can be in accordance with statute, as provided for under s 136 of the Law of Property Act 1925:

> Any absolute assignment by writing under the hand of the assignor (not purporting to be by way of charge only) of any debt or any other legal thing in action, of which express notice in writing has been given by the debtor, trustee or other person from whom the assignor would have been entitled to receive or claim such debt or thing in action, is effectual in law (subject to equities having priority over the right of the assignee) to pass and transfer from the date of such notice—
>
> (a) the legal right to such debt or thing in action;
> (b) all legal and other remedies for the same; and
> (c) the power to give a good discharge for the same without the concurrence of the assignor . . .

4.11 'Debt or other legal thing in action' as used in s 136 has been held to have wide application. 'Thing in action' or 'chose in action' refers to '. . . all personal rights of property which can only be claimed or enforced by action, and not by taking physical possession.'[20] Despite the phrase 'legal thing in action' being used in s 136 the phrase has been held to include equitable choses in action as well.[21] The point is, however, unlikely to have any practical effect.

[17] *Advanced Technology Structures Ltd v Cray Valley Products Ltd* (1992) 63 BLR 63, at 75.
[18] Ibid.
[19] *Norglen Ltd v Reeds Rains Prudential Ltd* [1999] 2 AC 1.
[20] *Torkington v Magee* [1902] 2 KB 427, 430 (reversed on other grounds [1903] 1 KB 644 (CA)).
[21] Ibid.; *Re Pain* [1919] 1 Ch 38, 44.

The essentials therefore for an assignment to comply with s 136 and be effective in law are: **4.12**

(a) The assignment must be in writing.[22]
(b) It must be absolute.
(c) Express notice in writing must be given to the person who owes the obligation to the assignor.

It should be noted that a statutory assignment does not require consideration.[23] Once the mandatory elements of a statutory assignment are in place, the beneficiary of the assignment can sue upon the obligation as if it were the assignor itself and give a good discharge. **4.13**

Assignments which do not come within the section are assignments which are: (a) not in writing; (b) conditional assignments[24] or assignments which are otherwise than absolute; and (c) assignments where express notice has not been given in writing. Of those three elements the last can be cured by the giving of a notice at any time. Until notice is given then the assignment is not effective as a statutory assignment and discharge cannot be given by the assignee. The essential benefit of the statutory assignment is that the assignee can sue in his or her own name. It is clear on its face that an assignment which is by way of a charge does not come within the section. However a question which does arise is whether any creation of a security interest, such as a pledge, can fall within s 136. The Court of Appeal considered this and decided that they did not.[25] In *Williams* part of a debt was assigned (or pledged) and it was held to be an equitable rather than a statutory assignment as full discharge could not be given in respect of the debt itself. **4.14**

However, under a building contract, it is possible to assign a specific debt which arises under that contract (eg the right to receipt of retention, or the right to receipt of monies due on a specific certificate) which might form part of the contract sum. In that event, it is a nice question whether the assignment is statutory (because it involves the absolute assignment of a specific debt) or equitable (because the assignment was of part of the contract sum). While the answer will depend upon the nature of the obligation to pay, it is submitted that if the contract involves interim certificates which are subject to review up to and including the issue of a final certificate, then the assignment of the right to receive retention is equitable. The reason for this is that although there is a right to interim payment, which creates periodic debts on interim certificates, the fundamental obligation of the employer is to pay the finally ascertained contract sum. Interim certificates in most standard forms can establish amounts due which are vulnerable to their being reviewed (and reduced) when the next certificate is issued. It would follow that whilst the certificate creates a debt, the debt is a payment which goes to discharge the ultimate debt, namely the obligation to pay the contract sum. However if the contract provides for unreviewable milestone **4.15**

[22] See also *Herkules Piling v Tilbury Construction Ltd* (1992) 61 BLR 107.
[23] *Re Westerton* [1919] 2 Ch 104.
[24] As in *Hughes v Pump House Hotel Ltd* [1902] 2 KB 190.
[25] *Williams v Atlantic Assurance Co Ltd* [1933] 1 KB 81.

payments then the assignment of one of them is the assignment of a single debt and this can be statutory.

Equitable assignment

4.16 An equitable assignment tends to arise where it is clear that it was intended to make an assignment but where that assignment does not comply with the s 136 formalities. An equitable assignment: (a) need not be in writing;[26] (b) can be otherwise than absolute (as for example a part of a debt); and (c) is not subject to notice in writing being given.

4.17 The difference in practice between an equitable and a statutory assignment is that an equitable assignee of a legal chose in action ought not as a matter of practice to sue in his or her name for the proceeds of the debt or the benefit of the legal right assigned without the assignor also being joined in the action. The assignor and the assignee therefore ought to be the parties which sue allowing the assignor to give a proper discharge of the debtor's obligations.[27] If the assignee sues in his own name, given that the right to sue has been transferred by the assignment to the assignee, and while for good order the assignor should be joined, in exceptional circumstances he need not be[28] and the presence of the assignor as a party will not be seen as essential to the cause of action. It follows that if limitation expires before the assignor is joined, this will not be fatal to the action brought by the assignor as the purpose of the assignor's presence in the action is for the sake of good order to permit a discharge to be given to the person sued.[29]

Assignment Subject to Equities

4.18 It is of venerable authority that an assignment ordinarily takes effect subject to equities.[30] A corollary of this rule is that an assignee cannot find himself in a better position than the assignor following assignment.[31] It follows that an assignment cannot abrogate an arbitra-

[26] Note that while an equitable assignment of a *legal* interest need not be in writing, an equitable assignment of an equitable interest must be. See s 53(1)(c) Law of Property Act 1925.

[27] See *Brandt's (William) Sons & Co v Dunlop Rubber Co Ltd* [1905] AC 454; *Performing Rights Society Ltd v London Theatre of Varieties Ltd* [1922] 1 KB 538, [1924] AC 1; *Walter & Sullivan Ltd v J Murphy & Sons Ltd* [1955] 2 QB 584; *Central Insurance Co Ltd v Seacalf Shipping Corp ('The Aiolos')* [1983] 2 Lloyd's Rep 24; *Weddell v JA Pearce & Major* [1988] Ch 26; and *Three Rivers DC v Bank of England* [1996] QB 292.

[28] See *Brandt's (William) Sons & Co v Dunlop Rubber Co Ltd* [1905] AC 454 and, even where objection is taken, the Court will not require joinder if there is no risk of a separate claim being brought. See *National Sporting Club Ltd v PCA Management Ltd* [2006] EWHC 3682 (QB).

[29] See *Central Insurance Co Ltd v Seacalf Shipping Corp ('The Aiolos')* [1983] 2 Lloyd's Rep 24; *Weddell v JA Pearce & Major* [1988] Ch 26; cf contrary dicta by Lord Denning MR in *Warner Brothers v Rollgreen Ltd* [1976] QB 430 and note its inconsistency with the above authorities and academically, see *Kloss*, 39 Conv (NS) 261; *Chitty on Contracts*, 30th edn, 2008, para 19-039; and Treitel, *The Law of Contract*, 8th edn, 1991, p 585, n 82; *Snell's Equity*, 29th edn, 1990, p 81; and the Law Commission's *Consultation Paper on Privity of Contracts, Contracts for the Benefit of Third Parties* (1991) LCCP No 121, p 50, n 99.

[30] See *Mangles v Dixon* (1852) 3 HLC 702 at 731. See also s 136 Law of Property Act 1925 itself as regards statutory assignments.

[31] This general rule can, however be excluded or modified by contract (see *William Pickersgill & Sons Ltd v London & Provincial etc, Ins Co Ltd* (1867) LR 2 Ch App 391, 397) or statute (eg Bills of Exchange Act 1882, s 38(2)).

tion clause in the relevant contract concerned[32] as an assignment of the cause of action preserves the limitations and incidents to that cause of action including the arbitration clause and other equities.

Accordingly if there is a set off by way of defence to an action, then that set off will operate as against the assignee (the assignee is in no better position than the assignor). Naturally that set off cannot become a counterclaim which demands money of the assignee because the assignee is not in contractual relations with the party against which the claim is made. **4.19**

Assignment When There Is Insolvency

See Chapter 17 on Insolvency (which addresses the operation of s 323 of the Insolvency Act 1986, r 4.90 of the Insolvency Rules 1986, and those mandatory provisions for a single sum to be due from one party to another upon an insolvency). For present purposes it is sufficient to note that separate causes of action relating to two or more parties, where there is a situation of mutual credits and debits, cannot be assigned upon an insolvency because the rule contemplates a single account for a net balance.[33] Insolvency set off is self executing and takes effect on the bankruptcy date. **4.20**

Assignment by the Contractor

This creates fewer difficulties than assignment by an employer of the benefit of its contract and the assignment of the right to receive payment under a contract is relatively straightforward. The contractor will have a debt and will assign that debt to another party. The only real questions are therefore whether that debt has accrued and whether the essential elements of the assignment have in fact been satisfied. If the essentials of an assignment have not been carried out, then the assignment will be ineffective. **4.21**

The most common form of equitable assignment in this context relates to retention monies[34] and clauses in standard forms, particularly the JCT forms. The JCT Standard Forms of sub-contract, which provide that the main contractor's interest in retention money is as fiduciary or trustee for the sub-contractor and that that money must be set aside by the main contractor for later payment to the sub-contractor, create an equitable assignment of the retention money. If the money is not set aside into a separate fund, there is no fund upon which the equitable assignment can bite and in that event the sub-contractor is an unsecured creditor. It is therefore essential that if a similar type of clause is inserted into a sub-contract as was the case in *Tout and Finch*, that the sub-contractor ensures that the retention money is in fact kept separate. **4.22**

---

[32] *Schiffahrtsgesellschaft Detlev von Appen GmbH v Voest Alpine Intertrading ('The Jay Bola')* [1997] 2 Lloyd's Rep 279 (CA); and *Through Transport Mutual Insurance Association v New India Assurance Co Ltd* [2005] EWHC 455 (Comm), [2005] 1 CLC 376, Moore Bick J.

[33] *Stein v Blake* [1996] AC 243 at 256.

[34] See *In re Tout and Finch* [1954] 1 WLR 178; *Rayack Construction Ltd v Lampeter Meat* (1979) 12 BLR 30; and *Re Arthur Sanders* (1981) 17 BLR 125.

4.23 A charge might be by way of an equitable assignment, whereas a charge cannot, by definition, create a legal assignment. If an assignment does create a charge on a book debt, then the charge must be registered otherwise it is void against the administrator or receiver or liquidator.[35]

Assignment by the Employer

4.24 There may be difficulties when an employer enters into a building contract and is not the party which will suffer damage in the event that the contract is broken by the builder. This is because under the normal rule an employer can recover its loss and damage by reason of a breach of contract, but cannot normally recover anyone else's loss.

4.25 This simple factual position has led to four cases, two in the House of Lords and two in the Court of Appeal, where problems arising out of this situation have had to be discussed. The essential problem is that if the employer under the building contract does not suffer the loss and cannot recover for that loss, then there is a black hole into which the builder's liabilities fall. Essentially the result would be that the builder could breach its contract with impunity and not have to pay any damages to its employer, because the employer has suffered no loss. These are described as the 'black hole' argument or the 'no loss' argument.

4.26 The four cases must therefore be looked at to see how the courts have dealt with the problem.

4.27 The first is *Linden Gardens Trust Ltd v Lenesta Sludge Disposals Ltd*.[36] That case dealt with two appeals in which the building contracts had prohibitions against assignment and thus purported assignments were ineffective. In each case the builder executed defective work in breach of contract and the case brought by the assignee therefore failed. However, in the second case the original assignor made a claim for substantial damages against the contractor and this was met with the argument that it had suffered no loss as it had parted with its interest in the property before there was any breach of the building contract. The House of Lords applied the principle set out by Lord Diplock in *The Albazero*:[37]

> The only way in which I find it possible to rationalise the rule in *Dunlop v Lambert* so that it may fit into the pattern of the English law is to treat it as an application of the principle, accepted also in relation to policies of insurance upon goods, that in a commercial contract concerning goods where it is in the contemplation of the parties that the proprietary interests in the goods may be transferred from one owner to another after the contract has been entered into and before the breach which causes loss or damage to the goods, an original party to the contract, if such be the intention of them both, is to be treated in law as having entered into the contract for the benefit of all persons who have or may acquire an interest in the goods before they are lost or damaged, and is entitled to recover by way of damages for breach of contract the actual loss sustained by those for whose benefit the contract is entered into.

---

[35] See s 395 of the Companies Act 1985 and *In re Cosslett* [1998] Ch 495.
[36] [1994] 1 AC 85.
[37] [1977] AC 774 at 847.

The effect of that was to allow the original contracting party to recover substantial damages because it was treated in law as having entered into the contract for the benefit of another. **4.28**

In *Darlington Borough Council v Wiltshier Northern Ltd*[38] a builder entered into two contracts with a finance company to build a recreation centre for the council which owned the site. The finance company assigned to the council all its causes of action against the builder. The council brought an action for damages for breach of contract against the builder and the judge at first instance held that the damages recoverable could be nominal only. The rationale was that the finance company had no proprietary interest in the property and had suffered no damage. The Court of Appeal held that even though the finance house had not suffered any loss, it could nevertheless recover on the same *Albazero* basis as was laid down in *Linden Gardens.* Dillon LJ held that the case was on all fours with *Linden Gardens* and relied on the following extract from Lord Browne-Wilkinson's speech in that case and quoted from Lord Browne-Wilkinson[39] as follows: **4.29**

> In my judgment the present case falls within the rationale of the exceptions to the general rule that a plaintiff can only recover damages for his own loss. The contract was for a large development of property which, to the knowledge of both Corporation and McAlpine, was going to be occupied, and possibly purchased, by third parties and not by Corporation itself. Therefore it could be foreseen that damage caused by a breach would cause loss to a later owner and not merely to the original contracting party, Corporation. As in contracts for the carriage of goods by land, there would be no automatic vesting in the occupier or owners of the property for the time being who sustained the loss of any right of suit against McAlpine. On the contrary, McAlpine had specifically contracted that the rights of action under the building contract *could* not without McAlpine's consent be transferred to third parties who became owners or occupiers and might suffer loss. In such a case, it seems to me proper, as in the case of the carriage of goods by land, to treat the parties as having entered into the contract on the footing that Corporation would be entitled to enforce contractual rights for the benefit of those who suffered from defective performance but who, under the terms of the contract, could not acquire any right to hold McAlpine liable for breach. It is truly a case in which the rule provides 'a remedy where no other would be available to a person sustaining loss which under a rational legal system ought to be compensated by the person who caused it.'

In *Alfred McAlpine Construction Ltd v Panatown Ltd*[40] a builder contracted to construct an office block and car park on a site owned by a sister company of the employer. In addition the builder entered into a duty of care deed with that sister company of the employer. The deed gave a direct remedy against the builder in respect of any failure to use reasonable skill and care and it was assignable. Defects were discovered and the employer went to arbitration seeking damages. The employer's difficulty was that it had suffered no loss. It sought to rely upon *Linden Gardens* and *Darlington* to recover substantial damages, and by a three to two decision the House of Lords held that the employer could recover nominal damages only because the contractual scheme was such that those cases did not apply. Having decided to create a duty of care deed, it was held to be contrary to the contractual intention that the employer should recover more than nominal damages. **4.30**

38 [1995] 1 WLR 68.
39 At 114–15.
40 [2001] 1 AC 518.

4.31 Finally the Court of Appeal in *Offer-Hoar and others v Larkstore Ltd and another (Technotrade Ltd, Part 20 defendant)* [2006] 1 WLR 2926 had to deal with a situation where a report prepared by Technotrade on the conduition of some land was prepared for the original owner. Larkstore purchased the site. The report was allegedly negligently prepared. The site suffered landslip damage which caused damage to adjoining owners who claimed against Larkstore. Larkstore then took an assignment of the rights and benefits of the report which had been prepared for Technotrade. The judge held that Larkstore, as assignee of the cause of action, could recover substantial damages even though the landslip causing the damage had occurred prior to the assignment.

4.32 Rix LJ reviewed the authorities and said:

> Underlying all these cases can be heard the drumbeat of a constant theme, which could possibly be described as ubi ius ibi remedium, the maxim that where there is a right there is a remedy; but it could also be said that the courts are anxious to see, if possible, that where a real loss has been caused by a real breach of contract, then there should if at all possible be a real remedy which directs recovery from the defendant towards the party which has suffered the loss. In the case of property development, where it is readily contemplated that a party which prepares the development will transfer the fruits of his work to one or more partners or successors, there is a particular need for some such solution.
>
> The courts have to work with the analytical tools which are to hand. But the essence of the matter is that the general principles which have been developed to ensure that claims are confined to victims (the rule that a party may only claim in respect of his own loss; the rule in favour of privity of contract) and that a wrongdoer should not be made to pay compensation which goes beyond his breach (the rule that an assignee may not recover more than his assignor could have recovered), rules which as far as they go are necessary and fundamental to good order and fairness in the litigation of claims, are not, if at all possible, to be allowed to become instruments of maladjustment and injustice. Thus the exception developed long ago in the carriage of goods context to allow a contracting party to recover damages against a carrier on behalf of another party to whom the goods in question are subsequently transferred has been brought into use in a modern situation where there is an equal need to find a solution which matches the commercial situation, and where no other solution had been found to be at hand. Of course, where a solution has been provided by statute, as where a contract of carriage of goods by sea is novated statutorily, as in the case of bills of lading, or where there are other solutions readily to hand (as in *The Albazero* [1977] AC 774 or in the *Panatown Ltd* case [2001] 1 AC 518), there may be no need, and thus it will be thought to be undesirable, to find an exception to general principle.

4.33 Accordingly the attitude of the courts is not friendly to those which it perceives as avoiding its obligations through the operation of chance when the result may be the creation of a legal black hole as a result of which a party who is considered as really responsible for losses would escape liability altogether. The tone of Rix LJ's judgment is quite clear on this topic. In the same case Mummery LJ's approach reaches the same result.

4.34 If an assignment is made, it may be said that the assignee can only recover the loss which the assignor suffered. If the assignor suffers no real loss because it is not the ultimate building owner, then it might be said that the assignee can be in no better position. However, the effect of these three cases leads to the conclusion that if the contractual circumstances are that the assignor, albeit itself suffering no loss, entered into the contract for the benefit of the assignee, then the assignee can recover its own losses in an action against the contractor since the assignor could itself have recovered substantial damages. But this is subject to a caveat that the contractual structure must permit the recovery of substantial damages by the assignor. If, as in *Panatown* it could not do so, then the assignee will be able to recover nominal damages only.

It is obvious that these cases lead to the need for great care in setting out the contractual structure so that there can be recovery of more than nominal damages in these very common situations. They point to the need for duty of care deeds which provide a proper indemnity to the party which is going to suffer the loss in the event of defects to the works causing loss to a non-contracting party. Sophisticated developers ought to have no difficulty in so arranging their affairs with proper specialist advice. **4.35**

Sub-Letting

This is no more than vicarious performance by the contractor. Many standard and bespoke forms prohibit sub-letting without the permission of the contract professional or the employer. However, since most contractors carry out their work entirely or nearly entirely with the use of sub-contractors, this is less insisted upon. **4.36**

Vicarious performance does not change the contractual relationship between main contractor and employer. The contractor will remain liable for defects and other defaults in the work contracted for. Sub-letting in many forms outwith the permission granted may make the contractor liable to forfeiture. There is little sanction for an employer when the prohibition against sub-letting is breached. There can be no claim for damages, because damage is not suffered unless the work is defective or late, and that relates to another breach altogether. The only remedy is forfeiture, but that may be thought of as a 'nuclear' option which is disproportionate to the loss or inconvenience suffered by the employer. **4.37**

Contracts (Rights Against Third Parties) Act 1999

Most of the cases relating to assignment referred to in this chapter pre-date the Contracts (Rights Against Third Parties) Act 1999. That Act was passed to deal with the perception that the law was defective because it did not allow third parties, who were not in contractual relations, to benefit from contracts designed for their benefit. The law's disallowance of such rights was ancient but enshrined by the House of Lords in *Dunlop Pneumatic Tyre Co Ltd v Selfridge*, albeit criticized academically and judicially ever since.[41] **4.38**

The House of Lords in *Linden Gardens* and the associated cases sought to permit recovery for parties whom the contract was designed to benefit. But that line of cases showed how hit and miss the position is. Whilst in some circumstances there may be a need to allow third parties to benefit, it is not obvious why this should be an axiom. After all, parties to a **4.39**

[41] [1915] AC 847. See for example Lord Diplock in *Swain v The Law Society* [1983] 1 AC 598 at 611 describing it as 'an anachronistic shortcoming that has for many years been regarded as a reproach to English private law.'

commercial bargain can ensure that their contractual structure is such that all who should benefit do in fact do so.[42]

4.40 Parliament passed the Contracts (Rights of Third Parties) Act 1999 in response to a Law Commission Report, and it came into force on 11 November 1999 and applies to contracts entered into after 11 May 2000. The scheme is simple. It permits third party rights to be enforced if the contract expressly provides for it or the contract purports to confer a benefit on that third party.[43] The third party must be referred to by name or by class and rely on an argument that he or she falls within the act by process of construction.[44] The Act gives the third party any remedy which he or she would have had were he a party to the contract. Such remedies will include a right to damages, specific performance, or an injunction. If the contract contains an arbitration clause, then the third party is bound by that.[45] In order to come within the Act therefore the claimant must show that the term of the contract relied on purports to confer a benefit on him. If the claimant succeeds in showing that, then he is entitled to enforce the term directly against the defendant unless the defendant persuades the court that the parties did not intend the term to be enforceable by him.[46]

4.41 In spite of the fact that the Act seems almost designed to assist in construction cases and developments where contracting developers may not own the property on which the structure is being built, development agreements now seem almost as a matter of routine to contract out of the effect of the Act.[47] However, the Act is playing an increasingly important role in a construction law context—and this trend is likely to continue.

[42] It is interesting that the charming story and result in *Beswick v Beswick* [1968] AC 58 did not need anything like the 1999 Act to permit justice to be done.

[43] See s 1 of the Contracts (Rights of Third Parties) Act 1999.

[44] *Avraamides v Colwill* [2006] EWCA Civ 1533.

[45] See s 8 of the Act. See *Through Transport Mutual Insurance Association v New India Assurance Co Ltd* [2005] 1 CLC 376.

[46] See *Laemthong International Lines Co Ltd v Abdullah Mohammed Fahem & Co* [2005] EWCA Civ 519 *per* Clarke LJ at [22].

[47] Pursuant to s 1(2) of the Act which states that the Act does not apply if on a proper construction of the Act the parties did not intend the term to be enforceable by a third party. The contracting out is usually by a clause saying that the Act is not intended to be of any effect under the agreement and no third party is intended to benefit. For a discussion of why this might be so (and the misconceptions within these reasons) see the article by Francis Ho, 'Third way' (2006) 150 SJ 5, 150–151.

Part II

LIABILITY

5

THE ROLES OF PROFESSIONALS

Background and Professionals Generally

The work of construction professionals is subject to the Housing Grants, Construction and Regeneration Act 1996 and therefore the implication of the statutory terms of that Act will apply to the professionals' contracts to which that Act is relevant. Consequently all written contracts are subject to the right to adjudication, the requirement of interim payment, and the exclusion of 'pay when paid' conditions. **5.01**

In traditional forms of contract the employer's professional is responsible for the design of the works and the contractor is responsible for their construction. These types of contract and the roles they create are considered first. The professional, be they architect, engineer, quantity surveyor, or project manager, shall be referred to here simply as the professional in spite of the comparative misnomer borne of the fact that, as has already been noted, contractors are as professional as the professionals. There is normally a lead consultant. However the lead consultant does not have to be an architect. More recently the role of the lead consultant had been described in the following terms by Coulson J: **5.02**

> The lead consultant may, but will not necessarily, be the architect; the discipline from which the lead consultant is drawn will depend on the type of project and the particular wishes of the employer. But, whichever professional is appointed to this role, he will usually be engaged by the employer pursuant to a written contract, the terms of which will require him to perform a wide range of duties. He must generally act as the employer's agent and do all that he reasonably can to implement the entirety of the employer's requirements. That may involve incorporating and integrating the specialist designs of others into one coherent whole, as well as the performance of an onerous administrative function . . . The lead consultant will usually agree, on behalf of the employer, the particular terms of the building contract appropriate to the project. This is important because it will usually be the lead consultant who will be responsible for administering that contract. In doing this he has to act as the employer's agent and, for certain functions, to adopt a quasi-arbitral role, balancing fairly the interests of both employer and contractor. Most of the standard forms of contract identify a wide range of tasks which the contract administrator must perform. It is therefore critical for the lead

consultant to ensure that the contract as between the employer and himself incorporates all the obligations which, in its terms, the standard form of building contract assumes have been imposed upon the lead consultant as contract administrator. . . . The lead consultant will usually have the necessary authority to make decisions and to issue all instructions on behalf of the employer. Indeed, it is impossible to administer a construction contract satisfactorily if the lead consultant does not have the power to issue such instructions.[1]

5.03 The professional's first duty is to gather up the employer's requirements with a view to the production of a design which is fit to go out to tender. In this section we will be dealing with the principal professional engaged by the employer. The professional must understand his limitations and ensure that he obtains appropriate assistance so that any gaps within his expertise are plugged. Traditionally it was the case that an architect was the 'leader of the design team', whose responsibility it was to put together a team of professionals to ensure that the employer's requirements were put into a proper design form. Thus they might assemble:

(a) a civil engineer to ensure that the underground work was properly designed;
(b) a structural engineer to ensure that the above ground structure can take the stresses and strains to be placed upon it;
(c) an electrical engineer to ensure that complicated electrics associated with modern buildings are properly dealt with;
(d) a mechanical engineer and a heating engineer to ensure that the heating and ventilating systems are properly designed.

5.04 Additional to those professionals there are specialist designers of specialist works. Curtain walling, windows, lifts, and other specialist items of construction must be integrated into the design. These specialist sub-contractors will not only design their works but they will also construct them. All their design must be integrated into the main body of the design which the other professionals are putting forward.

5.05 Ensuring that the design is integrated and that there are no clashes is more of a complicated business than ever. The wiring must go through routes which must not clash with the ventilating ducting, for example. There must be enough space in the roof area for all the elements which are required to be routed through it to fit together, to take another example. Over the last twenty years it has become apparent that architects are not necessarily the best suited professionals for this coordination role. The result of this has been the emergence of the project manager whose job is not necessarily creative in terms of design but is essential to the management of all the designers' work and their coordination. Allied with this has been the recognition that the management of a large and complicated site is not always required to be done by architects, many of whom consider themselves more creative than administrative.

5.06 All professionals are liable in damages to their employer for a breach of contract or negligence. A breach of the duty to take reasonable care can only be proved by expert evidence

[1] See paras 149–151 of Coulson J's judgment in *McGlinn v Waltham Contractors Ltd* [2007] EWHC 149 (TCC), [2008] Bus LR 233.

of another professional of like discipline which would be designed to show that the professional concerned and complained of had fallen below the standard of an ordinary competent professional.[2] The evidence of another professional in another, but similar, discipline will not do.[3]

The professional will not, save where there is an express term to that effect, guarantee a result. In the case of a defective structure therefore liability will only attach if the professional concerned has been negligent.[4] An error of judgment is not negligence.[5] **5.07**

Contract professionals which are employed by the building owner do not owe a duty of care to the contractor[6] in how they carry out their work or how they certify for work done.[7] The professional cannot serve two masters. It has been mooted that under-certification by a contract professional may give rise to a cause of action in the event that the contractor does not receive its due.[8] However there can be no proper rationale for such liability. If works are not certified properly, the contractor has a right to go to adjudication and arbitration in order to receive sums due.[9] There is no warrant for placing the professional in a legal relationship of proximity with the contractor which would allow direct recovery from the professional of sums which the contractor can obtain in adjudication or arbitration. **5.08**

---

2 *Bolam v Friern Hospital Management Committee* [1957] 1 WLR 583; and *Whitehouse v Jordan* [1981] 1 WLR 245.

3 *Investors in Industry v South Bedfordshire District Council* [1986] 1 All ER 787 at 808–9, a passage which did not find its way into other reports.

4 See *Bolam*, n 2 above, and *Greaves v Baynham Meikle* [1975] 1 WLR 582; and *Eckersley v Binnie* (1988) 18 Con LR 1 at 79–80.

5 *Saif Ali v Sydney Mitchell & Co* [1980] AC 198 at 220 *per* Lord Diplock.

6 Save as relating to the health and safety of employees and the duty will then be owed to the individual workmen on site.

7 The only case where this was argued at appellate level was in *Pacific Associates v Baxter* [1990] 1 QB 993 and the Court of Appeal rejected the argument. Prior to the appeal HH Judge James Fox-Andrews QC in *Michael Sallis v Calil* (1988) 4 Const LJ 125 held that such a duty may lie, but his reasoning does not accord with the reasoning of the Court of Appeal in *Pacific Associates*. That case was consolidated with the *Pacific Associates* case in the Court of Appeal but did not proceed.

8 See where this was mooted but not decided in *Costain Ltd v Bechtel Ltd* [2005] EWHC 1018 (TCC) particularly at [38]–[48] where Jackson J considered that the view of a duty being owed by the professional was arguable.

9 See Puchas LJ in *Pacific Associates v Baxter* [1990] 1 QB 993 at 1024 where he said:

> The position might well have been otherwise if G.C. 67 or some other provision for arbitration had not been included in the contract. The position of the engineer, if the contract so provided, might well be arbitral or quasi-arbitral: see *per* Lord Morris of Borth-y-Gest in *Sutcliffe v Thackrah* [1974] AC 727, 751G. On the other hand, he might well owe a duty to exercise skill and care to a contractor who relies on his valuations and certificates just as the person operating the weighbridge was found to be liable in *Glanzer v Shepard* (1922) 135 NE 275 or even to be 'shot at from both sides' as envisaged by Lord Salmon in *Arenson v Arenson* [1977] AC 405, 438. These are all exceptional cases. It will be rare in these days for a contract for engineering works of any substance in which a consulting engineer is appointed not to have an arbitration clause.

> Not only is it rare for there to be no arbitration clause, but when that case was decided there was no right to adjudication, which now applies to all construction contracts which are in writing or so evidenced. Furthermore, since that dictum the House of Lords overruled *Crouch* in *Beaufort Developments (NI) Ltd v Gilbert Ash (NI) Ltd* [1999] 1 AC 266.

5.09 It is often said that a contract professional has a duty to act fairly as between contractor and employer and apply his or her professional judgment impartially. This statement is true so far as it goes, in that the employer will be in breach of contract with the contractor if its agent does not act fairly and professionally.[10] This is because the employer has a duty to ensure that its agent acts fairly. But that is wholly different from the situation of a duty to act fairly being owed direct to the contractor by the contract professional. The contractual structure does not place the contract professional in contractual relations with the contractor, but instead it does with the employer.[11]

5.10 Contract professionals are entitled to payment of their fees. This ought to be pursuant to an express agreement setting out what payments are to be made and when. A contract which is subject to the Housing Grants, Construction and Regeneration Act 1996 will also have a right to periodic payment. Absent an agreement as to fees payable the professional will be entitled to a quantum meruit.

5.11 The contract professional will have a lien on his or her papers and documents as security for unpaid fees, just as solicitors do. There are no modern cases where this has been discussed. Subject to the lien the plans and papers in their possession are the property of the employer, save for the working papers, which remain the property of the contract professional.[12] However the copyright in the design, plans, and drawings, being an artistic work, will, absent any express terms of the contract of engagement, remain with the contract professional.[13] If, therefore, an employer wishes to use the design, plans, or drawings, then it ought to ensure that the contract of engagement transfers copyright to the employer, or, as is more usual, that the employer is granted a licence to use them. A professional may seek an injunction under s 96 of the Copyright, Designs and Patents Act 1988 to restrain a breach of copyright. But this is unlikely to succeed on an interim basis in case of a dispute because the balance of convenience will favour not granting the injunction and leaving the professional to recover damages set at the amount which would be chargeable as a licence for the use of the design documents.[14]

5.12 Subject to that introduction it is necessary to outline the respective roles of the principal professionals.

---

[10] See, for example, *Croudace v London Borough of Lambeth* (1986) 33 BLR 20.

[11] For discussion of the contractual structure in construction cases and its impact on duties of care see *Henderson v Merrett* [1995] 2 AC 145 *per* Lord Goff. But see also Neuberger LJ (now Lord Neuberger MR) in *Riyad Bank & Ors v Ahli United Bank (UK) Plc* [2006] EWCA Civ 780, [2007] PNLR 1, [2006] 2 Lloyd's Rep 292, [2006] 2 All ER (Comm) 777, [2006] 1 CLC 1007 on the topic of contractual structure and its interplay with the law of tort.

[12] *Leicester County Council v Michael Faraday and Partners* [1941] 2 KB 205; and *Chantrey Martin v Martin* [1953] 2 QB 286.

[13] See ss 1–4 and 65 of the Copyright, Designs and Patents Act 1988.

[14] See *Stovin-Bradford v Volpoint Properties Ltd* [1971] Ch 1007 at 1016–17, 1020, and 1022–3; *Hunter v Fitzroy Robinson* (1977) 10 BLR 88 at 94–5; and s 97 of the Copyright, Designs and Patents Act 1988.

The Architect

Unlike the other professionals, nobody can practice as an architect without being registered pursuant to statute.[15] Indeed it is an offence to describe yourself as an architect if you are not properly registered.[16] A firm of architects (a partnership or a company) can employ people who are not qualified as architects, but then the firm must be under the control or management of someone who is. There are exceptions to the full rigour of this, in that, for example, you can, if you wish, style yourself as a 'golf course architect',[17] but it is not the purpose of this chapter or any part of this book to delve into such niceties. It is enough to know that you would commit a crime if you style yourself as an architect and seek to practice as such. If you wish to consider such arcane pleasures as pursuing a hobby in that sphere, then it is worth reading the Architects Act 1997, or, better, taking legal advice. Furthermore if a contract is made with someone who describes themselves as an architect, then the 'architect' could not sue for fees. Any contract would be voided by the misrepresentation[18] and there would be no right to a quantum meruit since the 'work' would have been obtained by fraud. There is no English case where this has been considered. **5.13**

The normal rules of contract formation apply to all professional contracts, and therefore to architects as well. It is a little surprising how many agreements are of an informal nature. There is no requirement to practice on the RIBA's standard terms, although sometimes they will be incorporated by reference into a contract which is made by exchange of letters. **5.14**

The architect is often described as the leader of the design team[19] and, subject to the role of the specialist project manager, is responsible for assembling the design team, coordinating the professionals, and ensuring that the primary design is got out to the satisfaction of the employer. Then, once the design is completed the next stage is the initial estimation of cost (so that the employer will know whether the project is likely to be viable) and then the preparation of the tender documents. The contractors who are to tender are selected and the tender documents are put together. These will involve a specification and drawings and, if appropriate, Bills of Quantity. The latter are drawn up by quantity surveyors and are discussed separately. The drawings must not only set out the conceptual design but also provide for the design of other specialists, including mechanical and electrical engineers and specialist sub-contractors who provide their own design. It is an unusual design of a major structure which is only produced by the architect alone. The above duties may include the need to give specific warnings about price increases and inflation, or giving advice as to the net lettable area of the building, but these depend upon the terms of the **5.15**

---

15 s 20 of the Architects Act 1997.

16 s 21 of the Architects Act 1997.

17 See s 20(2) of the Architects Act 1997.

18 See misrepresentation and the Misrepresentation Act 1967.

19 Or the captain of the ship in *Kensington and Chelsea Health Authority v Wettern Composites* (1984) 31 BLR 57 at 84. This appellation must be open to doubt today in cases where a project manager is involved.

contract between architect and employer.[20] The architect then receives the tenders returned by the tenderers and must assess which is the most appropriate for the employer to accept. Evaluation of tenders will not only go to price but also to suitability, ability to complete to price, financial strength, and reputation. The lowest tender is usually the one which is accepted because tenderers should be selected on the basis that they are all as capable as one another of doing the job. This is not always the case and other factors sometimes will influence the decision of which tender is to be accepted. These may include familiarity with the contractor concerned and the contractor's perceived competence in any specialist area under consideration.[21] Following the decision on which tender is to be accepted, the architect's next, and possibly most important job, is to run the contract to the conclusion of the works.

5.16 The administration of the contract conditions will include:

(a) ensuring that the design is completed to the extent that it is not so prior to tender;
(b) issuing information to the contractor so that the works can be completed;
(c) awarding extensions of time, if appropriate;
(d) certifying the value of the work periodically (usually monthly);
(e) certifying when the works are practically completed; and
(f) certifying when all defects are remedied and the final certificate can be issued.

Perhaps the most important function of the architect, which is necessary to ensure that the above matters are properly dealt with, is to periodically inspect the works as they are progressed. This is often referred to as supervision but this is a misnomer. Under traditional forms of contract the architect does not supervise in the sense in which the term is normally understood. That function is carried out by the contractor's site personnel. Periodic inspection does not involve giving instructions to the contractor; rather it involves inspecting the work and its progress so that the above functions at (a)–(f) are properly carried out. Necessarily this process will involve checking whether the work has been properly done; in other words, whether the work is defective or not. If the work has not been properly done, then it does not fall for certification. As has already been noted, this process is part of the architect's responsibilities which are owed to the employer and are not for the benefit of the contractor.[22]

---

[20] As for example in *Nye Saunders v Alan E Bristow* (1987) 37 BLR 92 for inflation warnings; and *Gable House Estates v The Halpern Partnership* (1996) 48 Con LR 1.

[21] This is especially the case when dealing with old buildings. There are several contractors who have built up a well-deserved reputation in dealing with and restoring old buildings. If their tenders are greater than their competitors, it may be because they understand the risks involved with such structures better and also because they may carry greater overheads in the form of expertise than their competitors.

[22] See *Sutcliffe v Chippendale and Edmondson* (1971) 18 BLR 149, particularly at 162 where the judge identified the duties of the architect. That case was the first instance decision of Judge Stabb QC which led to the House of Lords decision in *Sutcliffe v Thackrah* [1974] AC 727. See also *Corfield v Grant* (1992) 29 Con LR 58; and *Consarc Design Ltd v Hutch Investments Ltd* (2002) 83 Con LR 36 where HH Judge Bowsher QC said: 'what is absolutely clear is that the architect does not guarantee that his inspection will reveal or prevent all defective work. . . .'; *George Fischer Holdings Ltd v Multi Design Consultants Ltd* (1998) 61 Con LR 85; *East Ham Corporation v Bernard Sunley* [1966] AC 406; and finally *McGlinn v Waltham Contractors Ltd* [2007] EWHC 149 (TCC), [2008] Bus LR 233 at paras 213–219.

The architect is, like any other professional, liable to his or her employer for any breach of contract and will, to their employer, owe a duty to take reasonable care in the performance of their duties.[23] There was discussion in the old cases of the architect acting in an arbitral or judicial role or a quasi arbitral or judicial role and being immune from suit accordingly. However that view has long gone. The rationale of the old cases seems to have surrounded the role of the architect in certifying sums due and a perception that his or her view was final. In nearly all construction contracts this is not the case, the architect's view is subject to review by an arbitrator. **5.17**

It has been held at first instance in the 1960s that an architect cannot delegate his or her duties.[24] This is too simple a proposition. No contract professional can delegate its legal duties, even if he or she does obtain vicarious performance. But architects are no longer the sole designer of work, if they ever were. Everyone expects that part of the job of the architect is to assemble a design team where specialists will play their part. It is therefore inevitable that the design will not be exclusively in the hands of the architect in all save the simplest jobs. Employers expect this. Therefore a sounder proposition would be that an architect will not escape liability for negligence if some of the work taken on by that architect is delegated to another—but the work taken on by an architect has a restricted meaning and does not include design work by specialists assembled as part of the design team for the benefit of the employer.[25] **5.18**

The Project Manager

The project manager's job is, as already stated, to manage and coordinate the various professionals' work and is a relatively new specialism whose principal members are quantity surveyors by training. In the era of management contracting and the proliferation of specialist expertise, project management is an obvious development in the evolution of the modern construction industry. Project managers arose out of the existing pool of expertise and have evolved from them. **5.19**

It would be unfair to say that project managers have arisen out of the failure of the architectural profession to properly manage as well as design construction works. But it is fair to say that management is a specialist expertise on its own and all architects cannot all be expected to have that ability. **5.20**

The construction industry is, therefore, still learning how to fit project management into the scheme of contractual undertakings. The Project Management Institute was founded in 1969 but the UK branch was only founded in 1995. It has defined project management as 'the art of directing and coordinating human and material resources throughout the life of a project by using modern management techniques to achieve predetermined objectives of scope, cost, time, quality and participation satisfaction.' **5.21**

---

23 *Sutcliffe v Thackrah* ibid.

24 *Moresk Cleaners v Hicks* [1966] 2 Lloyd's Rep 338.

25 For an application of this principle, see the decision in *London Borough of Merton v Lowe* (1981) 18 BLR 130 where the architect was not held liable for the defective design of a specialist nominated sub-contractor.

5.22 The above definition is slightly nebulous but it is helpful to look at it as being the basic objective of the project manager who is a construction professional. The essentials of project management in a construction context are:

(a) Specification of objectives as to what is to be built, the timeframe, and costs, and then choosing and assembling the design team.
(b) Ensuring resources are used to maximum effect through the use of proper planning and procurement techniques.
(c) Coordination and control of planning, design, estimating, contracting, and construction in the entire process.
(d) Developing effective communications and communication systems relevant to all involved in the project as well as effective and appropriate dispute resolution procedures.

5.23 Construction project managers get into the business via some other discipline, usually quantity surveying or management contracting. There is very little case law on the role of project managers for this reason. The two principal cases so far are first instance decisions only.[26] For these reasons it is too early to be definitive as to the scope of a project manager's role. However it can be said that it is primarily a coordinating role and the project manager is the hub of a wheel where the spokes are the employer, the contractor, the architect, and other professionals, and other specialist designers and sub-contractors—in other words the construction team. Good management may be the type of art which does not attract legal attention. However, this is unlikely and the lack of legal authority is more to do with the newness of the role than anything else. A poor project manager, who does not keep all the composite elements of the construction team informed of things which they need to do their job properly, is capable of causing enormous loss to his employer.

5.24 One aspect of the project manager's work is programming the work. This is a complicated business as will be seen when litigation programmers are discussed in Chapter 22, Litigation. But the basics of programming are simple. Construction of a structure involves successive and concurrent activities which are dependent upon one another periodically. To take a simple example: the construction of a house. First come the foundations, then the beams and slabs, then the internal skin of the external walls, then the internal walls, and so onward to first and second fix work. The foundations are the first job upon which all the others are dependent. The logic therefore requires the programmer to programme the foundations first and the beams and ground floor slab and walls thereafter. This is simple in concept because the example is simple. But the logic may change if, for example, the external walls are to be pre-fabricated. In that event the construction (but not the erection) of the external walls can be programmed simultaneously as opposed to successively. The programmer can therefore change the logic of the successive steps in construction to take account of the alteration in the mode of the construction of the walls. The programmer can, armed with the logic and construction periods for each concurrent and successive activity, create a

[26] The professional negligence cases *Pozzolanic Lytag v Bryan Hobson Associates* [1999] BLR 267 decided by Dyson J; and *Great Eastern Hotel v John Laing Construction Co Ltd* [2005] EWHC 181, (2005) 99 Con LR 45 (TCC) which was a decision by HH Judge Wilcox dealing with construction management.

programme which will allow the dates for the commencement and completion of each construction element to be ascertained.

Modern construction projects involve many different activities of a far more complicated nature than the simple example given above. It is the job of the programmer, and the project manager, to ensure that each of these many steps are appropriately programmed so that, in theory at least, it will be known when each activity will start and finish such that different trades will not get in each other's way and the worksite will be ready for the trades when they arrive on site at the programmed start. **5.25**

Engineers

The role of engineer is two-fold. First, there is the role of engineer as subordinate to the architect under a normal form of building contract. Secondly, there is the engineer as leader of the design team in his own right when he has similar duties to the architect—only as engineer under an engineering contract such as the ICE conditions. **5.26**

As a subordinate, the engineer's job is to carry out the instructions of the architect. This will normally involve checking, where appropriate, specialist design and doing such calculations as are appropriate such as to ensure that the structure will bear all the loads (dead load and applied load) which the structure is designed to bear. **5.27**

As the leader of the team, as already stated, the duties are similar to those of the architect. Under the ICE conditions the duties are of a more elaborate nature than those of the architect under the JCT conditions. In particular, the additional duties will involve checking and approving the contractor's method statements and programmes. The engineer is also more likely to be his own quantity surveyor and will be measuring and valuing the works, whereas an architect may have a quantity surveyor to carry out that role. **5.28**

Quantity Surveyors

Quantity surveyor is a specialism which is particularly English in character. Most other countries have their professional architects and engineers carry out the role of quantity surveyors. In the United Kingdom this is a specialism. Here the role begins with taking off the design from the drawings and drawing up bills of quantities which will measure the work which is to be done so that the work can be tendered for and valued by quantity. To use a crude example, in the construction of a house there will be so many square metres of brick walls on the external skin of brickwork; so many square metres of brickwork or blockwork on the internal skin and for the internal walls. That will produce a set of elements of structural work which is measurable—in this example by the metre. Then the tendering contractors will place a price per metre against each element. In this example it will be £x per metre of brickwork and £m per metre of blockwork. Thus the surveyor can come to site, at the time of ascertaining the value of the work, and measure what of each element of the work has been done and the price per metre can be applied to each total. Then when, for example, the employer requests the addition of further brickwork (by way of the construction of a garden wall or an additional house) the brickwork element of that extra work **5.29** can be valued by reference to the prices submitted. It will be important, however, to see exactly what costs are included in the unit rate. Are on site overheads included? Are off site overheads included? Are either to be measured and valued separately? Exactly what costs are included within the rate; is it simply materials and labour? Does it include plant? Does it include scaffolding? Does it include for the creation of working space? These are questions which must be addressed, usually by reference to criteria set out in the bills of quantity themselves and the standard method of measurement.

5.30 The above example given is crude but explains how bills of quantity work in general terms. Quantity surveyors also are known as construction cost consultants. Their expertise relates to the costing of work whether or not it is done as part of a contract under bills of quantities. Their expertise relates to all aspects of costings and cost control. One of the most crucial parts of their responsibilities relates to the ascertainment of a contractor's loss and expense should events happen which give rise to an entitlement to payment of such expenses. Loss and expense and its valuation is treated separately—see Chapter 14. It is the quantity surveyor's expertise which comes into play when these matters are to be considered. Since money is a function of time, quantity surveyors have become more and more involved with programming, as to which see the discussion of the project manager above. The matters attributed to project managers in relation to programming are also applicable to quantity surveyors, who, unsurprisingly, have become concerned with project management and, as has already been said, many project managers are quantity surveyors by training.

5.31 If the contract has a fluctuations clause which provides for increased payment to the contractor in the event of inflation of certain costs, then the quantity surveyor will be responsible for applying the formula to the value of the work done as the work progresses.

5.32 Whilst there has been a significant quantity of litigation concerning architects and engineers and therefore judicial consideration of their duties, there has been very little in relation to quantity surveyors. This may be because over-valuation by quantity surveyors leading to claims by employers is less likely with other professionals also involved. Whilst an architect or engineer may have no cross check on his work, a quantity surveyor will have the architect or engineer, as the case may be, cross checking the work. Additionally when it comes to matters of pure valuation, the work of the quantity surveyor is of a more objective nature and is not concerned with the quality of the work done.

6

TORT AND CONSTRUCTION LAW

The Common Law: Negligence

Introduction

The elements of the tort of negligence are: (1) the existence of a duty of care owed by the defendant to the claimant. The enquiry as to whether there is a duty of care is not carried out in the abstract but must consider whether this defendant owes this claimant a duty of care in relation to the type of damage sustained by the claimant; (2) a breach of that duty **6.01** of care by the defendant, ie a failure to satisfy the standard required by the law; (3) a sufficient causal link between the damage suffered by the claimant and the breach of duty by the defendant; (4) the damage suffered by the claimant must not be unforeseeable and thus too remote from the breach of duty by the defendant.

6.02 This section will focus on the first element of the tort of negligence, namely, the circumstances in which a defendant will be held to owe a duty of care to a claimant in respect of particular types of damage. This is necessary because, as is explained below, the law of tort in the last thirty years has undergone substantial change in relation to the circumstances in which contractors and construction professionals owe tortious duties of care. The position is by no means fixed today since the evolution of the law has been haphazard and dependent upon the judicial philosophies of different judges who had ascendancy at different periods during which the appellate based law was made.

6.03 In order to make the current position comprehensible, it is necessary to review the development of the law. However, before reviewing the authorities on the duty of care, the concept of 'pure economic loss' will be considered.

The characterization of loss

6.04 In one sense all loss is economic. Even in straightforward personal injury or property damage cases, the claimant will be seeking a monetary remedy. However, the courts have generally considered that different principles may be required when considering recovery in negligence in respect of loss which is not sufficiently tied either to personal injury or damage to property. Consequently, the proper characterization of damage is an important stage in the determination of a duty of care in a particular case.

6.05 Although the label 'pure economic loss' may be and has been applied to a wide variety of different types of case,[1] and a full discussion of them is beyond the scope of this book, three important types of loss will be considered.

6.06 First, there is the 'pure economic loss' which is not consequential upon any personal injury or physical damage to property belonging to the claimant[2] and is caused by negligently given advice and/or representations. This would include, in the construction context, for example, the advice given by a quantity surveyor to a contractor on a tender. If the quantity surveyor negligently advises too low a figure and the contractor wins the tender, the contractor may suffer a loss.

6.07 Secondly, there is the 'pure economic loss' which is consequential upon damage to the property of a third party. In *Spartan Steel and Alloys Ltd v Martin & Co (Contractors) Ltd*,[3] the plaintiffs operated a business manufacturing stainless steel alloys. The defendant negligently damaged an electrical cable under the road in which they were digging. The cable belonged to the electricity board but supplied the plaintiffs' factory. The damage caused a

---

[1] See, for example, the discussion by Oliver LJ in *Leigh and Sillivan Ltd v Aliakmon Ltd* [1985] 1 QB 350 at 377.

[2] Or in which he has a sufficient proprietary or possessory interest or personal injury.

[3] [1973] QB 27.

power failure which lasted for fourteen hours. Immediately the power supply failed, the plaintiffs had to pour molten metal out of their furnace to prevent the metal solidifying and damaging the furnace. As the plaintiffs could not keep the metal at the correct temperature and complete the 'melt', the metal depreciated in value by £368 and they lost a profit from the sale of the metal from that melt of £400. They could also have completed four further melts during the power cut and their loss of profit from those melts was £1,767. In an action for negligence against the defendants, the plaintiffs claimed all three sums as damages, a total of £2,535. The defendants admitted negligence and liability for the physical damage of £368 but denied liability for the plaintiffs' economic loss. The Court of Appeal held that the plaintiffs were entitled to recover the £400 as damages as the loss of profit from the melt was a foreseeable financial damage immediately consequential on the foreseeable physical damage to the metal, but they were not entitled to recover the loss of profit from the four melts due to the negligent interruption of the electricity supply.

Thirdly, there is the loss suffered by the owner when his property suffers from an inherent defect which reduces its value but causes no personal injury or damage to other property. It is this category which has provoked the most controversy, in particular in the construction field. **6.08**

There are a number of reasons why the courts have sought to treat claims for 'pure economic loss' in a different way to claims for damages for personal injury or damage to other property. **6.09**

First, there is a concern that economic loss can be suffered by a large variety of potential claimants following a single act or omission by the potential defendant. In the words of Cardozo CJ in *Ultramares Corp v Touche, Niven & Company*[4] 'the defendant would be exposed to a liability in an indeterminate amount for an indeterminate time to an indeterminate class.' This was a feature of the reasoning in *Caparo plc v Dickman*.[5] **6.10**

Secondly, there is a concern that permitting recovery of pure economic loss would trespass upon the proper province of contract. This was a feature of the reasoning of the House of Lords in *Murphy v Brentwood DC*.[6] **6.11**

Thirdly, there is a concern about how to identify the standard to which the building or other property should have been constructed, ie a feature of a building may only be a 'defect' in certain circumstances or subject to certain uses. Contracts provide objective standards against which performance can be measured. The duty of care in tort is unable to provide those standards. This was the concern of Lord Brandon in his dissenting speech in *Junior Books v Veitchi*.[7] **6.12**

[4] (1931) 255 NY 170.
[5] [1990] 2 AC 605 *per* Lord Bridge at 620H–621F.
[6] [1991] AC 398 *per* Lord Bridge at 468D–470C, 480F–481C.
[7] [1983] AC 520, [1983] 3 WLR 477, [1982] 3 All ER 210.

Historical overview

The modern genesis of the duty of care in relation to personal injury and damage to other property: Donoghue v Stevenson

6.13 The law of negligence initially evolved in order to deal with damage to persons or their property by reason of the negligent breach of a duty of care owed by one person against another. Prior to *Donoghue v Stevenson*,[8] the House of Lords had not accepted that any general principle could be used to determine when one person owed a duty of care to another to hold him harmless from personal injury or damage to property. Instead the courts had identified a number of particular types of relationship in relation to which a duty of care would be imposed. In *Donoghue v Stevenson*, the majority in the House of Lords held that there was a common principle underpinning the categories of relationship in which a duty of care had been found. Lord Atkin held:

> . . . in English law there must be, and is, some general conception of relations giving rise to a duty of care, of which the particular cases found in the books are but instances. . . . The rule that you are to love your neighbour becomes in law, you must not injure your neighbour; and the lawyer's question, Who is my neighbour? receives a restricted reply. You must take reasonable care to avoid acts or omissions which you can reasonably foresee would be likely to injure your neighbour. Who then, in law, is my neighbour? The answer seems to be—persons who are so closely and directly affected by my act that I ought reasonably to have them in contemplation as being so affected when I am directing my mind to the acts or omissions which are called in question.[9]

6.14 Lord Atkin's reliance[10] on cases such as *Heaven v Pender*[11] indicated that he considered that the principle was only applicable to cases in which the claimant sought to recover in respect of personal injury or damage to other property.

6.15 The principle in *Donoghue v Stevenson* was applied in the construction context by the Court of Appeal in *Clay v AJ Crump & Sons Ltd*[12] so as to render an architect liable for the personal injury suffered by an employee of a building contractor. The architect had negligently allowed the demolition contractor to leave a wall standing in a dangerous condition. The wall collapsed injuring the plaintiff.

6.16 At the macro level, the approach of the courts to duties of care in relation to claims for personal injury and damage to other property remained relatively stable until the adoption by the House of Lords of an expansive approach to duties of care in *Dorset Yacht Co Ltd v Home Office*,[13] which is discussed below.

---

[8] [1932] AC 562.

[9] [1932] AC 562 at 580.

[10] For example, at 580–1.

[11] (1883) 11 QBD 503. In *Heaven v Pender*, the plaintiff recovered damages for personal injury from the defendant. Brett MR defined the issue before the Court of Appeal as: 'the proper definition of the relation between two persons other than the relation established by contract, or fraud, which imposes on the one of them a duty towards the other to observe, with regard to the person or property of such other, such ordinary care or skill as may be necessary to prevent injury to his person or property;' (507–8).

[12] [1964] 1 QB 533, [1963] 3 WLR 866, [1963] 3 All ER 687, 4 BLR 80.

[13] [1970] AC 1004.

The modern genesis of the duty of care in relation to pure economic loss

In *Candler v Crane, Christmas & Co*,[14] the Court of Appeal refused to extend the principle in *Donoghue v Stevenson* to claims other than for property damage or personal injury. In that case, the plaintiff was considering the possibility of investing £2,000 in a limited liability company and wished to consider the company's accounts. The managing director of the company accordingly instructed the defendants, the company's accountants, to press on and complete them. The defendants were informed that the accounts were to be shown to the plaintiff, who they knew was a potential investor. The defendant prepared the accounts and showed them to and discussed them with the plaintiff who took a copy of them and submitted them to his own accountant for advice. As a result the plaintiff invested his money with the company. The accounts were in fact carelessly prepared, contained numerous false statements, and gave a wholly misleading picture of the state of the company, which was wound up within a year, the plaintiff losing the whole of his investment. The plaintiff sought damages in negligence from the defendants. **6.17**

Cohen and Asquith LJJ held that a false statement made carelessly, but not fraudulently, was not actionable in the absence of a relevant contractual or fiduciary relationship between the parties. In particular, Asquith LJ noted that the principle laid down by Lord Atkin in *Donoghue v Stevenson* had never been applied where the damage 'was not physical in its incidence to either person or property'.[15] **6.18**

Denning LJ dissenting held that, as accountants, the defendants owed a duty of care not only to their clients with whom they had contracted but to any third person to whom they showed the accounts or who they knew that their clients were going to show them when, to the knowledge of the accountants, that person would consider their accounts with a view to the investment of money or taking other action to his gain or detriment. Other professionals, such as surveyors, valuers, and analysts were under a similar duty. **6.19**

The correctness of the Court of Appeal decision in *Candler* was at issue in *Hedley Byrne v Heller & Partners Ltd*[16] in which the House of Lords was required to consider whether a claimant could recover for pure economic loss in the absence of a contract or fiduciary relationship. The plaintiffs had sought a reference for a customer from the defendants who were the customer's bankers. The defendants provided a favourable but carelessly prepared reference. The customer then went into liquidation and the plaintiff suffered loss. **6.20**

The House of Lords recognized that a person making a statement could owe a duty of care to the recipient (with whom he had no contract) to take reasonable care and that, in the event of breach, he could be liable for purely economic loss suffered by the recipient. **6.21**

Their Lordships differed as to the precise relevance of *Donoghue v Stevenson* to the issue before them and the relationship between the recovery of damages for physical loss and personal injury on the one hand and pure economic loss on the other. However, they all recognized that a control mechanism was required so as to limit the class of persons to **6.22**

---

14 [1951] 2 KB 164.

15 At 189.

16 [1964] AC 465.

whom a duty of care in respect of pure economic loss applied. Although different formulations of the control mechanism were provided, each focused on the nature of the relationship between the person making the statement and the person relying upon the statement. Lord Devlin said:[17]

> I think, therefore, that there is ample authority to justify your Lordships in saying now that the categories of special relationships which may give rise to a duty of care are not limited to contractual relationships of fiduciary duty, but include also relationships which in the words of Lord Shaw in *Nocton v Lord Ashburton* [1914] AC 932, 972 are 'equivalent to contract,' that is, where there is an assumption of responsibility in circumstances in which, but for the absence of consideration, there would be a contract.

6.23 Consequently, following *Hedley Byrne*, the common law provided for the recovery in negligence of damages for personal injury and physical damage on the one hand, and pure economic loss on the other. However, the tests to be applied when determining whether a duty of care existed in a particular situation were, apparently at least, quite different.[18]

Dorset Yacht Co Ltd v Home Office *and* Anns v Merton LBC

6.24 However, two different developments during the 1970s served to change the picture substantially. First, the House of Lords adopted an expansive approach to the principle set out in *Donoghue v Stevenson* and sought to assimilate the principles relating to recovery for economic loss with those for recovery for physical damage. Secondly, the Court of Appeal and the House of Lords categorized the damage suffered by a building owner caused by inherent defects in the building as physical damage and not pure economic loss.

6.25 The expansive approach to the principle in *Donoghue v Stevenson* and the assimilation of the different types of loss was first seen in *Dorset Yacht Co Ltd v Home Office*,[19] where Lord Reid stated:[20]

> In later years there has been a steady trend towards regarding the law of negligence as depending on principle so that, when a new point emerges, one should ask not whether it is covered by authority but whether recognized principles apply to it. *Donoghue v Stevenson* [1932] AC 562 may be regarded as milestone, and the well known passage in Lord Atkin's speech should I think be regarded as a statement of principle. It is not to be treated as if it were a statutory definition. It will require qualification in new circumstances. But I think that the time has come when we can and should say that it ought to apply unless there is some justification or valid explanation for its exclusion. For example, causing economic loss is a different matter; for one thing, it is often caused by deliberate action. Competition involves traders being entitled to damage their rivals' interests by promoting their own, and there is a long chapter of the law determining in what circumstances owners of land can and in what circumstances they may not use their proprietary rights so as to injure their neighbours. But where negligence is involved the tendency has been to apply principles analogous to those stated by Lord Atkin: cf *Hedley Byrne & Co Ltd v Heller & Partners Ltd.*

---

[17] At 528–9.

[18] It should be noted that in *Dorset Yacht Co v Home Office* [1970] AC 1004 at 1027, Lord Reid held, relying on *Hedley Byrne*, that in economic loss cases 'the tendency has been to apply principles analogous to those stated by Lord Atkin'.

[19] [1970] AC 1004.

[20] At 1026H–1027C.

This expansive approach was confirmed by the House of Lords in *Anns v Merton LBC*,[21] where Lord Wilberforce said:[22] **6.26**

> Through the trilogy of cases in this House—*Donoghue v Stevenson* [1932] AC 562, *Hedley Byrne & Co Ltd v Heller & Partners Ltd* [1964] AC 465, and *Dorset Yacht Co Ltd v Home Office* [1970] AC 1004, the position has now been reached that in order to establish that a duty of care arises in a particular situation, it is not necessary to bring the facts of that situation within those of previous situations in which a duty of care has been held to exist. Rather the question has to be approached in two stages. First one has to ask whether, as between the alleged wrongdoer and the person who suffered the damage there is a sufficient relationship of proximity or neighbourhood such that, in the reasonable contemplation of the former, carelessness on his part may be likely to cause damage to the latter—in which case a prima facie duty of care arises. Secondly, if the first question is answered affirmatively, it is necessary to consider whether there are any considerations which ought to negative, or to reduce or limit the scope of the duty or the class of person to whom it is owed or the damages to which a breach of it may give rise: see *Dorset Yacht* case [1970] AC 1004, *per* Lord Reid at p 1027.

The characterization of damage caused by an inherent defect as physical loss was first made by the Court of Appeal in *Dutton v Bognor Regis UDC*.[23] In that case Mrs Dutton sought to recover in respect of the damage to her property, caused by defective foundations, from the local authority which had approved those foundations. She was a subsequent purchaser of the property and claimed damages for the cost of repair, surveyors' fees, and loss of market value caused by the defective foundations. **6.27**

The key finding by Lord Denning MR in *Dutton* was that this type of loss was not economic loss, but physical damage to the house.[24] Sachs LJ held[25] that the question whether the loss was economic or physical damage was irrelevant. He considered that the key question was whether the damage was of a type that the proper exercise of the statutory power was designed to prevent. But if it was necessary to determine whether the loss was economic or physical damage, it was physical damage. Stamp LJ held[26] that the loss suffered was economic loss, but that an application of *Hedley Byrne* meant that the local authority (but not the builder) owed the plaintiff a duty in relation to economic loss. **6.28**

Both Lord Denning MR and Sachs LJ ignored the basic rule in respect of the liability of builders, namely that in negligence what was recoverable was damage to other property caused by the negligent act or omission rather than damage to the thing itself. To take a crude but simple example, if someone makes a defective fridge which does not work and sells the fridge to a shop which sells the fridge to a customer, then the manufacturer is not liable in negligence to the ultimate purchaser. The manufacturer is liable in contract to the shop, which is liable in contract to the customer, thereby creating a chain of liability. It is the contractual chain which creates recovery in the hands of the consumer. It is so with a **6.29**

---

21 [1978] AC 728.
22 At 751G–752C.
23 [1972] 1 QB 373, CA.
24 See Lord Denning MR at 396C–F.
25 403E–404E.
26 414D–415E.

house or other structure. Whilst the departure from this basic tenet of the law was pointed out to the Court of Appeal, it was impliedly rejected and liability attached.

6.30 Although *Dutton* was concerned only with the liability of the local authority, the Court of Appeal accepted that it would be wrong if the local authority was liable but the builder whose careless work was the direct cause of the defect was not, and held that a builder would be liable in respect of damage such as that claimed by Mrs Dutton.[27]

6.31 In 1977 the House of Lords upheld *Dutton* in *Anns v Merton LBC*.[28] Lord Wilberforce, who gave the majority speech, held that both a builder guilty of negligently constructing foundations[29] and a council guilty of negligently failing to identify the defective foundations would be liable in negligence to a subsequent purchaser of the building, not only in respect of any personal injury caused by the defects, but for the cost of restoring the dwelling to a condition in which it was no longer a danger to the health and safety of persons occupying the premises.

6.32 As to the nature of the damage, Lord Wilberforce held[30] that, at least in relation to the council, the key inquiry was not whether the loss was to be characterized as physical damage or economic loss, but whether it was reasonably foreseeable that particular damage would be caused by the breach of duty. However, if classification was required the damage was physical damage. Consequently, it was established that local authorities became liable in tort to subsequent owners if they negligently inspected foundations.

6.33 If, as established by *Dutton* and *Anns*, a local authority and builder were liable in negligence, then it followed that the relevant professional would also be liable (if fault was proved). Therefore, the paradigm defective structure case would involve claims against the builder, the architect or engineer or both, and the local authority. Most cases were settled (usually at the door of the court and expensively) but many fought, creating a sub-jurisprudence in the tort of negligence which arose out of the unanswered questions in *Anns*.

6.34 As the House of Lords pointed out in *D&F Estates*[31] *and Murphy v Brentwood DC*,[32] there was no obvious need for the decision in *Anns*. Through the Defective Premises Act 1972, Parliament had provided a statutory remedy to householders against builders and developers.

6.35 The effect of the *Anns* decision was that, contrary to the confidence of their Lordships as expressed in their speeches, floodgates opened and a stream of cases were brought in tort against local authorities arising out of damage to property because of negligent inspection by the building inspectors of either plans or construction work or both.[33]

---

27 See Lord Denning MR at 392F–394C; Sachs LJ at 401B–402C.
28 [1978] AC 728.
29 758G–759E.
30 759E–760A.
31 [1989] 1 AC 177 *per* Lord Bridge at 193B–195G.
32 [1991] AC 398 *per* Lord Keith at 472, *per* Lord Jauncey at 498D–E.
33 There were a number of reasons for this. First, that the action was possible and local authorities had deep pockets, therefore it became the convention to sue the local authority along with or instead of the builder and professionals, who may not be solvent (or insured) or even in existence by the time the action

The high water mark of the expansive approach was *Junior Books v Veitchi Co Ltd*[34] in which specialist nominated flooring sub-contractors were held to owe a duty of care in tort to the employer in relation to the cost of remedying defects in the floor which they had laid, despite the absence of any contractual relationship. However, *Junior Books* also contained the seeds of the retreat from *Anns*, in the form of the dissenting speech of Lord Brandon, who considered that *Donoghue v Stevenson* was confined to cases of physical damage. **6.36**

The retreat from Anns

In a series of cases[35] from the mid 1980s, the courts signalled the retreat from both the characterization of damage caused by inherent defects and the two-stage test for determining the existence of a duty of care established by *Anns*. But it was not until 1990 that the House of Lords decided formally to depart from its decision in *Anns*. **6.37**

In *Governors of the Peabody Donation Fund v Sir Lindsay Parkinson & Co Ltd*,[36] the House of Lords unanimously held that the plaintiff was not entitled to recover the cost of repairing defective drains from the local authority which had inspected and approved those drains. In so doing, the House of Lords cast doubt on both the expansive approach to the duty of care and the characterization of damage in *Anns*. In relation to the expansive approach to the duty of care, Lord Keith, with whom the other Law Lords agreed, sounded a note of caution[37] about treating the passage from the speech of Lord Wilberforce cited above as being of a 'definitive character' and emphasized the need to determine whether there existed in a particular case a relationship of sufficient proximity and whether it was 'just and reasonable' that a duty of care should exist between the plaintiff and the defendant. **6.38**

As to the characterization of damage, Lord Keith considered the plaintiff's claim was for economic loss.[38] He considered[39] that the plaintiff was not assisted by the decision in *Anns*, because the loss suffered by the plaintiff in *Anns* was not 'economic loss pure and simple' but 'expenditure necessary to avert injury to safety or health'. Consequently, *Anns* was limited to those cases in which the plaintiff was a person who may necessarily be expected to suffer injury to health or safety if the foundations of the property were inadequate. **6.39**

was commenced. Secondly, the two hot summers of 1975–6 meant that many houses on shallow foundations based upon clay began to subside meaning that there were a great many potential plaintiffs or claimants. Some of these houses had been built many years ago and indeed were constructed under the Public Health Act 1875, the predecessor of the 1936 Act.

[34] [1983] AC 520.

[35] In addition to the cases addressed in the text, the retreat from *Anns* was seen in cases such as *Leigh and Sillivan Ltd v Aliakmon Shipping Co Ltd* [1986] AC 785; *Yuen Kun Yeu v Attorney-General of Hong Kong* [1988] AC 175; *Rowling v Takaro Properties Ltd* [1988] AC 473; *Hill v Chief Constable of West Yorkshire* [1989] AC 53.

[36] [1985] 1 AC 210.

[37] 240F–241C.

[38] 239C–D.

[39] 242A–D and 244A–245B. It is worth noting that in *Junior Books v Veitchi* [1983] AC 520, Lord Keith held that the employer was entitled to recover the cost of remedying the defective floor only because it was necessary to do so to avert the economic loss caused by uneconomic working that the employer would otherwise suffer. Lord Keith specifically stated that he was not allowing recovery on the basis that there was a general liability in tort for the reduced value of defective premises or chattels (see 536E–537E).

6.40 A further significant step in the retreat from *Anns* was taken in *D&F Estates v Church Commissioners for England*.[40] The plaintiff had taken a lease of a property, which had been constructed between 1963 and 1965 by the third defendants as main contractors and the plaster work had been sub-contracted. In 1980 it was discovered that some of the plaster was loose and some fell down, damaging other property. The plaintiff carried out remedial work and then brought an action against, *inter alia*, the main contractors for the cost of that remedial work and the damage to the other property.

6.41 The House of Lords held that the cost of remedial work was pure economic loss and, as such, irrecoverable. Further, it was held that, in general, a main contractor was not liable in tort for damage to other property caused by the negligence of an apparently competent sub-contractor.

6.42 Lord Bridge held[41] that where either a chattel or real property contained a hidden defect which was discovered before it caused any personal injury or damage to other property, any claim by the owner against the person responsible for the defect was not covered by the principle in *Donoghue v Stevenson*. The loss suffered by the owner was economic loss and could only be recovered in contract. He considered that the position may be less straightforward in the case of complex structures or chattels, where it may be arguable that one element of the structure should be regarded for the purposes of liability in negligence as distinct from another element, so that damage to part of the structure caused by a defect in another part was damage to 'other property'.

6.43 Lord Oliver regarded *Anns* as having introduced, in relation to the construction of buildings, an 'entirely new type of product liability, if not, indeed, an entirely novel concept of the tort of negligence'.[42] He considered[43] that the decision in *Anns* was inconsistent with legal principle and was in any event limited directly to cases where the defect was one which threatened the health or safety of the occupants or third parties and (possibly) other property. It could not be extended to other cases of economic loss, recovery for which in negligence was governed by the principles in *Hedley Byrne*.

6.44 The House of Lords finally departed from *Anns*, and overruled *Dutton*, in *Murphy v Brentwood District Council*.[44] Lord Keith said[45] that *Anns*:

> . . . did not proceed upon any basis of established principle but introduced a new species of liability governed by a principle indeterminate in character but having the potentiality of covering a wide range of situations, involving chattels as well as real property, in which it had never hitherto been thought that the law of negligence had any proper place.

The House of Lords confirmed that damage to a building caused by an inherent defect was properly characterized as pure economic loss, even if it posed a danger to the health and

---

40 [1989] 1 AC 177.
41 206B–207C.
42 211B.
43 216F–217B.
44 [1991] 1 AC 398.
45 471B–C.

safety of the occupiers,[46] and as such was only recoverable in tort under the principle of *Hedley Byrne*.

As well as recategorizing the nature of the loss, the House of Lords made plain its disagreement with the two-stage test for establishing a duty of care set out in *Anns*. They approved[47] the 'incremental approach' adopted by the High Court of Australia in *Council of the Shire of Sutherland v Heyman*,[48] where Brennan J expressed his disagreement with the speech of Lord Wilberforce in *Anns* and said: **6.45**

> It is preferable, in my view, that the law should develop novel categories of negligence incrementally and by analogy with established categories, rather than by a massive extension of a prima facie duty of care restrained only by indefinable 'considerations which ought to negative, or to reduce or limit the scope of the duty or the class of person to whom it is owed'.

Murphy was directly concerned with the liability of the local authority who had approved plans. In *Department of the Environment v Thomas Bates & Sons Ltd*,[49] the House of Lords applied *Murphy* when deciding that a builder owed no duty of care in negligence to a subsequent owner in relation to damage arising from inherent defects. **6.46**

The demise of the two-stage test in *Anns* had in fact been confirmed by the House of Lords in *Smith v Eric Bush*,[50] decided shortly before *Murphy*. In *Smith*, a case concerning pure economic loss, the House of Lords set out[51] a three-stage test for establishing a duty of care. First, the plaintiff had to show that it was reasonably forseeable that if the defendant's advice was negligent the recipient was likely to suffer damage. Secondly, there had to be a sufficiently proximate relationship between the parties. Thirdly, it had to be just and reasonable to impose the duty. Lord Griffiths also reconsidered the decision in *Hedley Byrne* and stated that the 'voluntary assumption of responsibility test is not a helpful or realistic test of liability'. **6.47**

The three-stage test set out in *Smith* was itself qualified by the House of Lords in *Caparo Industries plc v Dickman*,[52] another case involving pure economic loss. Lord Bridge said:[53] **6.48**

> . . . in addition to the foreseeability of damage, necessary ingredients in any situation giving rise to a duty of care are that there should exist between the party owing the duty and the

[46] *Per* Lord Keith at 466H–468D; *per* Lord Bridge at 475A–476A; *per* Lord Oliver at 484A. The recategorization of inherent defects as pure economic loss continues to cause problems. In *Abbott v Will Gannon & Smith Ltd* [2005] EWCA Civ 198, the Court of Appeal held that *Pirelli General Cable Works Ltd v Oscar Faber & Partners* [1983] 2 AC 1 had not been expressly or impliedly overruled by *Murphy v Brentwood DC* and the court was bound to hold that a building owner's cause of action against his consulting engineer for negligent design accrued for limitation purposes when physical damage to the building first occurred. The Court of Appeal went on to hold that if it was not bound by Pirelli and the cause of action accrued when the claimant first suffered economic loss that was not when the work to the defective design was completed, but only when it manifested itself in a way that affected the market value. See further for a discussion of the *Pirelli* case Ch 15 paras 15.27–15.32.

[47] *Per* Lord Keith at 461C–D.

[48] (1985) 157 CLR 424 at 481 *per* Brennan J.

[49] [1991] 1 AC 499.

[50] [1990] 1 AC 831.

[51] *Per* Lord Griffiths at 864H–865A.

[52] [1990] 2 AC 605.

[53] 617H–618C.

party to whom it is owed a relationship characterized by the law as one of 'proximity' or 'neighbourhood' and that the situation should be one in which the court considers it fair, just and reasonable that the law should impose a duty of a given scope upon the one party for the benefit of the other. But it is implicit . . . that the concepts of proximity and fairness embodied in these additional ingredients are not susceptible of any such precise definition as would be necessary to give them utility as practical tests, but amount in effect to little more than convenient labels to attach to the features of different specific situations, which on a detailed examination of all the circumstances, the law recognizes pragmatically as giving rise to a duty of care of a given scope. Whilst recognizing, of course, the importance of the underlying general principles common to the whole field of negligence, I think the law has now moved in the direction of attaching greater significance to the more traditional categorization of distinct and recognizable situations as guides to the existence, the scope and the limits of the varied duties of care which the law imposes.

6.49 In both *Smith* and *Caparo*, recovery for economic loss in negligence was founded squarely upon *Hedley Byrne*, although doubts were expressed[54] as to the utility of the voluntary assumption of responsibility test advanced by some of the speeches in that case.

6.50 Although cases such as *Smith, Murphy*, and *Caparo* were concerned with recovery of damages for pure economic loss, the incremental approach to, and three-stage test for determining the existence and scope of, duties of care set out in those cases were also applicable to cases concerned with the recovery of damages for personal injury and physical damage.[55]

Henderson v Merrett

6.51 The law took a further turn in *Henderson v Merrett Syndicates Ltd*,[56] in which the House of Lords confirmed the existence of concurrent liability in contract and tort and established that the duty of care in relation to pure economic loss, founded upon the principle in *Hedley Byrne*, was distinct from the duty of care in relation to physical damage and personal injury.

6.52 As to the question of concurrent liability, the House of Lords confirmed that a person who owes a contractual duty of care to another may also owe that other a tortious duty of care, but only to the extent that the tortious duty of care is consistent with the contractual context. Lord Goff held:

> Yet the law of tort is the general law, out of which the parties can, if they wish, contract; . . . the same assumption of responsibility may, and frequently does, occur in a contractual context. Approached as a matter of principle, therefore, it is right to attribute to that assumption of responsibility, together with its concomitant reliance, a tortious liability, and then to enquire whether or not that liability is excluded by the contract because the latter is inconsistent with it.

[54] In *Smith*, Lord Griffiths at 862; in *Caparo*, Lord Roskill at 628.

[55] That is not to say that whenever a duty of care was held to exist in relation to personal injury, the duty would extend to pure economic loss. However, the three-stage test and the incremental approach which were adopted by the House of Lords in cases such as *Smith* and *Caparo* were also applied to cases involving personal injury and physical damage. See, for example, *Stovin v Wise* [1996] AC 923 *per* Lord Hoffmann at 949A–C; and *Marc Rich and Co AG v Bishop Rock Marine Co Ltd* [1996] AC 211.

[56] [1995] 2 AC 145.

. . . I am of the opinion that this House should now, if necessary develop the principle of assumption of responsibility as stated in *Hedley Byrne* to its logical conclusion so as to make clear that a tortious duty of care may arise not only in cases where the relevant services are rendered gratuitously, but also where they are rendered under a contract.[57]

And later: **6.53**

. . . an assumption of responsibility coupled with the concomitant reliance may give rise to a tortious duty of care irrespective of whether there is a contractual relationship between the parties, and in consequence, unless his contract precludes him from doing so, the plaintiff who has available to him concurrent remedies in contract and tort, may choose that remedy which appears to him to be the most advantageous.[58]

Although the House of Lords held that on the facts of the case the managing agents owed **6.54** a duty of care to the Indirect Names (ie the Names with whom they had no direct contractual relationship, but were part of a single contractual chain), Lord Goff held:[59]

I wish however to add that I strongly suspect that the situation which arises in the present case is most unusual; and that in many cases in which a contractual chain comparable to that in the present case is constructed it may well prove to be inconsistent with an assumption of responsibility which has the effect of, so to speak, short circuiting the contractual structure so put in place by the parties. It cannot be inferred from the present case that other sub-agents will be held directly liable to the agent's principal in tort. Let me take the analogy of the common case of an ordinary building contract, under which main contractors contract with the building owner for the construction of the relevant building and the main contractor subcontracts with sub-contractors or suppliers (often nominated by the building owner) for the performance of work or the supply of materials in accordance with standards and subject to terms established in the sub-contract. I put on one side cases in which the sub-contractor causes physical damage to the property of the building owner, where the claim does not depend on any assumption of responsibility by the sub-contractor to the building owner; though the subcontractor may be protected from liability by a contractual exemption clause authorized by the building owner. But if the sub-contracted work or materials do not in the result conform to the required standard, it will not ordinarily be open to the building owner to sue the sub-contractor direct under the *Hedley Byrne* principle, claiming damages from him on the basis that he has been negligent in relation to the performance of his functions. For there is generally no assumption of responsibility by the sub-contractor or supplier direct to the building owner, the parties having so structured their relationship that it is inconsistent with any such assumption of responsibility.

In relation to the recovery of pure economic loss in negligence, Lord Goff confirmed that **6.55** the founding principle was to be found in *Hedley Byrne* and held[60] that the governing principle underlying that case was the assumption of responsibility by the defendant combined with reliance by the plaintiff. In determining the governing principle in *Hedley Byrne*, neither Lord Goff nor Lord Browne-Wilkinson relied upon *Donoghue v Stevenson*, but instead referred to *Nocton v Lord Ashburton*,[61] thereby making it plain that the duty of care in relation to pure economic loss was not a species of *Donoghue v Stevenson* liability.

[57] 193B–C.
[58] 194D. See also, for example, *Pacific Associates Inc v Baxter* [1990] 1 QB 993.
[59] 195G–196C.
[60] At 180C–G.
[61] [1914] AC 932.

6.56 Nevertheless, Lord Goff also sought to explain the relationship between the 'assumption of responsibility' test and the three-stage test set out in *Smith* and *Caparo*. He said that once a case was identified as falling within the *Hedley Byrne* principle, there should be no need to embark upon any further enquiry whether it is 'fair, just and reasonable' to impose liability for economic loss.

6.57 Lord Goff held that in a context concerned with a liability which may arise under a contract or in a situation 'equivalent to contract', an objective test should be applied when asking whether the defendant had assumed responsibility to the plaintiff.[62] This is to be contrasted to the position of Lord Devlin in *Hedley Byrne* who said that the assumption of responsibility was not imposed by law but undertaken voluntarily.

6.58 It is important to note that although *Hedley Byrne* type liability had traditionally attached to the provision of information or advice, Lord Goff was clear that it could also attach to the provision of other services.[63]

Summary of the position following Henderson v Merrett

6.59 Following *Henderson v Merrett*, the law in relation to the existence and scope of duties of care could be summarized as follows.

6.60 In relation to claims for damages for personal injury and physical damage, the House of Lords had adopted two, apparently related, tests for determining the existence and scope of a duty of care. First, in *Murphy*, the House of Lords approved the 'incremental approach' proposed by Brennan J in *Heyman*. Secondly, in *Smith* and *Caparo*, the three-stage test (reasonable foreseeability, proximity, just and reasonable) was adopted.

6.61 In relation to claims for pure economic loss, the House of Lords in *Henderson v Merrett* had endorsed a third test (in addition to those applicable to cases of personal injury and physical damage), namely the modified 'assumption of responsibility' test established by *Hedley Byrne*. Further, *Henderson v Merrett* also established that the duty of care in relation to pure economic loss could arise not only in relation to negligent advice and information but also in relation to the negligent provision of services.

6.62 In *Murphy v Brentwood DC*, the House of Lords had reaffirmed that damage suffered by property caused by inherent defects was properly classified as pure economic loss.

Duties of care and pure economic loss: the relationship between the three approaches

6.63 As set out above, the House of Lords in *Smith, Murphy, Caparo*, and *Henderson* had approved three different approaches to determining the existence of duties of care in relation to pure economic loss. An important question, therefore, was the relationship between the different approaches. In *Henderson* itself, Lord Goff had sought to explain the relationship between the 'assumption of responsibility' test and the three-stage test by stating that, once

[62] See also *Williams v Natural Life Foods Ltd* [1998] 1 WLR 830 at 835; *Electra Private Equity Partners v KPMG Peat Marwick* [1999] Lloyd's Rep 670. In *White v Jones* [1995] 2 AC 207, Lord Browne-Wilkinson stated (at 273) that the 'assumption of responsibility' referred to the performance of the task and not legal liability.

[63] For example, at 193D.

it was established that a case fell within the *Hedley Byrne* principle, there was no need to consider further whether the imposition of a duty of care in relation to pure economic loss was 'just and reasonable'.

In *BCCI (Overseas) Ltd v Price Waterhouse*,[64] the Court of Appeal, having considered each of the three approaches to duties of care and pure economic loss adopted by the House of Lords, held that it may be useful to look at any new set of facts by using each of the three approaches in turn. If the facts were properly analysed and the policy considerations correctly evaluated each approach should yield the same result. **6.64**

A different approach was adopted by the Court of Appeal in *The Rebecca Elaine*,[65] in which the plaintiff sought to recover damages for pure economic loss allegedly caused by the defendant manufacturer's failure to warn about defects in an engine which had been fitted in a boat purchased by the plaintiff from a third party. In rejecting the claim, the Court of Appeal emphasized that, in a claim for pure economic loss, recovery was only permitted if the claim fell within the principle set out in *Hedley Byrne* which required the existence of a voluntary assumption of responsibility. Mummery LJ confirmed that there was no general duty in English law to take reasonable care to avoid inflicting financial loss on those whom it is reasonably foreseeable will suffer such loss as a consequence of careless acts or omissions. He emphasized that *Donoghue v Stevenson* was only concerned with claims for personal injury and damage to other property. **6.65**

The relationship between the three approaches to ascertaining the existence of a duty of care to prevent pure economic loss was reviewed by the House of Lords in *HMCE v Barclays Bank plc*.[66] The House of Lords emphasized the difficulty in identifying any single over-arching test by which a duty of care in relation to pure economic loss could be determined and identified the limits of each of the three approaches. **6.66**

Evolution of the Hedley Byrne *principle*

Although the speeches in *Hedley Byrne* and *Henderson v Merrett* focused upon the importance of reliance by the claimant on the defendant as the foundation of the assumption of responsibility, the House of Lords has since held that cases in which there is no reliance of the type usually required also fall within the *Hedley Byrne* principle. For example, in *White v Jones*,[67] the House of Lords held that a solicitor who negligently failed to carry out a testator's wishes thereby causing a loss to the intended beneficiaries owed a duty of care to those intended beneficiaries. The intended beneficiaries did not rely upon the negligent solicitor in the sense traditionally associated with the *Hedley Byrne* principle. Nevertheless the House of Lords stated that justice demanded that the intended beneficiaries should have a remedy in such a case. **6.67**

[64] [1998] PNLR 564. The Court of Appeal in *Parkinson v St James and Seacroft University Hospital NHS Trust* [2002] QB 266 approved the approach in *BCCI (Overseas) Ltd v Price Waterhouse* and in *Clerk & Lindsell*, 18th edn, paras 7–95 (see para 17). See also *Welton v North Cornwall District Council* [1997] 1 WLR 570 at 580E–F *per* Rose LJ.

[65] [1999] 2 Lloyd's Rep 1.

[66] [2006] UKHL 28.

[67] [1995] 2 AC 207, [1995] 2 WLR 187, [1995] 1 All ER 691. See also *Spring v Guardian Assurance plc* [1995] 2 AC 296, [1994] 3 WLR 354, [1994] 3 All ER 129.

6.68 *Williams v Natural Life Health Foods Ltd*[68] emphasized the need for a claimant to prove not only that he relied upon the assumption of responsibility by the defendant but that such reliance was reasonable.[69] The Court of Appeal emphasized that cases such as *White v Jones*,[70] in which the direct reliance traditionally required under *Hedley Byrne* was not found, were true exceptions in which the 'assumption of responsibility' approach was simply the best way of achieving justice.[71] It was implied that in most cases the requirement of direct reliance would be rigorously applied.

6.69 Lord Steyn[72] also provided guidance as to the practical application of what he called the 'extended *Hedley Byrne* principle':

> First, there is the approach to be adopted as to what may in law amount to an assumption of risk. This point was elucidated in *Henderson's* case by Lord Goff of Chieveley. He observed, at p 181:
>
> > especially in a context concerned with a liability which may arise under a contract or in a situation 'equivalent to contract', it must be expected that an objective test will be applied when asking the question whether, in any particular case, responsibility should be held to have been assumed by the defendant to the plaintiff . . .
>
> The touchstone of liability is not the state of mind of the defendant. An objective test means that the primary focus must be on things said or done by the defendant or on his behalf in dealings with the plaintiff. Obviously, the impact of what a defendant says or does must be judged in the light of the relevant contextual scene. Subject to this qualification the primary focus must be on exchanges (in which term I include statements and conduct) which cross the line between the defendant and the plaintiff.[73]

Complex structure theory

6.70 In *D&F Estates*,[74] Lord Bridge had raised the possibility that if one part of a complex structure caused damage to another part of the same structure, then the damage could be considered as 'damage to other property' and therefore fall within the principle of *Donoghue v Stevenson*.

6.71 This issue was considered by the House of Lords in *Murphy v Brentwood District Council*.[75] Lord Keith held[76] that it would be unrealistic to treat two parts of a building the whole of which had been constructed by the same contractor as different property. He contrasted this situation with the example of an electrical sub-contractor who carelessly installed wiring which later caused a fire which destroyed the building. In such a case, Lord Keith held, the sub-contractor may be liable. Lord Bridge also cast doubt[77] upon the complex structure theory. He considered that an important distinction was to be drawn between some part of

68 [1998] 1 WLR 830.
69 At 836E–837B.
70 [1995] 2 AC 207.
71 At 837C–G.
72 With whom Lords Goff, Hoffmann, Clyde, and Hutton agreed.
73 At 835E–H.
74 [1989] AC 177.
75 [1991] AC 398.
76 470C–G.
77 476B–479B.

a 'complex structure' which was said to be a 'danger' only because it did not perform its proper function in sustaining other parts and some distinct item incorporated in the structure which positively malfunctioned so as to inflict positive damage on the structure in which it is incorporated. He gave, as examples, a defective boiler or defective electrical installation. Lord Oliver, who had also referred to the complex structure theory in *D&F Estates*, said[78] that both he and Lord Bridge had not relied upon the complex structure theory with any enthusiasm, but as the only logical explanation of the categorization of the damage in *Anns* as 'material, physical damage'. Lord Jauncey[79] also considered that the complex structure theory had no application on the facts of *Murphy*, but may do where one integral component of the structure was built by a separate contractor and where a defect in such a component had caused damage to other parts of the structure. He gave as an example a steel frame erected by a specialist contractor which failed to give adequate support to floors or walls. He also stated that defects in ancillary equipment such as central heating boilers or electrical installations would be subject to the normal *Donoghue v Stevenson* principle if defects gave rise to damage to other parts of the building.

A number of questions arise from the speeches in *Murphy* in relation to the complex structure theory. First, it is not clear whether, in order to engage the complex structure theory, it is necessary that the two parts of the building were installed or constructed by different contractors. For example, what if the defective boiler or defective wiring is installed by same contractor who constructs the remainder of the building? Secondly, is it sufficient if the two parts are constructed by different contractors? If so, what if the foundations are constructed by one contractor and the superstructure by another? Thirdly, is it necessary in order to engage the complex structure theory that the defective element should cause positive damage, as opposed merely to fail to fulfil its function? **6.72**

Since *Murphy*, the complex structure theory has found little favour with the courts. For example, in *Tesco Stores Ltd v The Norman Hitchcox Partnership Ltd*[80] Judge Esyr Lewis QC, relying on Lord Bridge in Murphy, held[81] that defects in the parts of the structure intended to inhibit the spread of fire did not inflict positive damage on the remainder of the structure, so that there was no damage falling within the principle of *Donoghue v Stevenson*. **6.73**

Similarly in *Bellefield Computer Services Ltd v E Turner & Sons Ltd (No 1)*,[82] the Court of Appeal rejected an argument based upon the complex structure theory in relation to a compartment wall that failed to prevent the spread of fire at a dairy. The claimant who was a subsequent owner of the dairy had argued that the part of the building on the far side of the compartment wall from the seat of the fire was 'other property'. Despite misgivings, the Court of Appeal considered that it was bound by the decision in *Murphy* and held that the damage to the whole of the dairy was pure economic loss, and therefore outside the scope **6.74**

[78] 484E–F.
[79] 497B–E.
[80] (1997) 56 Con LR 42.
[81] At 167–8.
[82] [2000] BLR 97.

of the duty of care owed by the builder. The building was constructed as a single unit and purchased as a single unit by the claimant.

6.75 In *Samuel Payne v John Setchell Ltd*,[83] the defendant engineers had designed, inspected, and approved raft foundations upon which cottages were constructed. The foundations were defective and cracks appeared in the cottages. Judge Humphrey LLoyd QC held[84] that the complex structure exception was no longer tenable but considered that there may be circumstances in which it would be unreasonable not to afford a claimant a remedy for 'catastrophic damage caused by defective building work'.

Relevance of the contractual background

Introduction

6.76 In considering the relevance of the contractual background to a claim in negligence, it is necessary to examine both situations in which there is a contract between the claimant and the defendant and those in which there is no such contract but other relevant contractual relationships.

Contract between the claimant and the defendant

6.77 As set out above, where there is a contract between the claimant and the defendant, *Henderson v Merrett*[85] established that a tortious duty of care could co-exist with a contractual duty of care between the claimant and the defendant.

6.78 Further, where there is a contract between the claimant and the defendant, the scope and nature of the tortious duty of care will be informed and limited by the terms of the contract between the claimant and the defendant. For example, in *William Hill Organisation v Bernard Sunley*,[86] the plaintiff's claim under the contract between it and the defendant failed because, *inter alia*, the defendant had a good defence based on the contractual provision in relation to the final certificate. The Court of Appeal held that the plaintiff's claim in tort could not succeed because such a claim would be inconsistent with the contractual final certificate provision of the contract.

6.79 In *Greater Nottingham Co-operative Society Ltd v Cementation Piling and Foundations Ltd*,[87] the Court of Appeal held[88] that the direct contract entered into by an employer and a sub-contractor was inconsistent with any assumption of responsibility (and therefore liability

[83] [2002] BLR 489.
[84] At 511–12.
[85] [1995] 2 AC 145. See also *Hart Investments Ltd v Terence Fidler Partnership* [2007] EWHC 1058 (TCC), [2007] BLR 526, where it was held that an engineer employed by the claimant in relation to permanent works owed the claimant a duty of care in relation to the temporary works where he observed a state of the temporary works that was dangerous and causing immediate peril to the permanent works in respect of which he was employed. It was held that the engineer was under a contractual duty to warn and a concurrent duty in tort. It was also held that even if there had been no contractual duty there were special circumstances or justification for extending a duty to prevent economic loss.
[86] (1982) 22 BLR 1.
[87] [1990] 1 QB 71.
[88] *Per* Purchas LJ at 99C–100G; Woolf LJ at 106A–D.

in tort for pure economic loss) beyond that which had been undertaken in the direct contract.

No contract between the claimant and the defendant

Even where there is no direct contract between the claimant and the defendant, the overall contractual scheme in any case may well be relevant when determining the nature and scope of any duty of care. **6.80**

In *Henderson v Merrett*,[89] Lord Goff gave as an example of the type of case in which the contractual structure would usually preclude the existence of any duty of care the typical construction arrangement in which there is a chain of contracts with the employer contracting with the main contractor and the main contractor then contracting with a number of sub-contractors and suppliers. Lord Goff held that in such a situation it would be most unusual for there to be an assumption of responsibility by a sub-contractor to the employer so as to give rise to a duty of care. **6.81**

In *Pacific Associates v Baxter*,[90] the Court of Appeal held that the arbitration clause and a disclaimer clause in the contract between the contractor and the employer precluded the existence of any duty of care in respect of pure economic loss owed by the engineers appointed to administer a contract on behalf of the employer to the contractor. **6.82**

In *Norwich City Council v Harvey*,[91] the Court of Appeal held that the existence of a duty of care owed by a sub-contractor to an employer in relation to fire damage caused by the sub-contractor's employee was precluded by exemption and insurance clauses contained in the contract between the employer and the main contractor and a term in the sub-contract that the sub-contractor was bound to the 'same terms and conditions as those of the main contract'. The relevance of the contractual structure and the provisions in each contract in the chain in determining the existence and scope of a duty of care was emphasized by the House of Lords in *British Telecommunications plc v Thomson Ltd*.[92] In that case, the insurance obligations in the main contract provided that the employer, the main contractor, and nominated sub-contractors were to be joint insureds on the relevant policy. The defendant was a domestic sub-contractor and therefore fell outside the insurance scheme of the main **6.83**

---

[89] [1995] 2 AC 145. In *Galliford Try Infrastructure Ltd v Mott Macdonald Ltd* [2008] EWHC 1570 (TCC), (2009) 120 Con LR 1, [2009] PNLR 9, it was held that a consultant engineer engaged by a subsidiary of the employer owed no duty of care in relation to pure economic loss to the contractor engaged by the employer. The claimant and the defendant held discussions for the novation of the defendant's appointment to the claimant but irreconcilable differences remained as to the terms of the novation agreement. This pointed more to there being no duty of care than that there was a duty of care.

[90] [1990] QB 993, CA.

[91] [1989] 1 WLR 828.

[92] [1999] 1 WLR 9. See also *Hopewell Project Management Ltd v Ewbank Preece* [1998] 2 Lloyd's Rep 448, in which it was held that as engineers did not fall within the term 'contractors and sub-contractors', they were not joint insureds and therefore the insurance provisions of the main contract did not preclude the existence of a duty of care owed by the engineers to the employer. In *John F Hunt Demolition Ltd v Asme Engineering Ltd* [2007] EWHC 1507 (TCC), Judge Peter Coulson QC applied *Norwich City Council v Harvey (Paul Clarke)* [1989] 1 WLR 828, [1989] 1 All ER 1180, 45 BLR 14; and *British Telecommunications plc v James Thomson & (Engineers) Ltd* [1999] 1 WLR 9, [1999] 2 All ER 241, and found that the terms of the relevant main contract and sub-contract meant that no duty of care in tort was owed by the sub-contractor to the employer in respect of damage caused by fire.

contract. In these circumstances, the insurance obligations of the main contract reinforced rather than negatived the existence of a duty of care owed by the defendant to the claimant employer.

6.84 In *Bellefield (No 2)*,[93] the Court of Appeal held that the duty of care owed by an architect to a subsequent purchaser of a building in relation to damage to other property was limited by the terms of the architect's contract with its employer.

Contractors and the duty of care in relation to personal injury and damage to other property

6.85 A contractor will, in principle,[94] owe a duty of care in carrying out[95] his work in relation to personal injury and damage to other property to his employer, a subsequent owner of the property, and any third party who suffers personal injury or damage to other property.

6.86 For example, in *Bellefield Computer Services Ltd v E Turner & Sons Ltd (No 1)*,[96] the defendant builders had constructed a dairy, but had carelessly failed to build a compartment wall between a storage area and the rest of the dairy to the proper height. When a fire started in the storage area it spread to the rest of the dairy, because the compartment wall was not sufficiently high, causing damage not only to the dairy itself, but also other property. The Court of Appeal rejected an argument by the builders that they owed no duty of care in relation to the damage caused by the fire to other property within the dairy. The builders had argued that the substance of the complaint against them, that they had not built the compartment wall high enough, was a complaint of omission, and, in line with *Stovin v Wise*,[97] no liability should attach. However, the Court of Appeal held that *Stovin v Wise* should be distinguished, in particular because the builders had undertaken a contractual obligation to the original owners properly to build the compartment wall. Schiemann LJ accepted that the builders would have owed a duty in tort to the original owners, and the effect of finding that the builders owed a duty of care to the claimant in relation to property damage merely had 'the effect of substituting a different beneficiary of the builders' potential liability'. In accordance with this analysis, Schiemann LJ considered that the tortious liability of the builders to the subsequent owners would have been limited by any relevant contractual exclusion clause.

Construction professionals and the duty of care in relation to personal injury and damage to other property

6.87 A construction professional will, in principle,[98] owe a duty of care in relation to personal injury and damage to other property of his client, a subsequent owner of the property,

---

93 [2003] Lloyd's Rep PN 53.

94 ie subject to satisfying the criteria of reasonable foreseeability, proximity, and just and reasonableness.

95 As set out at para 6.122 to 6.123 below, the contractor will not normally be liable for the negligence of any independent sub-contractor.

96 [2000] BLR 97.

97 [1996] AC 923.

98 ie subject to satisfying the criteria of reasonable foreseeability, proximity, and just and reasonable. See for example, *Clay v AJ Crump & Sons Ltd* [1964] 1 QB 533.

and any third party who suffers personal injury or property damage as a result of his negligence.

In *Bellefield Computer Services Ltd v E Turner and Sons Ltd (No 2)*,[99] the Court of Appeal considered a contribution claim brought by a builder against the architects for the project in question. The builder's claim for damages was barred by s 14A of the Limitation Act 1980 and only the contribution claim remained. This claim depended on the nature of the duty of care, if any, owed by the architects to the subsequent building owners in relation to physical damage[100] suffered by the claimant. There was no contract between the architect and the original employer. **6.88**

At first instance, Forbes J held that the architects owed a duty of care in tort to a subsequent occupier of the building in respect of latent defects because there had been no reasonable possibility that the claimant would carry out an inspection which was likely to discover those defects before it went into occupation. **6.89**

The Court of Appeal upheld the judge's findings and reasoning. Potter LJ identified four principles concerning the existence and scope of an architect's duty to a subsequent owner or occupier: **6.90**

(a) An architect may, in appropriate circumstances, owe a duty of care in tort and be liable to a subsequent occupier of the building which the architect has designed and/or the construction of which he has supervised in respect of latent defects in the building of which there has been no reasonable possibility of inspection.

(b) The question whether a particular defect in a building comes within the scope of an architect's duty of care to a subsequent occupier will depend upon the original design and/or supervisory obligations of the architect in question. The architect will not owe a duty of care in respect of defects for which he never had any design or supervisory responsibility in the first place.

(c) If a dangerous defect arises as the result of a negligent omission on the part of the architect, he cannot excuse himself from liability on the grounds that he delegated the duty of design of the relevant part of the building works, unless he obtains the permission of his employer to do so.

(d) The detailed duties of an architect in relation to his design function depend upon the application of these general principles to the particular facts of the case, including any special terms agreed. The precise ambit of such duties will usually depend upon expert evidence from members of the profession as to what a competent, experienced architect would do in the circumstances.

[99] [2002] EWCA Civ 1823, [2003] Lloyd's Rep PN 53. See also *Sahib Foods Ltd v Paskin Kyriades Sands* [2003] EWHC 142 (TCC), [2003] PNLR 30, [2003] EWCA Civ 1832, [2003] PNLR 403.

[100] As set out above, in *Bellefield (No 1)*, the Court of Appeal had held that that the builder was liable for the physical damage but not the economic loss suffered by the owner.

Intermediate inspection

6.91 In *Donoghue v Stevenson*,[101] Lord Atkin stated that there would be no sufficient relationship of proximity and therefore no duty of care if the defect in the chattel was patent in the sense that it would be reasonably discoverable by an intermediate inspection which the manufacturer could reasonably expect would be made. The courts have applied this rule in a fact sensitive and flexible manner.

6.92 In *Clay v AJ Crump & Sons Ltd*,[102] the Court of Appeal rejected the argument that an architect owed no duty of care to the workman in relation to personal injury because of the opportunity for intermediate inspection by third parties of the wall which fell and caused the injury.[103]

6.93 Similarly in *Targett v Torfaen Borough Council*,[104] another personal injury case, the Court of Appeal held that knowledge or opportunity of inspection, did not by itself always negative the duty of care or break the chain of causation. It would only do so when it was reasonable to expect the plaintiff to remove or avoid the danger, and unreasonable for him to run the risk of being injured by the danger.

6.94 These cases contrast sharply with *Baxall Securities Ltd v Sheard Walshaw Partnership*,[105] a case involving damage to other property. The Sheard Walshaw Partnership ('SWP') were architects instructed to design light industrial units. Baxall Securities Ltd ('Baxall') wished to take a lease of one of the units and instructed a firm of surveyors to inspect the premises. This inspection revealed that there had been some water ingress and the surveyors recommended that the problem of ingress be remedied. In fact, there was (unknown to Baxall) a more fundamental defect. Baxall then entered into the lease and thereafter there were two serious floods, in May and September 1995, causing serious damage to Baxall's goods.

6.95 Baxall brought an action against SWP based upon breach of a tortious duty of care owed to Baxall as a subsequent occupier by SWP. The breach alleged was negligent design of the roof gutter system and failing to ensure that the detailed design carried out by the subcontractor had sufficient capacity to cope with the expected rainfall.

6.96 In order to recover, the claimant was required to show that the damage was caused by a latent defect. The Court of Appeal held that 'defect' meant the 'actual defect in the workmanship or design, not the danger presented by the defect'.[106]

6.97 The Court of Appeal held that a defect became patent not only if it was discovered prior to damage being suffered but if there had been a reasonable opportunity to discover the defect. David Steel J[107] held:

> In summary, I would put the matter in this way. Where in the normal course of events, a surveyor would be engaged in a survey of a building for a purchaser and, with the exercise

[101] [1932] AC 562.
[102] [1964] 1 QB 533, [1963] 3 WLR 866, [1963] 3 All ER 687, 4 BLR 80.
[103] See Ormerod LJ at 557.
[104] [1992] 3 All ER 27.
[105] [2002] BLR 100.
[106] Para 46.
[107] At para 53, with whom Hale and Brooke LJJ agreed.

of due diligence, that surveyor would have discovered a defect, that defect is patent whether or not a surveyor is in fact engaged and if engaged whether or not the surveyor performs his task competently. It follows that in my judgment, the Judge was right to conclude as follows:

> 108. In the present case, the precise defect had not been discovered before the floods, but there had been at least one previous flood evidenced by markings on the floor in the building. In my opinion, the duty of the defendants to the claimants must depend on the question, 'Was there a reasonable opportunity of inspection of the drainage system and discovering the defects before the floods?' If the claimants had a reasonable opportunity of inspecting the drainage system and discovering the defects before they suffered damage, it would not be fair just or reasonable to hold the defendants liable for that damage nor would it be right to say that there was a proximate relationship between the claimants and the defendants.
>
> 109. There was a reasonable opportunity of inspecting the building before the claimants took a lease. It would be normal procedure for any incoming tenant to have the building inspected by the surveyor, and that is what they did. Although the claimants received warnings from both the surveyors and from Mr Meiklejohn of a danger, they were not told what was the precise problem. The surveyors could, and in my view should, have told the claimants that there were no overflows, and that overflows should be provided. The cost of overflows was very small and if the claimants had been advised to install them then I cannot think that they would have failed to do so.
>
> 110. If Lambert Smith Hampton had been more assiduous in the performance of their duties, the claimants would have been expressly warned of the absence of overflows and the floods would not have occurred. To what extent is the claimants' claim affected by the acts of their professional advisers? Are they entitled to say, as they might in response to a defence of contributory negligence, that they took skilled advice and are entitled to rely on that advice? I do not think that that is the right approach.
>
> I do not think that it is fair just or reasonable that the extent of the liability of the defendants should depend on the assiduity of the surveyors instructed by the claimants. The claimants had the opportunity to discover the absence of overflows by reasonable inspection by professional advisers who might reasonably be expected to be instructed: whether that reasonable opportunity in fact revealed the defect is irrelevant. Because there was a reasonable opportunity to inspect, the defendants were not in a proximate relationship to the claimants so far as concerns defects which could have been discovered by that inspection, namely, the absence of overflows. But I repeat my previous finding that neither the claimants nor their surveyors could reasonably be expected to have discovered the underdesign of the drainage system.

Contractors and the duty of care in relation to pure economic loss

The judgment of the Court of Appeal in *Baxall Securities* must now be considered in the **6.98**
light of the decision of a differently constituted Court of Appeal in *Pearson Education Ltd v The Charter Partnership*.[108] In *Pearson Education*, the claimant brought a claim for damage to its property stored inside a warehouse, of which it was lessee, which had been designed by the defendant. The damage had been caused by a flood which occurred because the rainwater system designed by the defendant was inadequate. No survey of the warehouse was carried out on the occasion of any of the transfers of the lease which concluded with the

[108] [2007] BLR 324, [2007] EWCA Civ 130.

lease being transferred to the claimant. It was found that had a pre-purchase survey taken place it could not reasonably have been expected to identify the under-capacity of the rainwater system.

6.99 The Court of Appeal in *Pearson Education* held that there were two explanations of *Baxall Securities*. First, where it is reasonable to expect that an occupier will inspect a property before entering into occupation, no duty of care will be owed in respect of any defect that such an inspection should disclose. Second, where an occupier could reasonably have been expected in his own interests to carry out an inspection that would have revealed the defect, failure to carry out such an inspection, or to carry it out with reasonable skill and care, will break the chain of causation.

6.100 The Court of Appeal held that neither explanation of *Baxall Securities* afforded a defence to the architect in that case. In relation to the first principle, there was no defence because the architect's design shortcoming was truly latent. When the architect specified the capacity of the system there was no reason for it to expect that an inspection would be carried out that would reveal any error that they might make. In relation to the second principle, there was no defence because there was no reason why the claimant should have carried out any investigation of the adequacy of the rainwater system.

6.101 Finally, it should be noted that the Court of Appeal in *Pearson Education* doubted the correctness of the decision in *Baxall Securities*, whilst accepting that it could only be reversed by the House of Lords.

Introduction

6.102 In *Murphy v Brentwood District Council*,[109] Lord Bridge held:

> The House has already held in *D & F Estates* that a builder, in the absence of any contractual duty or of a special relationship of proximity introducing the *Hedley Byrne* principle of reliance, owes no duty of care in tort in respect of the quality of his work.[110]
>
> . . . There may, of course, be situations where, even in the absence of contract, there is a special relationship of proximity between builder and building owner which is sufficiently akin to contract to introduce the element of reliance so that the scope of the duty of care is wide enough to embrace purely economic loss. The decision in *Junior Books Ltd v Veitchi Co Ltd* [1983] 1 AC 520 can, I believe, only be understood on this basis.

6.103 Lord Bridge indicated that there could be two bases for imposing a *Hedley Byrne* type duty on a builder. First, it may arise because of the contractual relationship between the builder and the building owner. Secondly, it may arise (as in *Junior Books*) in the absence of any direct contractual relationship.

Contract between the builder and the building owner

6.104 There continues to be uncertainty as to the circumstances in which a builder will owe a duty of care in relation to pure economic loss to a person with whom he has a contractual relationship.

[109] [1991] 1 AC 398.
[110] At 480F–G.

In *Barclays Bank v Fairclough*,[111] the Court of Appeal held that a sub-sub-contractor, Trendleway, owed a duty of care in tort in relation to pure economic loss to the sub-contractor, Carne, to whom he owed a contractual duty of skill and care. As a consequence, Trendleway was liable for the loss suffered by Carne as a result of its negligence, namely the damages Carne had been required to pay to the main contractor, subject to a deduction for contributory negligence. **6.105**

Beldam LJ[112] relied upon Lord Goff's statement in *Henderson v Merrett*[113] that a duty of care in relation to economic loss could arise from the careless performance of services in holding:[114] 'A skilled contractor undertaking maintenance work to a building assumes a responsibility which invites reliance no less than the financial or other professional adviser does in undertaking his work.' **6.106**

There was no question of the duty of care short-circuiting the contractual structure, because of the contractual relationship between Carne and Trendleway, nor did it expose Trendleway to the risk of an unrestricted number of claims or to claims greater in extent than those for which it would have been liable for breach of the contractual obligations which it had undertaken. **6.107**

A number of issues were left unresolved by the judgment of Beldam LJ. First, was the concurrent duty of care dependent upon the existence of a contractual duty of care, or would a different type of contractual duty be sufficient to give rise to a tortious duty of care? Secondly, to what extent was it necessary to find that Trendleway was a specialist sub-contractor? Thirdly, was it relevant that the breach involved defective, as opposed to, say, late work? **6.108**

In *Storey v Charles Church Developments Ltd*,[115] the defendant contracted to design and build a house for the plaintiff. Some ten years after the completion of the house, structural faults appeared, and the plaintiff claimed damages against the defendant in tort, the contractual claim being time barred. It was common ground that the claim was for pure economic loss. **6.109**

Judge John Hicks QC held that the existence of a contract between the parties was not a bar to concurrent tortious liability but nor was it a sufficient condition for such liability. The key question was whether there had been a sufficient 'assumption of responsibility coupled with the concomitant reliance'. He held that a professional designer would usually owe his client a contractual duty of care coterminous with his contractual duty of care. He held that a builder who undertook contractual design obligations owed a tortious duty of care coterminous with his contractual design obligation. Judge John Hicks QC distinguished *Murphy* on the following grounds. First, the builder in *Murphy's* case was not the designer of the defective foundations. Secondly, Mr Murphy had no contract with the designers. Thirdly, Mr Murphy had no contract with the builders. It should be noted that in *Murphy* Lord **6.110**

[111] (1996) 76 BLR 1.
[112] With whom Sir Tasker Watkins and Nourse LJ agreed.
[113] [1995] 2 AC 145 at 180–1.
[114] At 24.
[115] (1996) 12 Const LJ 206, (1995) 73 Con LR 1.

Bridge held[116] that there may be situations where 'even in the absence of contract, there is a special relationship of proximity between builder and building owner which is sufficiently akin to contract to introduce the element of reliance so that the scope of the duty of care owed by the builder to the owner is wide enough to embrace purely economic loss.'

6.111 However, Judge John Hicks QC held[117] that:

> a line must be drawn somewhere if builders are not to be concurrently liable in tort for all their contractual duties, including workmanship, as well as design, and including those which amount to warranties as well as those which can be expressed in terms of a duty of care, and that there may be apparent anomalies between cases on each side of that line, but that is inherent in the recognition of concurrent duties in tort matching some, but not all, contractual obligations.

6.112 This case raised the important question of whether there can ever be an assumption of responsibility and concomitant reliance in relation to defective workmanship as opposed to defective design. If a client requires a builder to provide a building with a floor which can carry a particular load, why should the client's remedy be any different if the floor fails because of a design failure or from a workmanship failure?

6.113 In *Samuel Payne v John Setchell*,[118] the claimants were the purchasers of cottages which had been built for a Mrs Wright. Mrs Wright had contracted with the defendant civil and structural engineer to assist in the construction of the cottages. The defendant had designed raft foundations for the cottages, inspected the construction of those foundations, and certified that they were satisfactorily constructed and suitable for support of the cottages. It was later found that defects in the design of the foundations had led to structural damage to the cottages. The claimants argued that, until Mrs Wright sold the cottages to them, she had had a cause of action against the defendant for breach of his concurrent duty of care in tort in designing the foundations and certifying their suitability. It was common ground that if Mrs Wright had had such a cause of action, the claimants would have acquired an identical cause of action upon their acquisition of the cottages, under s 3 of the Latent Damage Act 1986.

6.114 Judge Humphrey LLoyd QC held that the defendant owed no concurrent duty of care in tort to Mrs Wright. He regarded Lord Keith's apparent approval of *Pirelli* in *Murphy* as 'mysterious' because, he said, liability for a negligent misstatement stems from a voluntary assumption of responsibility towards the claimant which creates the special relationship. It was thus generally not found when the parties' relationship is governed by contract, especially if there is anything other than the simplest arrangement. In essence, although Judge Humphrey LLoyd QC did not rule out *Hedley Byrne* type liability arising in a situation where there was a contractual relationship, far from being the foundation of such liability it pointed away from such liability.

[116] At 481D.
[117] At para 29.
[118] [2002] BLR 489.

Further, the learned judge regarded the decisions of the House of Lords in *Murphy* and *Department of the Environment v Thomas Bates & Son* as establishing that:[119] **6.115**

> as a matter of policy, any person undertaking work or services in the course of a construction process is ordinarily liable only for physical injury or for property damage other than to the building itself but is not liable for other losses—i.e. economic loss. If any liability for such economic loss is to arise it must be for other reasons, e.g. as a result of advice or statements made upon which reliance is placed in circumstances which create a relationship where there is in law to be an assumption of the responsibility for loss—i.e. within the principle of *Hedley Byrne v Heller* . . . In my judgment a designer is not liable in negligence to the client or to a subsequent purchaser for the cost of putting right a flaw in a design that the designer has produced that has not caused physical injury or damage, just as a contractor is not liable.

Accordingly, the judge held that the defendant did not owe to his client, Mrs Wright, any duty of care in tort to avoid causing economic loss. Since the claimants had suffered economic loss only (the cost of remedial works to cure the defects in the cottages' foundations), they acquired no useful cause of action upon acquisition of the cottages under s 3 of the Latent Damages Act 1986. **6.116**

In *Tesco v Costain*,[120] Judge Richard Seymour QC declined to follow the analysis of Judge Humphrey LLoyd QC in *Samuel Payne*. Judge Richard Seymour QC considered[121] that *Murphy* was not authority for the proposition that a builder could never owe a duty of care in relation to economic loss to a subsequent owner, but only that he could not do so in the absence of the necessary special relationship. He held: **6.117**

> If the position now is, as I consider that it is, that anyone who undertakes by contract to perform a service for another upon terms, express or implied, that the service will be performed with reasonable skill and care, owes a duty of care to like effect to the other contracting party or parties which extends to not causing economic loss, there seems to be no logical justification for making an exception in the case of a builder or the designer of a building.

Mirant-Asia Pacific Ltd v Ove Arup & Partners International Ltd[122] was a claim by an employer against an engineer ('Arup') in relation to the construction of a power station where there was a relevant contractual duty to exercise reasonable skill and care in providing the design. The facts of the case were in all material respects identical to those in *Samuel Payne v John Setchell*. However, Judge John Toulmin QC decided not to follow the judgment of Judge Humphrey LLoyd QC in that case. He held: **6.118**

> Applying an objective test and bearing in mind the questions raised by Sir Brian Neill in both the cited judgments [*McNaughton v Hicks Anderson & Co* [1991] 2 QB 113 and *BCCI v Waterhouse and Others* [1998] PNLR 564, CA], I conclude that in relation to [the claimant], Arup assumed a responsibility for economic loss following the reasoning derived from Lord Goff's judgment. In so far as I need to I respectfully disagree with the judgment of HHJ LLoyd QC in *Payne v Setchell* where he places a designer in the same position as a contractor. It seems to me that where the designer performs services of a professional or quasi-profes-

[119] At para 28.
[120] [2003] EWHC 1487 (TCC).
[121] At para 230.
[122] [2004] EWHC 1750 (TCC), (2005) 97 Con LR 1, (2005) PNLR 10.

sional nature it is in the same position as bankers, solicitors, surveyors, valuers and accountants.

.... It is not part of the reasoning of my finding, but in relation to the distinction between builder and designer it is right to note that a builder warrants that its works will be fit for purpose, see *Young and Martins v McManus Childs* (1968) 9 BLR 7 whereas a professional adviser warrants only that he will exercise reasonable skill and care.

6.119 In summary, Judge John Toulmin QC had similar qualms to the learned judge in *Storey*. He was content that a construction professional will owe a duty of care to the person with whom he has a contract extending to economic loss, but a builder will not.

No contract between builder and claimant

6.120 There appears to have been no reported case since *Junior Books v Veitchi*[123] in which a builder has been held to owe a duty of care in tort in relation to pure economic loss to a person with whom he did not have a contractual relationship. Although the House of Lords has not overruled *Junior Books*, it has been held to be confined to its own facts.[124]

Construction professionals and the duty of care in relation to pure economic loss

6.121 In the light of *Henderson v Merrett*,[125] a construction professional will usually owe a tortious duty of care in relation to pure economic loss to his client with whom he has contracted and to whom he owes a contractual duty of care. The tortious duty of care may not conflict with the contractual duty of care or any other term of the contract and will be limited by the terms of the contract which the professional and the client have entered into.

6.122 Further although the House of Lords in *Henderson v Merrett* also held that the Indirect Names were owed a duty of care in tort by the managing agents, despite the lack of a contractual relationship between the Indirect Names and the managing agents, Lord Goff stated that the case was unusual and in many cases where there was a similar contractual structure (such as the typical construction contractual structure), that contractual structure would preclude the assumption of responsibility by a sub-agent towards the agent's principal.

6.123 In *Smith v Eric S Bush*,[126] the House of Lords held that in certain circumstances, a valuer could owe a duty of care in tort to the mortgagor of a property even though the valuer and the mortgagor did not enter into a contract.

6.124 Until recently, it had been thought that a professional engaged by an employer could never owe a duty of care in relation to pure economic loss to a contractor.[127] However, in *J Jarvis*

---

123 [1983] AC 520.

124 See, for example, the speech of Lord Bridge in *D&F Estates Ltd v Church Commissioners for England* [1989] AC 177 at 202.

125 [1995] 2 AC 145.

126 [1990] 1 AC 831.

127 It was thought that *Pacific Associates v Baxter* [1990] 1 QB 993 was authority for this proposition.

& Sons v Castle Wharf Developments Ltd,[128] the Court of Appeal held, in the context of a claim by a contractor against a professional agent of the employer in relation to negligent misstatements made at the tender stage, that there was no reason in principle why the professional agent of the employer cannot become liable to a contractor for negligent misstatements by the agent to a contractor to induce the contractor to tender, if the contractor relies on those misstatements. Whether a duty of care in fact arises in any given situation must depend on all the circumstances.[129]

The courts have otherwise been reluctant to impose a *Hedley Byrne* type duty upon construction professionals in relation to those with whom the professional has no contractual relationship. **6.125**

In *Machin v Adams*,[130] the plaintiff purchased a property which was undergoing extension and conversion works. Under the terms of the purchase contract, completion of the purchase contract was to take place on 25 November 2002 or the issue of a final certificate of completion of the works. The plaintiff and the vendor were to appoint an architect to supervise the works and in due course issue a final certificate when they were completed to his satisfaction. On completion the vendors were to lend the plaintiff £100,000 towards the purchase price. The plaintiff could in certain circumstances require completion of the purchase contract notwithstanding non-completion of the works, retaining out of the purchase price a reasonable sum, as agreed or certified by the architect in respect of incomplete work. In fact, no architect was appointed under the terms of the purchase contract but the defendant was appointed by the vendors to inspect the building and provide a final certificate for them. The architect was not told of the terms of the purchase contract set out above. In the course of the building works, the defendant wrote a letter to the vendor providing his opinion of the cost and time to completion. The defendant must have known that the letter was intended for someone other than the vendor but he did not know the precise purpose. The Court of Appeal applied *Henderson v Merrett* and held that the defendant did not owe a duty of care to the plaintiff. Sir Brian Neill[131] stated: **6.126**

> I would start with the general proposition that if an adviser to one party to a transaction gives advice to that party his duty is prima facie to that party alone. But this general proposition has to be considered in the light of the recent authorities. First, it is clear that if the adviser to one party to a transaction makes a careless misrepresentation which that party communicates to the other party in circumstances which would make the first party liable for its accuracy to the other party, the adviser may be liable as his principal: cf Hobhouse LJ in *McCullagh v Lane Fox & Partners Ltd* (1995) 49 Con LR at 152–153. Second, the words 'assumption of responsibility' must be understood in the sense of a conscious assumption of responsibility for the task rather than a conscious assumption of legal liability to the plaintiff for the careful performance of the task (see Lord Browne-Wilkinson in *White v Jones* [1995] 1 All ER 691 at 715, [1995] 2 AC 207 at 273). The court must therefore examine all the circumstances of the case. Thus in my judgment, in order to establish liability against the adviser in a case such

[128] [2001] EWCA Civ 19.

[129] *Per* Peter Gibson LJ at [53].

[130] (1997) 59 Con LR 14, (1997) 84 BLR 79.

[131] With whom Morritt LJ agreed.

as the present the advisee must show some connecting thread between the task the adviser has undertaken to perform and the course of action upon which the advisee can be foreseen to be likely to embark.

Of course in some cases an adviser will provide information to his client in the knowledge that a third party will learn of and rely on that advice and in reliance on it will take an anticipated course of action. The adviser knows the purpose of the advice and the purpose for which it will be used by the third party. The knowledge may be actual or inferential. The mortgage cases fall into this category. But in order for the adviser to be liable to the advisee it seems to me that, if one applies Lord Oliver's criteria, he must have actual or inferential knowledge not only that the advice will be communicated to the advisee but also knowledge of the purpose for which the information is required by the advisee. It is only in such circumstances that there is room for a finding that the adviser has undertaken responsibility to the advisee in respect of some foreseeable loss.

6.127 In *Architype Projects Ltd v Dewhurst Macfarlane & Partners*,[132] the building owner claimed in relation to defects within the building against a sub-consultant of the architect with whom the building owner had contracted. Judge John Toulmin CMG QC held that he was only bound to follow *Junior Books* if the facts of the case were identical to those in *Junior Books*, which he found they were not.

Liability for the negligence of independent contractors

6.128 In general,[133] a person is not liable in tort for the negligence of his independent contractor.[134] Consequently, unless a contractor is guilty of carelessness in supervising or inspecting the work of an independent sub-contractor then loss and damage caused by the carelessness of that independent sub-contractor will not give rise to a claim in negligence against the contractor.

6.129 In *Bellefield (No 2)*, the Court of Appeal held, in relation to the duty of care owed by an architect to a subsequent purchaser in relation to damage to other property, that if a dangerous defect arises as the result of a negligent omission on the part of the architect, he cannot excuse himself from liability on the grounds that he delegated the duty of design of the relevant part of the building works, unless he obtained the permission of his employers to do so.

---

[132] [2003] EWHC 3341 (TCC), [2004] PNLR 38.

[133] The 'exceptions' to the general rule relate to non-delegable duties imposed upon the employer of the independent contractor. See *Clerk & Lindsell on Torts*, 19th edn, paras 6–52 to 6–67. In *Biffa Waste Services Ltd v Maschinenfabrik Ernst Hese GMBH* [2008] EWCA Civ 1238, the Court of Appeal considered the ambit of the rule that a defendant is liable for the negligence of his independent contractor where the activities of his independent contractor are 'ultra hazardous'. It held that the rule, which had been laid down in *Honeywill v Larkin* [1934] 1 KB 191, was unsatisfactory and should be kept as narrow as possible. The Court of Appeal also considered another exceptional basis for liability in negligence, namely the vicarious liability for the negligence of an employee of an independent contractor who becomes the defendant's borrowed employee (described in the authorities as its employee *pro hac vice* (for an occasion)).

[134] *D&F Estates v Church Commissioners* [1989] AC 177 *per* Lord Bridge at 208. See also *Tesco v Costain* [2003] EWHC 1487 (TCC), paras [224] to [225].

The Common Law: Nuisance

Although nuisance is notoriously difficult to define, its essence is 'a condition or activity which unduly interferes with the use or enjoyment of land.'[135] In both its public and private limbs, the concept of nuisance is perhaps less important than it once was, first because of the ever-widening reach of statute, and secondly because many of its underpinning principles have become assimilated with those of negligence.[136] The following discussion is limited solely to those features of nuisance of particular relevance to the construction field. **6.130**

By their nature, construction projects are likely to cause disturbance to those in the immediate vicinity, whether by way of noise or dust, or, more dramatically, by damage to the integrity of adjoining land, for instance, by the removal of supporting structures. Where the complaint is that the act complained of has interfered with enjoyment of land—for instance by reason of the noise generated—the courts have accepted that there are limits to the extent to which the inevitable disturbance caused by construction projects can found liability in nuisance; perhaps unsurprisingly, the key is that the interference must be unreasonable. **6.131**

It has long been established that: **6.132**

> when one is dealing with temporary operations, such as demolition and building, everybody has to put up with a certain amount of discomfort, because operations of that kind cannot be carried on at all without a certain amount of noise and a certain amount of dust. Therefore, the rule with regard to interference must be read subject to this qualification, and there can be no dispute about it, that, in respect of operations of this character, such as demolition and building, if they are reasonably carried on, and all proper and reasonable steps are taken to ensure that no undue inconvenience is caused to neighbours, whether from noise, dust, or other reasons, the neighbours must put up with it.[137]

However, the principle that harm caused by temporary building work is not generally actionable is confined to mere inconvenience and does not extend to physical damage to property.[138] **6.133**

---

[135] *Clerk & Lindsell*, 19th edn, para 20–01.

[136] As regards liability for the escape of substances the principles applicable are now to all intents and purposes the same whether the claim is framed in nuisance, negligence, or under the rule in *Rylands v Fletcher* (1868) LR 3 HL 330): to establish liability, it must be established that escape was caused by a failure by the defendant to exercise reasonable care in allowing the circumstances giving rise to the escape, or the escape itself, to arise or by a failure to exercise reasonable care to control or to stop the escape once it has started: *Goldman v Hargrave* [1967] 1 AC 645; *Leakey v National Trust for Places of Historic Interest or Natural Beauty* [1980] QB 485; and *Holbeck Hall Hotel Ltd v Scarborough BC* [2000] QB 836. For a recent review of negligence, nuisance (private and public), and *Rylands v Fletcher* see the liability judgment in relation to the Buncefield explosion: *Colour Quest Ltd v Total Downstream UK Plc* [2009] EWHC 540 (Comm), [2009] 2 Lloyd's Rep 1.

[137] *Andreae v Selfridge & Co Ltd* [1938] Ch 1, 5; cf *Matania v National Provincial Bank and Elevenist Syndicate Ltd* [1936] 2 All ER 633, where the nuisance stemming from temporary dust and noise caused by extensive building alterations was sufficiently substantial to be actionable.

[138] *Clerk & Lindsell*, 19th edn, para 20-15; *Clift v Welsh Office* [1999] 1 WLR 796.

6.134 It will also frequently be necessary in the construction law context properly to analyse the potential liabilities to third parties in nuisance of: (1) the employer, and (2) the contractor. As regards (2), the question is relatively straightforward. If the contractor is the wrongdoer, then he will be liable in nuisance. This liability extends to cover the erection of a building on another's land which itself constitutes a nuisance.[139] It should be noted that the contractor will remain liable in nuisance even after the building is handed over to the employer.

6.135 As regards (1), matters may well be more complex. The general principle is that, as in negligence, an employer is not vicariously liable for the acts of his independent contractor.[140] However, this principle is subject to exceptions, conveniently listed by Neill LJ in *Alcock v Wraith*[141] as cases:

(1) where the employer is under some statutory duty which he cannot delegate;
(2) where there is the withdrawal of support from neighbouring land;
(3) where there is an escape of fire;
(4) where there is an escape of substances, such as explosives, which have been brought on to the land and which are likely to do damage if they escape (the rule in *Rylands v Fletcher*[142]);
(5) where there are operations on the highway which may cause damage to persons using the highway;
(6) concerning non-delegable duties of an employer for the safety of his employees;
(7) concerning extra-hazardous acts.

Of these exceptions, (2) and (7) are likely to be of most relevance in construction cases.[143]

6.136 A *tour d'horizon* of the case law on the matter was conducted by Judge Anthony Thornton QC in *Johnson v BJW Property Developments Ltd.*[144] The position can be summarized as follows. For an employer to be vicariously liable in nuisance (or, indeed in negligence) for the acts of his independent contractor, he must have authorized, instructed, or adopted the activity of that contractor. The escape onto and damage to adjoining land of the substance in question, or the interference with the use of the adjoining land must also be foreseeable. If the relevant act was authorized and the escape or interference foreseeable, then it must be determined whether the circumstances were such as to give rise to vicarious liability. The test has been expressed in a variety of ways: for instance, that the activity causing the escape was one which gave rise to a special danger,[145] was of an extra-hazardous kind,[146] or was such to create a special risk of damage due to the difficulties inherent in

[139] *Thompson v Gibson* (1841) 7 M&W 456.
[140] *Salsbury v Woodland* [1970] 1 QB 324.
[141] (1991) 59 BLR 16.
[142] (1868) LR 3 HL 330.
[143] (2) must also be considered with reference to the Party Wall etc Act 1996, discussed further below at paras 6.144 to 6.151.
[144] [2002] 3 All ER 574.
[145] *Matania v National Provincial Bank and Elevenist Syndicate Ltd* [1936] 2 All ER 633.
[146] *Honeywill & Stein Ltd v Larkin Bros (London's Commercial Photographers) Ltd* [1934] 1 KB 191. In *Biffa Waste Services Ltd v Maschinenfabrik Ernst Hese GMBH* [2008] EWCA Civ 1238, [2009] BLR 1, (2009) 122 Con LR 1, [2009] PNLR 12, [2009] 3 WLR 324, the Court of Appeal held that the principle in the *Honeywill*

carrying out that activity.[147] Taken together, however, it appears that the test is variable: where the work in question is done away from the boundary or party wall, it needs to be extra-hazardous (*Matania* and *Honeywill*); where the work is carried out on or adjacent to the boundary or party wall, it need only create a special risk of damage (*Bower v Peate*[148] *and Alcock*) in order to cross the threshold.

Legislative provisions

Defective Premises Act 1972[149]

Section 1(1) of the Defective Premises Act 1972 provides that a person who takes on work for or in connection with the provision of a dwelling, whether the dwelling is provided by the erection or by the conversion or enlargement of a building, owes a duty to ensure that the work is done in a workmanlike or, as the case may be, professional manner, with proper materials and so that as regards that work the dwelling, when completed, will be fit for human habitation. By s 1(1)(a) and (b) respectively, that duty is owed both to the first owner of the dwelling and to any person who subsequently acquires a legal or equitable interest in it. **6.137**

The duty under s 1(1) is owed by all persons taking on work in connection with the dwelling, including the building owner, and, by s 1(4), any person who arranges for such work to be carried out in the course of a business which includes the provision of or arranging for the provision of dwellings or installations in dwellings. **6.138**

The duty under s 1(1) is additional to any others owed (s 6(2)) and cannot be restricted or excluded (s 6(3)). It covers both acts and omissions and a dwelling may be unfit for human habitation if it lacks some essential attribute when complete even if the problems caused by the lack of that attribute have not yet become apparent: *Andrews v Schooling*.[150] It is to be noted, however, that works 'in connection with the provision of a dwelling' are limited to the creation of a new dwelling, rather than rectification to an existing dwelling: *Jacobs v Moreton & Partners*.[151] The 1972 Act has not been the subject of extensive judicial consideration.[152] Moreover, it would appear that the duty imposed by s 1(1) is limited to preventing serious defects in construction, such as to make the dwelling unfit for habitation: *Alexander v Mercouris*[153] *and Thompson v Clive Alexander & Partners (a firm)*.[154] The Act does not operate, in other words, to impose liability in respect of defects which have arisen solely as a result of a failure to carry out work in a professional manner and with proper materials. **6.139**

---

case was anomalous and so unsatisfactory that its application should be kept as narrow as possible and should be truly exceptional.

[147] *Alcock v Wraith* (1991) 59 BLR 16.

[148] (1876) 1 QBD 321.

[149] For more detailed commentary on the provisions of the Act, see James, *Construction Law: Liability for the Construction of Defective Buildings*, 2nd edn, Palgrave, 2001, pp 103 ff.

[150] [1991] 1 WLR 783.

[151] (1994) 72 BLR 92.

[152] Guidance was given by the Court in *Bole v Huntsbuild Ltd* [2009] EWHC 483 (TCC)

[153] [1975] 1 WLR 1270.

[154] (1992) 59 BLR 81.

6.140 Section 1(5) of the 1972 Act provides that the cause of action in respect of a breach of the duty imposed by section 1 shall be deemed to accrue for limitation purposes at the time when the dwelling was completed. However, if, after that time, a person who has done work for or in connection with the provision of the dwelling does further work to rectify the work he has already done, any such cause of action in respect of that further work shall be deemed for those purposes to have accrued at the time when the further work was finished.[155]

6.141 Section 3(1) of the Act provides an important safeguard in respect of work done on premises. It provides that any duty of care owed because of work done on premises (including construction, repair, maintenance, or demolition) to persons who might reasonably be expected to be affected by defects created during such work, is not abated by the disposal of the premises by the person who owed that duty.

Building Act 1984 and associated regulations

6.142 Construction work must comply with regulations made under s 1(1) of the Building Act 1984. The most recent regulations are the Building Regulations 2000 (SI 2000/2531), as amended. The Regulations lay out a broad framework, with technical detail set down in approved documents.[156] For present purposes, key amongst the Regulations is reg 7, which provides that building work shall be carried out\(a) with adequate and proper materials which are: (1) appropriate for the circumstances in which they are used; (2) adequately mixed and prepared; and (3) applied, used, and fixed so as adequately to perform the functions for which they are designed; and (b) in a workmanlike manner.

6.143 It is to be noted, however, that the section (s 38(1)) providing for civil liability for breach of the Building Regulations has never been brought into force. Nor would it appear that an action would lie for breach of statutory duty for failure to comply with the Regulations.[157] The Act and the associated Regulations therefore do not provide a cause of action for third parties as against either contractor or employer. Moreover, since the decision in *Murphy*, analysed above, it would appear that building control authorities owe no duty to persons properly to enforce the Regulations.

Party Wall etc Act 1996

6.144 The Party Wall etc Act 1996 ('the 1996 Act') extends (with variations) the regime that previously applied to London alone under the London Building Acts (Amendment) Act 1939. It is a short but complex and not altogether satisfactory statute; discussion here will be limited only to those provisions directly relevant to construction matters.[158]

---

155 *Alderson v Beetham Organisation Ltd* [2003] EWCA Civ 408, [2003] 1 WLR 1686, [2003] BLR 217.

156 To be found at the time of writing at <http://www.planningportal.gov.uk>. Failure to comply with such approved documents does not establish civil or criminal liability, but can be relied on in any proceedings for evidential purposes tending to establish or disprove liability: Building Act 1984, s 7(1).

157 'Ex hypothesi there is nothing in the terms or purpose of the statutory provisions which supports the creation of a private law right of action for breach of statutory duty': *Murphy v Brentwood DC* [1990] 3 WLR 414 at 449H *per* Lord Oliver.

158 For a comprehensive survey, see Bickford-Smith, *Party Walls: Law and Practice*, 3rd edn, Jordans, September 2009.

For almost all practical purposes, the 1996 Act has replaced the common law applicable to party walls with a statutory framework governing the rights and obligations of owners and (to a lesser extent) occupiers of adjoining land. **6.145**

Broadly, party walls are boundary walls which belong to one or more structures (s 20). The substantive rights given to the owners of buildings by the 1996 Act are contained in ss 1 and 2; they are, however, conditional on the operation of the statutory machinery. The two sections are mutually exclusive; in broad terms, s 1 governs the situation where the party wall has not yet been built, while s 2 applies where a wall has been built. **6.146**

In both cases, it is necessary to serve the requisite notice on the owner of the adjoining land before starting work. In the case of notices served under s 3 in relation to works proposed under s 2, the adjoining owner may serve a counter-notice under s 4. Where consent is not given within fourteen days to either a notice or a counter-notice, then, by s 5, a dispute is deemed to have arisen, triggering the statutory dispute resolution mechanism contained in s 10 and described in more detail below. **6.147**

Section 6 of the 1996 Act contains important provisions (including a one-month notice provision) regarding excavation (including excavation for and erection of a building or a structure). The provisions vary depending on whether the proposed works are to take place within 3 metres or 6 metres of the land of the adjoining owner (deemed to be so adjoining by virtue of s 6(4)). The provisions restrict the rights of the owner of the land to conduct such excavations with a view to protecting buildings or structures belonging to adjoining owners from subsidence. It is important to note in this regard that, although s 6 sets in place a statutory framework governing excavations in proximity to adjoining land, s 6(10) provides that compliance with the regime does not relieve the building owner of any liability he may owe at common law to the adjoining owner or occupier. In certain circumstances, liability for withdrawal of support may arise in negligence: *Holbeck Hall Hotel Ltd v Scarborough Borough Council (No 2)*,[159] while the adjoining owner may also have obtained a right by prescription to have his building supported: *Rees v Skerrett*.[160] **6.148**

Section 7 of the 1996 Act requires compensation to be paid for loss and damage and places ancillary obligations upon a building owner exercising his rights under the Act. Importantly, any works carried out under the provisions of the Act must comply with statutory requirements and be executed in accordance with such plans, sections, and particulars as are either agreed between the owners or determined in accordance with the dispute resolution procedure provided for under s 10. The duties provided for under s 7 are statutory duties; breach is capable of giving rise to an action in tort.[161] Such an action is the only enforcement mechanism open to occupiers of adjoining land, as they are not able to rely directly on the mechanism provided for under s 10. **6.149**

Provisions are contained in s 8 as to rights of entry to allow works to be carried out in pursuance of the 1996 Act. **6.150**

---

[159] [2002] 2 WLR 1396.
[160] [2001] 3 EGLR 1.
[161] For a comprehensive survey, see Bickford-Smith and Sydenham, n 158 above.

6.151 Section 10, finally, provides a special dispute resolution procedure, modelled on that contained in s 55 of the London Building Acts (Amendment) Act 1939. Where a dispute arises or is deemed to have arisen between a building owner and an adjoining owner (but not, note, an adjoining occupier), then a mechanism is triggered for resolution of that dispute. In broad terms, this operates by the appointment of a surveyor, either directly by the parties by consent, or, where they cannot agree, by a surveyor appointed by surveyors appointed by the parties. By s 10(10), either the agreed surveyor or (in specified circumstances) the three surveyors or any two of them, can settle by award any matter in dispute connected with work to which the Act relates. The award by s 10(12) can determine: (1) the right to execute any work; (2) the time and manner of executing any work; and (3) any other matter arising out of or incidental to the dispute including the costs of making the award. A time-limited right of appeal to the County Court is given by s 10(17).

EC Legislation: Consumer Protection Act 1987[162]

6.152 The Consumer Protection Act 1987 ('CPA') implements the Product Liability Directive 1985.[163] The heart of Pt 1 of the Act is to be found in s 2(1), which provides that, 'subject to the following provisions of this Part, where any damage is caused wholly or partly by a defect in a product, every person to whom subsection (2) below applies shall be liable for the damage.'

6.153 The Act provides for strict liability: all that is required is proof of damage, which is defined in s 5(1) as death or personal injury or any loss or damage to any property, including land. The damage must exceed £275, and the Act excludes damage to the product itself (s 5(2)); moreover, it does not include damage to business property (s 5(3)).

6.154 Although the Act provides for strict liability, it also provides a number of defences in s 4. Perhaps the most important of these is found in s 4(1)(e), namely that the state of scientific and technical knowledge at the relevant time was not such that a producer of products of the same description as the product in question might be expected to have discovered the defect if it had existed in his products while they were under his control.

6.155 'Product' is defined in s 1(2) of the CPA 1987. It includes any goods or electricity and includes 'a product which is comprised in another product, whether by virtue of being a component part or raw material or otherwise.' 'Goods' are defined in s 45 as including 'substances, growing crops and things comprised in land by virtue of being attached to it, and any ship, aircraft or vehicle.'

6.156 By s 3(1), there is a defect in a product if the safety of the product is not such as persons generally are entitled to expect. By s 3(2), in analysing what persons are entitled to expect in relation to a product the circumstances to be taken into account include: (1) the manner in which, and purposes for which, the product has been marketed, its get-up, the use of any mark in relation to the product and any instructions for, or warnings with respect to, doing or refraining from doing anything with or in relation to the product; (2) what might

[162] For further discussion of the 1987 Act, see James, *Construction Law: Liability for the Construction of Defective Buildings*, 2nd edn, Palgrave, 2001, pp 260 ff.

[163] Directive 85/374/EEC [1985] OJ L210/29.

reasonably be expected to be done with or in relation to the product; and (3) the time when the product was supplied by its producer to another. Importantly, nothing in s 3(1) requires a defect to be inferred from the simple fact that a later product is safer than the product in question.

Section 2(2) defines the persons to whom liability can attach. They are: (1) the producer of the product; (2) any person who, by putting his name on the product or using a trademark or other distinguishing mark in relation to the product, has held himself out to be the producer of the product; and (3) any person who has imported the product into a Member State of the EU from a place outside the EU in order, in the course of any business of his, to supply it to another. The scope of s 2(2) is widened considerably by s 2(3), which provides that a person is deemed to be liable for damage if he is the supplier of the product (to the victim of the damage, the producer of the product in which the product in question is comprised, or any other person) and fails to comply within a reasonable period with a request from the victim of the damage to identify the actual producer of the product. **6.157**

Producer is, in turn, defined widely in s 1(2). For present purposes, the relevant categories of producer are: (1) a person who manufactured the product (s 1(2)(a)); and (2) a person who carried out an industrial process giving the product its essential characteristics, where that product has not been manufactured, won, or abstracted (s 1(2)(c)). This latter category would cover, for instance, a person engaged in the preparation of concrete, being a process giving the product its essential characteristics. **6.158**

Specifically in relation to construction law, it is important to note s 46(3) and (4). Section 46(3) provides that '. . . subject to subsection (4) below, the performance of any contract by the erection of any building or structure on any land or by the carrying out of any other building works shall be treated for the purposes of this Act as a supply of goods in so far as, but only in so far as, it involves the provision of any goods to any person by means of their incorporation into the building, structure or works.' Section 46(4) provides that '. . . references in this Act to supplying goods shall not include references to supplying goods comprised in land where the supply is effectuated by the creation or disposal of an interest in land.' **6.159**

Interpreting s 46(3) and (4), therefore, it would appear that the contractor will be liable to the employer under the provisions of the CPA 1987 in respect of defective products that the contractor has supplied and produced, but only where those products have actually been incorporated into the building.[164] If the building *itself* is defective, then the CPA 1987 is not apt to cover the situation; in specific situations, the DPA 1972 may impose duties on the contractor, as analysed above. **6.160**

One further point that must be noted is the situation pertaining where the building in question is erected by a builder on land that he owns. Where he then disposes of that property, then, by s 46(4), he will not be liable under the provisions of s 2(1) in respect of damage caused by any defective products incorporated into the building. It has been suggested **6.161**

[164] Mirroring the position obtaining as a result of the provisions in Pt 1 of the Supply of Goods and Services Act 1982.

by one commentator that the builder may be liable as 'producer' in the event that the completed building is defective.[165] However, this would appear to strain the meaning of the word 'product' even beyond the wide bounds given in s 1(2); moreover, where the building in question is a new dwelling, the builder will owe a duty to the purchaser under the DPA 1972.

EC Legislation: Construction Products Regulations 1991[166]

6.162 The Construction Products Regulations 1991 (SI 1991/1620, as amended by the Construction Products (Amendment) Regulations 1994 (SI 1994/3051)) implement the Construction Products Directive.[167] They will be referred to in the balance of this chapter as 'the 1991 Regulations'.

6.163 By reg 2(1), a 'construction product' is any product, other than a minor part product, which is produced for incorporation in a permanent manner in works. A minor part product is a construction product which is included in a list of products which play a minor part with respect to health and safety drawn up, managed, and revised periodically by the European Commission. This list is published in the OJ 'C' Series.

6.164 By reg 3(1), construction products, other than minor part products, shall have such characteristics that the works in which it is to be incorporated, assembled, applied, or installed can, if properly designed and built, satisfy the essential requirements when, where, and to the extent that such works are subject to regulations containing such requirements. The essential requirements are set down in Sch 2 to the Regulations, reproducing Annex I to the Construction Products Directive. The preamble to the Annex notes that the essential requirements must, subject to normal maintenance, be satisfied for an 'economically reasonable working life', and generally concern requirements which are foreseeable. The six essential requirements relate to:

(1) mechanical resistance and stability;
(2) safety in case of fire;
(3) hygiene, health, and the environment;
(4) safety in use;
(5) protection against noise; and
(6) energy economy and heat retention.

6.165 By reg 4(1), any product which bears the CE mark is presumed to satisfy the requirements of reg 3(1). Compliance with and enforcement of the (voluntary) CE marking scheme is therefore of considerable importance to the operation of the regime set in place by the 1991 Regulations.

6.166 It is important to note that the 1991 Regulations do not make provision in respect of civil liability—they are to be enforced by means of criminal sanction (see, for instance, reg 9(4)). Nor would it appear that an action would lie for breach of statutory duty in respect of

[165] See James, n 162 above, p 267.
[166] For discussion, see ibid. pp 270 ff.
[167] Directive 89/106/EEC [1989] OJ L40/12.

a breach of the Regulations. However, in appropriate circumstances, they place important flesh on the bones of duties owed by architects, engineers, and contractors. In respect of the former two, a failure to ensure that products compliant with the 1991 Regulations are used in a construction project would ground an allegation that there had been a failure to provide reasonable supervision. As between employer and contractor, the obligation to provide satisfactory quality materials implied into the contract by s 4(2)–(6) of the Supply of Goods and Services Act 1982 would arguably extend to the provision of construction products compliant with the Regulations. Similarly, a supplier of construction products would owe a duty to the contractor to provide regulation-compliant products for use in the building; a contractor would have a good Part 20 claim against a supplier in the event that the former was fixed with liability for damage arising out of unsatisfactory construction products.[168]

EC Legislation: General Product Safety Regulations 1994[169]

The General Product Safety Regulations 1994 (SI 1994/2328) implement the General Product Safety Directive 1992.[170] They are referred to in the balance of this chapter as 'the 1994 Regulations'. **6.167**

The 1994 Regulations are of limited relevance to construction matters given the scope of the 1991 Regulations, and also the fact that they do not cover products supplied for commercial use. They are, however, included here for the sake of completeness. **6.168**

By reg 7, no producer shall place a product on the market unless the product is a safe product. Producers are defined in reg 2(1) as: (1) the manufacturer of the product, when he is established in the EU, including any person presenting himself as the manufacturer by affixing to the product his name, trademark, or other distinctive mark, or the person who reconditions the product; (2), where the manufacturer is not established in the EU, either the manufacturer's representative in the EU or the importer (if the manufacture does not have a representative in the EU); and (3) other professionals in the supply chain, insofar as their activities may affect the safety properties of a product placed on the market. **6.169**

A product is defined in reg 2(1) as any product intended for consumers or likely to be used by consumers, supplied whether for consideration or not in the course of a commercial activity and whether new, used or reconditioned. Importantly, however, products used exclusively in a commercial context fall outside the scope of the Regulations. **6.170**

A safe product, by reg 2(1), means any product which, under normal or reasonably foreseeable conditions of use, including duration, does not present any risk or only the minimum risks compatible with the product's use, considered as acceptable and consistent with a high level of protection for the safety and health of persons, taking into account in particular: (1) the characteristics of the product, including its composition, packaging, instructions **6.171**

[168] See James, n 162 above, p 275 and cf *Young and Marten v McManus Childs* [1969] 1 AC 454 at 466 *per* Lord Reid, [1986] 2 All ER 1169 at 1172. Note, however, that the 1991 Regulations have yet to be the subject of any judicial consideration.

[169] For discussion, see James, n 162 above, pp 272 ff.

[170] Directive 92/59/EC [1992] OJ L228/24.

for assembly, and maintenance; (2) the effect on other products, where it is reasonably foreseeable that it will be used with other products; (3) the presentation of the product, the labelling, any instructions for its use and disposal, and any other indication or information provided by the producer; and (4) the categories of consumers at serious risk when using the product, in particular children. The fact that higher levels of safety may be obtained or other products presenting a lesser degree of risk may be available shall not of itself cause the product to be considered other than a safe product.

6.172 Regulation 10 sets down provisions for determining whether a product is safe. In broad terms, if it complies with specific UK health and safety rules, it is presumed to be safe; where there are no such rules applicable, then other criteria are to be used, including voluntary standards giving effect to European national standards and EU technical specifications.

6.173 Similarly to the position under the 1991 Regulations, professionals involved in the supervision of works should be fixed with a duty (as part of their duty to ensure reasonable supervision) to take reasonable care to ensure that products are used that are compliant with the 1994 Regulations. In the same vein, a contractor owes a duty to an employer to use products compliant with the Regulations (and a contractor would have grounds to bring Part 20 proceedings against the supplier of such unsatisfactory products).

7

EXEMPTION AND LIMITATION OF LIABILITY CLAUSES

Limitation of Liability Clauses in Construction Contracts and Professionals' Contracts

7.01 There are many types of clauses in construction contracts and contracts with professionals in which the contractor or the professional seeks to limit its liability for breach of contract and negligence. The clauses can be placed into the following categories:

(a) Clauses which exclude liability for defective performance altogether.
(b) Clauses which exclude liability for damages for breach of contract and substitute for it a liability to replace or repair.
(c) Clauses which seek to limit liability for defective performance to a fixed sum or a sum which is expressed as a percentage of the contract sum.
(d) Clauses which seek to limit liability to a sum which it is agreed that a contractor or professional (usually a professional) shall insure against.
(e) Clauses which provide for a 'net contribution' limitation of liability in which the liability of a professional shall be limited to that proportion of the employer's loss which relates to the professional's relative responsibility along with the contractor and other professionals.
(f) Clauses which exclude liability for loss of profits, loss of use, and other 'consequential losses'.
(g) Clauses which provide for an indemnity by the innocent party against loss caused by the contractor.

7.02 Some of the abovementioned categories, save for the first, appear in bespoke contracts as well as the current edition of the FIDIC form and the contracts issued by the Institution of Chemical Engineers and the Institution of Mechanical Engineers.

7.03 Clauses which exclude liability for defective performance altogether are rare in the sense that a total exclusion is unusual. However in high-value contracts they are common when they are allied with clauses which require the contractor to repair and replace any defective work.[1] Therefore the contractor will have the obligation to remedy defects and the liability for defective work is limited to that, just as the obligation of the contractor in respect of delay will be limited to liquidated damages, which may be nil and will certainly be much less in many cases than the loss the employer will suffer in the event of delay. Sometimes the damages will be limited to a fixed sum or at least a sum which represents a proportion of the contract sum. This is, again, inevitable, in high-value contracts where the

[1] As, for example in *Yorkshire Water Services Ltd v Taylor Woodrow Construction Northern Ltd* [2005] EWCA Civ 894 and see in particular paragraph [34] of the judgment of May LJ:

> Yorkshire Water started the proceedings on the wrong foot by formulating their claim as a claim for damages for breach of the contract. This disregarded clause 44.2 of the General Conditions which provided that, except in the case of termination of the contract under clause 41, Taylor Woodrow's liability to Yorkshire Water for any breach of contract should be limited to the expenses, charges, damages and reimbursements expressly provided in the contract. Yorkshire Water had to make their claim under the contract, not as a claim for damages for breach of it.

employer's loss as a result of delay or defects is likely to be as much if not greater than the contract sum.

How the Clauses Are Construed

Like the contract as a whole, the enforceability of the clause in question is assessed as at the time the contract was made (and not, for example, when the alleged breach occurs) and on the basis of the parties' actual and constructive knowledge at that time. There are two categories of clause: those at (a) to (f) above which are exclusion or limitation of liability clauses, and (g) an indemnity clause which is designed to have the same effect. However, they are construed differently. Exclusion clauses are now construed in the same way as any other clause of the contract, applying the ordinary meaning of the words which the parties have used. It used to be the case that courts would strain to find a way to make them inapplicable because of their perceived iniquity. However since *Photo Production Ltd v Securicor Transport Ltd*[2] this is no longer the case. The impact of the Unfair Contract Terms Act 1977 is that the courts can simply disregard exclusion clauses which are perceived as unfair and they no longer have to strain their meaning.[3] This is subject to the proviso (which is discussed below in relation to indemnity clauses) that clear words must be used to exclude the consequences of negligence. Indemnity clauses, which are designed to have the same effect as exclusion clauses provide that the contract breaker is to be indemnified by the innocent party against the consequences of the contract breaker's negligence or default. Their construction is subject to specific guidance in the dictum of Lord Morton in *Canada Steamship Lines Ltd v The King*:[4] **7.04**

> (1) If the clause contains language which expressly exempts the person in whose favour it is made (hereafter called 'the proferens') from the consequence of the negligence of his own servants, effect must be given to that provision. . . .
> (2) If there is no express reference to negligence, the court must consider whether the words used are wide enough, in their ordinary meaning, to cover negligence on the part of the servants of the proferens. If a doubt arises at this point, it must be resolved against the proferens . . .
> (3) If the words used are wide enough for the above purpose, the court must then consider whether 'the head of damage may be based on some ground other than that of negligence', to quote again Lord Greene in the *Alderslade*[5] case. The 'other ground' must not

[2] [1980] AC 827, [1980] 1 All ER 556. Exclusion and limitation clauses are therefore subject to the general presumption in favour of 'commercially sensible construction' without 'technical interpretations and undue emphasis on the niceties of language'; see generally *Mannai Investment Co Ltd v Eagle Star Life Ass Co Ltd* [1997] AC 749 at 771; and *Sirius International Insurance Co Ltd v FAI General Insurance Ltd* [2004] 1 WLR 3251 at [19].

[3] The inclination of the courts to disapply contract terms under the Unfair Contract Terms Act 1977 depends on the 'commerciality' of the contracting parties; *Watford Electronics v Sanderson* [2001] 1 All ER (Comm) 696; and *Granville Oil & Chemicals Ltd v Davis Turner & Co Ltd* [2003] EWCA Civ 570, where Tuckey LJ said at para [31]: 'The 1977 Act obviously plays a very important role in protecting vulnerable consumers from the effects of draconian contract terms. But I am less enthusiastic about its intrusion into contracts between commercial parties of equal bargaining strength, who should generally be considered capable of being able to make contracts of their choosing and expect to be bound by their terms.'

[4] [1952] AC 192, [1952] 1 All ER 305, [1952] 1 Lloyd's Rep 1.

[5] *Alderslade v Hendon Laundry Ltd* [1945] KB 189.

> be so fanciful or remote that the proferens cannot be supposed to have desired protection against it; but subject to this qualification, which is no doubt to be implied from Lord Greene's words, the existence of a possible head of damage other than that of negligence is fatal to the proferens even if the words used are prima facie wide enough to cover negligence on the part of his servants.

7.05 This three-stage test has been considered and applied ever since.[6] The third part of the test has been modified slightly in that in applying it the courts should not ask whether there is or might be a technical alternative head of legal liability and, if there is, immediately construe the clause as inapplicable to negligence. The facts at the time the contract was made and the background must be looked at and the court must ask what potential liabilities one to another did the parties apply their minds to or must be deemed to have so applied their minds.[7] There are quite a lot of dicta to the effect that it must not be treated like a statute. But equally such dicta normally accompany its application with a further explanation of its rationale.[8] The third test has been described as 'horn-brook

---

[6] See for example Judge Bowsher QC in *Stent Foundations Ltd v MJ Gleeson Group plc* [2001] BLR 134.

[7] *Lamport & Holt Lines Ltd v Coubro & Scrutton (M&I) Ltd ('The Raphael')* [1982] 2 Lloyd's Rep 42; and *Caledonia Ltd v Orbit Valve Co* [1994] 1 WLR 1515 at 1521–2, CA.

[8] See Hobhouse J in *Caledonia Ltd v Orbit Valve Co* [1994] 1 WLR 221 at 227 where he said:

> . . . the well known opinion of the Judicial Committee, delivered by Lord Morton of Henryton in *Canada Steamship Lines Ltd v The King* [1952] A.C. 192, . . . there laid down rules to guide the decision of a court upon the question whether a particular indemnity or exemption clause should be construed as covering the consequences of the negligence of the party seeking the indemnity or the exemption. As has been said on a number of occasions (see Salmon LJ in *Hollier v Rambler Motors (AMC) Ltd* [1972] 2 QB 71, 80; May LJ in *Lamport & Holt Lines Ltd v Coubro & Scrutton (M&I) Ltd, (The Raphael)* [1982] 2 Lloyd's Rep. 42, 49–50 and Viscount Dilhorne in *South Wales Switchgear* [1978] 1 WLR 165, 168), what Lord Morton said was no more than guidance, and every contract has to be construed having regard to its own terms and context and the circumstances in which it was made. The question remains one of the construction of the contract, applying the established principles of construction. These include the principle that the parties to a contract are not to be taken to have agreed that a party shall be relieved of the consequences of its negligence without the use of clear words showing that that was the intention of the contract. In *Walters v Whessoe Ltd* (1960) 6 BLR 23, 35, Devlin LJ said: 'The law therefore presumes that a man will not readily be granted an indemnity against a loss caused by his own negligence.' In *Smith v South Wales Switchgear Co Ltd* [1978] 1 WLR 165, 168, Viscount Dilhorne said:
>
> > While an indemnity clause may be regarded as the obverse of an exempting clause, when considering the meaning of such a clause one must, I think, regard it as even more inherently improbable that one party should agree to discharge the liability of the other for acts for which he is responsible.
>
> This principle overlaps with but is not the same as the 'contra proferentem' rule—that contractual provisions should prima facie be construed against the party who was responsible for the preparation of the contract and/or who is to benefit from the provision. In the present case the contract form was the plaintiffs' form, to which the defendants were asked to adhere, but the relevant provision, article 10(b), is a mutual provision which, as a matter of construction, is equally capable of benefiting either party and therefore does not call for an interpretation which is weighted in favour of or against either of the parties. The relevant principle is simply one which involves construing exemption and indemnity provisions as applying to a party's own negligence, only if that intent is made clear in the contract; this principle is equally capable of application whether the clause is mutual or unilateral. Lord Keith of Kinkel, in *South Wales Switchgear* case [1978] 1 WLR 165, 180, adopted two sentences from the judgments of the Court of Appeal in the *Whessoe* case, 6 BLR 23:
>
> > It is well established that indemnity will not lie in respect of loss due to a person's own negligence or that of his servants unless adequate and clear words are used or unless the indemnity could have no reasonable meaning or application unless so applied: *per* Sellers LJ, at p 33.

construction law'.[9] The wording which must be used to include within the scope of an indemnity a breach of contract or negligence must be clear and explicit. The term 'any and all liabilities' has been held not to be sufficiently clear to include within the indemnity a liability for a breach of contract.[10] It must be noted however that limitation of liability clauses may not be effective to limit or prevent recovery under the Civil Liability (Contribution) Act 1978 in cases of fraud.[11]

The Role of Insurance

All exclusion and indemnity clauses must be read with the notion of insurance in mind and, in particular, the insurance arrangements that were, objectively speaking, in the contemplation of the contracting parties at the time of the contract. A net contribution clause, for example, may well relate to a professional's insurance policy's limit of indemnity. Equally, when insurance decides who bears a particular risk, it seems to be anachronistic to apply the rules of construction in relation to indemnity and exemption clauses when everyone knows, or ought to know, that the true question in relation to professionals is 'who is to insure'. Insurance is, however, one factor and not the decisive factor when exclusion clauses come to be considered under the Unfair Contract Terms Act 1977.[12] Before considering the Act, insurance clauses and their effect must be considered. **7.06**

Common Insurance Clauses

The most common insurance clauses in construction contracts relate to the requirement that one or other of the parties is responsible for providing 'All Risks' cover. 'All Risks' cover is provided in standard insurance policies and is linked to the JCT Standard Forms of Contract which is normally defined as physical loss or damage to the work done and materials on site. **7.07**

The term 'All Risks' is a rather grander term than encompassing truly every risk. All Risks policies now exclude war and terrorism cover. Terrorism cover excluded in modern policies also excludes any action taken in controlling, preventing, or suppressing terrorism. Terrorism is often defined as not only the use of force but also the threat of the use of force and harm or damage to life or property including (but not limited to) harm or damage done by nuclear and/or chemical and/or biological and/or radiological means caused by a person or group of people in whole or in part for political, religious, ideological, or **7.08**

It is now well established that if a person obtains an indemnity against the consequences of certain acts, the indemnity is not to be construed so as to include the consequences of his own negligence unless those consequences are covered either expressly or by necessary implication: *per* Devlin LJ, at p 34.

9 *Caledonia Ltd v Orbit Valve Co* [1994] 1 WLR 1515 at 1523.

10 *Deepak Fertilisers and Petrochemical Corp v Davy McKee (London) Ltd* [1999] BLR 41 at 50, [1999] 1 Lloyd's Rep 387, [1999] 1 All ER (Comm) 69.

11 See *Nationwide Building Society v Dunlop Haywards and another* [2009] EWHC 254 (Comm).

12 See *Singer Co UK Ltd v Tees and Hartlepool Port Authority* [1988] 2 Lloyd's Rep 164 at 170; and *The Flamar Pride* [1990] 1 Lloyd's Rep 434 at 439.

similar purposes. Subject to that exemption, the purpose of the policies required to be taken out by either employer or contractor is to deal with any risk which is likely to give rise to damage to the works. Any risk does not, of course, include defective work by the contractor although it will include damage to the work caused by a peril such as fire or flood caused by carelessness on the part of the contractor. So if the contractor burns down the building, then that is an insurable event. However if the contractor executes defective work such as to cause the building to collapse, then that is not a risk covered.

Effect of Insurance Clauses on Duties of Care and Liabilities under the Contract

7.09 The insurance clauses will have different effects in different contracts, depending upon exactly what terms and conditions are agreed to. However in relation to the JCT Standard Forms there have been a number of authorities which make it clear that the insurance clause is for the benefit of both contractor and employer and is not an exclusion clause in the sense that such clauses are normally understood.

7.10 In *James Archdale & Co Ltd v Comservices Ltd*,[13] a contractor agreed to decorate the employer's premises under the JCT Standard Form's predecessor, the RIBA 1952 Edition, conditions. The conditions provided: (a) a standard indemnity clause in which the contractor agreed to indemnify the employer against 'any injury or damage whatsoever to any property, real or personal, in so far as such injury and damage arises out of or in the course of or by reason of the execution of the works . . .'[14]; and (b) an obligation on the employer to insure the existing structures and the works and that they shall 'be at the sole risk of the employer as regards loss or damage by fire'.[15] The inevitable happened and the contractor burned down the premises, or at least caused damage to it by fire. The employer sued the contractor under the indemnity and failed. It was held that the indemnity excluded from within its scope any damage caused by an insured peril and that the risk of fire was accordingly the employer's and he must insure against it and take his remedy against the insurance company. In 1986 the House of Lords upheld and affirmed that decision.[16] In each case the indemnity clause referred to and deferred to the insurance clause. In the first case the indemnity clause concluded with the words '. . . subject also as regards loss or damage by fire to the provisions contained in clause 15 [the insurance clause] of these conditions' and in the second case the indemnity clause began with the words 'Except for such loss or damage as is at the sole risk of the employer under . . . clause 20[C] of these conditions [the insurance clause] . . .' It was obvious, therefore, that the obligation to indemnity was

13 [1954] 1 WLR 459.

14 cl 14(b) of the current RIBA conditions.

15 cl 15(b) of the current RIBA conditions.

16 *Scottish Special Housing Association v Wimpey Construction UK Ltd* [1986] 1 WLR 995. In that case the equivalent indemnity provision was cl 18(2) and the insurance provision was cl 20[C].

subject to the displacement of risk and the obligation to insure under the insurance clause. That led Lord Keith in the second case to say:[17]

> I have found it impossible to resist the conclusion that it is intended that the employer shall bear the whole risk of damage by fire, including fire caused by the negligence of the contractor or that of sub-contractors. The exception introduced by the opening words of clause 18(2) must have the effect that certain damage caused by the negligence of the contractor or of sub-contractors, for which in the absence of these words the contractor would be liable, is not to result in liability on his part. The nature of such damage is to be found in clause 20[C], which refers in general terms to damage by fire to the existing structures. No sensible content can be found for the words of exception in clause 18(2) if they are not read as referring to damage of the nature described in clause 20[C]. Counsel for the association strove valiantly to indicate some such alternative content but was unable, in my view, to do so convincingly.
>
> A similar conclusion was arrived at by the Court of Appeal in England in *James Archdale & Co Ltd v Comservices Ltd*[18] upon the construction of similarly but not identically worded corresponding clauses in a predecessor of the standard form. I consider that case to have been correctly decided and to be indistinguishable from the present one.

This accords with good business sense. There is no point in having an obligation to insure agreed to be placed in the hands of one or other party if the party which insures then proceeds against the party which does not. The scheme is therefore clear. One party insures, both parties get the benefit of the insurance and under the JCT Forms the insurance money is applied to repair or reconstruction by way of a variation. This scheme has been upheld by the House of Lords in *Co-operative Retail Services Ltd v Taylor Young Partnership Ltd (Carillion Construction Ltd, Part 20 defendants)*.[19] In that case there was a JCT Standard Form used which applied clause 22A, which requires 'all risks' insurance to be taken out by the contractor. Architects and engineers were employed as were electrical sub-contractors. A fire damaged the works and the contractor repaired them pursuant to the terms of the contract requiring it to do so. The employer brought an action against the architects and engineers who brought contribution proceedings against the main contractors and their electrical sub-contractors. The House of Lords held that the contractual scheme provided for restoration of the damaged works through the insurance money; therefore, the contractor and its subcontractor were not liable to the employer in damages and accordingly could not be liable to contribution under the Civil Liability (Contribution) Act 1978.[20] **7.11**

Therefore, one could expect that similar clauses will produce similar results.[21] The application of the cases relating to insurance and the cases to the effect that a co-insured cannot be sued by way of a subrogated claim might be thought to ensure that the insurance clause is **7.12**

---

[17] At p 999 of the report.

[18] [1954] 1 WLR 459.

[19] [2002] 1 WLR 1419.

[20] Endorsing the reasons given by Brooke LJ in the Court of Appeal [2000] 2 All ER (Comm) 865.

[21] As the Court of Appeal did with the JCT Intermediate Form of Contract in *Scottish & Newcastle plc v GD Construction (St Albans) Ltd* [2003] BLR 131. See also however *London Borough of Barking and Dagenham v Stamford Asphalt Company Ltd* (1997) 82 BLR 25.

sufficient to insulate the contractor from liability in respect of the insured perils.[22] But it is not always so.[23] In *Surrey Heath Borough Council v Lovell Construction Ltd*,[24] a case with similar facts, the indemnity clause however did not have the exclusion within it taking out of the scope of the indemnity loss and damage which was the subject of insurance under clause 22. Therefore the contractor was liable to the employer under the indemnity provision. In *Surrey Heath* the omission looks at first sight like a drafting error but it was sufficient to ensure that on the true construction of the contract the insurance clause did not override the indemnity clause. The reasoning in *Surrey Heath* has been applied in 2008 by the Court of Appeal in *Tyco Fire & Integrated Solutions (UK) Ltd v Rolls-Royce Motor Cars Ltd*.[25] In that case it was argued that the insurance exclusion of liability in such cases amounted to a rule of law. This was not accepted by the Court of Appeal and it was held that it is necessary to look at the particular conditions of contract in each case to see if the principle of exclusion applies.[26] Although the scheme of the JCT insurance clauses in major works forms ought to be generally to the effect of the *Archdale, Scottish Special Housing Association*, and *Co-op* cases, without special care being taken in drafting, the intent may not be put into practice. It is clear, however, that if the draftsmanship of the indemnity and insurance clauses is as *per* the *Archdale, Scottish Special Housing Association*, and *Co-op* cases then the contractor and its sub-contractors will not be liable in contract or tort to the employer in respect of damage caused by the insured risks.[27] But if it is as *per Surrey Heath* and *Tyco* then there will be liability notwithstanding the insurance clause. The difficulty in getting the drafting right is illustrated by the difficult case of *The National Trust v Haden Young*,[28] known as the *Uppark* case, where a fire destroyed Uppark House, started by two employees of Haden Young, who admitted negligence but pleaded reliance upon clauses 6.2 and 6.3B of the JCT Minor Works Form (similar but not identical and shorter versions of the insurance and indemnity clauses dealt with in the *Archdale, Scottish Special Housing Association, Norwich City Council*, and *Co-op* cases). The Court of Appeal held Haden liable under the indemnity clause and received no protection from the insurance clause. It was held therefore that the insurance provision was insufficient to provide for an exclusion. You could have insurance for a peril which might be as a result of either no fault of the contractor or as a result of the fault of the contractor, fire being the prime example. The insurance provision would protect the employer and the contractor in the event that the fire was nobody's fault but it would not protect the contractor if it was the contractor's fault.

---

[22] See *Petrofina (UK) Ltd v Magnaload Ltd* [1984] QB 127, [1983] 3 All ER 35; and *Rowlands (Mark) Ltd v Berni Inns Ltd* [1986] QB 211, [1985] 3 All ER 473, CA; *Hopewell Project Management Ltd v Ewbank Preece Ltd* [1998] 1 Lloyd's Rep 448, 458.

[23] As it was not by the Court of Appeal with the bespoke contract in *Casson v Ostley PJ Ltd* [2003] BLR 147.

[24] (1990) 48 BLR 108.

[25] [2008] EWCA Civ 286, [2008] 1 CLC 625, [2008] Lloyd's Rep IR 617, [2008] 2 All ER (Comm) 584, [2008] BLR 285, 118 Con LR 25, [2008] 14 EG 100.

[26] Note too that the House of Lords in *Co-op*, ibid, expressly stated that no assistance would be obtained from the *Archdale* and *Scottish Special Housing Association* cases. See para [45] of the judgment of Lord Hope.

[27] See also *Norwich City Council v Harvey* [1989] 1 WLR 828.

[28] (1994) 72 BLR 1, CA; see also the first instance decision of Otton J (1993) 66 BLR 88.

The indemnity clause had to be excised for the insurance clause to provide for an exclusion.[29] The effect of that decision (as indeed in *Tyco* and *Surrey Heath)* is that under the JCT Minor Works Form, insurance against fire ought to be taken out by the contractor, even though the contract provides that the employer shall do so as well.

Unfair Contract Terms Act 1977: Generally

The Unfair Contract Terms Act 1977 ('the Act') imposes 'limits on the extent to which . . . civil liability for breach of contract, or for negligence or other breach of duty, can be avoided by means of contract terms and otherwise'.[30] **7.13**

The Act does not apply to contracts of insurance[31] and any contract 'so far as it relates' to the creation or transfer of an interest in land, or to the termination of such an interest.[32] **7.14**

The Act does not apply to international supply contracts and deals only with contracts having some substantial connection with part of the United Kingdom. The material provisions do not apply where a contract is governed by the law of part of the United Kingdom only by choice of the parties and would, apart from that choice, be governed by the law of another country outside the United Kingdom.[33] However, in order to frustrate devices for evading its provisions, the Act operates notwithstanding any contract term which applies or purports to apply the law of some country outside the United Kingdom, if: (a) the term appears to have been imposed wholly or mainly for the purpose of enabling the party imposing it to evade the operation of the Act; or (b) in the making of the contract one of the parties dealt as consumer, was habitually resident in the United Kingdom, and the essential steps necessary for the making of the contract were taken there.[34] **7.15**

A further restriction on devices for evading the provisions of the Act is contained in s 10, which deals with secondary contracts which may contain terms excluding or restricting liability. It provides that 'a person is not bound by any contract term prejudicing or taking away rights of his which arise under, or in connection with the performance of, another contract, so far as those rights extend to the enforcement of another's liability which this Part of this Act prevents that other from excluding or restricting.' **7.16**

The operation of almost all[35] the material provisions of the Act is dependent on the satisfaction of at least one of two preliminary definitions: 'business liability' and 'dealing as a consumer'. **7.17**

---

[29] See also Slade LJ's analysis in *Dorset County Council v Southern Felt Roofing Company Ltd* (1989) 48 BLR 96 at p 106.

[30] Preamble to the Act.

[31] Sch 1, para 1(a).

[32] Ibid, para 1(b).

[33] s 27(1).

[34] s 27(2).

[35] Except in so far as it deals with liability for breach of the obligations arising from s 12 of the Sale of Goods Act 1979 (seller's implied undertakings as to title, etc) or s 8 of the Supply of Goods (Implied Terms) Act 1973 (the corresponding provision in relation to hire-purchase). Section 6(4) of the Act is also concerned with non-business liability, whilst s 6(1) and (3) do not require an individual to deal as a consumer.

7.18 'Business liability', defined[36] as liability for breach of obligations or duties arising either: (a) from things done or to be done by a person in the course of a business; or (b) from the occupation of premises used for business purposes of the occupier.

7.19 The provisions which deal with liability arising in contract[37] (other than liability for negligence or arising from the sale or supply of goods) may be relied upon only by a party who deals as a consumer, or on the other party's written standard terms of business, when he makes the contract in question. The section dealing with indemnity clauses[38] and some of the provisions relating to sale and supply of goods, may be relied upon only by a party who deals as a consumer.

7.20 A party to a contract 'deals as a consumer' if: (a) he neither makes the contract in the course of a business nor holds himself out as doing so; and (b) the other party does make the contract in the course of business.[39] The concept of dealing as a consumer was considered by the Court of Appeal in *R & B Customs Brokers Co Ltd v United Dominions Trust Ltd*,[40] where a private company, which carried on business as shipping and freight forwarding agents, was treated as a consumer when it purchased a motor car for the personal and business use of one of its directors. The court considered that a party entering into a transaction which was integral to its business did not deal as a consumer; if a transaction was merely incidental to the business, a degree of regularity is required before it could be viewed as an integral part of the business carried on. This was applied in *The Chester Grosvenor Hotel Company Ltd v Alfred McAlpine Management Ltd*,[41] where the claimant sought to argue that it acted as a consumer when it retained the defendant to carry out works of alteration and refurbishment to its hotel. It was held that the claimant did not deal as a consumer since the relevant contracts were entered into as an integral part of its business by way of providing hotel facilities and were part of its established practice of refurbishing the hotel. Despite that case, there remains uncertainty as to the way in which a court would view a party which enters into a contract for purposes that are merely incidental to the core business but nevertheless related to it.[42] Whilst an individual who enters into an agreement for the construction,[43] renovation,[44] or extension of a single dwelling-house will almost certainly be dealing as a consumer, the position of a retailer which engages a contractor to construct its sales premises is less certain. Whilst it appears possible that a business which contracts for the construction of its only office building deals as a consumer, the position may be different if the contract is for a periodic refurbishment of the same building.

---

[36] s 1(3).

[37] s 3.

[38] s 4.

[39] s 12. In the case of a contract governed by the law of sale of goods or hire-purchase, or by s 7 of the Act there are additional requirements.

[40] [1988] 1 WLR 321.

[41] (1991) 56 BLR 115.

[42] See Law Commission Report No 292, 'Unfair Terms in Contracts', paras 3.20–3.22.

[43] For example *Moores v Yakeley Associates Ltd* (1998) 62 Con LR 76.

[44] See *Casson v Ostley PJ Ltd* [2003] BLR 147. Although the applicability of the Act in this case could have been a consequence of contracting on 'the other's written standard terms' it is not made clear in either judgment.

Negligence liability

Section 1(1) of the Act defines negligence as: (a) breach of a contractual obligation to take reasonable care or exercise reasonable skill in the performance of the contract; or (b) breach of a common law duty to take reasonable care or exercise reasonable skill; or (c) breach of the common duty of care imposed by the Occupiers' Liability Act 1957. Section 2 applies only to negligence so defined and clearly does not apply to provisions excluding or restricting strict liability. **7.21**

Section 2(1) provides that a person cannot, by reference to a contract term or notice, exclude or restrict his [business] liability for death or personal injury resulting from negligence to any person, whether a consumer or not. **7.22**

[Business] liability for loss and damage, other than death or personal injury, cannot be restricted or excluded except in so far as the relevant term or notice satisfies the 'requirement of reasonableness'.[45] As with s 2(1), this is not confined to cases where the other party deals as a consumer. In *Smith v Eric S Bush*,[46] the House of Lords held that a disclaimer purporting to exclude liability for the negligence of surveyors to buyers of dwelling-houses was a notice, within the meaning of the Act, and had to satisfy the requirement of reasonableness in order to be effective. Similarly, an estate agent's disclaimer as to representations made in the particulars of sale for a property worth £875,000 was required to be reasonable.[47] **7.23**

Section 2 catches only attempts to exclude or restrict the liability of the negligent party to the victim of negligence; it is not concerned with arrangements made between the negligent party and third parties. In *Philips Products v Hyland*,[48] a plant-hire agreement contained a clause making the plant owner's driver the servant or agent of the hirer, thereby purporting to exclude the liability of the owner for the driver's negligence; when the owner attempted to rely on the clause to avoid liability for damage caused to the hirer by the negligence of the driver, it was held to be subject to the requirement of reasonableness. However, in *Thompson v Lohan (Plant Hire)*[49] a similar clause in a plant-hire agreement was held not to be subject to s 2(1) when a third-party was killed by the negligence of the driver; the clause did not operate to exclude or restrict liability to the third-party victim but merely determined, as between the hirer and the owner, by whom that liability would be borne. **7.24**

Consumer contracts and standard form contracts

Section 3, which deals with clauses restricting or excluding liability for breach of contract or qualifying the performance of contractual obligations, applies as between contracting parties where one of them deals as consumer or on the other's written standard terms of business.[50] **7.25**

---

[45] s 2(2).
[46] [1990] AC 831, HL.
[47] *McCullagh v Lane Fox & Partners Ltd* [1996] 1 EGLR 35.
[48] [1987] 1 WLR 659, CA.
[49] [1987] 1 WLR 649, CA.
[50] s 3(1).

7.26 Whilst the Act gives no guidance on the meaning of 'the other's written standard terms of business', the cases suggest that the regularity with which the terms are employed by the party advancing them is of more importance than whether that party is author.[51] Where a party invariably contracts in the same written terms without material variation, those terms will become its 'standard form contract' or 'written standard terms of business'. However, it does not follow that because terms are not employed invariably, or without material variation, they cannot be standard terms. This leaves open the possibility that the definition is satisfied when an employer contracts on the terms of a JCT Form of Contract, or an architect is appointed under the terms of the RIBA Form of Agreement for the Appointment of an Architect.[52]

7.27 Where s 3 applies, the party contracting on their written standard terms or not dealing as a consumer, cannot by reference to any contract term, except in so far as that term satisfies the requirement of reasonableness: (a) exclude or restrict any liability in respect of his own breach;[53] (b) claim to be entitled to render a contractual performance substantially different from that which was reasonably expected of him;[54] (c) claim to be entitled in respect of the whole or any part of his contractual obligation, to render no performance at all.[55]

7.28 'Any liability', in the context of a party seeking to exclude or restrict liability for its own breach, extends to strict liability for breach of contract. In *The Chester Grosvenor Hotel Company* case, a clause in a contract between an employer and a management contractor[56] provided that, where the management contractor was liable to the employer for breach of the contract by reason of any act or omission of one of its sub-contractors, the employer was not entitled to recover from the management contractor a sum in excess of that which it had recovered from its sub-contractor. This clause was agreed to be subject to the requirement of reasonableness by virtue of s 3(2)(a) and was found to be reasonable.

7.29 Where a term purports to entitle a party to render a contractual performance substantially different from that which was reasonably expected of him, the operation of s 3(2)(b)(i) gives rise to difficult questions of fact and degree: what was the reasonable expectation of the other party and what constitutes a substantial difference? The contractual obligations

[51] The pre-Act case of *Tersons v Stevenage Development Corp* [1963] 1 Lloyd's Rep 333 has been cited as supporting the proposition that a standard form of contract, the ICE conditions, is not within that category (particularly Pearson LJ's dictum at 368). However that case did not have to decide the application of the 1977 Act and it is hard to see how if a contractor contracted habitually on the basis of such standard terms that they do not become their standard conditions. And see *RW Green Ltd v Cade Bros Farms* [1978] 1 Lloyd's Rep 602. There is not as yet a definitive decision on this point.

[52] In *British Fermentation Products Ltd v Compair Reavell Ltd* (1999) 66 Con LR 1, Judge Bowsher QC refused to lay down any general principle as to when or whether the Act applies in the generality of cases where use is made of model forms drafted by an outside body. However, he concluded that for the Act to apply to such model forms, the party who seeks to rely on the Act must prove that the form is at least usually used by the party in question.

[53] s 3(2)(a).

[54] s 3(2)(b)(i).

[55] s 3(2)(b)(ii).

[56] Made, so the judge held, on the construction manager's standard terms. The management contract in question did define McAlpine as 'Project Manager', and its sub-contractors as 'construction contractors'—but this terminology can cause the reader to lose sight of the fact that McAlpine was a main contractor, not a construction professional.

of the party seeking to rely on the term are not the relevant criterion; rather the term must be measured against the reasonable expectations of the other party.

Section 3(2)(b)(ii) is triggered where an agreement imposes a contractual obligation but purports to give the party contracting on their written standard terms or not dealing as a consumer a wide discretion as to whether to perform or to perform to the extent agreed. In contrast to s 3(2)(b)(i), the relevant criterion is not what the other party expects but the contractual obligations of the party seeking to rely on the term. However, if the term in question grants such a wide discretion that, on its true construction, there is no duty to perform at all then s 3(2)(b)(ii) will not apply since s 3 does not deal with clauses which exclude a duty.[57] **7.30**

Unreasonable indemnity clauses

Section 4 provides that a person dealing as consumer cannot by reference to any contract term be made to indemnify another person (whether a party to the contract or not) in respect of liability that may be incurred by the other for negligence or breach of contract, except in so far as the contract term satisfies the requirement of reasonableness. **7.31**

Section 4 applies whether the liability in question is: (a) directly that of the person to be indemnified or is incurred by him vicariously; or, (b) is to the person dealing as consumer or to someone else.[58] The important distinction in relation to s 4 is between a clause which purports to indemnify against liability to a third party and a clause which purports to indemnify against liability to the indemnifier himself. In the first situation the indemnity clause is not regarded as an exemption clause for the purposes of the Act since it does not exclude or restrict liability to the indemnifier.[59] Therefore, in the second situation, but not the first, the relevant clause may be subject to both s 4 and other sections of the Act. This is particularly significant since s 4 is, on its own, unlikely to be relevant to the majority of construction contracts because its application is limited to cases where one party deals as a consumer. **7.32**

Sale and supply of goods

Section 6(1) provides that liability for the breach of obligations arising from the undertakings as to title implied by statute[60] into contracts of sale or hire purchase cannot be excluded or restricted by reference to any contract term. **7.33**

Section 6(2) provides that, as against a consumer, liability for the breach of obligations arising from statutory implied terms[61] as to conformity of goods with description or sample, or as to their quality or fitness for a particular purpose, cannot be excluded or restricted by reference to any contract term. **7.34**

---

[57] s 13.

[58] s 4(2).

[59] *Thompson v Lohan (Plant Hire)* [1987] 1 WLR 649, CA.

[60] s 12 of the Sale of Goods Act 1979; s 8 of the Supply of Goods (Implied Terms) Act 1973.

[61] ss 13, 14, and 15 of the Sale of Goods Act 1979; ss 9, 10, and 11 of the Supply of Goods (Implied Terms) Act 1973.

7.35 Section 6(3) provides that, as against a person dealing otherwise than as consumer, liability for the breach of obligations arising from statutory implied terms[62] as to conformity of goods with description or sample, or as to their quality or fitness for a particular purpose, may be excluded or restricted by reference to a contract term, but only in so far as the term satisfies the requirement of reasonableness.

7.36 The liabilities referred to in s 6 are not limited to 'business liabilities'[63] as defined by s 1(3), but include those arising under any contract of sale of goods or hire-purchase agreement. However, it should be noted that the notion of business liability is incorporated into s 6(2) by the requirement that one party should deal as a consumer; this is only possible if the other party 'makes the contract in the course of a business'.[64] Further, the statutory implied terms as to quality or fitness for a purpose which are referred to in s 6(2) and (3) apply only where the seller or supplier acts in the course of business.[65]

7.37 Section 7 applies where possession or ownership of goods passes pursuant to contracts not governed by the law of sale of goods or hire purchase, such as contracts of hire or exchange. As against a person dealing as consumer, liability in respect of the goods' correspondence with description or sample, or their quality or fitness for any particular purpose, which arises as a consequence of a breach of an implied term, cannot be excluded or restricted by reference to a contract term.[66] As against a person dealing otherwise than as consumer, such liability can be excluded or restricted by reference to a term, but only in so far as the term satisfies the requirement of reasonableness.[67] Liability for breach of obligations arising under s 2 of the Supply of Goods and Services Act 1982[68] cannot be excluded or restricted by reference to a term.[69] Liability in respect of: (a) the right to transfer ownership of the goods, or give possession; or (b) the assurance of quiet possession to a person taking goods in pursuance of the contract, cannot (in a case to which s 2 of the Supply of Goods and Services Act 1982 does not apply) be excluded or restricted by reference to a term, except in so far as the term satisfies the requirement of reasonableness.[70]

7.38 The Act does not apply to contracts for the sale or supply of goods which are made between parties whose places of business (or, if they have none habitual residences) are in the territories of different states.[71]

---

62 ss 13, 14, and 15 of the Sale of Goods Act 1979; ss 9, 10, and 11 of the Supply of Goods (Implied Terms) Act 1973.

63 s 6(4).

64 s 12(1)(b).

65 s 14(2) and (3) of the Sale of Goods Act 1979; s 10(2) and (3) of the Supply of Goods (Implied Terms) Act 1973.

66 s 7(2).

67 s 7(3).

68 Implied terms about title etc in certain contracts for the transfer of the property in goods.

69 s 7(3A).

70 s 7(4).

71 s 26 and Sch 1. See recently *Trident Turboprop (Dublin) Ltd v First Flight Couriers Ltd* [2009] EWCA Civ 290.

The reasonableness test applied to section 3 exemption clauses

7.39 It is for the party claiming that a contract term or notice satisfies the requirement of reasonableness to show that it does.[72] Section 11(1) of the Act provides that 'in relation to a contract term, the requirement of reasonableness for the purposes of this Part of this Act . . . is that the term shall have been a fair and reasonable one to be included having regard to the circumstances which were, or ought reasonably to have been, known to or in the contemplation of the parties when the contract was made'. In relation to a non-contractual notice, the requirement of reasonableness is that it should be fair and reasonable to allow reliance on it, having regard to all the circumstances obtaining when the liability arose or (but for the notice) would have arisen.[73] Section 11(2) requires the court, for the purposes of ss 6 and 7 to have in mind the factors set out in Sch 2 to the Act. On authority Sch 2 should be regarded as being of general application, so far as practicable or appropriate, to the question of reasonableness and therefore to s 3.[74] Those factors as set out in Sch 2 are:

(a) The strength of the bargaining positions of the parties relative to each other taking into account (amongst other things) alternative means by which the consumer's requirements could have been met.
(b) Whether the customer received an inducement to agree to the term, or in accepting it had the opportunity of entering into a similar contract with other persons, but without having to accept a similar term.
(c) Whether the customer knew or ought reasonably to have known of the existence and extent of the term (having regard, amongst other things, to the custom of the trade and any previous course of dealing between the parties).
(d) Where the term excludes or restricts relevant liability if some condition is not complied with, whether it was reasonable at the time of the contract to expect that compliance with that condition would be practicable.
(e) Whether the goods were manufactured, processed, or adapted to the special order of the customer.

Categories of Exemption Clauses and the Reasonableness Test under the 1977 Act

7.40 The categories of exemption clause set out in paragraph 7.01 above will now be considered. Before doing so, it is important to bear in mind that the decision as to what is or is not unreasonable will be fundamentally a matter for the first instance judge.[75] Whilst courts do interfere, and have done so, with the assessment of the first instance judge on the question of reasonableness, this interference is discouraged and interference by the Court of Appeal

[72] s 11(5).

[73] s 11(3).

[74] *Stewart Gill Ltd v Horatio Myer & Co Ltd* [1992] QB 600 at 608G.

[75] For a recent demonstration of how the judge goes about the business of deciding whether any particular clause is or is not reasonable, see the decision of Ramsey J in *Lobster Group Ltd v Heidelberg Graphic Equipment Ltd & Anor* [2009] EWHC 1919 (TCC).

ought only to take place when the judge has misconstrued the clause, erred in principle or is otherwise plainly wrong.[76]

Clauses which exclude liability for defective performance altogether

7.41 Whilst such clauses can be upheld, in theory it is hard to see how in fact they can be and usually they are not.[77]

Clauses which exclude liability for damages for breach of contract and substitute for it a liability to replace or repair

7.42 These clauses are common, as already noted, and so far none of the standard or derived terms which do so have been held to be unreasonable. Parties of equal or almost equal bargaining power are generally left to make their own commercial assessment of risk.[78]

Clauses which seek to limit liability for defective performance to a fixed sum or a sum which is expressed as a percentage of the contract sum

7.43 The Act lays down guidelines for determining reasonableness in two specific situations: where a term limits a person's liability to a specific sum of money and where a term in a contract for the sale or supply of goods[79] is subject to the requirement of reasonableness. Where a contract term or notice, by reference to which a person seeks to restrict his liability to a specific sum, is subject to the requirement of reasonableness the court must have regard to: (a) the resources which he could expect to be available to him for the purpose of meeting the liability should it arise; and (b) how far it was open to him to cover himself by insurance.[80] In applying the second guideline the court considers the availability of insurance to the party seeking to restrict liability, rather than that party's actual insurance position.[81]

---

[76] *Regus (UK) Ltd v Epcot Solutions Ltd* [2008] EWCA Civ 361 at para [38]. Lord Bridge in *George Mitchell Ltd v Finney Lock Seeds* [1983] 2 AC 803 at 815–16 said:

> It would not be accurate to describe such a decision as an exercise of discretion. But a decision under any of the provisions referred to will have this in common with the exercise of a discretion, that, in having regard to the various matters to which the modified section 55(5) of the Act of 1979, or section 11 of the Act of 1977 direct attention, the court must entertain a whole range of considerations, put them in the scales on one side or the other, and decide at the end of the day on which side the balance comes down. There will sometimes be room for a legitimate difference of judicial opinion as to what the answer should be, where it will be impossible to say that one view is demonstrably wrong and the other demonstrably right. It must follow, in my view, that, when asked to review such a decision on appeal, the appellate court should treat the original decision with the utmost respect and refrain from interference with it unless satisfied that it proceeded upon some erroneous principle or was plainly and obviously wrong.

[77] See, for example, *Bacardi-Martini Beverages Ltd v Thomas Hardy Packaging* [2002] EWCA Civ 549; and *Fillite (Runcorn) Ltd v APV Pasilac Ltd* (1989) 49 BLR 27; *Edmund Murray v BSP International Foundations* (1992) 33 Con LR 1; *Sovereign Finance v Silver Crest* [1999] GCCR 2187.

[78] See, for example, *Yorkshire Water Services Ltd v Taylor Woodrow Construction Northern Ltd* [2005] EWCA Civ 894 where the point was not taken; and for a first instance upholding of the Institution of Mechanical Engineers Model Form C, see *British Fermentation Products Ltd v Compare Reavell Ltd* (1999) 66 Con LR 1 and the dictum of Judge Thayne Forbes QC cited in para 7.43.

[79] Subject to ss 6 or 7 of the Act.

[80] s 11(4).

[81] *Singer (UK) Ltd v Tees and Hartlepool Port Authority* [1988] 2 Lloyd's Rep 164.

In *Moores v Yakely Associates Ltd*,[82] the defendant architect was hired by a wealthy businessman to design a domestic bungalow. The contract for architectural services contained a term limiting the architect's liability for loss and damages to the sum of £250,000, which was based on his estimate of construction costs. These costs escalated sharply. The claimant later commenced proceedings for breach of contract, alleging that the defendant's standard of work had fallen below that required; the defendant argued that all the claims were limited to £250,000. The Court of Appeal held that the limitation was not unreasonable given that the fees which the defendant had expected to earn *at the time of the agreement*[83] were of the order of £20,000, less than one-tenth of the limitation, and that the claimant, who was a man of considerable means, had been in a stronger bargaining position than the defendant because of a severe recession in the construction industry at the time. The equality of bargaining power is often decisive. *In Salvage Association v CAP Financial Services Ltd* [1995] FSR 654 at 676, Judge Thayne Forbes QC said:

> Generally speaking, where a party well able to look after itself enters into a commercial contract and, with full knowledge of all relevant circumstances, willingly accepts the terms which provide for apportionment of the financial risks of the transaction, I think it is very likely that those terms will be held to be fair and reasonable.

This statement was approved by the Court of Appeal in *Watford Electronics v Sanderson CFL Ltd* [2001] 1 All ER (Comm) 696 at [63].

Contracts which seek to limit liability to a sum which it is agreed that a contractor or professional (usually a professional) shall insure against

These are a gloss upon the maximum sum terms previously discussed and will be treated similarly. If, however there has been extensive discussion as to the professional's upper limit of insurance and this is agreed to be the maximum sum due, even though the burden of proving reasonableness lies with the party asserting it, in an ordinary case, proof of reasonableness will be relatively easy in large arms-length transactions between major commercial parties. As between a small householder and an experienced professional, then this may be more difficult. In addition, the Unfair Terms in Consumer Contracts Regulations 1999 will apply to them. **7.44**

Contracts which provide for a 'net contribution' limitation of liability in which the liability of a professional shall be limited to that proportion of the employer's loss which relates to the professional's relative responsibility along with the contractor and other professionals

There is in the construction industry a lively debate between producer and consumer interests over whether professionals should have an unlimited liability or whether standard terms and conditions should contain within them a 'net contribution' clause. The purpose of the net contribution clause is to ensure that should one or more of the other professionals or the contractor be uninsured or insolvent or both, then the professional which is solvent and insured does not pick up the whole bill as it ordinarily would do. It is unlikely that such **7.45**

[82] [2001] BLR 322.

[83] That these fees later escalated by a multiple of fifteen was not relevant to the issue of reasonableness.

a clause will be held to be unreasonable because it deals primarily with solvency of co-contractors and tortfeasors and in large projects it is easy for employers to ensure that all the professionals are appropriately insured and to take a view as to the solvency of the main contractor or sub-contractors with which it has a direct warranty. It has been held that such clauses do not actually restrict liability but only make the negligent professional responsible for his or her damage.[84] This is a doubtful proposition. The rationale by the judge was that the applicable defendant was being held liable only for the consequences of its own breach of duty and were not held liable by the operation of the doctrine of joint and several liability for the breach of duty by other contractors. That reasoning is, it is suggested, defective. Joint and several liability is just that. To the extent that it is joint it is so because all defendants caused the relevant damage and thus *ex hypothesi* the party relying upon the net contribution clause caused the damage as much as the joint tortfeasor or contract breaker. However, in the same case it was held, it is submitted correctly, that if it was an exclusion clause, it would be fair and reasonable.[85]

Contracts which exclude liability for 'loss of profits, loss of use' and other 'consequential losses'

7.46 In non-consumer contracts, these clauses are commonplace in standard and bespoke forms and will usually be upheld.[86]

Sale or Supply of Goods and the Further Matters to Be Considered under the Reasonableness Test

7.47 Where a term in a contract for the sale or supply of goods is, pursuant to ss 6 or 7 of the Act, subject to the requirement of reasonableness, s 11(2) of and Sch 2 to the Act provide that regard is to be had to five further matters, in so far as they are relevant:

(a) the strength of the bargaining positions of the parties relative to each other;
(b) whether the customer received an inducement to agree to the term, or, in accepting it had an opportunity of entering into a similar contract with other persons, but without having to accept a similar term;
(c) whether the customer knew or ought reasonably to have known of the existence and extent of the term;
(d) where the term excludes or restricts any relevant liability if some condition is not complied with, whether it was reasonable at the time of the contract to expect that compliance with that condition would be practicable;
(e) whether the goods were manufactured, processed, or adapted to the special order of the customer.

---

84 *Langstane Housing Association Ltd v Riverside Construction Aberdeen Ltd & Ors* [2009] ScotCS CSOH_52.

85 Ibid. Furthermore, the judgment in that case gives a useful history of the role of net contribution clauses under the ACE conditions.

86 See eg *Regus (UK) Ltd v Epcot Solutions Ltd* [2008] EWCA Civ 361 at paras [39]–[43].

It is difficult to see why these guidelines should apply only in the context of contracts for the sale or supply of goods, and indeed the courts have treated them as being of general application and applied them, by analogy, to terms in other types of contract.[87]

The statutory guidelines are not exhaustive, even in relation to the situations covered by them. In *Smith v Eric S Bush*,[88] the House of Lords held that a disclaimer purporting to exclude liability for the negligence of surveyors to buyers of dwelling-houses was a notice, within the meaning of the Act, and had to satisfy the requirement of reasonableness in order to be effective. The disclaimer had been made by a building society on behalf of both itself and the surveyor carrying out the valuation. In deciding that the disclaimer did not satisfy the requirement, the House of Lords pointed to several factors: there was no equality of bargaining power, it was not reasonable to expect the buyers to commission their own structural survey since the houses were of modest value, and the surveyors could have insured against the risk at only modest expense. Whilst the same conclusion is likely to be reached in other cases which relate to modest domestic dwellings, a disclaimer might be reasonable if the valuation is a matter of great complexity or the value of the property is so high that insurance against professional liability is prohibitively expensive. Limiting or excluding liability for professional negligence in relation to representations concerning expensive domestic dwellings and industrial property is less likely to be viewed as unreasonable. In *McCullagh v Lane Fox & Partners Ltd*,[89] an estate agent's disclaimer as to representations relating to the floor area of the property, made in the particulars of sale for a property worth £875,000, was held to be reasonable; the purchaser in such a case could be expected to verify such information before completing the purchase. **7.48**

Effect of Repudiation

Where a contract term satisfies the requirement of reasonableness, effect may be given to it notwithstanding the termination of the contract by breach or by a party electing to treat it as repudiated.[90] Affirmation by a party entitled to treat a contract as repudiated does not of itself exclude the requirement of reasonableness in relation to any contract term.[91] **7.49**

General Terms and Severability

Where a clause contains more than one limitation or exemption, or is phrased in general terms, a court may hold one part effective and another part ineffective or that the clause as a whole is effective in certain circumstances and ineffective in others. The Act does not invalidate contractual terms, rather it states that certain liabilities cannot be excluded or restricted by reference to them. It is therefore possible that a clause rendered ineffective by the Act in so far as it purports to restrict a specific liability is effective in so far as it is relied **7.50**

[87] *Singer (UK) Ltd v Tees and Hartlepool Port Authority* [1988] 2 Lloyd's Rep 164.
[88] [1990] AC 831, HL.
[89] [1996] 1 EGLR 35.
[90] s 9(1).
[91] s 9(2).

upon in a situation not covered by the Act. Furthermore, where the Act provides that terms subject to the requirement of reasonableness are effective 'only in so far as', or ineffective 'except in so far as', they satisfy the requirement of reasonableness, a court may treat a clause which contains more than one exemption or limitation as severable and, having severed it, hold that one part meets the requirement and another is ineffective since it does not.

Effect of Different Types of Exemption Clause: Set Off

7.51 To the extent that the Act prevents the exclusion or restriction of liability it also prevents: (a) making the liability or its enforcement subject to restrictive or onerous conditions;[92] (b) excluding or restricting any right or remedy in respect of the liability, or subjecting a person to any prejudice in consequence of his pursuing any such right or remedy;[93] and (c) excluding or restricting rules of evidence or procedure.[94] Thus a clause in a standard form contract which excludes the remedy of set off where goods delivered are defective or incorrect is subject to the requirement of reasonableness in the same way as would be a clause excluding or restricting liability for defective or incorrect delivery.[95] However, a liquidated damages clause is unlikely to be subject to the Act since it may extend, as well as limit, liability. It is noteworthy that the Act explicitly provides that an agreement in writing to submit present or future differences to arbitration is not to be treated as excluding or restricting any liability.[96]

7.52 To the extent that ss 2 and 5 to 7 of the Act prevent the exclusion or restriction of liability they also prevent the exclusion or restriction of the relevant obligation or duty.[97] Thus a clause in a contract of employment which required the employee to work up to eighty-eight hours a week was held to be ineffective pursuant to s 2 because it operated to restrict or exempt the employer from liability for personal injury by restricting the scope of the duty of care to provide a safe system of work.[98] By contrast, a clause which makes the performance of a contractual obligation conditional on an event such as labour strikes is not caught by s 3(2)(b)(ii), since s 3 does not prevent the exclusion of a duty.

The Unfair Terms in Consumer Contracts Regulations 1999

Application

7.53 The Unfair Terms in Consumer Contracts Regulations 1999[99] ('the Regulations') give effect to Council Directive 93/13/EEC which is designed to promote the harmonization

---

[92] s 13(1)(a).
[93] s 13(1)(b).
[94] s 13(1)(c).
[95] *Stewart Gill Ltd v Horatio Myer & Co Ltd* [1992] QB 600.
[96] s 13(2).
[97] s 13(1).
[98] *Johnstone v Bloomsbury Health Authority* [1992] QB 333.
[99] SI 1999/2083.

of the laws of Member States so as to ensure that contracts with consumers do not include terms which are unfair to the consumer.[100]

The Regulations operate in parallel to the Act so that a term may be ineffective under one but valid under the other. They are of limited applicability to construction contracts since they are effective only where one party acts as a consumer[101] and apply only to terms which are not individually negotiated.[102] By contrast, s 3 of the Act—the provision most likely to be of relevance to construction contracts—applies where one party acts as a consumer or where the contract is on one party's written standard terms. The Regulations are likely to apply most frequently to private house building or works of renovation or modification. They can apply to contracts for the sale of interests in land.[103] Where they do apply, however, they are in one respect wider in scope than the Act: whereas the Act, with the exception of s 4,[104] deals only with clauses which exclude or restrict liability or a duty, the Regulations apply to clauses which confer rights upon the party relying on them. Such terms may, for example, confer rights which entitle a seller to increase his charges unfairly, or require a consumer to pay disproportionate compensation if he fails to fulfil his obligations.[105] **7.54**

The Regulations do not apply to terms which a contract is required, by statute, to contain[106] or terms which reflect the provisions or principles of international conventions to which European Member States or the Community are party.[107] **7.55**

The Regulations apply notwithstanding any contract term which applies or purports to apply the law of a non-Member State, if the contract has a close connection with the territory of the Member States.[108] **7.56**

Unfair terms and the Regulations

Regulation 7 provides: (a) that a seller or supplier shall ensure that any written term of a contract is expressed in plain, intelligible language; and (b) if there is doubt about the meaning of a written term, the interpretation which is most favourable to the consumer shall prevail. **7.57**

Regulation 8 provides that an unfair term in a contract concluded with a consumer by a seller or supplier shall not be binding on the consumer. However, the contract shall continue to bind the parties if it is capable of continuing in existence without the unfair term. **7.58**

A seller or supplier is defined as 'any natural or legal person who, in contracts covered by these Regulations, is acting for purposes relating to his trade, business or profession, **7.59**

---

[100] Fourth Recital to Dir 93/13/EEC [1993] OJ L95/29.

[101] reg 8(1).

[102] reg 5(1).

[103] See *London Borough of Newham v Khatun* [2004] 3 WLR 417, [2004] EWCA Civ 55; and Case C-237/02 *Freiburger Kommunalbauten GmbH Baugesellschaft KG v Hofstetter* [2004] 2 CMLR 13; and *Zealander & Zealander v Laing Homes Ltd* [2000] TCLR 724.

[104] Indemnity clauses.

[105] Examples taken from Sch 2 to the Regs.

[106] reg 4(2)(a).

[107] reg 4(2)(b).

[108] reg 9.

whether publicly owned or privately owned.'[109] A building contractor, architect, project manager, or surveyor may be suppliers for the purposes of the Regulations.[110] The reference to 'contracts covered by these Regulations' is not an allusion to a specific set of contracts but refers to contracts between a seller or supplier and a consumer which contains terms which have not been individually negotiated.

7.60 A consumer is 'any natural person who, in contracts covered by these Regulations, is acting for purposes which are outside his trade, business or profession'.[111] Whilst an individual who contracts for the building, renovation,[112] or modification of a private dwelling-house will almost certainly be a consumer for the purpose of the Regulations, an individual who contracts for the building of his own retail premises will not. The restriction of a consumer to a natural person[113] further diminishes the potential for the Regulations to apply to construction contracts. However, a 'legal' person may be a consumer even where the contract is of great value.[114]

7.61 Regulation 5(2) states that 'a term shall always be regarded as not having been individually negotiated where it has been drafted in advance and the consumer has therefore not been able to influence the substance of the term.' There is no requirement that the term be drafted by the seller or supplier, thus the Regulations would apply to the JCT Form of Building Contract,[115] or the RIBA Form of Agreement for the Appointment of an Architect. Where a specific term, or certain aspects of it, has been individually negotiated, the Regulations shall apply to the rest of the contract if an overall assessment of it indicates that it is a pre-formulated standard contract.[116] The burden of proving that a specific term was individually negotiated rests upon the seller or supplier who seeks to rely upon it.[117]

Unfairness and the Regulations

7.62 By reg 5(1), 'a contractual term which has not been individually negotiated shall be regarded as unfair if, contrary to the requirement of good faith, it causes a significant imbalance in the parties' rights and obligations arising under the contract, to the detriment of the consumer.' There are therefore two concepts which must be considered independently: 'a significant imbalance in the parties' rights'; and 'the requirement of good faith'. This is to be contrasted with the different test of reasonableness under the Act.

7.63 In determining whether there is a significant imbalance the court must take 'into account the nature of the goods or services for which the contract was concluded and by referring,

---

109 reg 3(1).

110 *Bryen & Langley Ltd v Martin Boston* [2005] EWCA Civ 973.

111 reg 3(1).

112 *Allen Wilson Shopfitters v Buckingham* [2005] EWHC 1165 (TCC); *Bryen & Langley Ltd v Martin Boston* [2005] EWCA Civ 973.

113 The one exception to this is for the purposes of the Arbitration Act 1996, where the Regs are extended by ss 89–91 to include legal persons in the case of consumer arbitration agreements where the amount claimed is £5,000 or less.

114 In *Standard Bank of London Ltd v Apostolakis* [2000] ILPr 766, the Regs were applied to a currency transaction worth $7 million.

115 eg *Bryen & Langley Ltd v Martin Boston* [2005] EWCA Civ 973, CA.

116 reg 5(3).

117 reg 5(4).

at the time of conclusion of the contract, to all the circumstances attending the conclusion of the contract and to all the other terms of the contract or of another contract on which it is dependent.'[118]

Among the factors which are relevant to the question of whether a party has acted contrary to the requirement of good faith are 'the strength of the bargaining positions of the parties, whether the consumer had an inducement to agree to the term and whether the goods or services were supplied to the special order of the consumer'.[119] The House of Lords has held[120] that the requirement of good faith is 'one of fair and open dealing', the seller or supplier is not permitted to 'take advantage of the consumer's necessity, indigence, lack of experience, unfamiliarity with the subject matter of the contract, weak bargaining position', and any terms must be expressed 'fully, clearly and legibly, containing no concealed pitfalls or traps'. Hence, whilst the Regulations do not specify that the term which is 'not individually negotiated' must be proposed by a party other than the consumer, in cases where the term has been proposed by the consumer it will be virtually impossible to demonstrate that the supplier has acted contrary to the requirement of good faith.[121] **7.64**

Schedule 2 to the Regulations contains an indicative and non-exhaustive list of the terms which may be regarded as unfair. Included in the list are terms which have the effect of: **7.65**

> excluding or hindering the consumer's right to take legal action or exercise any other legal remedy, particularly by requiring the consumer to take disputes exclusively to arbitration not covered by legal provisions, unduly restricting the evidence available to him or imposing on him a burden of proof which, according to the applicable law, should lie with another party to the contract.[122]

The effect of the Regulations on adjudication and arbitration provisions

In *Picardi v Cuniberti*,[123] Judge Toulmin CMG QC expressed the view[124] that an adjudication clause, contained in the RIBA conditions of engagement, would, had it been incorporated into a contract for the refurbishment of a domestic dwelling, have been unfair and therefore ineffective pursuant to the Regulations. He considered that the adjudication provisions were apt to cause a significant imbalance in the parties' rights and obligations, that requiring a consumer to follow the adjudication procedure would cause irrecoverable expenditure in either prosecuting or defending it and might hinder his right to take legal action. Further, he observed that, unless it was properly explained to the consumer, the fact that the adjudicator is to be a neutral, even if nominated by the architect's own professional body, may also give the appearance of unfairness. By contrast, in *Lovell Projects Ltd v Legg*,[125] Judge Moseley QC held that an adjudication clause caused no significant imbalance: it **7.66**

[118] reg 6(1).

[119] Dir 93/13/EEC, sixteenth recital.

[120] *Director General of Fair Trading v First National Bank plc* [2002] 1 AC 481.

[121] *Lovell Projects Ltd v Legg* [2003] 1 BLR 487; *Westminster Building Co Ltd v Beckingham* [2004] 1 BLR 265; *Bryen & Langley Ltd v Martin Boston* [2005] EWCA Civ 973, CA.

[122] Sch 2, para 1(q).

[123] [2003] 1 BLR 487.

[124] Obiter.

[125] [2003] 1 BLR 487.

applied equally to both contractor/supplier and employer/consumer. He also found that there could not have been a lack of good faith on the part of the contractor since it was not they who had introduced the adjudication clause. When the Court of Appeal was required to consider the issue of adjudication clauses in *Bryen & Langley Ltd v Martin Boston*,[126] it followed the reasoning of Judge Moseley QC, but only in so far as it held that an adjudication clause could not be unfair if it had been imposed by the consumer/employer, as was the position in the case, since there could be no question of bad faith on the part of the supplier/ contractor; the Court deliberately avoided the issue of whether such a clause created a significant imbalance between the parties. Therefore it might be said that the question of whether the adjudication provisions could be unfair is an open one. It is suggested that adjudication provisions do not fall foul of the Regulations. In *Domsalla v Dyason*[127] the adjudication procedure was described (rightly) as being a 'rapid, cheap and temporary legal process . . . conducted pursuant to minimum standards of fairness and impartiality . . . similar to the summary judgment procedures available in court but capable of being undertaken more speedily and economically.' The advantages of the adjudication procedure, and any associated disadvantages, apply to both parties. Whilst a claim for money by a contractor can be speedily enforced, so can a householder's defence be ascertained at less cost than an arbitration. This is all the more the case since historical evidence indicates that the majority of adjudications do not proceed on to arbitration or litigation.

7.67 The case law to date indicates that it may be only in exceptional circumstances that an arbitration clause be found to be an unfair term under the Regulations. In *Heifer International Inc v Christiansen & Ors*[128] the Technology and Construction Court declined to find that an arbitration clause referring a dispute between a Russian employer and a Danish architect to the Danish Building and Construction Arbitration Board in Copenhagen to be decided according to Danish law was unfair to the employer. This decision is consistent with the general policy of the court to enforce arbitration clauses, including those referring disputes arising in this country to foreign arbitration according to foreign law in a reputable jurisdiction. At the time of the first edition, only two arbitration clauses in construction contracts had been held to be unfair under the Regulations.[129] Since then Ramsey J in *Mylcrist Builders Ltd v G Buck* [2008] EWHC 2172 (TCC) held that an arbitration clause fell foul of the Regulations because contrary to the requirements of good faith it caused a significant imbalance in the parties' rights and obligations under the contract to the detriment of the 'consumer', applying *Director General of Fair Trading v First National Bank Plc* [2002] 1 AC 481. The mandatory stay under s 9 of the Arbitration Act 1997 meant that the clause did exclude or hinder a consumer's right to take legal action and that was a factor to be taken into account as was the low sum of money in dispute when set against the arbitrator's fees (£2,000) and the fact that the clause had not been properly brought to the consumer's attention. See also *Zealander & Zealander v Laing Homes* [2000] 2 TCLR 724. By contrast,

[126] [2005] EWCA Civ 973, CA.

[127] [2007] EWHC 1174 (TCC) at para [92]. Judge Toulmin QC did however hold at para [95] that on the facts the withholding notice provisions of the adjudication procedure were unfair.

[128] [2007] EWHC 3015 (TCC). The employer had the benefit of legal advice at the time of contract.

[129] *Zealander & Zealander v Laing Homes* [2000] 2 TCLR 724, HHJ Havery QC in relation to the NHBC Buildmark documentation; and *Munkenbeck & Marshall v Michael Harold* [2005] EWHC 1165.

HH Judge Toulmin QC in *Heifer International Inc v Christiansen* [2008] Bus LR D49 considered the fairness under the 1999 Regulations of a clause providing for Danish arbitration and held it to be fair. In *Mylcrist*, Ramsey J helpfully set out the following principles in deciding whether such a clause fell foul of the Regulations as follows:

> (1) A term is unfair if it causes a significant imbalance in the parties' rights and obligations under the contract to the detriment of the consumer in a manner or to an extent which is contrary to the requirement of good faith.
> (2) There is 'significant imbalance' if a term is so weighted in favour of the supplier as to tilt the parties' rights and obligations under the contract significantly in his favour.
> (3) The element of 'detriment to the consumer' makes clear that the Regulations are aimed at significant imbalance against the consumer, rather than the seller or supplier.
> (4) The requirement of good faith is one of fair and open dealing in which:
>   (a) Openness requires that the terms should be expressed fully, clearly and legibly, containing no concealed pitfalls or traps. Appropriate prominence should be given to terms which might operate disadvantageously to the consumer.
>   (b) Fair dealing requires that a supplier should not, whether deliberately or unconsciously, take advantage of the consumer's necessity, indigence, lack of experience, unfamiliarity with the subject matter of the contract, weak bargaining position or any other factor listed in or analogous to those listed in Schedule 2 to the 1994 Regulations (an inducement to the consumer to agree to the term, whether goods or services were sold or supplied at the special order of the consumer or whether the seller or supplier dealt fairly and equitably with the consumer). The supplier should deal fairly and equitably with the consumer.
> (5) Schedule 2 to the Regulations is best regarded as a check list of terms which must be regarded as potentially vulnerable to being unfair.
> (6) Useful approaches include:
>   (a) assessing the impact of an impugned term on the parties' rights and obligations by comparing the effect of the contract with the term and the effect it would have without it.
>   (b) considering the effect of the inclusion of the term on the substance or core of the transaction; whether if it were drawn to his attention the consumer would be likely to be surprised by it; whether the term is a standard term, not merely in similar non-negotiable consumer contracts, but in commercial contracts freely negotiated between parties acting on level terms and at arms' length; and whether, in such cases, the party adversely affected by the inclusion of the term or his lawyer might reasonably be expected to object to its inclusion and press for its deletion.
> (7) Where the consumer has imposed the term either by their own choice or a choice made by their professional agent then it is unlikely that there would be any lack of good faith or fair dealing with regard to the incorporation of the terms into the contract.

Assessment of the terms

The assessment of the fairness of a term will not touch upon the definition of the main subject matter of the contract[130] or the adequacy of the price paid for the goods **7.68**

[130] reg 6(2)(a) described as 'the essential bargain' in *Abbey National plc & Ors v Office of Fair Trading* [2009] EWCA Civ 116.

and services[131] as long as the relevant term is expressed in 'plain, intelligible language'. The Regulations are not a mechanism of price or quality control: they do not allow a court to investigate the adequacy of consideration and leave it for the parties to define the subject matter of the contract. If, however, a term is not expressed clearly and intelligibly and, on its true construction, provides for a price or subject matter which is not expected by the consumer, it may be unfair for lack of clarity.

Reform

7.69 In 2001, the Department of Trade and Industry asked the Law Commission to rewrite the law of unfair contract terms as a single regime, in a clearer and more accessible style, and to consider whether to extend the legislation to protect businesses, particularly small businesses. On 24 February 2005, the Commission published its final report,[132] together with a draft bill. There has been no parliamentary action on the draft bill.

7.70 The report recommends the replacement of both the Act and the Regulations with a single, unified regime to cover the whole of the United Kingdom. The report also recommends improved protection for small businesses; under the present regime, as described above, where a business contracts on the other party's standard written terms of business, it may challenge unfair exemption clauses, pursuant to the provisions of the Act, but cannot challenge other unfair clauses since it is not a consumer for the purposes of the Regulations.

7.71 The main provisions of the draft bill are summarized as follows: for consumer contracts, the draft bill: (a) extends to all the terms currently covered by the Regulations (not just exclusion clauses); (b) continues to hold that terms which limit liability for death or personal injury, or which exclude basic undertakings about the quality and fitness of goods, are ineffective; (c) includes negotiated clauses as well as standard clauses (as does the Act); and (d) states that in claims brought by consumers, the burden of proof lies on the business to show that the term is fair (again following the Act).

7.72 The draft bill includes special protection for the smallest businesses, referred to as 'micro businesses' and defined as businesses with nine or fewer staff. These would be able to challenge any standard term of the contract which has not been altered through negotiation, and is not the main subject matter of the contract or the price. There are to be a number of exceptions to the new small business protections: contracts for financial services, contracts over £500,000, and situations where an apparently small business is associated with other businesses, so that overall the group has more than nine employees, are all excluded. The exemptions currently in the Act are to continue to apply to the new controls on small business contracts, namely: the exclusion of contracts for land, intellectual property, or security interests or contracts to form or dissolve contracts.

---

[131] reg 6(2)(b).

[132] Law Commission Report No 292, *Unfair Terms in Contracts*. <http://www.lawcom.gov.uk/docs/lc292bill.pdf>.

8

MISTAKE, MISREPRESENTATION, AND FRUSTRATION

Common Mistake

A mistake can be simply defined as an erroneous belief. There are three types of mistake relevant to the law of contract: **8.01**

(a) In order for two parties to conclude a contract binding in law, they must agree the terms of the contract. Mistakes may prevent this mutuality of agreement that is necessary for the formation of a contract. An objective appraisal of the facts may reveal that no agreement has been reached as to the terms of the contract; in such circumstances no binding contract has ever been concluded.[1]

(b) A second type of mistake is where the parties erroneously spell out the contract in terms which fail to give effect to the antecedent agreement that they have reached. Such a mistake can result in rectification of the contract.[2]

(c) The third type of mistake is where both parties to a contract are agreed as to its express terms, but where both parties are under a common mistake of fact or law[3] which renders

---

[1] eg in *Raffles v Wichelhaus* (1864) 2 H&C 906, where the parties entered into a contract for the sale and purchase of a cargo to arrive '*ex Peerless* from Bombay'. The term was capable of applying equally to the cargo on board two different ships. The court accepted that parol evidence could be adduced to prove which shipment the parties had intended. Had they intended different shipments, no agreement necessary to form a contract would have been reached.

[2] It requires a strong case to persuade the court that something has gone wrong with the language. For a useful discussion of the principles applicable to this kind of mistake, and the availability of rectification, see the judgment of Lord Hoffmann in *Chartbrook Ltd v Persimmon Homes* [2009] UKHL 38, paras 14 ff.

[3] *Kleinwort Benson Ltd v Lincoln City Council* [1999] 2 AC 349; *Brennan v Bolt Burton* [2005] QB 303. The old law to the effect that the mistake had to be one of fact only is no longer applicable to the law of mistake. The mistake can be of law or fact or both.

the service to be provided under the contract, if it were to be performed according to its terms, essentially different from the performance which the parties contemplated. In such circumstances the contract is void.[4] The law of mistake in such cases is akin to the law of frustration, and the test is the same.[5] The test is narrow and arises out of the impossibility of the task agreed to. A contract will only be void for mistake if there was a common mistake: there has to be a common assumption as to the state of affairs, and the assumption cannot be brought about because of the fault of one of the parties. The voiding of contracts through common mistake is by operation of law and not through any implied term.[6]

8.02 Denning LJ in *Solle v Butcher*[7] propounded the theory of equitable mistake rendering a contract voidable, as opposed to void. This heresy has been extirpated.[8]

Unilateral Mistake

8.03 If a party has entered into a contract on the basis of a unilateral rather than a common mistake, the equitable remedies of rescission or rectification may be available.[9]

8.04 However those equitable remedies are exceptional, and will be available only where the conduct of the other party is tainted with fraud, sharp practice, or some other moral turpitude.[10] A simple unilateral mistake is not good enough without the turpitude: were it otherwise a non-mistaken and innocent party could find that a contract with which they were satisfied was rectified or rescinded at the behest of the mistaken party.[11] The type of sharp practice or moral turpitude which will bring about these remedies is not always easy to define with precision. It is not necessary, for example, for the 'guilty' party to have knowledge of the other's mistake. Stuart Smith LJ gave an example in *Commission for New Towns v Cooper (GB) Ltd*[12] where he said:

> I would hold that where A intends B to be mistaken as to the construction of the agreement, so conducts himself that he diverts B's attention from discovering the mistake by making false and misleading statements, and B in fact makes the very mistake that A intends, then notwithstanding that A does not actually know, but merely suspects, that B is mistaken, and it cannot be shown that the mistake was induced by any misrepresentation, rectification may be granted.[13]

[4] *Bell v Lever Brothers* [1932] AC 161.

[5] *Great Peace Shipping Co Ltd v Tsvliris Salvage Ltd* [2003] QB 679.

[6] Ibid.

[7] [1950] 1 KB 671.

[8] *Great Peace*, n 5 above.

[9] For rectification, see also the discussion in Ch 1.

[10] See *May v Platt* [1900] 1 Ch 616, 623; *Blay v Pollard and Morris* [1930] 1 KB 628, 633; *Torrance v Bolton* (1872) 8 LR Ch App 118, 124; *Taylor v Johnson* (1983) 151 CLR 422, 431; and *Thomas Bates and Son Ltd v Wyndham's (Lingerie) Ltd* [1981] 1 WLR 505, 515.

[11] *Olympia Sauna Shipping Co SA v Shinwa Kaiun Kaisha Ltd* [1985] 2 Lloyd's Rep 364, 371 *per* Bingham J.

[12] [1995] Ch 259 at 280.

[13] See also the discussion of this case in *George Wimpey UK Ltd v VI Construction Ltd* [2005] EWCA Civ 77 involving a contract for the sale of land.

The standard of proof is high, as it is in any normal rectification or rescission case. The more detailed the negotiations and the lengthier and more detailed the contract document, the more formidable the task it is to show that the agreement did not represent the parties' intentions.[14] **8.05**

Misrepresentation and the Misrepresentation Act 1967

Misrepresentation is related to the law of tort, and a misrepresentation will bring about a right in damages in the event of loss being caused. See Chapter 6 dealing with negligence and the law of tort. **8.06**

In 1967, Parliament passed the most obscure and badly drafted statute which it is possible to conceive in the Misrepresentation Act 1967 ('the 1967 Act'). **8.07**

The 1967 Act permits rescission of a contract where the contract has been entered into as a result of a misrepresentation, whether the representation was made innocently or fraudulently.[15] **8.08**

The 1967 Act permits damages to be payable in the event that a negligent or innocent misrepresentation causes someone to enter into a contract. Prior thereto, damages were only recoverable in cases of fraudulent misrepresentation. This was a clear injustice. The relevant section of the 1967 Act[16] is worded so as to permit such damages to be recoverable as if the representation were fraudulently made,[17] ie without the limitations of foreseeability and remoteness which a claim in negligence would have.[18] Furthermore, damages will not be reduced if something happens afterwards to reduce or mitigate the loss, although the ordinary rules of mitigation apply.[19] **8.09**

This provision is separate from the right to rescission. So if the right to rescission is lost because of an equitable defence (for example, affirmation or laches), the right to damages subsists under s 2(1). **8.10**

Section 2(2) of the 1967 Act gives the court a discretion to grant damages in lieu of rescission. This is akin to Lord Cairns' Act, the Chancery Amendment Act 1858, which permits the grant of damages in lieu of an injunction. For this section to be operative there must be a subsisting right to rescission, in contradistinction to the right to damages for misrepresentation under s 2(1) of the 1967 Act.[20] This is because you cannot have damages in lieu **8.11**

---

[14] *James Hay Pension Trustees Ltd v Kean Hird* [2005] EWHC 1093, Lawrence Collins J.

[15] See s 1 of the 1967 Act.

[16] s 2(1).

[17] *Royscot Trust Ltd v Rogerson* [1991] 2 QB 297.

[18] For the measure see *Doyle v Orlby (Ironmongers) Ltd* [1969] 2 QB 158; and *Clef Aquitaine Sarl v Laporte Materials (Barrow) Ltd* [2001] QB 488.

[19] *Smith New Court Securities Ltd v Citibank NA* [1997] AC 254 *per* Lord Steyn 'the Plaintiff is entitled to recover as damages a sum representing the financial loss flowing directly from his alteration of position under the inducement of the fraudulent representations of the defendants . . .' applying Bingham LJ in *Cemp Properties (UK) Ltd v Dentsply Reearch and Development Corp* [1991] EGLR 197 at 201.

[20] See *Floods of Queensferry Ltd v Shand Construction Ltd* [2000] BLR 81 applying *William Sindall plc v Cambridge City Council* [1994] 1 WLR 1016 and not following *Thomas Witter Ltd v TBP Industries Ltd*

of rescission if there is no right to rescind in the first place: it is not enough for the right to rescind to have subsisted in the past.[21] The position is therefore the same as that which obtains under the Lord Cairns' Act in relation to injunctions, where damages in lieu of an injunction are not recoverable in the event that there is no right to an injunction. Accordingly the remedy of an injunction (and therefore damages in lieu) is only available if there is a subsisting right to an injunction and equitable defences such as laches, acquiescence, and estoppel are not good.[22]

8.12 The measure of damages recoverable under s 2(2) of the 1967 Act is not the same as the fraudulent measure recoverable under s 2(1): under s 2(2) the court has a broad discretion to do what is equitable.[23]

8.13 Parties cannot contract out of the 1967 Act unless the relevant term passes the test of reasonableness under s 11 of the Unfair Contract Terms Act 1977.[24] See the discussion of reasonableness and the 1977 Act in Chapter 7. However if it is an international supply contract to which s 26 of the Unfair Contract Terms Act 1977 applies, then the exclusion clause will not have to pass the reasonableness test.[25]

Rescission

8.14 This term is often used in the context of the acceptance of a repudiation by one or other party to a contract which brings the contract to an end. It is, however, an equitable remedy which enables one or other party to rescind the contract on the grounds of misrepresentation. If a contracting party has been induced into entering into a contract because of a misrepresentation by the other party, then the innocent party has the right to rescind. Under the 1967 Act, the innocent party also has the right to recover damages. The contract can also be rescinded if the contract has been obtained by the use of fraud or bribery.[26] Most standard forms have a clause which gives an express right to rescind in such circumstances. Rescission has the effect of unravelling the whole transaction. A builder who is entitled to rescind, therefore, will be entitled to an award which will put him or her into the position he or she would have been in but for the misrepresentation.[27] If work has been done this may well involve a quantum meruit, unlike in the repudiation cases where this is open to doubt.[28] The factors giving rise to a right to rescind a construction contract are not common, and there are no modern cases where rescission of a construction contract has been discussed. In practical terms, if a misrepresentation has been made and a construction contract has been partly performed, then damages under s 2(1) or (2) of the 1967 Act will

[1996] 2 All ER 573. See also *Zanzibar v British Aerospace (Lancaster House) Ltd* [2000] 1 WLR 2333; and *UCB Corporate Services Ltd v Thompson* [2005] EWCA Civ 225.

21 For academic discussion see Prof Hugh Beale's note in (1995) 111 LQR 386.

22 *Jaggard v Sawyer* [1995] 1 WLR 269, see Millett LJ at 287.

23 *Floods of Queensferry Ltd v Shand Construction Ltd* [2000] BLR 81 at 93, para 31.

24 See s 3 of the 1967 Act.

25 See *Trident Turboprop (Dublin) Ltd v First Flight Couriers Ltd* [2009] EWCA Civ 290.

26 *Panama & South Pacific Telegraph Co v India Rubber Co* (1875) LR 10 Ch 515.

27 See s 2(1) of the Misrepresentation Act 1967 and *Doyle v Orlby (Ironmongers) Ltd* [1969] 2 QB 158.

28 See the discussion of *Lodder v Slowey* in Ch 12.

normally be sufficient. Many bespoke and standard forms of contract have express terms which exclude claims for damages for misrepresentation and also have entire agreement clauses which state that the agreement was entered into without reference to any prior representations.[29] The most common form of misrepresentation complained about relates to ground conditions, with the contractor complaining that boreholes and other site investigation materials which were supplied by the employer were misleading. Since many engineering contracts will have a remedy for adverse physical conditions (under clause 12 of the ICE conditions or equivalent) or will have terms which exclude the employer from liability for reliance on such information, the need for rescission as a remedy by a contractor in those circumstances is not great.

Rescission is an equitable remedy, and all equitable defences are available to a claim for rescission, such as laches (delay), lack of clean hands on the part of the claimant, and impossibility of restitutio in integrum (restoring the parties to their original position). **8.15**

Frustration

Lord Radcliffe in *Davis Contractors v Fareham UDC*[30] stated: **8.16**

> [F]rustration occurs whenever the law recognizes that without default of either party a contractual obligation has become incapable of being performed because the circumstances in which performance is called for would render it a thing radically different from that which was undertaken by the contract. Non haec in foedera veni. It was not this that I promised to do.

A further definition in the same case by Lord Simon was as follows:[31] **8.17**

> Frustration of a contract takes place when there supervenes an event (without default of either party and for which the contract makes no sufficient provision) which so significantly changes the nature (not merely the expense or onerousness) of the outstanding contractual rights and/or obligations from what the parties could reasonably have contemplated at the time of its execution that it would be unjust to hold them to the literal sense of its stipulations in the new circumstances; in such case the law declares both parties to be discharged from further performance.

The elements of the test for frustration as set out in *Davis v Fareham* are: **8.18**

(a) The cause of the event is not the default of either party. Therefore there is no such thing as self-induced frustration.[32]

---

[29] See, for example, *Deepak Fertilisers and Petrochemical Corp v Davy McKee (London) Ltd* [1999] BLR 41, [1999] 1 Lloyd's Rep 387, [1999] 1 All ER (Comm) 69.

[30] [1956] AC 696 at 729.

[31] At 700.

[32] On this point see the decision of the House of Lords in *Paal Wilson & Co A/S v Partenreederie Hannah Blumenthal (The Hannah Blumenthal)* [1983] 1 All ER 234, where it was held that an arbitration agreement was not frustrated by long delay (over seven years), since this delay occurred through the fault of the parties. As Lord Brandon stated, the frustrating event 'must have occurred without either the fault or default of either party to the contract' (at para 44).

(b) The performance of the contract must have become something radically different from that which the parties contemplated when it was concluded. An event which makes the performance of the contract uneconomic does not frustrate it. There is no such concept as 'commercial frustration'.

(c) There must be no sufficient provision for the event in the contract. Where a contract has made provision for a certain event, the occurrence of such an event will not frustrate the contract, even if the occurrence of the same event would have frustrated an otherwise identical contract which made no such provision.

This formulation of the test for frustration in *Davis v Fareham* has been applied by the House of Lords subsequently.[33]

8.19 The event must supervene the contract. A frustrating event cannot precede the contract formation because then the task undertaken will be exactly in accordance with what the contract provides, rather than something different.[34] Frustration is not impossibility, although supervening impossibility will be frustration.[35]

8.20 Frustration is rare in construction cases and there have not been any recent cases where frustration has been found to have taken place. It cannot include dealing with the consequences of hazards such as fire or flood, since those are normally (or can easily be) provided for and are insurable events. Therefore the burning down of the works is not a frustrating event. However if the site where the works are to be performed disappears into the sea, that may be a frustrating event.[36] That would be because there would then be no place or site where the works could be done. Thus the destruction of a flying freehold will be a frustrating event.

8.21 In *Gamerco SA v ICM/Fair Warning Agency*,[37] Garland J found that a contract for the performance of a rock concert (by Guns N' Roses) was frustrated when the authorities banned the use of the stadium because it was discovered to have high alumina cement in its construction. Since the contract was discharged by frustration, the promoters were entitled to recover their advance payment of $412,500 pursuant to s 1(2) of the Law Reform (Frustrated Contracts) Act 1943 ('the 1943 Act'). If, however, the contract had not specified the particular stadium in which the group was to play, the non-availability of one stadium would not have been a frustrating event. The 1943 Act allows for the court to permit recovery for any benefit received pursuant to a frustrated contract, including any advance payments made. The recovery of payment or the ordering of a payment for a benefit is discretionary and will be ordered if, having regard to all the circumstances, it is just to do so.

---

[33] *National Carriers Ltd v Panalpina (Northern) Ltd* [1981] AC 675; *Pioneer Shipping Ltd v BTP Tioxide (The Nema)* [1982] AC 724; *Paal Wilson & Co A/S v Partenreederie Hannah Blumenthal* [1983] 1 AC 854.

[34] Although the doctrine of common mistake may cause the contract to be void.
See *Great Peace Shipping Ltd v Tsavliris Salvage Ltd* [2003] QB 679 at 703, para 76; and *Brennan v Bolton Burdon* [2005] QB 303 at 310.

[35] For a discussion of the law of frustration, see *Great Peace Shipping Ltd v Tsavliris Salvage Ltd* [2003] QB 679 at 700–3, paras 62–73; for this proposition see particularly para 63.

[36] As was assumed to be in the case of the hotel falling into the sea. But in that case there was no judicial finding of frustration *Holbeck Hall Hotel v Scarborough Borough Council* (1997) 57 Con LR 113 at 152.

[37] [1995] 1 WLR 1226, [1995] CLC 536.

The 1943 Act dealt with the inflexibility of the old law, which made the loss lie where it fell. The 1943 Act is not concerned with frustration itself, but with its consequences; it creates statutory remedies, enabling the court to award restitution in respect of benefits conferred under contracts thereafter frustrated. Because there is little case law where frustration has been found, there is little guidance on how the 1943 Act is to be applied. Perhaps the most useful guidance comes from the decision of Robert Goff J in *BP Exploration Co (Libya) v Hunt (No 2)*,[38] the first case decided under the 1943 Act. In that case Robert Goff J held that: **8.22**

(1) the fundamental principle underlying the Act itself, is prevention of the unjust enrichment of either party to the contract at the other's expense;
(2) the 1943 Act is not designed to apportion loss between parties;
(3) the 1943 Act is not designed to put the parties in the position that they would have been had the contract been performed;
(4) although s 1(2) and (3) are concerned with restitution in respect of different types of benefit, it is right to construe the two subsections as flowing from the same basic principle and therefore, so far as their different subject matters permit, to achieve consistency between them;
(5) an award under the 1943 Act may rescue the claimant from an unprofitable bargain, especially where a claim is made under s 1(2), where the award is generally speaking an award for repayment of money paid to the defendant pursuant to the contract. Where an award is made under this section, the court has a discretion to make allowance for expenses incurred by the defendant.
(6) Where the claim is made under s 1(3), the court may award to the plaintiff such sum, not greater than the value of such benefit, as it considers just having regard to all the circumstances of the case. There is nothing in s 1(3) which necessarily limits an award to the contract consideration, but this may be highly relevant to the assessment of the just sum to be awarded. The 'benefit' conferred, however, is not the actual services provided, but rather the end product of such services. Robert Goff J gave the example of a building contract frustrated by a fire which destroys both the building and a substantial amount of work already done by the claimant. He noted that whilst it may be thought just to award the claimant a sum assessed on a quantum meruit basis in respect of the work he has done, the effect of s 1(3)(b) will be to reduce the award to nil, because of the effect, in relation to the defendant's benefit, of the circumstances giving rise to the frustration of the contract. He saw it as 'quite plain' that, in s 1(3)(b), the word 'benefit' is intended to refer to the actual improvement to the building, because that is what will be affected by the frustrating event; the subsection therefore contemplates that, in such a case, the benefit is the end product of the plaintiff's services, not the services themselves. The value of the benefit serves to fix a ceiling on the value of the award.
(7) The value of the benefit is to be measured on the date of frustration;
(8) The 1943 Act is silent as to the relevance of the contractual terms to the assessment of a 'just sum'. The courts are, however, likely to take such terms into account: in many cases it will not be just to impose upon the defendant an obligation to make restitution at a level higher than the contract rate.[39]

[38] [1979] 1 WLR 783 at 799.
[39] [1979] 1 WLR 783 at 799–800.

9

GUARANTEE AND INDEMNITY

Guarantee and Indemnity Contrasted

9.01 A guarantee[1] has no definitive judicial definition,[2] however it is a contract in which the guarantor[3] agrees to be answerable for the debt or default of another to a third party promisee.[4] The essential characteristics of it are that the primary obligation of a third party is underwritten by the guarantor. The guarantor becomes answerable for the faults of the third party.[5] An indemnity is different in that the indemnitor undertakes a primary liability to another party to indemnify the other party against a specific event, which may or may not involve the act or default of a third party.

9.02 The two overlap and a guarantee is a sub-species of an indemnity, but it is important to note the distinction because of the operation of the Statute of Frauds 1677, s 4 which provides:

> No action shall be brought . . . whereby to charge the defendant upon any special promise to answer for the debt default or miscarriages of another person[6]. . . unless the agreement upon

[1] May also be referred to as a 'bond'.

[2] Although s 4 of the Statute of Frauds 1677 gives a similar definition for the purposes of prescribing the type of guarantee contract which must be in writing before it is enforceable. It thus becomes somewhat circular in that it can be said that a guarantee is not enforceable unless it is in writing but that a contract which falls within the section and must be in writing is a guarantee. See also the Holmes J definition above and the definitions in *Re Coney* [1938] 2 All ER 127 at 130–1.

[3] May also be referred to as the 'surety', 'bondsman' or 'obligor'.

[4] May also be referred to as the 'creditor', 'obligee' or 'beneficiary'. In building contracts this will often be the building owner or if the bond is in favour of a main contractor, securing the obligations of a sub-contractor, then the main contractor.

[5] The 'third party' whose obligations are guaranteed under the bond will generally be the main contractor or the sub-contractor who may often be referred to as 'the debtor'.

[6] ie, a guarantee.

which such action shall be brought or some memorandum or note thereof shall be in writing and signed by the party to be charged therewith or some other person thereunto by him lawfully authorized.

Therefore any guarantee must be in writing and signed if it is to be enforceable.[7] Whilst there may be complaint that this requirement allows the unscrupulous to evade their obligations, in construction cases, the 1677 Act creates no difficulty. It has been on the books long enough for all who deal in such matters to know that if they want their guarantee to be enforceable, then they must ensure that it is in writing. The mischief of the statute is somewhat outdated today but it remains a potential trap for those involved in informal arrangements.[8] **9.03**

The statute does not apply to indemnities, only to guarantees. Therefore there often has to be some analysis of the obligation to see whether it falls within the section and must be evidenced in writing in accordance with the statute. In most construction contracts and documents associated with them, this is academic because the whole suite of contracts[9] will be in writing. But less sophisticated arrangements will not necessarily be so. **9.04**

The distinction is best understood by the characteristic of a guarantee being that it involves the answering to the defaults of another and does not involve answering to the defaults of the guarantor itself if it is a co-contractor. So a parent company will be a guarantor and subject to the statute of frauds if it simply guarantees the performance of its subsidiary, but it will not be so if it undertakes to perform the primary obligation jointly with its subsidiary. **9.05**

Guarantees are often used in domestic construction projects to secure the performance of the contractor[10] and are usually provided in a sum expressed as a percentage of the contract sum.[11] In engineering and particularly in international construction contracts, it is not uncommon for an employer to provide an advance payment to the contractor to fund the mobilization of the project. In such circumstances, in addition to a performance bond, the contractor will also be required to provide a guarantee for the repayment of the advance payment.[12] **9.06**

In construction contracts the most frequently used form of indemnity is that which is contained in sub-contracts where the sub-contractor agrees to indemnify the main contractor **9.07**

---

[7] See *Actionstrength Ltd v International Glass Engineering SpA* [2003] 2 AC 541 where the House of Lords held that an oral contract of guarantee was unenforceable for want of signed writing. Moreover the oral promise to pay could not give rise to an estoppel so as to prevent the promisor from relying on s 4 of the Statute of Frauds 1677 to defeat the claimant's claim.

[8] See the *Actionstrength* case ibid.

[9] Generally there will be a minimum of 3 contracts: (1) the principal/underlying building contract between the owner and main contractor; (2) the agreement between the contractor and the bondsman; and (3) the bond under which the bondsman agrees to guarantee the obligations of the contractor.

[10] The contractor will either provide an express indemnity to the bondsman that it will duly perform its obligations under the principal contract, alternatively such an indemnity will be implied. The real risk which has to be weighed by the bondsman therefore is the potential insolvency of the contractor which would prevent it from both performing its obligation under the principal contract and from meeting its liability to the bondsman under its indemnity.

[11] A fact which gives rise to continuing doubts about whether the limited financial protection they afford is sufficient to justify the costs of their provision which obviously impact on the project costs.

[12] See for example the FIDIC conditions of Contract for Construction 1999, cl 14.2.

against any loss it may suffer by reason of any default in carrying out the sub-contract works—usually the obligation to pay damages to the employer and prolongation and disruption costs of the contractor. The need for the indemnity is in order to avoid the difficulties which come about when it takes some time for the losses caused by the sub-contractor to be known about (for example defects to the employer's building) and the employer might sue the contractor just within the limitation period but too late for the contractor to sue the sub-contractor.

9.08 There are two types of bond, commonly referred to as a 'conditional' bond and an 'on demand' bond. In the case of a conditional bond, in order to make a call on the bond, the beneficiary must establish both the default of the contractor and also prove damage.[13] In the case of an 'on demand' bond, the beneficiary may make a call on the bond without proof of default by the contractor and, absent fraud, the only restriction is that the call must be made in strict compliance with the terms of the bond. On demand bonds have, on occasions, been described pejoratively as simply 'discounts' from the contract price.[14] The most frequently used guarantee is the bond in traditional form, which is a conditional bond.

Common Forms of Bond: Their History

9.09 In spite of the strictures of the House of Lords which are set out in footnote 13 to this chapter the most common form of bond which is used today is in substantially the same form as used in the 1930s, and was arcane then. The form is short enough to be able to be set out in full:

> *Amount £500,000*
>
> Know all men by these presents that Messrs X plc (hereinafter called the Contractors) and Y plc (hereinafter called the Surety) are held and firmly bound to Z plc (hereinafter called the Owners) well and truly to be made and done the said Contractors and the said Surety bind themselves and their successors jointly and severally firmly by these presents:
>
> Signed Sealed and Dated this 12th day of October 2009
>
> Whereas the Contractors have entered into a certain written contract with the Owners bearing date 12th May 2009 (hereinafter referred to as the contract) for the execution of certain works of demolition and the construction completion and maintenance of certain new works in the construction completion and maintenance of an extension of the Building and a new entrance thereto and all other works in clause 1 of the contract mentioned (all of which works are more particularly shown described or referred to in the contract) subject to and in accordance with the provisions of the said clause 1 which contract with all its provisions covenants conditions and obligations is hereby made a part of the above written Bond to all intents and purposes as though the contract had been incorporated herein. Now therefore the condition of the above written Bond is such that if the Contractors shall execute construct and maintain the said recited works subject to and in accordance with the provisions of the said clause 1 of

[13] Generally there is no requirement for the beneficiary first to make or prove his claim against the contractor or even to join him as co-defendant to any claim made against the bondsman. However, in the event of a claim, the bondsman may decide to join the contractor.

[14] *Edward Owen Engineering Ltd v Barclays Bank International Ltd* [1978] QB 159, at 170, *per* Lord Denning MR.

> the contract and shall in all things whatsoever observe perform fulfil and keep all and singular the clauses and conditions agreements matters and things which on the part of the Contractors are and ought to be observed performed fulfilled and kept by them according to the contract then the above written Bond shall be null and void otherwise it shall remain in full force and virtue Provided always and it is hereby agreed and declared that the Surety shall not be released or discharged from the above written Bond by any agreement which may either with or without the assent or notwithstanding the dissent of the Surety be made between the Contractors and the Owners or between the Contractors and the engineers in the contract mentioned for any alteration in or to the said works or the contract or any forbearance by the Owners or by the said engineers whether as to payment time performance or otherwise howsoever . . .

The instrument is a three-way contract between the bondsman, the contractor and the owner/beneficiary. It involves the answering of the bondsman for the default of the contractor and it provides also for the bondsman to be a joint contractor to this extent, namely it places upon the bondsman the obligation of ensuring that the contractor carries out its obligations as well as giving the indemnity against the consequences of the contractor not doing so, at least up to the value of the bond. **9.10**

That the form is commonly still used today can be explained by its well known form and its simplicity and shortness. It takes up only one side of paper; it is familiar to professional bondsmen and contractors and the arcane nature of the language has not created any significant mischief beyond the provocation of occasional judicial exasperation.[15] Attempts to replace the form have foundered upon the length and complication of any proposed replacement. In fact a performance bond of this type could be more simply expressed as follows: **9.11**

> We, the Bondsman, agree to guarantee the performance of X Contractor plc under their contract with Y Employer plc [dated] for the construction of the Abbey Municipal Swimming Pool and to pay Y Employer plc, the Employer under that contract, all damages which may be proved by them as being recoverable from X plc for breach of contract up to a maximum of £500,000. It is also agreed that the Bondsman shall not be discharged from its obligations under this bond by any indulgence by Y plc towards X plc or any alteration in the obligations of X Contractor plc to Y Employer plc by contract or otherwise.

There is contained in each form a short indulgence clause of the type which appears at the end of the standard form of bond ensuring that the bondsman remains liable under the bond notwithstanding any variation of the contractor's responsibilities under the principal **9.12**

---

[15] In particular *Trade Indemnity Co v Workington Harbour and Dock Board* [1937] AC 1 at 17 *per* Lord Atkin:

> I may be allowed to remark that it is difficult to understand why business men persist in entering upon considerable obligations in old fashioned forms of contract which do not adequately express the true transaction. The traditional form of marine policy is perhaps past praying for; but why insurance of credits or contracts, if insurance is intended, or guarantees of the same, if guarantees are intended, should not be expressed in appropriate language, passes comprehension. It is certainly not the fault of lawyers.

Trafalgar *House Construction (Regions) Ltd v General Surety & Guarantee Co Ltd* [1996] AC 199 *per* Lord Jauncey at 209: 'I find great difficulty in understanding the desire of commercial men to embody so simple an obligation in a document which is quite unnecessarily lengthy, which obfuscates its true purpose and which is likely to give rise to unnecessary arguments and litigation as to its meaning.'

contract or other indulgence by the employer. The reason for the need for that type of clause is dealt with next.

The Rule in *Holme v Brunskill*[16]

9.13 This rule is of Court of Appeal authority only[17] and is long-standing albeit that it has attracted judicial criticisms. It provides that if the contractor and the employer make any alteration of the contractor's duties under the principal contract, for the better or for worse, then the bondsman is discharged altogether. The reasons for this rule are twofold: first, that if the primary obligations are altered, then that is not what the bondsman has underwritten; and, secondly, that if any indulgence is given to the contractor, then that may irredeemably prejudice the bondsman. An example of this would be a surety being discharged by the premature payment of a debtor.[18]

9.14 The first reason is easily understood and creates no difficulty. The bondsman ought only to underwrite the obligation it has underwritten and no more. The second creates difficulties in that the obligation of the contractor is not changed and, far from it, is forgiven or ameliorated. For example the employer may agree to release retention early to the contractor so as to enable it to fund the completion of the works in circumstances where the contractor's ability to do so otherwise is in doubt. In that event the bondsman will be discharged because of that indulgence. But the rationale behind the rule is that the bondsman's exposure is increased by that indulgence because in the event that the contractor goes bankrupt, then the obligation of the bondsman in respect of the cost of remedial works will be increased by the amount of the retention surrendered because, as a result of the indulgence, it is no longer available to contribute to the costs of the remedial works.

9.15 Whilst the theory is impeccable, it is rigid and may operate unfairly. An employer may release retention on the basis that this will assist the cash flow of the contractor so that defects are remedied to the ultimate benefit of the bondsman. There is no real justice in that operating as a discharge. Or the amount of retention may be small in relation to the overall failures of the obligor. Retention may dwarf the cost of remedying defects which are subsequently found.

9.16 In *Holme v Brunskill*, the Court of Appeal held that the bondsman would be relieved from liability unless it was evident, without enquiry, that the alteration was insubstantial or that it could not be otherwise than beneficial to the bondsman. If the insubstantial nature of the change was not self evident, the court would not look into the impact of the change and it was for the bondsman to determine whether or not he wished to remain liable. In more recent cases the impact of the rule has been softened by the courts development of this

[16] (1878) 3 QBD 495.

[17] Though recently applied by the Court of Appeal in *Lloyds TSB Bank plc v Norman Hayward* [2005] EWCA Civ 466. In that case though there was no discussion of the ambit of the rule or its applicability in the modern world, cf *Wardens Commonalty of the Mystery of the Mercers v New Hampshire Insurance* [1992] 2 Lloyd's Rep 365 where the distaste for the rule and its application is obvious.

[18] See *St Microelectronics NV v Condor Insurance Ltd* [2006] EWHC 977 (Comm).

exception and it has been held that less significant indulgences or even breaches of contract by the beneficiary will not discharge the bondsman unless the breaches are substantial and the term breached is 'embodied' and amounts to a condition of the bond.[19] Accordingly, a merely technical breach which causes no damage or prejudice to the bondsman will be insufficient to release him from his obligations under the bond. That said, it is obviously sensible for parties to ensure that the bond includes a clause which obviates any discussion of the character of any variation or indulgence granted by the owner to the contractor.

There is, of course, no reason why the law should not have developed a 'pro tanto' rule whereby the prejudice to the bondsman by the indulgence is measured and decreases the bondsman's obligations in proportion to the amount of that prejudice. Thus, if retention of £50,000 is released prematurely but defects to the amount of £500,000 are eventually discovered, then the bondsman's obligations could be reduced to the extent that the £50,000 release caused real prejudice. That might be nil if the amount of the bond is small and dwarfed by the value of the outstanding defects or it may be £50,000 if the value of the bond is £500,000 or more. Alternatively, it may be nil if the early retention release is several years before the defects are discovered. Some States in the United States, New York for example, have developed and operate a pro tanto rule. **9.17**

Most bonds contain an indulgence clause to deal with this problem and so there should be no commercial mischief to the commercially aware. But it is surprising how many bonds still do not contain such an indulgence clause, with the result that the beneficiary under the bond is exposed to a significant commercial risk. **9.18**

Other jurisdictions have also developed a pro tanto rule and there is no reason why the courts of this country should not do the same. However, it is only open to them to do so at House of Lords level; it is not open to the Court of Appeal to do so. Given the ability of the parties to avoid this pitfall with an indulgence clause, it may never be needed with proper draftsmanship. **9.19**

On Demand Bonds and Conditional Bonds: *Trafalgar House v General Surety*

As already noted,[20] a conditional bond is so called because, in the event of default by the contractor, the claimant beneficiary under the bond must plead and prove breach of contract by the contractor and prove damages as well. Conversely, an 'on demand' or 'unconditional' bond, which is in form and substance no different from a standard letter of credit,[21] **9.20**

[19] *Wardens Commonalty of the Mystery of the Mercers v New Hampshire Insurance* [1992] 2 Lloyd's Rep 365. In this case the Court of Appeal held that the bond was not a performance guarantee but a guarantee of the obligation to repay the advance payment on the contract which had been created by a side letter. In addition, however, the Court of Appeal held that the owner's breach of the building contract [the date for possession had been put back by agreement and then further delayed by the owner] was insubstantial and not a condition of the bond; *National Westminster Bank v Riley* [1986] BCLC 268, CA followed.

[20] See para 9.08 above.

[21] See *Edward Owen Engineering Ltd v Barclays Bank International Ltd* [1978] QB 159 at 169. See also *Intrano Ltd v Notis Shipping Corp* [1981] 2 Lloyd's Rep 256.

entitles the beneficiary to demand and receive payment without inquiry as to whether or not there was in fact default on the part of the Contractor. It is the equivalent of a letter of credit in that provided the documents are presented in proper form, it must be paid by the bondsman. Any dispute between the owner and contractor over whether the call on the bond was justified must be settled between themselves.

9.21 The business purpose behind them is to ensure that the beneficiary does not have to go to the trouble of proving default before it can receive the money payable under the bond. If, for example, the beneficiary is likely to need the cash to remedy the defects to the works caused by the obligor, then a simple call upon the bond will allow for that payment to be made. On demand bonds are not often used in domestic construction contracts (outside of the petroleum and power industries) but are a standard feature of many international projects.

9.22 The issuer of an 'on demand' bond is usually a bank or finance house whereas the issuer of the traditional type of performance bond, known as a 'conditional bond' is usually a professional surety company.

9.23 It is not always straightforward to decide which is which. The Court of Appeal in one case held that a traditional form of conditional bond in fact created on demand obligations upon the bondsman and it took the decision of the House of Lords to put that error right.[22] The courts place the burden on the claimant to prove that a bond is of an 'on demand' nature.[23] It would follow that in cases of genuine doubt the courts would fall back on a presumption that the bond was not on-demand.

9.24 It is at the margins, of course, where that difficulty arises. The traditional conditional bond exampled above should give rise to no difficulty in construction because it has the authority of the House of Lords which gives it the appropriate categorization. Therefore the beneficiary under the bond must prove default by the contractor and loss on its behalf before a claim is admitted. Equally a standard bond in the form of a Letter of Credit from a bank will create no difficulty. It will require payment on demand.

9.25 The margins of the argument come with careless drafting, or perhaps deliberately ambiguous drafting. If, for example the bond provides that payment must be made upon the beneficiary's 'first written demand' then, all other things being equal, the bondsman must pay the amount demanded without query or inquiry.[24] But if the bond is expressed as follows 'if the contractor is in default such as to cause damage to the employer in any

---

[22] *Trafalgar House Construction (Regions) Ltd v General Surety & Guarantee Co Ltd* [1996] AC 199.

[23] *Marubeni Hong Kong Ltd v Mongolian Government* [2005] 1 WLR 2497 applied in *IIG Capital LLC v Van Der Merwe & Anor* [2008] EWCA Civ 542 *per* Waller LJ para 30. In that case the judge and the Court of Appeal held that the presumption was rebutted.

[24] See *Esal Commodities v Oriental Credit* [1985] 2 Lloyd's Rep 546; *IE Contractors v Lloyds Bank* [1990] 2 Lloyd's Rep 496; and *Gold Coast Ltd v Caja de Ahorros del Mediterraneo* [2001] EWCA Civ 1806, [2002] 1 Lloyd's Rep 617:

> Where in international transactions a bond or guarantee is expressed to be payable upon demand, in the absence of clear words indicating that liability under it is conditional upon the existence of a liability or the part of the account party in connection with the underlying transaction, the guarantee is intended and should be construed as an independent guarantee entitling the beneficiary to payment

amount, then the employer may demand that the bondsman shall pay the amount of its loss', then it is likely that the bond is in fact conditional in nature because the demand is expressed to be subject to a condition precedent which must be subject to proof. But if the words are 'and the bondsman shall pay the amount demanded upon the first written demand', then that points to the bond in fact being of an on demand nature. The existence of ambiguous bond wording is a surprising thing to the lawyer because it is in fact very easy for the draftsman to make the intention clear. Clarity is of great importance not only because of the uncertainty legally but also because an on demand bond is usually more expensive to secure than a conditional one.

Illustrative of this is the case where the Court of Appeal has recently held that an on demand letter from a government was not an on demand bond. The express wording contained in the letter was 'the undersigned Ministry . . . unconditionally pledges to pay to you upon your first simple demand all amounts payable under the Agreement if not paid when the same becomes due'. One reason that assisted the Court of Appeal in construing the obligation as conditional was that the giver of the letter was a government rather than a bank.[25] The decision is best explainable on the basis that the words were not clear enough to create an on demand obligation (especially given the words 'all amounts payable' in the letter). It is hard to see how words should mean one thing when uttered by a bank and another when uttered by a government. However the law as applied by the Court of Appeal in that case suggests that the identity of the party creating the instrument is important. Were the same words to have emanated from a bank, then the Court of Appeal would have decided the case differently and the term 'first written demand' would have triggered an unconditional obligation to pay. The case, however, is illustrative of how wording which might be designed to create an on demand obligation, can be accompanied by wording which makes the obligation to pay conditional. If that happens, then there is a real possibility that the instrument will be construed as being of a conditional nature. **9.26**

Cross Guarantees

Whilst bonds may be of a conditional nature, the normal industry practice is for the contractor who has to procure the bond to secure the performance of his obligations to give the bondsman a cross guarantee of an on demand nature.[26] The obligations contained in the cross guarantee are usually of two overlapping types. The first obligation is the obvious one, to pay the amount of loss suffered by the bondsman. But that obligation is normally expressed to be of an on demand nature. So it will require the contractor to pay the bondsman 'upon their first written demand the amount sought by the beneficiary', so all that is required for the bondsman to recover from the obligor is a demand of it from the beneficiary. But the second obligation normally contained in the cross guarantee is that the obli- **9.27**

simply against an appropriately worded demand accompanied by such other documents (if any) as the guarantee may require.

[25] *Marubeni Hong Kong and South China Ltd v The Mongolian Government* [2005] EWCA Civ 395.

[26] See, for example, *Technical & General Guarantee Co Ltd v Mark Patterson*, L Henderson QC (Ch D, 12 February 2003).

gor pay to the bondsman on demand any sum, up to the value of the bond as security for any payment it might make to the beneficiary. These two obligations are then normally swept up by a clause which states that the contractor may not question the validity of any payment made to the beneficiary and that such a payment, if made, is conclusive evidence of its validity.

9.28 These clauses may be perceived as onerous, but they ensure that the bondsman's risks are kept as low as possible with the result that the premium paid to the bondsman is kept low as a result. These cross guarantees have created little litigation, and such that there is, is usually the result of the impecuniosity of the contractor.[27]

9.29 Once an on demand bond has been paid, that is not the end of the matter so far as the parties are concerned. The on demand nature of the call gives rise to the risk that the amount paid under the bond exceeds the amount of loss which the beneficiary has in fact suffered. This risk is particularly likely in the case of defective work where its full extent may not be known or where the beneficiary is already holding a substantial retention against the background of a fall in construction prices so that remedial work costs are, contrary to the normal position, cheaper than the original tender costs. In that event the beneficiary must account to the bondsman for the balance. This is a right of the bondsman in quasi contract or restitution.[28]

The Intervention of the Courts to Prevent Calls on Bonds in the United Kingdom and Internationally

9.30 The courts have always been anxious to ensure that the on demand nature of bonds is not compromised by judicial intervention.[29] Thus the courts will not unless there is a clear case of fraud, prevent by injunction either a call being made upon the on demand bond or instrument or a payment being made by the bondsman.[30] The courts apply this maxim even if the injunction is sought in support of an arbitral process.[31]

---

[27] As indeed was the case ibid.

[28] See the discussion in *Trafalgar House Construction (Regions) Ltd v General Surety & Guarantee Co Ltd* [1996] AC 199.

[29] *Edward Owen Engineering Ltd v Barclays Bank International Ltd* [1978] QB 159. See also *Intrano Ltd v Notis Shipping Corp* [1981] 2 Lloyd's Rep 256 where Donaldson LJ said 'Thrombosis will occur if, unless fraud is involved, the courts intervene and thereby disturb the mercantile practice of treating rights thereunder as being the equivalent of cash in hand'. See also *Group Josi Re v Walbrook Insurance Co Ltd* [1996] 1 WLR 1152, [1996] 1 Lloyd's Rep 345; and *Harbottle (Mercantile) Ltd v National Westminster Bank* [1978] QB 146. Where a seriously arguable case of fraud is shown, note that *Group Josi* casts doubt upon the correctness of the decision of the majority of the Court of Appeal in *Themehelp Ltd v West* [1996] QB 84. See also May LJ in *Sirius International Insurance Corp v FAI General Insurance Co Ltd* [2003] 1 WLR 2214 at para 31; and Rix J in *Czarnikow-Rionda Sugar Trading Inc v Standard Bank London Ltd* [1999] 2 Lloyd's Rep 187 at 202.

[30] *Group Josi Re (formerly Groupe Josi Rèassurance SA) v Walbrook Insurance Co Ltd* [1996] 1 WLR 1152 at 1160–2.

[31] *Permasteelisa Japan KK v Bouyguesstroi & Anor* [2007] EWHC 3508 (TCC). See also *Sabmiller Africa BV v Tanzania Breweries Ltd* [2009] EWHC 2140 (Comm).

For a call or payment to be prevented by injunction it is not enough to make an allegation of fraud and neither is it enough that there is a suspicion of fraud.[32] It must be clear and incontrovertible before there will be an injunction granted.[33] Where there are issues to be tried an injunction will not lie.[34] **9.31**

This rule is of worldwide scope. Nearly all other jurisdictions operate similarly. Whilst fraud may in fact be committed, if it is not clear, then a fraudulent call will not be prevented. It might be thought that this leads to unfairness. It does, but it is a risk of unfairness which is bargained for commercially and the consequence of any different legal rule being operative would be that an on demand bond would be de facto a conditional bond, contrary to the commercial purpose of the instrument. Given that the commercial world prefers certainty, that objective is best achieved by the rules in relation to on demand bonds being immutable. **9.32**

It was once thought that there was a distinction between an injunction to restrain a call upon a bank and an injunction to restrain payment by the bank. This is not correct and it is hard to see how the difference ever gained currency.[35] **9.33**

The exception to the fraud rule may be when otherwise than fraudulently, a claim is made in breach of an express term as suggested in *Sirius International Insurance Corp v FAI General Insurance Co Ltd*[36] which was a decision of the Court of Appeal that dealt with the autonomy **9.34**

---

[32] *Kvaerner John Brown Ltd v Midland Bank plc* [1998] CLC 446.

[33] This proposition is supported by many cases.

In *Turkiye Is Bankasi AS v Bank of China* [1996] 2 Lloyd's Rep 611 at 616 Waller J said: '... He may show an arguable case that the demand is not honest, but that is not sufficient. He must also establish that "the only realistic inference is that the demands were fraudulent."'

This approach and decision were upheld by the Court of Appeal; see [1998] 1 Lloyd's Rep 250.

In *Czarnikow-Rionda v Standard Bank* [1999] 2 Lloyd's Rep 187 at 202 Rix J said:

> However, the fact that the claimant gets the benefit of a lower standard of proof for the purposes of a pre-trial hearing, places on the Court, as I believe the cases demonstrate, an additional requirement to be careful in its discretion not to upset what is in effect a strong presumption in favour of the fulfilment of the independent banking commitments.

In *Solo Industries v Canara Bank* [2001] 1 WLR 1800 at 1813 Mance LJ said: '... the court should be careful not to allow too extensive a dilution of the presumption in favour of the fulfilment of independent banking commitments.'

In *Banque Saudi Fransi* [2007] 2 Lloyd's Rep 47 at 55 Pill LJ, whilst accepting that the test in CPR Part 24.2 applied, said, at para 34, that the task of showing a real prospect of proving that the beneficiary could not honestly have believed in the validity of its demands was 'a high hurdle, as the authorities in my judgment recognise.'

The above cases were applied in *Enka Insaat Ve Sanayi AS v Banca Popolare Dell'alto Adige SPA* [2009] EWHC 2410 (Comm).

See also *Balfour Beatty Civil Engineering v Technical and General Guarantee Co Ltd* (2000) 68 Con LR 180, [2000] CLC 252.

[34] *TTI Team Telecom International Ltd v Hutchison 3G UK Ltd* (TCC, 23 January 2003); although in that case the judge permitted an interim injunction to lie until the final hearing when the application was dismissed. It is suggested that the approach was wrong and interim injunctions or interim undertakings should not be given or asked for.

[35] *Bolivinter Oil SA v Chase Manhattan Bank NA* [1984] 1 Lloyd's Rep 256 *per* Donaldson MR at 256. See also *Deutsche Rückversicherung AG v Walbrook Insurance Co Ltd* [1995] 1 WLR 1017 where Phillips J examined and set out the authorities at 1030.

[36] [2003] 1 WLR 2214.

principle in such documents; but that issue was not dealt with by the House of Lords. In that case it is possible that contrary to the general rule an injunction might lie against the beneficiary rather than the bank. However it is submitted that such a position would be inconsistent with the law requiring that fraud be the sole reason for preventing payments out of such instruments and all such documents issued by banks cannot be the subject of any restraint save in the case of fraud.

9.35 The court will, in an appropriate case, impose a freezing injunction upon the fruits of a performance guarantee.[37]

9.36 The approach of courts outside the United Kingdom is broadly similar. Most United States jurisdictions apply the same principles and will not restrain calls on on-demand instruments. However the Federal and State judges before whom the application may come might not be as familiar with the jurisdiction as they ought to be and first instance injunctions have been granted in several cases only to be overturned on appeal. This is much like the position in the United Kingdom twenty years ago when a fortunate injunction was taken from a Bear Garden judge. In other Commonwealth or former Commonwealth jurisdictions such as Singapore and Hong Kong, there are similar rules to those in the United Kingdom.[38]

[37] *Intraco Ltd v Notis Shipping Corp ('The Bhoja Trader')* [1981] 2 Lloyd's Rep 256 at 258.
[38] See for example *Rajaram v Ganesh* [1995] SLR 159.

Part III

PERFORMANCE

10

CERTIFICATES AND THE RIGHT TO PAYMENT

Simple Contracts; Lump Sum Contracts; Substantial Performance; Non-completion

Simple oral contracts will be examined first. **10.01**

The traditional simple contract required completion of the work before payment was **10.02**
made. There is no logic to this in that there is no theoretical reason why the law should not have required payment before performance. But since *Cutter v Powell*,[1] the case of the unfortunate seaman's widow who was not paid his wages because he did not complete the voyage, the legal theory has been that pay succeeds performance. The theory caused the description of building contracts to be 'lump sum contracts' in which the builder was paid a lump sum upon completion of his (always a male then) obligations. The gross inequity of that situation caused the emergence of the doctrine of substantial performance, whereby the builder would receive payment if the work was substantially completed, even if not absolutely complete in all particulars.[2] The position can now be shortly stated: under a simple oral contract a builder will be entitled to receive the contract price when the work is substantially completed, even if there are a few defects to the work. If the builder throws up the contract before substantial completion, then he or she will not be entitled to any money at all.

[1] (1795) 6 Term Rep 320, [1775–1802] All ER Rep 159.

[2] *Hoenig v Isaacs* [1952] 2 All ER 176 applied to a decorator; and *Hutchinson v Harris* (1978) 10 BLR 19 applied to an architect.

10.03 In practical terms, under a simple oral contract, the courts will be fairly generous in their interpretation of what is substantial completion if the position is that the employer obtains a substantial benefit from the works. The courts will not be anxious to permit an employer to receive a substantial benefit from executed works without payment.

10.04 However simple oral contracts for works of any substantial nature are rare and uncommercial. No builder will undertake to execute work of any substance without: (a) a written agreement, however informally executed; or (b) an express right to stage payments as the work progresses. An oral agreement to stage payments will be interpreted according to the words used, which means in the light of the oral evidence given at trial. Such a hazardous venture is not recommended nor is it common. Indeed it is proposed that oral contracts come under the purview of the Housing Grants, Construction and Regeneration Act 1996. At the time of writing it is not known whether these proposals will be brought into force. See paragraphs 1.110 and 1.111 above.

10.05 A written agreement carries with it the right to interim payments. Either the right is agreed expressly in the contract itself or it is implied into the agreement by the operation of ss 109–110 of the Housing Grants, Construction and Regeneration Act 1996 and dealt with in Chapter 1. Thus all agreements now make provision for interim payments as and when the work progresses. Each interim payment is usually an estimate of the work done and is not supposed to be a precise measure, because precision is at the end of the job.[3]

Certificates and Payment

10.06 Most construction contracts make provision for interim payments by way of certificate or approval of the contract professional or the employer's agent. The contract professional must normally make an assessment of the value of the work and certify a payment, usually monthly. The contract will normally provide a set of objective criteria by which the work is to be valued (Bills of Quantities, Schedules of Rates, and other like documents) and this

---

[3] See Hobhouse J in *Secretary of State for Transport v Birse-Farr Joint Venture* (1993) 62 BLR 36 at 53:

> Certification may be a complex exercise involving an exercise of judgment and an investigation and assessment of potentially complex and voluminous material. An assessment by an engineer of the appropriate interim payment may have a margin of error either way. It may be subsequently established that it was too generous to the contractor just as it may subsequently be established that the contractor was entitled to more. Further the sum certified may be made up from a large number of constituent figures, some of which may likewise be assessed favourably to one party or the other. It may be that a contractor can say under a certain heading he did not have certified as high a figure as can later be seen to be appropriate but that under another heading he has to accept that the figure certified can be shown to have been an over-certification. At the interim stage it cannot always be a wholly exact exercise. It must include an element of assessment or judgment. Its purpose is not to produce a final determination of the remuneration to which the contractor is entitled but is to provide a fair system of monthly progress payments to be made to the contractor. The strict twenty-eight day period in sub-clause (2) can be contrasted with the three months allowed in sub-clause (3) 'after receipt of this final account and of all further information reasonably required for its verification.'

See also *Henry Boot Construction Ltd v Alstom Combined Cycles Ltd* [2005] EWCA Civ 814, [2005] 3 All ER 932.

will enable the contract professional to exercise his or her judgment against those criteria in arriving at a sum which should be certified and paid. Thus if the contract professional certifies an insufficient or excessive amount, then an adjudicator is able to step in and remedy the default. Furthermore if there is an arbitration clause, then the arbitrator is able to step in and do likewise, as is the court.[4]

The presence of objective criteria to govern how valuations are carried out is the touchstone of a sensible commercial arrangement. **10.07**

However most contracts provide an element of discretion in the certifier, whether he or she be a contract professional or the employer's employee. The question then is what happens in the event of a non-approval or a non-certification by the contract professional or employee and where there is no mechanism of review by an arbitrator of the certifier's discretion. Before this discussion, it should be said that most contracts provide for review by an arbitrator or the court. There are clauses which give contractual discretion to the certifier which is expressed to be his or hers alone and not subject to review. To be effective and to displace a court or arbitrator clear words must be used and such an intention is rarely to be inferred.[5] Under the contract, that must be given effect to, but the question remains as to what can be done when the certifier, whose discretion is all, goes wrong. **10.08**

Non-approval or Certification by the Contract Professional or Employee of the Developer or Contractor and Wrong Certification

The ancient authorities do not speak with one voice but modern authorities do. Under the old authorities there was a line of cases which suggested that if the contract professional did not approve or certify then the contractor would have no cause of action because its right was dependent upon a certificate. Equally there was a line which suggested that non-certification would cause the condition precedent of certification or approval to be dropped. The modern authorities are to the effect that any discretion must be properly exercised and thus the court can step in when certificates or approvals are being withheld. Because the modern authority is not all based in construction law, the old authorities must be examined as well. **10.09**

The classic case which embodies the confusion stated above is *Dallman v King*.[6] In that case Mr Dallman agreed to spend £200 of his own money building a kitchen and executing other work in premises he had taken from the defendant charity. The contract stipulated that 'such erection and alteration or repairs to be inspected and approved of by the said William King . . . And it is agreed that the said William Dallman shall be allowed the sum of £200 towards such erections alterations or repairs and shall be at liberty to retain the **10.10**

[4] See *Beaufort Developments (NI) Ltd v Gilbert Ash (NI) Ltd* [1999] 1 AC 266 applying *Robins v Goddard* [1905] 1 KB 294 and overruling the eccentric decision of the Court of Appeal in *Northern Regional Health Authority v Derek Crouch Construction Co Ltd* [1984] QB 644.

[5] See for example *London & Regional (St George's Court) Ltd v Ministry of Defence and Secretary of State for Defence* [2008] EWCA Civ 1212 para 25.

[6] 4 Bing (NC) 105, SC 5 Scott 382, 3 Hodges 283, and 7 LJCP 6.

same out of the first year's rent . . .' Mr King did not approve the work. The question was whether, given that the work was done properly, the non-approval of Mr King precluded Mr Dallman from his £200 credit. The decision of the Court of Common Pleas, Tindall CJ and Vaughan, Bosanquet, and Coltman JJ was that Mr Dallman had his credit without the need for Mr King's approval. The decision has been cited ever since for the words of Tindall CJ: 'It never could have been intended that he should be allowed to capriciously withhold his approval; that would have been a condition which would go to the destruction of the thing granted, and if so, according to the well known rule, the thing granted would pass discharged of the condition.'

10.11 However, the case ought to be viewed with caution as the rationale behind any need to dispense with approval or certification. The decision was based as a matter of construction of that particular contract because the need for approval was in a separate sentence from the payment of the £200. Therefore, the payment of the £200 was conditional upon the work being done and the approval was merely the means whereby Mr King came to be informed of the work having been done. The case distinguished a decision of Tindall CJ (four years previously) in *Morgan v Birnie*[7] in which he decided that where a builder was to be paid on an architect's certificate that certificate was a condition precedent to payment. That case provided that the architect was to be the sole arbitrator in deciding rates and prices and indeed all disputes. However the dictum of Tindall CJ in *Dallman* was not necessary for the decision. The other members of the court decided the matter on the point of construction because they could not accept that it would be right that Mr King could capriciously refuse his approval on the words used in the agreement. Bosanquet J said that the case was different from *Morgan v Birnie* because in that case the architect was sole arbitrator and therefore his approval was indeed a condition precedent. Nevertheless, in *Eaglesham v McMaster*[8] Lord Reading CJ, and in *Dunlop & Ranken Ltd v Hendall Steel Structures Ltd*[9] the Court of Appeal, whilst not having *Dallman* or indeed any other pertinent case cited to them[10] decided that the issue of a certificate was a condition precedent to any payment.

10.12 The modern view is that non-certification does not mean that there is no right to recovery. In *Panamena Europea Navigacion (Compania Limitada) v Frederick Leyland & Co Ltd*,[11] a ship-repairing contract, the contractors were absolved from obtaining a certificate because the certifier refused to give a certificate until he received information which he was not entitled to receive on the true construction of the contract. Lord Thankerton then said at 435:

> Dr Telfer declined to proceed with the matter unless he was provided with the information to which on his erroneous view of the contract, he held himself entitled; in this view the appellants concurred, and this position was maintained up to and after the issue of the writ. This means that an illegitimate condition precedent to any consideration of the granting of

---

[7] 9 Bing 672 and SC 3 Moo & Sc 76.

[8] [1920] 2 KB 169.

[9] [1957] 1 WLR 1102.

[10] Particularly *Panamena Europea Navigacion (Compania Limitada) v Frederick Leyland & Co Ltd (J Russell & Co)* [1947] AC 428 discussed below.

[11] Ibid.

a certificate was insisted on by Dr Telfer and by the appellants. It is almost unnecessary to cite authority to establish that such conduct on the appellants' part absolved the respondents from the necessity of obtaining such a certificate, and that the respondents are entitled to recover the amount claimed in the action. In *Hotham v East India Co*,[12] Ashhurst J, in delivering the judgment of the court, said:

> It is unnecessary to say whether the clause relative to the certificate be a condition precedent or not; for granting it to be a condition precedent, yet the plaintiffs having taken all proper steps to obtain the certificate, and it being rendered impossible to be performed by the neglect and default of the company's agents, which the jury have found to be the case, it is equal to performance. If it were necessary to cite any case for this, which is evident from common sense, it was so held in Rolle's Abridgement 445, and many other books.

It will be sufficient to add two short passages from the well-known case of *Roberts v Bury Improvement Commissioners*.[13] Blackburn J says:[14] 'It is a principle very well established at common law, that no person can take advantage of the non-fulfilment of a condition the performance of which has been hindered by himself...'

In that case the certifier was an employee of the employer rather than an independent contract professional. It used to be thought that architects and other professional certifiers had a quasi arbitral role which rendered them immune from suit. That has not been the case since 1974,[15] so there is no reason why the contract professional's non-feasance should, as against the contractor, be treated any differently from the non-feasance of the person directly employed by the employer under a building contract. **10.13**

That leaves the question of some certification or approval by the contract professional but which falls outside what is reasonable. The position has now been reached that in commercial contracts discretion must, in the absence of clear words to the contrary, be exercised honestly and in good faith and cannot be exercised capriciously. This is by way of an implied term. The current law was summarized by Potter LJ in *Horkulak v Cantor Fitzgerald International*:[16] **10.14**

---

12 (1787) 1 TR 638 at 645.

13 (1870) LR 5 CP 310.

14 Ibid, 326.

15 *Sutcliffe v Thackrah* [1974] AC 727, see Lord Morris at 751:

> The decision in *Chambers v Goldthorpe* [1901] 1 KB 624, has not hitherto come directly before this House for consideration. In *Hickman & Co v Roberts* [1913] AC 229 there are passages in the speeches which indicate that when the architect (pursuant to the clauses of the contract in that case) gave his certificate he was in a quasi-judicial position of an arbitrator, and in *RB Burden Ltd v Swansea Corporation* [1957] 1 WLR 1167 Lord Radcliffe, referring to the decision in *Chambers v Goldthorpe*, said that it was an established principle of law that in granting a final certificate under a building contract the architect acts in an arbitral capacity. But in neither of those two cases was the authority of *Chambers v Goldthorpe* directly challenged. Now that it is I consider for the reasons which I have set out that it should be overruled.

16 [2005] ICR 402 at 414, paras 26–8. See also particularly the cases cited by Mance LJ in *Gan Insurance v Tai Ping Insurance Co Ltd (No 2)* [2001] 2 All ER (Comm) 299 in relation to the exercise of a contractual discretion. He cited the need to act: 'in good faith and after consideration and on the basis of facts giving rise to the particular claim, and not with reference to considerations wholly extraneous to the subject matter of the particular reinsurance or arbitrarily...' See also Lord Cooke in *Equitable Life Assurance Society v Hyman* [2002] 1 AC 408 at 460 '... no legal discretion, however widely worded ... can be exercised for purposes

So far as commercial contracts are concerned, it has been rightly said that: 'the authorities do not justify any automatic implication, whenever a contractual provision exists putting one party at the mercy of another's exercise of discretion. It all depends on the circumstances . . .': *Gan Insurance v Tai Ping Insurance Co Ltd (No 2)* [2001] 2 All ER (Comm) 299, 322, para 62, *per* Mance LJ. However, a number of authorities collected and considered in the *Tai Ping* case at pp 322–323, para 62 make clear the willingness of the court to read into a discretion the requirement or implied term that: 'Not only must the discretion be exercised honestly and in good faith but, having regard to the provisions of the contract by which it is conferred, it must not be exercised arbitrarily, capriciously or unreasonably,' see *Abu Dhabi National Tanker Co v Product Star Shipping Ltd (The Product Star) (No 2)* [1993] 1 Lloyd's Rep 397, 404, *per* Leggatt LJ. In that passage, Leggatt LJ properly observed that any analogy of approach with that adopted in the judicial control of administrative action must be applied with caution to the assessment of whether a contractual discretion has been properly exercised. However, in the *Gan Insurance* case, p 323, para 64, Mance LJ was not in doubt that 'what was proscribed was unreasonableness in the sense of conduct or a decision to which no reasonable person having the relevant discretion could have subscribed'.

10.15 The Court of Appeal took a similar approach in *Paragon Finance plc v Nash*,[17] which was a case involving the setting of variable mortgage interest rates. The right to set the rates was not unfettered. In that case the term implied was that rates would not be set in a *Wednesbury*[18] unreasonable sense. This borrows a public law concept but only as a matter of contractual interpretation rather than duty (public or private). In *Socimer International Bank Ltd v Standard Bank London Ltd* [2008] EWCA Civ 116, [2008] 1 Lloyd's Rep 558, [2008] Bus LR 1304, Rix LJ summarized the position, in relation to the exercise of a contractual discretion, as follows: (paragraph 66)

> It is plain from these authorities that a decision-maker's discretion will be limited, as a matter of necessary implication, by concepts of honesty, good faith, and genuineness, and the need for the absence of arbitrariness, capriciousness, perversity and irrationality. The concern is that the discretion should not be abused. Reasonableness and unreasonableness are also concepts deployed in this context, but only in a sense analogous to Wednesbury unreasonableness, not in the sense in which that expression is used when speaking of the duty to take reasonable care, or when otherwise deploying entirely objective criteria: as for instance when there might be an implication of a term requiring the fixing of a reasonable price, or a reasonable time. In the latter class of case, the concept of reasonableness is intended to be entirely mutual and thus guided by objective criteria.[19]

10.16 There are therefore limits to the assault on discretion which can be made against a certifier. Naturally if the arbitrator can open up and revise the decisions of the certifier, then the ultimate discretion will lie with the arbitrator or the court. But if the discretion lies with the contract professional or the employer, then the limits upon the discretion are similar in concept to the limits on the discretion of public authorities as suggested by Rix LJ above. This is applicable even when the discretion is expressed to be 'absolute'. There is, however,

contrary to those of the instrument by which it was conferred . . .' See also *Secretary of State for Transport v Birse-Farr Joint Venture* (1993) 62 BLR 36 (Hobhouse J).

[17] [2002] 1 WLR 685.

[18] *Associated Provincial Picture Houses Ltd v Wednesbury Corp* [1948] 1 KB 223, [1947] 2 All ER 680.

[19] See also *Lymington Marina Ltd v MacNamara* [2007] EWCA Civ 181, [2007] 2 All ER Comm 825 *per* Arden LJ para 37and Pill LJ para 69.

nothing to stop a sufficiently tightly worded clause vesting in the certifier a final say, however capricious that may be. But such a clause would be unattractively uncommercial and hard to draft.

Schedules of Rates; Bills of Quantities; Milestone Payments

The objective criteria by which works are to be valued and interim payments are made are usually set out in a value document which sets out how the work is to be measured and valued as the work progresses. Schedules of Rates or Bills of Quantities provide the contract professional and the contractor with the means whereby the work can be progressively valued. They are the most popular methods of valuation, not only because they permit variations to be valued but also because they will allow for different areas of work and different trades to be valued at the many different stages which they have reached. Milestone payments depend upon particular stages which the whole work has reached. In some cases, for example roads or the fabrication of oil platforms, this may be relatively easy because particular stages of completed work are easy to identity (in the case of a platform, for example, load out and sailaway; in the case of a road the completion of a particular stage at set chainages). But in most construction contracts, with work being done by many different trades simultaneously and reaching different stages of completion at different times, then milestones are not appropriate. Milestone payments may, however, be the complete remedy the contractor has in respect of the work done, in that the contract will provide for a series of milestone payments with no further assessment or re-assessment of the value of the work because the contractor obtains its complete remedy with each milestone.[20] **10.17**

The Final Account; Certification of the Final Account; Effect of Final Certificates; Discharge from Further Performance by Final Certificate

Subject to the milestone exception of paragraph 10.16 interim certificates are usually made as approximations of the value of work done at the point the valuation is made and the certificate is issued. Necessarily there is no precision because the work is ongoing. At the end of the work then the final contract sum must be ascertained. **10.18**

The JCT Standard Forms have made provision for the finality of such certificates in that some wording has, over the years, been expressly or unintentionally designed to discharge the contractor from all actions for breach of contract unless a notice of arbitration has been served within a specified time.[21] The mechanism of discharge has been one of evidence, namely that the certificate has been conclusive evidence that the work has been carried out and completed in accordance with the terms and conditions of the contract, and therefore strictly speaking does not operate as an exclusion clause. The current edition of the JCT **10.19**

[20] *Birse Construction Ltd v McCormick (UK) Ltd* [2005] EWCA Civ 940.

[21] See *Hosier & Dickinson v Kaye* [1971] 1 WLR 146; and *Crown Estates Commissioners v John Mowlem & Co Ltd* (1994) 40 Con LR 36.

contract is not designed to create that kind of finality. However bespoke contracts sometimes do so, particular when matters such as performance tests are the touchstone of completion.[22]

Set Off

10.20 There was a brief heresy in which the Court of Appeal held that certificates issued by architects had some special character and operated as cash.[23] The practical reason for the heresy was that cash flow was the lifeblood of the construction industry. However the House of Lords in *Modern Engineering v Gilbert Ash*[24] decided that the law of set off would apply to certificates issued under standard forms of building contract as much as any other contract and there is a rule of construction that clear words are needed to rebut the presumption that a contracting party does not intend to abandon any remedies for breach of the contract arising by operation of law.[25] Those remedies include a defence by way of set off against sums due. Set off is an equitable remedy in which a cross claim arising out of the same transaction or closely associated with the original claim acts as a defence to the claim to the extent of the value of the cross claim.[26]

10.21 Abatement is the common law equivalent of set off whereby the price is reduced by the amount by which the value of the work done is reduced.[27] Legal set off is of mutual debts through the Statutes of Set-off, which provided that if there were mutual debts between a plaintiff and a defendant then one debt might be set against the other. But both claims have to be liquidated debts or money demands which could be ascertained with certainty at the time of pleading. Legal set off is rarely used because equitable set off covers the same ground and does not require the liquidation of the debt.

---

[22] As in *Deepak Fertilisers and Petrochemical Corp v Davy McKee (London) Ltd* [1999] BLR 41, [1999] 1 Lloyd's Rep 387, [1999] 1 All ER (Comm) 69.

[23] *Dawnays Ltd v FG Minter Ltd* and *Trollope and Colls Ltd* [1971] 1 WLR 1205, [1971] 2 All ER 1389.

[24] [1974] AC 689.

[25] For an example of a case where clear words have been used to contract out of set off, see *Coca-Cola Financial Corp v Finsat International Ltd* [1998] QB 43.

[26] For the best description of set off and the species of set off, see the seminal judgment of Morris LJ in *Hanak v Green* [1958] 2 QB 9 as well as *Modern Engineering v Gilbert Ash* itself, n 24 above.

[27] See, for example, *Mondel v Steel* (1841) 8 M&W 858, (1976)1 BLR 106. See also *Hutchinson v Harris* (1978) 10 BLR 19; *Acsim (Southern) Ltd v Danish Contracting and Development Ltd* (1989) 47 BLR 55; and *A Cameron Ltd v John Mowlem and Co Ltd* (1990) 52 BLR 24.

11

TIME FOR COMPLETION

Dates for Completion

All construction contracts provide for a time by which the contract works are to be completed. If there is no express term of the contract which provides for a completion date, then the law will imply a reasonable time for their completion. What is a reasonable time will depend upon all the circumstances of the case. There has been no case law which assists in working out what a reasonable time is. All we are left with is the bald, and some may think unhelpful, statement that a reasonable date is what is reasonable in all the circumstances. That means that if no date is provided for it is nearly impossible to work out what a reasonable rate of progress for the works is. When disputes arise as to whether a contractor has defaulted by failing to complete within a reasonable time, the court or arbitrator will be faced with little or no help in deciding the point. It will take a brave employer to allege that a reasonable time for completion has passed unless the delay has been gross. In practical terms an employer in that situation ought to take periodic photographs of the site such as to be able to show that the progress has been so bad that it is clearly unreasonable that the works have not already been completed. We will look at this question again when we come to consider the position when a fixed date for completion has been discharged by operation of law. **11.01**

When a fixed date for completion has been agreed, the simple position is that the contractor has until that date to complete and in default of completion by that date the contractor must pay damages to the employer for the breach of contract. Such simple situations are sufficiently rare as to be virtually nonexistent in practice with the growth of contracts which provide for extensions to the contract period and for liquidated and ascertained damages. **11.02**

11.03 In the 19th century it was frequently the case that contracts provided for fixed dates for completion of the works and liquidated and ascertained damages in the event of default. The bargaining position of contractors was relatively poor and so many contractors signed contracts which provided for onerous contract periods and large liquidated damages.

11.04 The fixed date for completion would become almost impossible for the contractor to keep because of delaying events occurring, some of them being within the control of the employer. In that event the courts would discharge the fixed date for completion and pronounce that the contractor no longer had to complete by that fixed date but instead would have an obligation to complete within a reasonable time. The rationale behind this was that the courts would not allow the contracts to become an instrument of commercial oppression.

11.05 The response to this was that employers made provision within their contracts for the fixed date for completion to be re-fixed in the light of delaying events, particularly those which were the responsibility of the employer. The objective of the employer was to ensure that its right to recover liquidated damages remained in place. A fixed date for completion, from which liquidated damages could be calculated, remained in place under this scheme. Thus it was that extension of time clauses were put into construction contracts for the benefit of the employer and not, as is often thought, for the benefit of the contractor.

Extension of Time Clauses: An Introduction

11.06 The courts were as suspicious of the extension of time clauses as they were of the original fixed date for completion. Therefore they read the clauses narrowly and against the employer (to use the Latin term *contra proferentem*). The clauses were for the benefit of the employer to preserve a right to recover liquidated damages.

11.07 It followed that, if an employer caused delay to the contractor and that delaying event was not strictly within the extension of time clause, then the courts would hold that, because there was no proper mechanism to adjust the completion date, the fixed date for completion was discharged and the employer was left to prove its actual losses suffered during the period between when there was a reasonable date for completion and actual completion. This requirement for strictness took two forms. First, that if a delaying event caused by the employer was not precisely within the meaning of an event prescribed by the clause, then the extension of time clause would cease to operate because the fixed time for completion was discharged. Secondly, if the extension of time clause was too general, perhaps providing for an extension 'for any reason whatsoever as is reasonable', then again it was not specific enough to save the fixed date for completion and with it the right to charge liquidated damages. The theory was that if a fixed date for completion was discharged and a reasonable date was substituted, then liquidated damages could not apply to an unfixed and reasonable date.

11.08 It is still the law today that extension of time clauses are construed strictly against the employer. There have been recent decisions of the Privy Council,[1] the Court of Appeal,[2]

[1] *Philips v Attorney General of Hong Kong* (1993) 61 BLR 41.
[2] *Murray v Leisureplay plc* [2005] EWCA Civ 963.

and at first instance[3] which suggest that the time for this approach has passed. Indeed there is no commercial or social reason why this should remain the law particularly when the employer and the contractor will be of equal or almost equal bargaining power and when, as now, liquidated damages are often used to limit a contractor's liability for delay. The reasons for the development of this doctrine have effectively gone, but its consequences remain, much like an appendix.

Therefore it is prudent for any employer which wishes to have a provision for the payment of liquidated damages to ensure that all possible causes of delay to the contractor's progress which might be caused by the employer should be expressly catered for in an extension of time clause. A relatively recent decision involved a fixed date for completion being discharged because the standard form of building contract made no express provision for delay caused by the employer failing to give the contractor possession of the whole of the site.[4] That case was in the JCT Standard Form for the day and was very specific, in accordance with good practice, in the reasons for delay which would result in the payment of liquidated damages. But the failure to cater for lack of possession of the site was fatal to the claim for liquidated damages.[5] **11.09**

The distaste that the courts had for liquidated damages clauses was accompanied by a refusal to enforce provisions for liquidated damages which the courts saw as oppressive. Again this was a function of the unequal bargaining power between employer and contractor in the nineteenth century. Indeed it could be said that it was emblematic of the unequal bargaining power. It is impossible to see how any commercial organization would agree to pay damages in excess of an employer's losses for any reason, let alone for the reason of delay to completion unless the bargaining power is grossly one sided. **11.10**

The touchstone of what was an oppressive and unenforceable rate of liquidated damages was if the damages were in excess of what would be or was a genuine pre-estimate of the likely loss which the employer would suffer in the event of delay. If liquidated damages are in excess of a genuine pre-estimate, then they are unenforceable even if in fact the losses suffered approximate or exceed the liquidated rate. The employer will have to prove its actual losses. But the commercial reason for liquidated damages clauses was and remains that, in many cases, it is difficult to quantity the losses suffered because of delay to the works and if the damages can be agreed in advance then the parties know exactly where they are. That commercial reason remains to this day. **11.11**

---

[3] *Alfred McAlpine Capital Projects v Tilebox Ltd* [2005] EWHC 281 (TCC); *Multiplex Constructions (UK) Ltd v Honeywell Control Systems Ltd (No 2)* [2007] BLR 195, 111 Con LR 78, [2007] CILL 2458; and *Steria Ltd v Sigma Wireless Communications Ltd* [2008] BLR 79, 118 Con LR 177, [2008] CILL 2544 (discussed below).

[4] *Bramall & Ogden Ltd v Sheffield City Council* (1983) 29 BLR 73. That was a case dealing with sectional completion. See a case where a similar argument was unsuccessful in striking down the clause as a penalty: *Liberty Merican Ltd v Dyball Construction Ltd* [2008] EWHC 2617 (TCC). In *Avoncroft Construction Ltd v Sharba Homes (CN) Ltd* (2008) 119 Con LR 130; (2008) 26 EG 118 at paras [6]–[7] HHJ Kirkham held that there was no requirement for the contractor to have acted blamelessly in order to rely upon the principle in *Bramall* since any such defence was a legal and not an equitable one.

[5] *Rapid Building Group Ltd v Ealing Family Housing Association Ltd* (1984) 29 BLR 5; applying *Peak Construction (Liverpool) Ltd v McKinney Foundations Ltd* (1970) 1 BLR 111, (1969) 69 LGR 1.

11.12 This is particularly the case with non-commercial buildings. A municipal swimming pool will not cause the authority building it financial loss if it is constructed late. But it will suffer social costs in the loss of enjoyment of its taxpayers who are paying for the pool. Liquidated damages are designed to quantify losses such as those in a way which means that the local authority gets a fair recompense for the contractor's default.

11.13 Finally it is to be noted that liquidated damages are the employer's complete remedy for delay to the works. If a contract provides for the payment of liquidated damages, those damages cannot be added to if the employer suffers greater losses than anticipated. Because the employer has liquidated all its damages, it is taken to have received full and complete compensation for all its delay losses when it recovers liquidated damages.

11.14 The legal rules which can be ascertained from the above discussion are:

(a) Liquidated damages can only be recovered if they are a genuine pre-estimate of the likely loss to be suffered by the employer in the event of delay.

(b) The matter is judged by reference to the pre-estimation rather than the outturn.

(c) Liquidated damages are the employer's complete and sole remedy for delay. No further or other delay related damages can be claimed beyond the liquidated damages recoverable. All losses of a delay nature are assumed to be liquidated in the agreed damages.[6]

(d) Extension of time clauses are construed strictly against the employer so that if a delaying event occurs, which is the fault or responsibility of the employer, and which is not catered for in the extension of time clause, then the extension of time clause will cease to operate, the fixed date for completion will be discharged, and the contractor will have a reasonable time for the completion of the works.

(e) Once the fixed date is discharged, liquidated damages are also discharged and the employer must prove its actual losses.

(f) Obversely, once it is shown that the contractor has failed to complete by the fixed date for completion, the employer does not have to prove its actual losses arising from the delay, it can recover the liquidated damages provided for by the contract.

Effect of Discharge of Liquidated Damages and Recovery of Actual Losses

11.15 Some legal problems remain unresolved by authority. The principal one is what happens if the liquidated damages provision is discharged by the employer's default and the employer's actual losses by reason of the delay are greater than the losses provided for in the liquidated damages clause.

11.16 This problem has been discussed but not decided in one case. The textbooks which discuss it all agree that the employer cannot recover damages at a greater rate than the liquidated damages allowed for in the contract. The reasoning is two fold. The first is that by setting

[6] *Peak Construction (Liverpool) Ltd v McKinney Foundations Ltd* (1970) 1 BLR 111, (1969) 69 LGR 1.

out in the contract what the liquidated damages are, the employer is representing what its damages are and cannot resile from that representation. The second reason is an appropriation of the legal maxim (which is really a canon of contractual construction) that you cannot take advantage of your own wrong. It does offend a sense of justice that an employer recovers damages for delay at a greater rate than that set out in the contract simply because the employer itself has defaulted.[7]

It is hard to fault either of those two reasons. So far no employer has argued in an English court that it is entitled to recover damages at a greater rate than the liquidated damages rate because of its own defaults in discharging the provision for liquidated damages. Whilst both sets of reasoning easily deal with an employer's subsequent default, only the first deals with the position where an employer's liquidated damages rate was a penalty because it was not a genuine pre-estimate of its likely loss. It might be a surprisingly bold employer which argues that its provision for liquidated damages are a penalty and seeks to recover a greater sum. So far this has not been attempted. **11.17**

The area where such an argument might be attempted is where liquidated damages are in fact a limitation of liability clause. But the two reasons given above are likely to be more than sufficient to deal with such an argument. **11.18**

The fact remains that the law relating to penalties, extensions of time, and liquidated damages is rooted in different social times which are irrelevant to today's construction market and a less technical approach is now the modern law. **11.19**

The above discussion can be looked at with the consideration of the modern law relating to penalties. **11.20**

The Law Relating to Penalties

Penalty clauses are not enforced by the courts. However the burden of proof is upon the party who is sued upon the clause to prove that it was a penalty clause.[8] **11.21**

The definition of a penalty was set out by Lord Dunedin in the House of Lords in *Dunlop Pneumatic Tyre Co v New Garage and Motor Co Ltd*:[9] **11.22**

> 1. Though the parties to a contract who use the words 'penalty' or 'liquidated damages' may prima facie be supposed to mean what they say, yet the expression used is not conclusive. The Court must find out whether the payment stipulated is in truth a penalty or liquidated damages. This doctrine may be said to be found passim in nearly every case.

[7] In *Cellulose Acetate Silk v Widnes Foundry* [1933] AC 20 at 26 Lord Atkin left open the question whether a clause in a contract, which fixed a single sum as payable upon breach of a number of different terms of the contract, some of which breaches may cause only little damage but others damage greater than the stipulated sum, would be treated as imposing a limit on the damages recoverable. Thus it would operate to reduce the damages which would otherwise be recoverable at common law. This approach, it is suggested, would be appropriate in this situation. But note Lloyd LJ in *Rapid Building Group Ltd v Ealing Family Housing Association* (1984) 29 BLR 5 where this point was left open.

[8] *Robophone Facilities Ltd v Blank* [1966] 1 WLR 1428 at 1447 *per* Diplock LJ.

[9] [1915] AC 79.

2. The essence of a penalty is a payment of money stipulated as in terrorem of the offending party; the essence of liquidated damages is a genuine covenanted pre-estimate of damage (*Clydebank Engineering and Shipbuilding Co v Don Jose Ramos Yzquierdo y Castaneda* [1905] AC 6).
3. The question whether a sum stipulated is a penalty or liquidated damages is a question of construction to be decided upon the terms and inherent circumstances of each particular contract, judged of as at the time of the making of the contract, not as at the time of the breach (*Public Works Commissioner v Hills* [1906] AC 368; and *Webster v Bosanquet* [1912] AC 394).
4. To assist this task of construction various tests have been suggested, which if applicable to the case under consideration may prove helpful, or even conclusive. Such are:
   (a) It will be held to be a penalty if the sum stipulated for is extravagant and unconscionable in amount in comparison with the greatest loss that could conceivably be proved to have followed from the breach (illustration given by Lord Halsbury in *Clydebank Case*).
   (b) It will be held to be a penalty if the breach consists only in not paying a sum of money, and the sum stipulated is a sum greater than the sum which ought to have been paid (*Kemble v Farren* 6 Bing 141). This, though one of the most ancient instances, is truly a corollary to the last test. Whether it had its historical origin in the doctrine of the common law that when A promised to pay B a sum of money on a certain day and did not do so, B could only recover the sum with, in certain cases, interest, but could never recover further damages for non-timeous payment, or whether it was a survival of the time when equity reformed unconscionable bargains merely because they were unconscionable—a subject which much exercised Jessel MR in *Wallis v Smith* 21 Ch D 243—is probably more interesting than material.
   (c) There is a presumption (but no more) that it is a penalty when 'a single lump sum is made payable by way of compensation, on the occurrence of one or more or all of several events, some of which may occasion serious and others but trifling damage' (Lord Watson in *Lord Elphinstone v Monkland Iron and Coal Co* 11 App Cas 332).

On the other hand:

   (d) It is no obstacle to the sum stipulated being a genuine pre-estimate of damage, that the consequences of the breach are such as to make precise pre-estimation almost an impossibility. On the contrary, that is just the situation when it is probable that pre-estimated damage was the true bargain between the parties (*Clydebank Case*, Lord Halsbury; *Webster v Bosanquet*, Lord Mersey).[10]

11.23 This definitive statement has not been altered by recent authority,[11] but it has been explained and considered and applied.[12] For example, it is insufficient for a party seeking to establish that a liquidated damages provision is a penalty to identify situations where the application of the provision could result in a larger sum being recovered by the injured party than his

[10] See also *Clydebank Engineering and Ship Building Co Ltd v Don Jose Ramos Yzquierdo Y Castaneda* [1905] AC 6 at 11; *Robophone Facilities Ltd v Blank* [1966] 1 WLR 428 at 1447B–D; and *General Trading Co v Richmond Corporation Ltd* [2008] EWHC 1479 (Comm), (2008) 2 Lloyd's Rep 475 at para 130 *per* Beatson J, relying on the above references.

[11] See *Jeancharm Ltd (trading as Beaver International) v Barnet Football Club Ltd* (2003) 92 Con LR 26 in which it was held that the decision of the Privy Council in *Philips v Attorney General of Hong Kong* (1993) 61 BLR 41 did not alter the statement by Lord Dunedin in any way.

[12] For example, by the Court of Appeal in *Lansat Shipping Co Ltd v Glencore Grain BV* [2009] EWCA Civ 855.

actual loss.[13] The fact that greater loss can be recovered under a liquidated damages provision than at common law may lead to the conclusion that the provision is a penalty, but it may not. It depends on the circumstances.[14]

Whether a provision is to be treated as a penalty is a matter of construction to be resolved by asking whether at the time the contract was made the predominant contractual function was to deter a party from breaking the contract or to compensate for a breach.[15] Therefore, a particular clause which might in some circumstances give greater damages than at common law, may be commercially justifiable provided its dominant purpose was not to deter the other party from a breach.[16] Factors which militate against a clause being a penalty include the importance to the parties both of knowing what will be the financial consequences to them of a breach of contract.[17] Another large factor militating against a clause **11.24**

---

[13] *Phillips v Attorney General of Hong Kong* (1993) 61 BLR 41. See particularly at 58–9 where Lord Woolf said:

> Except possibly in the case of situations where one of the parties to the contract is able to dominate the other as to the choice of the terms of a contract, it will normally be insufficient to establish that a provision is objectionably penal to identify situations where the application of the provision could result in a larger sum being recovered by the injured party than his actual loss. Even in such situations so long as the sum payable in the event of non-compliance with the contract is not extravagant, having regard to the range of losses that it could reasonably be anticipated it would have to cover at the time that the contract was made, it can still be a genuine pre-estimate of the loss that would be suffered and so a perfectly valid liquidated damage provision. The use in argument of unlikely illustrations should therefore not assist a party to defeat a provision as to liquidated damages. As the Law Commission stated in Working Paper No 61 (page 30): 'The fact that in certain circumstances a party to a contract might derive a benefit in excess of his loss does not . . . outweigh the very definite practical advantages of the present rule upholding a genuine estimate, formed at the time the contract was made of the probable loss'. A difficulty can arise where the range of possible loss is broad. Where it should be obvious that, in relation to part of the range, the liquidated damages are totally out of proportion to certain of the losses which may be incurred, the failure to make special provision for those losses may result in the 'liquidated damages' not being recoverable. (See the decision of the Court of Appeal on very special facts in *Ariston SRL v Charly Records Limited*, The Independent, 13 April 1990.) However, the court has to be careful not to set too stringent a standard and bear in mind that what the parties have agreed should normally be upheld. Any other approach will lead to undesirable uncertainty, especially in commercial contracts.

[14] See *Murray v Leisureplay plc* [2005] EWCA Civ 963 at para [49].

[15] See *Lordsvale Finance plc v Bank of Zambia* [1996] QB 752 at 762G; and *Murray v Leisureplay plc* at para [106]. See also *Cine Bes Filmcilik Ve Yapim Click v United International Pictures* [2003] EWCA Civ 1699 particularly Mance LJ.

[16] *Murray v Leisureplay plc*, above, para [106(v)]; and *Cine Bes Filmcilik Ve Yapim Click v United International Pictures*, n 15 above.

[17] See *Robophone Facilities v Blank* [1966] 1 WLR 1426 where Diplock LJ in a dictum approved by Clarke LJ in *Murray v Leisureplay plc*, n 15 above, said at 1447:

> Nevertheless the courts would be doing an ill turn to those whom the rule about 'penalty clauses' is designed to protect if they were to apply it so as to make it impracticable for parties to agree at the time when they enter into a contract upon a fair and easily ascertainable sum to become payable by one party to another as compensation for the loss which the latter will sustain as a consequence of its breach. It is good business sense that parties to a contract should know what will be the financial consequences to them of a breach on their part, for circumstances may arise when further performance of the contract may involve them in loss. And the more difficult it is likely to be to prove and assess the loss which a party will suffer in the event of a breach, the greater the advantages to both parties of fixing by the terms of the contract itself an easily ascertainable sum to be paid in that event. Not only does it enable the parties to know in advance what their position will be if a breach occurs and so avoid litigation at all, but if litigation cannot be avoided, it eliminates what may be the very heavy legal

being a penalty is the expense which is avoided in litigation where actual loss does not have to be proved and the parties know where they stand.[18] At the same time, the provisions of the contract must be capable of generating with certainty the liquidated damages which flow from an identified or particular breach by the contractor. If they do not do so, then the clause will be a penalty.[19]

11.25 It is important to bear in mind the requirement of unconscionability (see Lord Dunedin's definition at (a) in paragraph 11.22 above).[20] The two alternatives of the clause being a penalty or a genuine pre-estimate of loss have no middle ground. Therefore, if the court cannot with confidence say that the clause is a deterrent, then it is forced back upon finding it to be a genuine pre-estimate of loss.[21] For example, in *Liberty Mercian Ltd v Dean & Dyball Construction Ltd*[22] Coulson J held that, construing a sectional completion agreement as a whole, it was not unfair for liquidated damages to be levied in respect of multiple sections of the work where delay to an initial section caused knock-on delays to other sections.

11.26 However, a contract is not unconscionable simply if it is put on a 'take it or leave it' basis. In the *Dunlop* case itself a liquidated damages clause was upheld by the House of Lords notwithstanding its status as such (sometimes referred to as a contract of adhesion).

11.27 There are few modern reported cases in construction contracts where the amount of liquidated damages have been held to be penal.[23] The cases where penalties are more readily found relate to employment contracts, where, for example large sums are payable to the employee if he or she is discharged (and even those have a heavy bias in favour of their enforcement) and hire-purchase contracts where finance houses have preyed on the poor. The cases do refer to inequality of bargaining power as being possibly relevant, which is not surprising since the doctrine relating to penalties arose out of inequality of bargaining power. Such reference is usually to the effect that the parties are of equal or almost equal

costs of proving the loss actually sustained which would have to be paid by the unsuccessful party. The court should not be astute to descry a 'penalty clause' in every provision of a contract which stipulates a sum to be payable by one party to the other in the event of a breach by the former.

See also *Steria Ltd v Sigma Wireless Communications Ltd* [2008] BLR 79, 118 Con LR 177, [2008] CILL 2544 at para [106].

[18] See *Kemble v Farren* (1829) 6 Bing 141 at 148 approved by Buxton LJ (para [115]); and Clarke LJ (para [106(vi)]) in *Murray v Leisureplay plc* [2005] EWCA Civ 963; and Diplock LJ in *Robophone* cited in the immediately preceding footnote.

[19] *Braes of Doune Windfarm (Scotland) Ltd v Alfred McAlpine Business Services Ltd* [2008] EWHC 426 (TCC).

[20] See also Lord Woolf in *Philips v Attorney General of Hong Kong* (1993) 61 BLR 41.

[21] *Murray v Leisureplay plc* [2005] EWCA Civ 963 *per* Buxton LJ at para [111].

[22] [2009] BLR 29, [2009] CILL 2648, [2009] 6 EG 102, [2009] 1 EGLR 1, [2008] EWHC 2617 (TCC).

[23] See, for example, where a penalty has been found *Commission of Public Works v Hills* [1906] AC 368. In *Lansat Shipping Co Limited v Glencore Grain BV* [2009] EWCA Civ 855 a clause in a charterparty was held to be a penalty because the primary purpose of the clause was to deter the charterers from breaching their obligations. The case, particularly at paras 31–43, provides a useful summary of the authorities in relation to penalties.

bargaining power.[24] But it must be doubted that inequality of bargaining power will be found in construction contracts in any but the rarest and most unusual of cases. Nevertheless where the intention is deterrence the court will not flinch from finding a clause to be penal.[25]

An argument that a liquidated damages provision was, in reality, a penalty arose in *Braes of Doune Windfarm (Scotland) Ltd v Alfred McAlpine Business Services Ltd*[26] Akenhead J considered an appeal from an arbitrator who had found that the provisions of the contract were not capable of generating with certainty liquidated damages flowing from an identified breach by the contractor, and that therefore the relevant clause should not be enforced. The learned judge concluded that not only was the arbitrator not obviously wrong but that his decision was ultimately right. The problem in that case was that, because of how it was drafted, the liquidated damages clause could well impose a liquidated damages liability on the contractor in respect of delays to individual wind turbines which were caused by another contractor altogether. **11.28**

In *Alfred McAlpine Capital Projects v Tilebox Ltd* Jackson J summarized the modern law in three propositions:[27] **11.29**

> There seem to be two strands in the authorities. In some cases judges consider whether there is an unconscionable or extravagant disproportion between the damages stipulated in the contract and the true amount of damages likely to be suffered. In other cases the courts consider whether the level of damages stipulated was reasonable. [Counsel] submits, and I accept, that these two strands can be reconciled. In my view, a pre-estimate of damages does not have to be right in order to be reasonable. There must be a substantial discrepancy between the level of damages stipulated in the contract and the level of damages which is likely to be suffered before it can be said that the agreed pre-estimate is unreasonable.
>
> Although many authorities use or echo the phrase 'genuine pre-estimate', the test does not turn upon the genuineness or honesty of the party or parties who made the pre-estimate. The test is primarily an objective one, even though the court has some regard to the thought processes of the parties at the time of contracting.
>
> Because the rule about penalties is an anomaly within the law of contract, the courts are predisposed, where possible, to uphold contractual terms which fix the level of damages for breach. This predisposition is even stronger in the case of commercial contracts freely entered into between parties of comparable bargaining power.

Whilst there must be a substantial discrepancy between the level of damages stipulated in the contract and the level of damage which is likely to be suffered before it can be said that the agreed pre-estimate is unreasonable, it must also be borne in mind that this is only a guide, and does not necessarily always provide the answer by itself. The question is a broad **11.30**

[24] As indeed Jackson J did in *Alfred McAlpine Capital Projects v Tilebox Ltd* [2005] EWHC 281 (TCC) cited in para 11.29.

[25] See for example *Lansat* cited in n 23.

[26] [2008] EWHC 426 (TCC), [2008] 1 Lloyd's Rep 608, [2008] 1 CLC 487, [2008] BLR 321, [2008] 2 All ER (Comm) 493, [2008] Bus LR D137.

[27] [2005] EWHC 281 (TCC) at para [48].

and general one and in commercial contracts the courts will exercise great caution before striking down a clause as penal.[28]

Extension of Time Clauses; Reasons for Extension of Time Clauses and the *Contra Proferentem* Rule in Relation to Liquidated Damages

11.31 As indicated in the introductory section of this chapter, extension of time clauses were designed to ensure that the fixed date for completion was adjusted rather than discharged so that the provisions for liquidated damages remain. This principle was restated[29] in *Dodd v Churton*[30] where Lord Esher MR said:

> The principle is laid down in Comyns' Digest, condition L(6), that, where one party to a contract is prevented from performing it by the act of the other, he is not liable in law for that default; and, accordingly, a well recognised rule has been established in cases of this kind, beginning with *Holme v Guppy* ((1838) 3 M & W 387), to the effect that, if the building owner has ordered extra work beyond that specified by the original contract which has necessarily increased the time requisite for finishing the work, he is thereby disentitled to claim the penalties for non-completion provided for by the contract.

11.32 This principle has been applied in *Peak Construction (Liverpool) Ltd v McKinney Foundations Ltd*[31] and therefore must be stated to be the law as it stands at present. The logic is that the fixed date can be re-fixed by the express terms of the contract, but if the employer causes delay and the delaying event does not fall within the extension of time clause, then the fixed date is discharged. However there is no logic to this process. There is no reason why the courts cannot fix a date which is a reasonable date for completion having regard to all the circumstances and liquidated damages should not be payable from that reasonable date. When *Peak* is cited in the higher courts today, the proposition of discharge is a source of surprise.[32] There is no reason why, as suggested by Lloyd LJ in *Rapid Building Group Ltd v Ealing Family Housing Association Ltd*,[33] where there are two separate and distinct causes of delay with two separate causes and where the dispute relates to one of those causes, it would not be just and convenient to be able to claim liquidated damages in relation to the other period. The doctrine of discharge as set out in *Dodd v Churton*[34] is a blunt instrument, much like the doctrine of discharge in cases of material variation of the underlying contract in contracts of guarantee.[35] It may have worked perfectly well when there were good reasons why the courts were predisposed to release parties from unconscionable bargains and fixed dates for completion were in the nineteenth century part of the unconscionable

[28] As emphasized by Buxton LJ in *Murray* at [111] and by Jackson J in *Tilebox* at [48.3]—see also *Steria Ltd v Sigma Wireless Communications Ltd* [2008] BLR 79, 118 Con LR 177, [2008] CILL 2544 at para [100].

[29] Substantially taken from *Holme v Guppy* (1838) 3 M&W 387.

[30] [1897] 1 QB 562 at 566.

[31] (1970) 1 BLR 111, (1969) 69 LGR 1.

[32] As in Peak itself and in *Rapid Building Group Ltd v Ealing Family Housing Association Ltd* (1984) 29 BLR 5 at 19 *per* Lloyd LJ.

[33] (1984) 29 BLR 5 at 19.

[34] [1897] 1 QB 562.

[35] See the discussion of *Holme v Brunskill* and the *Mercers* case in Ch 9.

bargain which liquidated damages often embodied. But today, when liquidated damages and extension of time clauses are designed to provide commercial parties with certainty, it is very odd that the courts will seek to abandon the measure of loss which the parties have pre-agreed for a measure of loss which requires all the difficulties and expense of proof. It might be said that the answer lies in the hands of the parties, namely that they can make bargains which allow for an extension of time in every case of employer's delay. But this may be an insufficient response because of the *contra proferentem* rule historically being applied to extension of time clauses[36] to make general phrases such as 'other causes beyond the contractor's control' insufficiently specific to inxclude employer's defaults.[37] Whatever the position as to this, it would in any case be wrong for the *contra proferentem* rule to be applied outside its normal remit—which is of course confined to contractual terms that are ambiguous. (See further as to this below.)

Given that the interpretation and application of liquidated damages clauses are matters of construction and given that the reasons for the suspicion of extension of time clauses has gone and given the general words used in many standard and bespoke contracts which provide for the grant of extensions of time (such as the JCT Minor Works Form), it is likely that, should the courts have to consider the *contra proferentem* rule as applied to extension of time clauses, they will either abandon that rule in this context (either expressly or practically) or at the very least ensure that it is kept within its proper limits—ie in situations of ambiguity. It seems unlikely that the courts will refuse to hold that an act of the employer in, say, refusing possession of the site, is something which falls outside the general words of a short form of extension of time clause which gives rise to an extension of time for causes beyond the contractor's control. The courts will want to uphold the commercial bargain and not permit an outdated approach to the construction of these clauses to frustrate them. **11.33**

Salmon LJ summarized the law in *Peak Construction (Liverpool) Ltd v McKinney Foundations Ltd*:[38] **11.34**

> The liquidated damages and extension of time clauses in printed forms of contract must be construed strictly *contra proferentem*. If the employer wishes to recover liquidated damages for failure by the contractors to complete on time in spite of the fact that some of the delay is due to the employer's own fault or breach of contract, then the extension of time clause should provide, expressly or by necessary inference, for an extension on account of such a fault or breach on the part of the employer . . .

One answer to this—which we would advocate—would be express abrogation of the *contra proferentem* rule in this context. An alternative is a more benevolent construction of the words 'necessary inference' permitting more general words to sweep up employer's breaches and thus allowing the extension of time clause to work as intended. In practice most arbitrators apply the extension of time clauses as they were intended to be used and do not set **11.35**

[36] *Peak Construction (Liverpool) Ltd v McKinney Foundations Ltd* (1970) 1 BLR 111, (1969) 69 LGR 1 particularly Salmon LJ at BLR 121.

[37] *Holme v Guppy* (1838) 3 M&W 387; *Wells v Army & Navy Co-operative Society* (1902) LT 764; and the discussion in *Amalgamated Building Contractors v Waltham Holy Cross Urban District Council* [1952] 2 All ER 452.

[38] (1970) 1 BLR 111 at 121.

time 'at large' with all the inconvenience and commercial uncertainty which would attend such an act.

11.36 In the meantime, the courts have already reinforced the limitations upon the application of the rule in this context. In the recent case of *Multiplex Constructions (UK) Ltd v Honeywell Control Systems Ltd (No 2)*[39] Jackson J reviewed the authorities in this area and derived three propositions:

> (i) Actions by the employer which are perfectly legitimate under a construction contract may still be characterized as prevention [ie acts which it has prevented the contractor from performing and which cannot be relied upon to hold the contractor to a specified completion date], if those actions cause delay beyond the contractual completion date.
> (ii) Acts of prevention by an employer do not set time at large, if the contract provides for extension of time in respect of those events.
> (iii) In so far as the extension of time clause is ambiguous, it should be construed in favour of the contractor.

11.37 The first point here is that Jackson J was clearly restating the familiar principle referred to above that it is only in the event of ambiguity that the *contra proferentem* rule should apply. He went on however to observe that his last proposition must be treated with care,[40] since what was important was that in the case of any ambiguity the term was construed in such a way as to validate the instrument and confer rights on both parties to the contract. Thus, where an extension of time clause was ambiguous, the court should lean in favour of a construction which permits the contractor to recover appropriate extensions of time in respect of events causing delay. Such an approach accords with the principle of construction that in cases of ambiguity the courts should lean towards a construction which validates rather than invalidates the instrument.

11.38 This approach was approved in *Steria Ltd v Sigma Wireless Communications Ltd*.[41] In that case HHJ Stephen Davies agreed with Jackson J that it was only in the event of genuine ambiguity that the *contra proferentem* rule should have any application. This was because the extension of time provision conferred benefits on both parties; in particular it enabled a contractor to recover reasonable extensions of time whilst still maintaining the contractually agreed structure of a specified time for completion (together, in the majority of cases, with the contractual certainty of agreed liquidated damages, as opposed to uncertain unliquidated damages).[42] Consequently he held that, where there was a genuine ambiguity as to whether or not notification was a condition precedent to a time extension, then the notification should not be construed as being a condition precedent, since such a provision operates for the benefit of only one party, ie the employer, and operates to deprive the other party (the contractor) of rights which he would otherwise enjoy under the contract.[43]

---

39 [2007] BLR 195, 111 Con LR 78, [2007] CILL 2458 at [56].
40 Ibid at para [57].
41 [2008] BLR 79, 118 Con LR 177, [2008] CILL 2544.
42 Ibid at para [89].
43 At para [89].

Whilst these recent first instance decisions fall short of an express abrogation of the *contra proferentem* rule, nevertheless it is apparent that its scope is in practice being limited in this particular context. **11.39**

The Grant of Extensions of Time and Its Effect

Under bespoke and standard forms of contract it is the employer's agent, usually the contract professional who is empowered to grant extensions of time to the contractor. Under some bespoke forms it is the employer itself, through its own employee, who is responsible for the task. **11.40**

It used to be that employers would be slow to grant extensions of time. The tactic was often to ask for further information and when that was supplied ask for more before an extension of time was properly considered. Now there is a right to adjudication, the right to an extension of time is granted more readily as an adjudicator can provide swift redress. Previously standard and bespoke contracts only gave the right to arbitration when the works were completed or alleged to have been completed. This would mean that the contractor would not know whether its claims were to be honoured or not, with the result that it would be placed in a dilemma, whether to accelerate the work to catch up or whether to trust the legal or arbitral process long after the works were completed. As noted elsewhere, the temptation was to seek to say that the failure to grant an extension of time was a deemed instruction to accelerate the works. This plea has never succeeded in the English courts. There is less commercial reason than before for a plea of constructive acceleration now that contractors have the right to an adjudication. There is no reason for the finding of constructive acceleration when the contractor can have a swift adjudication on the merits. Furthermore, even before adjudication there was no juridical reason for a development of construction law to find constructive acceleration. Each of the parties would be uncertain as to the outcome and the contractor would have agreed its remedy and the timing of its remedy. It would lead to too much further uncertainty if a failure to grant an extension of time might give rise to unspecified and unprovided for rights to additional payment. **11.41**

Once an extension of time is granted, any liquidated damages which accrued prior to that grant become repayable forthwith. Naturally the employer can seek to arbitrate and reverse the extension of time granted. In practice this rarely happens. Under JCT Standard Forms, before liquidated damages are deductible, the contract professional must issue a certificate to the effect that the works ought to have been reasonably completed by a set date following which liquidated damages can be deducted.[44] The purpose of this certificate is to prevent precipitate deduction of liquidated damages. **11.42**

The grant of an extension of time does not mean that the contractor is entitled to payment of its prolongation or other costs arising out of the delay. It is often said that 'money carrying' extensions of time lead inexorably to payment of loss and expense. It may be that this happens in practice. But there is no axiom that this is so. See also the discussion of loss and expense in Chapter 14. **11.43**

[44] *JF Finnegan Ltd v Community Housing Association Ltd* (1995) 77 BLR 22.

Part IV

BREACH, TERMINATION, AND DAMAGES

12

DAMAGES FOR BREACH OF CONTRACT

Introduction

The position in English law is that the normal rules of contract law apply to construction contracts in relation to damages which are recoverable for breach of contract. No special rules apply, though construction contracts provide a multitude of special circumstances which allow the law to be examined in some detail. First, however, it is necessary to see the basic law of recoverable damages for breach of contract and that involves a set of rules which go back to the famous and ancient case of *Hadley v Baxendale*,[1] which is one of the very few cases from the 19th century still commonly cited and relied upon alone today. However, two further cases from the 20th century, which have considered and explained *Hadley v Baxendale* and which are often considered together with it, are *Victoria Laundry (Windsor) Ltd v Newman Industries Ltd*[2] *and C Czarnikow Ltd v Koufos Ltd*.[3] **12.01**

[1] (1854) 9 Ex 341.
[2] [1949] 2 KB 538.
[3] [1969] 1 AC 350.

The Rule in *Hadley v Baxendale*

12.02 The rule in *Hadley v Baxendale* was stated by a unanimous court as follows:

> We think the proper rule in such a case as the present is this: Where two parties have made a contract which one of them has broken, the damages which the other party ought to receive in respect of such breach of contract should be such as may fairly and reasonably be considered either arising naturally, ie, according to the usual course of things, from such breach of contract itself, or such as may reasonably be supposed to have been in the contemplation of both parties, at the time they made the contract, as the probable result of the breach of it.

12.03 The recoverable loss is therefore divided into two categories. First, 'the damages which the other party ought to receive in respect of such breach of contract should be such as may fairly and reasonably be considered either arising naturally, ie, according to the usual course of things, from such breach of contract'. This might be described as the damages which flow in the natural way from the breach and is referred to in judgments and textbooks as the 'first limb' of the rule in *Hadley v Baxendale*. Secondly there are those 'such as may reasonably be supposed to have been in the contemplation of both parties, at the time they made the contract, as the probable result of the breach of it'. This is described as the second limb of the rule in *Hadley v Baxendale* and relates to special circumstances known to the parties at the time the contract was made. In *Hadley v Baxendale* it was put in this way at 354–5:

> . . . if the special circumstances under which the contract was actually made were communicated by the plaintiffs to the defendants, and thus known to both parties, the damages resulting from the breach of such a contract, which they would reasonably contemplate, would be the amount of injury which would ordinarily follow from a breach of contract under the special circumstances so known and communicated.

12.04 The special circumstances are often used to argue for a greater loss than would ordinarily be contemplated. But in fact a lesser loss as a result of special knowledge may be all that is recoverable by reference to pre-contract imputed knowledge.[4]

12.05 The objective of the law of damages is related to the underlying rule of the common law in that 'where a party sustains a loss by reason of a breach of contract, he is, so far as money can do it, to be placed in the same situation with respect to damages, as if the contract had been performed'.[5] A pragmatic view must be taken in the assessment of damages[6] so that the purpose, which is compensatory rather than punitive, is achieved.[7]

---

[4] See Devlin J in *Biggin & Co Ltd v Permanite Co Ltd* [1951] 1 KB 422 at 435–6 where he said: 'Damages which arise under the so-called 'second rule' in *Hadley v Baxendale*, (1854) 9 Exch 341, are sometimes referred to as if they were an increased sum which the plaintiff could obtain if he could show 'special circumstances', or as if the rule embodied a measure of damage specially beneficial to the plaintiff which he could invoke if he fulfilled the necessary conditions. It is, no doubt, true that it generally operates in favour of a plaintiff rather than against him, but I think that it is capable of doing either.' Quoted with approval by Otton LJ in *Bence Graphics Ltd v Fasson Ltd* [1998] QB 87 at 99. See also Lord Pearce in *C Czarnikow Ltd v Koufos Ltd* [1969] 1 AC 350 at 416.

[5] See Parke B in *Robinson v Harman* (1848) 1 Exch 850 at 855 and *Sally Wertheim v Chicoutimi Pulp Co* [1911] AC 301 at 307 *per* Lord Atkinson.

[6] *Golden Straight Corporation v Nippon Yusen Kubishka Kaisha (The Golden Victory)* [2007] UKHL 12.

[7] See the discussion relating to liquidated damages and penalties in Ch 11.

That underlying rule may lead to unfair results and thus the rule in *Hadley v Baxendale* is designed to ensure that there is a proper ambit of recoverable damages and that improbable and unpredictable results are not recovered. **12.06**

The two limbs lead to questions of foreseeability and the concept that if damage is foreseeable then it is recoverable and if not then it is not. However recovery is not simply governed by foreseeability. A party may not be liable for foreseeable losses if they are not of the type or kind for which he can be treated as having assumed responsibility.[8] As is shown below, foreseeability is a dangerous concept if taken too far from the words in *Hadley v Baxendale* itself. But it is a useful shorthand. Nevertheless before considering foreseeability, there should be an overarching and prior consideration of the question of whether the contract breaker can be treated as having assumed responsibility for that type of loss claimed.[9] Foreseeability depends upon the knowledge possessed by the parties, or the contract breaker, and it is of two types, actual knowledge or imputed knowledge. Imputed knowledge is that which is taken to be known by a reasonable person and which arises out of the first limb of the rule in *Hadley v Baxendale*. Actual knowledge is that which the contract breaker possesses outside the ordinary course of things and is of such a kind which informs the contract breaker that a breach would be liable to cause more loss than would occur in the ordinary course of things.[10] **12.07**

Foreseeability (and thus whether the loss falls within one or other limb) is to be judged at the time the contract is entered into.[11] **12.08**

Language in analysing the first and second limb

Whilst the law has been re-stated and reformulated in subsequent cases, judges return to *Hadley v Baxendale* when analysing what is recoverable pursuant to a breach of a construction contract. Indeed there have been explanations of the first rule in *Hadley v Baxendale* which tried to define what should 'fairly and reasonably be considered either arising naturally, ie, according to the usual course of things, from such breach of contract itself' and it led to terms like 'liable to result' (acceptable to some but not to Lord Reid, see footnote on reasonable foreseeability below), 'a real danger' (acceptable), and 'on the cards' (unacceptable)[12] being considered, but it must be doubted whether they improved upon the language adopted in *Hadley v Baxendale*. The term 'reasonable foreseeability' is sometimes used in cases and textbooks but should not be.[13] Whilst there is a danger of being too **12.09**

[8] *Transfield Shipping Inc v Mercator Shipping Inc* [2008] UKHL 48 [2009] 1 AC 61, [2008] 2 Lloyd's Rep 275, [2008] 2 CLC 1, [2008] 4 All ER 159, [2008] Bus LR 1395, [2008] 2 All ER (Comm) 753, Lord Walker in para 69, Lady Hale at para 92 and Lord Hoffmann at para 12.

[9] See Lord Hope in *Transfield* ibid in para 31 'Assumption of responsibility, which forms the basis of the law of remoteness of damage in contract, is determined by more than what at the time of the contract was reasonably foreseeable.'

[10] *Victoria Laundry (Windsor) Ltd v Newman Industries Ltd* [1949] 2 KB 538 at 539.

[11] See *Hadley v Baxendale* itself and *Victoria Laundry (Windsor) v Newman Industries Ltd*, above, at 539.

[12] See the denigration of the term in *C Czarnikow Ltd v Koufos Ltd* [1969] 1 AC 350.

[13] Ibid *per* Lord Reid at 389 where he said:

> But what is said to create a 'landmark' is the statement of principles by Asquith L.J. This does to some extent go beyond the older authorities and in so far as it does so, I do not agree with it.

purist, it is submitted that the language of *Hadley v Baxendale* ought to be looked at in each case when recoverability is considered and euphemism may assist in arriving at a conclusion as to whether or not the first limb is satisfied, but ought not to do more than assist and when such recourse is made it should be borne in mind that in the past when people have done so, it has often led to error. However this is subject to one proviso, namely that the term 'directly' should yield to the reasonable contemplation of the parties.[14]

12.10 Analysis of the second limb of *Hadley v Baxendale* is relatively easy. What has to be decided is exactly what was known at the time of the contract. Whilst that may involve an extensive fact-finding exercise, in the end from the point of view of legal analysis that is all that has to be done. So, for example, if it was known at the time a contract between a developer and contractor was entered into that the contractor was going to enter into a sub-contract which would mean that special and large damages would be suffered by the contractor in the event of a breach of contract by the developer, then in the event of a breach of contract by the employer causing that special and large loss that special and large loss will be recoverable. That will not be recoverable under the first limb but rather the second.

12.11 It has been stated by Lord Diplock in many cases[15] that the obligation to pay damages for breach of contract is a 'secondary obligation' in contra-distinction to the primary obligation

---

> In paragraph (2) it is said that the plaintiff is entitled to recover 'such part of the loss actually resulting as was at the time of the contract reasonably foreseeable as liable to result from the breach.' To bring in reasonable foreseeability appears to me to be confusing measure of damages in contract with measure of damages in tort. A great many extremely unlikely results are reasonably foreseeable: it is true that Lord Asquith may have meant foreseeable as a likely result, and if that is all he meant I would not object further than to say that I think that the phrase is liable to be misunderstood. For the same reason I would take exception to the phrase 'liable to result' in paragraph (5). Liable is a very vague word but I think that one would usually say that when a person foresees a very improbable result he foresees that it is liable to happen.

14 Ibid and *Bence Graphics Ltd v Fasson Ltd* [1998] QB 87 at 93.

15 See for example *Photo Production Ltd v Securicor Transport Ltd* [1980] AC 827 at 849:

> Every failure to perform a primary obligation is a breach of contract. The secondary obligation on the part of the contract breaker to which it gives rise by implication of the common law is to pay monetary compensation to the other party for the loss sustained by him in consequence of the breach; but, with two exceptions, the primary obligations of both parties so far as they have not yet been fully performed remain unchanged. This secondary obligation to pay compensation (damages) for non-performance of primary obligations I will call the 'general secondary obligation'. It applies in the cases of the two exceptions as well.
>
> The exceptions are: (1) Where the event resulting from the failure by one party to perform a primary obligation has the effect of depriving the other party of substantially the whole benefit which it was the intention of the parties that he should obtain from the contract, the party not in default may elect to put an end to all primary obligations of both parties remaining unperformed. (If the expression 'fundamental breach' is to be retained, it should, in the interests of clarity, be confined to this exception). (2) Where the contracting parties have agreed, whether by express words or by implication of law, that any failure by one party to perform a particular primary obligation ('condition' in the nomenclature of the Sale of Goods Act 1893), irrespective of the gravity of the event that has in fact resulted from the breach, shall entitle the other party to elect to put an end to all primary obligations of both parties remaining unperformed. (In the interests of clarity, the nomenclature of the Sale of Goods Act 1893, 'breach of condition', should be reserved for this exception).
>
> Where such an election is made (a) there is substituted by implication of law for the primary obligations of the party in default which remain unperformed a secondary obligation to pay

to perform the contract properly. That formulation is sometimes helpful to use when analysing the obligations of the parties in the event of a breach.

Employer's Remedies for the Contractor's Breaches of Contract

The contractor's two responsibilities are to deliver up a structure to specification within the time prescribed. Thus there are two differing types of damages payable in the event of a breach of contract: damages for breach of the obligation to build to specification, and damages for breach of the obligation to deliver up to time. In relation to the latter obligation, see Chapter 11 dealing with liquidated damages and within it the term damages 'at large'. **12.12**

In relation to damages for breach of the obligation to build to specification, the courts have looked at two separate measures of damage. One is the cost of executing remedial works and the other is the diminution in value of the structure by reason of the defect. The most frequently cited and applied measure of damage is the cost of remedial works. It is the basis upon which Scott Schedules are ordered and produced in the TCC.[16] **12.13**

But it may be that the measure of loss is the amount by which the value of the completed structure is diminished by the breach. That is because it is not always either reasonable or sensible for an employer to remedy the defect, particularly if the defect has no effect at all on the value of the delivered up structure. The classic example of this is the swimming pool case where the House of Lords refused the claimant the cost of deepening the pool in circumstances where it was plainly unreasonable to do so.[17] The facts in each case must be looked at to see whether it is fair and reasonable for the loss to be measured in reinstatement value terms. In the case of the defective swimming pool, where the owner never intended or never could reasonably have been taken to have intended to do the remedial work, then it was easy for the House of Lords to hold the owner to some smaller measure of loss than reinstatement value. **12.14**

The modern law can be set out in the following propositions which were accepted by the Court of Appeal in *Southampton Container Terminals Ltd v Schiffahrtsgesellschaft 'Hansa Australia' mgH and Co*,[18] a case dealing with the loss of a crane in an accident:[19] **12.15**

(1) The question of damages is largely one of fact, subject only to the most general guiding principles. The relevant principle in the law of tort is *restitutio in integrum* (that the claimant should be made whole again).

monetary compensation to the other party for the loss sustained by him in consequence of their non-performance in the future and (b) the unperformed primary obligations of that other party are discharged. This secondary obligation is additional to the general secondary obligation; I will call it 'the anticipatory secondary obligation.

[16] See *McGlinn v Waltham Contractors Ltd* [2007] EWHC 149 (TCC), [2008] Bus LR 233 paras 787–9. That case is one of the classic Scott Schedule cases.

[17] See *Ruxley Electronics and Construction v Forsyth* [1996] AC 344 at 100; and *Birse Construction Ltd v Eastern Telegraph Co Ltd* (2004) 47 EG 164.

[18] [2001] 2 Lloyd's Rep 275.

[19] Submissions made by Peter Gross QC (now Gross J).

(2) Common to the law of tort and contract are the principles that a loss is only recoverable if: (i) it is caused by the breach of duty or contract complained of; and (ii) it could not have been mitigated.[20]

(3) Damages are only recoverable if reasonable. If damages are unreasonable then they were not caused by the breach or should have been mitigated, and are irrecoverable.

(4) In particular, where replacement costs exceed market value, they will only be recoverable if shown to be reasonable.

(5) Whether or not a claimant has an intention to replace the item is relevant to the question of reasonableness.

(6) Whether the dispute arises in a commercial context, or whether the dispute involves questions of personal preference is relevant to the question of reasonableness.

(7) The essential question is as follows: what loss did the claimant really suffer? This is a question of fact and degree. If it is unreasonable in a particular case to award the cost of re-instatement it must be because the loss sustained does not extend to the need to re-instate.

12.16 In the same case Clarke LJ, now Lord Clarke, set out the following five propositions, which must now be regarded as authoritative on this topic:

(1) The claimant will ordinarily be entitled to the diminution in value of the property unless the reasonable claimant would have reinstated the land at less cost.

(2) The claimant who has in fact reinstated the property will ordinarily be entitled to recover the reasonable cost of doing so, even if the cost is greater than the diminution in value, unless he has acted unreasonably in reinstating the property.

(3) Where the claimant has not in fact yet reinstated the property, (subject to 4 and 5 below) he will ordinarily be entitled to recover the reasonable cost of reasonable reinstatement, even if it is greater than the diminution in value.

(4) In assessing what is the reasonable cost of reasonable reinstatement, the court will consider whether the amount awarded is objectively fair; that is fair to both parties. In particular, the court will not award a sum which is out of proportion to the benefit conferred on the claimant.

(5) In assessing what steps it is reasonable to take by way of reinstatement, the court will take account of the cost of the reinstatement. Thus it may not be reasonable fully to reinstate the property because the cost of doing so may not be justified. All will depend on the circumstances of the particular case.[21]

[20] For recognition in the cases of the close relationship between causation and mitigation, see eg *Radford v De Frobeville* [1977] 1 WLR 1262 at 1272 *per* Oliver J; *Dodd Properties v Canterbury City Council* [1980] 1 WLR 433 at 453 and 456; *The Borag* [1981] 1 WLR 274 at 281 and 285 *per* Templeman LJ; and *The Alecos M* [1991] 1 Lloyd's Rep 120 at 125 *per* Neill LJ.

[21] In formulating these propositions Lord Clarke relied upon the following cases from which the five propositions are distilled: *Lodge Holes Colliery Co Ltd v Wednesbury Corp* [1908] AC 323; *The Susquehanna* [1926] AC 655; *Hutchison v Davidson* 1945 SC 395; *Hollebone v Midhurst and Fernhurst Builders* [1968] 1 Lloyd's Rep 38; *Harbutt's 'Plasticine' Ltd v Wayne Tank and Pump Co Ltd* [1970] 1 QB 447; *Hole & Son (Sayers Common) v Harrisons of Thurnscoe* [1973] 1 Lloyd's Rep 345; *CR Taylor (Wholesale) Ltd v Hepworths*

The above propositions of Lord Clarke originally related to trespass to land, but were applied to the destruction of a crane,[22] and the same principles were applied to the destruction of an archive where it was held that the measure of damage was the cost of restoration and replacement rather than the market value. See *Aerospace Publishing Ltd and another v Thames Water Utilities Ltd* [2005] EWHC 2987 (QB) and [2007] EWCA Civ 3. They apply to construction cases. The courts have moved towards a single set of principles which govern all damages claims in contract and tort. **12.17**

The problem with the diminution in value test is that it takes no account of the fact that it is the specific structure which is sought by the employer, not some different alternative which many would think to be as good. But that problem is taken account of at the consideration of the reasonableness test. It must be noted that in *Ruxley* the reinstatement cost was far greater than the diminution in value. It is necessary for the court to stand back a little and make an assessment of what is reasonable in each case. In *Farmer Giles Ltd v Wessex Water Authority*[23] Russell LJ put it this way: **12.18**

> In reality, therefore, in making that award, I take the view that the judge has done no more than reflect in the damages the true diminution in value of this site in the circumstances of this case caused by the destruction of the building. The award, particularly when contrasted with the cost of full reinstatement, in my judgment, also passes the test of reasonableness. I add that test of reasonableness because the authorities to which we have been referred indicate that reasonableness always has to be taken into account. The judge must stand back, when he has done his arithmetic, and ask himself whether the figure achieved by his finding is fair both to the plaintiff and the defendants.

An example of this standing back is *R&M Stansfield Enterprises Ltd v AXA Insurance UK plc* [2006] EWCA Civ 881 which dealt with the termination of a salvage contract. The Court of Appeal held that the benefits accruing to the innocent party as a result of the termination must be included in the calculation of loss. **12.19**

Ltd [1977] 1 WLR 659; *Dodd Properties (Kent) Ltd v Canterbury City Council* [1980] 1 WLR 433; *Ward v Cannock Chase District Council* [1986] Ch 546; *Farmer Giles Ltd v Wessex Water Authority* [1990] 1 EGLR 177; *Dominion Mosaics and Tile Co v Trafalgar Trucking Co* [1990] 2 All ER 246; *Jordan v Norfolk County Council* [1994] 1 WLR 1353; and *Ruxley Electronics Ltd v Forsyth* [1996] 1 AC 344. His dicta have been applied by Coulson J in *McGlinn v Waltham Contractors Ltd* [2007] EWHC 149 (TCC), [2008] Bus LR 233 paras 790–794.

[22] With a significant caveat that they are, in the case of chattel destruction, to be read with the following propositions set out by Holland J in that case and quoted with approval by Lord Clarke:

(1) On proof of the tortious destruction of a chattel, the owner is prima facie entitled to damages reflecting the market value of the chattel 'as is'.
(2) He is so entitled whether or not he intends to obtain a replacement.
(3) The market or resale value is to be assessed on the evidence, there being no standard measure applicable to all circumstances.
(4) If the claimant intends to replace the chattel, and if the market or resale value as assessed is inadequate for that purpose, then the higher replacement value may, in the event, be the appropriate measure of damages.
(5) When and if replacement value is claimed, the claimant can only succeed to the extent that the claim is reasonable; that is, that it reflects reasonable mitigation of its loss.
(6) The claim will ordinarily be reasonable if it is reasonable to replace the chattel and the cost of replacement is reasonable.

[23] [1990] 1 EGLR 177 at 179.

12.20 Sometimes one of the consequences of breach of contract is that the innocent party's staff and management are required to divert time to investigating and mitigating the loss caused. In such circumstances, damages will be awarded if the claimant properly establishes the fact and extent of diversion of staff time and that the diversion caused significant disruption to its business. Although strictly speaking the claim should be cast in terms of loss of revenue caused by the diversion and disruption, ordinarily it will be reasonable for the court to infer from the disruption that but for the breach, staff would have applied the diverted time to activities which would, directly or indirectly, have generated revenue for the claimant in an amount at least equal to the costs of employing them during that time; *Aerospace Publishing Ltd v Thames Water Utilities Ltd* [2007] EWCA Civ 3.[24] See also the discussion of management time in Chapter 14.

Damages claimed for defects remedied during the defects liability period

12.21 This is in a different category to defects which emerge after the contractor has completed and the defects liability period is over. If construction work is carried out under a contract which provides for the contractor to make good defects during a defects liability period, then it is not open to the employer to remedy the defects itself during that period. The contractor, in those circumstances, has not only the duty but also the right to remedy the defects. If the employer goes ahead and remedies them itself, then the damages it can recover are limited to the amount that it would have cost the builder to remedy the defects.[25] This is hard to reconcile with the above authority. The contract breaker has not caused the employer's loss; the employer has done so by remedying defects which it is the right of the contractor to remedy. Therefore as a matter of causation it might be said that the builder has not caused the remedial work costs to be incurred. So perhaps the situation can be explained by some sort of restitution to the employer of the saving (in terms of the cost of remedying defects) made by the contractor. Whatever the jurisprudential theory, in the end practical justice is done because the builder is saved the cost of remedying the defective work and has breached its contract, in spite of the theory or heresy which is next discussed.

The temporary disconformity theory or heresy

12.22 There is a theory, born of the speech of Lord Diplock in *Kaye v Hosier & Dickinson*,[26] that all defects which occur prior to practical completion are not breaches of contract at all. The relevant passage in his speech which gives rise to that theory is as follows:

> During the construction period it may, and generally will, occur that from time to time some part of the works done by the contractor does not initially conform with the terms of the contract either because it is not in accordance with the contract drawings or the contract bills or because the quality of the workmanship or materials is below the standard required by condition 6 (1). The contract places upon the contractor the obligation to comply with any

---

24 See also *Bridge UK.Com Ltd v Abbey Pynford plc* [2007] EWHC 728 (TCC); and *Al-Rawas v Pegasus Energy Ltd* [2008] EWHC 617 (QB) for an application of these principles.

25 See *Pearce & High Ltd v John P Baxter* [1999] BLR 101.

26 [1972] 1 WLR 146 at 165.

> instructions of the architect to remedy any temporary disconformity with the requirements of the contract. If it is remedied no loss is sustained by the employer unless the time taken to remedy it results in practical completion being delayed beyond the date of completion designated in the contract. In this event the only loss caused is that the employer is kept out of the use of his building beyond the date on which it was agreed that it should be ready for use. For such delay liquidated damages at an agreed rate are payable under condition 22 of the contract.
>
> Upon a legalistic analysis it might be argued that temporary disconformity of any part of the works with the requirements of the contract even though remedied before the end of the agreed construction period constituted a breach of contract for which nominal damages would be recoverable. I do not think that makes business sense. Provided that the contractor puts it right timeously I do not think that the parties intended that any temporary disconformity should of itself amount to a breach of contract by the contractor.

It is to be noted that this is obiter dicta in a dissenting speech. The 'temporary disconformity' theory has been argued in several cases but it has never yet given rise to a finding that the contractor is not in breach of contract when it has executed defective work. Nevertheless there is no good reason why the contractor should be liable in damages so long as it has the right to remedy the defect and practical completion has not been reached. In that event the employer's remedy is to require that the contractor remedy the defect, not to claim damages. Sometimes a middle course has been argued, namely that if the employer remedies the defective work, then it can recover only the cost which the contractor would have incurred in remedying the defect. That analysis does produce broad justice and has been held by the Court of Appeal to be correct in cases where the employer remedies defective work after practical completion,[27] but it does not deal with the point that the employer has no right to remedy the defect and the contract gives it its remedy. As a matter of causation, the expenditure of the employer in remedying the defect is not caused by the execution of the defective work but is caused by the employer by-passing the contractual machinery in place for such remedial work. **12.23**

Most standard and bespoke contracts have an express provision within them which permits the employer or the employer's professional or agent to instruct the removal of defective work and its remedy along with the right to remedy the defect if it is not put right within a given period. The existence of that right does not alter the above analysis. It is merely an additional right in the hands of the employer which is specific. If the employer instructs the removal of defective work, then it can operate the clause and recover the cost of remedying the defect. But it cannot simply go ahead and remedy the defective work, so long as the contractor has the right and obligation to do so, and claim damages from the contractor. **12.24**

---

[27] See *Pearce & High Ltd v John P Baxter* [1999] BLR 101.

General damages

12.25 In a commercial context there is no claim for general damages for breach of contract against a contractor.[28] However in a domestic context where the object of the agreement may be to enhance the pleasure of the householder, then damages will be recoverable.[29] The level of damages will be modest. Ramsey J in *Eiles v London Borough of Southwark* [2006] EWHC 1411 directed himself in a domestic case as follows:

> The approach of the courts to the question of general damages for such vexation, distress and worry is to provide compensation which is 'not excessive, but modest' and which 'may not be very substantial' per Lord Denning MR and Oliver LJ in *Perry v Sidney Phillips* [1982] 1 WLR 1297. In *Watts v Morrow* [1991] 1 WLR 1421 Bingham LJ said that general damages were recoverable but were limited, generally, to damages for physical inconvenience and discomfort and mental suffering directly related to that.

12.26 In that case he awarded £2,250 where the homeowner suffered the physical inconvenience and discomfort of having a house which was cracking and having the house invaded by various people investigating the cracks. She had to move out of the house for a period of a year, causing the inconvenience of moving out, moving back, settling into the alternative temporary accommodation, and moving back to a house which had been the subject of the remedial work. She did not have to suffer the level of inconvenience and discomfort that would have occurred if she had remained in the house. She did, however, continue to visit to keep an eye on the work being carried out to her home.

12.27 Coulson J in *Iggleden v Fairfield New Homes* [2007] EWHC 1573 (TCC) in a 'mid range' case awarded £750 for each year during which disruption and inconvenience was suffered. In *Glen Haysman v Mrs Roger Films Ltd* [2008] EWHC 2494 (QB) the Deputy Judge awarded £1000 in a similar type of mid range case.

Where the employer may not suffer all the loss

12.28 Because the employer may not be the party which suffers the loss there is the potential for the builder to escape liability for its breach of contract. However the courts will be astute to avoid this if they can and will permit, in the right circumstances (where the loss of the non-contracting party is in the contemplation of the contracting parties), the employer to recover not only its own loss but also the loss of the party which suffers it.[30]

---

[28] *Johnson v Unisys Ltd* [2003] AC 518 *per* Lord Millett at para 70.

[29] See two articles by Kim Franklin *Damages for Heartache: The Award of General Damages for Inconvenience and Distress in Building Cases* (1988) 4 Const LJ 264 and *More Heartache: A Review of the Award of General Damages in Building Cases* (1992) 8 Const LJ 318. *Watts v Morrow* [1991] 1 WLR 1421; *Farley v Skinner* [2002] 2 AC 732.

[30] See *Linden Gardens Ltd v Lenesta Sludge Disposal Ltd* [1994] AC 85 and the cases identified in Ch 4 on Assignment and Sub-Letting dealing with the legal black hole. See for an application of this principle in an ordinary contractual setting *John Harris Partnership v Groveworld* (1999) 75 Con LR 7. See also the overt distaste for a legal black hole in *McLaren Murdoch & Hamilton Ltd v Abercromby Motor Group Ltd* (2002) 100 Con LR 63.

Damages against professionals

The above principles in relation to *Hadley v Baxendale* apply when defective performance by professionals leads to defects to the completed structure.[31] 12.29

However, in surveyor's negligence cases, where the surveyor inspects prior to purchase and in reliance on the surveyor's report the building is bought, the measure of loss recoverable by a disappointed purchaser from its surveyor is the difference in value between the value of the premises as described and as they should have been described.[32] That measure has produced litigation because it has often been thought that the two measures produce radically different results with the remedial work costs being higher than the diminution in value. But it is hard to see how this need be so. If the structure requires to be underpinned because the foundations are failing, then most valuers would say that its value is diminished by at least the cost of underpinning. 12.30

As against the contractor there is no right to general damages for disappointment in a commercial context. But in the case of a householder's contract with a professional as with the contractor, where the object of the contract is to buy pleasure, relaxation, or peace of mind then general damages are recoverable.[33] 12.31

Where loss is suffered by a party who is not the contracting party, then the same considerations as to recoverability apply to professionals as to contractors.[34] 12.32

Contractor's Remedies for Employer's Breach of Contract

Most standard and bespoke forms of contract provide an obligation upon the employer to provide the contractor with the information which it needs to execute and complete the works. Failure to provide information which is needed and which it is only in the employer's gift is, if there is no express term to cover it, a breach of the *Mackay v Dick* implied term of the contract discussed above under implied terms. Obviously the contractor cannot build without the information needed to do so. 12.33

Aside from the obligation to provide information, and the obligation (where applicable) to nominate sub-contractors,[35] the principal obligation of the employer is to pay the contractor in accordance with the terms of the contract. 12.34

The obligation to make payment has only one remedy, namely an action for sums due. Interest is payable pursuant to statute (see s 35A of the Supreme Court Act 1981 (which permits simple interest) and s 49 of the Arbitration Act 1996 (which permits compound interest)). Interest under section 35A is discretionary and only simple interest can be 12.35

[31] See *Sutcliffe v Thackrah* [1974] AC 727; and *Sutcliffe v Chippendale & Edmondson* (1977) 18 BLR 149.

[32] *Phillips v Ward* [1956] 1 WLR 471; and *Perry v Sidney Phillips* [1982] 1 WLR 1297.

[33] See *Farley v Skinner* [2002] 2 AC 732, cf *Watts v Morrow* [1991] 1 WLR 1421. *Knott v Bolton* (1995) 11 Const LJ 375 overruled. See also *Hobbs v London and South Western Railway Co* (1875) LR 10 QB 111; and *Ruxley Electronics and Construction Ltd v Forsyth* [1996] AC 344.

[34] *John Harris Partnership v Groveworld* (1999) 75 Con LR 7.

[35] See *Bickerton v North West Metropolitan Regional Hospital Board* [1970] 1 WLR 607.

awarded, whilst compound interest can now be awarded as a result of the decision of the House of Lords in *Sempra Metals Limited v IRC* [2008] 1 AC 561, that is only in respect of the calculation of unjust enrichment caused by the payment of money to which the defendant in that case was not entitled.[36] Interest under s 35A is usually granted at either the normal commercial rate payable generally or at the normal commercial rate payable by the successful party.[37] Successful claimants are well advised to ensure that they are able to demonstrate the rate of interest they actually pay their banks if they wish to recover more than a normal commercial rate. This often gets forgotten in the excitement of victory, but a change of only 1 percent in the interest awarded can be significant and with a credit crunch interest rates paid by different commercial entities can vary enormously. It should be noted that if claims are not formulated with proper speed, the contractor may not recover interest for the whole period it is out of its money.[38] Interest may also be payable at rates provided for under the contract. However, a contractual rate of interest must be no greater than the rate which would reflect the loss the contractor would be expected to suffer by reason of non-payment at the time the contract was entered into. The law relating to penalties, which is applicable to liquidated and ascertained damages, is applicable also to contractual rates of interest. It is, however, relatively rare for a rate of interest to be contained within the contract documents. There is no further remedy available to the contractor if it has suffered further losses as a result of the failure to make payment to time.

12.36 Contractor's claims for loss and expense are considered under loss and expense and are not dealt with separately here. The elements of loss and expense which are set out in Chapter 14 are claimable as loss and expense. This will include finance charges costs of keeping plant and machinery idle as falling under the first limb of *Hadley v Baxendale*.[39]

How the Remedies Inter-relate to Claims under the Contract

12.37 The provision of late information by the employer or its professional is usually catered for in standard forms. In JCT Standard Forms late information is provided for as a ground for recovery of loss and expense. This will involve the certification and payment of loss and expense pursuant to certificates issued. Such recovery is normally expressed to be without prejudice to the other rights and remedies of the contractor. Even if it was not so expressed, it would require clear words in the contract before the contractor's rights to recover damages for breach of contract were excluded by reason of the contractual remedy. Some standard forms do set out those clear words by saying that recovery under the contract pursuant to express terms are both sides' exclusive remedies and therefore recovery must be pursuant to the express terms of the contract.

---

[36] For a useful discussion of this topic and the distinction, see the judgment of Patten LJ in *Benedetti and another v Naguib Onsi Naguib Sawiris and others* [2009] EWHC 1806 (Ch) particularly paras 8–10.

[37] *Benedetti and another v Naguib Onsi Naguib Sawiris and others* [2009] EWHC 1806 (Ch) at para 11 which in that case was 1% above Euribor.

[38] *Socimer International Bank Ltd v Standard Bank London Ltd* [2006] EWHC 2896 (Comm) Gloster J. The authorities are summarized in relation to this issue in paras 2–10 of the judgment.

[39] See *FG Minter Ltd v Welsh Health Technical Services Organisation* (1980) 13 BLR 1.

It is common for contractors in litigation or adjudication or arbitration to claim loss and expense both as damages for breach of contract and as a claim for sums due under the contract in the alternative. **12.38**

Recovery under the contract may be subject to provisions of notice. **12.39**

Recovery of Loss Following a Settlement: The Rule in *Biggin & Co Ltd v Permanite*

Where there is a claim against a contractor which has a claim over against a sub-contractor it will often happen that the contractor will wish to settle the claim and then make a claim over against the sub-contractor. This situation will be replicated in cases where, for example, the professional and the contractor are claimed against and one wishes to settle with the claimant and the other does not. The question arises as to the status of the settlement. Can it be applied as being the measure of the loss of the main contractor? **12.40**

Ramsey J in *Siemens Building Technologies FE Ltd v Supershield Ltd* [2009] EWHC 927 (TCC) examined the authorities on the question and formulated the following propositions from them: **12.41**

(1) For C to be liable to A in respect of A's liability to B which was the subject of a settlement it is not necessary for A to prove on the balance of probabilities that A was or would have been liable to B or that A was or would have been liable for the amount of the settlement.
(2) For C to be liable to A in respect of the settlement, A must show that the specified eventuality (in the case of an indemnity given by C to A) or the breach of contract (in the case of a breach of contract between C and A) has caused the loss incurred in satisfying the settlement in the manner set out in the indemnity or as required for causation of damages and that the loss was within the loss covered by the indemnity or the damages were not too remote.
(3) Unless the claim is of sufficient strength reasonably to justify a settlement and the amount paid in settlement is reasonable having regard to the strength of the claim, it cannot be shown that the loss has been caused by the relevant eventuality or breach of contract. In assessing the strength of the claim, unless the claim is so weak that no reasonable party would take it sufficiently seriously to negotiate any settlement involving payment, it cannot be said that the loss attributable to a reasonable settlement was not caused by the eventuality or the breach.
(4) In general if, when a party is in breach of contract, a claim by a third party is in the reasonable contemplation of the parties as a probable result of the breach, then it will generally also be in the reasonable contemplation of the parties that there might be a reasonable settlement of any such claim by the other party.
(5) The test of whether the amount paid in settlement was reasonable is whether the settlement was, in all the circumstances, within the range of settlements which reasonable people in the position of the settling party might have made. Such circumstances will generally include:
    (a) the strength of the claim;
    (b) whether the settlement was the result of legal advice;
    (c) the uncertainties and expenses of litigation;
    (d) the benefits of settling the case rather than disputing it.
(6) the question of whether a settlement was reasonable is to be assessed at the date of the settlement when necessarily the issues between A and B remained unresolved.

It is fair to say that a number of judges have recently sought to distill the law on this topic[40] and all their analyses repay close scrutiny but the above principles are the latest (and probably most helpful) authority on the subject.

12.42 *Biggin & Co Ltd v Permanite Ltd* was a case which involved the plaintiffs buying defective goods from the defendants for resale to the Dutch government. The government's claim against the plaintiff went to arbitration and was settled on legal advice for £43,000 plus costs. The plaintiff then sought to recover this sum from the defendants. The Court of Appeal held that the settlement was relevant (the judge below did not) and, on the facts, reasonable with the result that the plaintiff recovered the amount paid pursuant to the settlement. Somervell LJ at 321 said:

> I think that the judge here was wrong in regarding the settlement as wholly irrelevant. I think, though it is not conclusive, that the fact that it is admittedly an upper limit would lead to the conclusion that, if reasonable, it should be taken as the measure. The result of the judge's conclusion is that the plaintiffs must prove their damages strictly to an extent to show that they equal or exceed 43,000l.; and that if that involves, as it would here, a very complicated and expensive inquiry, still that has to be done. The law, in my opinion, encourages reasonable settlements, particularly where, as here, strict proof would be a very expensive matter. The question, in my opinion, is: what evidence is necessary to establish reasonableness? I think it relevant to prove that the settlement was made under advice legally taken. The client himself could do that, but I do not think that the advisers would normally be relevant or admissible witnesses. I say 'normally'. It may be that in special cases they might be. The plaintiff must, I think, lead evidence, which can be cross-examined to, as to facts which the witnesses themselves prove and as to what would probably be proved if, as here, the arbitration had proceeded, so that the court can come to a conclusion whether or not the sum paid was reasonable. The defendant may, by cross-examination, as was done here, seek to show—and perhaps successfully show—that it was not reasonable. He may do so, or call evidence which leads to the same conclusion. He might in some cases show that some vital matter had been overlooked . . . [B]ut if there is evidence at the end of the matter of the kind which I have indicated, on which the court can come to the conclusion that this was a reasonable settlement in the circumstances, then I think that it should be the measure. Parties, Bowen, LJ, said, have been held to contemplate litigation in the sort of circumstances which have arisen here. It would, I think, be unfortunate if they were not also held to contemplate reasonable settlements in the type of circumstances which have arisen here.

12.43 The claimant must prove a liability on the part of the defendant. It is not enough to simply adduce evidence of a settlement.[41] But if liability is established and the compromise is

---

[40] For example HH Judge Anthony Thornton QC in *Bovis Lend Lease Ltd v RD Fire Protection Ltd* [2003] EWHC 939 (TCC) paras 69–80; HH Judge Coulson QC in *John F Hunt Demolition Ltd v ASME Engineering Ltd* [2008] 1 All ER 180 [2007] EWHC 3159 (TCC); and HH Judge John Toulmin QC in *Mirant Asia-Pacific Construction (Hong Kong) Ltd v Ove Arup and Partners International Ltd & Anor* [2007] EWHC 918 (TCC) para 41 onwards. See also *Fairfield Mabey Ltd v Shell UK Ltd* (1988) 45 BLR 113; *Oxford University Press v John Stedman Group* (1990) 34 Con LR 1; *DSL Group Ltd v Unisys International Services Ltd* (1994) 67 BLR 117; *P&O Developments Ltd v The Guys and St Thomas' Hospital NHS Trust* [1999] BLR 3; *The Royal Brompton Hospital NHS Trust v Frederick Alexander Hammond* [1999] BLR 163; *Ascon Contracting Ltd v Alfred McAlpine Construction Isle of Man Ltd* (1999) 66 Con LR 119. It is also helpful to remember that Evans LJ in *Dubai Aluminium Co Ltd v Salaam* [2001] QB 113 at 132 referred to 'The common law rule that a settlement payment may be recoverable when it is made reasonably to compromise the claims in the action (*Biggin & Co Ltd v Permanite Ltd*).

[41] See *Fletcher and Stewart Ltd v Peter Jay and Partners* (1976) 17 BLR 38.

reasonable, then the settlement sum is recoverable.[42] In construction cases the principle has been considered and applied at first instance in several cases.[43] There has been some debate as to whether or not the principle is a question of the measure of damage or of mitigation.[44] But from the *Biggin* case itself it is plainly a matter of measure. However *Biggin* does refer to the concept of the upper limit. It is clear from the language that the reference in *Biggin* to the settlement being an 'upper limit' means no more than that, whether as a matter of mitigation or causation, there could be no question of Biggin recovering an amount in excess of the settlement.[45]

It is prudent when entering into a settlement, to make proper allocation of settlement damages to the acts or omissions of any putative defendant against whom recovery will be sought. When there is no allocation, there is a danger that the court will not have the material to assess the quantum of a claim which has been compromised along with several others.[46] However, given that a settlement allocation, where a main contractor is settling with an employer in respect of defective work by more than one sub-contractor, is always open to being coloured as self serving, it is difficult to predict the weight which will be given to it. For example if there are two sub-contractors responsible for two different defects and one sub-contractor is solvent and the other is not, the temptation to allocate the larger part of the settlement sum to the solvent sub-contractor's defects will be too strong for many to resist. The employer, when presented with an allocation of the global sum which it is prepared to take in settlement of its claim, will have little interest in the allocation of damages from one defect to another. In those circumstances the employer is only interested in the global sum and not interested in which sub-contractor will have to make what restitution to the main contractor. Therefore it is hard to see the probative value of the allocation in those circumstances. **12.44**

Discharge from Further Performance by Breach; Fundamental Breach

All contracts have their terms divided into three categories: (a) warranties, (b) conditions, and (c) innominate terms. **12.45**

A warranty is a term of a contract where the remedy of the innocent party is damages only or an action by an arbitrator or court remedying the default. The term will be a warranty because the obligation is not sufficiently important to the running of the contract to permit **12.46**

[42] *Comyn Ching Co Ltd v Oriental Tube Co Ltd* (1979) 17 BLR 47. See also *Jones v Williams* (1841) M&W 473; and *Fisher v Val de Travers* (1876) 45 LJNS 479.

[43] See *Fairfield Mabey Ltd v Shell UK Ltd* (1988) 45 BLR 113; *Oxford University Press v John Stedman Group* (1990) 34 Con LR 1; *DSL Group Ltd v Unisys International Services Ltd* (1994) 67 BLR 117; *P&O Developments Ltd v The Guys and St Thomas' Hospital NHS Trust* [1999] BLR 3; *The Royal Brompton Hospital NHS Trust v Frederick Alexander Hammond* [1999] BLR 163; *Ascon Contracting Ltd v Alfred McAlpine Construction Isle of Man Ltd* (1999) 66 Con LR 119.

[44] Judge Hicks QC in *The Royal Brompton Hospital NHS Trust v Frederick Alexander Hammond* [1999] BLR 163 treated the matter as one of the measure of damages when the matter was debated before him.

[45] See the decision of Peter Gross QC in *Total Liban SA v Vitol Energy SA* [2001] QB 643 at 663.

[46] As, for example, was the case in *Bovis Lend Lease Ltd v RD Fire Protection Ltd* [2003] EWHC 939 (TCC) Judge Thornton QC.

any other remedy. In construction contracts, these may be the minority of contractual terms because all breaches of contract are capable of having a profound effect on the other party's ability to perform it. An example of such warranties would be the obligation to issue a 'Final Certificate'. The remedy in the hands of the contractor is to ask an arbitrator to do so. The contractor's ability to execute the works is unaffected by the breach and always would be so unaffected. The parties can agree that a term of the contract which might be regarded as a condition shall be treated as a warranty.[47]

12.47 A condition of the contract is a term which is (or is agreed to be) so fundamental to the contract's operation that without it the other party cannot carry out its obligations or its ability to do so is profoundly and seriously affected. This can be a term which is fundamental because, by its nature, a breach of the term will deprive the other contracting party of the benefit of the contract. It can also be a term which the parties expressly in their contract designate as a condition even if the effect of a breach is minor.[48] A breach of such a term would, the law holds, amount to a repudiation of the contract by the guilty party which the innocent party can accept. In the event of the breach of a condition, the innocent party has the right to either: (a) if practicable, claim damages for breach of contract; or (b) accept the breach as being a repudiation and bring the contract to an end. If the innocent party does not accept the repudiation, then it will be taken to have affirmed the contract and the right to accept the repudiation will be lost unless there is a fresh breach.

12.48 An innominate term is the most common. It is a term the breach of which does not necessarily have a profound effect upon the ability of the other party to perform its obligations, but, depending upon its seriousness may do so. In the event of the serious breach the innocent party's rights are the same as if there had been a breach of a condition. The test of whether the breach entitles the innocent party to bring the contract to an end is whether the innocent party has, by the breach, been deprived of substantially the whole benefit of the contract.[49]

---

[47] See *Hong Kong Fir Shipping Co Ltd v Kawasaki Kisen Kaisha Ltd* [1962] 2 QB 26 *per* Diplock LJ at 69 and the cases cited in n 50 below.

[48] There is nothing to prevent the parties from so agreeing as they can agree that what would ordinarily be thought of as a condition is in fact a warranty. See *Lombard North Central plc v Butterworth* [1987] QB 527; and *Bettini v Gye* (1876) 1 QBD 183 at 187; *Hong Kong Fir Shipping Co Ltd v Kawasaki Kisen Kaisha Ltd* [1962] 2 QB 26 at 70; *Financings Ltd v Baldock* [1963] 2 QB 104; *Photo Production Ltd v Securicor Transport Ltd* [1980] AC 827 at 849E; *Bunge Corp, New York v Tradax Export SA, Panama* [1981] 1 WLR 711 at 715, 719.

[49] See *Hong Kong Fir Shipping Co Ltd v Kawasaki Kisen Kaisha Ltd* [1962] 2 QB 26 where at 70 Diplock LJ said:

> There are, however, many contractual undertakings of a more complex character which cannot be categorized as being 'conditions' or 'warranties', if the late nineteenth-century meaning adopted in the Sale of Goods Act, 1893, and used by Bowen LJ in *Bentsen v Taylor, Sons & Co* be given to those terms. Of such undertakings all that can be predicated is that some breaches will and others will not give rise to an event which will deprive the party not in default of substantially the whole benefit which it was intended that he should obtain from the contract; and the legal consequences of a breach of such an undertaking, unless provided for expressly in the contract, depend upon the nature of the event to which the breach gives rise and do not follow automatically from a prior classification of the undertaking as a 'condition' or a 'warranty'. For instance, . . . [the] breach of an undertaking by a shipowner to sail with all possible dispatch to a named port does not necessarily relieve the charterer

In the event that an employer or contractor either breaches a condition of the contract or breaches an innominate term in such a serious way as to have the effect of being a breach of a condition of the contract, then the innocent party can choose its remedy, as stated above: claim damages or bring the contract to an end. **12.49**

The innocent party must elect its remedy. It cannot blow hot and cold. If it does not accept the breach of contract as a repudiation within a reasonable time, then it will be taken to have affirmed the contract and its remedy will be damages only. **12.50**

However, once the breach of contract is accepted as a repudiation, the employer's remedy will simply be damages for breach of contract. If the innocent party is the contractor, then it may have a further choice to make in relation to the damages which it can claim. It is often stated that the contractor can elect between two remedies: (a) the contract value of the work carried out plus loss of profit on work not carried out, and (b) a quantum meruit for all the work done to date. **12.51**

The authority for the proposition that the contractor has this choice is the decision of the Privy Council in *Lodder v Slowey*.[50] However that case is a very thin basis upon which to found the suggestion that the contractor can abandon its rates and prices for work done and recover a quantum meruit. The statement that this is a possibility is casually stated and not truly central to the decision and it does not appear that the contrary was seriously argued. Strictly speaking it is not binding in England since it is a Privy Council decision relating to a case from New Zealand.[51] There is extensive discussion in the authorities about whether a quantum meruit is claimable in these circumstances and one possible position being that the innocent party can claim a quantum meruit but that is capped by the contract rates on the basis that those rates indicated the parties' agreement as to the value of those works and thus the quantum meruit.[52] **12.52**

of further performance of his obligation under the charterparty, but if the breach is so prolonged that the contemplated voyage is frustrated it does have this effect.

[50] [1904] AC 442.

[51] Note Cooke J (later Lord Cooke) in *Canterbury Pipe Lines v Christ Church Drainage* (1979) 16 BLR 76 at 100 in the New Zealand Court of Appeal declining to follow it (and distinguishing it). There is a Scottish case which states, obiter, that a quantum meruit can be recovered: see *ERDC Construction Ltd v HM Love & Co* (1994) 70 BLR 67 at 88, but that was a case where a contractor claimed a quantum meruit where there was no contractual discharge and therefore, unsurprisingly, the contractor failed to recover a quantum meruit. See also *Morris-Knudsen Co Inc v British Columbian Hydro and Power Authority* (1978) 85 DLR (3rd) 186.

[52] *Taylor v Motability Finance Ltd* [2004] EWHC 2619 (QBD (Comm)); *GEC Marconi Systems Pty Ltd v BHP Information Technology Pty Ltd* (2003) FCA 50 (Fed Ct (Aus) (Full Ct)); *Newton Woodhouse v Trevor Toys Ltd* (Unreported, 20 December 2001) (CA); *Renard Constructions (ME) Pty Ltd v Minister for Public Works* (1992) 26 NSWLR 234 (CA (NSW)); *Lusty Architects v Finsbury Securities* (1991) 58 BLR 66 (CA (Civ Div)); *Jennings Construction Ltd v QH & M Birt Pty Ltd* (Unreported, 16 December 1988) (Sup Ct (NSW)); *Morrison-Knudsen Co v British Columbia Hydro and Power Authority* (1978) 7 Const LJ 227 (CA (BC)); *Heyman v Darwins Ltd* [1942] AC 356 (HL); *Luxor (Eastbourne) Ltd v Cooper* [1941] AC 108 (HL); *Chandler Bros Ltd v Boswell* [1936] 3 All ER 179; *Segur v Franklin* (1934) SR (NSW) 67; *Lodder v Slowey* [1904] AC 442 (PC (NZ)); *Ankeny v Clark* 148 US 345 (1893); *United States v Behan* 110 US 338 (1884); *Appleby v Myers* (1866–67) LR 2 CP 651 (Ex Chamber); *Bartholomew v Markwick* (1864) 15 CB (NS) 711; *Harrison v James* (1862) 7 H&N 804; *Prickett v Badger* (1856) 1 CB (NS) 296; *Ranger v Great Western Railway* (1854) 5 HL Cas 72; *De Bernardy v Harding* (1853) 8 Exch 822; *Planche v Colbourn* (1831) 8 Bing 14. See the impressive article by Julian Bailey in Construction Law Journal on this topic which cites the cases in this footnote: (2006) 22 Const LJ 4, 217–240.

12.53 In the event that the contractor has under-tendered, a repudiation by the employer would, if remedy (b) was open to it, release the contractor altogether from its bad bargain and give it a remedy which is out of all proportion to the loss the breach of contract has caused it. After all, the payment of the contract value of the work done to date plus loss of profit on work not carried out compensates the contractor in full for its losses.

12.54 Given that the case is of Privy Council authority only, it must be doubtful that it would stand scrutiny in a proper appellate setting as being authority for the proposition that the quantum meruit remedy is open to the contractor. The only jurisprudential basis for the quantum meruit remedy is that, since the employer has repudiated the contract, then it should not be able to rely upon the terms of the contract for the purposes of valuing the contractor's work. But that has no logic to it because the remedy which is normal in all other cases (aside from construction contracts) when there has been a repudiation is damages for breach of contract. If an innocent party has a claim for breach of contract, then a quantum meruit is scarcely the same remedy. It makes no sense for a remedy in restitution to be given in cases where the complaint is for a breach of contract.

12.55 This position is open to review by the English courts. There has been no case where what is perceived to be the rule in *Lodder v Slowey* has been applied by the courts of England to give a contractor a quantum meruit. Until there is review by the courts, it is still possible to argue that the innocent contractor can make the election for a quantum meruit and such arguments are made in arbitrations with success.

Impecuniosity and the Additional Damage That Causes

12.56 *Liesbosch Dredger (Owners of) v Owners of SS Edison ('The Liesbosch')*[53] is a famous and now overruled case which decided (or appeared to decide) that if an impecunious innocent party suffered loss through the tort or breach of contract of another, then the additional loss and damage caused by that impecuniosity was not recoverable. The case itself involved the loss of a dredger, the owners of which could not afford to replace it. Instead they hired another dredger for a period until they were able to purchase a replacement. The hire cost was held not to be recoverable. Lord Wright at 460 said:

> The respondents' tortious act involved the physical loss of the dredger . . . But the appellants' actual loss in so far as it was due to their impecuniosity arose from that impecuniosity as a separate and concurrent cause, extraneous to and distinct in character from the tort; the impecuniosity was not traceable to the respondents' acts, and in my opinion was outside the legal purview of the consequences of these acts . . . In the varied web of affairs, the law must abstract some consequences as relevant, not perhaps on grounds of pure logic but simply for practical reasons. In the present case if the appellants' financial embarrassment is to be regarded as a consequence of the respondents' tort, I think it is too remote, but I prefer to regard it as an independent cause, though its operative effect was conditioned by the loss of the dredger.

[53] [1931] P 230, [1932] P 52, CA, [1933] AC 449, HL(E).

The case was never satisfactory since it departed from the principle that you must take your victim as you find him, a principle which is applicable to contract and tort. That dictum of Lord Wright ran contrary to Lord Collins' statement in *Clippens Oil Co Ltd v Edinburgh and District Water Trustees*[54] that the wrongdoer must take his victim talem qualem (as he finds him), adding that: 'if the position of the latter is aggravated because he is without the means of mitigating it, so much the worse for the wrongdoer, who has got to be answerable for the consequences flowing from his tortious act.' **12.57**

The Liesbosch came in for much academic criticism[55] and was followed reluctantly.[56] In construction cases it received its first death blow in *Dodd Properties (Kent) Ltd v Canterbury City Council*[57] which was a case involving damage to a garage building which was damaged by pile driving. The garage could not be repaired immediately because of financial stringency. At 1970 prices the cost would have been £11,375 but in 1978 the cost was £30,327. The trial judge applied the 'breach date' rule, that damages were to be assessed at the date of the breach and applied the *Liesbosch* test. However the Court of Appeal reversed the decision, casting doubt on *Liesbosch* and referring to the above quoted dictum of Lord Collins in *Clippen.* From that date there was an uneasy position in construction cases with claimants relying upon *Dodd* as giving them damages assessed as at the date repairs to defective structures were reasonably carried out, but pleaders were anxious to say that the delay in doing the work had nothing to do with impecuniosity but perhaps with financial stringency. To most people, save perhaps a good damages lawyer, the distinction between the two categories is nonexistent. The Privy Council in a case involving damage to a householder's roof by a factory emitting corrosive pollution so as to damage it[58] permitted recovery of the cost of repair at the date the householder could raise the funds to do so where repair costs had gone up from $135 per square foot to $600 per square foot. The distaste for *Liesbosch* in Lord Slynn's Opinion given in the Privy Council was palpable, but the case was not overruled. It was distinguished on the grounds that the extra hire in *Liesbosch* was an additional claim which was too remote, whereas in the case that the Privy Council was dealing with the additional repair costs were part of a single element of recoverable loss. The combined effect of that case and *Dodd* might be said to have eliminated *Liesbosch* as a factor in construction litigation. But it would always be there as a factor to be considered commercially in any negotiation. However the House of Lords in a case dealing with car hire by an impecunious victim of a collision decided unanimously that *Liesbosch* should be departed from.[59] The case was not formally overruled.[60] That might be considered as no more than judicial good manners. But the effect of the case is to ensure that *Liesbosch* arguments will **12.58**

---

[54] [1907] AC 291 at 303.

[55] For example by IN Duncan Wallace in 'Costs of Repairs: Date for Assessment' (1980) 96 LQR 341 and Coote, 'Damages, *The Liesbosch*, and Impecuniosity' [2001] CLJ 511.

[56] For example in *Cia Financiera 'Soleada' SA v Hamoor Tanker Corpn Inc* [1981] 1 WLR 274; and *Ramwade Ltd v WJ Emson & Co Ltd* [1987] RTR 72.

[57] [1980] 1 WLR 433, [1980] 1 All ER 928.

[58] *Alcoa Minerals of Jamaica Inc v Broderick* [2002] 1 AC 371.

[59] *Lagden v O Connor* [2004] 1 AC 1067.

[60] See Lord Hope at 1088, para 62 that the rule should now be departed from; Lord Walker at 1099, para 102 that Lord Wright's distinction between mitigation of damage and measure of damage was unhelpful; and Lord Nichols at 1073, para 8 agreeing with Lords Walker and Hope that the remarks of Lord Wright must be

be ineffective. A contract breaker or tortfeasor cannot now allege that an impecunious claimant cannot recover the increased cost of repair as a result of inflation between the time that the damage was done until the time that the defects or damage were actually repaired, unless the delay can be shown to be unreasonable. Victims of a tortfeasor or a contract breaker must mitigate their damage, however. Nothing in any of the cases casts doubt on that rule. But if defects to a structure result from a breach of contract, the contract breaker cannot plead that reasonable delay in carrying out remedial work which increases the cost of repair causes the resulting increased cost to be disallowed.

New for Old

12.59 When defects to a structure are repaired, it is inevitable that the repaired structure will not be exactly as the old one was. It might be better in the sense that it is more durable or more aesthetically pleasing than the unrepaired structure. The question arises as to what 'betterment' must be credited to the damages recoverable. The law has evolved to the following position:

(a) If a structure is repaired or replaced as a result of a defect brought about by a contract breaker or tortfeasor, then the prima facie position is that the innocent party is entitled to replacement or repair costs.

(b) If, however, the repair or replacement is to a higher standard than strictly necessary to deal with the breach of duty, then the excess cost beyond that which is necessary to deal with the breach of duty will not be recoverable.

(c) If, however, necessary repair or replacement means that the innocent party gets back a better structure than he or she had in the first place, then no credit should be given for the improvement because otherwise the innocent party would be forced to spend money on improvement which he or she should not be required to do.

12.60 These principles were spelt out by the Court of Appeal in *Harbutt's 'Plasticine' Ltd v Wayne Tank and Pump Co Ltd*.[61] In that case a factory, which was in an old mill, was destroyed by fire as a result of defects in the design of equipment supplied by the defendants. A new factory had to be built by the claimants if they were to continue their business of making plasticine because they were not allowed to rebuild the old mill. The defendants said that their recovery should be limited to the difference in the value of the old mill before and after the fire and that the claimants should not be allowed the cost of replacing it with a new building. This argument was rejected. Lord Denning MR said:[62]

> If a second-hand car is destroyed, the owner only gets its value; because he can go into the market and get another second-hand car to replace it. He cannot charge the other party with the cost of replacing it with a new car. But when this mill was destroyed, the plasticine company had no choice. They were bound to replace it as soon as they could, not only to keep

now regarded as overtaken by events agreed with by Lord Slynn (author of the judgment of the Privy Council in *Alcoa*) at 1073, para 10.

[61] [1970] 1 QB 447 and approved by Lord Hope in the House of Lords in *Lagden v O Connor*, ibid.

[62] At 468.

> their business going, but also to mitigate the loss of profit (for which they would be able to charge to defendants). They replaced it in the only possible way, without adding any extras. I think they should be allowed the cost of replacement. True it is that they got new for old; but I do not think the wrongdoer can diminish the claim on that account. If they had added extra accommodation or made extra improvements, they would have to give credit. But that is not this case.

Widgery LJ said:[63]

> It was clear in the present case that it was reasonable for the plaintiffs to rebuild their factory, because there was no other way in which they could carry on their business and retain their labour force. The plaintiffs rebuilt their factory to a substantially different design, and if this had involved expenditure beyond the cost of replacing the old, the difference might not have been recoverable, but there is no suggestion of this here.
>
> Nor do I accept that the plaintiffs must give credit under the heading of 'betterment' for the fact that their new factory is modern in design and materials. To do so would be the equivalent of forcing the plaintiffs to invest their money in the modernizing of their plant which might be highly inconvenient for them.

Cross LJ said:[64]

> I can well understand that if the plaintiffs in rebuilding the factory with a different and more convenient lay-out had spent more money than they would have spent had they rebuilt it according to the old plan, the defendants would have been entitled to claim that the excess should be deducted in calculating the damages.

New for old therefore may have to be paid for by the contract breaker. But this is no more than another aspect of the rule that you must take your victim as you find him. More recently this doctrine has been applied by the High Court in *Glen Haysman v Mrs Rogers Films Ltd* [2008] EWHC 2494 (QBD). In that case a homeowner's driveway was damaged by a film crew on a location shoot. The film company admitted a breach of the agreement with the homeowner in which the film company was allowed to bring machinery onto the property and film there. The film company pleaded that the recoverable damages should have been discounted to reflect an improvement or betterment of the driveway once it was re-instated.[65] The Deputy Judge applied *Harbutt's Plasticine* and accepted the claimant's submission that, to the extent that there might have been some improvement in the appearance of the driveway, this was an inevitable consequence of the repairs and therefore a new for old deduction was not appropriate. **12.61**

Where conventional or identifiable financial damage cannot be proved

Sometimes the court will consider that the compensation which is the claimant's due cannot be measured (or cannot be measured solely) by reference to identifiable financial loss. In such cases it will nevertheless award damages. This is comparatively rare in construction cases since financial loss is normally easily ascertainable. But the best example may involve trespass by a builder on an adjoining owner's land as did Brightman J in *Wrotham Park* **12.62**

[63] At 473.
[64] At 476.
[65] See para 15 of the judgment.

Estate Co Ltd v *Parkside Homes Ltd* [1974] 1 WLR 798. That case is an example only. Others may be the use of an architect's design or an extinguishment of a right to light.[66] The principle is applicable in all cases and has been applied many times since.[67] Such damages remain compensatory in character[68] but the law is in a state of development on this topic and it is impossible to say that a restitutionary remedy will not be granted in the right case.[69]

[66] *Forsyth-Grant v Allen & Anor* [2008] EWCA Civ 505, [2008] Env LR 41.

[67] *Attorney General v Blake* [2001] 1 AC 268; *Experience Hendrix LLC v PPX Enterprises Inc* [2003] EWCA Civ 323; and *WWF-World Wide Fund for Nature & Anor v World Wrestling Federation Entertainment Inc* [2007] EWCA Civ 286.

[68] *WWF-World Wide Fund for Nature & Anor v World Wrestling Federation Entertainment Inc* [2007] EWCA Civ 286; *Forsyth-Grant v Allen & Anor* [2008] EWCA Civ 505, [2008] Env LR 41.

[69] See particularly Toulson LJ in *Forsyth-Grant* ibid.

13

DETERMINATION OR FORFEITURE CLAUSES

Types of Forfeiture Clause

Forfeiture clauses, discussed briefly in Chapter 2, are designed to provide a contractually regulated ending to the obligations of the parties under the contract. A forfeiture clause may allow determination for cause or determination without cause. **13.01**

Determination for Cause

Determination for cause will involve the default of one of the parties. That default need not necessarily involve a breach of contract but usually will involve a course of conduct or a position which will, if it continues, result in a breach of contract. It is not, for example, a breach of contract for a contractor or an employer to go into administration or receivership.[1] Neither is it a breach of contract for a contractor to proceed with the works with the lack of regularity and diligence which would enable it to complete the works to time. Both, however, are events which commonly permit the other party to determine the contractor's employment 'for cause'. **13.02**

The clause may make the giving of notice a condition precedent to determination. Most clauses permit[2] or require[3] an automatic determination in the event of administration, receivership, or other similar event. However other defaults, such as a failure to make payment on the part of the employer or a failure to proceed regularly and diligently on the part **13.03**

[1] As for example insolvency. See *Perar v General Surety and Guarantee Co* (1994) 66 BLR 72.
[2] See for example cl 8.5 of the JCT Standard Building Contract with Approximate Quantities 2005.
[3] See for example the JCT SFBC 1998.

of the contractor, do not normally entitle a party to determine unless a valid notice of default is given.

13.04 Sometimes the notice of default will precede by a number of days an indefeasible right to determine. For example, clause 64 of the ICE conditions provides that if the engineer certifies that the contractor has committed one of five defaults then, upon seven days' notice being given in writing, the employer can determine the contract. Other forms require a notice to be given and default to be persisted with or repeated. For example, clause 8.4 of the JCT Intermediate Building Contract 2005 permits the employer to terminate the contractor's employment if, following the issue by the architect/contract administrator of a notice specifying a relevant default, the contractor continues the specified default for a period of 14 days. The employer may also terminate if the contractor does not continue for a period of 14 days but later repeats the default specified in the notice.[4] Thus under the JCT Form, the first notice given under the clause acts as a warning to the contractor or employer which, if heeded, avoids the determination.

13.05 The justifications for determination of the contractor's employment by the employer are usually:[5]

(a) Suspension of the works by the contractor.
(b) A failure to proceed 'regularly and diligently' with the works.
(c) Failing to remedy defective work after a notice requiring removal of that work.
(d) Insolvency or the appointment of a Receiver.

13.06 The justifications for determination on the part of the contractor are usually related to non-payment of certificates and sometimes interference with the certification process.[6]

Determination without Cause

13.07 Contracts sometimes give the employer the right to determine without the fault of the contractor. In such circumstances, the contract will usually provide for the contractor to be paid for work done and may also prohibit the employer from re-letting the contract works. An example is condition 15.5 of the FIDIC Form for works designed by the employer,[7] which provides that the employer may determine the contract at any time by giving notice 28 days before the determination is to take effect. Condition 15.5 requires the contractor to be paid for work carried out and for any costs relating to the aborted works. It may not be relied upon if the employer is to execute the works himself or to arrange for their execution by another contractor.

[4] For a helpful analysis of the operation of such clauses see *Reinwood Ltd v L Brown & Sons (No 2)* (2008) 121 Con LR 1.

[5] See, for example, the JCT Standard Building Contract with Approximate Quantities 2005, cl 8.4.

[6] See, for example, the JCT Intermediate Building Contract 2005 Edition, cl 8.9

[7] FIDIC General Conditions for Building and Engineering Works Designed by the Employer 1999.

Strict Compliance or *Contra Proferentem* Rule

All such clauses are read against the party relying upon them, or strictly. This is often a proposition cited when the determining party has served notice to determine out of time; namely, either too late or too soon. The authority relating to strict construction is old.[8] Whilst the principle appears redundant in the modern commercial context, it is still good law.[9] **13.08**

The determination provisions, and particularly those relating to timetable, must be followed to the letter, otherwise the innocent party risks becoming the guilty one through inadvertently repudiating the contract themselves. This is because if a determination is not properly executed in accordance with the contract, logically it must be a repudiation of the contract,[10] unless it can itself be held to be an acceptance of a repudiation of the contract by the other party. The difficulty with that approach, in practice, may be that the other side's defaults do not themselves amount to a repudiation and therefore there is no repudiation to accept. **13.09**

This will lead to injustice in many cases where insufficient care is taken. The most frequent mistake which is made is of failing to count the number of days between the first notice (of warning) and the second notice (of determination) and as a result delivering the letter which is designed to determine the contract a day early or a day late. A notice which is given one day early or a notice which is one day too late is not a notice which is given in accordance with the terms of the contract. **13.10**

Often determination provisions will state that the notice should be given by registered post or recorded delivery. There is authority to the effect that this requirement is not essential and that so long as the notice has been received at the appropriate moment by the party receiving it then that is sufficient.[11] Registration and recording of post does often ensure that the determining party knows that its notice has been received and it may also dispose of any arguments which may be raised as to the timing of the receipt of the notice, which has long been a source of potential dispute. **13.11**

Given that the consequence of a wrongful determination is that the party which is in default may manage to turn the legal tables and allege that the other party has in fact repudiated the contract, it is all the more necessary that care is taken to ensure that the second notice is given at the correct time and that all notices are in proper form. If a defaulting contractor is able to allege that its employer has repudiated the contract, then there is the risk that it can recover a quantum meruit for its work. This is a viable risk, given that one **13.12**

[8] *Roberts v Bury Commissioners* (1870) LR 5 CP 310 at 326.

[9] For a recent case where the principle was applied, see *Robin Ellis v Vinexsa International* [2003] BLR 373, 93 Con LR 92.

[10] For an example of a court seeking to get away from such a formalistic approach, see *Canterbury Pipe Lines v Christchurch Drainage* (1979) 16 BLR 76.

[11] *Goodwin & Sons v Fawcett* (1965) 195 EG 27; and *JM Hill v London Borough of Camden* (1980) 18 BLR 35.

reason why contractors may not be proceeding diligently would be that the tender rates were insufficient.

13.13 One possible, and frequently used, way round the problems of an over-technical forfeiture clause, in cases where the breach specified by the clause also constitutes a repudiatory breach, is to seek to accept the contractor's repudiation of the contract at the same time as operating the forfeiture clause. This has a great deal to recommend it in terms of a belt-and-braces exercise. But it has the conceptual difficulty that it involves the determining party seeking to affirm the contract (by operating its machinery) whilst at the same time seeking to put that very contract to an end. It is unlikely that such a course would succeed and indeed it has been held not to be possible.[12]

13.14 It might be thought that this discussion is academic since all determinors can be careful to ensure that the contractual machinery is properly operated. Therefore, if the time passes when the second and determining notice can be given, all that is needed to be done is to serve a first notice again and restart the process. But this is not always so easy to do as it involves delaying the process and delay may be an expensive option. The only clear solution for a contractor or employer is, at the time a determination is thought about as a serious option, that they should with their legal advisors plot exactly the form, substance, and timing of all notices which have to be given.

Repudiation

13.15 An acceptance of a repudiation is an altogether different process to a determination. It involves an assessment that the other party has repudiated its contract (or fundamentally broken it) and then deciding to accept the repudiation. Not all breaches of contract are repudiations. The law divides breaches of contract into three types:

(a) breach of a condition of the contract, which is a serious breach entitling the innocent party to accept the resulting repudiation and bring the contract to an end;

(b) breach of a warranty, being a breach of a term which only entitles the innocent party to a claim in damages; and

(c) breach of an 'innominate term', being a term which if breached may involve a repudiation or may not depending upon the seriousness of the breach itself.[13]

13.16 Poor workmanship is the most frequent example of a contractor's repudiation. But the breach must be serious. However an accumulation of relatively small breaches may be sufficiently serious to justify calling them, taken together, repudiatory.[14] In the case of an

---

[12] By the Court of Appeal in Tanzania in *Mvita Construction Co v Tanzania Harbours* (1988) 46 BLR 19. But note the discussion in *Stocznia Gdynia SA v Gearbulk Holdings Ltd* [2009] EWCA Civ 75 particularly at paras 43–45. See also *Stocznia Gdanska SA v Latvian Shipping Co and Others* [2001] 1 Lloyd's Rep 537, [2002] 2 Lloyd's Rep 436, CA; and *Dalkia Utilities Services plc v Celtech International Ltd* [2006] 1 Lloyd's Rep 599.

[13] *Hong Kong Fir v Kawasaki Kisen Kaisha* [1961] 2 QB 26.

[14] See *Sutcliffe v Chippendale & Edmondson* (1971) 18 BLR 157 at 161 where Judge Stabb QC said:

> . . . the contractors' manifest inability to comply with the completion date requirements, the nature and number of complaints from sub-contractors and [the architect's] own admission that in

employer which wishes to get rid of a contractor which is going too slowly, the assessment of whether there has been a repudiation requires some care. The obligation to complete to time is not a condition of the contract unless it is made to be so by a notice being given to make time of the essence.[15] Therefore before the employer can accept what is perceived as a repudiation, it should serve the notice which makes time of the essence. The notice ought to be reasonable and give the contractor a chance to perform the contract properly.

The contractor's options when the employer is in default of its obligations to make payment are greater than before. Usually the forfeiture clause will permit a determination if payment is not made within a certain period. **13.17**

A failure to make payment is also a breach of an 'innominate term', namely a term of the contract the breach of which may be fundamental or may not be depending on the circumstances. In addition to the right to determine and the right to accept a repudiation, the contractor also has the express power under s 112 of the Housing Grants, Construction and Regeneration Act 1996 to suspend work. That right, provided by statute is of great practical value. It used to be the case that non-payment by the employer would give the contractor no right to suspend working. Therefore it was bound to carry on with the work even though it knew that there was a real risk that the employer would be unable to pay for the work. It gave the contractor a real dilemma. If it suspended work, it ran the risk that the nearly insolvent employer would seek to determine the contractor's employment for failure **13.18**

> May and June the quality of work was deteriorating and the number of defects was multiplying, many of which he had tried unsuccessfully to have put right, all point to the truth of the complainant's expressed view that the contractors had neither the ability, competence or the will by this time to complete the work in the manner required by the contract...

See also Hale LJ's comments in n 17 below.

[15] See *United Scientific Holdings v Burnley Council* [1978] AC 904 at 937 *per* Viscount Dilhorne:

In *Parkin v Thorold* (1852) 16 Beav 59, 65 Sir John Romilly MR said that under the doctrine of a court of equity:

> time is held to be of the essence of the contract... only in cases of direct stipulation, or of necessary implication. The cases of direct stipulation are, where the parties to the contract introduce a clause expressly stating, that time is to be of the essence of the contract. The implication that time was of the essence of the contract is derived from the circumstances of the case... It is needless to refer to the authorities, which are numerous, to support these propositions.

and *per* Lord Salmon at 958:

> the law is correctly summarized in the following passage from *Halsbury's Laws of England*, 4th ed, vol 9, para 481: The modern law, in the case of contracts of all types, may be summarized as follows. Time will not be considered to be of the essence unless: (1) the parties expressly stipulate that conditions as to time must be strictly complied with; or (2) the nature of the subject matter of the contract or the surrounding circumstances show that time should be considered to be of the essence; or (3) a party who has been subject to unreasonable delay gives notice to the party in default making time of the essence.

But see May LJ in *Shawton Engineering Ltd v DGP International Ltd (t/a Design Group Partnership) & Anor* [2005] EWCA Civ 1359 at para 76 where he said:

> I accept that, even if time is not of the essence, it is theoretically possible for a party to show that another party's delay is so profound as to be repudiatory. But what has to be shown is, not mere breach, but a breach of such gravity as to deprive the other party of substantially the whole benefit which it was the intention of the parties that they should obtain from the contract.

See also Denning LJ in *Rickards v Oppenheim* [1950] 1 KB 616 at 624.

to proceed regularly and diligently. But if it continued working, then it ran the risk that the insolvent employer would not pay for that work as well as the work for which it was already in default. Now that there is the express right under statute to suspend work, that dilemma is much reduced and perhaps for most practical purposes, eliminated. In addition there is the right to adjudication given by the 1996 Act. It must therefore be doubted that a simple and single refusal to pay, even a large sum of money, will amount to a repudiation of the contract by the employer. However it has been held that persistent late or non payment of a cynical kind will amount to a repudiation.[16] Persistent late or non payment can be looked at cumulatively with a view to the court saying that unacceptable past performance is an indicator of future unacceptable performance entitling the innocent party to treat itself as discharged.[17] The time when the contractor had the dilemmas whether to continue is now past. A contractor may suspend work and then adjudicate and thereby obtain relief. There are virtually no recorded cases where non-payment has, of itself, been held to amount to a repudiation of the contract by the employer. This is partly because most standard forms of contract make express provision for the remedy of determination in the event of non-payment and therefore the pressure on the contractor to accept a repudiation is absent. More often the problems came not from non-payment but from interference by the employer in the certification process and under-certification. The effects of this are also mitigated by the provisions permitting adjudication. Therefore, cases which suggest that interference with the certification process constitutes repudiation by the employer must be treated with caution.

Remedies in the Event of Forfeiture

13.19 Once there is a determination pursuant to the express terms of the contract, the rights and remedies of the parties are those which are expressly set out in the contract. Usually the clauses provide, in the case of a determination for cause, that the innocent party can recover such sums as it would normally recover in the event of a repudiation. But it is not always so. Some determination clauses are limited in the remedy they give to the determining party. The remedy given by the determination clause is complete in the sense that you can recover what the clause allows you to but no more and no less.[18]

---

16 *Alan Auld Associates Ltd v Rick Pollard Associates & Anor* [2008] EWCA Civ 655.

17 See *Rice (T/A The Garden Guardian) v Great Yarmouth Borough Council* [2000] All ER (D) 902 Hale LJ at paras 36 and 38:

> The question for the court (and indeed the contracting parties) in a case like this is whether the cumulative effect of the breaches of contract complained of is so serious as to justify the innocent party bringing the contract to a premature end. . . . These contracts are like building contracts in that the accumulation of past breaches is relevant, not only for its own sake, but also for what it shows about the future. In my view, the judge was right to ask himself whether the cumulative breaches were such as to justify an inference that the contractor would continue to deliver a substandard performance.

18 *Thomas Feather v Keighley Corp* (1953) 52 LGR 30. But see cases in fn 12 above.

A forfeiture clause may give the employer the right, following determination, to withhold interim payments which are due to the contractor but have not yet been paid.[19] Such a clause has been held by the House of Lords to be compatible with the provisions of Part II of the Housing Grants, Construction and Regeneration Act 1996.[20] Where the clause states the right may only be exercised in respect of interim payments which have been 'reasonably' withheld, the question of whether it was reasonable to withhold will usually turn on whether there was a valid withholding notice.[21] **13.20**

It is therefore essential for the determining party to consider the recovery provision of the forfeiture clause before deciding whether to go down the route of using it. It may be that the better practical course is to accept a repudiation, assuming that it is open to the determining party to do so. For a discussion of the recovery in the case when repudiation is accepted, see Chapter 12. **13.21**

Liens, Bills of Sale, and Vesting Clauses in Relation to Plant and Machinery; Fixed or Floating Charges and Registration

Under the ICE and FIDIC conditions of contract, a determination for cause by the employer gives it the right to use the contractor's plant and machinery which is on the site for the purposes of completing the works. That is a contractual right which is untrammelled. It gives the employer the equivalent of a lien on the plant and machinery.[22] A lien is a right to retain possession of the goods of another until claims are satisfied. It depends upon possession and lasts only as long as possession is retained. **13.22**

Forfeiture clauses which give the employer the right to sell the contractor's plant and apply the proceeds towards the discharge of the contractor's liabilities to the employer may give rise to the creation of a floating charge over that plant and machinery. If it should do so, then the charge should be registered, otherwise the charge is void against any liquidator, receiver, or administrator and the right to sell falls away.[23] The essence of the status of a charge, like a mortgage on a house, is that the plant is appropriated to the discharge of the debt or other obligation of the contractor and any excess upon sale is repayable payable to the contractor. **13.23**

Such rights may also create a Bill of Sale under the Bills of Sale Acts.[24] There is old Court of Appeal authority to the effect that the Bills of Sale Acts do not apply in engineering contracts. But that authority is of doubtful weight. Such a Bill of Sale, if indeed one is created **13.24**

[19] See, for example, cl 27.6.5.1 of the JCT Standard Form of Building Contract with Contractor's Design (1998).

[20] *Melville Dundas v George Wimpey UK* [2007] UKHL 18, 112 Con LR 1.

[21] *Pierce Design v Johnston* (2007) 115 Con LR 110.

[22] *Cosslett v Bridgend* [1998] Ch 495.

[23] This has been held to be the case under the ICE conditions, Fifth Edition. See *Cosslett v Bridgend* [1998] Ch 495.

[24] Expressly left open by Lord Scott in the House of Lords in *Smith v Bridgend* [2002] 1 AC 336 at para 64. There is 19th century Court of Appeal authority to the effect that the Bills of Sale Act 1878 and the Bills of Sale Act (1878) Amendment Act 1882 do not apply to construction contracts and forfeiture clauses, but there is no reason in principle why they should not do so. However as in *Smith* the point is likely to be academic.

by the clause, must also be registered to be effective against the liquidator, receiver, or administrator. The ICE conditions have been modified in the seventh edition to take away the element of charge or sale in the forfeiture clause; however, such provisions still exist in the FIDIC Forms and often exist in bespoke contracts.

Penalties

13.25 For a discussion of the modern law in relation to penalties, see Chapter 11. Any provision in a forfeiture clause which provides for the payment of a sum which is more than the employer is going to suffer by reason of the determination is a penalty and unenforceable.[25] Forfeiture clauses which make such a provision are rare and only appear in bespoke contracts. Such penalties are easy to spot because they will plainly not be a genuine pre-estimate of the likely loss to be suffered by the employer in the event of the default of the contractor. Liquidated damages and other such matters are irrelevant to forfeiture clauses. In the event of such a clause being a penalty, the courts will permit the employer to recover its actual loss.

Relief against Forfeiture

13.26 This is a concept appropriate to leases. The court has power in its equitable jurisdiction to prevent the determination of a lease and grant relief against its forfeiture. The law relating to such relief is outside the scope of this book.

13.27 It may be suggested that the courts have power to grant relief against the forfeiture of a construction contract. The closest that the courts came to this was in *Hounslow v Twickenham Garden Development*[26] where an injunction was refused to an employer seeking to enter its site. That decision of Megarry J has rightly come in for a great deal of academic criticism. A construction contract does not grant any interest in land to the contractor, which only has a licence to occupy the site for the purposes of the construction contract. It must follow that the law relating to relief against the forfeiture of parties which have rights in land is wholly inapposite. Furthermore a construction contract is a commercial asset of a contractor and damages are an adequate remedy for the contractor should the employer wrongfully determine the contractor's employment.[27] It may be that the term 'forfeiture clause' misleads practitioners who are used to dealing with leases into thinking that the two are similar. Whatever the position, it is noteworthy that the *Hounslow* case has not been followed in the United Kingdom. One reason may be that contractors prefer to sue for damages when they have had their contracts wrongly determined rather than take the strangely uncommercial decision to require the employer to continue to employ them through the expedient of refusing them access to their own property.

---

[25] *Ranger v Great Western Railway* (1854) 5 HLC 72.

[26] [1971] Ch 233.

[27] *Ericsson AB v EADS Defence and Security Systems Ltd* [2009] EWHC 2598 (TCC).

Part V

CLAIMS

14

LOSS AND EXPENSE

The Right to Order Variations; the Distinction between Variations under the Contract and a Variation of the Contract

14.01 Most standard forms and bespoke forms of contract provide a mechanism whereby the employer can order changes to the work to be done. This is usually defined as a variation to the quantity or quality of the work to be done, by way of addition, omission, or variation.

14.02 The power to order varied work, once enshrined in the contract, becomes a right to do so. This right is usually used frequently by employers who have often not had all their requirements sorted out at the time of the letting of the contract. It is a rare contract which proceeds from commencement to completion without the employer invoking its right to order variations to the work.

14.03 In design and build contracts, where the contractor is responsible for the design of the works, the scope for variations should be less than in a traditional form. Instead of being termed a variation, a change under a design and build contract is called a 'Change to the

Employer's Requirements'. This is because in contradistinction to a traditional contract where the employer may change small matters of design, under a design and build the change will usually be of a more far-reaching nature.

14.04 A variation ordered under the terms of a building or engineering contract does not change or vary the contract itself. That is because the contract gives the employer the power to order variations and therefore the instruction to vary the work is an instruction issued under and envisaged by the contract. In taking on the contract, the contractor will have taken on the obligation to execute varied work.

14.05 Without a variation clause empowering the employer to instruct changes to the works, the employer would indeed have to negotiate a fresh contract with the contractor before a change could be made. This is because in that event, the contractor would have the right as well as the duty to execute the works in accordance with the contract documents.

14.06 There are two limitations upon the employer's right to vary the work pursuant to a normal variations clause. The first is that the change ordered cannot alter the essential characteristic of the contract itself.[1] The second is that the employer may not order the omission of work simply in order to instruct another contractor to do the same or almost the same work. Each must be looked at.

14.07 So far as the first is concerned, it is a matter of fact and degree as to whether the instruction goes beyond a change which is envisaged by the contract and becomes asking for something different to that which was expected and contracted for. That said, it is rarely argued by contractors that the variation instructed takes the work outside what the employer is entitled to ask for under the variations clause. As with all matters of fact and degree it is easier to demonstrate at the extremes than it is at the margin. So, for example, a contract to build a beach house will have implicit within the variation clause a right to change the colour of the external paintwork. It will not have the right to order that the contractor build a lighthouse alongside. Matters which may be at the margin on the wrong side of the line would be an instruction to build another house next door. On the right side of the line would be an instruction to build another storey to the house. It is impossible to be more than illustrative in defining what can and cannot be done. Each case will have to be looked at on its own facts. But the essential question for a judge or arbitrator to ask in deciding whether the variation was within the power of the employer to order is: 'is this change the kind of change which the parties reasonably expected could be ordered at the time the contract was entered into?'

14.08 The second limitation is easier to define and understand. Implicit in a building contract is the right of the contractor to carry out and complete the whole of the work. That means that the employer cannot, without agreement, abridge that right save pursuant to the variations clause of the contract. A variation which involves a change in what the employer wants built is a valid change. So, for example, if there is a contract to build a housing estate of fifty houses and the employer decides that it only wants forty-five built, then it is within the employer's power to vary the work. But if the employer wishes fifty houses to be built

[1] See the discussion in Ch 3 concerning quantum meruit, para 3.01.

but has been able to find another contractor to build the five houses at a cheaper rate than the contractor has agreed to build them for, then there is no true change in what the employer wants. In that event the contractor has two potential remedies. The first is to seek to recover as damages for breach of contract the losses it has suffered by reason of the omission of the five houses. The second remedy, which may not be very commercial, is to seek a declaration from a court or arbitrator that it has the right to construct the five houses and that the variation instruction is null and void and not issued under the contract conditions. In most cases the first remedy will be the one sought since all the contractor will lose from the non-construction of the five is profit. But that may not necessarily be so if the project is prestigious and the emblem of building the whole estate is of great value (albeit partly non-economic).

Contractor's Claims under the Contract; The Most Common Types of Claim

Mostly claims for sums due under the contract are dealt with non-contentiously. They will involve claims such as: **14.09**

(a) claims for increased costs under the fluctuations clause of the contract;
(b) claims for the valuation of additional or varied work;
(c) claims for an extension of time for the completion of the work;
(d) claims for loss and expense arising out of late instructions or variations or pursuant to the power granted under some other clause of the contract which gives rise to a right to payment of loss and expense; and
(e) claims for the issue of a certificate of practical completion or a final certificate.

Claims for increased costs are usually mechanical in nature. The contractual formulae are used to ascertain what additional payments are due to the contractor. These payments are a function of the increases in the cost of labour, plant, and materials which are measured contractually by reference to formulae, often the Baxter index. Little more need be said, except to note that the right to receive these payments arises irrespective of the pace at which the work is done. It is not open to the employer, where there is a provision for the payment of liquidated damages for delay, to plead that the work should have been completed by the time the price indices increased and that accordingly no extra payment should be made. Since liquidated damages are a complete remedy for the delay to the completion of the works, those damages include the additional cost to the employer payable to the contractor by reason of the increased costs which the employer must pay pursuant to the fluctuations clause.[2] **14.10**

Claims for additional and varied work are probably where the most common contentious claims are made. The contract will usually have a formula for the valuation of this work. The best known formula is that which states that work of a similar character and carried out under similar conditions to ordinary contract work shall be valued at the rates and prices **14.11**

[2] *Peak Construction (Liverpool) Ltd v McKinney Foundations Ltd* (1970) 1 BLR 111, (1969) 69 LGR 1.

stipulated in the contract. That presupposes an almost mechanical application of rates and prices to the work done. So, for example, if a fifty metre wall is extended by twenty-five metres, then one would expect the additional meterage to be valued pro rata by reference to the contract rates. Such valuations are normal and non-contentious. However, the problems occur when the work is not of a similar character or carried out under similar conditions as the contract work and also when the additional work causes delay and disruption to the contractor. So far as the delay and disruption is concerned, that is considered under the heading of loss and expense. So far as the work being of dissimilar character is concerned, that is normally prescribed as being recoverable pursuant to a fair valuation. If it is not recoverable as a fair valuation, then it may be recoverable as loss and expense.

14.12 Claims for the issue of certificates of practical completion and final certificates are usually concerned with the time that liquidated damages are supposed to stop (practical completion) and the final ascertainment of the contract sum (the final certificate). Works are practically complete when they are fit for normal use, not when the works are completed and all defects are remedied.[3] The defects liability period is the time when small defects, which do not prevent beneficial occupation of the structure, are remedied. It is a matter of judgment by the contract professional as to when practical completion has taken place since many different people may have justifiable but differing views as to when that has taken place. The major source of claim relates to loss and expense and extensions of time, which is looked at next.

Loss and Expense and Extensions of Time

14.13 Loss and expense claims are the most common type of contractor's claims which are made in construction litigation, arbitration, or adjudication, aside from the related question of extensions of time.

14.14 Since they are related, both are considered together. They often bring up the same questions which appear in cases of damages for breach of contract. However they make up a separate topic.

14.15 These claims are contractual and because of their special nature are dealt with separately from the topic of damages claims arising from breaches of contract, statutory duty, and tort.

---

[3] *Jarvis v Westminster Corporation* [1969] 1 WLR 1448, at 1458 D to F: *per* Salmon LJ

> . . . the words 'practically completed' and 'practical completion' are used in clauses 15 and 16 of the contract. I take these words to mean completion for all practical purposes, that is to say, for the purpose of allowing the employers to take possession of the works and use them as intended. If completion in clause 21 meant completion down to the last detail, however trivial and unimportant, then clause 22 would be a penalty clause and as such unenforceable. . . . .

Jarvis v Westminster Corporation [1970] 1 WLR 637 at 646 E to F *per* Viscount Dilhorne:

> One would normally say that a task is practically completed when it was almost but not entirely finished, but 'practical completion' suggests that that is not the intended meaning and that what is meant is the completion of all construction work that has to be done.

Applied by Warren J in *Menolly Investments 3 Sarl v Cerep Sarl (Rev 1)* [2009] EWHC 516 (Ch), (2009) 125 Con LR 75 paras 61–66.

It may be thought surprising that there is very little case law dealing with these topics and therefore there is little judicial guidance given as to how such claims are proved and how loss is calculated. **14.16**

The classic contractual route for the claiming of loss and expense by contractors is pursuant to express terms of the contracts issued by the JCT, and formerly by RIBA. **14.17**

In order to provide a framework for discussion, the JCT model principally will be used in this Chapter. Other forms will be referred to very briefly. However the JCT Form is the paradigm form whereby classical loss and expense claims are made in the industry and many other forms of contract will either follow the same scheme in outline if not in detail or the principles applicable to this discussion will be, with adaptations, relevant to them. **14.18**

Under the JCT Form the contractor is entitled to claim an extension of time where specific events obtain. These can be divided into the following categories: **14.19**

(a) Events which are the result of the acts or omissions of the employer. The provision of late information to the contractor, for example, or the issue of instructions requiring the variation of the works to be done as another example. These might be described as employer induced events.

(b) Events which are the result of matters which neither the employer nor the contractor could reasonably be expected to be responsible for without express contractual allocation of risk. Examples of this would be force majeure, terrorism, and events which are normally the subject of insurance.

(c) Events which might ordinarily be thought to be contractor's risk items but which the contract provides shall, so far as the time for completion is concerned, not be at the contractor's risk. Examples of this are: (i) failure to obtain materials or labour owing to shortage; (ii) strikes or lock outs; (iii) delay by a nominated sub-contractor.[4]

The grant of an extension of time has one direct effect, namely the avoidance on the part of the contractor of the obligation to pay liquidated and ascertained damages for the period of delay caused by the event concerned. The other effect is that it relieves the contractor of the obligation to complete by the original date for completion and creates a fresh completion obligation to complete at a later date. **14.20**

The above-mentioned types of event allocate risk to either the employer or the contractor so far as time is concerned. They do not directly affect loss and expense, but do so indirectly. **14.21**

So far as loss and expense is concerned there is only one category of event which gives rise to a claim and that is the type of event which is in the employer's sphere. The most common of these are the provision of late instructions or the issue of variations by the employer's agent, the architect. **14.22**

The clauses provide that upon a timeous notice being given by the contractor, of the event concerned the employer's agent shall ascertain and certify for payment the loss and expense which is incurred by the contractor. **14.23**

[4] Though not in the 2005 edition.

14.24 The recoverable losses which a contractor is likely to suffer are:

(a) Prolongation costs; the costs of on-and-off site overheads and the increased costs to the contractor resulting from cost inflation, commonly known as fluctuations.
(b) Disruption costs; the costs of inefficient working and the increased cost to the contractor of executing the work at a different time or in a different sequence to that anticipated (except cost inflation or fluctuation costs which come under category (a)).
(c) Finance charges incurred (including compound interest) as a result of the event giving rise to the claim.[5]
(d) Additional management time expended in dealing with the relevant event.[6]
(e) Damages or costs payable to third parties.

14.25 Logically it is appropriate to consider the principles governing the grant of an extension of time before coming to consider loss and expense, which must ordinarily be considered in the context of the period for completion of the works.

The Principles Which Govern the Grant of an Extension of Time

14.26 Upon being given notice by the contractor of a delaying event (which is to be given whenever it becomes reasonably apparent that the progress of the works is being delayed or is likely to be delayed) and particulars thereof, the employer's agent must then form an opinion as to whether an event has occurred which gives the contractor the right to an extension of time and completion of the works is likely to be delayed beyond the completion date by it.

14.27 If the employer's agent forms that opinion, he or she must give an extension of time 'as he then estimates to be fair and reasonable'.

14.28 The clause provides for further opinions to be formed and further extensions of time to be granted and it also provides for the fixing of an earlier completion date in certain circumstances, but the essential steps remain the same namely: (a) a notice is given by the contractor of an event which has caused delay; (b) an opinion is formed by the employer's agent; and (c) an extension of time which is fair and reasonable is granted.

14.29 There is a proviso that the contractor 'shall constantly use his best endeavours to prevent delay in the progress of the Works, howsoever caused, and to prevent the completion of the Works being delayed beyond the Completion Date . . .'

14.30 The essential first step however is the happening of the delaying event which means that 'the completion of the Works is likely to be delayed thereby beyond the Completion Date'.

---

[5] *FG Minter Ltd v Welsh Health Technical Services Organisation* (1980) 13 BLR 1; *Rees & Kirby Ltd v Swansea City Council* (1985) 30 BLR 1. These cases, along with *Wadsworth v Lydell* [1981] 1 WLR 598; *Department of Environment for Northern Ireland v Farrans* (1981) 19 BLR 1; and *Farrans (Construction) v Dunfermline DC* (1988) Const LJ 314 entrench finance charges as a special head of loss which is recoverable. The correct view must be that finance charges, or interest, are recoverable as sums due under a contract. Interest as damages for breach of contract may be more difficult to establish. See *President of India v La Pintada Compania Navigacion SA* [1985] AC 104; and *Lips Maritime Corp v President of India* [1988] AC 395.

[6] See para 14.96 where this is discussed.

It is the occurrence of that event which leads to the obligation to grant a fair and reasonable extension of time.

The Principles Which Govern the Certification and Payment of Loss and Expense

The certification and payment of loss and expense follows a series of specific steps: **14.31**

(a) The possession of the site or the regular progress of the works is or is likely to become affected by an event which the contract provides gives rise to an entitlement to loss and expense (an entitlement event).

(b) As soon as the entitlement event becomes apparent or should reasonably have become apparent, the contractor makes a written application to the employer's agent to that effect.

(c) The employer's agent either himself ascertains or he instructs the quantity surveyor to ascertain the loss and expense and that is certified and paid as appropriate.

There are two essential differences between the provisions as regards the grant of an extension of time and the certification and payment of loss and expense. The first is that the giving of a timeous notice is stated to be a condition precedent to the payment of loss and expense whereas it is not stated expressly to be so, so far as the grant of an extension of time is concerned. The second is that the grant of the extension of time is to be so far as is fair and reasonable, whereas loss and expense is a pure ascertainment process. **14.32**

In addition the payment of loss and expense is expressly stated to be without prejudice to any other rights and remedies which the contractor may have. Therefore the contractor can, as an alternative or in addition, make a claim for damages for breach of contract in respect of the same events. **14.33**

Common Problems in the Examination of Extensions of Time and Loss and Expense

Remarkably there is much academic and professional debate about how extensions of time and loss and expense are to be assessed but in practice there has been little case law. The Society of Construction Law has, however, produced a protocol: *The Society of Construction Law Delay and Disruption Protocol*.[7] This is designed to produce a common outlook in the construction industry dealing with delay and disruption. The Protocol has not caused controversy to cease, but it has brought into focus the problems inherent in analysis of extensions of time and loss and expense. **14.34**

Specific problems and debates which recur include: **14.35**

(a) Concurrent delay and how it is to be treated. Allied with this is the test to be applied to causation.

[7] October 2000, reprinted October 2004.

(b) The role of computer programs and experts in making assessments in arbitration and litigation.
(c) How the contractor's 'float' is to be dealt with.
(d) How are late variations to be assessed, on a net or gross basis?
(e) Global claims and how losses are to be apportioned.
(f) Head office overheads and how they are to be assessed and calculated.

14.36 Headings (a) to (d) are common to extensions of time and loss and expense. Headings (e) and (f) relate to loss and expense only.

Concurrent Delay and Causation: Extensions of Time

14.37 What happens when there are two events which cause delay at the same time, one which entitles the contractor to an extension of time and one which does not? In that event there is a substantial body of opinion which states that the contractor is entitled to an extension of time but does not receive loss and expense.[8] This seems appropriate and just in that the contractor pays no liquidated damages and neither does it receive loss and expense. Loss lies where it is.

14.38 However it is probably useful to look at examples to see how this might work out in practice and what it actually means.

14.39 For example with the construction of a simple house, where the employer is late in giving dimensional information to the contractor such that it cannot commence the construction of the external walls until January, when the work ought to have begun in the previous November. Then one can, on the face of it, see that the contractor would be entitled to at least one month's extension of time and prolongation costs of the same amount, on the basis that without the external walls no progress could have been made on the house for the period of delay. But what if, concurrently, the contractor failed to order the bricks on time so that in practice it could not have commenced the construction of the external walls until January in any event?

14.40 Is this a concurrent delay caused by a late instruction and a failure to properly procure materials? Or is a more sophisticated analysis required?

---

[8] Protocol, paras 9 and 10 of the 'Core Principles' and p 15 Guidance, section 1; para 1.4.1 and paras 1.4.6 and 1.4.7. See also *Henry Boot Construction (UK) Ltd v Malmaison Hotel (Manchester) Ltd* [1999] All ER 118, 70 Con LR 32, Dyson J at para 13 where it was agreed by Counsel:

> . . . if no work is possible on site for a week not only because of exceptionally inclement weather (a relevant event), but also because the contractor has a shortage of labour (not a relevant event), and if the failure to work during that week is likely to delay the works beyond the completion date by one week, then if he considers it fair and reasonable to do so, the architect is required to grant an extension of time of one week. He cannot refuse to do so on the grounds that the delay would have occurred in any event by reason of the shortage of labour.

Applied by Judge Stephen Davies in the TCC in Manchester in *Steria Ltd v Sigma Wireless Communications Ltd* [2007] EWHC 3454 (TCC).

So far as an extension of time is concerned, the architect may take the view that it is fair and reasonable to grant an extension of time for the late instruction. The architect has a wide discretion in relation to the grant of such extensions as are fair and reasonable. The architect may think that it would be wrong for the contractor to pay liquidated damages in respect of a delay which would have been inevitable. However if the contractor was in no position to commence the work until January, how can it be said that the late instruction caused any delay at all? Equally if the contractor could not have started the work until January because it did not have the information, how can it be said that late procurement caused delay? **14.41**

Often extension of time problems are discussed as if they were problems of concurrency when in fact the true question relates to causation in law. One can ask the twin questions here: did the late information cause delay; or did the late ordering of the brickwork cause delay? If the answer is that both caused the delay, then an extension of time will follow because it has been shown that the late information did cause delay. **14.42**

This approach would be in line with the law as declared by Devlin J in *Heskell v Continental Express*[9] relating to breaches of contract.[10] It is that if there are two concurrent causes of loss of equal efficacy and one is a breach and one is not, then there can be recovery in full for damage sustained by the breach.[11] In this example there is recovery in full for the extension of time. **14.43**

---

[9] [1950] 1 All ER 1033.

[10] See also Steyn J in *Banque Keyser SA v Skandia (UK) Insurance* [1990] QB 665 at 717, accepted by the Court of Appeal at 813A to 814C. Steyn J at 717 said:

> There is no more difficult area of our law than causation. Scientific precepts and philosophical notions are frequently invoked. Ultimately, it seems to me, a judge is on safer ground if he puts his trust in precedent, or, in its absence, in common sense. That is the approach adopted by Devlin J. in *Heskell v Continental Express Ltd* [1950] 1 All ER 1033. He said, at p 1047:
>
> > Where the wrong is a tort, it is clearly settled that the wrongdoer cannot excuse himself by pointing to another cause. It is enough that the tort should be a cause and it is unnecessary to evaluate competing causes and ascertain which of them is dominant...
>
> To the best of my knowledge this proposition has not been questioned since that decision. It is relevant, of course, to the plaintiffs' case based on a breach of the common law duty of care.

[11] See also *Wayne Tank Co v Employers Liability* [1974] QB 57 at 73; *JJ Lloyd Instruments v North Star Insurance* [1987] 1 Lloyd's Rep 32; and *Fairfield Mabey v Shell UK Ltd* (1989) 45 BLR 113, 27 Con LR 1, [1989] 1 All ER 576.

The most recent affirmation of the *Heskell* principle is by the Court of Appeal in *Holladay v East Kent Hospitals NHS Trust* [2003] EWCA Civ 1696 at [32] *per* Scott Baker LJ where he said:

> The critical question it seems to me is not what might or might not have happened in different circumstances, but whether the respondent's breach of duty caused the appellant to be arrested. I accept the appellant's argument that the respondent's breach of duty does not have to be the sole cause of the arrest, or even the dominant cause, as long as it was a material cause: see *Bonnington Castings v Wardlaw* [1956] AC 613 at 621. Mr Geoffrey Cox QC for the appellant relied on a passage in the judgment of Devlin J in *Heskell v Continental Express Limited* [1950] 1 All ER 1033, 1047C:
>
> > Where the wrong is a tort, it is clearly settled that the wrongdoer cannot excuse himself by pointing to another cause. It is enough that the tort should be a cause and it is unnecessary to evaluate competing causes and ascertain which of them is dominant: see *Minister of Pensions v Chennell* [1947] KB 250 per Denning J.
>
> Any doubt that the rule is the same in contract was dispelled by *County Limited v Girozentrale Securities* [1996] 3 All ER 834.

14.44 It has been suggested that in circumstances such as this, under the JCT Standard Form 1998 Edition, the architect could exercise his discretion and not grant the extension of time on the grounds of fairness and reasonableness.[12] It would follow from that approach, if correct, that the architect could make an apportionment between the two delaying events and grant a proportionate extension of time which reflected what the architect thought as an appropriate split of responsibility. However in the example just given, this would not be logical.

14.45 It is submitted that the appropriate formula for use in this situation is as follows: When there are two concurrent causes of delay, one which gives rise to a right to an extension of time and the other does not, then the task of the court or arbitrator, in the case of a dispute, is to ascertain the real and effective cause[13] of delay by way of factual inquiry. Where the delaying events are of equally potency, then the extension of time would follow because both events are effectively dominant together.[14] Common sense is an essential ingredient of the factual inquiry,[15] and the essence of the inquiry is to discover the reality rather than the theoretical position.[16] The fact that two delaying events took place side by side does not mean that they are of the same causative potency, one may be of no potency at all. If the

---

[12] John Marrin QC, 'Concurrent Delay', paper given to a meeting of the Society of Construction Law (5 February 2002).

[13] The term 'dominant' cause was used by Lord Dunedin in *Leyland Shipping Co Ltd* [1918] AC 350 at 363. However the term 'dominant, real or efficient' is used in *Yorkshire Dale Steamship Co v Ministry of War Transport* [1942] AC 691 at 706 (Lord Wright); *Gray v Barr* [1971] 2 QB 554 at 567 the term used is 'dominant and effective' (Lord Denning MR); *Wayne Tank and Pump Co Ltd v Employers Liability Assurance Corporation Ltd* [1974] QB 57 at 65–6 (Lord Denning MR) 'effective and dominant.' Glidewell LJ used the term 'an effective or dominant cause of the loss' in *Galoo v Bright Grahame Murray* [1994] 1 WLR 1360 at 1375G. May LJ used the term 'the effective, dominant cause of the relevant loss' in *Lisk Carew v Birmingham City Council* (CA, 23 April 2004) at para [29]. The use of 'dominant', interchangeably with 'effective', is useful when simple causation is to be analysed. However when the analysis relates to competing causes, the term 'dominant' is no longer appropriate.

[14] See, for example, *Henry Boot Construction (UK) Ltd v Malmaison Hotel (Manchester) Ltd* [1999] All ER 118; *Steria Ltd v Sigma Wireless Communications Ltd* [2007] EWHC 3454 (TCC).

[15] See *Monarch Steamship Co Ltd v Karlshamns Oljefabriker AB* [1949] AC 196 *per* Lord Wright at 228: 'Causation is a mental concept, generally based on inference or induction from uniformity of sequence as between two events that there is a causal connection between them . . . The common law, however, is not concerned with philosophic speculation, but is only concerned with ordinary everyday life and thoughts and expressions . . .'. See also Glidewell LJ in *Galoo v Bright Grahame Murray* [1994] 1 WLR 1360 at 1374–5: 'How does the court decide whether the breach of duty was the cause of the loss or merely the occasion for the loss? . . . The answer in the end is "By the application of the court's common sense."'

[16] Thus in this context the term 'real' is to be preferred to 'dominant'. But see Clarke LJ (now Lord Clarke of Stone-cum-Ebony) in *100 Old Broad Street Ltd v Sidley* [1999] EWCA Civ 1286 where he quotes from *Chitty on Contracts*, 27th edn, para 26-015 which said 'The courts have avoided laying down any formal tests for causation: they have relied on commonsense to guide decisions as to whether a breach of contract is a sufficiently substantial cause of the plaintiff's loss. (It need not be the sole cause).' And he then went on to say 'The authorities quoted for these propositions include *Monarch Steamship Co Ltd v Karlshamns Oljefabriker* [1949] AC 370 and *Quinn v Burch Brother Builders Limited* [1966] 2 QB 370 together with *Galoo Limited v Bright Grahame Murray* [1994] 1 WLR 1360. To this I would add, as the speeches in the *Monarch Steamship case* made clear, that the breach must be held to have been an "effective" or "dominant" cause of the plaintiff's loss. The test is equally sound whether the claim be in contract, or in tort.'

cause which gives rise to the extension of time is of significant causative potency in fact, then the extension of time is to be granted for the delaying event. If the delaying event cannot be shown to be of causative potency, then it is to be disregarded.

Thus, in our example, the court or arbitrator may have to inquire whether the late ordering of materials was influenced by the fact that the instructions were late; whether the late instructions were given in the knowledge that the contractor could not proceed any earlier so in fact the instructions were not given at a date which was later than the contractor's readiness to receive them. Many arbitrations proceed to ascertain whether a piece of late information really caused delay, or whether the information was late by reference to an early programme or by reference to a specific request in writing which asked for the information long before it was needed. **14.46**

Much analysis, in articles and textbooks, of examples such as this, proceeds on the assumption that there is true concurrency. In fact in most cases a deeper analysis of the facts demonstrates that this is not the case and what is thought of as concurrent delay is in reality nothing of the sort. **14.47**

The example given in this section of possible concurrent delay illustrates the difficulty of the problems of concurrency in their pure form. The fact finder must ascertain whether there is actual delay in each case, applying the test: what is the real and effective cause of delay as opposed to a concurrent event which might in other circumstances have caused delay but did not do so in this case? **14.48**

To adapt the above example, if the contractor was asking for and not receiving the dimensional information and calculated that it need not order the materials until it received the information, then the 'late' material order was in fact nothing of the kind. It was merely postponed in reaction to events and was not a delaying event at all. On the other hand the contractor might have had difficulty in finding the appropriate supplier and have been incapable of putting in the order before it actually did; in that event there is a concurrency of delay. **14.49**

The permutations are endless, even with a basic example such as the one which has been chosen. All require factual inquiry and simple recitation of programme dates by reference to a computer-generated network will not assist save in a very limited way. **14.50**

Construction contracts are unique in dealing with such problems. Most commercial contracts deal with serial and sequential operations where the fact finding exercise is relatively simple. If a ship is bound from Port A to Port B and thence onwards to Port C, it is easy to draw the inference that a delay in the journey from Port A to Port B will produce some (albeit not always pro rata) delay to the arrival at Port C. How much delay is involved will be easy to ascertain. However in construction contracts, this is not the case, there are sequential operations but they are often concurrent with many others and the interdependence of one operation upon another is not always static. **14.51**

It is surprising how little authority there is on this topic. Apart from the discussion in the *Malmaison* case cited above, there is nothing of note. **14.52**

Concurrent Delay and Causation: Loss and Expense

14.53 The same principles apply to loss and expense, save that concurrency of an event which would bring with it an entitlement to loss and expense and one which would not usually mean that to the extent that there is concurrency, there is no entitlement to prolongation loss and expense.

14.54 The lack of entitlement to loss and expense for prolongation arises because the contractor will not be able to prove that its additional period on site was brought about by the event complained of.[17]

14.55 The most common difficulty which arises in relation to concurrent delays relates to what are described as 'global claims' or rolled up claims.

Programs and Computer Analysis

14.56 Most substantial claims made in court for an extension of time and time-related loss and expense are supported by experts' reports. This has probably led to an over-emphasis upon computer programs which are used to model delay claims on the basis of the critical path for the works. The term 'critical path' has been defined as 'the sequence of activities through a project network from start to finish, the sum of whose durations determines the overall Project duration', see BS.6079–2.2000 Part 2, 2.41.[18] The program is often taken as a starting point and then updated as the works progress and extensions of time are granted[19] and computer software is used for analysis.

14.57 This leads to the giving of programming experts' reports, in litigation or arbitration, which are designed to prove by computer analysis that certain events led inexorably to an entitlement to an extension of time. Frequently these computer analyses have arrived at an extension of time which is far beyond the amount of delay which has actually taken place; and often the experts' reports which set out the entitlement are full of lengthy and incomprehensible appendices which reproduce bar charts and networks with a doubtful grip on the reality of the job.

14.58 The decision of Judge Wilcox in *Skanska Construction UK Ltd v Egger (Barony) Ltd*[20] emphasizes the difficulties which these reports throw up. In that case the judge's remarks upon an expert's 'sophisticated impact analysis' was that it was 'only as good as the data put in. The court cannot have confidence as to the completeness and quality of the input into this complex and rushed computer project . . .' Many of the judge's strictures, which were made

---

[17] For a logical analysis see Protocol: p 7 Core Principles, item 10 and Guidance Notes Guidance, section 1, para 1.10 and particularly 1.10.4. See also *Henry Boot Construction (UK) Ltd v Malmaison Hotel (Manchester) Ltd* [1999] All ER 118, 70 Con LR 32 *per* Dyson J at para 13, cited above.

[18] *Mirant Asia-Pacific Construction (Hong Kong) Ltd v Ove Arup And Partners International Ltd* [2008] Bus LR D1, [2007] EWHC 918 (TCC) para 121.

[19] Protocol: Guidance, section 2 and see also Core principles, p 7, item 11 'Accurate identification of float and concurrency is only possible with the benefit of a proper programme, properly updated.'

[20] [2004] EWHC 1748 (TCC) at para [419].

in that case, regarding the duties of programming experts such as these to make themselves familiar with the facts of the case, demonstrate the difficulties which occur when the computer programs take over from a proper factual analysis of what actually happened and when. In another case the same judge said of an expert that he:

> carefully constructed three critical paths starting at the beginning of the project in July 1997 when the milestone dates did not exist. At best they are theoretical constructs identified retrospectively once the project was completed. These are not paths identified by either party during the project itself. His theory that the second critical path did not run through the Infill Block collided with reality on site . . .[21]

The experience of Judge Wilcox in *Skanska* has not been shared by all the TCC judges. Judge Toulmin QC in in *Mirant Asia-Pacific Construction (Hong Kong) Ltd v Ove Arup And Partners International Ltd* [2008] Bus LR D1, [2007] EWHC 918 (TCC) found programming analysis very helpful, see particularly paragraphs 123–136 of his judgment. A programming expert, who gives a critical path analysis, may be able to assist in backing up the factual analysis and in showing that the logic of the program which the contractor was building to was appropriate. But it cannot be a substitute for an understanding of how the activities in fact inter-linked or how the original program's logic in relation to the activities on site could be altered to produce a different result. **14.59**

The critical path analysis undertaken by programming experts has often been shown to be flawed in adjudications and arbitrations. In litigation it has been in the *Egger*[22] case and the Great Eastern[23] case. The reason, it is submitted, is that they have been allowed to take centre stage in delay analysis as opposed to providing a supporting role. Furthermore, there can be misunderstandings as to criticality of parts of the works. In *Mirant Asia-Pacific* Judge Toulmin QC said in paragraph 127 of his judgment: 'It is of course possible to affect the critical path by giving some activities a longer time than is necessary, thus keeping them on the critical path. This protects other activities from being on the critical path.' It is respectfully suggested that statement is in error. A contractor may assign more time to an activity than needed on a programme. That does not put the activity on the critical path. It makes it appear so for as long as the excessive length of the particular activity is taken at face value. To assign an excessive period to an activity on a programme does not 'protect' the other activities from being on the critical path, it merely masks their presence there. It is suggested that the true critical path can only be ascertained by inputting the real period for an activity. The excessive period has the potential to mislead an onlooker (be it judge or engineer) into thinking that the critical path is other than what in fact it is. The judge's statement illustrates the danger of too much reliance on computer generated documents for the purpose of ascertaining delay and criticality. If the programme contains an inaccurate period or periods for activities, then any analysis which relies upon those periods is likely to be flawed. **14.60**

---

[21] Judge Wilcox in *Great Eastern Hotel Co Ltd v John Laing Construction Ltd* [2005] EWHC 181, 99 Con LR 45 at paras [306]–[308].

[22] See para 14.58.

[23] See n 18 above.

14.61 The reason for the programming experts taking centre stage are not hard to find. It is time consuming and expensive, in relation to an average construction project which has gone wrong, to ascertain the facts on the ground, and on top of that the witnesses interviewed may not be objective or reliable, whereas a computer program produces a neat report of apparently impeccable logic and is therefore appears to be attractive and reliable.

14.62 The correct approach is to examine the facts by reference to the witnesses who attended the site at the time with the aid of the site records. Thus, in our example given above, the reality is addressed directly. Why was the information late? Why did the contractor not order the materials? Could the job have taken on a different sequence in order to avoid the delay—in programming terms: could the logic be adjusted so as to avoid delay?

14.63 If the information was given when it was (and theoretically late) because the employer knew that the contractor had procurement difficulties and was therefore unhurried, then it might be difficult to say that in fact the lateness of the information actually caused delay. A computer program cannot ascertain such a situation. If the contractor did not order the materials because it knew that they would not be needed when originally required in view of the lateness of the information coming from the employer, then equally it will be difficult to say that the late procurement caused any delay. If critical 'interior' work could have been done without the external walls being in place then it might be that no critical delay was caused at all. Given, under the JCT Standard Form, that the contractor has the obligation to prevent delay, the alteration of sequence can obviate many delays which appear on a critical path. The truth is that many jobs have many potential critical paths and if one or more are overlooked by a planning expert, the result is that the expert evidence is likely to be unhelpful.

14.64 What of loss and expense? Here the question is different to that relating to extensions of time; it is whether the late instruction caused the contractor to suffer loss and expense? The answer to that question in the example given above is likely to be no, because the contractor may not be able to say that it was on site in respect of this period for a day longer than it would have been in any event, given its inability to procure materials timeously. However the further analysis identified above in relation to extensions of time may, after all, give rise to an assessment that loss and expense was caused.

Ownership of Float

14.65 A common matter of discussion amongst programmers is what is described as the ownership of float. 'Float' is an additional period allowed for in a contractor's programme beyond that which is needed for any activity before the next critical activity. In the Protocol it is defined as 'The time available for an activity in addition to its planned duration'.[24]

14.66 Again the question is best illustrated by example: The contractor programmes the construction of foundations to take eight weeks, from say week 10 to week 18, but the following activity is programmed to start after week 22, then there is a four-week period of float.

[24] Protocol: Definitions p 56.

The reason why this might be contained in a programme is to allow for contingencies. The foundations sub-contractor might over-run and this allows for a period of four weeks before the next critical activity takes place so a delay to the foundations of four weeks ought not to create a delay to the completion of the works. **14.67**

The question of ownership of the float therefore relates to the question of whether the employer may take advantage of that four weeks' delay at no cost to itself in terms of prolongation of the job. This can happen in a number of ways. The architect might not give instructions as to the type of brick to be used until one week before the construction of the external walls was to commence. Strictly speaking, using this example, the architect has issued the instruction three weeks after the foundations were completed. Is the contractor entitled to an extension of time for those three weeks? The common sense answer will be no, provided the contractor did not intend to commence the brickwork for a further week. But what if the contractor built in that four weeks of float as a safety valve which was to be used if appropriate, but if not, was to be passed on to the next activity. For example, what if the contractor decided that if the foundations were built by week 18, then the external walls would start then, and there would be a four week float before the commencement of the next critical activity? The response of the employer might be that there is still no entitlement to an extension of time because the works could still be completed to time even though four weeks of float is lost in the beginning. **14.68**

That is a defensible position to take. But then, what if the brickwork sub-contractor is four weeks late in completing its work and that puts back the roofing work by four weeks and the critical path lies from foundations to walls to roof before splitting into several alternative critical paths? **14.69**

In that event the contractor can complain that if the information as to the brickwork was not late, then even though the brickwork sub-contractor was slow, the project would not have been late because the brickwork subcontractor's lateness would have been absorbed by the early start. The employer could say that the problem was brought about by the slow progress of the brickwork sub-contractor. Had the brickwork sub-contractor not progressed so slowly the project would still have completed to time as more than sufficient time was allowed for the brickwork activity. **14.70**

It is surprising that such problems, which have been put in a basic way but which occur in reality frequently, have not brought about any judicial decision. The basic question: 'Who owns the float?' relates to such problems. **14.71**

There are three possible answers: (a) the contractor; (b) the employer; and (c) the project. **14.72**

If it is the contractor who owns the float, then the employer is liable for an extension of time and loss and expense for the four weeks' delay caused by the late instruction. If it is the employer who owns the float, then the contractor is liable for the delay caused by the errant sub-contractor. If it is the project, then the party which uses the float first has the benefit of it. The Protocol suggests that generally the project owns the float.[25] The same is the **14.73**

[25] See Protocol: Core principles, p 6, item 7 and Guidance Notes, Guidance, section 1, para 1.3.1.

position in the United States where there has been judicial decision on the topic.[26] The position in England and Wales may be the same on the basis that the first event which uses up float has no causative effect and events which do not have any effect on critical activity ought not to be weighed in the balance against either the contractor or the employer.

14.74 However it seems instinctively unfair, to use our example, that the contractor, which makes provision for things to go wrong, should have to pay for a delay caused by its own fault when, had the employer not been in default, the employer's delay could ordinarily have been absorbed into the programme with no ill effects.

14.75 Under the JCT Standard Form the position is not straightforward so far as the contractor's position is concerned as regards late information because the obligation to give instructions is related specifically to the necessity of the contractor to receive them.[27] It can be argued in some cases that, having regard to the available float, the employer did not issue late instructions because it was not necessary for the contractor to receive the instructions at a time which would have preserved the float for the contractor.

14.76 Float can also be looked at broadly, as, for example, when the contractor plans to complete the work early. The question arises as to what happens to extensions of time and loss and expense when the employer causes a delaying event such as a late instruction or a variation. In that event the Protocol states that the contractor is not entitled to an extension of time if the completion date is not itself threatened, but is entitled to loss and expense.[28]

14.77 This appears correct because although the completion date is not affected by the late instruction, the contractor, in this scenario, ought to be able to show that it has suffered the expense of completing the work at a time later than it otherwise would.

14.78 The terms of the contract are crucial to how this is worked out. The contractor cannot rely upon an implied term that the contractor can receive instructions so as to be able to complete in accordance with its programme, when that programme shows an early completion.[29] Only an express term of the contract can allow payment of loss and expense and an early completion date may not make it necessary for the contractor to receive 'late' information which, in theory, will allow the contractor to complete early.

14.79 The Protocol suggests that this issue should be addressed in each construction contract.[30] It is impossible to criticize this suggestion since it tries to get the parties to deal expressly with a problem which construction planners have had to grapple with for a long time. However the true answer must be that the float is not owned by anyone and that the contract professional must weigh all the factors, including the float in order to arrive at a fair and reasonable extension of time and a proper assessment of what loss and expense is payable.

---

[26] *Weaver-Bailey Contractors Inc v United States* (1990) 19 Ct Cl 474.

[27] See cl 5.4 of the 1998 JCT Standard Form. See also *Neodox v The Borough of Swinton and Pendlebury* (1958) 5 BLR 39 *per* Diplock J.

[28] Protocol Guidance, Guidance Notes, section 1, para 1.12.

[29] See *Glenlion Construction Ltd v Guinness Trust* (1987) 39 BLR 89, 11 Con LR 126, (1988) 4 Const LJ 39 (TCC), Judge Fox-Andrews QC; see also the cases on implied terms.

[30] Protocol Guidance, Guidance Notes, section 1, para 1.12.4.

How are Late Variations to Be Assessed: On a Net or Gross Basis?

For many years claims were made by contractors for extensions of time in circumstances which defied common sense. If, for example, a contractor was ten weeks in delay and the employer issued a variation instruction shortly before actual completion, then the contractor would argue that it was entitled to a full extension of time on the basis that the contractor could not have completed the work as varied on any date earlier than an appropriate period after the late instruction was issued. The argument would be that the ten weeks of culpable delay was wiped out by the late variation. **14.80**

Allied with this argument was a further one relating to 'time at large'. If the 'late' instruction was issued after the completion date it would be argued that this meant that the contractor was released from the fixed date for completion and time was 'at large'.[31] **14.81**

The first instance decision of Colman J in *Balfour Beatty Building Ltd v Chestermount Properties Ltd*[32] has given both these arguments their quietus. Although the case is first instance only, it has been adopted as correct by lawyers in the TCC and in arbitrations and has not been queried in any legal decision since. **14.82**

So far as the 'net' versus 'gross' assessment of an extension of time argument is concerned, which was decided by that case the best summary is by Dyson J in *Henry Boot Construction (UK) Ltd v Malmaison Hotel (Manchester) Ltd*:[33] **14.83**

> First it is agreed that the analysis of Colman J in *Balfour Beatty Building Ltd v Chestermount Properties Ltd* (1993) 32 Con LR 139 should be applied. In his valuable interpretation of cl 25 of this form of contract, Colman J. said, inter alia, that the purpose of the power to grant an extension of time under cl 25.3 was to fix the period of time by which the period of time available for completion ought to be extended having regard to the incidence of the relevant events, measured by the standard of what is fair and reasonable. The completion date as adjusted was not the date by which the contractor ought to have achieved practical completion, but the end of the total number of working days starting from the date of possession within which the contractor ought fairly and reasonably to have completed the works.

That statement of the law as to how extensions of time are to be looked at under a JCT Standard Form but applicable to nearly every other form of contract, bespoke or otherwise, as agreed by the parties and endorsed by Dyson J, disposes of both the argument that an extension of time should be gross and measured from the time the instruction was given and also deals with the 'time at large' point. **14.84**

The absence of any further authority evidences the acceptance by the profession of this legal position and the fact that contrary arguments are unlikely in any context where there is the potential for judicial oversight. **14.85**

---

[31] See discussion of time at large in Ch 7.

[32] (1993) 62 BLR 1.

[33] 70 Con LR 32 at para 12.

How Loss and Expense Is Calculated and the Essentials for Pleading and Putting Forward a Claim for Loss and Expense

Calculation of loss and expense

14.86 Loss and expense will normally have the elements identified in paragraph 14.24 above.

Prolongation costs

14.87 These are on- and off-site overheads and fluctuations. Fluctuations have already been dealt with in paragraph 14.10 above.

14.88 On-site overheads are relatively easy to calculate. The hire of plant and site establishment can be broken down into appropriate weekly costs and then applied to the prolongation period. Sometimes contractors will present their claim on the basis of their tendered rates for on-site overheads. The true view is that the contractor's actual costs must be examined as they are the true cost to the contractor of staying on site for a period longer than it otherwise would. If the plant is hired, then there is little difficulty. If it is owned by the contractor, then the argument will be made that nothing has been lost and the contractor has not proved that element of claim because the plant would not have been used elsewhere anyway.[34] Whether or not that is correct is a question of fact. But it is suggested that it ought not to be too difficult for a contractor to be able to show that, just as with off-site overheads, it has lost the opportunity to put the plant to use and accordingly has lost whatever a court or arbitrator assesses as a reasonable amount of lost hire or alternative value.

14.89 Off-site overheads, or head office overheads, are a source of much academic and professional controversy but little judicial decision. This category of loss is in contradistinction to the additional management costs which might be incurred to deal with the event complained of dealt with below. This is a loss which is incurred by the contractor sustained by its inability to generate income through construction or other works. For example in relation to a claim for loss of profit, it must be shown that on a balance of probability there was profit capable of being earned elsewhere and there was no change in the market thereafter affecting profitability of the work. In relation to head office overheads it must be shown that the contractor was unable to deploy resources elsewhere and had no possibility of recovering them accordingly but for the employer's default.

This requires the fact finder to be persuaded that but for the event complained of the contractor could have generated income. That is a matter of proof. In a normal case this may not be of difficulty. However many contractor's claims are met with the response that the contractor cannot prove any lost jobs. Such a response by an employer is easy to make and it is usually impossible for the contractor to show that it has lost any specific income generating job. The true view is that the contractor will have overheads to fund for any particular period of operation. Few contractors will be able to point to specific contracts lost and it is suggested that this head of loss is not to be turned aside on the basis that a specific

[34] See for example *Alfred McAlpine Homes North Ltd v Property and Land Contractors Ltd* (1995) 76 BLR 59.

contract cannot be found to have been lost. The court or arbitrator will usually be able to infer on the basis of all the evidence before him the fact that, for example, profitable work could have been got elsewhere even if no specific job can be pointed to. In those circumstances therefore a formula calculation is appropriate.

The formulae which are in general use seek to ascertain lost profit and overhead related turnover by reference to a percentage of a putative turnover lost in a specific period of delay. There are two: (1) the Hudson formula,[35] and (2) the Eichleay formula. **14.90**

The first formula takes a percentage, best called the 'Contract Percentage'. This percentage is the percentage of overhead and profit applicable to the contractor's turnover. That can be the tendered percentage or actual overhead and profit percentage. Each has disadvantages. The tendered percentage may be optimistic. The actual percentage will be affected by the delay to the works themselves and so will not necessarily be appropriate. The formula is as follows: **14.91**

$$\frac{\text{contract percentage} \times \text{contract sum} \times \text{period of delay}}{\text{contract period}}$$

It can be seen that it operates a unitary method to an overhead and profit percentage to be applied to the contract sum. If the fact finder is in doubt as to the contract percentage to apply, one method might be to take the average of the overhead and profit percentage of the contractor for the previous three years.[36] That has the advantage of taking an average figure which is thereby neither inflated nor unduly low. Most arbitrators and adjudicators use the Hudson formula which, for all its imperfections, is likely to produce a fair result provided appropriate proof of loss is established. It has been pointed out that the formula assumes that there has been no change in the contract price and when there is an increase in the value of the work then the contractor can recover its overheads in that way and the reduced earning may not in fact happen.[37] The criticism is fair insofar as it requires that the contractor must be able to show lost overhead recovery and for that purpose the contractor's records must be examined to see whether in fact there is in the applicable year some loss of overhead recovery. But the danger of carrying this too far is that the reality that what is ought to be ascertained cannot normally be divined in any other way and an increase in turnover does not mean that there has been no lost contribution to overheads and profit. A contractor may have a good recovery in any given year, but that does not mean that it has not lost any further recovery. In the end the formula is a means of working out loss and expense in this category by way of inference from the primary information in the form of **14.92**

[35] Recognized by Akenhead J in *VGC Construction Ltd v Jackson Civil Engineering Ltd* [2008] EWHC 2082 (TCC) as being . . . 'a claim for time-related overheads and profits on the so-called "Hudson Formula" basis. This Formula is well known throughout the industry and involves a simple calculation in effect to work out what the time-related head office overhead and profit contribution is or should be related on a weekly basis to a period of over-run.'

[36] As Lord Hamilton did in *Beechwood Development Company (Scotland) Ltd v Stuart Mitchell* [2001] CILL 1727, [2001] ScotCS 30. Also by Sir William Stabb QC in *JF Finnegan Ltd v Sheffield City Council* (1988) 43 BLR 124 at 134–6.

[37] See the commentary in the Building Law Reports, vol 28, to *Ellis-Don v Parking Authority of Toronto* (1978) 28 BLR 99 at 102.

inputs into the equation. It is not a means to avoid proof, but is a recognition that 'proof' in the classic sense for example by way of the provision of invoices etc is not usually possible and therefore the court or arbitrator is invited to infer the loss and expense.

14.93 The Eichleay formula is used in the US Court of Claims and is designed to be a more precise way of assessing loss. It is designed to ascertain a weekly overhead contribution which can then be multiplied by the period of delay to arrive at a gross sum claimed. It gets there by dividing the delayed contractor's invoices in the contract period by all invoices of the contractor in the contract period and multiplying the result by the total fixed overheads of the contractor in the contract period to arrive at a figure called the 'required overhead contribution in the contract period'. That required overhead contribution is then divided by the contract period to arrive at the weekly overhead contribution. This formula has not yet been applied in the TCC. It suffers from two main faults. First, it relies upon invoices, which in many cases will only reflect a percentage of the contractor's working activities, and secondly it does not ascertain the profit percentage lost.

Disruption costs

14.94 This is often the most contentious area of claim by contractors. Whilst the on- and off-site overheads can be calculated using data and formulae by reference to a period of delay, disruption costs are far more wide ranging. Similarly with regard to inefficient usage of plant, it is usually possible to ascertain the productivity of certain plant that a surveyor would expect the plant to normally achieve against what it actually achieved, giving credit for disrupting events for which the employer is not responsible. The basic problem with any disruption claim relates to labour and it is that it is difficult to prove that the contractor's tendered labour costs would have been sufficient had the works not been disrupted. Therefore, a comparison of actual labour costs with the tendered costs will always suffer from the attack of employers to the effect that the contractor has not proved 'cause and effect'. However if the contractor has proved disruption as a fact, then it will be a matter of inference that labour costs have increased as a result. Ascertainment then becomes a matter of: (1) examining the tendered provision against standard productivity norms as well as some historical examination of the contractor's labour productivity in order to arrive at a labour and productivity value to apply to the tendered work; then (2) ascertaining what disrupting events have taken place for which the contractor is contractually responsible; then (3) ascertaining that disrupting events have taken place which are the employer's responsibility; and finally (4) making an assessment of the effect of (2) and (3) on the increase in labour cost. This is subjective and however any claim is formulated, it can be subject to attack. In the end an adjudicator or arbitrator must make an assessment based on experience and regard for all the factors.

Finance charges

14.95 Two Court of Appeal decisions in the 1980s[38] decided that recoverable loss and expense by a contractor would include finance charges and that finance charges will include

[38] *FG Minter Ltd v Welsh Health Technical Services Organisation* (1980) 13 BLR 1; *Rees & Kirby Ltd v Swansea City Council* (1985) 30 BLR 1.

compound interest payable by the contractor. There are now formulae and spreadsheets which can assess a contractor's finance charges and the assessment might now be more mechanical than formerly. The justice of causing an employer to compensate a contractor for the costs which it has actually incurred is obvious.

Management time

The employees of a commercial business are taken to generate revenue which is at least equal to the cost of employing them. Recovery under this head is different from overhead recovery and relates to specific management time of the contractor's senior staff. **14.96**

In *Tate and Lyle Food and Distribution Ltd v Greater London Council* [1982] 1 WLR 149 the defendants were liable to the claimants for having failed to dredge silt which they had caused to be accumulated when constructing new piers for the Woolwich ferry and which had obstructed the claimants' use of their barge moorings. The result had been that the claimants themselves had had to dredge the silt and, as part of their claim, they claimed managerial and supervisory expenses referable to that operation. In rejecting a claim for management time Forbes J said at 152 E–G: **14.97**

> I have no doubt that the expenditure of managerial time in remedying an actionable wrong done to a trading concern can properly form the subject matter of a head of special damage. In a case such as this it would be wholly unrealistic to assume that no such additional managerial time was in fact expended. I would also accept that it must be extremely difficult to quantify. But modern office arrangements permit of the recording of the time spent by managerial staff on particular projects. I do not believe that it would have been impossible for the plaintiffs in this case to have kept some record to show the extent to which their trading routine was disturbed by the necessity for continual dredging sessions.

The claim therefore failed only because there had not been properly demonstrated the amount of management time diverted and the extent to which the trading routine had been disturbed by it.

In *Standard Chartered Bank v Pakistan National Shipping Corp* [2001] EWCA Civ 55, [2001] CLC 825 as part of a claimant's attempt to mitigate its loss it sent one of its officers to Vietnam. The bank claimed an amount equal to his salary during the two trips he made there. Potter LJ cited the above passage in the judgment of Forbes J and continued: **14.98**

> It does not seem to me that either the passage quoted or the circumstances relating to the claim in *Tate and Lyle* justify the recovery of the proportion of Mr Griffiths' salary claimed in this case. No doubt it was true, as the judge stated, that, in visiting Vietnam, Mr Griffiths was engaged in an unusual task. However it is not suggested that his trip abroad, as an employee engaged in the business of [the bank] and in respect of whose responsibilities his salary was in any event payable, led to any significant disruption in [the bank's] business or any loss of profit or increased expenditure on [the bank's] part . . . In certain situations, involving particular types of trading concern, such a claim may be appropriate. In particular, building contractors who, by reason of delay, suffer increased costs attributable to a particular job which costs are irrecoverable elsewhere, may claim for a proportion of their fixed overheads (including head office salaries) as part of their claim for consequential loss. However, that is not this case. There is no suggestion that the business of [the bank], or the system of charging upon which its profits depend, were in any way adversely affected by the diversion of Mr Griffiths to Vietnam.

14.99 In *Horace Holman Group Ltd v Sherwood International Group Ltd* [2001] All ER (D) 83 (Nov)[39] the defendants were liable to the claimants for having failed to provide them with an adequate software package. As part of their damages, the claimants claimed time wasted by their directors and staff in struggling with the inadequacies of the software provided by the defendants. In allowing this claim HH Judge Bowsher QC, sitting in the Technology and Construction Court said at paragraph 73: 'I cannot and do not say, in the absence of records there is to be no recovery.' The judge rejected the submission that there were distinctions between an income-producing employee and a 'back office' employee and between short periods and long periods of diverted time. He continued: 'In all cases, the claimants were paying for time which was to be a benefit to them and they lost the benefit of that time.'

14.100 In *Admiral Management Services Ltd v Para-Protect Europe Ltd* [2002] EWHC 233 (Ch), [2002] 1 WLR 2722 Stanley Burnton J decided that while in theory a claim for damages for loss of revenue might be available to the claimants, there was no evidence to justify it in that claim and he disallowed it.

14.101 In *R + V Versicherung AG v Risk Insurance and Reinsurance Solutions SA* [2006] EWHC 42 (Comm), [2006] All ER (D) 209 (Jan) Gloster J looked at the above decisions, observing that she had some difficulty with the reasoning of Stanley Burnton J in *Admiral Management Services* and then she said:

> In my judgment, as a matter of principle, such head of loss (ie the cost of wasted staff time spent on the investigation and/or mitigation of the tort) is recoverable, notwithstanding that no additional expenditure 'loss', or loss of revenue or profit can be shown. However, this is subject to the proviso that it has to be demonstrated with sufficient certainty that the wasted time was indeed spent on investigating and/or mitigating the relevant tort; i.e. that the expenditure was directly attributable to the tort . . . This is perhaps simply another way of putting what Potter LJ said in *Standard Chartered*, namely that to be able to recover one has to show some significant disruption to the business; in other words that staff have been significantly diverted from their usual activities. Otherwise the alleged wasted expenditure on wages cannot be said to be 'directly attributable' to the tort.

14.102 In *Aerospace Publishing v Thames Water* [2007] EWCA Civ 3 the Court of Appeal examined these authorities and concluded that they established the following propositions:

(a) The fact and, if so, the extent of the diversion of staff time have to be properly established and, if in that regard evidence which it would have been reasonable for the claimant to adduce is not adduced, he is at risk of a finding that they have not been established.

(b) The claimant also has to establish that the diversion caused significant disruption to its business.

[39] On 7 February 2002, Dyson LJ refused permission to appeal against Judge Bowsher's judgment: [2002] EWCA Civ 170. This decision was applied by the same judge in *Sam Business Systems Ltd v Hedley and Co* [2002] EWHC 2733.

(c) Even though it may well be that strictly the claim should be cast in terms of a loss of revenue attributable to the diversion of staff time, nevertheless in the ordinary case, and unless the defendant can establish the contrary, it is reasonable for the court to infer from the disruption that, had their time not been thus diverted, staff would have applied it to activities which would, directly or indirectly, have generated revenue for the claimant in an amount at least equal to the costs of employing them during that time.[40]

For this head of claim to be recoverable good records must be kept demonstrating that the people involved spent time on the relevant project and what they did must be clearly ascertainable. It will always be hard to show that a particular senior employee is doing work attributable to an event for which the employer is liable as opposed to a neutral event or an event for which the contractor is liable. Timesheets must be task specific. If they are, then the claim is likely to be recoverable in principle. **14.103**

Damages or costs payable to third parties

Settlement claims over are discussed in paras 12.40 et seq. In the pre-2005 JCT forms of contract and sub-contract there was provision for a name borrowing arbitration by the sub-contractor which was granted the right to the benefits of the rights and remedies of the main contractor. So an arbitration against the employer by both the main contractor on its own account and with the sub-contractor seeking its own recovery was a common scenario. In that event the main contractor was best advised to keep the sub-contractor's claim in the pleadings very clearly separate in terms of quantum. **14.104**

Loss and Expense and Rolled Up Claims

There are three essentials to plead in any construction case: (1) duty, (2) breach, and (3) damage. But overarching all three is linkage between the three. Pleading all three is normally no problem in the theoretical sense. But without the linkage between the three a claim will fail. **14.105**

Saying that a claim will fail in those circumstances does not mean that, in cases where cause and effect is difficult to establish, there will be no recovery. There has been a great deal of confusion on this subject with employers' representatives taking the line that the contractor who is so disrupted by numerous variations cannot prove its case because it cannot attribute cause and effect, and contractors' representatives taking the line that it is all so confusing that they do not have to try. **14.106**

[40] This dictum has been applied in *Al-Rawas v Pegasus Energy Ltd & Ors* [2008] EWHC 617 (QB); *Glen Haysman v Mrs Rogers Films Ltd* [2008] EWHC 2494 (QB); and *Cheltenham Borough Council v Laird* [2009] EWHC (QB) 1253. See also *Nationwide Building Society v Dunlop Haywards and another* [2009] EWHC 254 (Comm) where Clarke J gave recovery of wasted staff time in investigating a fraud.

14.107 There are now numerous cases which deal with global claims.[41] The position from these cases can be summarized in the following propositions:

(a) When a contractor can show cause and effect from specific variations or other events for which the employer is responsible, then it ought to do so because any other course will result in a prolonged forensic exercise examining the adequacy of the pleadings.

(b) The case must be put forward in a way which will allow the employer to know the case it has to meet.

(c) If a 'composite' claim or 'rolled up' claim is put forward, then all duplication ought to be excluded from the pleading of the quantum of the claim.

(d) The employer cannot non suit a contractor if the particulars given permit it to know the case it has to meet.

(e) The contractor cannot succeed in a rolled-up claim if a significant and unascertainable portion of the delay and disruption has been caused by events for which the contractor is responsible. However that rule must not be too rigorously applied. An element of common sense must enter into it.[42]

---

[41] *Wharf Properties v Eric Cumine Associates* (1991) 52 BLR 1, 29 Con LR 84 (Privy Council); and *J Crosby & Sons v Portland UDC* (1967) 5 BLR 121 (Donaldson J); and see also *ICI v Bovis Construction Ltd* (1992) 32 Con LR 90; *British Airways Pension Trustees Ltd v Sir Robert MacAlpine & Sons Ltd* (1994) 72 BLR 26, 45 Con LR 1; and *Bernhard's Rugby Landscapes Ltd v Stockley Park Consortium Ltd* (1997) 82 BLR 39; *John Doyle Construction v Laing Management (Scotland) Ltd* [2004] BLR 295; the Supreme Court of South Africa in *Imprefed (Pty) Ltd v National Transport Commission* (1993) (3) SA 94; and the US Court of Claims in *Wunderlich Contracting v United States* (1965) 351 F2nd 956; and *Boyajian v United States* 423 F 2d 1231. See also *John Holland Construction & Engineering Pty Ltd v Kvaerner RJ Brown Pty Ltd* (1996) 82 BLR 81.

[42] See particularly Lord Drummond Young in *John Doyle Construction v Laing Management (Scotland) Ltd* [2004] BLR 295 where he said:

> The logic of a global claim demands, however, that all the events which contribute to causing the global loss be events for which the defender is liable. If the causal events include events for which the defender bears no liability, the effect of upholding the global claim is to impose on the defender a liability which, in part, is not legally his. That is unjustified. A global claim, as such, must therefore fail if any material contribution to the causation of the global loss is made by a factor or factors for which the defender bears no legal liability . . . The point has on occasion been expressed in terms of a requirement that the pursuer should not himself have been responsible for any factor contributing materially to the global loss, but it is in my view clearly more accurate to say that there must be no material causative factor for which the defender is not liable.
>
> Advancing a claim for loss and expense in global form is therefore a risky enterprise. Failure to prove that a particular event for which the defender was liable played a part in causing the global loss will not have any adverse effect on the claim, provided the remaining events for which the defender was liable are proved to have caused the global loss. On the other hand, proof that an event played a material part in causing the global loss, combined with failure to prove that that event was one for which the defender was responsible, will undermine the logic of the global claim. Moreover, the defender may set out to prove that, in addition to the factors for which he is liable founded on by the pursuer, a material contribution to the causation of the global loss has been made by another factor or other factors for which he has no liability. If he succeeds in proving that, again the global claim will be undermined.
>
> The rigour of that analysis is in my view mitigated by two considerations. The first of these is that while, in the circumstances outlined, the global claim as such will fail, it does not follow that no claim will succeed. The fact that the pursuer has been driven (or chosen) to advance a global claim because of the difficulty of relating each causative event to an individual sum of loss or expense does not mean that after evidence has been led it will remain impossible to attribute individual sums of loss or

(f) If the case of the contractor is such that specific attribution of loss and expense to specific events is impossible, then the court will permit an aggregation of causes to be attributed to certain loss and expense so long as the attribution is comprehensible and credible.

Therefore the essentials are: (a) to ensure that the delay and disruption which is pleaded and claimed is not mixed in with matters which do not give rise to a claim; and (b) to see that the judge, arbitrator, or adjudicator can see how the events complained of and giving rise to a claim caused loss. One vice which appears in many claims is that a list of variations is amalgamated into a schedule which is then cross referred to two further schedules which allege: (1) delay, and (2) disruption. What has to be done is to explain how variation A caused delay and how it caused disruption. A variation, of itself, does not do so. A change in the colour of the paint for a fence is unlikely to cause either. However it might if it means that work has to be re-done, or stopped whilst the fresh paint is sourced. Similarly 100 paint colour changes may not cause any delay or disruption. The sheer number of variations is not, of itself, sufficient indicia of delay and disruption. What is required is narrative and this is best done by reference to those who were on site and can attest to what happened. Expert evidence may be used. as Ramsey J said in *London Underground Ltd v Citylink Telecommunications Ltd*:[43] 'The essence of a global claim is that, whilst the breaches and the relief claimed are specified, the question of causation linking the breaches and the relief claimed is based substantially on inference, usually derived from factual and expert evidence.' **14.108**

Therefore, in preparing any claim, the claimant's advisors should concentrate on narrative and explanation and should advise that the project team is not disbanded until that has been provided. **14.109**

There have of late been cases where the law as to global claims has been sought to be stated and applied.[44] The most recent is the decision of Judge Wilcox in *Maersk Oil UK Ltd v Dresser-Rand (UK) Ltd* [2007] EWHC 752 (TCC). At paragraph 685 of a massive judgment he said: **14.110**

> Useful guidance in relation to such claims is to be found in *Laing Management (Scotland) Ltd v John Doyle Construction Ltd* [2004] BLR 295
>
> John Doyle were works contractors employed by management contractors Laing under an Amended Scottish Works Contract Form. The project was the construction of new corporate headquarters for Scottish Widows. John Doyle claimed a 22-week extension of time and

expense to individual causative events. The point is illustrated in certain of the American cases. The global claim may fail, but there may be in the evidence a sufficient basis to find causal connections between individual losses and individual events, or to make a rational apportionment of part of the global loss to the causative events for which the defender has been held responsible.

The second factor mitigating the rigour of the logic of global claims is that causation must be treated as a common sense matter . . . That is particularly important, in my view, where averments are made attributing, for example, the same period of delay to more than one cause.

[43] 114 Con LR 1, [2006] EWHC 388 (Comm), [2006] 1 All ER (Comm) 529, [2006] 2 Lloyd's 1, [2007] BLR 391, [2007] EWHC 1749 (TCC), [2007] 2 All ER (Comm) 694.

[44] For example, Judge Toulmin QC in *Mirant Asia-Pacific Construction (Hong Kong) Ltd v Ove Arup And Partners International Ltd* [2008] Bus LR D1, [2007] EWHC 918 (TCC) at para 45.

£4.8 million for loss and expense caused by delay and disruption. Laing argued that since John Doyle had made a global claim, if any of the loss and expense claimed was not caused by them then the whole claim must fail.

It was held that provided a contractor is able to establish that all of the events on which he relies are the responsibility of the employer, in law, it is not necessary for him to prove causal links between individual events and particular heads of loss.

Where the loss was caused both by matters for which the employer is responsible, and also by matters for which he is not responsible, then for the claim to succeed the contractor must establish that those events for which the employer is responsible, are the proximate or dominant cause of the loss. The proximate or dominant cause established by proven fact and commonsense analysis will be treated as the operative cause and the person responsible for it will be responsible for the whole loss.

In the event that the contractor cannot identify the dominant cause of the loss, it may be possible to apportion the loss between the causes for which the employer is responsible and other causes. In such a case it is obviously necessary that the event or events for which the employer is responsible should be a material cause of the loss. Provided that condition is met, the apportionment of loss between the different causes is possible in an appropriate case.

At page 303 column 1 Lord Maclean said:

> Such a procedure may be appropriate in a case where the causes of the loss are truly concurrent in the sense that both operate together at the same time to produce a single consequence . . ..
>
> Apportionment in this way on a time basis is relatively straightforward in cases that involve any delay. Where disruption to contractors work is involved matters become more complex. Nevertheless, we are of the opinion that apportionment will frequently be possible in such cases, according to the relative importance of the various causative events in producing loss. Whether it is possible will clearly depend on the assessment made by the judge or arbiter, who must of course approach it on a wholly objective basis. It may be said that such an approach produces a somewhat rough and ready result. This procedure does not however seemed to us to be fundamentally different in nature from that used in relation to contributory negligence or contribution among the joint wrongdoers. Moreover, the alternative to such an approach is the strict view that, if a contractor sustained a loss, caused partly by events for which the employer is responsible and partly by other events he cannot recover anything because he cannot demonstrate that the whole of the lost is the responsibility of the employer. That would deny him a remedy even if the conduct of the employer or the architect is plainly culpable . . . It seems to us that in such cases the contractor should be able to recover for part of his loss and expense, and we are not persuaded that the practical difficulties of carrying out the exercise should prevent him from doing so.
>
> An apportionment procedure of this nature has been used with apparent success in the United States in cases before the Court of Claims. Thus in *Lichter v Mellon-Stewart Company* 305 F2nd 216 (1962) the plaintiff's total cost claim on one contract was rejected on the ground that a substantial amount of their loss was the consequence of factors other than breaches of contract by the defendants. Nevertheless the Court of Claims allowed a claim based on another contract between the same parties to succeed in part and its decision was upheld about United States Court of Appeals for the Third circuit. The Court of Claims had held that part of the Plaintiff's extra cost on this contract was attributable to the fault of the defendant and part was attributable to the other non-compensable factors . . . . . .
>
> The important point that emerge from this decision are, first, that the Federal Courts in the United States are willing to undertake an apportionment exercise and, secondly,

> that any such apportionment must be based on the evidence and carried out on the basis that is reasonable in all the circumstances. In our opinion, a similar procedure should be available in Scots law. We stress, however, that the allocation must be based on the evidence, and that under Scottish procedure the evidence must be based on a foundation in the pleadings.

[Judge Wilcox concluded]

> If an apportionment, based on the evidence is possible, albeit difficult it would be manifestly unjust to deny a remedy, where there are plain contractual breaches by the defendant.

The judge held in that case that it was not possible on the basis of the evidence to identify a dominant cause for the cost claimed and it failed accordingly.

15

LIMITATION OF ACTION

Rationale

15.01 The prosecution of stale claims led Parliament to pass acts of limitation which were designed to prevent the bringing of claims after a certain period and thus eliminate the injustice of stale claims. The first of these was the Limitation Act 1623, and the current statute is the Limitation Act 1980, as amended by the Latent Damage Act 1986. Limitation of action is statutory and is a matter of policy. In *Rhone-Poulenc Rorer International Holdings Inc & Anor v Yeda Research and Development Co Ltd* [2006] EWHC 160 (Ch) Lewison J said:

> In domestic law limitation periods have a number of different policy justifications. The principal strands are:
>
> i) *Evidentiary reasons.* Evidence may depend on the recollection of witnesses, which deteriorates over time. It may depend on the preservation of written records which may be lost or destroyed.
>
> ii) *The interests of defendants.* Defendants may have changed their position or conducted their businesses in the belief that a claim would not be made. It is for this reason that limitation statutes have been described as 'acts of peace' or 'statutes of repose'. People should be free to get on with their lives or businesses without the threat of stale claims being made.
>
> iii) *Encouragement of claimants.* A limitation period also encourages claimants to bring their claims promptly and not, in the old phrase, 'to sleep on their rights'.

iv) *The public interest.*The public interest has always been concerned that litigation should be brought within a reasonable time. This enables cases to be dealt with properly and justly. Moreover the public interest requires the principle of legal certainty.

The statute does not extinguish the cause of action, it only bars the remedy.[1] The distinction is unimportant in most or nearly all cases but it does have two effects. The first is that the Act must be pleaded by a defendant before it has any effect. If a cause of action which is statute barred is pleaded, then the court cannot dismiss it unless the defendant to it pleads the Act. Once it has been pleaded, however, the claimant has the burden of showing that his claim accrued within the period. The second effect relates to cases with a foreign element. The Limitation Act is procedural in nature rather than substantive. **15.02**

Time stops under the Limitation Act 1980 when proceedings are commenced. In court proceedings this is when a claim form is issued. In arbitration this is governed by the parties' agreement or if there is no agreement when one party serves upon the other a notice requiring him or her to submit the matter to arbitration.[2] The action brought in court may be capable of amendment outside the limitation period (see amendments below) but an arbitration notice (most frequently a notice to concur in the appointment of an arbitrator) must comprehensively set out all matters in dispute because matters outside the ambit of the arbitration notice may not be brought into the arbitration because they are outside the scope of the arbitration. **15.03**

Limitation Period for Contracts under Seal

When a contract is under seal the limitation period is twelve years from the date on which the cause of action accrued.[3] A cause of action is not defined under the Act but authorities from earlier legislation show that it means '*that which makes action possible*';[4] it has also been famously defined by Diplock LJ in his judgment in *Letang v Cooper*[5] as: 'simply a factual situation the existence of which entitles one person to obtain from the court a remedy against another person'.[6] **15.04**

[1] For an example where the distinction does matter see *Laroche v Spirit of Adventure (UK) Ltd* [2009] 3 WLR 351, paras 69–74.

[2] See the Arbitration Act 1996, ss 12(5), 13, and 14.

[3] Limitation Act 1980, s 8.

[4] Preston and Newsom, *Limitation of Actions*, 3rd edn, p 4 citing Lord Dunedin in *Board of Trade v Cayzer, Irvine & Co Ltd* [1927] AC 610 at 617.

[5] [1965] 1 QB 232 at 242–3.

[6] In *Read v Brown* (1889) 22 QBD 128 (a case on procedure only regarding whether a cause of action arose in the City of London and thus could be tried there) Lord Esher MR, citing *Cooke v Gill* (1873) Law Rep 8 CP 107, defined a cause of action as follows: 'every fact which it would be necessary for the plaintiff to prove, if traversed, in order to support his right to the judgment of the Court.' Lopes LJ at 133 expressly approved Lord Esher's definition. Fry LJ had a slightly different approach in saying, 'Everything which, if not proved, gives the defendant an immediate right to judgment, must be part of the cause of action.' Lord Diplock's definition is the one used today, but clearly it comes from Lord Esher's, albeit that the case was not cited in *Letang* in either the argument or the judgments. The definition was expanded upon slightly by Lord Esher in *Coburn v Colledge* [1897] 1 QB 702 at 706–7, 'The language I used obviously means this: the plaintiff in order to make out a cause of action must assert certain facts which, if traversed, he would be put to prove.'

15.05 In construction contracts the factual situation arises in two ways. First, when sums become due under the contract and, secondly, when the contract is breached and damages become payable as a result (known as the 'breach date rule'). This will be looked at in more detail below.

15.06 It is not necessary for the contract to be literally sealed with sealing wax or some other physical seal. All that is necessary is that the contract was so signed that it demonstrates an intention by the parties that the contract should be considered under seal. The words 'signed as a deed' alongside the place for the signature of a deed will normally be sufficient. If there is a place also for signature for the contract to be signed 'under hand' and that is left blank, then the irresistible inference will be that the parties intended that the contract be under seal if they affix their signature where the words 'signed as a deed' or 'as a deed' appear.

15.07 Aside from the period of limitation there are no different rules relating to limitation to those which obtain under a simple contract.

Limitation Period for Contracts Not under Seal or Simple Contracts

15.08 The limitation period in contract is six years from the date the cause of action accrued.[7] The term cause of action is as defined and described above in relation to contracts under seal.

15.09 The essential point is therefore to decide when the cause of action accrued, in other words when the factual situation arose which gave rise to the right to a remedy. This section discusses the remedy of claims for sums due and claims for damages for breach of contract.

Claims for sums due: contractors and professionals

15.10 The contract will determine when and how sums become due. When work is done and the contract makes no provision for when payment is due, then payment will become due when the work is completed.[8] The right to recovery is at the earliest time when the action could be brought.[9]

15.11 When work is done and the contract does make provision for payment, then the right to recovery of sums due under the contract will be dictated by the express terms of the contract.

15.12 Most construction contracts provide that payment shall be made periodically upon certificate. Therefore a cause of action accrues in contract each time a certificate becomes due and owing. However if a certificate is not issued at all or is issued later than it should, that is a breach of contract and a cause of action accrues when the certificate should have been

[7] Limitation Act 1980, s 5.

[8] See *Cutter v Powell* (1795) 2 Smiths Leading Cases 1; *Coburn v Colledge* [1897] 1 QB 702.

[9] See *per* Lindley LJ in *Reeves v Butcher* [1891] 2 QB 509 '. . . the cause of action arises at the time when the debt could first have been recovered by action. The right to bring an action may arise on various events; but it has always been held that the statute runs from the earliest time at which an action could be brought.'

issued.[10] Therefore a more complete statement of the position is that a cause of action accrues when a certificate is issued or when a certificate should have been issued.

That position obtains serially and periodically up to and including the Final Certificate or date for final payment[11]. The cause of action upon interim certificates is not the same as the cause of action in relation to a final certificate because the nature of the interim certificates is rough and ready and prone to either error or approximation. The purpose of interim certificates is to provide a fair interim payment as the works progress.[12] The final certificate and payment is a far more complex and precise task. Thus in respect of any piece of work done, there is likely to arise a cause of action at the time the certificate applicable to the work is issued or should have been issued and then again at the time of the final certificate. **15.13**

[10] *Henry Boot Construction Ltd v Alstom Combined Cycles Ltd* [2005] EWCA Civ 814, [2005] 3 All ER 932.

[11] For an academic criticism of this analysis, see Wendy MacLaughlan, 'Courts and Construction: Accrual of a Cause of Action for Payment in Engineering Contracts', (2006) 22 Const LJ 7, 445–460.

[12] See ibid at [55]–[58] and [72] *per* Dyson LJ particularly at [56] where he said:

> In my view, the cause of action in respect of an Engineer's failure to include a sum in an interim certificate is not the same as the cause of action in respect of the failure to include a sum in the final certificate, even if the two sums happen to be the same. This is because interim certificates are no more than provisional estimates of the sum to which Boot is entitled by way of instalment payments. The Engineer is required to certify the amount which on the basis of Boot's monthly statement in his opinion is due 'on account of clause 60(1)(a) and (d)'. Clause 60(1)(a) requires the monthly statement to show the 'estimated contract value of the Permanent Works'. Clause 60(1)(d) requires the statement to show the 'estimated amounts to which the Contractor considers himself entitled in connection with all other matters . . .' (ie other claims). The Engineer is, therefore, required to decide, on the basis of estimates, what in his opinion is due as a payment 'on account'. The contract does not state what supporting documentation *if any* Boot must put forward in support of its estimates. Boot is not even obliged to claim the full estimated values or amounts, because the whole of clause 60(1) is subject to the qualification: 'unless in the opinion of the Contractor such values and amounts together will not justify the issue of an interim certificate'. These words may do no more than reflect the language of clause 60(3). More importantly, the Engineer is not required to carry out a detailed and accurate valuation each month. His obligation is no more than to certify the amount which in his opinion is due *on account of* clause 60(1)(a) and (d). This is of particular relevance in relation to claims for extra expense pursuant to various clauses of the contract, where the calculation of Boot's true entitlement can be very complicated and time-consuming, and where it is often impossible to form an accurate view at the interim stage of the extra expense that will finally be payable.

Dyson LJ at paragraph 56 was applied by the Court of Appeal in *London & Regional (St George's Court) Ltd v Ministry and Defence* [2008] EWCA Civ 1212. See also Hobhouse J in *Secretary of State for Transport v Birse-Farr Joint Venture* (1993) 62 BLR 36 at 53:

> Certification may be a complex exercise involving an exercise of judgment and an investigation and assessment of potentially complex and voluminous material. An assessment by an engineer of the appropriate interim payment may have a margin of error either way. It may be subsequently established that it was too generous to the contractor just as it may subsequently be established that the contractor was entitled to more. Further the sum certified may be made up from a large number of constituent figures, some of which may likewise be assessed favourably to one party or the other. It may be that a contractor can say under a certain heading he did not have certified as high a figure as can later be seen to be appropriate but that under another heading he has to accept that the figure certified can be shown to have been an over-certification. At the interim stage it cannot always be a wholly exact exercise. It must include an element of assessment or judgment. Its purpose is not to produce a final determination of the remuneration to which the contractor is entitled but is to provide a fair system of monthly progress payments to be made to the contractor. The strict twenty-eight day period in sub-clause (2) can be contrasted with the three months allowed in sub-clause (3) 'after receipt of this final account and of all further information reasonably required for its verification'.

15.14 When payment is due pursuant to milestones or particular stages of the work being completed, rather than certificates, then the position is likely to be different and there will be no right to revisit and reclaim at the end. In that event, the sums are likely to be due when the stage of the works dictating payment is reached,[13] and no further cause of action is likely to arise thereafter. If that is the case, then the cause of action arises once in respect of any item of work or claim, namely at the time the sum should have been paid. In that event when there is no recurring right to review of previous payments or claims, then a cause of action in respect of an early milestone or payment may accrue at the beginning of the job and that cause of action will become statute barred six years from then. It is axiomatic that each contract must be looked at and construed on its own merits to see whether there is a single cause of action or a series of obligations to certify the work right up to completion and a final obligation to review all claims at the end with the final certificate and payment. Most standard forms of building and engineering contract adopt the latter model and therefore the analysis of Dyson LJ in the *Alstom* case will be applicable.[14]

15.15 Such considerations concerning when the right to payment accrues apply to professionals, such as architects, engineers, surveyors, and project managers. The right to payment accrues when the contract provides for payment to be made or, if no provision is made, upon completion. With professionals, however, payment is normally due periodically upon completion of various stages of the work, rather than upon certification.

Claims for damages for breach of contract: contractors

15.16 The cause of action accrues upon breach. There is a breach of contract when a certificate is not issued when it should have been. There is a breach of contract when defective work is done. However the obligation of the contractor is to deliver up a completed building in accordance with the contract. Accordingly, even if the contractor is in breach of contract by the execution of defective work, a fresh cause of action accrues at the date of practical completion or handover. At the moment the contractor has purported to hand over to the employer a structure which is not in accordance with the contract, it is in breach of contract.

15.17 The date the breach is discovered or is discoverable is irrelevant.[15]

Claims for damages for breach of contract: professionals

15.18 Professionals are normally responsible for design and for contract administration. Each responsibility must be treated separately.

15.19 In relation to design, the breach of contract takes place when the defective design is carried out. However it has been held that a professional will sometimes, but not invariably, have

[13] See, for example, *Birse Construction Ltd v McCormick (UK) Ltd* [2005] EWCA Civ 940.

[14] Though there are of course exceptions: see *London & Regional (St George's Court) Ltd v Ministry of Defence and Secretary of State for Defence* [2009] BLR 20.

[15] See *Bagot v Stevens Scanlan & Co* [1966] 1 QB 197.

an obligation to review its design until practical completion of the works has been reached. See *Brickfield Properties v Newton*[16] *per* Sachs LJ:

> Where there are found in completed buildings serious defects of the type here under review the facts relating to design, execution and superintendence are inextricably entangled until such time as the court succeeds in elucidating the position through evidence. The design has inevitably to be closely examined even if the only claim relates to superintendence, and all the more so if the designs are, as is alleged here, experimental or such as need amplification as the construction progresses. The architect is under a continuing duty to check that his design will work in practice and to correct any errors which may emerge.

The duty to review the design has been held by first instance authority to extend throughout the life of the construction process itself.[17] Other, perhaps more persuasive, first instance authority is less supportive of that position. See the decision of Dyson J in *New Islington and Hackney Housing Association v Pollard Thomas and Edwards Ltd*:[18] **15.20**

> But it is necessary to consider the scope of that duty in a little more detail. What does the duty to review the design entail? In what circumstances will an architect be in breach of that duty? I find it convenient to consider an example. Let us suppose that an architect is engaged on the standard RIBA Conditions of Engagement to provide the full service (as PTE were in the present case), including administering a building contract in a standard JCT form of contract. Suppose that he designs the foundations of a building (a large office block), the foundations are constructed in accordance with his design, and several years later, practical completion is achieved. Let us further suppose that the design of the foundations is defective and one which no reasonably competent architect would have produced: in other words, the architect was negligent. There can be no doubt that the architect commits a breach of contract when he completes the design and gives instructions to the contractor to construct the foundations in accordance with it. But in what sense and to what extent is the architect under a duty to review his negligent design once the foundations have been designed and constructed?
>
> In my view, in the absence of an express term or express instructions, he is not under a duty specifically to review the design of the foundations, unless something occurs to make it necessary, or at least prudent, for a reasonably competent architect to do so. For example, a specific duty might arise if, before completion, the inadequacy of the foundations causes the building to show signs of distress; or if the architect reads an article which shows that the materials that he has specified for the foundations are not fit for their purpose; or if he learns from some other source that the design is dangerous. In such circumstances, I am in no doubt that the architect would be under a duty to review the design, and, if necessary, issue variation instructions to the contractor to remedy the problem. But in the absence of some reason such as this, I do not think that an architect who has designed and supervised the construction of foundations is thereafter under an obligation to review his design.
>
> I do not accept that in every case where an architect has negligently introduced a defective design into a building, he is also by the same token in breach of a continuing contractual

---

16 [1971] 1 WLR 873.

17 See *Chelmsford DC v TJ Evers* (1984) 25 BLR 99 *per* Judge David Smout QC at 106; *Equitable Debenture Assets Corp Ltd v William Moss Group Ltd* (1984) Con LR 1 *per* Judge John Newey QC at 24; *London Borough of Merton v Lowe* (1982) 18 BLR 130 *per* Judge William Stabb QC; *Victoria University of Manchester v Hugh Wilson* 2 Con LR 43 *per* Judge John Newey QC at 78; See also *Mirant-Asia Pacific Ltd v Oapil* [2004] EWHC 1750 (TCC) Judge Toulmin QC at 494. For a case where it was held to last beyond practical completion see *University of Glasgow v William Whitfield* (1988) 42 BLR 66 at 78 *per* Judge Peter Bowsher QC, but that was pursuant to an express term of the contract to do so.

18 [2001] BLR 74 at paras 15 et seq.

obligation to review his design. In *Midland Bank v Hett, Stubbs & Kemp* [1979] Ch 384, 403C, Oliver J said:

> It is not seriously arguable that a solicitor who or whose firm has acted negligently comes under a continuing duty to take care to remind himself of the negligence of which, ex hypothesi, he is unaware.

In my view, that observation is as apt to apply to an architect as it is to a solicitor. The position is quite different where the architect (or solicitor) knows, or ought to know, of his earlier negligence. When that occurs, then he may well be under a contractual obligation to review his earlier performance, and advise his client honestly and competently of his opinion. Whether he is in fact under such a duty when he has actual or constructive knowledge of his earlier breach of contract will depend on whether the contract is still being performed. If the contract has been discharged (for whatever reason), then the professional person may be under a duty in tort to advise his client of his earlier breach of contract, but it is difficult to see how he can be under any contractual duty to do so[19].

15.21 The point has also been addressed more recently by Ramsey J in *Oxford Architects Partnership v Cheltenham Ladies College*:[20]

> In terms of a cause of action for breach of contract it is sometimes said that contractors and Architects owe a continuing contractual duty up to at least Practical Completion. There is, however, in my judgment, a distinction to be drawn between the position of the contractor and the position of a professional such as an architect.
>
> The position of a contractor of course depends on the terms of the contract but generally there is an obligation to 'carry out and complete' the works. Thus, there will be a cause of action for a failure properly to complete the work by the date for completion. In those circumstances a cause of action will accrue right up to Practical Completion if the contractor fails to complete the works properly . . . There may then, depending on the defects liability provisions in the contract, be a further cause of action after Practical Completion.
>
> In the case of an architect the duty, again, depends on the terms of the engagement. If an architect merely carries out a design which is issued to a contractor for construction then it is difficult for a continuing duty to arise during the period of construction. Where, however, the architect's engagement includes services during the period of construction, there may be a continuing duty . . .
>
> . . . under clause 3.1.4 of the conditions of Engagement it provides:
>
>> Except in an emergency the Architect shall make no material alteration of addition to or omission from the approved design during construction without the knowledge or consent of the Client.
>
> In these circumstances it seems to me that the appropriate obligation on the Architects in relation to the period of construction is that identified by Dyson J as he then was in *New Islington and Hackney Housing Association Limited v Pollard Thomas and Edward Limited* [2001] BLR 74 at page 79 where he said in relation to the Architect's continuing duty to review his design:
>
>> In my judgment, the duty does not require the Architect to review any particular aspect of the design that he has already completed unless he has good reason for so doing.

[19] Applied by Judge Richard Seymour QC in *Tesco Stores Ltd v Costain Construction Ltd* [2003] EWHC 1487 (TCC) see para 272 of the judgment.

[20] [2007] 1 BLR 293

> What is a good reason must be determined objectively, and the standard is set by reference to what a reasonably competent Architect would do in the circumstances.
>
> In terms of a breach of contract, if an architect has good reason to review any aspect of the design then a breach will occur if the architect fails to review the design or if the architect reviews the design but fails to do so properly in accordance with the terms of engagement . . .
>
> There is, it seems, some confusion as to the meaning of a continuing breach in this context and in particular the date when a cause of action accrues for a breach of that duty. If an architect produces a design which is in breach of the terms of engagement and issues it to the contractor who constructs the work to that design, then a breach of contract will occur when, for instance, the architect under stages F to G 'prepares production drawings' or under stage J 'provides production information' as required by the building contract, a cause of action will then accrue based upon that breach of contract.
>
> The continuing duty does not, however, give rise to a single and continually accruing cause of action. Rather, a different cause of action accrues at various stages. Thus, the cause of action for a failure properly to review the design is a different cause of action from a failure to provide a proper design in the first place. The causes of action will therefore accrue on different dates.

It is submitted, then, that in contract, unless there is an express term of the contract which requires a review of the design or there is some other circumstance which means that the design or the relevant aspect of it requires review, there is no contractual obligation to revisit the design of the works. Therefore if, for example, defective foundations are designed and then covered up, there will be a breach of contract at the time the foundations are defectively designed and there may be a further breach when they should have been inspected. But there is no continuing duty to review the foundation design, in contract. It is to be noted that Sachs LJ referred to an obligation to correct any errors which may emerge. That is far from a duty to revisit design when there is nothing to cause the architect or professional to do so. **15.22**

In relation to defective supervision or inspection by the professional, the breach of contract takes place when the defective inspection or supervision takes place. There is no warrant for the suggestion that the professional has committed a breach of contract at the time of practical completion in the same way as a contractor will have done. **15.23**

Limitation Period in Tort

The limitation period in tort is six years from when the cause of action accrued and the cause of action accrues when the damage takes place. **15.24**

The essential question is therefore 'what is damage?' This is a question which has been troublesome for many years and has been discussed at length in many authorities. The recent highest authority is the House of Lords in *Law Society v Sephton*,[21] which was a case involving when the Law Society suffered damage when a solicitor failed to spot the misappropriation of funds. The speeches in that case repay close study since they examine, again, the concept of damage in tort, when it occurs and when actionable loss is suffered. **15.25**

[21] [2006] 2 AC 543.

That case involved considerations of contingent liabilities and whether they constitute damage. The decision of the House of Lords was that a contingent liability did not constitute damage. Damage did not occur until the contingency is fulfilled and the loss becomes actual. When the damage is personal injury caused as a result of defective design or construction, then it is usually[22] easy to ascertain when loss occurs. However that is more the province of personal injury lawyers than construction lawyers. It is enough to note that if someone is injured through a defect to a structure, then the cause of action will accrue when the injury takes place rather than when the defect to the building was created. The duty of care sounds in *Donoghue v Stevenson*[23] and is to the effect that the builder owes a duty to 'use reasonable care to prevent damage to persons whom he may reasonably expect to be affected by his work'.[24]

15.26 Damage to the structure itself as a result of careless building does not, as has been seen (see Chapter 6), give rise to liability on the part of the builder.[25] *Classic Donoghue v Stevenson* liability is not forthcoming in relation to the structure itself as against the builder. Therefore the limitation position as regards the builder is an irrelevance. There is no liability.

15.27 However the position as regards professionals is more complicated. The professional's liability has been analysed by the House of Lords in *Murphy v Brentwood District Council*[26] *as springing not from Donoghue v Stevenson* but from *Hedley Byrne*.[27] In classic *Donoghue v Stevenson* liability the damage takes place when there is damage to the building itself, whether or not it is detectable. It was upon this basis, applying *Cartledge v E Jopling*,[28] that the House of Lords decided *Pirelli General Cable Works v Oscar Faber and Partners*.[29] In that

[22] Save for the asbestos cases like *Rothwell v Chemical & Insulating Co Ltd and another* [2007] UKHL 39, [2007] 1 AC 876.

[23] [1932] AC 562.

[24] See *Billings v Riden* [1958] AC 240. See also Lord Keith in *Murphy v Brentwood* [1991] AC 398 at 464.

[25] See *Murphy v Brentwood District Council* [1991] 1 AC 398; and *D&F Estates v Church Commissioners of England and Wales* [1989] AC 177; and *Bellefield Computer Services Ltd v E Turner & Sons* [2000] BLR 97.

[26] [1991] 1 AC 398 at 466.

[27] But note Dunn LJ in the Court of Appeal in *Forster v Outred* [1982] 1 WLR 86 where he said the opposite:

> Speaking for myself I do not find the cases on physical and material damage very helpful. This is a case of economic loss. In *Ministry of Housing and Local Government v Sharp* [1970] 2 QB 223, 278. Salmon LJ said:
>
> > It is true that in *Donoghue v Stevenson* [1932] AC 562 it was physical injury that was to be foreseen as a result of the failure to take reasonable care whereas in the present case it is financial loss. But this no longer matters, and it is now well established that quite apart from any contractual or fiduciary relationship, a man may owe a duty of care in what he writes or says just as much as in what he does: see *Hedley Byrne & Co Ltd v Heller & Partners Ltd* [1964] AC 465. No doubt in our criminal law, injury to the person is or should be regarded as more serious than damage to property and punished accordingly. So far, however, as the law of negligence relating to civil actions is concerned, the existence of a duty to take reasonable care no longer depends upon whether it is physical injury or financial loss which can reasonably be foreseen as a result of a failure to take such care.
>
> I approach this case on the basis that it is sufficient that it is financial loss that should be foreseen, and I would hold that in cases of financial or economic loss the damage crystallizes and the cause of action is complete at the date when the plaintiff, in reliance on negligent advice, acts to his detriment.

[28] [1963] AC 758.

[29] [1983] AC 1.

case a firm of consulting engineers were sued in tort in respect of defects to a chimney which caused undetectable damage. Damage in the legal sense occurred when the damage to the chimney occurred, even though it was not detectable (in the sense that cracks appeared at the top of a very high chimney where inspection did not normally take place). This state of affairs was felt to be unsatisfactory with the result that the Latent Damage Act 1986 was passed (see discussion of the Act below). The analysis in *Pirelli* was that a defect to the work was separate from the damage it caused and it was upon, and only upon, the damage to the structure (in this case cracking to a chimney) occurring that the cause of action accrued. That is wholly in accord with a *Donoghue v Stevenson* liability. The arguments in *Pirelli* assumed that the individual claimant personally suffered damage as a result of physical damage to its building. It was thus not surprising that the House of Lords applied *Cartledge v E Jopling*, and applied a liability much like in a case involving damage to goods.[30]

However, once the duty of a professional is classified as a *Hedley Byrne* liability, then the question of what is damage must be answered in a different way. This is because *Hedley Byrne* liability does not link damage to the structure in quite the same way as *Donoghue v Stevenson* does. In *Hedley Byrne* the elements of the cause of action are: (a) duty; (b) breach of duty; (c) reliance upon the professional in breach of duty; and (d) damage resulting therefrom. In *Hedley Byrne* itself the complaint related to a negligent credit reference and the case was often cited solely as one relating to negligent misstatement. However, the ambit of the cause of action is wider. The breach of duty was the giving of the negligent reference and the damage was caused by the reliance upon that negligent reference. In construction cases involving defective structures, where the professional is negligent the employer relies upon the professional and if there is either defective design or defective supervision the employer suffers damage upon reliance upon the professional which means that they have a less valuable structure than they otherwise would. If that analysis is accepted as being correct, then the cause of action accrues when the employer receives a defective structure which is less valuable than it otherwise should be. That analysis runs contrary to the *Pirelli* analysis in the House of Lords which decides that the employer will not suffer loss until the defect to the structure causes damage to it. The Latent Damage Act 1986 was, unsurprisingly, drafted on the basis that the *Pirelli* analysis correctly stated the law.[31] **15.28**

In the case of a defective structure applying *Donoghue v Stevenson*, the owner suffers damage when the structure suffers damage, usually in the form of cracking or some other deterioration. However in the case of professional negligence against a professional applying *Hedley Byrne*, damage will occur when there is reliance upon the advice given by the professional causing a detriment to the claimant. **15.29**

This is an unsatisfactory state of the law and has been recognized as such by the courts. There have been departures from *Pirelli* in Canada[32] and New Zealand,[33] the Privy Council **15.30**

[30] See, for example, *Homburg Houtimport BV v Agrosin Private Ltd* [2003] UKHL 12, [2004] 1 AC 715.

[31] See particularly s 3 of the Act and s 1, which inserted ss 14A(6)(a) and 14B(1)(b) into the Limitation Act 1980.

[32] *Kamloops (City) v Neilson* [1984] 2 SCR 2.

[33] *Invercargill City Council v Hamlin* [1996] AC 624.

in the latter case going so far as to describe the decision in *Pirelli* as unfortunate, and noting judicial and academic criticism of the judgment with apparent approval. The Court of Appeal in *Abbott v Will Gannon & Smith*[34] decided that the *Pirelli* analysis had to be followed by them. In that case it was alleged that engineers designed remedial work to a bay window in 1995; the work was completed in 1997; cracking because the remedial work was inadequate was discovered in 1999; and the claim against the engineers in contract and tort was commenced in 2003. The claim in contract was statute barred. If the cause of action accrued in tort when the work was completed in 1997, then the claim was time barred. If, however, it was when physical damage first occurred in 1999, then it probably was not. The Court of Appeal held that it was bound by *Pirelli* and held that no damage occurred until 1999 and so the matter was not time barred. It also held that even if the cause of action occurred when economic loss was suffered, then damage occurred when the cracks occurred because prior to then the building had not lost value.[35]

15.31 The second part of the decision, namely that damage occurred when the cracks occurred as a matter of economic loss, is, with respect, flawed. If liability in tort is to be analysed along *Pirelli* lines, then it is not economic loss but damage to property which is the touchstone. If, however, it is economic loss and *Hedley Byrne* liability then it is impossible to see how in *Pirelli* the cause of action did not accrue when the employer relied upon the advice of its engineers by having the defectively design executed. The reliance basis of the liability in *Hedley Byrne* has been set out in several cases and the logic of them is impeccable.[36]

15.32 The decision, as an application of the doctrine of precedent, is correct insofar as it places the proper analysis of the law in the hands of the House of Lords. But the statement by Tuckey LJ in paragraph 17: 'So what is the current state of the law of England? With three House of Lords cases to guide us it ought to be possible to give a clear answer to this question, but I regret that I feel unable to do so with any confidence' is apt and indicative of the position that the question of how damage is categorized in defective structure cases in tort has to be considered by the House of Lords, given the current incoherence. The recent cases of *Law Society v Sephton* [2006] 2 AC 543 and *Rothwell v Chemical & Insulating*

---

[34] [2005] EWCA Civ 198.

[35] Note however that Lord Mance in *Law Society v Sephton* [2006] 2 AC 543 at para 56 referred to *Pirelli* without adverse comment. However, that case was dealing with contingent liabilities and *Pirelli* was not actually examined.

[36] *Forster v Outred* [1982] 1 WLR 86; *UBAF Ltd v European American Banking Corp* [1984] QB 713; *DW Moore v Ferrier* [1988] 1 WLR 267; *Baker v Ollard & Bentley* (1982) 126 SJ 593; *Knapp v Ecclesiastical Insurance Group plc* [1998] Lloyd's Rep IR 390. See particularly Hobhouse LJ in *Knapp* who said 'From these authorities it can be seen that the cause of action can accrue and the plaintiff can have suffered damage once he has acted upon the relevant advice "to his detriment" and failed to get that to which he was entitled. He is less well off than he would have been if the defendant had not been negligent.' See also *Iron Trade Mutual Insurance Co Ltd v JK Buckenham Ltd* [1990] 1 All ER 808, a decision of Kenneth Rokison QC; and *Bell v Peter Browne & Co* [1990] 2 QB 495.The House of Lords has considered negligent advice damage in respect of valuation in *Nykredit Mortgage Bank plc v Edward Erdman Group Ltd (No 2)* [1997] 1 WLR 1627, followed by the Court of Appeal in *Byrne v Hall Pain and Foster* (a firm) [1999] 1 WLR 1849. In an unrelated area of the law, the House of Lords in *West Bromwich Building Society v Wilkinson* [2005] UKHL 144 decided that the right to recover the shortfall of a mortgage advance from the mortgagors was from the date when the mortgagors defaulted in the monthly instalment payments. This is consistent with the approach in *Forster*. See also *Law Society v Sephton* [2006] 2 AC 543 which approved *Forster* but made it clear that the loss was not a contingent liability but was actual loss.

Co Ltd and another [2007] UKHL 39, [2007] 1 AC 876 do not specifically deal with defective structure cases.

The position at present however is that a professional's duty of care will be decided on *Hedley Byrne* principles but damage will be held to have taken place when the structure itself suffers damage. **15.33**

This leads to the problem of what happens if no actual damage (as one would ordinarily understand it) is suffered, but a defect is noticed prior to damage to the structure being made manifest. For example, what is the position if an engineer prescribes a concrete mix which will cause flaking of the concrete and structural damage in an aggressive environment after, say, six years and the employer reads an article concerning this type of concrete mix after four years. Does the employer have a cause of action when he reads the article? The *Pirelli* analysis would say 'no' because the employer has discovered a defect but no damage has occurred. What if the defective mix is contained in the specification which is handed out to all possible purchasers of the building? Does damage occur when the information goes to possible purchasers on the basis that at that moment the building is less valuable than it otherwise would be if the correct mix had been used? The current state of the law provides no clear answer to these questions, which may be surprising given the amount of case law which has considered limitation of action in negligence. **15.34**

The law has been pragmatic in dealing with such questions. Three examples suffice. First, *Dove v Banham Patent Locks*[37] where a defective gate or lock permitted thieves to enter. The damage took place at the time of the burglary. Secondly, *Tozer Kemsley & Millbourne (Holdings) Ltd v J Jarvis & Sons Ltd*[38] where a defective air conditioning plant was held to cause damage when the defective plant was handed over in a defective state. Thirdly, *London Congregational Union Inc v Harriss and Harriss*[39] where there were defective drains which led to a flood and damage was held to have taken place when the flood occurred.[40] Until the House of Lords deals with the problem of reliance loss in defective structures, courts of first instance and the Court of Appeal can be expected to be pragmatic rather than principled in dealing with questions of damage. **15.35**

It is submitted that the House of Lords will have a completely clear field in deciding where the law should go given the difficulties and lack of intellectual coherence outlined in this chapter.[41] The correct view, it is suggested, is that *Pirelli* must be regarded as wrongly decided and overruled on the basis that it proceeded on an incorrect assumption as to the nature of the duty owed by the engineers. That incorrect assumption was borne of the fact that the parties were arguing about latent damage and therefore never got around to a proper analysis of the nature of the duty owed. Since a professional's liability is in *Hedley Byrne* the employer suffers damage when it relies to its detriment, which is usually when it **15.36**

[37] [1983] 1 WLR 1436.
[38] (1983) 1 Const LJ 79.
[39] [1988] 1 All ER 15.
[40] But note the powerful and persuasive dissenting judgment of Sir Denys Buckley.
[41] See also comments by the Court of Appeal in *Law Society v Sephton & Co* [2004] EWCA Civ 1627, [2005] QB 1013 at [56]–[58].

has constructed a structure which is less valuable than otherwise it might be as a result of the defective design. At that point, the building owner suffers damage of a relevant and measurable kind.[42] If the employer manages to sell the building before the defect is discovered, then it mitigates its damage. Otherwise the damage takes place when reliance loss is suffered, which is normally at the latest at practical completion, but it will usually be earlier because, in reliance on the defective design, the employer is constructing a less valuable structure than otherwise it would have and so loss takes place on the date of reliance on the design when the defective design is incorporated into the works. Waller LJ in *Spencer v Secretary of State for Work and Pensions* [2009] 2 WLR 593 put the law in summary as follows:

> I would draw the following conclusions from the authorities including the *Sephton & Co* case and the *Nykredit* case. The facts may demonstrate that no measurable damage has been suffered at the date when negligent advice has been given or negligent failure has occurred and, as the *Sephton & Co* case itself demonstrated, that will be so where damage is totally contingent. But if it can be shown that a claimant is worse off in terms that can be measured financially at the date of receipt of the advice or the negligent failure, the cause of action will accrue on that date, even though accurate measurement of damage would be difficult and some of the damage may still be contingent. In particular, if the allegation is of a failure to provide a term in a contract or a failure to provide an effective insurance policy, the cause of action will accrue on receipt of the negligently drafted contract or receipt of the ineffective policy because, as at that date, the claimant has received something of less value and has thus suffered loss.

15.37 Therefore if, for example, defective foundations mean that the building which is placed on top of them is less valuable than otherwise it might be as it is built, then loss will occur when defective foundations are built upon and the asset the building owner has is worth less than it should be. If this leads to time running out before damage is discovered, then the Latent Damage Act 1986 may apply and rescue the employer. The fact that the damage is not discovered for a period ought now as a matter of legal policy to be as irrelevant as in the solicitor reliance cases like *Forster v Outred* because legislation has expressly dealt with that injustice.

Latent Damage: The New Three-Year Period and the Fifteen-Year 'Long Stop'

15.38 Given that the current state of the law is that damage occurs when there is deterioration to the structure which is often undetectable, the potential injustice of a potential claimant never discovering that it has a cause of action until limitation has expired, led Parliament to pass the Latent Damage Act 1986 which now forms part of the Limitation Act 1980 at ss 14A, 14B, and 28A with an additional provision in relation to successive owners in s 3 of the Act. Following the Law Commission Report on Latent Damage,[43] Parliament had to

---

[42] See Lord Mance's analysis in *Law Society v Sephton* [2006] 2 AC 543 particularly at paras 60, 67–70, and 78.

[43] Law Reform Committee's 24th Report, *Latent Damage* (Cmnd 9390). See also the Law Reform Committee's 21st Report in 1977, *Final Report on Limitation of Actions* (Cmnd 6923).

consider the injustice set out in *Pirelli* by which a cause of action may accrue at a point in time when it is neither discovered nor discoverable, and balance against that potential injustice the reasonable entitlement of professionals to a degree of certainty in their professional dealings, and to protection from having to defend potentially very stale claims of which they might have very little recollection.

Parliament legislated for two dates in respect of damages for negligence not involving personal injuries. **15.39**

The first of these dates is the date at which the cause of action accrued. There are two possible alternatives when determining this date, of which the later in time is applicable. The first date is six years from the date the cause of action accrued (s 14A(4)(a)). The second is three years from the 'starting date' (s 14A(4)(b)). The starting date is the earliest date on which the claimant or any person in whom the cause of action was vested prior to the claimant first had both the knowledge required for the bringing an action for damages in respect of the relevant damage and a right to bring such an action.[44] The term 'knowledge required for bringing an action' has a special meaning under the statute. Knowledge is of the material facts about the damage in respect of which the damages are claimed and the material facts are such facts about the damage which would lead a reasonable person who had suffered such damage to consider it sufficiently serious to justify instituting proceedings. It is also knowledge that the damage was attributable in whole or in part to the act or omission which is alleged to constitute negligence, and knowledge of the identity of the defendant and any other relevant person.[45] Essentially, knowledge occurs when both the damage and the potential defendant (or other relevant party) have been or might reasonably have been discovered. Put another way, it is knowing with sufficient confidence to justify embarking on the preliminaries to the issue of a writ.[46] Knowledge that the damage is attributable in whole or in part to the potential defendant's acts or omissions means knowledge on broad terms of the facts on which the claim is based and knowing that there is a real possibility that the acts or omissions of the potential defendant have been a cause of the damage.[47] The burden of proof in these cases is upon the claimant[48] and it is important to remember that it is the acts or omissions of a defendant which are relevant; knowledge that the acts or omissions are negligent is irrelevant.[49] **15.40**

It is questionable whether these statutory provisions apply to claims for breaches of a duty of care arising out of a contract. There is authority to suggest that the Act does not extend **15.41**

[44] See Limitation Act 1980, s 14A(5). For an examination of the knowledge required for the purposes of this special time limit, see *Haward v Fawcetts* [2006] 1 WLR 683. Applied in *Harris Springs Ltd v Howes* [2007] EWHC 3271 (TCC)

[45] See Limitation Act 1980, s 14A(6)–(8).

[46] *Haward v Fawcetts* [2006] 1 WLR 683. Applied in *Harris Springs Ltd v Howes* [2007] EWHC 3271 (TCC); and *Shore v Sedgwick Financial Services Ltd & Ors* [2007] EWHC 2509 (QB).

[47] *Haward v Fawcetts* [2006] 1 WLR 683.

[48] See *Haward* and, for a discussion of the importance of pleading and proving this, see *3M United Kingdom Plc & Anor v Linklaters & Paines (A Firm)* [2006] EWCA Civ 530.

[49] *Haward v Fawcetts* [2006] 1 WLR 683.

to such claims.[50] If correct, the Act will not apply where concurrent duties are assumed in contract and tort and only cases of pure tort will be caught by the section. The contrary argument may be that *Pirelli* itself, which brought about the legislation, was a concurrent duty case.

15.42 The second date legislated for by Parliament is what has been described as a 'long stop' date which is fifteen years from the date (or the last of the dates) on which there occurred any act or omission which is alleged to constitute negligence and to which the damage in respect of which damages are claimed is alleged to be attributable. The long stop provision bars the right even if the cause of action has not accrued or the starting date has not arrived. Therefore after the long stop date any action is barred even if the damage could not reasonably have been discovered.

Limitation Periods for Guarantee and Indemnity

15.43 A guarantee is a contract and the normal rules relating to contracts apply. Six years if under hand and twelve years if under seal. The period runs from the date when the guarantor is in breach. This is at the same time as the primary obligor because the guarantor guarantees the primary obligor's obligations and is accordingly in default when the obligor is in default.

15.44 An indemnity is also a contract and the six- and twelve-year periods apply. But the cause of action accrues when the party indemnified suffers the loss which is the subject of the indemnity. If the indemnity is given by a sub-contractor, the main contractor will not have suffered loss until it is either held liable to the employer or agrees liability to the employer.[51] Therefore if, for example, a main contractor executes a contract under seal, it can be sued up to twelve years after the breach of contract. If the employer sues on the last day of the limitation period and it then takes three years of contested proceedings before the main contractor agrees liability or is held liable by judgment or award by an arbitrator, then the main contractor will (if the indemnity is given pursuant to a contract under hand) have six years from the time of the agreement or award or judgment to bring an action against the sub-contractor. If the sub-contract is under seal, then that six-year period becomes twelve years.

15.45 Thus indemnities given by sub-contractors give rise to the possibility of an action being brought against the sub-contractor a very long time after the breach of sub-contract complained of and profoundly longer than the normal six- or twelve-year limitation period.

[50] *Iron Trades Mutual Insurance Co Ltd v JK Buckenham Ltd* [1990] 1 All ER 808; and *Société Commerciale de Reassurance v ERAS* [1992] 2 All ER 82.

[51] See *County and District Properties v Jenner* [1976] 2 Lloyd's Rep 728; *Green & Silley Weir Ltd v British Railways Board* [1985] 1 WLR 570, (1980) 17 BLR 94; *Telfair Shipping Corp v Inersea Carriers* [1985] 1 WLR 553. However it is a question of construction in each case therefore, cf *Morgan Grenfell Development Capital Syndications Ltd v Arrows Autosports Ltd* [2004] EWHC 1015 (Ch) Lindsay J at [43].

The Separate Cause of Action in an Arbitration Award

An arbitration agreement involves an implied promise by the parties that they will honour any arbitration award. This implied promise means that the award produces a separate cause of action additional to any breach of contract which has given rise to the award itself. Therefore, if an award is issued, then the six- or twelve-year period will begin to run from the moment the award is made.[52] The original cause of action which brings about the award may be extinguished, but the award survives for a fresh limitation period.[53] This state of affairs, from the authorities identified above, has been more recently summarized by Mocatta J in *The 'Rosarino'*[54] where he said: 'In my judgment by entering into a charter-party with an arbitration clause . . . the parties agreed *ipso facto* to meet any award that might be made thereunder'. **15.46**

The position was reiterated, with the Limitation Acts expressly in mind, by Otton J in *Agromet Motoimport v Maulden Engineering*:[55] **15.47**

> I am satisfied that I can and should adopt the approach of Mustill & Boyd. In my judgment, the action on the award and the action to enforce an award is an independent cause of action. It is distinct from and in no way entangled with the original contract or the breach occurring from it, as reflected in the award. I have come to the conclusion that there is nothing repugnant in implying such a term into the contract. Indeed, in argument Mr Newman conceded that such an implied term could be read into such a contract. In my view, therefore, there is such an implied term that an award will be honoured when it is made. That implied term is, of course, in the original agreement of 2 September 1972, and the implied term continues, that if the award is not honoured, there is then a breach of that implied term and time begins to run for the purposes of the Limitation Act.

The reasoning in *Agromet* was approved by the Court of Appeal in *International Bulk Shipping & Services Ltd v Minerals & Metals Trading Corp of India* [1996] 1 All ER 1017. Addressing the issue of the point at which time starts to run for the purposes of enforcing an arbitration award, Evans LJ reiterated that '*conceptually the claim arises under a contractual undertaking to honour the award*' and that the six-year limitation period in that case began '*whenever the claimants became entitled to enforce the Awards*'. **15.48**

Actions on a Judgment

The position is governed by s 24(1) of the Limitation Act 1980 which provides that the limitation period for an action to enforce a judgment is six years from the date the judgment becomes enforceable. Section 38(1) ('Interpretation') states that an action, 'unless the **15.49**

---

[52] See Limitation Act 1980, s 7.

[53] See *Doleman v Ossett* [1912] 3 KB 357 *per* Fletcher Moulton at 267; and *Bremer Oeltransport GMBH v Drewry* [1935] 1 KB 753, (1933) 45 Ll L Rep 133 particularly *per* Slesser LJ at 764–5; cf *Hoole Urban District Council v Fidelity and Deposit Company of Maryland* [1916] 1 KB 24 *per* Bailhache J [1916] 2 KB 568 (CA), probably wrongly decided and inconsistent with all the other authorities.

[54] [1973] 1 Lloyd's Rep 21 at 25.

[55] [1985] 1 WLR 762 at 772.

context otherwise requires' includes 'any proceeding in a court of law, including an ecclesiastical court'. This does not prevent the execution of a judgment after six years but only bars fresh claims to enforce the judgment.[56] It does not apply to bankruptcy or winding up proceedings based on judgment debts.[57]

The Limitation Period for a Contribution

15.50 Under the Civil Liability (Contribution) Act 1978, contract breakers and tortfeasors may seek a contribution from another party who is also liable for the same damage. In the case of a claim by an employer for damages for breach of contract, the employer may choose to bring a claim against the professionals as well as the contractor. If it does so then it is usual for the defendants to the action to bring their claim for contribution at the same time as the multi-party claim is being brought. It will involve a saving of time and cost if this is done and the judge who hears the action is in a good position to decide where the relative blame lies and therefore how contribution should be apportioned. But this may not always be the case. If the professionals and the contractor are sued under contracts which contain an arbitration clause, then one or more of the defendants may choose to invoke the arbitration agreement. In that event the employer may have to sue or arbitrate separately.

15.51 A party's right to bring a claim for a contribution is two years from the date that the party seeking the contribution is held liable to the claimant.[58] That means that strictly speaking a claim for a contribution which is made in an action might be made before the cause of action in contribution itself accrues because liability will not be (or may not be) established until the end of the action. In the first edition of this book it was said that this technical position has never been the source of difficulty. That cannot any longer be regarded as correct. In *Aer Lingus v Gildacroft Ltd and Sentinel Lifts Ltd* [2006] EWCA Civ 4, the Court of Appeal had to decide whether the judgment spoken of by the statute includes a judgment on liability only for damages to be assessed or whether a judgment on quantum as well as liability is required to start time running. The Court held (allowing the appeal) that time runs from a judgment or award which ascertains the quantum and not merely the existence of the tortfeasor's liability. The judgment of the Court of Appeal makes it clear that the language of the statute is by no means clear.

The Defective Premises Act 1972

15.52 A cause of action accrues when the dwelling is completed.[59] However, if after completion a person who has done work for or in connection with the provision of the dwelling does further work to rectify the work he has already done, a cause of action in respect of that further work accrues when it is finished.

[56] *Lowsley v Forbes* [1999] AC 329.
[57] *Ridgeway Motors (Isleworth) Ltd v ALTS Ltd* [2005] 1 WLR 2871.
[58] Limitation Act 1980, s 10.
[59] Defective Premises Act 1972, s 1(5).

Actions for Sums Recoverable Pursuant to Other Statutes

The limitation period is six years from the date the cause of action accrued.[60] When the cause of action accrues depends upon the statutory provision which must be carefully considered in each case.[61] Actions arising under the Civil Liability (Contribution) Act 1978 are unaffected by this provision.[62] **15.53**

Concealed Fraud and the Extension of the Limitation Period

The Limitation Acts have always provided that time does not run if the breach of duty is subject to concealment by the contract breaker or tortfeasor. As the primary purpose of the Limitation Acts was and is to prevent the injustice which comes about through stale claims being litigated, it follows that there can be no injustice if the reason the claim has become stale is through some deliberate act of concealment or fraud on the part of the person in breach. The Limitation Acts have, throughout the years, used different wording. The current statute is s 32 of the Limitation Act 1980 (as amended by the Consumer Protection Act 1987) and for construction cases the relevant sub-sections are (1) and (2) which provide as follows: **15.54**

> 32(1) . . . where in the case of any action for which a period of limitation is prescribed by this Act, either—
>
> (a) the action is based on the fraud of the defendant; or
> (b) any fact relevant to the plaintiff's right of action has been deliberately concealed from him by the defendant; or
>
> . . .
>
> the period of limitation shall not begin to run until the plaintiff has discovered the fraud, concealment or mistake (as the case may be) or could with reasonable diligence have discovered it . . .
>
> 32(2) For the purposes of subsection (1) above, deliberate commission of a breach of duty in circumstances in which it is unlikely to be discovered for some time amounts to deliberate concealment of the facts involved in that breach of duty.

The concealment does not have to happen at the same time as the breach of duty itself. If there is concealment subsequent to the cause of action, time is stopped from running until the concealment either is discovered or ought reasonably to have been discovered.[63] **15.55**

For a brief period there was a legal heresy to the effect that negligent advice fell within s 32(2) because it was always deliberate and ipso facto the breach would not be found **15.56**

---

[60] Limitation Act 1980, s 9(1).

[61] See for example *Monckton v Payne* [1899] 2 QB 603; *Central Electricity Board v Halifax Corp* [1963] AC 785; *Sevcon Ltd v Lucas CAV Ltd* [1986] 1 WLR 462; *Yorkshire Electricity Board v British Telecommunications plc* [1986] 1 WLR 1029; and *BCL Old Coal Ltd & Others v BASF SE (formerly BASF AG) & Others* [2009] EWCA Civ 434, *The Times*, 9 July 2009.

[62] Limitation Act 1980, s 9(2).

[63] *Sheldon v RHM Outhwaite (Underwriting Agencies) Ltd* [1996] 1 AC 102.

for some time. That has now been laid to rest.[64] Lord Millett summarized the position neatly:[65]

> Given that section 32(2) is (or at least may be) required to cover cases of nondisclosure rather than active concealment, the reason for limiting it to the deliberate commission of a breach of duty becomes clear. It is only where the defendant is aware of his own deliberate wrongdoing that it is appropriate to penalize him for failing to disclose it.
>
> In my opinion, section 32 deprives a defendant of a limitation defence in two situations: (i) where he takes active steps to conceal his own breach of duty after he has become aware of it; and (ii) where he is guilty of deliberate wrongdoing and conceals or fails to disclose it in circumstances where it is unlikely to be discovered for some time. But it does not deprive a defendant of a limitation defence where he is charged with negligence if, being unaware of his error or that he has failed to take proper care, there has been nothing for him to disclose.

15.57 Therefore the covering up of bad work will come within the section if the intention is to conceal.[66] Being unaware that design was defective and issuing it[67] and covering up poor work in the normal course of construction does not.[68]

15.58 Unethical builders or workmen have featured in several cases where poor work has been covered up deliberately or concealed.[69] Each should be looked at on its facts with attention paid to the concealment. The essential question for the purposes of the statute is: 'Has there been a deliberate concealment or failure to disclose?'

Amendments to Court Actions

15.59 As has already been discussed, an amendment to a claim in an arbitration may not be made if it is outside the scope of the arbitration notice or notice to concur. That is because the notice to concur or arbitration notice will define the dispute referred and therefore the terms of reference of the arbitrator.

15.60 Therefore the amendment of claims outside the limitation period in arbitration need not be considered.

15.61 However the position in court is prescribed by statute and the Civil Procedure Rules (CPR). Section 35 of the Limitation Act 1980 provides:

> (1) For the purposes of this Act, any new claim made in the course of any action shall be deemed to be a separate action and to have been commenced—
> (a) in the case of a new claim made in or by way of third party proceedings, on the date on which those proceedings were commenced; and
> (b) in the case of any other new claim on the same date as the original action.
>
> (2) In this section a new claim means...
> (a) the addition or substitution of a new cause of action;...

---

64 *Cave v Robinson Jarvis and Rolf* [2003] 1 AC 384, [2002] UKHL 18.
65 In paras 24 and 25 of the speeches.
66 See, for example, *Applegate v Moss* [1971] 1 QB 406; *King v Victor Parsons* [1973] 1 WLR 29.
67 *Kaliszewska v John Clague and Partners* (1984) 5 Con LR 62.
68 *William Hill v Bernard Sunley* (1982) 22 BLR 1.
69 See, for example, *Gray v TP Bennett & Son* (1987) 15 BLR 1.

(3) Except as provided by . . . rules of court, neither the High Court nor any county court shall allow a new claim within subsection (1)(b) above . . . to be made in the course of any action after the expiry of any time limit under this Act which would affect a new action to enforce that claim. . . .
(4) Rules of court may provide for allowing a new claim to which subsection (3) above applies to be made as there mentioned, but only if the conditions specified in subsection (5) below are satisfied, and subject to any further restrictions the rules may impose.
(5) The conditions referred to in subsection (4) above are the following—
 (a) in the case of a claim involving a new cause of action, if the new cause of action arises out of the same facts or substantially the same facts as are already in issue on any claim previously made in the original action;. . .

Section 35(3) refers to 'rules of court', the relevant rules being CPR 17.4(1) and (2)[70] and, in respect of the addition of new parties, CPR 19.5. CPR 17.4(1) and (2) provide that where a party applies to amend its statement of case and the limitation period has expired then: **15.62**

> The court may allow an amendment whose effect will be to add or substitute a new claim, but only if the new claim arises out of the same facts or substantially the same facts as a claim in respect of which the party applying for permission has already claimed a remedy in the proceedings.

The essential point is that new claims may not be made outside the limitation period unless they come within the provisos of the CPR and s 35 of the Limitation Act 1980. If the limitation period has expired and the provisions do not apply, then the court simply has no power to allow the amendment.[71] The cases which deal with these provisions are not always easy to reconcile. In respect of three of the leading cases, for example, it has been held that an amendment will be allowed when the pre-existing action involves a claim in negligence against an architect for defective design and the amendment adds a claim for poor supervision of construction.[72] Whilst both negligent actions relate to the same building, they may take place years after each other and take place at radically different periods of time. On the other hand in another case, claims involving an air conditioning system will not admit an amendment relating to defective blockwork.[73] The question of whether the claim arises out of the same facts or substantially the same facts has been held by the courts and to be a matter of fact and degree[74] and therefore it must be a matter of judgment for the judge in each **15.63**

[70] CPR 17(3) and (4) also permit, respectively, amendment in order to correct a genuine mistake as to the name of a party and amendment to alter the capacity in which a party claims if the new capacity is one which that party had when proceedings started or has since acquired.

[71] *Welsh Development Agency v Redpath Dorman Long Ltd* [1994] 1 WLR 1409; and *Bajwa v Furini* [2004] 1 WLR 412.

[72] *Brickfield Properties Ltd v Newton* [1971] 1 WLR 862.

[73] *Steamship Mutual Underwriting v Trollope & Colls* (1986) 33 BLR 77 explaining *Idyll v Dinnerman* (1971) 1 Const LJ 294, a case which at first sight went the other way but where the unsuccessful submission was made that different defective bricks produce different causes of action. Applied in *British Airways Plc v Apogee Enterprises Inc* [2007] EWHC 93 (TCC).

[74] *Steamship Mutual Underwriting v Trollope & Colls* (1986) 33 BLR 77 at 98 *per* May LJ: 'I do not think one can look only to the duty on a party, but one must look also to the nature and extent of the breach relied upon, as well as to the nature and extent of the damage complained of in deciding whether, as a matter of degree, a new cause of action is sought to be relied upon.' See also *Del Grosso v Payne & Payne (A Firm)* [2007] EWCA Civ 340, applying *Welsh Development Agency v Redpath Dorman Long Ltd* [1994] 1 WLR 1409.

individual case. Thus in a third case it was held that an existing claim relating to negligent acts may not arise out of the same facts as a negligent misstatement.[75]

15.64 In *Secretary of State for Transport v Pell Frischmann Consultants Ltd* [2006] EWHC 2909 (TCC) Jackson J set out the following propositions which, it is suggested, accurately state the law in relation to construction projects:

(i) If the claimant asserts a duty which was not previously pleaded and alleges a breach of such duty, this usually amounts to a new claim.
(ii) If the claimant alleges a different breach of some previously pleaded duty, it will be a question of fact and degree whether that constitutes a new claim.
(iii) In the case of a construction project, if the claimant alleges breach of a previously pleaded duty causing damage to a different element of the building, that will generally amount to a new claim.
(iv) When considering whether one claim arises out of 'substantially the same facts' as a previous claim, it is necessary to examine the extent to which the facts of the first claim and the facts of the second claim overlap, and the extent to which they diverge. It will then be a matter of impression whether the test is satisfied.

15.65 The exercise, be it described as a balancing exercise or an exercise of judgment as a matter of impression is very much one for the first instance judge.[76] However recent authority suggests that the analysis must be scrupulous in relation to new facts. The Court of Appeal in *The Society of Lloyd's v Henderson* [2008] 1 WLR 2255 said at paragraph 53:

> Before us, it was argued that a new claim sufficiently 'arises out of' the same facts as an existing claim if there is a sufficient nexus between the old and the new claim, in the sense that some or a substantial part of the facts relied on to promote the new claim were relied on to promote the old claim. That takes far too broad an approach to the rule, which it effectively rewrites. The new claim does not arise out of the facts on which the old claim was based if, in order to prove it, new facts have to be added. That is why this court has said that the basic test is whether the plea introduces new facts: *Goode v Martin* [2002] 1 WLR 1828, paragraph 42.

15.66 In a construction context, in *Seele Austria GmbH & Co Kg v Tokio Marine Europe Insurance Ltd* [2009] EWHC 2066 (TCC) at paragraph 55, Christopher Clarke J said: 'In my judgment a claim in respect of 26 brick clad windows on the South and West elevations does not arise out of the same or substantially the same facts as a claim in respect of 18 stone clad windows on the North and East elevations.'

15.67 The leading construction case on this topic is the decision of the Court of Appeal in *Welsh Development Agency v Redpath Dorman Long Ltd*.[77] This decided that on the facts of the case some amendments would be allowed and others would not. An amendment of the claim against consulting engineers was allowed adding identical claims in respect of ten further

---

[75] *Hydrocarbons GB v Cammell Laird Shipbuilders* (1991) 25 Con LR 131.

[76] See *Society of Lloyd's v Laws* [2003] EWCA Civ 1887, where in commenting on comparison between the old and the new claims required by CPR r 17.4(2) the court said at para 52:

> That balancing exercise was very much a matter for the judge and, in our view, this court should not interfere with his conclusion unless he erred in principle or was plainly wrong. We should not simply substitute our view for his, even if we would have reached a different conclusion if sitting at first instance.

[77] [1994] 1 WLR 1409.

factory units on an estate when the claim was commenced in respect of two only. However an amendment was disallowed adding a claim for negligent misstatement to a claim for ordinary professional negligence. The decision was unremarkable and was a classic question of fact and degree by the trial judge which the Court of Appeal refused to interfere with. It is a useful case to study in order to see the approach to amendments in defective structure construction cases. The question of whether the new claim arises out of the same facts is a matter of impression or essentially a reasoned qualitative judgment which must be derived from a reasoned assessment of the relevant factors.[78]

Unless the new claim falls within the exceptions provided by the rules of court, leave to amend cannot be given if the limitation period expired by the time the application was heard even if the application was issued before the expiry of the limitation period. The key date therefore is the hearing, not the issue of the application to amend.[79] This is counter-intuitive but correct on the reading of the rules. **15.68**

The effect of these provisions may summarized as follows: **15.69**

(a) A new cause of action is a 'new claim' within s 35. The court may not allow a new claim to be made after the expiry of a limitation period which would affect a new action to enforce such a claim unless the new cause arises out of the same or substantially the same facts as those already in issue in the action.
(b) If the court does allow a new claim to be made by amendment, it shall be deemed to have been commenced when the writ in the original action was issued.
(c) If, when the court is considering whether to allow an amendment, the limitation period for a new action has not expired, the claim is not caught by s 35(3). CPR 17 then permits the court to allow the amendment without any particular requirement being satisfied.
(d) However, if the court has power to allow an amendment, whether this be because the period of limitation for a new action has not expired or because the claim is based on substantially the same facts as those already in issue in the action, the court retains a discretion whether or not to allow the amendment.[80]

In many defective structure cases the expiry of the limitation period will not be clear as it will not always be easy to ascertain when damage occurred and therefore whether the case falls within s 35 or not. This ought to present no particular problems when the argument centres around whether the conditions subsist which permit the exercise of discretion (for example if the new claim arises out of the same facts). But if those conditions do not subsist, then leave to amend ought not to be granted and the claimant ought to be left to commence fresh proceedings, which can then be ordered to be tried at the same time as the original proceedings. If that course is taken, then the question of limitation is not decided at an **15.70**

[78] See *Chantry Vellacott v The Convergence Group plc* [2005] EWCA Civ 20 at [104] and [115]; and *Paragon Finance Plc v DB Thakerar & Co* [1999] 1 All ER 400 at 418, in which Millett LJ suggested that it might be appropriate to describe it as essentially a matter of impression in borderline cases, but in others it must be a question of analysis. See also *Aldi Stores Ltd v Holmes Building* [2003] EWCA Civ 1882.

[79] *Welsh Development Agency v Redpath Dorman Long Ltd* [1994] 1 WLR 1409.

[80] See ibid.

interim stage but upon the full facts after a trial, which is a far more satisfactory position than on an application made in the middle of the action. The claimant is not prejudiced in this situation because if it is not statute barred, then it will be found to be not so after a trial.

15.71 The burden of proof is on the party seeking the amendment that it is just and equitable to allow the amendment.[81]

15.72 A defence and counterclaim, being by way of set off may be added to the defence notwithstanding that it is outside the limitation period.[82]

15.73 A new party can be added by amendment if the addition of the new party is necessary for the determination of the action.[83] After joinder, which takes place when the new party is served with process, the new party is, by s 35(1), deemed to have been added on the date the proceedings were commenced.

[81] See *Hancock Shipping Company v Kawasaki Heavy Industries* [1992] 1 WLR 1025; and *Woodspring District Council v Venn* (1985) 5 Const LR 54.

[82] *JFS (UK) Ltd v Dwr Cymru Cyf* [1999] BLR 17.

[83] See Limitation Act 1980, s 35(5)(b).

16

INJUNCTIONS AND SPECIFIC PERFORMANCE

Dealing with a Breach of Contract; Specific Performance or Mandatory Injunctions

The power to grant an injunction derives from s 37 of the Supreme Court Act 1981 and the court's inherent jurisdiction.[1] Specific performance is an order of the court which tells a contracting party to perform its promises. In contracts for the sale of land an order for specific performance will encompass, if necessary, the empowerment of an officer of the court to perform the obligation in the event that the party which has contracted to sell then refuses to do so. **16.01**

Specific performance is, however, to all intents and purposes a sub-species of an injunction. It is an order which is made in the event of a breach of covenant by the party against whom it is sought. An injunction is the court ordering a party to either refrain from doing something (a negative injunction) or to do something (a positive injunction). The former is the most common and least controversial. All the court has to do is order that a party refrains from doing something. Then if the injunction is breached it is relatively easy to ascertain whether or not the injunction has been broken. However, with a positive injunction the position is more difficult. **16.02**

[1] See *Fourie v Le Roux* [2007] UKHL 1, [2007] 1 WLR 320 para 25.

16.03 There is old authority to the effect that the courts will not compel the building of houses.[2] But this is too simple a proposition. The reason why an order for specific performance or a mandatory injunction can be granted is because damages are not an adequate remedy for the innocent party. The adequacy of damages is the touchstone. If damages are an adequate remedy for the innocent party, then he or she will be left to that remedy. The equitable remedies of injunctions and specific performance came into being because the common law remedy of damages was not a sufficient remedy.[3] Thus in *Costain Ltd & Ors v Bechtel Ltd & Anor* [2005] EWHC 1018 (TCC) Jackson J held that it would be a wrong exercise of the court's discretion, at the interim stage, to use the machinery of an injunction to correct any shortcomings in the certification process. Damages would be an adequate remedy.

16.04 Once it is established that damages would not be an adequate remedy then the court will have a discretion as to whether or not to grant the remedy.

16.05 The factors which will be considered in deciding whether or not to grant specific performance or an injunction if damages are not an adequate remedy are:

(a) Does the order require supervision by the court? If so, then that is a reason why an order will not be made. In building contracts, that is an exaggerated factor because the works themselves will not often require supervision, although the result may require inspection.[4]

(b) Is what has to be done defined so that exactly what has to be done is easily understood and the court will be able to know whether or not its order has been complied with? If the order is not capable of precision, then it will not be granted.[5]

(c) Is the defendant in possession of the land on which the work is to be done? If so, then the remedy is more likely to be granted. This was a common factor in cases involving building leases or other transactions where a developer undertook to erect structures on its land thus allowing an amenity to the other party.[6]

(d) Has the application been made promptly and if not would it be inequitable to grant an injunction?[7]

16.06 Orders for the specific performance of covenants to repair under leases[8] and under tenancies of social housing are granted as a matter of routine in the county court.[9] This is because the tenant will be in no position financially to do the work.

---

[2] *Wilkinson v Clements* (1872) LR 8 Ch 96 *per* Mellish LJ.

[3] *Co-operative Insurance Society Ltd v Argyll Stores (Holdings) Ltd* [1998] AC 1. See also for example *Société des Industries Metallurgiques SA v Bronx Engineering Co Ltd* [1975] 1 Lloyd's Rep 465.

[4] For a discussion of supervision see *CH Giles & Co Ltd v Morris* [1972] 1 WLR 307 at 318–19.

[5] *Wolverhampton Corp v Emmons* [1901] 1 KB 515.

[6] See for example *Carpenter Estates Ltd v Davies* [1940] Ch 160. But on analysis this is a sub-species of the requirement that the party seeking the order cannot be compensated in damages. Obviously if the claimant is in possession of the land, then he or she is capable of doing the work and then claiming damages.

[7] *Shaw v Applegate* [1977] 1 WLR 970; *Gafford v Graham* (1997) 77 P & CR 73.

[8] See *Price v Strange* [1978] 1 Ch 337; see also *Jeune v Queens Cross Properties Ltd* [1974] Ch 97.

[9] Pursuant to express statutory power under the Housing Acts.

The cases where a building contract is specifically enforced must be very rare because of the adequacy of the remedy of damages, which means that the courts will not get to consider the question of discretion. It has been suggested that an order may be made in the case of specialized work where no alternative contractor is available to do the work. If such an order is to be made, it will be necessary to show that an alternative is impossible. The mere fact that a certain type of work cannot be sourced from elsewhere only means that the structure will have to be varied. It is therefore doubtful whether a situation will arise where such an order will lie. Specific performance's first cousin is the mandatory injunction. In *Redland Bricks Ltd v Morris*,[10] Lord Upjohn set out the following criteria governing the grant of a mandatory injunction: **16.07**

> It is in this field that the undoubted jurisdiction of equity to grant a mandatory injunction, that is an injunction ordering the defendant to carry out positive works, finds its main expression, though of course it is equally applicable to many other cases. Thus, to take the simplest example, if the defendant, the owner of land, including a metalled road over which the plaintiff has a right of way, ploughs up that land so that it is no longer usable, no doubt a mandatory injunction will go to restore it; damages are not a sufficient remedy, for the plaintiff has no right to go upon the defendant's land to remake his right of way. . . . The grant of a mandatory injunction is, of course, entirely discretionary and unlike a negative injunction can never be 'as of course'. Every case must depend essentially upon its own particular circumstances. Any general principles for its application can only be laid down in the most general terms:
>
> 1. A mandatory injunction can only be granted where the plaintiff shows a very strong probability upon the facts that grave damage will accrue to him in the future. . . . It is a jurisdiction to be exercised sparingly and with caution but in the proper case unhesitatingly.
> 2. Damages will not be a sufficient or adequate remedy if such damage does happen. This is only the application of a general principle of equity; it has nothing to do with Lord Cairns' Act or *Shelfer's* case [1895] 1 Ch. 287.
> 3. Unlike the case where a negative injunction is granted to prevent the continuance or recurrence of a wrongful act the question of the cost to the defendant to do works to prevent or lessen the likelihood of a future apprehended wrong must be an element to be taken into account:
>   (a) where the defendant has acted without regard to his neighbour's rights, or has tried to steal a march on him or has tried to evade the jurisdiction of the court or, to sum it up, has acted wantonly and quite unreasonably in relation to his neighbour he may be ordered to repair his wanton and unreasonable acts by doing positive work to restore the status quo even if the expense to him is out of all proportion to the advantage thereby accruing to the plaintiff. As illustrative of this see *Woodhouse v Newry Navigation Co* [1898] 1 IR 161;
>   (b) but where the defendant has acted reasonably, though in the event wrongly, the cost of remedying by positive action his earlier activities is most important for two reasons. First, because no legal wrong has yet occurred (for which he has not been recompensed at law and in equity) and, in spite of gloomy expert opinion, may never occur or possibly only upon a much smaller scale than anticipated. Secondly. because if ultimately heavy damage does occur the plaintiff is in no way prejudiced for he has his action at law and all his consequential remedies in equity.
> 4. If in the exercise of its discretion the court decides that it is a proper case to grant a mandatory injunction, then the court must be careful to see that the defendant knows exactly

[10] [1970] AC 652 at 665–7.

in fact what he has to do and this means not as a matter of law but as a matter of fact, so that in carrying out an order he can give his contractors the proper instructions.

. . . There may be some cases where, . . . the defendant can be ordered 'to restore the right of way to its former condition'. This is so simple as to require no further elucidation in the court order. But in anything more complicated the court must in fairness to the defendant tell him what he has to do, though it may well be by reference to plans prepared by some surveyor, . . . The principle is summed up by Maugham LJ, in *Fishenden v Higgs & Hill Ltd* (1935) 153 LT 128, 142:

> I should like to observe, in the first place, that I think a mandatory injunction, except in very exceptional circumstances, ought to be granted in such terms that the person against whom it is granted ought to know exactly what he has to do.

16.08 Those criteria are as applicable to specific performance and it is obvious from the stringent criteria set out that they will be very rare in construction cases.

Injunctions, Namely Orders Otherwise than Seeking to Enforce a Contract and 'Damages in Lieu'

16.09 As with specific performance, if damages are an adequate remedy, then an injunction will not lie. As has been seen, mandatory injunctions are rare in building cases and most injunctions sought involve restraining someone from a course of action.

16.10 A trespass cannot be compensated for by an award of damages and therefore a nuisance will give rise to an injunction; for example an order stopping a crane from oversailing the claimant's land,[11] or preventing a party from causing damage to an adjoining property.[12] Where there is an interference and an injunction would ordinarily lie, the court may instead grant damages in lieu. That would be pursuant to a discretion to do so granted by the Chancery Amendment Act 1858 which provides in s 2:

> In all cases in which the Court of Chancery has jurisdiction to entertain an application for an injunction against a breach of any covenant, contract, or agreement, or against the commission or continuance of any wrongful act, or for the specific performance of any covenant, contract, or agreement, it shall be lawful for the same court, if it shall think fit, to award damages to the party injured, either in addition to or in substitution for such injunction or specific performance, and such damages may be assessed in such manner as the court shall direct.

16.11 It is important to note that the damages which lie in lieu arise out of the fact that an injunction will lie rather than not. This may be confusing conceptually because an injunction will not lie when damages are an adequate remedy. However the courts do now have the flexibility to grant damages in lieu of an injunction. But such damages ought not to be granted, save with the utmost caution, if the effect is to permit the trespasser to purchase a right

[11] *Anchor Brewhouse Developments Ltd v Berkeley House (Docklands Developments) Ltd* (1987) 38 BLR 82, overruling *Woollerton and Wilson v Richard Costain* [1970] 1 WLR 411 where damages were awarded in lieu pursuant to the Lord Cairns Act, which is the Chancery Amendment Act 1858 (21 & 22 Victoria), ch 27.

[12] As, for example, the garden wall in *Udal & Anor v Dutton & Anor* [2007] EWHC 2862 (TCC).

in property.[13] But if the injury to the claimant's rights is small, capable of being estimated in money, and can be compensated for by a small payment and it would be oppressive to grant an injunction then damages in substitution for an injunction will be granted.[14] But the court will not award nominal damages in lieu of an injunction to restrain a continuing trespass.[15] An injunction has been granted to require an occupier of a property to install appropriate floor coverings in order to prevent a noise nuisance in an adjoining property.[16]

An injunction was once refused to an employer which wanted access to its site; such refusal had the effect of restraining forfeiture of a building contract.[17] But the correctness of that decision must be doubted. There is no reason why a builder cannot obtain an adequate remedy in damages. A building contract is a commercial enterprise in which the builder receives money in exchange for work. If the contract is determined then the builder has its rights in damages. Even if it is known that the employer is not entitled to determine but is determined to do so, there is no reason why damages are not an adequate remedy for the builder.[18] But if denial of access to the site causes non monetary loss which cannot be compensated in damages, then an injunction will lie.[19] **16.12**

Interim Injunctions

Interim injunctions (formerly known as interlocutory injunctions) are injunctions obtained pending the trial of an action. Positive interim injunctions are rare, save for landlord's repairing obligation cases. Negative interim injunctions restraining a party from doing something, are common, but rare in building cases. They normally relate to nuisance or trespass by builders arising out of the operation of their works. Most interim injunction cases which come before the courts have nothing to do with construction contracts and are more to do with either attempting to restrain the publication of confidential information or seeking to enforce restrictive covenants in employment contracts. **16.13**

The purpose of a negative interim injunction is to preserve the position pending a decision on the merits.[20] The purpose of an interim mandatory injunction is similar but the **16.14**

---

[13] *Jaggard v Sawyer* [1995] 1 WLR 269 *per* Millett LJ at 286 where he makes it clear that the jurisdiction is nevertheless there to do so.

[14] *Shelfer v City of London Electric Lighting Co* [1895] 1 Ch 287 at 323 *per* AL Smith LJ, applied in *Young & Co's Brewery plc v Gordon* by David Donaldson QC (4 May 2001) (Lawtel).

[15] See *Donald Nelson v John Nicholson, The Independent*, 22 January 2000; and *Jaggard v Sawyer* [1995] 1 WLR 269. See also *Attorney General v Blake* [2001] 1 AC 268; *Experience Hendrix LLC v PPX Enterprises Inc* [2003] EWCA Civ 323; and *WWF- World Wide Fund for Nature & Anor v World Wrestling Federation Entertainment Inc* [2007] EWCA Civ 286.

[16] *Stannard v Charles Pitcher Ltd* [2002] BLR 441.

[17] *Hounslow v Twickenham Garden Developments* [1971] Ch 233.

[18] *Wiltshier Construction v Parker Developments* (1996) 13 Const LJ 129.

[19] *Bath and North East Somerset District Council v Mowlem Plc* [2004] EWCA Civ 115, [2005] 100 Con LR 1, [2004] BLR 153.

[20] See *Attorney General v Punch Ltd* [2003] AC 1046 *per* Lord Hoffmann at para 74:

> In the case of a negative interlocutory injunction, the purpose which the court is intending to fulfil is ordinarily to preserve the existing position pending a decision on the merits. It does not involve any decision as to whether any particular act falling within the prohibition would, or even arguably

preservation of the position relates to the irreparable harm which one party will suffer if the injunction is not granted.

Negative interim injunctions

16.15 The principles relating to negative interim injunctions are those set out in *American Cyanamid v Ethicon*[21] in which it was decided that first there had to be a serious question to be tried, and secondly the balance of convenience had to favour the granting of the injunction. The serious question to be tried relates to whether a permanent injunction could lie. In that regard all the factors relating to whether an injunction would lie apply. A serious question to be tried is not as high a test as formerly when the test was a strong prima facie case. Then the balance of convenience balances out the likelihood of irreversible damage on either side if the injunction is granted.[22] The touchstone for the judge is 'what if I am wrong, where does the most irreversible damage lie'.[23] So if irreversible damage is done by not granting the injunction and if the party injuncted can be compensated in damages if the injunction turns out to have been wrongly granted, then the balance of convenience will favour the granting of the injunction. Similarly if the party against whom the injunction is sought will suffer irreversible damage if it is granted, then the court is involved in a balancing exercise. The claimant must give to the court an undertaking that it will pay to the defendant any damages it might suffer in the event that the grant of the injunction turns out to be wrongful.

---

would, infringe the plaintiff's rights. It is true that a judge will not make such an order unless he considers that, in the words of Lord Diplock in *American Cyanamid Co v Ethicon Ltd* [1975] AC 396, there is a serious question to be tried. But otherwise the decision to grant or withhold such an order is usually based upon discretionary considerations of the balance of convenience. As Lord Diplock said in the *Cyanamid* case, at p 406:

> the decision whether or not to grant an interlocutory injunction has to be taken at a time when ex hypothesi the existence of the right or the violation of it, or both, is uncertain and will remain uncertain until final judgment is given in the action.

[21] [1975] AC 396.

[22] See Lord Diplock ibid at 406:

> The object of the interlocutory injunction is to protect the plaintiff against injury by violation of his right for which he could not be adequately compensated in damages recoverable in the action if the uncertainty were resolved in his favour at the trial; but the plaintiff's need for such protection must be weighed against the corresponding need of the defendant to be protected against injury resulting from his having been prevented from exercising his own legal rights for which he could not be adequately compensated under the plaintiff's undertaking in damages if the uncertainty were resolved in the defendant's favour at the trial. The court must weigh one need against another and determine where 'the balance of convenience' lies.

[23] See *Films Rover Ltd v Cannon Film Sales Ltd* [1987] 1 WLR 670 at 680 *per* Hoffmann J:

> The principal dilemma about the grant of interlocutory injunctions, whether prohibitory or mandatory, is that there is by definition a risk that the court may make the 'wrong' decision, in the sense of granting an injunction to a party who fails to establish his right at the trial (or would fail if there was a trial) or alternatively, in failing to grant an injunction to a party who succeeds (or would succeed) at trial. A fundamental principle is therefore that the court should take whichever course appears to carry the lower risk of injustice if it should turn out to have been 'wrong' in the sense I have described.

It is sometimes tempting for parties to have long affidavits or witness statements which set out the merits.[24] This is usually only forensic hot air. The court is not concerned with who is right at this stage, only with whether there is a serious question to be tried. **16.16**

It is also important to remember that the question to be tried is whether a permanent injunction will lie. It is not whether any party is in breach of its obligations or has behaved badly. **16.17**

Injunction to restrain the advertisement and presentation of a winding up petition

It is thought sometimes that it is appropriate to present a winding up petition in order to cause or pressure a party in dispute to pay sums claimed. This is wrong. An injunction will lie to prevent the advertisement and presentation of such a petition in the event that there is a bona fide dispute.[25] Such a step will also lead to an order for costs to be paid on an indemnity basis and the dismissal of the petition.[26] If there is a dispute as to payment of sums, then the proper course is to seek judgment or an arbitral award for the sums claimed and only present a winding up petition when it is clear beyond peradventure that there is no dispute as to the sums claimed and a winding up petition is appropriate because the other party is in fact insolvent. **16.18**

Injunction to prevent the call of an on-demand bond

An injunction will not lie to prevent the call of an on-demand bond save in the case of clear fraud. See Chapter 9 paragraphs 9.30–9.35 **16.19**

An injunction to prevent a public authority from entering into a contract when the tendering procedure does not comply with the Public Contracts Regulations 2006

Whilst it might be thought that the unsuccessful contractor has only lost money by losing the contract, this is incorrect and an injunction can lie.[27] **16.20**

Positive interim injunctions

There is a greater reluctance to grant a positive injunction on an interim basis than a negative one.[28] This reluctance stems from the fact that the order requires the defendant to do something rather than merely desist. Therefore, it has been said that a high degree of assurance **16.21**

---

[24] See Lord Diplock in *American Cyanamid v Ethicon* at 406:

> My Lords, when an application for an interlocutory injunction to restrain a defendant from doing acts alleged to be in violation of the plaintiff's legal right is made upon contested facts, the decision whether or not to grant an interlocutory injunction has to be taken at a time when ex hypothesi the existence of the right or the violation of it, or both, is uncertain and will remain uncertain until final judgment is given in the action.

[25] See *Penwith District Council v VP Developments Ltd* [2005] BCC 393; *Favermead v FPD Savills Ltd* [2005] EWHC 626 (Ch); *James Dolman & Co Ltd v Eric Pedley* [2004] BCC 504.

[26] *In The Matter of Sat-Elite Ltd*, 17 November 2003, Evans Lombe J; *In Re A Company (No 006798 of 1995)* [1996] 1 WLR 491, [1996] 2 All ER 417, [1996] BCC 395, [1997] 2 BCLC 48, 49 Con LR 39.

[27] *Lettings International Ltd v London Borough of Newham* [2007] EWCA Civ 1522.

[28] *Locabail International Finance Ltd v Agroexport* [1986] 1 WLR 657.

that the claimant will succeed is necessary.[29] But *American Cyanamid* applies equally to mandatory or positive injunctions as negative ones.[30] The reluctance arises because of the balance of convenience and the greater risk of injustice in a normal case which comes about because of requiring a party to do a thing as opposed to restraining that party.[31] The cases discuss these injunctions normally in the context of landlord's covenants to repair and whether specific performance will be granted to enforce such covenants on an interlocutory or interim basis.[32] In construction cases injunctions are most likely to lie where a contractor or sub-contractor wishes to require that retention money is paid into a separate account or that its materials on site are kept separate and marked until paid for,[33] or where drawings or other intellectual property is sought to be obtained by an order that it be handed over.[34] Frequently the balance of convenience in favour of the grant of such injunctions will be related to the defendant's inability to pay damages to the solvent claimant which would, absent such an injunction, suffer grave financial loss.[35] Sometimes insolvent or nearly insolvent contractors will see an opportunity to hold on to documents which are valuable to either an employer or main contractor and actually belong to that contractor or employer. Injunctions for delivery up are useful counters to that opportunism. The TCC can act quickly in such circumstances.

[29] See Megarry J in *Shepherd Homes Ltd v Sandham* [1971] Ch 340 at 351 approved in *Locabail* ibid at 664.

[30] *Zockoll Group Ltd v Mercury Communications Ltd* [1998] FSR 354.

[31] *Films Rover Ltd v Cannon Film Sales Ltd* [1987] 1 WLR 670 at 680 *per* Hoffmann J:

> The passage quoted from Megarry J in *Shepherd Homes Ltd v Sandham* [1971] Ch. 340, 351, qualified as it was by the words 'in a normal case', was plainly intended as a guideline rather than an independent principle. It is another way of saying that the features which justify describing an injunction as 'mandatory' will usually also have the consequence of creating a greater risk of injustice if it is granted rather than withheld at the interlocutory stage unless the court feels a 'high degree of assurance' that the plaintiff would be able to establish his right at a trial. . .

[32] For example, in *Peninsular Maritime Ltd v Padseal Ltd* [1981] 2 EGLR 43 there was an interlocutory mandatory injunction to use best endeavours to put a lift into working order.

[33] For retention, see *In re Tout and Finch* [1954] 1 WLR 178; *Rayack Construction Ltd v Lampeter Meat* (1979) 12 BLR 30; and *Re Arthur Sanders* (1981) 17 BLR 125. For retention of title in the seller and the need to keep goods separate where title is sought to be retained in property pending payment, see *Aluminium Industrie Vaassen v Romalpa Aluminium* [1976] 1 WLR 676, and the discussion of those clauses in Ch 17, Bankruptcy and Insolvency.

[34] See, for example, *Psychometric Services Ltd v Merant International* [2002] FSR 8; and for a construction case Coulson J in *AMEC Group Ltd v Universal Steels (Scotland) Ltd* [2009] EWHC 560 (TCC): in that case the injunction was to deliver up to AMEC Quality Assurance documentation in 15 separate categories relating to four jetty piles manufactured by USSL for AMEC.

[35] As in *AMEC* ibid.

17

BANKRUPTCY AND INSOLVENCY

Insolvency

Bankruptcy and insolvency are governed by the Insolvency Act 1986, as amended. As a result of the *Cork* report and the White Paper, *A Revisited Framework for Insolvency Law* (Cmnd 9175, 1984), insolvency laws have been subject to reform such as to create mechanisms which are designed to seek to rescue the insolvent party through the appointment of administrators and administrative receivers. The 1986 Act also put into one statute the law relating to the insolvency of companies (formerly dealt with under the Companies Acts) and individuals (formerly dealt with under the Bankruptcy Acts), and in 444 sections and 14 Schedules amended and consolidated the law of insolvency. For the purposes of this chapter, the question of insolvency will not be examined beyond the happening of an act of bankruptcy or insolvency and there will be no discussion of the duties of administrators, receivers, and liquidators. Such is the province of specialist literature. Insolvency involves either the inability of a party to be able to pay their debts as they arise or when the value of the party's liabilities exceed their assets. **17.01**

Effect of Contractor's Insolvency and Employer's Insolvency

Bankruptcy and insolvency are comparatively frequent events in the construction industry and the general law and the standard forms make provision for them. Overlying the standard terms, however, is the scheme of the insolvency legislation which is that the assets of the insolvent party must be applied pari passu in the discharge of the insolvent's debts and there must be no priority as between unsecured creditors. Thus there must not be **17.02** transfers at an undervalue or attempts to ensure that one creditor is favoured over another. The courts will be rigorous to ensure that the contractual arrangements between parties do not violate the scheme of the Insolvency Act 1986 and the rights of the other creditors.[1]

17.03 Most standard and bespoke forms make express provision for what is to happen in the event of an act of insolvency of either the employer or the contractor. The construction contract is automatically determined as at the date of the act of insolvency and each party is left to its remedies under the contract.

17.04 This normally creates no difficulty and the process of the administration of the contractual machinery begins at that stage. There are, however some special terms and conditions which need separate attention. They are:

(a) The effect of clauses which entitle the employer to use and then sell the contractor's plant and equipment.

(b) Clauses which seek to retain the title in goods in the vendor until payment (sometimes called *Romalpa* clauses).

(c) Clauses which require retention to be placed unto a special account.

(d) Clauses which permit direct payment of sub-contractors.

The Effect of Clauses Which Entitle the Employer to Use and Then Sell the Contractor's Plant and Equipment

17.05 Forfeiture clauses such as those contained in the ICE conditions (fifth edition)[2] will allow the employer to use the contractor's plant and equipment in order to complete the works. This right of usage is unaffected by the insolvency of the contractor.[3] No right or security interest in the plant and machinery is created and therefore the employer's right to use the plant and equipment stands.

17.06 If, however, there is added to the right to use the plant and equipment a right to sell the same and apply the proceeds to the debt of the contractor to the employer, then that creates a security interest in the plant and equipment which is in the form of a floating charge which must be registered. Failing registration the charge is unenforceable against the liquidator, administrator, or receiver or administrative receiver.[4] Therefore if an employer

---

[1] See, for example, the decision of the House of Lords in *British Eagle International Airlines Ltd v Compagnie Nationale Air France* [1975] 1 WLR 758, [1975] 2 All ER 390 where the House of Lords effectively struck down the clearing system of cross debts between airlines.

[2] The seventh edition has been modified to exclude the right to sell plant and apply the proceeds to the sums owed by the contractor to the employer.

[3] *Re Cosslett (Contractors) Ltd* [1998] Ch 495; *Smith (Administrator of Cosslett (Contractors) Ltd) v Bridgend County Borough Council* [2002] 1 AC 336.

[4] Under s 395 of the Companies Act 1985, which is repealed by s 1295 of the Companies Act 2006 with effect from 1 October 2009. The position is governed by s 874 of the Companies Act 2006: a charge will be void against a liquidator, administrator, or creditor unless s 860 (requirements to register a company charge) is complied with; ss 878 and 889 govern the position in Scotland.

wishes, by its contract, to cause the contractor to create a security interest in plant and machinery or other materials, then the charge thereby created must be registered.[5] This advice is slightly complicated by the fact that fixed charges do not have to be registered. However it is not always easy to decide whether a charge is fixed or floating. There has been a great deal of case law on the topic, and aside from footnoting the leading cases, the scope of this chapter will not be lengthened by a detailed analysis of what charge or security interest is floating and what is fixed. However, it is important to note that whether a charge is fixed or floating will not be determined by the language which is used. A fixed charge may be created in a deed with the title 'Floating Charge' and vice versa. What is important is the nature of the rights created. If the charge is ambulatory, then it will be a floating charge and thus registrable.[6]

Clauses Which Seek to Cause the Vendor to Retain the Title in Goods Until Payment

This must be prefaced by the proposition that where all the circumstances point to the passing of property at a particular moment then a party who wishes to say that it did not do so must be able to rely upon an express term of the contract to contradict the proposition that property has passed.[7] The famous case of Aluminium Industrie Vaassen BV v Romalpa Aluminium Ltd 8 involved a retention of title clause and the Court of Appeal had to deal with its effect in the case of a bankruptcy. The purpose of a retention of title clause is to give the seller of goods the security of being able to repossess them rather than prove in the liquidation of the company and receive a payment which is only a fraction of what is owed by the insolvent buyer. There were two types of such clause which that case had to consider. The first was a straightforward retention of title clause, in other words a clause in a contract of supply of goods which stated that until payment for the goods was made in full then the title to the goods remained with the seller and did not pass to the buyer. The second was a more difficult clause which was to the effect that if the goods were put into a manufacturing process, then the manufactured goods would become the property of the seller. **17.07**

The first clause is known as a *Romalpa* clause or a retention of title clause. This type of clause has been recognized as being of long standing in contracts for the sale of goods **17.08**

[5] For an erudite commentary on this topic, see the Law Commission paper *Registration Of Security Interests: Company Charges And Property Other Than Land (A Consultation Paper)* [2002] 164 EWLC 6 (14 June 2002).

[6] See *In re Spectrum Plus Ltd (in liquidation)* [2005] UKHL 41, [2005] 4 All ER 209; where the House of Lords applied *Agnew v Comr of Inland Revenue* [2000] 1 NZLR 223, [2001] UKPC 28, [2001] 2 AC 710, [2001] 3 WLR 454; and overruled *In re New Bullas Trading Ltd* [1994] 1 BCLC 485, CA; and *Siebe Gorman & Co Ltd v Barclays Bank Ltd* [1979] 2 Lloyd's Rep 142. See also *In re Yorkshire Woolcombers Association Ltd* [1903] 2 Ch 284, Farwell J and CA, sub nom *Illingworth v Houldsworth* [1904] AC 355. See also the application of the above cases in *In the Matter of Beam Tube Products Ltd* [2006] EWHC 486 (Ch), [2006] BCC 615, [2007] 2 BCLC 732.

[7] *Anonima Petroli Italiana SpA v Marlucidez Armadora SA* [1991] 2 Lloyd's Rep 337 at 342–3.

[8] [1976] 1 WLR 676, [1976] 2 All ER 552.

and has been enshrined in effect by the Sales of Goods Acts. The Sale of Goods Act 1979 provides for them as follows:

> Where there is a contract for the sale of specific goods or where goods are subsequently appropriated to the contract, the seller may, by the terms of the contract or appropriation, reserve the right of disposal of the goods until certain conditions are fulfilled; and in such a case, notwithstanding the delivery of the goods to the buyer, or to a carrier or other bailee or custodier for the purpose of transmission to the buyer, the property in the goods does not pass to the buyer until the conditions imposed by the seller are fulfilled.

17.09 A *Romalpa* clause will not create any rights in the goods in the buyer until the payment has been made. The clause does not create a charge because a charge could only be created by the buyer in favour of the seller as a security interest if the buyer had an interest in the goods, which, ex hypothesi, it does not.[9]

17.10 In one such case[10] the clause said:

> However, the ownership of the material shall remain with the seller, which reserves the right to dispose of the material until payment in full for all the material has been received by it in accordance with the terms of this contract or until such time as the buyer sells the material to its customers by way of bona fide sale at full market value.

It was upheld as keeping the title to the goods in the hands of the seller.

17.11 Therefore suppliers can use such clauses to keep the legal title in their goods until payment. However the better practice is to ensure that their goods are kept separate from the rest of the contractor's goods and materials on site and the better drafted clauses require the contractor not only to keep the goods separate, but also to ensure that they are marked or otherwise delineated as being the property of the seller. Otherwise the danger is that there will be a fight with the liquidator or receiver over who owns what. In the same case there was a further clause which provided 'If such payment is overdue in whole or in part the seller may (without prejudice to any of its other rights) recover or resell the material or any of it and may enter upon the buyer's premises by its servants or agents for that purpose.' Such a clause is a prudent one to have along with the requirement that the goods be kept separate.

17.12 Once the goods have been sold on, then the abovementioned clause does not have the effect of keeping title with the seller. Similar problems apply in construction contracts when the goods are incorporated into the works. Assuming that the goods are bricks or roof tiles, for example, at the point that they are incorporated into the works, then they become part of the freehold and therefore the property of the freeholder. At that point the property in the goods is transferred and two debts are created: (a) a book debt owed by the employer to the

[9] See *Clough Mill Ltd v Martin* [1985] 1 WLR 111; and *In re Peachdart Ltd* [1984] 1 Ch 131, [1983] 3 WLR 878, [1983] 3 All ER 204. As N Strauss QC sitting as a Deputy High Court Judge in *Moriarty & Anor v Various Customers of BA Peters plc (in administration)* [2008] EWHC 2203 (Ch) pointed out: 'It is perhaps surprising that decisions on this issue should sometimes turn on the linguistic skill of the draftsman of a provision which is intended to give the seller a security interest in the proceeds of sub-sales until he has been paid in full, and it may be that, but for the original decision in the *Romalpa* case, all provisions of this kind would be treated as unregistered charges . . .'

[10] *Clough Mill*, ibid.

main contractor, and (b) a book debt by the main contractor to the supplier. The retention of title clause will purport to create a charge on those book debts and will be void as against the liquidator, receiver, etc if not registered.[11]

The second type of clause, namely to the effect that if the goods are put into a manufacturing process, the manufactured goods would become the property of the seller, is problematical. This clause will be ineffective when the goods are incorporated into the works, as discussed above. But the effect of such a clause will be important in contracts of supply, as, for example, when goods are incorporated into a product in a fabricator's workshop. Then the law goes in two directions. On the one hand you have the simple proposition that if that is what the parties agree, then why should not such a clause have the effect which it says it has? On the other hand if goods are intermingled in a manufacturing or construction process, then the goods are no longer identifiable.[12] The first proposition is to be preferred, provided the goods are not incorporated into the freehold.[13] Most retention of title clauses do not go so far as to say that the property in the manufactured goods is transferred back to the seller. If there is then a conventional retention of title clause, then, depending upon its exact terms, the clause will create a charge on the manufactured goods which must be registered or it will be void against the liquidator. **17.13**

Clauses Which Require Retention to be Placed into a Special Account

If a contract provides that retention shall be kept in a separate account and held in trust for the contractor,[14] then that clause is effective to keep the beneficial interest in the retention in the hands of the contractor. However if in fact the money is not actually kept in a separate account, then the clause will not be of assistance to the contractor because the retention **17.14**

[11] Under s 395 of the Companies Act 1985, replaced by s 860 of the Companies Act 2006 from 1 October 2009. See n 4 above. See *In re Bond Worth Ltd* [1980] Ch 228, [1979] 3 WLR 629, [1979] 3 All ER 919; and *Borden (UK) Ltd v Scottish Timber Products Ltd* [1981] Ch 25, [1979] 3 WLR 672, [1979] 3 All ER 961, CA.

[12] See the dictum of Staughton J in *Hendy Lennox (Industrial Engines) Ltd v Grahame Puttick Ltd* [1984] 2 All ER 152 at 159 where he said: 'Alteration and accession: According to Bracton and Blackstone when a thing is changed into a different species, as by making wine, oil, bread or malt out of the grapes, olives, wheat or barley of another, the operator becomes the new owner thereof and is only liable (in damages) to the former proprietor for the value of the materials he has so converted.' But that statement does not deal with an express clause which transfers the property in the transformed goods to the seller.

[13] See Lord Oliver in *Clough Mill*, n 9 above, at 124:

> English law has developed no very sophisticated system for determining title in cases where indistinguishable goods are mixed or become combined in a newly manufactured article and, to adopt the words of Lord Moulton in *Sandeman & Sons v Tyzack & Branfoot Steamship Co* [1913] AC 680, 695, 'the whole matter is far from being within the domain of settled law'; and though, like Sir John Donaldson MR, I prefer to reserve my opinion, I am not sure that I see any reason in principle why the original legal title in a newly manufactured article composed of materials belonging to A and B should not lie where A and B have agreed that it shall lie . . .

See also *Hendy Lennox (Industrial Engines) Ltd v Grahame Puttick Ltd* [1984] 1 WLR 485; and *In re Andrabell Ltd* [1984] 3 All ER 407 for authority for the proposition that where property is incorporated into some other article, then provided it keeps its separate identity, the title will remain with the seller under a conventional *Romalpa* clause.

[14] For the purposes of this discussion the contractor is in the same position vis-à-vis the employer as the sub-contractor is vis-à-vis the contractor.

money will form part of the assets of the employer which cannot be separately traced.[15] That is why an injunction will lie to require payment into the separate account if that is what the contract provides.[16] Without the injunction the contractor would suffer irreparable loss in that in the event of an insolvency it would have to prove in the liquidation and receive a dividend as opposed to the money which ought to be held in trust for it.

Clauses Which Permit Direct Payment of Sub-contractors

17.15 One area of real difficulty is the provision within standard forms for the direct payment of sub-contractors by the employer. These terms usually involve the following process: (a) the contractor fails to pay the sub-contractor; (b) the sub-contractor gives notice of that non-payment to the employer or its agent; and (c) at the next time for payment the employer may pay the sub-contractor direct.

17.16 This process means that the normal chain of liability, from employer to con-tractor to sub-contractor is short circuited. The problem with this is that the work executed by the sub-contractor will create a right to payment by the contractor and the contractor is then in possession of a right to payment by the employer. That right to payment in the hands of the contractor as against the employer is an asset in the hands of the contractor which, upon an insolvency, must be applied pari passu to all the creditors. A direct payment by the employer permits the sub-contractor to recover in full as opposed to receiving a dividend and the policy of the law is to disallow that.[17]

17.17 Although the standard forms usually expressly provide that the provisions for direct payment become inapplicable upon the happening of an insolvency, that does not really deal with the difficulty. A non-payment of sub-contractors is one of the early warning signs that an insolvency is on the cards and is, if not imminent, not very far away. An employer which pays a sub-contractor direct is, in these circumstances, in danger that the eventual liquidator of the main contractor will seek to recover the value of the sub-contract work from the employer or at least bring the value of that work into account when it comes to ascertaining the credits and debits of the insolvent's assets. The danger of paying twice is significant. So far there has been no decided case which deals with this potential difficulty. It is probably because in most cases the liquidator's claim for the money is, in practice, off set by the greater debt owed by the contractor to the employer by reason of the determination.

Insolvency Set Off

17.18 Section 323 of the Insolvency Act 1986 provides for an account of mutual credits and mutual debts or other mutual dealings between a bankrupt and creditor. This is a re-enactment of s 31 of the Bankruptcy Act 1914, which has been repealed by this section.

---

[15] See *In re Tout and Finch* [1954] 1 WLR 178; *Rayack Construction Ltd v Lampeter Meat* (1979) 12 BLR 30; and *Re Arthur Sanders* (1981) 17 BLR 125.

[16] As discussed under Injunctions in Ch 16.

[17] *In re McLaughlin & Harvey plc (in liquidation)* (1996) 86 BLR 1.

Rule 4.90 of the Insolvency Rules 1986 is to like effect; headed 'Mutual Credit and Set-off', it provides:

> (1) This rule applies where, before the company goes into liquidation there have been mutual credits, mutual debts or other mutual dealings between the company and any creditor of the company proving or claiming to prove for a debt in the liquidation.
>
> (2) An account shall be taken of what is due from each party to the other in respect of the mutual dealings, and the sums due from one party shall be set off against the sums due from the other. . . .
>
> (3) Only the balance (if any) of the account is provable in the liquidation. Alternatively (as the case may be) the amount shall be paid to the liquidator as part of the assets.

The section applies to bankruptcy and the rule applies to corporate insolvency. Both are mandatory and cannot be excluded by prior agreement between the parties. Insolvency set off is self executing and takes effect on the bankruptcy date.[18] If therefore there is litigation between two parties, each with a claim against the other, then the litigation is merely part of the process of retrospective calculation for the purposes of ascertaining the net balance due from one party to another.[19] The term 'mutual dealings' is wide and does not mean simply dealings arising out of the same transaction. Therefore a claim by a subcontractor for sums due against a contractor can be met by a claim by the contractor for sums due to it under another contract. **17.19**

Separate causes of action cannot be assigned upon an insolvency because the rule contemplates a single account for a net balance.[20] **17.20**

Therefore if there is a claim by a builder for sums due pursuant to an adjudicator's decision and a cross claim by the employer, the temporary finality of the decision is only effective insofar as it states that there is a claim. But against that claim may be a set off of the counterclaim.[21] **17.21**

[18] *Halesowen Presswork & Assemblies Ltd v National Westminster Bank Ltd* [1972] AC 785, [1972] 1 All ER 641, HL.

[19] *Stein v Blake* [1996] AC 243 at 255.

[20] Ibid at 256. *Melville Dundas Limited (in receivership) and others v George Wimpey UK Limited and others* [2007] Bus LR 1182, [2007] BLR 257, [2007] UKHL 18, [2007] 1 WLR 1136, [2007] 3 All ER 889 *per* Lord Hoffmann at para 13 'any creditor who owes a debt to an insolvent company, no matter how long overdue, may set off that debt in full against his own claim in the liquidation.'

[21] *Bouygues (UK) Ltd v Dahl-Jensen (UK) Ltd* [2001] 1 All ER (Comm) 1041, [2000] BLR 522, CA. Cited with approval in *Melville Dundas Ltd (in receivership) and others v George Wimpey UK Ltd and others* [2007] Bus LR 1182, [2007] BLR 257, [2007] UKHL 18, [2007] 1 WLR 1136, [2007] 3 All ER 889.

Part VI

PROCUREMENT, COMPETITION IN THE EUROPEAN COMMUNITY, AND PARTNERING

18

PROCUREMENT AND THE EUROPEAN COMMUNITY

Overview

Lord Denning vividly described European law as being like an incoming tide, flowing into the estuaries and up the rivers of English jurisprudence.[1] In the field of public procurement the tidal surge has been particularly strong, and this is not surprising. Fundamental to the economic aims of the so-called European project has been to establish a single, integrated market, stripped of national barriers, with free movement of people, services, goods, and capital.[2] The inevitable propensity of concerns from one nation state to favour tenders from concerns of the same nationality is an obstacle to that policy, and has proved difficult to eradicate. National governments, with an eye to domestic politics, have been particularly inclined to prefer local contractors. Hence there has been considerable European intervention. **18.01**

[1] *HP Bulmer Ltd v J Bollinger SA* [1974] Ch 401 at 422.

[2] The 'four freedoms'—see Titles I and III of Pt 3 of the Consolidated Version of the Treaty Establishing the European Community (as renumbered after the Amsterdam Treaty, and afterwards cited as the EC Treaty in the text), [2002] OJ C325, now proposed as 'fundamental freedoms' in the Treaty Establishing a Constitution for Europe, [2004] OJ C310.

The aim has been to promote transparency and eliminate discrimination, so as to increase competition and open up a truly continental market.

18.02 European activity in this area has its roots in the EC Treaty itself: Art 12 forbids discrimination on grounds of nationality; Art 28 forbids restrictions on imports 'and all measures having equivalent effect'; Art 43 allows the nationals of any Member State to establish themselves in business in any other Member State, and to trade on the same terms as its own citizens; and Art 49 contains similar provisions in respect of the supply of services. These underpinning principles of European law will apply to almost any act of public procurement, even if the transaction in question is outside the scope of the specific legislation in respect of procurement.

18.03 An illustration of the effect of these provisions, and how they may act to prevent indirect as well as direct discrimination, is *Commission v Ireland*, the *Dundalk Water* case.[3] Inviting tenders for a water project, Dundalk Council required the piping to be certified as complying with an Irish technical standard. It refused to consider whether piping certified by equivalent international bodies met the same criteria. This was found to be discriminatory, as it favoured domestic suppliers.

18.04 These overarching principles of European law have been supplemented in detail, so far as procurement is concerned, by a series of directives, which have themselves been subject to amendment on several occasions. In March 2004, the legislation was overhauled and consolidated by two new directives, one regulating public sector procurement and the other regulating procurement in the utilities sectors ('the Directives').[4] Directive 2004/18/EC co-ordinates the procedures for the award of public works contracts, public supply contracts and public services contracts ('the Public Sector Directive'). Directive 2004/17/EC co-ordinates the procurement procedures of entities operating in the water, energy, transport, and postal services sectors ('the Utilities Directive'). The Directives were implemented by the Public Contracts Regulations 2006 and the Utilities Contracts Regulations 2006 respectively (collectively, 'the Procurement Regulations').

Threshold Values

18.05 The provisions of the Procurement Regulations only apply to transactions which exceed certain financial thresholds. The thresholds as originally enacted in the Directives were overhauled by Regulation 1422/2007 EC in order to bring the regime in line with the EC's conclusion of the Uruguay Round of the General Agreement on Tariffs and Trade (GATT) negotiations, as given effect to by Decision 94/800/EC. The threshold for works contracts now stands at €5.15 million.[5] The threshold values for supplies and services contracts are

[3] 45/87 [1988] ECR 4929, [1989] 1 CMLR 225.

[4] Directive 2004/18/EC of 31 March 2004 on the co-ordination of procedures for the award of public works contracts, public supply contracts and public services contracts ('the Public Sector Directive'). Directive 2004/17/EC of 31 March 2004 co-ordinating the procurement procedures of entities operating in the water, energy, transport and postal services sectors ('the Utilities Directive').

[5] Art 56, as amended by Regulation 1422/2007 Art 2(3).

much lower than those for works—for certain types of transaction they are as low as €133,000. Where a contract could be characterized as having either works or services elements, this may give rise to difficulty in knowing whether the Directives apply: a transaction might exceed the threshold for regulation if categorized as being for services rather than for works. The approach taken by the European Court of Justice has been to identify the main aim of a project, and classify it accordingly. Thus in *Gestion Hotelera Internacional SA v Communidadd Autonoma de Canarias*,[6] the European Court of Justice found that a contract providing for the renovation of a hotel was a services contract, as the building work was incidental to the main aim of the project, which was to run a hotel and casino resort. This conclusion was reached even though a significant proportion of the contract price was attributable to the building work.

Determining the value of public works will, of course, often be complicated by the way in **18.06** which a transaction is structured. Procurement may take place in several stages, and arrangements may be made for related works in separate packages. The Public Contracts Regulations provide (by reg 8(7)) that the value of the works is the total consideration payable, net of VAT. The total consideration is to be calculated in accordance with reg 8, and is to take into account (where appropriate) any form of option; renewal of the contract; prize or payment awarded by the contracting authority to the economic operator; insurance premiums; and fees, commission, interest, and other remuneration payable for banking, other financial, and design services.[7] There are a number of further rules relating to aggregation:

(1) Where proposed works, or a single requirement for goods and services, are met by the entering into of a number of contracts for the works or to fulfil the requirement, the consideration under the separate contracts is to be aggregated for the purposes of Public Contracts Regulations (reg 8(11)).[8]

(2) Where a public authority enters into a series of contracts, or a renewable contract in order to fulfil a requirement for goods or services required over a period of time, the estimated value of the contract for the purposes of the thresholds is to be calculated by aggregating the value of contracts which have similar characteristics *and* are for the same types of goods and services during the last financial year or twelve-month period immediately preceding the relevant time, and by adjusting that amount to take into account expected changes; *or* by estimating the aggregate value of the consideration payable under contracts which have similar characteristics *and* are for the same type of goods or services during:

    (a) In the case of public supply contracts, the period of twelve months from the first date of the delivery of the goods to be purchased or hired, or in the case of public services contracts, from the first date on which the services will be performed; or

    (b) The financial year if that is longer than twelve months. (reg 8(13) and (14)).

[6] C-331/92 [1994] ECR I-1329.

[7] reg 8(8).

[8] However, this is subject to an exception where a particular contract is both valued at less than €80,000 for a public services or public supply contract or €1 million for a public works contract, *and* is less than 20% of the aggregate value of all the contracts.

(3) The division of transactions into separate contracts in order to avoid the EC procurement regime is specifically forbidden. (reg 8(19))

18.07 The European Court of Justice considered the question of avoidance in *Commission v France*, the *Sydev* case.[9] In that case it was held that individual contracts with specific municipalities participating in a provincial street lighting scheme should have been aggregated and treated as a single work. It was thought significant that the invitations for tenders had been made at the same time, that the contract notices were similar, and that a body representative of the various municipalities had initiated and coordinated the contracts within a single geographical area. This willingness to aggregate contracts has significant implications for large works contracts. For example, the ECJ itself suggested that the provisions of the Directive should be applied to a road project, even though it might cross the territory of several local authorities each having administrative responsibility for a particular section of the highway, if the aggregate value of the work on the road as a whole exceeded the threshold level.[10] Anything else, it thought, would 'thwart' the objective of the European legislation.

Excluded Contracts

18.08 Certain types of contract are excluded from regulation by the Directives. These include contracts relating to telecommunications;[11] contracts which are classified as secret or where the performance of the contract must be accompanied by special security measures in accordance with law or the national interest; contracts where Art 296 of the EC Treaty applies (another provision relating to national security); certain contracts entered into in accordance with international agreements with non-EC States, or which relate to the stationing of troops; contracts which are to be entered into in accordance with the contract award procedures of organizations to which only States or international organizations belong; contracts for the acquisition of land; contracts related to the acquisition, development, or production of programme material intended for broadcasting; contracts for arbitration or conciliation services; contracts for financial services in connection with the issue, purchase, sale, or transfer of securities or other financial instruments; contracts for central bank services; contracts of employment or service; and contracts for research and development services not intended wholly to benefit the contracting authority and not wholly paid for by that authority.[12]

9 C-96/98 [2000] ECR I-8315.

10 Cf para 45 of the judgment.

11 Not previously excluded, this new exclusion reflects the liberalization and privatization of the telecommunications sector in most EC Member States.

12 See reg 6 of the Public Contracts Regulations and regs 6–9 of the Utilities Contracts Regulations.

Equality of Tenderers

In *Commission v Denmark*,[13] the European Court of Justice identified 'the principle of equal treatment of tenderers' as being 'at the very heart' of the Public Works Directive.[14] The legislation aims to give that principle effect by making provisions for contract documents, for the procedures leading up to the award of contracts, and for the criteria to be employed in the award of contracts. Here follows a summary of its key features.[15] **18.09**

Technical Specifications

The potential for national technical specifications to be used as a means of preferring domestic contractors has already been noted.[16] Hence, the Public Contracts Regulations impose particular requirements in this respect.[17] Where a contracting authority wishes to lay down technical specifications which must be met by the services, goods, or works to be supplied or performed under the contract, it must specify them in the contract documents.[18] The specifications must afford equal access to contractors, and must not have the effect of creating unjustified obstacles to competition. **18.10**

Subject to mandatory UK technical requirements (but only to the extent that those requirements are compatible with EC Treaty obligations), technical specifications must be defined by reference to technical specifications in the following order of preference:[19] **18.11**

(i) British standards transposing European standards;
(ii) European technical approvals;
(iii) common technical specifications;
(iv) international standards; or
(v) other technical reference systems established by the European standardization bodies.[20]

---

[13] C-243/89 [1993] ECR I-3353.

[14] Para 33. The decision related to an earlier Directive, 71/305/EEC, but the same view would obviously be taken of its successors.

[15] The focus will be on the provisions relating to public works, as now governed by the Public Sector Directive. In general, the regime for utilities is more flexible. A full survey of the two regimes is impossible in a general work, given the massive size of the Directives, and of the domestic legislation proposed to give them effect. It must also be recalled in what follows that the terms of the domestic legislation are unfinalized at the time of writing—see above.

[16] See the discussion of the *Dundalk Water* case above.

[17] See reg 9 of the Public Contracts Regulations.

[18] ie the invitation to tender for or to negotiate the contract, the descriptive document (if any), the proposed conditions of contract, the specifications or descriptions of the goods, services, work or works required and of the materials or goods to be used, and all supplementary documents.

[19] However, technical specifications may also be defined in terms of performance or functional requirements, so long as the requirements are sufficiently precise to allow a contractor to determine the subject of the contract and to allow the contracting authority to award the contract. It is possible to use these two ways of stating technical specifications in combination: see reg 9(7) of the Public Contracts Regulations.

[20] reg 9(6)(a) of the Public Contracts Regulations.

Only in the absence of such technical specifications may purely British standards, technical approvals, or technical specifications be referred to. Reference to such standards must, however, be accompanied by the words 'or equivalent'.[21]

18.12 Even where technical specifications are set in this way, tenders must not be rejected on grounds of non-compliance if the tenderer can prove by any appropriate means that the solutions it proposes in its tender satisfy the requirements of those technical specifications in an equivalent manner.[22] Equally, where technical specifications are defined in terms of performance or functional requirements, a tender which complies with European or international technical standards, approvals, or specifications may not be rejected if the contractor proves by any appropriate means that those specifications address the prescribed performance or functional requirements.[23] The certificates of recognized laboratories or standards bodies established in other Member States must be accepted when considering whether prescribed technical specifications have been met.[24]

18.13 Technical specifications which refer to materials or goods of a specific make, source, or process, or to trademarks, patents, types, origin, or means of production, must not be made if that would have the effect of favouring or eliminating particular tenderers.[25] This rule will, exceptionally, not apply where the subject of the contract makes the use of such references indispensable or if the subject of the contract cannot otherwise be described. However, even in this situation the specification must state that equivalents are acceptable.[26]

Procedures for Seeking Tenders

18.14 There are four principal procedures for seeking tenders—the open, the restricted, the negotiated, and competitive dialogue. This latter procedure was introduced by the Public Sector Directive, as implemented by the Public Contracts Regulations, and is available only in the case of 'particularly complex' contracts.[27] There is no equivalent provision for utilities.

Open procedure

18.15 Where the open procedure is adopted, any interested contractor may tender for the project in question. A key requirement of the anti-discrimination regime is that contracts be advertised outside of the domestic media. Hence, the contracting authority must publicize its intention to seek tenders in the *Official Journal of the European Communities* (OJ) as soon as possible after that intention is formed, and must not advertise domestically beforehand.[28] Contractors must be given at least fifty-two days from the despatch of the notice to submit their tenders. Slightly shorter periods are allowed if the notice is transmitted to the

[21] Ibid, reg 9(6)(b).
[22] Ibid, reg 9(10).
[23] Ibid, reg 9(11).
[24] Ibid, reg 9(14) and (15).
[25] Ibid, reg 9(16).
[26] Ibid, reg 9(17).
[27] Cf Arts 28 and 29 of the Public Sector Directive; reg 18 of the Public Contracts Regulations.
[28] See reg 15 of the Public Contracts Regulations. The form which the notice in the OJ must take is prescribed by Annex VIIA of the Public Sector Directive.

Journal electronically (a reduction of seven days),[29] and if the contract documents are available for unrestricted inspection on the internet (a reduction of five days).[30]

An alternative procedure allows contracting authorities to publicize their intention to seek tenders in the future by a 'prior information notice' to the Commission, which should be given at the beginning of their financial year.[31] If this course is taken, there remains an obligation to publish a notice when tenders are sought, but the time allowed for the submission of tenders may be reduced, usually to thirty-six days, but exceptionally to twenty-two days.[32] **18.16**

If contract documents are not available online, they must be sent to any contractor who applies for them in good time, within six days of its request. Reasonable requests for further information made within good time must also be answered within six days. If these time limits are not observed, the time for submitting tenders must be extended. An extension must also be granted if it is necessary for the contractor to carry out site inspections or similar.[33] **18.17**

Restricted procedure

Under the restricted procedure, only pre-selected contractors may be invited to tender for the contract. Once again, there are strict requirements for notice to be given, with a requirement to advertise at the outset of the process in the *Official Journal*.[34] Contractors must usually be given at least thirty-seven days to apply for pre-selection (again, there are shorter time periods if electronic methods of communication are used).[35] **18.18**

Contractors may be excluded from the short lists for tendering only on limited grounds, either of their personal standing (for example because they have been convicted of certain offences or are insolvent),[36] or because they fail to satisfy requirements as to economic and financial standing,[37] or technical or professional ability.[38] However, the number of contractors invited to tender may be further restricted, so long as objective and non-discriminatory criteria, which have been stated in the contract notice, are applied.[39] The minimum number of contractors to be selected must also be stated in the notice, and it must be sufficient to ensure genuine competition and in any event exceed five. However, fewer than that number may be shortlisted if insufficient applicants meet the requisite criteria.[40] **18.19**

Once a short list has been made, written invitations to tender must be sent simultaneously to each contractor selected, either enclosing the contract documents, or indicating where **18.20**

---

[29] reg 15(5) of the Public Contracts Regulations.
[30] reg 15(6) of the Public Contracts Regulations.
[31] Art 35 of the Public Sector Directive. Provisions for the form of the notice are found in Annex VIII.
[32] Ibid, Art 38.
[33] See generally reg 15 of the Public Contracts Regulations.
[34] reg 16(2) of the Public Contracts Regulations.
[35] See Art 38 of the Public Sector Directive; reg 16(3) and (5) of the Public Contracts Regulations.
[36] See Art 45 of the Public Sector Directive.
[37] See Art 47 of the Public Sector Directive; reg 16(7)(b)(i) of the Public Contracts Regulations.
[38] See Art 48 of the Public Sector Directive; reg 16(7)(b)(ii) of the Public Contracts Regulations.
[39] reg 16(9)(a) of the Public Contracts Regulations.
[40] reg 16(9)(b) of the Public Contracts Regulations.

they can be obtained (for example, from a website).[41] The invitation must specify the date, address, and language for tenders, the information which will be required, and the weighting of the criteria which will be used to select the contractor.

18.21 There are usually at least forty days to tender, although this is again subject to truncation where electronic means of communication are employed, or where a prior information notice has been published. As with the open procedure, there are obligations to meet reasonable requests for further information within a reasonable time, and extensions must be granted where this is necessary.[42]

Negotiated procedure

18.22 The negotiated procedure is adopted where the contracting authority negotiates the terms of the contract with one or more contractors which it has selected. Public sector authorities may use it only in exceptional circumstances (utilities are not so restricted).[43] Where the contract has been advertised by publication of a contract notice, these include where use of the open or restricted procedure has been discontinued after only irregular tenders were received or tenders only from contractors legitimately excluded (eg on grounds of personal or financial standing);[44] where the nature of the contract does not, exceptionally, permit prior overall pricing or, in the case of a contract of services, where the nature of the services to be provided cannot be established with sufficient precision to permit selection by the open or restricted procedures; or where the contract is not profit making and is contemplated only for research and development purposes.[45]

18.23 The negotiated procedure may be used without advertisement where no, or no suitable, tenders or applications were submitted in response to an open or restricted procedure (provided that the initial conditions of contract are not substantially altered and that a report is sent to the Commission if it requests); when, for technical, artistic, or intellectual property reasons, the contract may be awarded only to a particular contractor; or in cases of extreme and unforeseen urgency (the urgency not being the fault of the contracting authority).[46] Where the contract is for additional works or services necessary for the performance of an existing contract and which were not foreseen when the initial contract was made, the contracting authority may, without advertisement, negotiate the supplemental contract

---

[41] See Art 40 of the Public Sector Directive; reg 16(13) of the Public Contracts Regulations.

[42] See generally reg 16 of the Public Contracts Regulations.

[43] See Art 30 of the Public Sector Directive and reg 13 of the Public Contracts Regulations for use of the negotiated procedure with prior publication of a contract notice, and Art 31 and reg 14 for its use without such prior publication.

[44] Even here, however, the need to advertise is lost if every contractor who unsuccessfully tendered or applied under the abortive open or restricted procedure is invited to negotiate.

[45] This is commonly relied on by HM Treasury in PFI projects. The Commission has expressed some discontent at the Treasury's readiness to resort to the negotiated procedure on these supposedly exceptional grounds: cf Press Release IP/00/869 (27 July 2000). However, nothing has come of their objections, and Richards J gave short shrift to a similar complaint in *R (Kathro) v Rhondda Cynon Taff County BC* [2001] EWHC 4527 (Admin). See A Brown, 'The Impact of the New Public Procurement Directive on Large Public Infrastructure Projects' (2004) 13 PPLR 160 at 163–4.

[46] As one would expect, the tendency of the ECJ has been to interpret this entitlement narrowly: C-24/91 *Commission v Spain* [1992] ECR I-1989; C-107/92 *Commission v Italy* [1993] ECR I-4655.

directly with the original contractor, so long as the aggregate value of any supplemental contracts does not exceed 50 percent of the value of the original contract. Likewise, where a contract consists of the repetition of similar works to those in a contract previously awarded by an open or restricted procedure, and where the new works are in conformity with the basic project for which the original contract was made, a negotiated procedure may be held with the original contractor so long as: (a) this possibility was stated when the original contract was put up for tender, and (b) the procedure for the award of the subsequent contract is used within three years of the conclusion of the original contract.

Where an advertisement is necessary, the procedure to be adopted to select contractors for negotiation is similar to that for selecting contractors for invitations to tender under the restricted process: hence, for example, there must be a notice in the *Official Journal*, to which responses are usually invited within not less than thirty-seven days.[47] Again, contractors may only be excluded from the process on limited grounds of personal standing or failure to satisfy minimum financial, technical, or professional standards. Shortlisted contractors must be selected according to published and non-discriminatory criteria, and (unless insufficient suitable contractors apply), the shortlist for negotiation should not be fewer than three. As with the restricted procedure, shortlisted contractors must all be contacted simultaneously, and provided with necessary information. There are, however, no time limits for the duration of the negotiations which follow. **18.24**

The manner in which negotiations are to be conducted is not specified in detail, although usual European law principles of non-discrimination and transparency are obviously expected of the contracting authority. Indeed, the need for equality of treatment, and particularly the non-discriminatory provision of information, is stated in the Public Sector Directive.[48] The Public Sector Directive also confirms that negotiations may take place in stages, and contractors can be excluded as the process continues; this must be stated in advance in the contract notice or specifications however.[49] Where possible, moreover, the contracting authority must ensure that the number of contractors to be invited to negotiate the contract at the final stage is sufficient to ensure genuine competition, and contractors may be eliminated only in accordance with the authority's stated award criteria.[50] **18.25**

Competitive dialogue

As noted, the new competitive dialogue procedure applies only to 'particularly complex' public sector projects. Before considering what exactly that means, it is useful to summarize some of the features of the new process,[51] which are as follows (the similarities with the negotiated procedure will be clear): **18.26**

1. The contracting authority publishes a contract notice in the *Official Journal* setting out its needs and requirements.[52]

---

[47] See reg 17 of the Public Contracts Regulations.
[48] Art 30(3). See also reg 17(21) of the Public Contracts Regulations.
[49] Art 30(4); reg 17(22) of the Public Contracts Regulations.
[50] Art 44(4).
[51] Art 29.
[52] Public Contracts Regulations, reg 18(4).

2. A shortlist of at least three candidates is selected, subject to the same requirements as for a shortlist for the negotiated process.[53]
3. The authority invites those selected to participate in a 'dialogue',[54] 'the aim of which shall be to identify and define the means best suited to satisfying its needs'[55] during which it 'may discuss all aspects of the contract with the participants selected'[56] and in which it 'shall ensure equality of treatment among all participants'[57] and 'not reveal to the other participants solutions proposed or other confidential information communicated by a participant without that participant's agreement'.[58]
4. In the course of the dialogue, the contracting authority may specify that payments be made to the participants in respect of expenses incurred participating in the competitive procedure.[59]
5. The procedure may take place in 'successive stages in order to reduce the number of solutions to be discussed during the dialogue stage by applying the award criteria in the contract notice or the descriptive document.'[60] However, where the competitive dialogue procedure is to take place in successive stages, the contracting authority must ensure that a sufficient number of economic operators are invited to participate in the final stage to ensure genuine competition.[61]
6. The dialogue continues until the contracting authority 'can identify one or more solutions, if necessary after comparing them, which are capable of meeting its needs'.[62]
7. The authority informs the candidates that the dialogue is concluded, and invites them to submit their final tenders on the basis of the solution or solutions presented and specified during the dialogue.[63]
8. The tenders must contain all the elements required and necessary for the performance of the project, and the authority may request that they be 'clarified, specified and fine-tuned'. However, such clarification, specification, fine-tuning or additional information may not involve changes to the basic features of the tender or the call for tender, variations which are likely to distort competition or have a discriminatory effect.[64]
9. The authority must assess the tenders received on the basis of the award criteria laid down in the contract notice or specifications, and must choose the most economically advantageous.[65]
10. The authority may request the successful tenderer to clarify aspects of the tender or confirm commitments contained in the tender provided this does not have the effect

[53] Ibid, reg 18(10)–(12).
[54] Ibid, reg 18(16).
[55] Ibid, reg 18(20).
[56] Ibid, reg 18(21)(a).
[57] Ibid, reg 18(21)(b).
[58] Ibid, reg 18(21)(c).
[59] Ibid, reg 18(29).
[60] Ibid, reg 18(22).
[61] Ibid, reg 18(23).
[62] Ibid, reg 18(24).
[63] Ibid, reg 18(25).
[64] Ibid, reg 18(26).
[65] Ibid, reg 18(27).

of modifying substantial aspects of the tender or of the call for tender and does not risk distorting competition or causing discrimination.[66]

The aim of this process is to allow greater flexibility where 'contracting authorities which carry out particularly complex projects . . . find it objectively impossible to define the means of satisfying their needs or of assessing what the market can offer in the way of technical solutions and/or financial/legal solutions', and this is without 'any fault on their part'.[67] A 'particularly complex' contract is therefore defined as one where the contracting authority is 'not objectively able to: (a) define the technical means . . . capable of satisfying their needs or objectives; or (b) specify the legal and/or financial make-up of a project or both.'[68] Examples given[69] are 'important integrated transport infrastructure projects, large computer networks or projects involving complex and structured financing the financial and legal make-up of which cannot be defined in advance'. The point has been well made that these provisions are both general and nebulous.[70] Moreover, there is considerable overlap with the existing negotiated procedure, and the boundaries between them may also be subject to debate. Hence, this new procedure has had a mixed reception from lawyers, and it is possible that its precise scope will need to be worked out in the future through the courts.[71] **18.27**

Other Innovations in the New Directives

The Directives also contain a number of provisions intended to modernize procurement, and more readily to embrace standard form arrangements. So far as the latter is concerned, contractors may enter 'framework agreements' with contracting authorities to establish the terms governing contracts to be awarded during a given period, in particular with regard to price and, where appropriate, the quantity envisaged. Such agreements may only themselves be concluded after contractors have been selected in accordance with the Directives' procedures (open, restricted, etc), may not usually exceed four years in duration, may not be significantly amended, and must not be used in such a way as to prevent, restrict, or distort competition. They are apt to allow repeated business to be put out to tender only to parties to a framework agreement, dispensing with the need for notice to be given in respect of each contract to the world at large.[72] **18.28**

The modernizing agenda of the Directives also manifests itself in their attempt to encompass electronic methods of procurement. There is provision for 'dynamic purchasing systems', defined as a 'completely electronic process for making commonly used purchases, the characteristics of which, as generally available on the market, meet the requirements of **18.29**

[66] Ibid, reg 18(28).

[67] Recital 31 to the Public Sector Directive.

[68] Reg 18(1) of the Public Contracts Regulations.

[69] Recital 31 to the Public Sector Directive.

[70] Brown, n 45 above, pp 170–172.

[71] Brown, ibid; R Williams, 'The New Procurement Directives of the EU' (2004) 13 PPLR 153; S Treumer, 'Competitive Dialogue', ibid, 178.

[72] Art 32 of the Public Sector Directive/reg 19 of the Public Contracts Regulations; Art 14 of the Utilities Directive/reg 18 of the Utilities Contracts Regulations.

the contracting authority'.[73] Such arrangements must be limited in duration (usually four years) and open throughout their validity to any contractor which satisfies the selection criteria and has submitted a tender that complies with the specification.[74] The purchasing system allows the contracting authority to issue a contract notice stating the specifications of the desired purchases, allowing contractors full and unrestricted access to the specifications by electronic means. Formal provision is also made for tenders to be awarded by electronic auctions, 'a repetitive process involving an electronic device for the presentation of new prices, revised downwards, and/ or new values concerning certain elements of tenders, which occurs after an initial full evaluation of the tenders, enabling them to be ranked using automatic evaluation methods'.[75]

Criteria for Awarding Contracts

18.30 Having completed both the selection process (in the case of a non-open procedure) and the tendering process, the essential question arises as to the basis on which the contract itself may be awarded.[76] Only two are permitted: either the lowest price, or 'the most economically advantageous.'[77] So far as the latter is concerned, an indicative list of the criteria the contracting authority may have in mind is offered: 'quality, price, technical merit, aesthetic and functional characteristics, environmental characteristics, running costs, cost-effectiveness, after-sales service and technical assistance, delivery date and delivery period or period of completion'. The actual criteria must be specified in the contract notice or documents, which must also state the relative weighting given to each of the criteria in question. The weightings can be expressed by providing for a range with an appropriate maximum spread. If weighting is not possible 'for demonstrable reasons', the contracting authority must instead indicate the applicable criteria in the contract notice or documents, and list them in descending order of importance.[78] The overriding principle, as expressed by the ECJ, is that contracting authorities must disclose 'all the elements to be taken into account by the contracting authority in identifying the economically most advantageous offer, and their

---

[73] Art 1(6) of the Public Sector Directive.

[74] Detailed rules, including as to notice, are set out ibid Art 33 and Art 15 of the Utilities Directive; time limits are set down in Arts 38 and 45 of the respective Directives. Since this relates primarily to supply of goods, rather to construction contractors, the full details will not be set out here.

[75] Art 1(7) of the Public Sector Directive. Detailed provisions are made ibid Art 54 and in Art 56 of the Utilities Directive. See also S Arrowsmith, 'Electronic Reverse Auctions under the New EC Procurement Directives' (2005) 14 PPLR 203.

[76] See Art 53 of the Public Sector Directive and Art 55 of the Utilities Directive.

[77] reg 30(1) of the Public Contracts Regulations; reg 30(1) of the Utilities Contracts Regulations. The former is the 'default' basis—*Harmon v House of Commons* (2000) 72 Con LR 21. A tender may be rejected if the price is abnormally low, so long as the tenderer is first permitted to explain how it has been reached: Case 76/81 *SA Tranportoute et Travaux v Minister of Public Works* [1982] ECR 417.

[78] For a discussion of the extent to which secondary policies not inconsistent with fundamental aims of the legislation like equality of tenders may be taken into account in the specification of relevant criteria, cf J Arnould, 'Secondary Policies in Public Procurement: the Innovations of the New Directives' (2004) 13 PPLR 227. The Public Contracts Regulations place some emphasis on the need for modern social concerns, such as providing for disabled access, to be taken into account in public procurement projects (see reg 9(3)).

relative importance'.[79] This disclosure obligation is a 'necessary consequence of the overarching principles of transparency and equal treatment',[80] and its application does not depend on the consequences of non-compliance. A claim can be brought for breach of the regulations by claimants who '*risk suffering . . . loss or damage*' (Art 47(6) of the Public Sector Directive).

Once a contract is awarded, notice must be given in the *Official Journal* within forty-eight days, in a specified form.[81] **18.31**

The contracting authority must also, as soon as possible after the decision has been made, inform any contractor which took part in the tender or selection to tender for the award of the contract, or of any decision to abandon or to recommence the contract award procedure. The contracting authority must give reasons for its decision and provide them in writing if so requested. The contractor must also be informed of the reasons why it was unsuccessful. Where it submitted an admissible tender, the contractor must also be informed of the characteristics and relative advantages of the successful tender and the name of the successful contractor. The contracting authority must keep detailed records of the contract process, and must, on request, send a report to the Commission in which such details are made available. These requirements (to which there are certain exceptions)[82] are designed to assist in the policing of the procurement system, and to ensure that interested parties will have the necessary information to allow them to perceive, and act on, any improper behaviour. **18.32**

Post-award Variations

While it is almost inevitable, in the case of complex projects, that there will be post-tender discussions with the successful party, problems may arise if negotiations lead to a significant alteration in the terms of the tender. This may be unfair to other bidders, being contrary to basic principles of transparency and equality of treatment.[83] Both the courts and the Commission have shown themselves flexible to the inevitable developments that may take place in negotiations in very large contracts; however, there would undoubtedly come a point at which a significant change in specification agreed between contracting authority and contractor after tenders had been submitted would be held to be a breach of the obligations which the legislation creates.[84] The ECJ considered the issue as to when an amendment to a public contract constitutes a new contract, and thus requires a fresh procurement exercise, in *Pressetext Nachrichtenagentur GmbH v Austria*. It concluded that, for amendments to cause a new procurement to arise, they have to be 'materially different in character from the original contract . . . such as to demonstrate the intention of the parties to renegotiate **18.33**

[79] See the *Emm G Lianakis* case [2008] All ER (D) 172 (Jan), para 36, citing the *Universale-Bau* [2002] ECR I-11617 and *ATI EAC* cases.

[80] *Risk Management Partners* [2008] EWHC 1094 (Admin) at para 87.

[81] Cf Annex VIIA of the Public Sector Directive for the form which such a notice is to take.

[82] See reg 32(13) of the Public Contracts Regulations.

[83] Cf C-87/94 *Commission v Belgium* [1996] ECR I-2043.

[84] The limited right to agree to variations post-tender is, in the context of competitive dialogue, now explicitly provided for in the legislation: see item 10 of the characteristics of that procedure set out in para 18.26 above.

the essential terms of the contract.' The ECJ further provided guidance as to what may constitute a 'material change', seeing that a change may be material if it serves to:

(1) introduce conditions which, had they been part of the initial award procedure, would have allowed for the admission of tenderers other than those initially admitted or would have allowed for the acceptance of a tender other than the one initially accepted;
(2) extend the scope of the contract considerably to encompass services not initially covered by the initial award procedure;
(3) change the economic balance of the contract in favour of the contractor in a manner not provided for in the terms of the initial contract.

Enforcement, Penalties, and Remedies

18.34 The Commission, either of its own initiative or on the complaint of a third party, may bring enforcement proceedings in the European Court, under Art 186 of the EC Treaty. A direct right of action in domestic courts was, however, also introduced, by the Remedies Directives.[85] These directives have been replaced by a consolidated Remedies Directive, Directive 2007/66/EC which had to be implemented in the UK by December 2009.[86] The Remedies Directive does not address transitional issues. The OGC's current preference was to apply the new rules only to those procurement processes beginning after the implementation date, and the Commission confirmed that this position was acceptable.

18.35 The UK Office of Government Commerce (OGC) published its second consultation on the implementation of this consolidated Remedies Directive. As part of this consultation, draft amendments to the Public Contracts Regulations have been published. Such amendments are to be applied by analogy to the Utilities Contracts Regulations. A subsequent period of consultation ended on 24 July 2009, for the draft amendments to be laid before Parliament.

18.36 Under the resulting Public Contracts (Amendment) Regulations 2009, the courts will be required to make a declaration of ineffectiveness for any public sector or utility contract awarded in breach of certain EU public procurement rules.[87] Those in breach of those rules will also be liable to financial penalties. This position represents a departure from the position under the current Regulations, where the remedy for a contractor who has suffered, or risks suffering, loss by virtue of a breach of the duty in question is restricted to bringing a damages action once a contract has been awarded in breach of the Regulations.

18.37 The proposed amendments do not vary the existing three-month time limit for the bringing of proceedings. Time starts to run for these purposes from the date when grounds for starting the proceedings first arose.[88] The Court has a discretion to extend this time limit

[85] Directives 89/995/EEC and 92/12/EEC.
[86] See Public Contracts (Amendment) Regulations 2009 (SI 2009/2992).
[87] Reg 47J(a).
[88] Reg 47D(2).

where it considers 'that there is a good reason for doing so.'[89] Special time limits apply for seeking a declaration of ineffectiveness.[90] Where a relevant contract award notice has been published in the Official Journal, such proceedings must be brought within thirty days of the date following the date the notice was published.[91] In all other cases, the relevant time limit is six months from the day following the day the contract was entered into.[92]

Where such proceedings are brought, the court may issue an injunction (even against the Crown) or otherwise make an interim order suspending the award procedure or the implementation of any decision taken by the contracting authority. In considering the power to make an interim order in *Lion Apparel Systems v Firebuy Ltd* [2007] EWHC 2179 (Ch), Morgan J held the applicable principles to be those laid down in *American Cyanamid*. Powers of final relief include the setting aside of decisions of the contracting authority (so long as no contract has actually been entered into), and (in cases where loss is suffered) the award of damages.[93] Where a contracting authority has failed to comply with its obligations of equality, transparency or objectivity, there is no scope for a margin of appreciation.[94] Where, however, a contracting authority decides on 'matters of judgment or assessment' it does have a margin of appreciation, and the court will only overturn its decision where the authority has committed a 'manifest error', ie 'where an error has clearly been made'.[95] **18.38**

Where the Court is satisfied that a decision or action taken by a contracting authority was in breach of their duty owed to economic operators, and before a contract has been entered into, the Court has the power to order that the decision or action be set aside; order that the contracting authority amend any document; and award damages to an economic operator which has suffered loss in consequence of the breach.[96] **18.39**

Competition: The European Community

Competition law may impact on the construction industry in the same way as it may on any other, in its concern to prevent the development of cartels, price-fixing, restrictions on tenders, and abuses of dominant positions. It too is now heavily permeated by European principles, with their particular concern to prevent anti-competitive behaviour creating barriers between inter-State commerce and thereby impeding the development of the single market. Moreover, the main domestic provisions (the Competition Act 1998 and the Enterprise Act 2002) have in large part been designed to bring domestic law in line with the European. Competition is now a well recognized area of specialty for commercial **18.40**

[89] Reg 47D(4).
[90] Reg 47E.
[91] Reg 47E(2)(a).
[92] Reg 47E(2)(b).
[93] See reg 47 of the Public Contracts Regulations.
[94] *Firebuy* at para 36.
[95] Ibid, at 38.
[96] Reg 47I.

lawyers, and in a work of this scope it is impossible to give anything other than a broad overview.[97]

The EC Treaty

18.41 The foundation of the European law is once again the EC Treaty. Article 81 forbids 'as incompatible with the common market all agreements . . . decisions . . . and concerted practices which may affect trade between Member States and which have as their object or effect the prevention, restriction or distortion of competition within the common market'. Specific examples are given of measures that are offensive, including those which:

- directly or indirectly fix purchase or selling prices or any other trading conditions;
- limit or control production, markets, technical development, or investment;
- share markets or sources of supply;
- apply dissimilar conditions to equivalent transactions with other trading parties, thereby placing them at a competitive disadvantage;
- make the conclusion of contracts subject to acceptance by the other parties of supplementary obligations which have no connection with the subject of such contracts.

18.42 Any agreement or decision which breaches Art 81 is declared to be automatically void: Art 81(2). However, Art 81(3) provides that in certain circumstances agreements may be granted an exemption from the general prohibition. Exempt agreements must contribute to improving the production or distribution of goods, or to promoting technical or economic progress. Any restrictions they impose must be indispensable to their aims, and must not be such as to create the possibility that competition will be eliminated in respect of the products in question. Consumers must be allowed 'a fair share of the resulting benefit.' Such exemptions may be granted either individually or *en bloc*. Individual exemptions have been granted to companies collaborating in joint ventures to develop new technologies. Various block exemptions have also been granted: a commonly encountered example is that which allows the selective distribution and servicing of motor vehicles by preferred dealerships.[98]

18.43 Article 82 of the EC Treaty prohibits abuses of a dominant position 'within the common market or a substantial part of it', insofar as it may affect trade between Member States; again, examples are given which include the imposition of unfair prices or trading conditions; limiting production, markets or development to the prejudice of consumers; applying dissimilar conditions to equivalent transactions with other trading parties so as to place them at a competitive disadvantage; and making contracts subject to acceptance by the other parties of supplementary obligations which have no connection with the subject of such contracts.

18.44 Articles 81 and 82 have direct effect: that is, they create rights which can be enforced in the courts of the Member States. The articles following (Arts 84–86), inter alia, empower the

[97] The leading specialist work is P Roth (ed), *Bellamy & Child: European Law of Competition*, 5th edn, London, 2001.

[98] Reg 1475/95.

Commission to legislate to prevent abuses, and to take action against Member States which do not give the measures effect.

The requirement that trade between Member States be affected for European law to be engaged has been widely construed by the courts. It is sufficient for there to be a potentially adverse effect, whether direct or indirect.[99] Although it has been suggested that few agreements between construction undertakings will impact on trade between Member States, given the similarity within EC and domestic competition law, this apparent limitation on the reach of EC principles may not be of great practical importance. **18.45**

Application of Article 81

The prohibition in Art 81 applies to practices as well as agreements: there is no need for a formal contract, and the focus of the court will be on economic effect and not legal form.[100] The exchange of information between competitors, leading to the informal coordination of their market activities, will amount to a 'concerted practice' with an adverse effect upon the market.[101] In practice, this may be difficult to prove: the modification of prices by one competitor in short succession to a similar modification by another may simply rank as an intelligent response to a change in the market. However, coordination may be inferred, especially where there is evidence of meetings between the competitors in question.[102] **18.46**

The article relates to arrangements between two or more 'undertakings'—it cannot impact on unilateral actions. An 'undertaking' means almost any legal entity undertaking an economic activity, and not necessarily for profit. It includes, for example, trade associations, and pricing indications given by such associations to their members may be caught by Art 81, even if merely advisory. However, employees are not considered to be undertakings.[103] In general, subsidiary companies are not treated as separate undertakings from their parents (so that an agreement between parent and subsidiary would not be caught by Art 81, as it would be treated as a unilateral arrangement). This is not an invariable rule, and a subsidiary that was managed in a different way from its parent, and pursed different market aims, might be treated as a distinct undertaking.[104] **18.47**

The arrangements to which Art 81 may apply are often sub-divided into 'horizontal agreements' and 'vertical agreements'. The former refers to arrangements made between competitors at the same level in the market, in an attempt to manipulate price, supply, or similar. To take a current example, were leading public schools found to have been collaborating in setting their fees, as natural competitors they would have entered into a horizontal agreement. A vertical agreement refers to arrangements between undertakings occupying different positions in the market, which may impact upon third parties. An agreement to **18.48**

[99] 123/83 *Bureau International Interprofessional du Cognac v Clair* [1985] ECR 391; 8/72 *Vereniging van Cementhandelaren v Commission* [1972] ECR 1977; 22/78 *Hugin Kassaregister AB v Commission* [1978] 3 CMLR 345.

[100] Cf 56/65 *La Technique Minière v Maschinenbau Ulm GmbH* [1966] 1 CMLR 357 at 374.

[101] 40–48/73 et al *Suiker Unie v Commission* [1976] 1 CMLR 295.

[102] See, for example, T-202/98 et al *Tate & Lyle plc et al v Commission* [2001] 5 CMLR 22.

[103] *Suiker Unie v Commission*, n 101 above.

[104] See C-73/95 *Viho Europe BV v Commission* [1997] ECR I-5457; 30/87 *Bodson v Pompes Funèbres* [1988] ECR 2479.

supply an essential commodity only to one manufacturer would be an example of such an arrangement, giving that manufacturer an advantage over would-be competitors.

18.49 The prohibition in the article is engaged by any arrangement that has the 'object or effect' of preventing, restricting, or distorting competition. 'Object' or 'effect' are alternative, not cumulative.[105] The Commission has taken action against undertakings that entered into agreements designed to restrict competition which in the result were unsuccessful: object alone suffices.[106] Whether or not the object of an agreement is to restrict competition will be determined objectively, with reference to the subject matter of the agreement. It does not depend on subjective intent. In any event, if the effect of an agreement is to restrict (etc) competition, Art 81 will be engaged regardless of the parties' objectives.

18.50 Agreements which merely have a small and incidental effect on competition will not, however, be captured by the prohibition in Art 81 (this principle is often referred to as 'appreciability', because it refers to situations where the impact on the market is unappreciable). In *Franz Völk v Ets Vervaecke SPRL*,[107] a distribution agreement relating to just 600 washing machines manufactured by a concern with a negligible market share was allowed to survive Art 81, even though it would undoubtedly have been offensive had the scale of the enterprise been greater. This decision was followed by a *Notice on Agreements of Minor Importance* from the Commission, which has been updated on a number of occasions. The most recent[108] indicates that horizontal agreements between competitors who between them have less than a 10 percent market share will be regarded as de minimis. In the case of vertical agreements, the threshold is 15 percent. Where the agreement is not readily categorized, the threshold is as for horizontal agreements. A different regime applies where there is a network of small agreements between undertakings. So long as no individual agreement involves parties which together have a market share exceeding 5 percent, the arrangement as a whole will usually avoid prohibition if its aggregate market share is less than 30 percent. However, certain agreements will never attract a de minimis exception, for example those which fix prices for sales to third parties, which limit output or sales and, in the case of vertical agreements, agreements with territorial protection provisions.

Application of Article 82

18.51 Article 82 applies to the abuse of a dominant position. In contrast to US anti-trust law, European law is not concerned with the attainment of dominance per se. However, the law imposes on dominant undertakings 'a special responsibility not to allow [their] conduct to impair genuine undistorted competition in the Common Market.'[109] This means that conduct which might be permissible in a modest concern is unacceptable in an undertaking which enjoys a dominant position.

---

105 *La Technique Minière v Maschinenbau Ulm GmbH* [1966] 1 CMLR 357.

106 eg 86/398 *Polypropylene* [1986] OJ L230/1: '[t]he fact that in practice the cartelization of the market was incomplete and did not entirely exclude the operation of competitive forces does not preclude application of article 81.'

107 5/69 [1969] CMLR 273.

108 [2001] OJ C368/13.

109 322/81 *Nederlandsche Banden-Industrie Michelin NV v Commission* [1985] 1 CMLR 282, para 57.

As to the question of what constitutes a dominant position, in *United Brands Co v Commission*[110] it was said to be 'a position of economic strength enjoyed by an undertaking which enables it to prevent effective competition being maintained on the relevant market by giving it the power to behave to an appreciable extent independently of its competitors, customers and ultimately of its consumers.' It is not necessary for there to be no competition at all, but a feature of dominance will be that the dominant undertaking has an 'appreciable influence' over the conditions in which competition develops, and can act 'largely in disregard of it'.[111] In practice, the question of whether a concern is dominant will be determined by the product sector and geographical area in which it trades, its market share, its competitors' market share, and the existence of other obstacles to the development of competition. The last would include, for example, the existence of very high start up costs, and any inhibition on the development of competitor products resulting from intellectual property rights.[112] It is also possible for dominance to arise collectively, when two or more undertakings act in concert in a particular market so that, in effect, they trade as a collective entity.[113] **18.52**

For Art 82 to be engaged at all, the dominance must, of course, be exerted over at least a 'substantial part' of the market, and the abuse in question must be one that potentially affects trade between Member States. With the recent expansion of the EC, the question of what amounts to a 'substantial part' is perhaps moot. The traditional view, however, is that the market in any given Member State must qualify for the description, and in 1997 the Commission took the view that the market for sugar in the Irish Republic (which amounted to just 1.2 percent of the EC total) amounted to a 'substantial part' of the market.[114] In another case, southern Germany (pre-unification) was held to constitute (so far as sugar was concerned) a substantial part of the common market.[115] **18.53**

Article 82 gives a number of examples of what amounts to abusive behaviour (see above). Others may include: **18.54**

- Exclusive purchasing agreements—an obvious device by which the entry of new competitors might be stifled.[116]
- Tied sales—a requirement to buy another product with the 'dominant product' whether needed or not is an obvious instance of a dominant position being exploited.[117]
- Discounts and rebates—where the dominant undertaking pays its customers for their loyalty, the practical effect may be the same as an exclusive purchasing agreement.[118]

---

[110] 27/76 [1978] 1 CMLR 429. The quotation is from para 65.

[111] 85/76 *Hoffmann-La Roche & Co AG v Commission* [1979] ECR 461.

[112] An example is given by the Commission's investigation into that universal *bête noire* the tetra pak. It was found to have 91.8% market share, benefiting from the use by food and drink manufacturers of machines specially designed for its product, and from patent protection: *Tetra Pak I* 88/501 [1988] OJ L272/27.

[113] C-395/96 et al *Compagnie Maritime Belge SA v Commission* [2000] 4 CMLR 1076.

[114] 97/264 [1997] OJ L258/1.

[115] 40–48/73 et al *Co-operative Vereniging 'SuikerUnie' UA v Commission* [1976] 1 CMLR 295, para 448.

[116] eg *Hoffmann-La Roche & Co AG v Commission* [1979] ECR 461.

[117] eg *Eurofix-Bauco/Hilti* 88/138 [1988] OJ L65/19.

[118] Again considered in *Hoffmann-La Roche*. See also, for a construction related example, *British Plasterboard Industries* 89/22 [1989] OJ L10/50.

- Refusal to contract—the refusal of a dominant undertaking to enter into contractual relations with another concern may obviously act to stifle competition.

Enforcement, penalties, and remedies

18.55 The Commission is the EC institution responsible for enforcing Arts 81 and 82, and may investigate any apparent breach.[119] It may act of its own motion, or in response to complaints made to it either by undertakings who believe themselves to have suffered from anti-competitive behaviour, or by the public at large. If an investigation is made, the Commission has substantial powers. For example, it may take statements, examine all books and other business records and take copies, demand oral explanations from employees and officers of the undertaking, and may enter any premises, land or vehicles in the course of its investigation.[120] A failure to cooperate is punishable by a fine of up to 1% of turnover.[121] If the Commission finds that there has been anti-competitive behaviour, it may impose a fine of up to 10 percent of turnover.[122] It will also order the termination of the arrangement in question, and may 'impose . . . any behavioural or structural remedies which are proportionate to the infringement committed and necessary to bring the infringement effectively to an end'.[123] Such measures may also be put in place as interim remedies after the Commission concludes that there is a prima facie case of infringement, so long as the situation is urgent, with a 'risk of serious and irreparable damage to competition'.[124] In addition to financial penalties, the Commission may place special reporting requirements on an undertaking, so that its behaviour may be monitored in the future. Appeals against the Commission's findings and penalties may be made to the European Court of Justice.

18.56 Articles 81 and 82 have always been treated as having direct effect. Regulation 1/2003 now provides in terms that agreements contrary to the articles are prohibited, 'no prior decision to that effect being required.' Hence, Arts 81 and 82 may be relied upon in the domestic courts, and enforced by domestic competition authorities. Thus, an English court could, without more, find an agreement to be illegal and without effect by virtue of a breach of Art 81 (subject to the doctrine of severance, which is recognized by EC as well as common law). Moreover, breaches of the articles also give rise to a cause of action pursuable in the domestic courts: *Garden Cottage Foods Ltd v Milk Marketing Board*.[125] Hence, a third party who suffered loss by reason of anti-competitive action could seek damages. However, no such case is known to have proceeded to trial in the United Kingdom, *Garden Cottage Foods* being merely an interlocutory decision. While it was once thought that a party to an illegal arrangement under Art 81 had no right of action against the other undertakings concerned,[126]

[119] Reg 17/62 [1962] OJ 13/204, now replaced by Reg 1/2003 [2003] OJ L1/1.
[120] Arts 17–21 of Reg 1/2003.
[121] Ibid, Art 23.
[122] Ibid.
[123] Ibid, Art 7.
[124] Ibid, Art 8.
[125] [1984] AC 130 HL at 141.
[126] *Gibbs Mew plc v Gemmell* [1998] Eu LR 588, CA.

the European Court of Justice has held that this should not be the case where the party bringing the action is less at fault for the infringement in question.[127]

UK Competition Law

Chapters I and II of the Competition Act are closely related to Arts 81 and 82 of the EC Treaty, and indeed the harmonization of domestic and European law was one of the principal aims of the Act's introduction. Section 60 of the Act contains an explicit statement to this effect, and an injunction to the domestic courts where possible to interpret it in accordance with EC law, and to 'have regard' to decisions and statements of the Commission. **18.57**

The Chapter I prohibition

The central provision of Ch I of the Act is a general prohibition which, save for its domestic emphasis, clearly has an Art 81 provenance. Known as the Ch I prohibition, it states: **18.58**

> Subject to section 3 [excluded agreements], agreements between undertakings, decisions by associations of undertakings or concerted practices which—
> (a) may affect trade within the United Kingdom, and
> (b) have as their object or effect the prevention, restriction or distortion of competition within the United Kingdom,
>
> are prohibited unless they are exempt in accordance with the provisions of this Part.

There is then an illustrative list of offensive transactions: **18.59**

> [the prohibition] applies, in particular, to agreements, decisions or practices which—
> (a) directly or indirectly fix purchase or selling prices or any other trading conditions;
> (b) limit or control production, markets, technical development or investment;
> (c) share markets or sources of supply;
> (d) apply dissimilar conditions to equivalent transactions with other trading parties, thereby placing them at a competitive disadvantage;
> (e) make the conclusion of contracts subject to acceptance by the other parties of supplementary obligations which, by their nature or according to commercial usage, have no connection with the subject of such contracts.

Any agreement or decision which is prohibited by the general prohibition is declared to be void.

The principal addition which this makes to the Art 81 regime is that it removes the need to show that trade between Member States might be affected. Otherwise, most of what has already been said about Art 81 may be applied to the Ch I prohibition. As with the EC law, alleged anti-competitive behaviour must have an appreciable effect on the market in order to engage the Act. Certain 'small agreements' are exempt from financial penalties under the Act.[128] The Office of Fair Trading's view on 'appreciability' is seemingly more indulgent **18.60**

---

[127] C-453/99 *Courage Ltd v Crehan* [2001] ECR I-6297.

[128] ss 39 and 40, in conjunction with the Competition Act 1998 (Small Agreements and Conduct of Minor Significance) Regulations 2000. So far as the Ch I prohibition is concerned, a small agreement is one where the combined turnover of all parties does not exceed £20m in the business year ending in the calendar year preceding the one in which the infringement took place. So far as the Ch II prohibition is concerned (see

than the Commission's—in its published guidelines on the Ch I prohibition,[129] it states that in general an agreement will not have an appreciable effect if the combined market share of the participating undertakings does not exceed 25 percent. However, as with the Commission, certain infringements are mortal, not venial: price fixing and resale price maintenance is likely to be considered unacceptable, even when practised by small concerns.

18.61 Some categories of agreement are exempt from the Ch I prohibition,[130] and as with the Commission, it is open to the Secretary of State to grant exclusions in certain circumstances.[131] Section 9 of the Act exempts agreements which contribute to improving production or distribution or promoting technical or economic progress, so long as they allow consumers 'a fair share of the resulting benefit' and impose only such restrictions as are 'indispensable' to attaining those objectives. Such agreements must not, however, afford the participating undertakings the possibility of eliminating competition in respect of a substantial part of the products in question. This is very similar to the exemption under Art 81(3) of the EC Treaty, but it applies to such agreements automatically, not by the granting of an exemption. It is for the undertakings who wish to take advantage of the provision to prove that it applies.

The Chapter II prohibition

18.62 Chapter II of the Act contains a further prohibition, this time implementing Art 82 in the domestic law. Section 18 of the Act provides:

> (1) Subject to section 19 [excluded cases], any conduct on the part of one or more undertakings which amounts to the abuse of a dominant position in a market is prohibited if it may affect trade within the United Kingdom.
>
> (2) Conduct may, in particular, constitute such an abuse if it consists in—
> (a) directly or indirectly imposing unfair purchase or selling prices or other unfair trading conditions;
> (b) limiting production, markets or technical development to the prejudice of consumers;
> (c) applying dissimilar conditions to equivalent transactions with other trading parties, thereby placing them at a competitive disadvantage;
> (d) making the conclusion of contracts subject to acceptance by the other parties of supplementary obligations which, by their nature or according to commercial usage, have no connection with the subject of the contracts.
>
> (3) In this section—
> 'dominant position' means a dominant position within the United Kingdom; and 'the United Kingdom' means the United Kingdom or any part of it.

18.63 As with the Ch I prohibition, the central addition which the Act makes to the European law is that the prohibited abuse does not need to effect trade between nation states. Otherwise,

below), 'conduct of minor significance' is conduct by an undertaking whose turnover does not exceed £50m for the business year ending in the calendar year preceding the one in which the infringement took place.

[129] OFT 401.

[130] s 3 of the Act, and Schs 1–3. None of the exceptions are of particular interest from a construction law perspective, although some (such as the exception for mergers, which are regulated by the Enterprise Act 2002) could of course be relevant to construction companies from other perspectives.

[131] Previously most vertical agreements, other than those fixing resale prices, were excluded. However, the material statutory instrument was repealed on 1 May 2005.

the approach to dominance and what constitutes an abuse of it will be similar to that in EC law. The reference in the Act to dominance within the United Kingdom or 'any part of it' should not be taken as a reference to a constituent *national* part of the Kingdom such as Wales or Scotland: it may refer to smaller geographical areas. Hence, for example, investigations have been made into abuses conducted by newspapers published only in Aberdeen.[132]

As to what constitutes dominance, the Office of Fair Trading has stated that it is 'unlikely' an undertaking will be individually dominant if its market share is less than 40 percent, although the exceptional weakness of competitors in a particular market might lead to a different conclusion.[133] As with EC law, it would be possible for two or more undertakings acting monolithically to achieve a collective dominance within the compass of the Act. So far as what constitutes abuse is concerned, there is little to add to the European discussion. It will be noted that the list of illustrative examples in s 18(2) is the same as those in Art 82 of the EC Treaty.[134] **18.64**

Enforcement and penalties

The OFT is responsible for investigating breaches of the Act, and like the Commission it has wide powers with which to do so, including requiring the production of (non-privileged) documents,[135] the entry to premises (on notice, without a judicial warrant; without notice, with a judicial warrant),[136] and the taking of statements. Again, like the Commission, it may order remedial steps, giving 'such directions as it considers appropriate to bring the infringement to an end',[137] and impose interim measures in cases of urgency, so long as that is necessary either to prevent serious and irreparable damage or to protect the public interest.[138] In cases of intentional or negligent infringement of either prohibition, the OFT may impose fines of up to 10 percent of worldwide turnover.[139] A right of appeal lies to the Competition Appeals Tribunal (created by the Enterprise Act 2002), with a further appeal lying, with permission, to the Court of Appeal. **18.65**

Failing to cooperate with an OFT investigation may give rise to criminal liability. The knowing or reckless destruction or falsification of documents or giving of false or misleading information give rise to 'either way' offences under ss 43 and 44 of the Act, punishable by a fine on summary conviction of up to £5,000 or, on indictment, by an unlimited fine, up to two years in prison, or both. A lesser offence of failing to cooperate (eg by not admitting the OFT to premises) is punishable by a fine only.[140] **18.66**

[132] *Aberdeen Journals Ltd v DG of Fair Trading* [2002] Comp AR 167, CAT.
[133] *Assessment of Market Power* (OFT 415), para 2.12.
[134] Again, there is an OFT Guidance Paper, *Assessment of Conduct*.
[135] Competition Act 1998, ss 26 and 30.
[136] Ibid, ss 27 and 28.
[137] Ibid, ss 32 and 33.
[138] Ibid, s 35.
[139] s 36 together with the Competition Act 1998 (Determination of Turnover for Penalties) (Amendment) Order 2004 (SI 2004/1259). For the exemption of 'small agreements' from such penalties, see the discussion of the Ch I prohibition above.
[140] s 42.

18.67 The Enterprise Act 2002 creates a new criminal offence, known as the cartel offence. On indictment, it may be punished by up to five years' imprisonment, an unlimited fine, or both. It provides that an individual is guilty of an offence if he dishonestly agrees with one or more other persons that undertakings will participate in one or more of the following anti-competitive arrangements: price fixing (direct or indirect); limitation of production or supply; market sharing; or bid rigging.[141] It is difficult to overstate how radical a development this is in domestic competition law. As one authority has observed, the last time that cartel activity in England was subjected to a criminal sanction was in the reign of King Edward VI (1547–53). The penalty then alighted upon was a spell in the pillory and the loss of an ear.[142]

Corruption and Bribery

18.68 Much of the law relating to bribery and corruption is connected to election rigging and jury tampering, and is therefore of no direct relevance to the subject of this work. In a construction context, corruption is most likely to arise where a public or corporate officer either solicits or is offered a bribe in order to advance the commercial interests of a construction undertaking. The law differs slightly depending on whether the 'target' of the bribe is a public or private official.

Public officials only

18.69 At common law, it is an offence for anyone in 'an office of trust concerning the public' to misbehave in that office.[143] This is capable of encompassing many acts of corruption, one example being where a local authority manager used workmen employed at public expense to work on his partner's house.[144] It is also an offence to bribe a public official, or for a public official to accept a bribe. These common law offences are punishable by a fine, imprisonment, or both.

18.70 Under the Public Bodies Corrupt Practices Act 1889, it is an offence for any person corruptly to solicit or receive 'any gift, loan, fee, reward or advantage whatever as an inducement to, or reward for, or otherwise on account of any member, officer, or servant of a public body . . . doing or forbearing to do anything in respect of any matter or transaction whatsoever, actual or proposed, in which the said public body is concerned'. Similarly, an offence is committed by anyone who corruptly gives, promises or offers such a bribe or other inducement.[145] The penalty on summary conviction is a fine of up to £5,000, six months' imprisonment, or both. On conviction on indictment, there may be an unlimited fine, up to seven years' imprisonment, or both.[146] A 'public body' includes 'local and public authorities of all descriptions'.[147] Hence, it is an offence to offer bribes to employees of

---

141 Enterprise Act 2002, ss 228–90.

142 See M Furse, *Competition Law of the EC & UK*, 4th edn, Oxford, 2004, p 210.

143 *R v Bembridge* (1783) 3 Doug 327.

144 *R v Bowden* [1996] 1 WLR 98, CA.

145 Public Bodies Corrupt Practices Act 1889, s 1.

146 s 2.

147 s 7, as supplemented by s 4(2) of the Prevention of Corruption Act 1916.

organizations as diverse as an NHS trust or the National Rivers Authority. However, the Crown is not a public body within the meaning of the Act.[148] The statutory definition requires the bribe to be solicited or offered 'corruptly'. The precise meaning of this word is not clear—one definition is unhelpfully circular ('purposely doing an act which the law forbids as tending to corrupt'[149])—but it doubtless ensures that behaviour that is warranted by long custom and usage (the giving of reasonable entertainment or modest tips) does not lead to conviction. No prosecution may be brought under the Act without the Attorney-General's consent in any event.

Corruption involving officers and employees of the Crown is governed by the Prevention of Corruption Act 1906. This is also the principal statute addressing the corruption of private officials, and will be addressed in detail in the next section. **18.71**

Section 2 of the Prevention of Corruption Act 1916 creates a presumption of corruption in certain situations where public or Crown employees are concerned. If it is proved 'that any money, gift, or other consideration has been paid or given to or received' by such an employee 'from a person, or agent of a person, holding or seeking to obtain a contract from His Majesty or any Government Department or public body, the money, gift, or consideration shall be deemed to have been paid or given and received corruptly . . . unless the contrary is proved.' The fact that this applies only where a contract is being sought or has been obtained means that there is no presumption where, for example, the alleged bribe is directed at some non-contractual purpose such as seeking planning permission.[150] This presumption applies to prosecutions under both the 1889 and 1906 Acts. **18.72**

Private and public officials

Section 1 of the Prevention of Corruption Act 1906 creates offences which may be committed in cases involving private or public officials, including Crown officials. It provides that an offence will be committed: **18.73**

> If any agent corruptly accepts or obtains, or agrees to accept or attempts to obtain, from any person, for himself or for any other person, any gift or consideration as an inducement or reward for doing or forbearing to do, or for having after the passing of this Act done or forborne to do, any act in relation to his principal's affairs or business, or for showing or forbearing to show favour or disfavour to any person in relation to his principal's affairs or business; or
>
> If any person corruptly gives or agrees to give or offers any gift or consideration to any agent as an inducement or reward for doing or forbearing to do, or for having after the passing of this Act done or forborne to do, any act in relation to his principal's affairs or business, or for showing or forbearing to show favour or disfavour to any person in relation to his principal's affairs or business; or
>
> If any person knowingly gives to any agent, or if any agent knowingly uses with intent to deceive his principal, any receipt, account, or other document in respect of which the principal is interested, and which contains any statement which is false or erroneous or defective in any material particular, and which to his knowledge is intended to mislead the principal.

[148] *R v Natji* [2002] 1 WLR 2337, CA.
[149] *R v Wellburn* (1979) 69 Cr App R 254, CA.
[150] *R v Dickinson* (1948) 33 Cr App R 5, CA.

18.74 An 'agent' within the meaning of the Act includes an employee or commercial agent, any person serving under a public body, and any person serving under the Crown or local government. Hence, an offence will be committed when a bribe is offered to or solicited by a private person, if the intent is to influence that person's principal or employer.

18.75 The penalties under the 1906 Act are the same as those under the 1889 Act, and once again prosecutions may be brought only with the Attorney General's consent.

18.76 A conviction for corruption excludes a person from tendering to the public sector under the Public Sector Procurement Directive.[151]

Crimes committed abroad

18.77 The Anti-terrorism, Crime and Security Act 2001 extended both the 1889 and 1906 Acts to encompass bribery offences committed outside the United Kingdom. The common law offence of bribery is similarly enlarged. Hence, it is an offence for any UK national, or company incorporated under the law of any part of the United Kingdom, either to offer or to receive bribes (or the equivalent) outside the United Kingdom if the same action would amount to an offence within the United Kingdom.[152] Proceedings for the offence may be brought in the UK criminal courts. The penalties for the offences remain the same, but the limited presumption of corruption which may arise under the 1916 Act does not apply to international offences.

Law reform

18.78 In January 1998, the Law Commission published a report arguing that the existing law as to corruption was 'obscure, complex, inconsistent and insufficiently comprehensive'.[153] A draft bill was published after the 2002 Queen's Speech. The bill was criticized in committee,[154] and has not since progressed. The Law Commission tried again and in November 2008 produced its final report on bribery proposing the creation of new general offences of bribery and one specific offence of bribing a foreign public official. A Bribery Bill is now under scrutiny in Parliament.[155] There is an international drive to tighten and standardize laws against corruption, which has been led by the UN, EC, the Council of Europe and, in particular, the OECD. The extension of the existing UK offences to bribery committed abroad brought by the Anti-terrorism, Crime and Security Act 2001 resulted from initiatives by the OECD. Hence, further reform is possible. However, so far as the categories of corruption referred to in this section are concerned, it seems unlikely that there will be fundamental changes of approach. Proposed modifications in those respects are directed largely at form. The use of the language of principal and agency in the 1906 Act, for example, is thought to be confusing to modern juries. It is likely to be overhauled, but it appears probable that the essence of the offences will remain the same.

---

151 Art 45(1)(c) of Directive 2004/22/EC.

152 Anti-terrorism, Crime and Security Act 2001, s 109.

153 *Legislating the Criminal Code: Corruption* (Report No 48).

154 Report, 17 July 2003, HL Paper 157; HC Paper 705.

155 As before it is coming in for criticism, principally that it does not go far enough.

19

PARTNERING

How the Term 'Partnering' Is Understood

The term 'partnering' describes an open method of working with a view to avoiding disputes by adding value and ensuring a collaborative approach from all the parties involved in it. One definition of partnering can be found in J Bennett and S Pearce *Partnering in the Construction Industry*: **19.01**

> Project Partnering is a set of actions taken by the work teams that form a project team to help them cooperate in improving their joint performance. Specific actions are agreed by the project team taking account of the project's key characteristics, and their own experience and normal performance. The choice of actions is guided by a structured discussion of mutual objectives, decision-making process, performance improvements and feedback.[1]

Conversely, another commentator characterizes partnering as something more akin to 'a fuzzy set of relationships dependent on altruistic behaviour and the voluntary compromise of commercial rights and obligations' which can appear fragile or even imprudent when held up in a harsh commercial light.[2] Others say that partnering is best considered as a set of collaborative processes[3]. It is said that such processes emphasize the importance of **19.02**

[1] J Bennett and S Pearce, *Partnering in the Construction Industry: a Code of Practice for Strategic Collaborative Working*, Butterworths-Heinemann, Oxford, 2006.

[2] D Mosey, *Early Contractor Involvement in Building Procurement*, Wiley-Blackwell, Chichester, 2009.

[3] J Barlow, M Cohen, A Jashpara, and Y Simpson, *Towards Positive Partnering: Revealing the Realities for the Construction Industry*, The Policy Press, Bristol, 1997.

common goals and raise such questions as how such goals are agreed upon, at what level are they specified, and how they are articulated.[4] The use of such words suggests that the term 'partnering' is not a precise word as it is normally understood in the construction industry. The lack of precision is reflected in the fundamental question which hovers over a great many partnering decisions: is the partnering agreement to be binding or non-binding?

Background to Partnering

19.03 The implementation of partnering in the construction industry has been patchy, with varying degrees of success and, in many instances, its adoption is more an exception that the norm according to one commentator.[5] However, the concept has become more prominent through endorsement by many construction industry reports, most notably *Constructing the Team* by Sir Michael Latham; *Achieving Excellence* by Sir John Egan and *Modernising Construction* by the OGC and the National Audit Report. The Latham Report stated that the employer should be at the core of the construction process. The use of teamwork and co-operation was advocated to improve employer satisfaction. The report said:

> Partnering includes the concepts of teamwork between supplier and client, and of continuous improvement. It requires openness between the parties, ready acceptance of new ideas, trust and perceived mutual benefit . . . We are confident that partnering can bring significant benefits by improving quality and timeliness of completion whilst reducing costs.

19.04 Upon the publication of the Latham Report, the Construction Industry Board ('CIB') was established with the ultimate aim of implementing Latham's fifty-three recommendations.

Disputes under Partnering Agreements

The basis of the contract

19.05 The main philosophy that underlies partnering is to reduce the adversarial and litigious culture that exists in the construction industry, and to resolve problems jointly and informally through more effective forms of inter-firm collaboration.[6]

19.06 Allied to these principles is the importance of having in place a proper legal framework at the outset. In *Alstom Signalling Ltd (trading as Alstom Transport Solutions) v Jarvis Facilities Ltd* [2004] EWHC 1232 (TCC) the court had to consider whether a contract contained a pain/gain sharing arrangement (where the parties had not agreed the precise detail of that arrangement). If the contract did contain such a provision, the court would then determine its terms where details of that arrangement were unclear and had not been finalized. Neither Alstom nor Jarvis disputed that they were in contract with each other, but they never

[4] D Jones, D Savage, and R Westgate, *Partnering and Collaborative Working: Law and Industry Practice*, Hammonds LLP, London, 2003.

[5] F Phua, 'When is Construction Partnering likely to happen? An empirical examination of the role of institutional norms', (2006) 24 CME, 6, June 2006, 615–24.

[6] M Latham, *Constructing the Team*. Final Report of the Government/Industry Review of Procurement and Contractual Arrangements in the UK Construction Industry, HMSO, London, 1994.

formally executed the contract in question, which was the IChemE Form of Sub-Contract for Process Plants ('the Yellow Book'), and the contract was performed. Mr Recorder Reese QC granted a declaration that there was an agreement that the contract would contain a pain/gain share provision but the parties had not reached agreement on the precise detail of that arrangement so the court could resolve the difference between the parties as to what was a fair and reasonable mechanism by which the parties would share any pain/gain.

It is to be noted that the court in that case had to consider many of the cases dealing with certainty of terms and agreements to agree which are discussed in Chapter 1 of this book and it is far from obvious that every judge would find the binding agreement which was the subject matter of the declaration, though it is suggested that the conclusion was correct in that case. **19.07**

Even though relationships can flourish without a contract nevertheless it has been suggested that a construction project should have a contract that reflects the commercial wishes of the parties.[7] The dangers where this did not happen were demonstrated in *Birse Construction Ltd v St David Ltd (No 1)* [2000] BLR 57. No formal contract document was signed between the parties and instead they had had signed a non-binding charter. HH Judge Humphrey LLoyd QC, while stating that the charter was 'clearly not binding', nevertheless noted the importance of its terms as they were 'clearly intended to provide the standards by which the parties were to conduct themselves and against which their conduct and attitude were to be measured.' The judge said: 'People who have agreed to proceed on the basis of mutual co-operation and trust, are hardly likely at the same time to adopt a rigid attitude as to formation of a contract.'[8] **19.08**

The issue of good faith in non-binding agreements

Whilst it is generally recognized that the concept of good faith has no place in English law,[9] the *Birse* case appeared to suggest a change in the way in which the courts would view such an obligation in a non-binding context. However, since *Birse* the case law in this area has not developed significantly and it is not possible to predict the way in which the court will treat such agreements in the future. One distinguished commentator[10] has noted that the reluctance of the English Courts to entertain such notions in partnering cases rests on two principal factors. Firstly, the introduction of such a concept is apparently irreconcilable with freedom of contract: once the parties have agreed the terms of their bargain there is no necessity to imply such a term. Secondly, a duty of 'good faith' is vague and not easily capable of definition. **19.09**

[7] Speechly Bircham LLP, *Could Partnering Sound the Death Knell for Confrontational Contracting?*, 8 July 2003, <http://www.speechlys.com/public/documents/confrontational%20 contracting%20seminar.pdf>.

[8] On appeal, the Court of Appeal did not address any of the matters pertaining to partnering agreements.

[9] See Vinelott, J in *London Borough of Merton v Stanley Hugh Leach Ltd* (1986) 32 BLR 51, in which he said: 'Although the courts will imply a duty to do whatever was necessary in order to enable a contract to be carried out, the requirement of good faith has not been incorporated into English law . . .'

[10] N Jefford, '*Soft Obligations' in Construction Law: Duties of Good Faith and Co-operation.* Keating Chambers, London, 2005.

19.10 Another case which concerns a partnering relationship without a contractual agreement to govern it is *Baird Textile Holdings Ltd v Marks & Spencer Plc* [2001] EWCA Civ 274. In *Baird*, the facts centred on a long-term relationship which afforded no protection whatsoever to a supplier when the employer, Marks & Spencer suddenly decided to terminate their contract. Baird claimed that Marks & Spencer had deliberately refrained from entering into a contract because it could achieve greater flexibility without one. Baird had been a supplier to Marks & Spencer for thirty years. Baird brought a claim against Marks & Spencer on the basis that, given the length of the relationship that had gone before it, it was entitled to three years' notice. The Court of Appeal rejected Baird's claims. In doing so, the courts noted that it: (i) will only imply a contract by reason of the conduct of the parties if it is necessary to do so; (ii) would be fatal to the implication of a contract that the parties would or might have acted as they did without any such contract. To do otherwise would be to make a bargain for the parties rather than seeking to enforce a bargain which they themselves had made.

The issue of good faith in binding agreements

19.11 The contracting parties in a partnering arrangement may choose to incorporate a binding agreement which includes an express obligation on the parties to act in good faith. Given the absence of a uniform statutory or common law definition of good faith under English law it could be said that it is unclear how such obligations would be interpreted. However, it is suggested that a non-binding agreement remains just that, whatever the moral imperatives otherwise.

19.12 The law in this area is less developed than that of other jurisdictions. However, in certain circumstances, the obligation to use good faith may act as a fetter on the authority's rights to terminate the contract; as was found in *Timeload v British Telecommunications plc* [1995] EMLR 459.

19.13 No matter which form of partnering agreement is eventually selected, it is clearly important to ensure that all the terms of partnering arrangement are agreed, recorded in writing, and that the final agreement is signed by the parties at the outset. This is suggested despite initial views to the contrary fashioned by some at the beginning of the partnering revolution. For example, it was said, perhaps a little optimistically:

> The Task Forces wishes to see . . . an end to reliance on contracts. Effective partnering does not rest on contract. Contracts can add significantly to the cost of a project and often add no value to the client. If the relationship between contractor and employer is soundly based and the parties recognize their mutual interdependence, then formal contract documents should gradually become absolute.[11]

[11] J Egan, *Rethinking Construction*. Report of the Construction Task Force published by Department of the Environment, Transport and Regions, London, 1998.

Many proponents[12] of partnering recognize the importance of a contract and this can been seen in the PPC 2000 Guidance Notes, which state: **19.14**

> A number of myths have built around partnering, including firstly the suggestion that you can partner with no contract at all . . . it is difficult to find any public or private sector organization that does not need to know the terms on which it commits or expects to receive a substantial sum of money.

The Notes also draw attention to the CIC Guide to Project Team Partnering, which states: **19.15**

> There has been much talk in the press and comments from industry experts that may suggest that partnering is not all about contracts. That is partly true. What partnering is not about is for each party to use contracts to strengthen their position at the expense of others and the overall project. It is not about the conflict and the lengthy and complex disputes that are frequently the result when parties are bound by intrinsically adversarial contracts.

Public procurement issues

There has been a rapid growth in the use of framework agreements by public bodies. Their popularity is most likely attributable to the fact that initial selection and award procedures can be dispensed with in any subsequent contract award thereby offering significant savings in time and money to those using them. A framework agreement (in a public procurement context): **19.16**

> Establishes the terms (in particular in relation to price and, as appropriate, quality) under which the economic operator will enter into one or more contracts with a contracting authority in the period during which the Framework Agreement applies. (reg 2 of the Public Contracts Regulations 2006—'the Regulations').

There has been a significant increase in the use of public procurement law to challenge the award of contracts by public bodies. The implementation of the Remedies Directive (2007/66/EC) ('the Directive') in the United Kingdom from 20 December 2009 is likely to facilitate more challenges. From then, the remedies available to unsuccessful bidders will include the new remedy of 'ineffectiveness'. The Office of Government Commerce ('OGC') confirmed on 30 April 2009 that 'ineffectiveness' will mean the relevant contract will be terminated with effect from the date of the relevant court order. **19.17**

The two cases of *Henry Bros (Magherafelt) Ltd and others v Department of Education for Northern Ireland (No 2)* [2008] NIQB 105, and *McLaughlin and Harvey Ltd v Department of Finance and Personnel (No 2)* [2008] NIQB 91 related to framework agreements. **19.18**

In *Henry Bros*, the Court ruled that the reliance on 'fee percentages' in the framework agreement in this case did not allow the contracting authority to judge the most economically advantageous or lowest price bid. It ruled that a discussion of rates and costs would still need to take place after an individual contract under the framework was granted. **19.19**

---

[12] J Critchlow, *Capital Projects in the Education Sector Partnering—Threats and Opportunities*, 12 October 2004,<http://www.fenwickelliott.co.uk/files/Contract%205%20-%20Partnering%20Threats%20and%20Opportunities.pdf>.

19.20 In *McLaughlin*, the contracting authority had used a methodology (in particular regarding weightings) that would have affected the presentation of tenders but had not been disclosed to bidders prior to the submission of their tenders and therefore breached the requirement of transparency.

19.21 In both cases the applications for interim relief were unsuccessful because of the balance of convenience between the parties. However, in *Lettings International Ltd v London Borough of Newham* [2007] EWCA Civ 1522 the balance of convenience did allow for interim relief and it is suggested that the courts in England and Wales will follow the *Lettings* approach when it comes to considering the balance of convenience. Were it otherwise, the Regulations may have little practical commercial force except in retrospect.

Contractual or non-contractual partnering?

19.22 When entering into a partnering arrangement, an important choice has to be made as to whether cooperation and transparency should have binding effect under the given agreement or whether they should simply remain a statement of aspiration.[13] For an example of a failure to make a firm choice between these two alternatives, see *BP Exploration Operating Co Ltd v Kvaerner Oil Field Products Ltd* [2004] EWHC 999 (see particularly paras 47, 54, 57, and 95). In that case reliance upon still live negotiations for a partnering relationship were not accepted as relevant matrix evidence for the interpretation of the obligations of parties as to obligations to procure insurance.[14]

Contractual, Non-Contractual, and Hybrid Partnering Arrangements

Contractual partnering

19.23 When a party agrees to enter into a contractual partnering arrangement it will expressly be agreeing that its conduct will be governed by a contractual framework that is intended to create legally enforceable rights and obligations. There are many forms that can be used to facilitate this objective, each of which has differing provisions. The relevant forms can be summarized as follows:

- PPC 2000 ACA Standard Form of Contract for Project Partnering and the SPC 2000 Specialist Contract.
- ICE Partnering Addendum to the ICE forms of contract (2003).
- The NEC/EEC and IChemE (partnering type agreements).
- NEC Partnering Option X12 (2001).

[13] D Jones, *Partnering, Prime Contracting and other Collaborative Working*—a paper for the International Federation of Consulting Engineers (FIDIC), 2004.

[14] Since that case was decided there has been judgment in *Chartbrook Ltd v Persimmon Homes Ltd and others* [2009] UKHL 38 endorsing the rule of law that negotiations are not matrix evidence.

Non-contractual partnering

The non-contractual (non-binding) partnering route envisages the operation of a charter that in hierarchical terms operates (notionally) over the top of the head or main contract. It seeks to influence rather than mandate certain behaviours and is no more than an expression of intent between two parties to behave in accordance with a set of agreed principles. The form that adopts this approach in practice is the JCT Partnering Charter/JCT Practice Note 4 (2001) for use with JCT contract forms. **19.24**

The hybrids

The JCT Constructing Excellence Contract 2006 adopts a contractual approach but conversely requires the inclusion of a non-partnering charter. The NEC 3 Option X12 might also be classed as contractual even though it does not create contractually binding links between the parties to it. **19.25**

Benefits in Partnering

There is a widely held view that partnering has the potential to bring consistently better results than the more traditional approach.[15] Benefits from partnering could be: **19.26**

- Reduced exposure to litigation.
- Improved project outcomes in terms of cost, time and quality.
- Lower administrative and legal costs.
- Increased opportunity for innovation and value engineering.
- Increased chances of financial success.[16]

Summary

The problem for 'partnering' as a concept is that on one level it decries binding contractual commitments and on the other it leaves open the potential for one or other party to be bereft of remedy when things go wrong. Whilst a non-contractual approach can, with the right parties, benefit employer and contractor, parties may be exposed if either employer or contractor does not operate in the spirit of the partnership as regards any specific project. If the relationship is long term, however, and importance is not attached to a specific project rather than to a long term relationship encompassing many contracts, then the benefits of dispute avoidance may be great. **19.27**

[15] C Skeggs, 'Project Partnering in the International Construction Industry', [2003] ICL Rev 456.

[16] See Construction Industry Institute Australia, *Partnering: Models for Success*, Research Report No 8, Sydney, CIIA, 1996.

Part VII

DISPUTES

20

MEDIATION AND ALTERNATIVE DISPUTE RESOLUTION

Mediation and ADR: The Movement for Change and ADR

Dissatisfaction with the legal and arbitration process is dealt with in Chapter 22 dealing with litigation. One of the more far reaching responses to that dissatisfaction has been the emergence of mediation, 'a flexible process conducted confidentially in which a neutral person actively assists parties in working towards a negotiated agreement of a dispute or difference, with the parties in ultimate control of the decision to settle and the terms of resolution.'[1] The glossary to the CPR defines ADR as a 'collective description of methods of resolving disputes otherwise than through the normal trial process'.[2] As was made clear by the Court of Appeal in *Halsey* (see below) ADR should be treated as including all forms of negotiation, including 'round table meetings' (without mediators), correspondence (whether Part 36 or otherwise)—and presumably even telephone negotiations. **20.01**

[1] The definition now adopted by CEDR, one of the leading promoters of the mediation process.

[2] It has also been described as a process of structured negotiation, which is carried out with the assistance of a neutral third party, the mediator, with all the advantages of conventional negotiation but the involvement of the mediator makes the negotiation much more effective.

20.02 Since it took hold in the mid 1990s, mediation has been very successful, especially in the commercial and construction field. It is now considered an integral part of the legal process. Lord Clarke of Stone-cum-Ebony MR has given two lectures which emphasize this point.[3]

20.03 The estimate routinely cited by mediation organizations is that 80% of disputes which come to mediation are settled during the mediation itself. However, precise statistics are difficult to come by, especially as many mediations which appear on their face to fail are followed by resolution by settlement in the period immediately following the mediation.

20.04 It is impossible to say exactly why mediation has been so successful, but four factors can perhaps be identified. The first is the dissatisfaction with the alternatives of litigation or arbitration, which are cost heavy and uncertain in outcome. The second is that many disputes contain within them, on each side, the perception that there has been a failure of the other party to understand the opposing point of view; the process of mediation permits that to be addressed. Thirdly, given that litigation and arbitration in construction cases is a commercial enterprise, the object of which is to obtain a monetary or other financial advantage, mediation provides a forum in which the commercial decision makers can assess their case against that of the other side or sides. In assessing their case, it is of considerable importance that the decision makers can also draw on the resources of an independent third party—the mediator—to assist them in coming to a view. Finally, mediation can provide a broader range of outcomes, some of which may be much more suited to the parties' commercial objectives (which may well include a desire for an ongoing commercial relationship) than a black and white finding of damages due or the like.

20.05 The courts now encourage mediation as a matter of routine and they also do so even if one party is resolutely against the idea.[4]

20.06 Before examining mediation in more detail, it is worth pausing for a moment to consider two other forms of ADR with which the success of mediation can be compared. One is 'pendulum arbitration' where the arbitrator must choose between the two sides' expressed positions and cannot make any alternative award. It is all or nothing for one or other side. It has the advantage of preventing parties taking extreme positions, something otherwise commonly adopted in the hopes of each party that the dispute can be negotiated to a middle ground moved in the direction of their extreme position. Pendulum arbitration eliminates the advantages of this approach. However, it has the significant disadvantage that neither side can see the other's true position. The dispute then becomes a game of poker, with the risk that neither party will see the other's true position before it is too late.

20.07 Adjudication is dealt with elsewhere. It has emerged as one of the biggest forms of ADR, allowing an independent third party to look at the dispute between the parties and come to a conclusion. The commercial and legal arms of the parties can then ask themselves

---

[3] <http://www.judiciary.gov.uk/docs/speeches/mr_mediation_conference_may08.pdf> and <http://www.judiciary.gov.uk/docs/speeches/mr-littleton-chambers-080609.pdf>. See also Lord Phillips CJ's lecture on mediation <http://www.judiciary.gov.uk/docs/speeches/lcj_adr_india_290308.pdf>.

[4] *Cable and Wireless v IBM United Kingdom Ltd* [2002] EWHC 2059 *per* Colman J; and *Shirayama Shokusan Co Ltd v Danovo Ltd* [2004] BLR 207 *per* Blackburne J.

whether a significantly different conclusion would be come to by a judge or arbitrator. Of course at the margin they will. But in most cases they do not and litigation, if there is any, will centre around the enforcement of the adjudicator's decision rather than the commencement of proceedings afresh on the merits.

Mediation is distinguishable from adjudication primarily because the mediator does not come to a decision. Nor does he or she normally give an opinion directly on the merits of the case, even if many mediators are able to give, in private, a forceful assessment of the merits—a reality check—which almost comes to the same thing. As is emphasized in the training literature produced by mediation organizations, the hallmark of a good mediator is the ability to provide such a reality check without stepping over the line into judgment. **20.08**

The appointment of the mediator can be crucial to the success or failure of the mediation. If the mediation is to be successful, all parties must be satisfied that the mediator is sufficiently skilled; those skills, importantly, must not include just technical competence (whether in the law or in construction), but also the requisite prowess in the process of mediation. It is to be noted also that it is not necessary that the mediator in fact be a lawyer; indeed there may be occasions when technical expertise is of more value than a professional background in the law. Mediators can either be chosen through an ADR organization such as CEDR or independently; either way, there is now a cadre of specialist mediators, many of them being drawn from the ranks of construction professionals who were involved in litigation and arbitration. **20.09**

It may on occasion be of assistance to choose two mediators, one of whom has particular expertise in the specialist field of the dispute, and one more known for their ability, for instance, to corral powerful personalities into agreement. It is also increasingly common practice for there to be an assistant mediator (in other words a trained mediator attending, in part, to gain experience): it goes without saying that the parties must consent for such a person to attend.[5] **20.10**

What Is Involved in a Mediation

The appointment of the mediator will invariably involve signature of a mediation agreement by the parties, setting the ground rules for the mediation. ADR bodies such as CEDR have standard mediation agreements. While they may vary, a heavy emphasis is always—and rightly—placed on the confidential nature of mediation, as discussed further below. The parties must also agree between them how long must be set aside for the mediation. Most disputes will take only one day, but in a significant number of cases more is required. It is, however, very rare for a mediation to last longer than three days. **20.11**

Once the mediator is appointed the parties must acquaint him/her with the nature of the dispute. This normally takes place through each party exchanging statements of their position: 'position statements'. Such position statements do not normally take the form of pleadings, since the process is not legal, nor are they drafted on the basis of the acrimonious **20.12**

[5] Note that it is CEDR policy that an assistant mediator attend every mediation held under its auspices.

correspondence which may have gone before. The best position statements are short (about ten pages) and set out how and why the party concerned feels that its position in the dispute is correct, why its claim is valid or the other side's claim is misconceived, and, sometimes, what the resolution of the dispute will mean to the party concerned.

20.13 The parties should, if possible, agree a 'mediator's bundle' which should contain the position statements and the key documents. The ideal mediator's bundle should be no larger than a lever arch file. The mediator will not have time to absorb more than that prior to the date of the mediation meeting. Key documents will include the contract, or, if the contract is a lengthy document, relevant extracts from the contract, any experts' reports which have been obtained, and any key correspondence which will assist in understanding the dispute. The correspondence included should be carefully selected so as only to cover those items central to the crystallization of the dispute. Exactly what are key documents will vary in each case. The parties' advisors should be astute to ensure that the documentation is kept to a minimum. The temptation will be to provide the mediator with as much as possible to convince him/her of the rightness of the cause. This is a mistake. The purpose of the statement and bundle is not to do that, but to enable him/her to see where the parties' positions are and to be clear as to exactly the nature of the dispute.

'Rules' of Mediation; 'Without Prejudice'; Effect on Agreements and Admissions; Expert Evidence; Other Statements, Threats etc; the Position of the Mediator

20.14 The central and essential part of a mediation is that it is a confidential process both internally and externally. Nothing which is said in a mediation can be repeated outside it and nothing which is said to the mediator by one of the parties can be repeated by him/her to the other in the mediation itself without the permission of the party confiding the information. Confidentiality has to be complete otherwise the process becomes impossible. It is only if the parties are able to speak freely to the mediator that s/he is in a position to see their respective current real positions and is able to judge where they might end up.

20.15 Allied to the confidentiality of mediations is their without prejudice status, which protects admissions in mediations from being allowed in as evidence in any subsequent hearing. This is carefully protected by the courts.[6]

Evidence of the negotiations

20.16 Can a court hear evidence of without prejudice negotiations in a mediation in order to determine whether or not there was a concluded agreement to settle at or after the mediation? This question arose in *Brown v Rice and Patel and ADR Group* [2007] EWHC 625 (Ch), Mr Stuart Isaacs QC (sitting as a Deputy Judge of the High Court). (The ADR Group had intervened in the action in order to make representations.)

[6] Note, however, that the court will have access to *offers* to mediate in correspondence marked 'without prejudice save as to costs' when considering costs orders at the conclusion of a trial: *Reed Executive v Reed Business Information* [2004] 1 WLR 3026.

The judge had no difficulty in deciding that the court had to look at without prejudice material in order to decide whether or not there was a concluded agreement to settle. That was on the basis of the existing without prejudice rules, including the exception to it, and not any special category of mediation privilege. He also decided that an agreement to settle, which was not reduced to writing and signed by, or on behalf of, the parties, could not be a binding agreement by reason of the term of the Mediation Agreement providing otherwise (and because the agreement was incomplete). **20.17**

Finally, he said, albeit *obiter*, that an offer left open for acceptance after the mediation day and then accepted was 'as much made in the mediation as if it were made at the hearing itself.' **20.18**

It is worth noting that in so doing he did not accept any of the following:

- ADR Group's submission that nothing said or done in preparation for, at, or in consequence of the mediation can ever be used outside the mediation;
- ADR Group's further submission that the court should not readily open up without prejudice discussions to the scrutiny of the court. Where there is an exception to the rule, the judge said the court must look at the necessary material;
- The submission that the Mediation Agreement prevented the court looking at events relevant to its decision on the matters within the exception to the without prejudice rule. That was contended for by reason of clause 1.4 of the Mediation Agreement which provided the settlement would not be binding unless it was in writing and signed by the parties It had been argued that this removed the exception to the without prejudice rule.

Experts

One important question that can arise is whether an agreement between experts that is made in the context of a mediation is to be treated as without prejudice or is to take effect as an open agreement in the litigation by reason of the Civil Procedure Rules (CPR). **20.19**

In *Aird and Another v Prime Meridian Ltd* [2006] EWCA Civ 1866, the Court of Appeal overturned the decision of HHJ Coulson QC as to this, holding that an open joint statement which was agreed and signed by expert witnesses was a document prepared for the court under the rules of the court (CPR r 35.12) and was not privileged from production even if one of the purposes for its preparation was use in a mediation which was itself conducted under the cloak of without prejudice. **20.20**

Furthermore, the court stressed that orders should not be made in court proceedings that experts produce a privileged statement for use in mediation. The court has no power to order how parties conduct a mediation, arising out of the position that the court has no power to order parties to mediate. **20.21**

It will be noted that the determining factors for the Court of Appeal in this case were that the experts' agreed statement was not (ultimately) marked without prejudice, and was found by the CA to have been made pursuant to r 35.12, rather than for any more limited purpose. However, this case provides a salutary lesson as to what (particularly in the field of **20.22** evidence) may and—more importantly—may not attract mediation privilege. The ruling is likely if anything to inhibit free expression except within the mediation process itself.

Other statements; threats etc

20.23 Similar issues can arise in the context of statements made during the course of without prejudice discussions at mediations. In the related cases of *Venture Investment Placement Ltd v Hall* [2005] EWHC 1227 (Ch) and *Hall and Another v Pertemps Group Ltd and Another* [2005] EWHC 3110 (Ch), the issue arose as to in what circumstances, if any, something said during mediation, alleged to amount to threats, might be outside the cloak of without prejudice and confidentiality created by: (i) the mediation agreement, and (ii) Part 36 negotiations, and if so what the costs consequences might be.

20.24 Bearing in mind the starting point that the mediation proceedings have to be guarded with great care as to confidentiality, the issue was whether what took place was properly within the scope of the mediation privilege or was it within the 'unambiguous impropriety exception' as explained by Walker LJ in *Unilever v Proctor and Gamble* [2000] 1 WLR 2436.

20.25 The first ruling was to the effect that such issues could not be decided at an interlocutory hearing, so an injunction to restrain disclosure was ordered, pending trial or further order.

20.26 Ultimately at trial Lewison J resolved the following as distinct preliminary issues:

(i) In the ordinary case, public policy will prevent a court inquiring into what went on in a mediation conducted on a without prejudice basis. However, where, as here, there had been satellite litigation in which one party alleged that threats had been made by the other party during the mediation and that other party denied that the threats had been made, without prejudice protection was lost by reason of a mutual waiver in the satellite litigation. Accordingly, the court decided it could look at what took place in deciding what costs order to make.

(ii) The circumstances in relation to the withdrawal of a Part 36 Offer were different because that went wider than had been pleaded in the satellite litigation. So, it was covered by the without prejudice cloak and there had been no waiver by the parties. Accordingly, the court would not take the circumstances in relation to the withdrawal of the Part 36 Offer into account in deciding costs.

20.27 More recently, issues relating to 'Without Prejudice' principles and privilege have been considered in detail by Ramsey J in *Farm Assist Ltd (In Liquidation) v The Secretary of State for the Environment, Food and Rural Affairs (No 2)*.[7] Here the judge was concerned with the position of mediator herself, against whom a witness summons had been issued which she was seeking to have set aside.

20.28 DEFRA had brought proceedings against Farm Assist Ltd (FAL) in relation to its handling of the foot and mouth epidemic in 2001. The parties attended a mediation. It was successful and a settlement agreement was entered into. FAL then brought proceedings against

---

[7] [2009] EWHC 1002 (TCC), (2009) 125 Con LR 154.

DEFRA seeking to set aside the settlement agreement on the basis that it had been entered into under economic duress at the mediation.

Neither party objected to the mediator being called to give evidence about the mediation (including her private conversations with the parties). The court directed that the parties should jointly write to the mediator to find out if she had retained any notes or documents from the mediation, whether she had any factual (or other) recollection of the mediation, and to invite her to disclose to the parties such notes or documentation she may have retained. **20.29**

The mediator responded by saying that, as the mediation had occurred many years ago, she had very little factual recollection of the events. Furthermore, her files contained no personal notes. She refused to meet DEFRA lawyers who wanted to take a witness statement. So DEFRA served a witness summons on the mediator seeking her attendance at the trial. The mediator applied to set aside the witness summons. **20.30**

The Mediation Agreement contained a number of standard form terms. Clause 6 provided that each party was deemed to agree to the confidentiality provisions in the Mediation Procedure. The Mediation Procedure included terms providing that: **20.31**

(i) all communications relating to and at the mediation would be without prejudice;

(ii) all information produced for, or arising in relation to, the mediation would be kept confidential;

(iii) all documents produced for, or arising in relation to, the mediation would be privileged and not be admissible as evidence or discoverable in any litigation or arbitration connected with the Dispute (as defined) unless they would be admissible or discoverable in any event; and

(iv) none of the parties to the Mediation Agreement would call the mediator as a witness in any litigation or arbitration '*in relation to the Dispute*' and the mediator would not voluntarily act in any such capacity without the written agreement of all the parties.

Ramsey J dismissed the mediator's application and held that she should give evidence in response to the witness summons. He summarized the position as to confidentiality, privilege and the without prejudice principle as follows:

(1) Confidentiality: The proceedings are confidential both as between the parties and as between the parties and the mediator. As a result even if the parties agree that matters can be referred to outside the mediation, the mediator can enforce the confidentiality provision. The court will generally uphold that confidentiality but where it is necessary in the interests of justice for evidence to be given of confidential matters, the Courts will order or permit that evidence to be given or produced.

(2) Without Prejudice Privilege: The proceedings are covered by without prejudice privilege. This is a privilege which exists as between the parties and is not a privilege of the mediator. The parties can waive that privilege.

(3) Other Privileges: If another privilege attaches to documents which are produced by a party and shown to a mediator, that party retains that privilege and it is not waived by disclosure to the mediator or by waiver of the without prejudice privilege.

He concluded that, in the interests of justice, the mediator should give evidence as to what was said and done in the mediation as:

(i) the issue in these proceedings was whether the settlement agreement was procured by economic duress during the mediation. The allegations concerned what was said and done in the mediation and this would require evidence of what FAL said had been done by the mediator. That evidence formed a central part of FAL's case and the mediator's evidence was necessary for the court to determine properly what was said and done:

(ii) the fact that the mediator had no recollection of the mediation did not prevent her from giving evidence. A witness summons would not be set aside because the witness says they cannot recall matters. Frequently memories are jogged and recollections come to mind when documents were shown to witnesses and they had the opportunity to focus, in context, on events some years earlier;

(iii) calling the mediator to give evidence would not be contrary to the express terms of the Mediation Agreement. The parties' agreement not to call the mediator as a witness '*in relation to the Dispute*' was limited to litigation or arbitration in relation to the underlying dispute. Dispute was a defined term that related to the work performed by FAL during the foot and mouth epidemic in 2001 and did not extend to a dispute as to whether the settlement agreement was entered into under duress.

(iv) Even if the wording of the Mediation Procedure did apply to this case, that would not in itself lead to the witness summons being set aside. Rather, it would be a factor for the court to take into account in deciding whether, in the interests of justice, the mediator should be called as a witness;

a. by agreeing that the mediator could be called as a witness, the parties had waived any without prejudice privilege in the mediation and since it was their privilege, they were entitled to do so; and

b. the mediator had a right to rely on the confidentiality provision in the Mediation Agreement, but this case was an exception, where the interests of justice lay strongly in favour of evidence being given of what was said and done at the mediation.

20.32 Most mediation agreements contain provisions seeking to restrict the parties' ability to call the mediator as a witness in subsequent disputes. This judgment may result in mediators inserting stricter provisions into their mediation agreements restricting the extent to which the parties may call them as a witness even though it appears from the judgment that even express words dealing with the situation that arose here would not have led the judge to a different conclusion. It would have been just one factor to take into account when applying the interests of justice test.

20.33 There are no rules which govern how a mediation is to take place. In particular, there is nothing like the CPR or a set of arbitration rules which prescribe the process. However, most mediations take similar forms with occasional variations to suit the circumstances. The following provides a broad outline of a typical mediation.

How Mediations Are Run; Opening Statements; Responses; Without Prejudice Meetings

20.34 At the first stage the parties gather. It is an obvious, but important, point that the client who attends must have authority to settle the matter under mediation, or have access to the decision-maker (a point which can create practical difficulties where the mediation lasts into the small hours). Others who might attend are their lawyers (solicitors and sometimes counsel), and experts. Much depends upon what is in issue. If there are important legal disputes, then counsel may be useful. Sometimes counsel is used for the presentation, but the process is not a legal one and even if counsel is instructed in the case it is by no means necessary that s/he attend.

Each party should have a room in which it can house all the principals. In addition to a room for each party there should be a room where all the participants in the mediation can be. It is also prudent to have at least one breakout room where the mediator can arrange face-to-face meetings separate from and concurrent with other mediation meetings. For example the mediator may wish to put the experts together whilst s/he meets with the parties. The reason for this may be that the experts are thought by the mediator to be capable of coming to an agreement on an aspect of the case only if they are left alone together without the rest of the team. **20.35**

In a mediation which is going to last more than one day a small room for the mediator him/herself may be useful. In day-long mediations the room where all the parties meet or the separate breakout room may be sufficient. The parties' advisors should think carefully about the number of rooms which are needed and their relative comfort. For much of the day or days set aside the participants will be left alone whilst the mediator sees the other side. If the room is uncomfortable, overheated, or oppressive, tempers may fray and patience may wear thin. **20.36**

At the start of the day the mediator will meet the parties in their own rooms and will judge whether it would be appropriate to have a plenary session at which all attend and statements can be made. Most mediations start with this plenary session. Each party is invited to make an opening statement of about five minutes in length. The purpose of this is to allow each side to vent its feelings in front of the other. It is also a useful opportunity to create goodwill by the making of any necessary or appropriate apology or the like: very often the real reasons for a dispute having become intractable have more to do with feelings than with logic, and one of the advantages of this initial a face-to-face meeting is that it can maximize the chances of getting rid of unnecessary obstacles to resolution of that sort. If the mediation statement is properly written, there should be no need for a lengthy address concerning the merits of the case. Importantly, even if it is drafted by the lawyers, an opening statement will be rather different to an opening before a court: it should acknowledge, for instance, the commercial realities underlying the stance adopted by the party on whose behalf it is given, as well as the perceived position of the other side. **20.37**

On occasion, these plenary sessions come close to being counter-productive: some people find it too much to have to listen to the other side's point of view, and they lose their composure. Sometimes, of course, this is play acting. But the session is normally valuable for the mediator who can assess the dynamics of the two groups—s/he can, for instance, identify those individuals who cannot stand one another and should be kept apart and those who might be able to do business together. **20.38**

Following the plenary session the parties will retire to their rooms and the mediator will then meet with the parties separately or arrange special meetings between different groups of people. The scale and nature of the meetings can vary widely, but in all of them, the mediator tries to discover how s/he can move the parties closer together. **20.39**

20.40 In the private meetings, the mediator will sometimes be relaying information or offers/counter-offers from the other side, sometimes s/he will be probing each party's views of their own and the other side's positions and case, and sometimes will simply be allowing the parties to vent their feelings. On occasion, plenary meetings may be reconvened to allow matters to be raised in open session; alternatively, meetings may be arranged between particular individuals or particular categories of individuals.

The Role of the Mediator; The Role of Legal Advisors and Professionals

20.41 During the course of the mediation, the experts and the legal advisors will have to take their role from the mediator, in particular as to their involvement in specific meetings. It is the hands-on nature of the role of the mediator which makes their (perceived) expertise all the more important: the more confidence the parties have in the mediator, the more willing they will be to accept his/her decisions as to the course of the mediation.

20.42 One practical problem often encountered is that the mediator finds that s/he has to spend more time with one party than another. S/he should, so far as possible, give the party which s/he is leaving alone a time estimate of when she may return and /she should normally update the waiting party if that changes. Where there are co-mediators or an assistant mediator, it can be of assistance for the mediators to split on occasion so that parties are not left on their own for too long.

20.43 Often mediations will go on into the night and the early hours. Frequently an agreement is reached near the time when the mediation clock is running out. Discussion and mediation may fill the time available, but perhaps the best explanation for the last minute nature of many agreements is simply that the approach of the eleventh minute of the eleventh hour concentrates the minds of the parties, who by then will have invested so much in the process. If a mediation is 'unsuccessful' in that an agreement is not reached on the day, it is often reached between the parties (with or without the help of the mediator) in the two weeks which follow the mediation. The process of discussion and the will to put an end to the dispute will have taken on a momentum which neither party wishes to stop, however intransigent they are at the start of the process. It is to try to ensure this momentum is maintained that skilled mediators are astute to ensure that parties at the end of a 'failed' mediation crystallize their thoughts as to the way ahead before leaving the building. Sometimes this can be in a final plenary meeting; sometimes—for instance where tempers have frayed—it would be more appropriate to do this in private.

Settlement

20.44 If an agreement is reached orally, no competent mediator will allow the process to stop until the agreement is put into writing and signed by the parties. Oral agreements have a way of unravelling as the 'deal improvers' get thinking and as misunderstandings grow up spontaneously in the soil of the parties' distrust.

Costs; *Dunnett* Mediations; Effect of Not Mediating Upon Any Costs Applications Made at the End of Arbitration or Litigation; *Cowl*; the Costs of the Mediation itself

Mediation is now part of the legal landscape,[8] the courts provide venues for mediation,[9] and no construction dispute is likely to come to court unless it has been tried.[10] Even mediations which both parties see as a waste of time at the outset are often successful. **20.45**

There may be cases where mediation is not suitable. For example the Commercial Court Working Party on ADR wrote in 1999: **20.46**

> The Working Party believes that there are many cases within the range of Commercial Court work which do not lend themselves to ADR procedures. The most obvious kind is where the parties wish the court to determine issues of law or construction which may be essential to the future trading relations of the parties, as under an on-going long term contract, or where the issues are generally important for those participating in a particular trade or market. There may also be issues which involve allegations of fraud or other commercially disreputable conduct against an individual or group which most probably could not be successfully mediated.

Those factors outlined will be rare in most construction cases, but the points dealing with questions of construction in long-term relationships will be present in PFI contracts. **20.47**

---

[8] CPR r 1.4(1) obliges the court to further the overriding objective of enabling the court to deal with cases justly by actively managing cases, and r 1.4(2)(e) defines 'active case management' as including 'encouraging the parties to use an alternative dispute resolution procedure if the court considers that appropriate and facilitating the use of such procedure'. Rule 26.4(1) provides: 'a party may, when filing the completed allocation questionnaire, make a written request for the proceedings to be stayed while the parties try to settle the case by alternative dispute resolution or other means.' The virtues of mediation in suitable cases are also recognized in the Chancery Guide (paras 17.1 and 17.3), the Queen's Bench Guide (para 6.6), the Admiralty and Commercial Court Guide (para D8.8), and the Technology and Construction Court Guide (para 6.4). Judges in the Commercial Court routinely make 'ADR orders' in the form set out in Appendix 7 to the Admiralty and Commercial Court Guide. In March 2001, the Lord Chancellor announced an 'ADR pledge' by which all government departments and agencies made a number of commitments including that: 'Alternative dispute resolution will be considered and used in *all* suitable cases wherever the other party accepts it.'

[9] See, for example, the Central London County Court scheme.

[10] Note particularly the stringent comments of Ward LJ, agreed with by Rix LJ in *Burchell v Mr and Mrs Bullard* [2005] EWCA Civ 358 in which he described a small building dispute between householder and builder (very much of the garden variety): '. . . a small building dispute is *par excellence* the kind of dispute which, as the recorder found, lends itself to adr . . .' The substance of these comments was repeated by him in *Khudados v Hayden & Ors* [2007] EWCA Civ 466 (3 April 2007) in which he said:

> But I add this footnote: this litigation fills me with real gloom and despair. Its first expedited hearing was listed for two to three days. It is now listed for 25 days. Like Topsy it has grown and I cannot but wonder whether it has grown out of all proportion. It strikes me yet again as the kind of case which should not return to this court before the parties have undertaken some mediation and the Court of Appeal's mediation service is here to assist in that endeavour. This case cries out for that to be attempted before further steps are taken and I urge it upon the parties with all the force that I can.

Sedley LJ agreeing said 'I respectfully associate myself strongly with his final indication that mediation is a course which even, or perhaps particularly at this late stage, the parties ought to attempt.'

20.48 The timing of a mediation may be crucial. The parties' legal advisors will have to judge when it is likely to be most beneficial. It is not always the case that mediation should be begun prior to the commencement of court or arbitration proceedings. Sometimes in disputes which appear intractable, it is better for the parties to set out their case in pleadings and even instruct experts to report before a mediation begins. In addition to giving an intransigent party a taste of the legal process the steps of exchange of pleadings and reports may crystallize the dispute in a way which even a pre-action protocol exchange of correspondence does not. It is a matter for the judgment of the parties' advisors as to the right time to try mediation. Obviously mediation ought not to be delayed until too much in costs has been expended. Today, with the less confrontational approach to litigation, it ought to be possible for solicitors for the parties to pre-agree when mediation is likely to be tried to best effect.

20.49 The TCC and the Commercial Court judges encourage mediation, but the courts do not any longer push the parties unwillingly to mediation.[11] Often judges will ask whether it has been considered and if the legal advisors are competent and reputable, the judge will usually defer to the parties' advisors as to the use of mediation and its timing in any particular case. This having been said, this area has been the subject of some sustained scrutiny by the higher courts, analysed in the following paragraphs.

20.50 In *R (Cowl) v Plymouth City Council*,[12] a public law case, Lord Woolf made it clear that litigation was a last resort, and that ADR should be encouraged. In *Dunnett v Railtrack plc*,[13] the Court of Appeal denied a successful party its costs because it had not countenanced mediation. The twin effect of those decisions was that it was thought by the legal profession that mediation had to be tried in all cases because if it was not their client would suffer in costs even if successful.

20.51 Thus it was that there sprang up what mediators called 'Done It' mediations—a (rather bad) play on the word *Dunnett*. These were mediations where one or both parties went through the motions of a mediation so that they did not risk an adverse costs order against them. The spate was further increased following the decision of Lightman J in *Hurst v Leeming*.[14]

20.52 Many mediators have complained about the futility of some mediations which were entered into simply to avoid any potential costs awards being made against the party which thought it had a watertight case.

20.53 The trend towards such 'Done It' mediations has been substantially abated by the effect of the decision of the Court of Appeal in the conjoined appeals of *Halsey v Milton Keynes General NHS Trust* and *Steel v Joy*.[15] In *Halsey*, the Court of Appeal decided that, in order

[11] Indeed such a course would run contrary to Art 6 of the Convention for the Protection of Human Rights and Fundamental Freedoms. See *Halsey v Milton Keynes General NHS Trust* and *Steel v Joy* [2004] 1 WLR 3002 at para 9; and *Deweer v Belgium* (1980) 2 EHRR 439 at para 49.

[12] [2002] 1 WLR 803.

[13] [2002] 1 WLR 2434.

[14] [2003] 1 Lloyd's Rep 379.

[15] [2004] 1 WLR 3002.

to deprive a successful party of some or all of its costs on the ground that the party had refused to agree to ADR, the unsuccessful party bore the burden of showing that the refusal had been unjustified. Dyson LJ, giving the judgment of the Court, identified several factors that would go into the equation (albeit stressing, at 3009, that 'in many cases no single factor will be decisive, and that [the] factors should not be regarded as an exhaustive check list'). The factors[16] are: (1) the nature of the dispute; in other words: does the dispute lend itself to mediation or is it of the type which is unsuitable for reasons given by the Commercial Court Working Party?; (2) the merits of the case; if a party has a 'watertight' case and reasonably believes that it does so, that may be sufficient justification for the refusal of mediation; however borderline cases, even if won in the end, may not; (3) the extent to which other settlement methods had been attempted; this may obtain if one party takes a realistic view of the case and the other party does not, nevertheless mediation often succeeds when other attempts to settle have failed; (4) whether the costs of the ADR would have been disproportionately high; this is rare in construction cases because they are disproportionately costs heavy, nevertheless a day in mediation will be as expensive as a day in court; (5) whether any delay in setting up and attending the ADR would have been prejudicial, for example putting off the trial date; and (6) whether the ADR had a reasonable prospect of success viewed from the point of view of the successful party.[17]

At this point consideration must be given to the burden of proof. In *Halsey* Dyson LJ[18] placed the burden of proof on the unsuccessful party to litigation to show that there was a reasonable prospect that the mediation would have been successful: **20.54**

> In our judgment, it would not be right to stigmatize as unreasonable a refusal by the successful party to agree to a mediation unless he showed that a mediation had no reasonable prospect of success. That would be to tip the scales too heavily against the right of a successful party to refuse a mediation and insist on an adjudication of the dispute by the court. It seems to us that a fairer balance is struck if the burden is placed on the unsuccessful party to show that there was a reasonable prospect that mediation would have been successful. This is not an unduly onerous burden to discharge: he does not have to prove that a mediation would *in fact* have succeeded. It is significantly easier for the unsuccessful party to prove that there was a reasonable prospect that a mediation would have succeeded than for the successful party to prove the contrary . . .
>
> Where a successful party refuses to agree to ADR despite the court's encouragement, that is a factor which the court will take into account when deciding whether his refusal was unreasonable. The court's encouragement may take different forms. The stronger the encouragement,

---

[16] Identified at 3009D–E and expanded at 3009F–13F.

[17] This decision has come in for some criticism. See, for example, Arthur Marriott QC's address at the Annual General Meeting of the London Branch of the Chartered Institute of Arbitrators on 27 April 2005 reprinted in [2005] 71 Arbitration 228, where at 229 Mr Marriott said:

> The rules oblige the courts to encourage and use ADR and to impose sanctions for refusal or abuse . . . This is a classic instance of a superior court not understanding and not being told, of how to give effect to the fundamental policy issues of attempting to reduce cost and to improve access to justice . . . There is a real need to review *Halsey* and to overturn it. I hope a case or series of cases can be reviewed by the Lord Chief Justice and heads of division or by the House of Lords. Failing that we shall need to legislate.

[18] At 3012.

the easier it will be for the unsuccessful party to discharge the burden of showing that the successful party's refusal was unreasonable.

20.55 Although *Halsey* makes it clear that 'Done It' mediations may well become a thing of the past, it is nonetheless clear that careful consideration must be given to any reasonable offer to mediate.[19] As Ward LJ held in April 2005, explicitly applying *Halsey*, '[t]he profession can no longer with impunity shrug aside reasonable requests to mediate'.[20] In *Burchell*,[21] Ward LJ was even prepared to accept that an unreasonable refusal to mediate can be taken into account where it took place before the start of formal proceedings.[22]

20.56 ADR can be no substitute for litigation in the final analysis because parties cannot be made to settle. ADR is a process which ought to go side by side with litigation. Most construction disputes are capable of settlement with the use of mediation. Nevertheless the courts recognize that some disputes cannot settle. This may be because of the intransigence of one or other party, or it may be because the parties actually want a definitive result or because it is one of those rare cases where the costs of litigation are in proportion to the amounts in issue.

20.57 The Court of Appeal has also made it clear in *Daniels v Commissioner of Police for the Metropolis* [2005] EWCA Civ 1312 that the principle that a refusal to negotiate will not, if it is in the circumstances reasonable, result in a costs penalty (as laid down in *Halsey*) are not confined to mediation but apply to ADR generally, and thus include a refusal to be drawn into negotiations in correspondence or involving CPR Part 36.

20.58 However the fact remains that the costs of construction disputes in court or arbitration are out of scale with what is in issue in most cases. In 'small' cases involving loss and expense or defects or both where the amount in issue is up to £100,000, or even more, this is all the more stark a position. For a paradigm case of a householder in a dispute with a domestic builder the case of *Burchell*[23] is instructive. The figures in the case deserve examination. The claim by the builder was for £18,318.45. The counterclaim by the householder was £100,815.34. The claim succeeded in full and the counterclaim to the extent of £14,373.15 making the builder the net winner in the litigation of £5,000. The builder's costs were £65,000 plus a further £27,500 of costs incurred by the builder's sub-contractor and the householder's costs were £70,000. That makes a total costs expenditure in the action of over £160,000 which is so far out of scale to the £5,000 net recovery to be scandalous. Nevertheless this is prosaically common. The TCC tries to manage this situation with case management.[24] But with the best will in the world, these figures may be ameliorated only.

---

[19] Note too the criticism of *Halsey* by the then Master of the Rolls in his May 2008 lecture <http://www.judiciary.gov.uk/docs/speeches/mr_mediation_conference_may08.pdf>. See also the discussion in Ch 22 and para 22.39.

[20] *Burchell v (1) Mr Bullard (2) Mrs Bullard* [2005] EWCA Civ 358, [2005] BLR 330 at para [43].

[21] A low value (but relatively high cost) construction claim, said by Ward LJ (at [27]) to be '*par excellence* the kind of dispute . . . which lends itself to ADR.'

[22] Para [43] of the judgment. Note also para 5.4 of the TCC's pre-action protocol for construction and engineering disputes which expressly requires the parties to consider at a pre-action meeting whether some form of ADR would be more suitable than litigation.

[23] *Burchell v (1) Mr Bullard (2) Mrs Bullard*, n 20 above.

[24] See, for example, the use of a single joint expert.

No sane person would litigate over these figures. Unfortunately it is often too late before they become apparent and when they do much will have been spent. Any sensible legal advisor therefore will make strenuous efforts to try to keep their client out of court and away from litigation like this. Mediation may not be a universal panacea. Sometimes litigation is inevitable, but mediation has a good record in ensuring that a scandal like that excessive cost expenditure either does not occur or is minimized. Any small builder or householder should examine the figures in the *Burchell* case set out above before contemplating litigation. Any legal advisor is negligent not to warn his or her client of them. The absurdity of costs in larger litigation cannot be ignored either, but they are more likely to be in scale with appropriate assessment by costs judges.[25]

The encouragement to mediate has continued, both from the Court of Appeal and from the Rules Committee. The Court of Appeal has emphasized the limitations on the principles set out in *Halsey* and *Daniels* (ie that it is only likely to be in exceptional cases that a refusal to negotiate will be reasonable), and affirmed its commitment in the general case to ADR, and mediation in particular, as the most suitable way to resolve disputes. In both *Dr SR Burne v 'A'* (an allegation of professional negligence) and *IDA Ltd and Others v University of Southampton and Others* (about who owns a patent) it recommended mediation to the parties in unequivocal terms. **20.59**

In *Dr SR Burne v 'A'* [2006] EWCA Civ 24, the court was concerned with allegations of professional negligence against a doctor, who is a GP. 'A' had suffered a heart attack and brain damage as a small child as a result of a blockage in a ventriculo-peritoneal shunt (a device that continuously drains fluid from the brain cavity) that had been fitted when he was 9 weeks old. The Judge at first instance had put the agreed expert evidence found for 'A' on the basis of agreed expert evidence. The Court of Appeal ordered a re-trial but Lord Justice Sedley, in giving the judgment of the Court, said: **20.60**

> I do, however, feel very strongly that this is a case which must be referred to alternative dispute resolution before it is restored for the re-trial. Both parties should take stock of the strengths but also the weaknesses of their respective cases which are now plain for all to see and I hope mediation will bring a swift conclusion to a tragic event.

In *IDA Ltd and Others v University of Southampton and Others* [2006] EWCA Civ 145, the court was concerned with costly litigation over who owned a patent. In allowing the appeal, Lord Justice Jacob, in giving the decision of the court, said: **20.61**

> Parties to these disputes should realise, that if fully fought, they can be protracted, very, very expensive and emotionally draining. On top of that, very often development or exploitation of the invention under dispute will be stultified by the dead hand of unresolved litigation. That may be the case here: there has not yet been any exploitation by either side, some 8 years after the original PCT application. It will often be better to settle early for a smaller share than you think you are entitled to—a small share of large exploitation is better than a large share of none or little.

[25] Nevertheless, see *Henry Boot Construction Ltd v Alstom Combined Cycles* [2005] 3 All ER 932 at paras [87]–[90] where a contractor submitted to the Court of Appeal a bill of costs of £350,000 for a two-day appeal on a point of pure law.

> This sort of dispute is particularly apt for early mediation. Such mediation could well go beyond conventional mediation (where the mediator facilitates a consensual agreement). I have in mind the process called 'med-arb' where a 'mediator' trusted by both sides is given the authority to decide the terms of a binding settlement agreement. Such a power in effect already exists in the Comptroller once he has found a case of entitlement (see s 8(2)). But by then it will probably be far too late.

20.62 It has been pointed out in an article by David Cornes, a very experienced mediator, that arbitration followed by mediation (rather than the other way round) would be more efficacious in such situations, since it would avoid the potential difficulties as to how a mediator deals with information that he has gleaned from a mediation when he comes to give an arbitration award. Arb-med can be set up so that the arbitration award is put in an envelope on terms that it is not to be opened unless and until the subsequent mediation fails to produce an agreed settlement.

20.63 The 'Practice Direction—Protocol', part of the Civil Procedure Rules, was recently amended as to mediation so as to give more emphasis to pre-action attempts to settle, including mediation, providing the court with a power to require evidence as to what alternative means of settlement were considered and to have regard to such matters when considering costs. This Practice Direction applies to the pre-action protocols which have been approved by the Head of Civil Justice. So, for example, it applies (paragraph 1.1) to the Construction and Engineering Disputes Protocol.

20.64 It has always been a requirement (paragraph 4.3(f)) that the claimant's Pre-Action Protocol letter should state that the claimant wishes to enter into mediation (or some other form of ADR). However, paragraph 4.7 has been re-written as follows:

> The parties should consider whether some form of alternative dispute resolution procedure would be more suitable than litigation, and if so, endeavour to agree which form to adopt. Both the Claimant and Defendant may be required by the Court to provide evidence that alternative means of resolving their dispute were considered. The Courts take the view that litigation should be a last resort, and that claims should not be issued prematurely when a settlement is still actively being explored. Parties are warned that if the protocol is not followed (including this paragraph) then the Court must have regard to such conduct when determining costs.
>
> It is not practicable in this protocol to address in detail how the parties might decide which method to adopt to resolve their particular dispute. However, summarized below are some of the options for resolving disputes without litigation:
>
> - Discussion and negotiation.
> - Early neutral evaluation by an independent third party (for example, a lawyer experienced in that field or an individual experienced in the subject matter of the claim).
> - Mediation—a form of facilitated negotiation assisted by an independent neutral party.

The costs of the mediation

20.65 The parties will normally provide expressly within their mediation agreement as to who should bear the costs of the mediation.

20.66 However the question has arisen as to whether the costs involved in any mediation might (like, for example, those of complying with a Pre-Action Protocol) be recoverable as costs *'incidental to'* any subsequent proceedings pursuant to s 51 of the Supreme Court Act 1981.

This was considered in relation to the costs of a pre-action mediation by Coulson J in *Lobster Group Ltd v Heidelberg Graphic Equipment Ltd.*[26] He said: **20.67**

> I do not believe that the costs of a separate pre-action mediation can ordinarily be described as 'costs of and incidental to the proceedings'. On the contrary, it seems to me clear that they are not. They are the costs incurred in pursuing a valid method of alternative dispute resolution. Those costs were incurred in a form of dispute resolution which had no connection to these proceedings, and which here took place 2.5 years before the proceedings even started. Both the course of the mediation itself and the reasons for its unsuccessful outcome are privileged matters, known only to the parties. As a matter of general principle, therefore, I do not believe that the costs incurred in respect of such a procedure are recoverable under s 51.

In that case the judge was plainly influenced by the fact that the parties had agreed to bear their own costs of the mediation. It would be a breach of that agreement to allow one party to recover from the other their costs of the mediation. **20.68**

He did acknowledge however that there could be specific exceptions to the general rule that pre-action mediation costs are not recoverable in subsequent proceedings. He gave the example of an expert's report provided for the mediation and then subsequently used in the court proceedings. Such a report could be recoverable as costs, but only if in some way, the parties had agreed, despite their agreement that the costs of the mediation would be shared, that those specific costs could be the subject of any subsequent application. **20.69**

A similar question arose in *National Westminster Bank Plc v Thomas James Feeney and Linda Catherine Feeney* [2006] EWHC 90066 (Costs), Master Campbell, and on appeal to Mr Justice Eady sitting with Assessors in May 2007 (Unreported). The issue there was whether, where a Tomlin Order settling proceedings provided for costs to be assessed, the costs of mediation were recoverable as part thereof. **20.70**

The answer is that both the legal costs and the mediation fees are, as a matter of principle, costs that can be within the definition of recoverable costs under the CPR. However, in this case, the terms of the mediation agreement prevented their recovery. **20.71**

In *Roundstone Nurseries Ltd v Stephenson Holdings Ltd*[27] the judge concluded that the mediation was an integral part of the parties attempt to comply with the Pre-Action Protocol process. In that case there had been no agreement as to the costs of the mediation. In these circumstances the judge held that he did have jurisdiction to make an order in respect of the costs of the mediation which had been cancelled at short notice by the defendant. He ordered the defendant to pay the claimant's costs thrown away in connection with this mediation as: **20.72**

> (i) the mediation was an agreed part of the Pre-Action Protocol process. The defendant was therefore obliged to participate in it;

[26] [2008] EWHC 413 (TCC).
[27] [2009] EWHC 1431 (TCC).

(ii) without the mediation, there was no way in which the Pre-Action Protocol requirement for a without prejudice meeting between claimant and defendant could be fulfilled;

(iii) the mediation had been arranged before there was any question of inviting a third party who was possibly liable instead of the defendant, to participate, and should have gone ahead even without such third party's involvement; and

(iv) the third party did not participate in the mediation because of the late service of the defendant's expert's report. It was not unreasonable for them to take that line. It was not then for the defendant unilaterally to cancel what would have been the completion of an eight-month Pre-Action Protocol process.

21

ARBITRATION

The Arbitration Agreement

Before you can have an arbitration, you must have an agreement to arbitrate. An arbitration agreement is an agreement to submit to arbitration present or future disputes;[1] those disputes can be of any nature and do not have to be contractual.[2] There have been oral agreements to arbitrate, but they are so rare as to be for practical purposes non-existent. This is all the more the case now that the Arbitration Act 1996, which governs all arbitration in England and Wales, provides that an arbitration agreements is made in writing if: **21.01**

(a) it is made in writing; or
(b) it is made by an exchange of communications in writing; or
(c) it is evidenced in writing; or
(d) it is not made in writing but is made by reference to terms which are in writing; or
(e) it is made orally and the parties exchange written submissions in arbitration or legal proceedings and an arbitration agreement is alleged by one party and not denied by the other party.[3]

[1] A dispute will include any difference between the parties. See s 82(1) of the Arbitration Act 1996.
[2] Arbitration Act 1996, s 6.
[3] s 5.

21.02 Therefore, if parties agree to arbitrate by an exchange of email correspondence or by letter, or even if they make an oral agreement to arbitrate in accordance with a standard form of agreement (such as, for example, Article 7 of the JCT Minor Works Building Contract 2005) then they will have made an agreement in writing for the purposes of the Arbitration Act 1996.

21.03 An arbitration agreement is of a special nature. Unless otherwise specifically agreed, it is not discharged by death[4] and survives the repudiation, frustration or discharge of the agreement with which it is concerned.[5] It is treated as distinct and separable from the substantive agreement of the parties; the arbitration agreement itself would have to be directly impeached for it to be invalid.[6]

21.04 The first step for any employer, contractor, or sub-contractor, when involved in a dispute, is to check and see whether the dispute is governed by an arbitration agreement. Whilst this is usually contained within a clause which forms part of the agreement, it can, as has been seen, be incorporated by reference. The most common incorporation by reference is when parties agree that a particular set of standard conditions shall govern their agreement.[7] This may not be at first sight obviously done; references to standard conditions can be cryptic and elusive. For example, one exchange of correspondence had a sub-contractor and a main contractor agree that their agreement should be governed by the main contract and it was held that in fact they agreed to be bound by the standard form of subcontract.[8] It is easy to know if your contract has an arbitration clause when it is incorporated within the body of the contract; in that event there is little scope for argument. But incorporation by reference may be more difficult to ascertain.[9] Whilst general words will normally be sufficient to incorporate by reference standard terms to be found elsewhere, a stricter approach requiring express reference to an arbitration clause may be applied by the courts where two contracts are involved; for example, where a sub contract refers to the terms of a main contract.[10]

21.05 Many construction disputes will have been subject to adjudication. Adjudication is dealt with in Chapter 23. This chapter is concerned only with arbitration. The only relevance of adjudication is that it has been a step which may have been taken and which provides a temporary finality to the dispute referred to it. The finality is subject to a later decision of an arbitrator or court which has the responsibility to finally decide any dispute between the parties.

---

4 s 8.

5 *Heyman v Darwins* [1942] AC 356 and s 7 of the Arbitration Act 1996.

6 s 7 and *Fiona Trust & Holdings Corporation and Ors v Privalov and Ors* [2008] 1 Lloyd's Rep 254, [2007] UKHL 40.

7 See, for example, *Brightside v Mitchell Construction* [1975] 2 Lloyd's Rep 493.

8 See ibid.

9 An example is provided by *Crest Nicholson (Eastern) Ltd v Western & Another* [2008] All ER (D) 249, in which an arbitration clause contained in the NHBC scheme was held inapplicable to a dispute between developer and purchaser.

10 Guidance to the approach of the courts was provided in *Sea Trade v Hellenic Mutual War Risks (The 'Athena')* [2006] 2 CLC 710.

Enforcement of the Arbitration Agreement

In many jurisdictions, if a dispute is subject to an arbitration agreement the courts will not hear the case should one of the parties refer to it in their defence. The court will simply decline jurisdiction and refuse to hear the case save for the purposes of dismissing it. In England and Wales, if a court action is begun then the defendant may apply for a stay of the action[11] and the stay will be granted, unless the applicant has taken a step in the proceedings to answer the substantive claim or the arbitration agreement is null and void or otherwise incapable of being performed. **21.06**

Therefore, the courts will, if asked, stop any court action which relates to a dispute which is subject to an arbitration agreement. The courts are likely to be generous in construing the scope of an arbitrator's jurisdiction, assuming that the parties will have intended, unless they have provided to the contrary, that any dispute arising from their relationship, including disputes over the validity of their agreement itself, should be covered by their agreement to arbitrate.[12] As seen in Chapter 23 dealing with adjudication, the fact that there is no true dispute between the parties because the position taken by the other side is palpably wrong is not a reason for the matter not to be referred to arbitration.[13] **21.07**

The Arbitration Act 1996: The Theory of Party Autonomy

The Arbitration Act 1996 came into force on 31 January 1997 after just over one year's consideration by Parliament following a long consultation process through the Departmental Advisory Committee on Arbitration Law under the chairmanship of Lord Mustill, Lord Steyn, and then Lord Saville. The Act, remarkably for any piece of legislation, is readable and readily understood. It applies to all arbitrations commenced after 31 January 1997, and therefore it will apply to all arbitrations relevant to the reader of this work. **21.08**

The Arbitration Act 1996 was designed to provide a new departure for arbitration, which was perceived as being too much subject to court interference by many arcane extensions of the supervisory role of the courts. The courts' ability to correct arbitration awards and hear appeals meant that the United Kingdom and London was losing its central role as a seat of arbitration because the parties to commercial contracts felt that the court's potential interference in their affairs was too much of a risk. They would rather go elsewhere and have the arbitration award achieve finality. **21.09**

The countervailing tension was that, if a party wished his or her case decided by arbitration, that did not mean that the party did not want their case decided according to law, and if there was no right of appeal against an arbitral award which was made contrary to law, then there was a real risk that the party would not have its legal rights decided properly at all. The **21.10**

[11] Arbitration Act 1996, s 9(4).

[12] See *Fiona Trust & Holdings Corporation and Ors v Privalov and Ors* [2008] 1 Lloyd's Rep 254, [2007] UKHL 40.

[13] See *Halki Shipping v Sopex Oils Ltd* [1998] 1 Lloyd's Rep 465, [1998] 1 WLR 726; and *Hayter v Nelson* [1990] 2 Lloyd's Rep 265.

rejoinder to that was, of course, that if you choose arbitration then you choose to run the risk that the arbitrator will fall into error, just as courts sometimes do.

21.11 The Arbitration Act 1996, therefore, regulated and restricted the right of appeal against an arbitrator's award. It also provided for the courts to provide support to the arbitration or the arbitrator. Section 1 of the Arbitration Act sets out the overarching scheme of the Act, namely, that the parties are free to agree how their dispute is to be resolved (subject to public interest safeguards) with the objective of arbitration being to obtain a fair resolution of the dispute by an impartial tribunal without unnecessary delay and expense, and the court will not intervene except in the limited cases which the Act expressly permits, on the basis that court interference should be kept to a minimum. The parties are masters of their own fate, the philosophy being that the parties have by contract agreed arbitration and with it comes the plusses and minuses which go with it. Had the parties wished to have a court determination, then they could have agreed to that. So the Act has set the parties on a course whereby they embark upon an arbitration and, save for certain mandatory rules, it leaves them and the arbitrator to adopt their own procedures. Over a decade after the Act was brought into force, it is widely thought that this intended balance has been met and the international standing of London as an arbitral centre has been restored.

Arbitration Procedure

21.12 The first matter for consideration is whether the arbitration is subject to a particular set of rules or whether the arbitration and the procedure is left in the hands of the arbitrator and the parties. More and more arbitration clauses provide that the arbitration should be subject to specific rules. This is a trend which has accelerated over the last twenty years. Now common standard forms, such as the JCT and the ICE, each provide that the arbitration is subject to rules.

The nominating body and choice of arbitrator

21.13 Other arbitrations, often erroneously called 'ad hoc' arbitrations, stem from arbitration agreements which are contained within a set of contract conditions which in turn provide only that all disputes shall be referred to the arbitration of a person to be agreed between the parties or failing an agreement someone nominated by a particular nominating body. The most popular nominating bodies for domestic (non-international) arbitrations are:

(a) the Royal Institution of Chartered Surveyors;
(b) the Royal Institute of British Architects;
(c) the Institution of Civil Engineers; and
(d) the Chartered Institute of Arbitrators.

21.14 Once a dispute is established, the dissatisfied party should serve a notice specifying the dispute and inviting the other side to agree their choice of arbitrator.[14] It is distressingly

[14] Commonly called a 'Notice of Dispute' or a notice 'to concur in the appointment of an arbitrator', s 14. See *Taylor Woodrow v RMD Quickform Ltd* [2008] BLR 383, 118 Con LR 57, Ramsey J, for a case concerning the requirements of such notice in the context of limitation protection.

frequent that the arbitrator who is proposed by one side is automatically not agreed by the other. It is hard to get disputants out of that mind set. Automatically the proposed arbitrator will be thought of as unacceptable to the other side. This has led to a dangerous trend, whereby a party's second choice is proposed in the first round of correspondence secure in the knowledge that the person concerned will not be agreed, and then in the second round the first choice will be proposed. The trend now is for arbitrators to be named as part of a panel in the arbitration clause (which is now often contained in a schedule to the agreement) and the arbitration agreement will then permit one of those to be chosen from the panel when a dispute arises. This has spilled over from panels of experts in expert determination clauses and dispute resolution panels. There is also a growing practice for parties to interview jointly potential arbitrators.[15]

Agreement of terms with the arbitrator on his or her fees

Once an arbitrator has been agreed, or nominated by the nominating body, the next stage is for the parties to agree terms with the arbitrator. Normally an experienced arbitrator will have a standard form of agreement and a fixed hourly rate which he or she will propose and the other parties will agree to. If no agreement is made, then the Arbitration Act 1996 provides that the arbitrator shall recover his or her reasonable fees and that the parties should be jointly and severally liable; in the event of a dispute the court can decide what is reasonable.[16] This is a rarely-used facility since fees are normally agreed in advance. If the parties find the proposed fee structure exorbitant, then they can seek to agree another arbitrator or ask the nominating body to re-nominate. Normally if parties raise their concerns about the arbitrator's fees and charges with the proposed arbitrator (particularly the often difficult subject of cancellation charges), they are answered satisfactorily for all concerned. **21.15**

Once there is agreement with the arbitrator, which is usually a tripartite agreement (but if one of the parties is truly recalcitrant there will have to be a bilateral agreement between the arbitrator and the party seeking arbitration) then it is up to the arbitrator to set the arbitration procedure in motion. **21.16**

The key questions influencing procedure

No two arbitrations are the same; but there are common questions in construction cases as follows: What is the nature of the dispute (defects; loss and expense; a combination)? How can that dispute be best and economically decided? Should there be a face-to-face hearing like a court hearing? Should witnesses give oral evidence and if so should it be under oath? Should there be disclosure of documents? Should there be a site visit? If the arbitration concerns whether the work is defective, can the arbitrator make a decision on that question without hearing from expert witnesses? How will the parties be represented? Does the arbitrator need expert assistance? **21.17**

The arbitrator will normally wish to hear from the parties either at a preliminary meeting or on paper as to their procedural preferences. As part of his or her decision the arbitrator **21.18**

[15] There is written guidance on this from the Chartered Institute of Arbitrators.

[16] s 28.

will have to decide the answer to the questions listed above, and each is briefly considered below.

21.19 What is the nature of the dispute (defects or loss and expense for example) is the fundamental question, allied with it is the question of size. If the case involves small sums of money, then that will dictate how extensive or thorough the procedural steps will have to be. A case involving a small sum of money should be conducted with an eye on proportionality and that factor will impinge on the questions which follow. Most arbitrations will involve one of three types of dispute: first, (a) defective work, leading to three questions: (i) is the work in accordance with the contract?; (ii) what remedial work is required if it is not?; and (iii) what is the cost of that remedial work?; second, (b) valuation, leading to two basic questions: (i) is the dispute relating to measurement of the work?; and (ii) is the dispute relating to loss and expense and how it is ascertained?; third, (c) is the dispute relating to extension of time claimed? Many cases involve a combination of these questions, but, equally, many relate to only one. The nature of the dispute will assist the arbitrator (and the parties) to decide how best the matter can be decided and what procedural steps need to take place in order to bring the matter to a conclusion.

The expertise of the arbitrator

21.20 If the case relates to defects and the arbitrator has expertise in the area of work about which the dispute relates, then the arbitrator ought to consider going to the site, examining the work and seeing whether, with the help of the parties, he or she can make a decision there and then as to whether or not the work is contract compliant. This is especially important in small domestic arbitrations between a small builder and a householder. Otherwise the costs of experts on both sides can get out of control, or at least become disproportionate to the matter in issue. It is surprising how often parties appoint their own experts and then ask the arbitrator to adjudicate between the two, when one of the strengths of arbitration is that the arbitrator could be the expert. There are real advantages to having experienced trade arbitrators.[17] However the expertise of the arbitrator will not permit him to arrive at a conclusion which has not been canvassed before him by the parties. He must conduct himself fairly and allow the parties to put their case before him rather than come up with his own conclusions based upon his own special knowledge or expertise without reference to the parties; see serious irregularity and unfairness below.[18]

Techniques for reducing cost and directions to be given by the arbitrator

21.21 Hearings are expensive and procedures which include disclosure are very expensive. The arbitrator will have to consider whether disclosure of documents is needed and if so how widespread it should be. There is no automatic right to disclosure and arbitrators should consider whether a fair resolution of the dispute requires disclosure. Similarly, with regard

[17] *Bremer Handelsgesellschaft mbH v Ets Soules et Cie* [1985] 1 Lloyd's Rep 160 *per* Mustill J at pp 164–165.

[18] *Fox v Wellfair Ltd* [1981] 2 Lloyd's Rep 514; and *London Underground Ltd v Citylink Telecommunications Ltd Rev 1* [2007] EWHC 1749 (TCC), 114 Con LR 1, [2006] 1 All ER (Comm) 529, [2006] 2 Lloyd's Rep 1, [2007] BLR 391, [2007] 2 All ER (Comm) 694 and the cases cited therein. See also *Zermalt Holdings SA v Nu-Life Upholstery Repairs Ltd* [1985] 2 EGLR 14 at 15 *per* Bingham J.

to an oral hearing, it may not necessarily be that one is necessary and an arbitrator may fairly be able to come to a conclusion on documents alone without witnesses attending for cross examination.[19]

Following appointment, the arbitrator may suggest directions for the conduct of the arbitration up to and including an award. Alternatively he or she may give directions up to the close of pleadings or statements of case in order to see the size of the dispute and whether any preliminary issues ought to taken which might unlock the arbitration. **21.22**

There are many types of directions which the arbitrator might give. The following three are the most common. The first is a standard set of directions up to and including a full hearing; the second, directions up to and including a decision being made on the papers without a hearing; and, thirdly, directions leading to an interim award on a preliminary issue which may give the parties an opportunity to settle their differences in the light of the arbitrator's decision on the preliminary issue. **21.23**

The effect of adjudication and mediation's arrival on the scene is that before an arbitration gets under way the issues between the parties are much better known than in the past. This has the effect of ensuring that the parties are more creative as to how the arbitrator can help them with such steps as preliminary issues or programmed hearings which mean that the era of long hearings which go on for months, if not over, has moved into a period where arbitrations are quicker. **21.24**

The effect of adjudication is twofold. First, the parties may have already had an adjudication prior to arbitration and therefore are better informed about the dispute and its pinch points. This is a reflection of adjudication. But the second is that arbitrators who act as adjudicators are more pro-active in seeking a speedier process. Thus, they are starting to import techniques refined in adjudication in arbitration in order to speed up the traditionally laborious process. Examples of this would be orders that written submissions be only of a certain size (say no more than twenty pages), another would be that witness statements have a precis of the main points in them before the main body of the witness statement is set out, and another that each party present to the arbitrator a mini bundle of no more than 100 pages of what they believe to be the key documents in the case. In the case of the mini bundle, it is surprising how often each side's views as to the important documents coincide and often the arbitrator can make his or her own amalgamation of the two mini bundles to be no more than 150 pages. None of these techniques are designed to prevent the parties from presenting their full case. But one of the vices of arbitration (and litigation) is the fact that both parties' legal advisors do not have the confidence to winnow out documents so that bundles for hearings reach absurdly large sizes of many lever arch files. The solicitor who prepares the bundle will often not have either the seniority or the confidence to omit a document from the bundle because of the catastrophic consequences of a seemingly innocuous document being in fact described by either the client or the senior counsel in the case as being a key document. Equally there is a tendency amongst counsel to keep options **21.25**

---

[19] See also the Construction Industry Model Arbitration Rules which expressly give to the arbitrator such time-saving techniques.

open and that too may cause the bundle to be bulkier than necessary. But if there is a limit to what goes in a 'core bundle' or 'key bundle', then there has to be rigorous selection and that process gives confidence to the selector of the mini bundle. Just as the large bundles of lever arch files result in relatively few documents being referred to at a hearing, so a hearing where a mini bundle has been made will often have comparatively few additional documents referred to.

Quality of Arbitrators: The Pool of Talent

21.26 Concerns as to quality have receded and transferred more to adjudication. But it is of some concern that many arbitrators (and adjudicators) are not as able as they might be. One problem is that nominating bodies have tended to nominate uncritically the same relatively weak arbitrators. In arbitrations of size the legal advisors will often take steps to try and avoid the vice referred to above and agree an arbitrator of repute. The fact remains that training of arbitrators is essential so that elementary mistakes are avoided, as, for example, one arbitrator (albeit some time ago) producing an interim award before hearing both sides of the case.[20] Or more recently an arbitrator contacting witnesses directly but failing to inform the parties to an arbitration that he had done so and failing to keep or disclose a record of what they had said.[21] There are many who want to be arbitrators and there are many training courses run for them, but there is a relatively limited band of those equipped by judgment, training, and experience to shine in that sphere.

Sealed Offers

21.27 Just as with litigation, the costs of an arbitration will often become as significant, if not more significant, than the amounts in issue. Also one party will usually know that it will be a paying party. It is just a question of how much. Aside from winning, the only tried and tested way for a party to avoid having to pay the costs of the arbitration (both sides costs including lawyers and experts, the arbitrator's fees, the costs of the room where the hearing is held, and the costs of a shorthand writer) is to make a well judged sealed offer. The offer will represent a proposal to settle the proceedings on certain terms. A sealed offer is not, as formerly, an offer which is kept sealed and put in an envelope for the arbitrator to open after having made the award. Rather it is an offer to settle which will be shown to the arbitrator after the award on liability and quantum (but not costs) has been made. That means that the award must be an interim award, because a final award must deal with everything, including costs.

21.28 The essentials of a sealed offer are:

(a) It must be clear;
(b) It must be an offer to settle part of or the whole of the claims which, if accepted, does so;
(c) It must deal with any counterclaim in the sense that it must be made clear that the counterclaim is or is not to be taken into account;

[20] See *Miskin v Modern Engineering* (1980) 15 BLR 82, CA.

[21] In that case *Norbrook Laboratories Ltd v A Tank and Moulson Chemplant Ltd* [2006] EWHC 1055 (Comm), as in *Miskin*, the arbitrator was removed for misconduct.

(d) It must deal with interest, making it clear whether the sum offered includes interest or whether interest is payable on top and in the latter case at what rate and for what period;
(e) It must deal with costs;
(f) It must set the time during which it can be accepted;
(g) It must deal with VAT.

There is no rule setting out the form of a sealed offer which will cause an arbitrator to award the costs of an arbitration to the maker of the offer. However, the following template is suggested as a starting point for consideration. **21.29**

Draft Sealed Offer

This is an open letter written on the basis that it shall not be shown to the arbitrator until all matters of liability and quantum have been dealt with and the arbitrator comes to deal with the question of costs. The intention of this letter is to approximate, as nearly as possible, CPR Part 36 Offer.

The Employer offers to pay the Contractor the sum of £xxx, inclusive of interest and VAT, in full and final satisfaction of its claims made in this arbitration.

The offer includes for settlement of all claims by the Contractor for payment of certified sums, for damages for breach of contract and for reimbursement for loss and/or expense arising out of the contract for the construction of the building and takes into account the Employer's claim against the Contractor for liquidated and ascertained damages.

Save for those costs which are subject to the existing orders of the Arbitrator, the Employer also offers to pay the Contractor's reasonable costs of the arbitration, such costs to be assessed if not agreed, and to pay the costs of the Arbitrator and the arbitration. Both sets of costs are those reasonably incurred up to the date of acceptance of this offer. Costs which are subject to pre-existing orders of the Arbitrator shall remain payable in accordance with those orders.

This offer is open for acceptance for a period of 21 days after the date of this letter. After the expiry of the 21 days, it may still be accepted on the following modified terms:

1. The payment of the Contractor's costs and those of the Arbitrator relates only to those costs incurred up to and including the date of this letter.
2. The Contractor bears its own costs incurred after the date of this letter and also pays the costs of the Arbitrator and the arbitration incurred after the date of this letter.
3. The Contractor pay the Employer's costs incurred after the date of this letter, such costs to be assessed if not agreed.
4. The Employer may set off against the sum payable by it, the costs which the Contractor is liable to pay under sub-paragraph 3 above.

Provided always that the offer cannot be accepted after 21 days if the Employer considers that anything has happened which would, were this offer a CPR Part 36 Offer, cause it to have proper grounds for opposing an application to take the money out of court. The Employer shall be sole judge of whether those grounds exist.

This offer expressly excludes any claim which the Employer may have in respect of defective work which is not the subject matter of this arbitration.

We believe this letter to be clear and to approximate so far as can be a CPR Part 36 Offer. If, however, there is anything which is not understood by you or you consider that anything within it is less favourable to you than a CPR Part 36 Offer please let us know forthwith, we will consider what you have to say.

Kindly acknowledge receipt of this letter.

21.30 The above draft has within it all the elements at (a) to (g) above but with the additional provision that the offer is capable of acceptance outside the time for acceptance specified on condition that the costs incurred by the offeror after the making of the offer are paid. It follows, as far as possible, the provisions of a CPR Part 36 Offer but has added to it the provision that if any part of the offer is not understood then the other side can ask. That last provision does not form part of the essentials of a sealed offer. However it is useful in cases where the other side may have a difficult client which will not accept the offer, encouraging its legal advisors to argue that the offer is ambiguous or otherwise ineffective. The tactical idea in those circumstances is to sow the seeds of doubt in the minds of the offeror (not at the time the offer is made but later when negotiations take place) that the sealed offer is not after all effective, not because the amount offered is not enough, but because of some real or perceived ambiguity in the wording of the offer. In simple cases for the offer of money for a claim, this tactic ought not to be effective because the offer can be a straightforward money offer for an identified claim. But sometimes the case is more complicated than that, as, for example, when there is a claim for defective work and the employer wishes to ensure that it keeps open its entitlement to claim for latent defects which strictly speaking do not form part of the arbitration. The act of making that particular point clear may cause the other side to alight on a real or imagined ambiguity to the detriment of the integrity of the sealed offer. Whilst a clear sealed offer ought to survive examination and quibbling, it is as well to have the invitation there in the letter just in case there is something of substance in any question arising out of the sealed offer. It is easier to deal with such points at the time, rather than after more costs have been incurred.

21.31 The sealed offer procedure may extend, by analogy with developments under CPR 36, to offers made before the arbitration has begun and also to offers made by claimants aimed at providing the protection of an order for indemnity costs.

Costs

21.32 The costs of an arbitration comprise the tribunal's fees and expenses, the fees and expenses of any institution concerned with the arbitration (such as the ICC), and the legal, expert, and other costs of each party.[22]

21.33 The parties can agree how the costs can be dealt with. But if they do not agree, then the arbitrator must, in the exercise of his or her discretion,[23] decide how they are to be awarded and must do so on the principle that the costs follow the 'event'[24] unless the circumstances

[22] Arbitration Act 1996, s 59.

[23] Which is a wide discretion within the bounds of the principles contained in s 61. CIMAR r 13 expressly provides that '*Subject to any agreement between the parties, the arbitrator has the widest discretion in awarding which party should bear what proportion of the costs of the arbitration*'.

[24] Characterization of the '*event*' can raise difficult questions particularly where an arbitration award results in a large corrective payment following temporary payment following an adjudication. This issue arose in *Treasure v Dawes* [2009] EWHC 1932 (TCC), Manchester *per* HHJ Stephen Davies, in which the '*event*' was treated as the outcome of the arbitration despite this resulting in a large corrective payment to the unsuccessful party.

make it appear to the arbitrator that some other order should be made.[25] There is no statutory guidance as to what those circumstances are, but the rule mirrors CPR r 44.3(1) and therefore the principles which are applied to litigation will also apply to arbitration.[26] The circumstances which will call for a departure from the normal order are many but the principal ones are:

(a) If the winner has recovered less than has been offered either openly or in a sealed offer.

(b) If the winner has prolonged the hearing or reference unnecessarily or otherwise caused the costs of the arbitration to be larger than they ought. This is normally by prolix pleadings and witness statements or indiscriminate disclosure but it can also be by the introduction of expert evidence which in the end is found to be either irrelevant or capable of agreement.

(c) If the winner has lost sub-issues which have taken a significant proportion of the time spent in the arbitration.

The arbitrator has power to cap costs unless the parties otherwise agree that he or she should not do so.[27] The costs cap can apply to the whole of the arbitration or to specific parts of it. But costs capping must take place sufficiently in advance of the incurring of the costs so that the parties can take the cap into account when they make their expenditure decisions. This is an important power because it allows arbitrators to give the parties a sense of proportion. Some arbitrators may be a little zealous in applying costs capping. But the safety valve is the fact that the parties can agree that there shall be no costs cap. The arbitral community is split on the question of costs capping. Some apply it invariably and some never. **21.34**

Appeals

Whilst the court's jurisdiction under the Arbitration Act cannot be excluded in domestic arbitrations, for example, by the arbitration agreement saying that the award shall be final, binding and conclusive,[28] there are severe restrictions on the right of appeal from an arbitrator to the court and from the first instance judge to the Court of Appeal. Restrictions on appeals began with the Arbitration Act 1979 and the guidelines by Lord Diplock in *The Nema*[29] *and The Antaios*.[30] However, those guidelines have been subject to statutory change and now s 69 of the Arbitration Act 1996 applies. **21.35**

Section 69(1) allows an appeal on a question of law. Questions of law only found an appeal and you cannot appeal on a question of fact dressed up as a question of law.[31] But an appeal **21.36**

[25] Arbitration Act 1996, s 61(2).

[26] If an arbitrator is to depart from the normal rule, then he or she should give the parties a hearing, otherwise an important part of natural justice will not be honoured. See *Gbangbola v Smith & Sherriff Ltd* [1998] 3 All ER 730.

[27] Arbitration Act 1996, s 65.

[28] *Shell Egypt West Manzala GMBH and others v Dana Gas Egypt Ltd* [2009] EWHC 2097 (Comm).

[29] [1982] AC 724.

[30] [1985] AC 191 at 203–4.

[31] *Penwith District Council v VP Developments Ltd* [2007] EWHC 2544 (TCC) *per* Akenhead J.

can only be launched if either the parties agree or the court gives leave.[32] In standard forms such as the JCT there is a pre-agreed right of appeal. However, if there is no agreement then leave is required. The strict guidelines setting out the criteria upon which leave will be granted are set out in s 69(3) of the Arbitration Act 1996.

21.37 The hurdle for getting permission to appeal is that the question of law must be one which could substantially affect the rights of one or more of the parties to the arbitration agreement. Therefore, appeals on marginalia or academic points will not be allowed.

21.38 Once that hurdle is passed, there are two sets of circumstances for consideration before leave can be given. The first is that the decision is 'obviously wrong'. That means that the court must look at the decision and be clear in its own mind that it is wrong. It is not enough to look at it and think that the decision might be wrong. If the decision is obviously wrong, then leave is given to appeal to the High Court. That is going to be to the Commercial Court in all save construction cases where the natural forum is the TCC and the normal practice is for the matter to be transferred to the TCC if it has not already been commenced there.

21.39 If the decision is not obviously wrong, then leave can only be granted if: (a) the question is one of general public importance, and (b) the decision of the arbitral tribunal is 'at least open to serious doubt'.

21.40 Whether the matter is of public importance goes wider than whether the question relates to a standard form of contract. It might relate to a standard form and not be of public importance or it might relate to a bespoke term and be of public importance. Whether the decision is 'at least open to serious doubt' is a relaxation of the requirements set out in *The Nema* and *The Antaios* and the door to an appeal is opened a little more widely than those guidelines stated and is a broader test.[33] A minority view of a dissenting arbitrator has been held by the Court of Appeal to be grounds for passing the test (provided the question is of public importance).[34] There is literature in which eminent writers ask about the use of dissents of arbitrators in commercial arbitration.[35] Whatever the merits of that debate, clearly a dissent will make it more likely that the court will think the decision open to doubt.

21.41 Finally there is a sweeping up hurdle set out in s 69(3)(d) that it must be just and proper to ask the court for a determination in spite of the agreement that the parties would resolve their differences by arbitration.

21.42 The application for leave to appeal to the High Court is a paper exercise but reasons must be given to enable the losing party to understand why the judge reached the decision he or

[32] s 69(2).

[33] See *CMA CGM SA v Beteiligungs-Kommanditgesellschaft MS 'Northern Pioneer' Schiffahrtgesellschaft mbH & Co* [2003] 1 WLR 1015, [2003] 1 Lloyd's Rep 212 at paras 11 and 60.

[34] See ibid. In that case an eminent arbitrator in a panel of three came to a different view to the other two, and the Court of Appeal stated at para 64: 'The difference of view between the experienced arbitrators in this case provides, of itself, ground for contending that the decision of the majority is "at least open to doubt".'

[35] See, for example, the excellent article entitled 'Dangerous Dissents' by Alan Redfern in (2005) 71 Arbitration 200, which was based on his Masters Lecture of the Worshipful Company of Arbitrators. In that article he states that he believes they have little part to play in commercial arbitration.

she did;[36] but it is nevertheless a paper exercise and there ought not to be much if any scope for elaborate submissions. Any application for leave must normally be capable of being read and digested by the judge in half an hour.[37] First instance judges are often tempted to allow the paper exercise to expand to submission and counter submission; however this ought to be (and probably in the light of recent authority now will be) resisted.

Challenges on the ground of serious irregularity

Consistently with the Act's policy of supporting arbitration, the courts are able to interfere with the arbitral process only in cases of serious procedural irregularity affecting the tribunal, the proceedings or the award. Section 68 covers a number of possibilities including a tribunal exceeding its powers, failing to comply with its duty under section 33, failing to conduct the proceedings on the basis of a procedure agreed by the parties, and failure to deal with all the issues that were put to it. In order to interfere, a court must also consider that the serious irregularity has caused, or will cause, substantial injustice to the applicant.[38] **21.43**

The House of Lords has limited the scope of any challenge on the ground that the tribunal exceeded its powers.[39] However the other grounds of challenge continue to provide a basis for the courts to uphold challenges in appropriate cases where an award is clearly flawed or the arbitrator has treated one or other party unfairly resulting in a substantial injustice.[40] The courts will not be technical and on the look-out for irregularity. Only substantial injustice will do.[41] **21.44**

Misconduct and Bias

The cornerstone of an arbitration is that the arbitrator is fair and impartial and unbiased. Bias can show itself by the kind of misconduct referred to in paragraph 20.26 above. Bias is assumed if the arbitrator has an interest in the outcome of the case. Bias is established if on examination of all the relevant circumstances the court concludes that there was a real danger (or possibility) of bias.[42] In two passages in *R v Gough* [1993] AC 646 at 661 and 670 Lord Goff of Chieveley set out the test as follows: **21.45**

> In my opinion, if, in the circumstances of the case (as ascertained by the court), it appears that there was a real likelihood, in the sense of a real possibility, of bias on the part of a justice or other member of an inferior tribunal, justice requires that the decision should not be allowed

---

[36] See *North Range Shipping Ltd v Seatrans Shipping Corp* [2002] 1 WLR 2397, [2002] 2 Lloyd's Rep 1.

[37] See *CMA CGM SA*, n 33 above, at para 23.

[38] *F Ltd v M Ltd* [2009] EWHC 275 (TCC); *Groundshire v VHE* [2001] BLR 395.

[39] *Lesotho Islands Development Authority v Impreglio SpA and Ors* [2006] 1 AC 221.

[40] See, for example, *Metropolitan Realisations v Atmore Investments* [2008] All ER (D) 304; *Van der Giessen De Noord Shipbuilding Division BV v Imtech Marine and Offshore BV* [2008] All ER (D) 284; *London Underground Ltd v Citylink Telecommunications Ltd Rev 1* [2007] EWHC 1749 (TCC), 114 Con LR 1, [2006] 1 All ER (Comm) 529, [2006] 2 Lloyd's Rep 1, [2007] BLR 391, [2007] 2 All ER (Comm) 694.

[41] See *F Ltd v M Ltd* [2009] EWHC 275 (TCC); and *Groundshire v VHE* [2001] BLR 395. See also a pre-1997 Act case *Zermalt Holdings SA v Nu-Life Upholstery Repairs Ltd* [1985] 2 EGLR 14 but it is to be noted that as stated by Judge Bowsher QC in *Groundshire*, the post-1997 Act cases are the ones to look at.

[42] *Locabail (UK) Ltd v Bayfield Properties Ltd & Anor* [1999] EWCA Civ 3004, [2000] QB 451, [2000] 1 All ER 65.

to stand. I am by no means persuaded that, in its original form, the real likelihood test required that any more rigorous criterion should be applied. Furthermore the test as so stated gives sufficient effect, in cases of apparent bias, to the principle that justice must manifestly be seen to be done, and it is unnecessary, in my opinion, to have recourse to a test based on mere suspicion, or even reasonable suspicion, for that purpose. (at 661)

In conclusion, I wish to express my understanding of the law as follows. I think it possible, and desirable, that the same test should be applicable in all cases of apparent bias, whether concerned with justices or members of other inferior tribunals, or with jurors, or with arbitrators. Likewise I consider that, in cases concerned with jurors, the same test should be applied by a judge to whose attention the possibility of bias on the part of a juror has been drawn in the course of a trial, and by the Court of Appeal when it considers such a question on appeal. Furthermore I think it unnecessary, in formulating the appropriate test, to require that the court should look at the matter through the eyes of a reasonable man, because the court in cases such as these personifies the reasonable man; and in any event the court has first to ascertain the relevant circumstances from the available evidence, knowledge of which would not necessarily be available to an observer in court at the relevant time. Finally, for the avoidance of doubt, I prefer to state the test in terms of real danger rather than real likelihood, to ensure that the court is thinking in terms of possibility rather than probability of bias. Accordingly, having ascertained the relevant circumstances, the court should ask itself whether, having regard to those circumstances, there was a real danger of bias on the part of the relevant member of the tribunal in question, in the sense that he might unfairly regard (or have unfairly regarded) with favour, or disfavour, the case of a party to the issue under consideration by him. . . .

21.46 There is more and more arbitral and court jurisprudence regarding bias of arbitrators.[43] For example the question arises as to whether a barrister in the same chambers as an arbitrator is liable for removal on the grounds that he or she could be expected to be biased.[44] Whatever

[43] See cases cited by Khawar Qureshi in his article in (2006) 156 NLJ 7223, 744–745 namely *Smith v Kvaerner Cementation Foundations Ltd* [2006] EWCA Civ 242; *AWG Group Ltd (formerly Anglian Water Plc) v Morrison* [2006] EWCA Civ 6, [2006] 1 WLR 1163, [2006] 1 All ER 967; *ASM Shipping Ltd of India v TTMI Ltd of England* [2005] EWHC 2238 (Comm), (2006) 1 Lloyd's Rep 375; *Petroships Pte Ltd of Singapore v Petec Trading & Investment Corp of Vietnam (The Petro Ranger)* [2001] 2 Lloyd's Rep 348; *AT&T Corp v Saudi Cable Co* [2000] 2 All ER (Comm) 625; *Laker Airways Inc v FLS Aerospace Ltd* [2000] 1 WLR 113; *Porter v Magill* [2001] UKHL 67, [2002] 2 AC 357 (HL).

[44] In England that would appear at the moment to be settled by Rix J's decision in *Laker Airways Inc v FLS Aerospace Ltd* [2000] 1 WLR 113 in which he held that there was no such bias and quoted the decision of the Paris Court of Appeal in *Kuwait Foreign Trading Contract & Investment Co. v Icori Estero SpA* (Unreported), 29 June 1991:

> Considering that the independence of an arbitrator is essential to his jurisdictional function and must be exclusively arm's length, in particular, of the parties; considering that the circumstances arising to challenge this independence must characterize, by the existence of physical or intellectual ties with one of the parties to the proceedings, a situation of such a nature as would affect the judgment of that arbitrator by the appearance of a definite risk of predisposition towards one party to the arbitration; considering that the arbitrator's duty to inform the parties in order that they may exercise their right of objection must be determined with regard to both the significance of the situation in question and of its impact on the arbitrator's judgment; considering that, in this instance, it is apparent from elements of the dispute (especially from the translation of an extract from the work 'International Commercial Litigation'); from the memorandum drafted by the London firm of solicitors, Masons, for the appellant, and from the statements of Mr. Anthony Scrivener, barrister and Edward Singleton, solicitor) that the function of a barrister is essentially carried out independently and that belonging to a set of chambers of barristers—an age-old institution unique to the British system—is characterized, in essence, by the sharing of premises and support staff, without necessarily creating

the position in England, it may not be straightforward in other jurisdictions.[45] However it must be doubted whether the view that an arbitrator in the same chambers as counsel before him escapes the inference of bias would prevail today. Barristers' chambers market themselves collectively more and more and project themselves as a 'brand', giving rise to a perception that they are in business together as quasi partners rather than simply sole practitioners sharing accommodation and administrative staff.

The International Chamber of Commerce (the ICC) and ICC Arbitrations

The world's most representative dispute resolution body is the ICC International Court of Arbitration. It is not a court in the true sense of the word. Rather, it is a body which sees that the rules of arbitration of the ICC are observed. The Court is assisted by a Secretariat comprising seventy full-time staff based in Paris, but with an operational branch office in Hong Kong. It is also supported by a number of national committees involving the ninety countries where its members are located. **21.47**

Under the ICC Rules the arbitral tribunal consists of one or more arbitrators. When only one arbitrator is designated he or she is appointed by the Court unless the parties otherwise agree. Where three arbitrators are required, each party nominates an arbitrator, and the third arbitrator, who chairs the tribunal, is either nominated by agreement by the parties or the co-arbitrators or appointed by the Court. When the parties are unable to agree on the number of arbitrators the ICC Rules provide for the Court to appoint a single arbitrator unless it appears to the Court that the appointment of three arbitrators is justified. **21.48**

The course of arbitration under the ICC Rules is in general terms as follows: **21.49**

(i) the claimant submits a Request for Arbitration to the Secretariat of the ICC. This sets out details of the parties, a description of the nature and circumstances giving rise to

any professional ties, which would imply—such as in the French Law Society for example—common interests or a certain economic or intellectual dependence between various members of a set of chambers who are, between themselves, frequently called, usually because of the specialism of the chambers, to plead against each other or to participate in arbitral tribunals before which other members of the same set of chambers may appear in the capacity as counsel; considering that if no other objective element existed in the current case which would have affected the independence of the president of the arbitral tribunal; from the single fact that a member of the same set of chambers acted as counsel to one of the parties, this situation creates no tie of dependence between the arbitrator and that particular party to the arbitration proceedings; considering that if it is established that the appellant and its counsel were not aware of the fact that the president of the arbitral tribunal belonged to the same set of chambers as counsel to the respondent, the appellant cannot have grounds against the arbitrator for not having revealed a situation which was not, in itself, of such a nature as would affect his judgment and could exist without affecting the exercise of his jurisdictional function; considering that, in these circumstances, it is not established in this instance that the arbitrator referred to in the appeal had not disposed of the independent spirit necessary for the exercise of his jurisdictional power; these being the conditions necessary to invalidate the constitution of the arbitral tribunal or to constitute an infringement of the principle of equality of the parties and respect of the rights of defence; that the first two grounds are without basis.

[45] See the article by Michael Hwang SC, 'Arbitrators and Barristers in the Same Chambers—An Unsuccessful Challenge' (2005) 6 BLI 2, 235.

the underlying dispute, the relief sought, as well as the claimant's views on the place and language of the arbitration and the applicable law of the case. The request should be accompanied by a copy of relevant agreements and, in particular, the arbitration agreement as well as payment of US$ 2,500 as an advance on administrative expenses;

(ii) upon receipt of the Request the Secretariat
  (a) normally asks the claimant to pay a provisional advance intended to cover an advance on the costs of the arbitration until the Terms of Reference have been drawn up;

  and
  (b) asks the other party/parties to the arbitration to file an Answer (and any counterclaim) within thirty days. This period may be extended by the Secretariat;

(iii) following receipt of the advance on costs and the Answer, the Secretariat submits a file of relevant documentation to the arbitral tribunal. The latter is required to draw up Terms of Reference within two months of its receipt, having regard to such documentation and any submissions made by the parties. The Terms of Reference will usually set out the full names of the parties and the arbitrator(s), the place of arbitration, a summary of the parties' claims, details concerning the applicable procedural rules and a provisional timetable for the conduct of the case. They will also set out the issues to be determined unless the arbitrator(s) should consider this to be inappropriate. If one of the parties should refuse to take part in drawing up, or signing, the Terms of Reference, the latter are submitted to the Court for approval;

(iv) the parties and the tribunal will decide the rules of procedure to be followed, subject to any applicable mandatory provisions. They can decide, for example, on whether and to what extent there should be discovery and/or whether there is to be cross-examination of witnesses;

(vi) in the absence of agreement between the parties the tribunal decides the language(s) of the proceedings and the law applicable to the merits of the case;

(v) the tribunal establishes the facts of the case, probably by way of an oral hearing. After the parties have had a reasonable opportunity to put their cases, it draws up a draft award;

(vi) the ICC Court scrutinizes the draft award. It does not interfere with the substance of any decision made but it may draw to the attention of the arbitrator(s) any gaps or the need for modifications in form, without interfering with any substantive decision made in so doing. Once approved the award is signed by the arbitrator(s) and released to the parties by the Secretariat.

21.50 The objective of the ICC Rules is that an award will be made within six months from the signature of approval of the Terms of Reference, although this period may be extended by the Court.

21.51 If the parties so agree, the arbitral tribunal can act as *amicable compositeur* or decide the case without regard to any particular law but on the basis of *ex aequo et bono*.

21.52 Arbitration under the ICC Rules has a number of unique features:

(i) the drawing up by the arbitral tribunal of Terms of Reference. Their preparation serves to bring the arbitrators and the parties together at a relatively early stage of the proceedings and such a get together may lead to the early resolution of the underlying dispute in consequence;

(ii) the scrutiny given by the Court to the proposed award. Such scrutiny is intended to ensure that the award is of a high standard and less open to challenge in a national court than it might otherwise be;

(iii) the basis of remuneration of the arbitral tribunal. An arbitrator plays no part in fixing his fees and expenses. Rather, the remuneration of the tribunal is fixed at the end of the proceedings by the Court, by reference to a scale of charges attached to the ICC Rules (where monetary relief is sought). The scale takes account of the amount in dispute but, in fixing the actual fees payable, the Court will also take account of the diligence of the arbitrator(s), time spent, the speed at which the case has proceeded; and its complexity. The basis of remuneration is intended to encourage the efficient handling of cases.

The powers of the Court are also worthy of note. By way of example, where the appointment of an arbitrator is challenged at the outset, because of, say, lack of independence, the Court can decide on the challenge, and its decision is final. Furthermore, before a file is transmitted to the arbitral tribunal, the Court may be called upon to decide whether it is prima facie satisfied that an agreement to arbitrate under the ICC Rules exists and, if it is not, the Court can stop the reference in its tracks at that stage. If the place of arbitration has not been agreed by the parties it is fixed by the Court, normally in a 'neutral' country. **21.53**

22

LITIGATION

Construction Litigation and the TCC Guide

Construction litigation, that is the litigation of claims by or against contractors, sub-contractors, and construction professionals, usually takes place in the Technology and Construction Court (TCC) which is a branch of the Queen's Bench Division of the High Court. Small claims, being those litigated in the county courts, will not be separately examined in this chapter. The TCC has now published the second edition of the Technology and Construction Court Guide (TCC Guide) dated October 2007 and it is reproduced in full as Appendix 5 to this volume. This chapter must be read with the Guide. The Guide is an up-to-date statement of the practices of the court setting out what must be done, and when.[1] **22.01**

The types of cases tried by the TCC are numerous and are not limited to cases involving construction and engineering. Nevertheless, the core disputes which are tried in the TCC and are looked at in this chapter are: **22.02**

(a) Claims for damages for breach of contract or negligence arising out of work not being carried out in accordance with the contract or negligently done work. The most frequent type relates to defective work which leads to a failure in the building requiring expensive remedial work. The most frequent types of claim will be by the building owner against the contractor or the contractor against the sub-contractor.
(b) Allied with the claim against the builder is a claim by the building owner against the professional. Most frequently this involves poor inspection or supervision of construction or poor design.
(c) Claims by contractors or sub-contractors for payment of sums due under the contract.
(d) Claims for enforcement of decisions by adjudicators.
(e) Applications under the Arbitration Act 1996.

Pre-Action Protocols and the Mischief They Seek to Avoid

Pre-action Protocols: The TCC Protocol

Prior to the commencement of an action the courts now require that steps be taken to try to avoid litigation and therefore steps to be taken before proceedings are commenced. This involves compliance with pre-action protocols, which are therefore considered first before litigation itself. **22.03**

[1] TCC litigation must of course take place within the ambit of the regime applicable to civil litigation in general. That regime is currently under review. On 3 November 2008, the then Master of the Rolls, Lord Clarke of Stone-cum-Ebony, announced that Jackson LJ had been appointed to conduct a fundamental review into the costs of civil litigation. The review commenced in early 2009 and an interim report entitled *Review of Civil Litigation Costs: Preliminary Report* (the Jackson Report) was published in May 2009. The ensuing consultation process is underway at the time of writing. Vol 1, ch 34 deals specifically with the TCC. Where comments on areas of litigation are made in the Jackson report which specifically relate to the TCC, they are referred to in the relevant part of the chapter dealing with those areas. Other potential areas of reform are addressed at the end of this chapter.

22.04 There are two protocols and a Practice Direction which are relevant to consider:

(a) The Practice Direction—Pre-Action Conduct, which came into force on 6 April 2009 (the Pre-Action Practice Direction);
(b) The protocol in respect of proceedings in the TCC, formally known as the Pre-Action Protocol for the Construction and Engineering Disputes (the TCC Protocol); and
(c) The Pre-Action Protocol for Professional Negligence (the Professional Negligence Protocol).

With the exception of claims which involve the enforcement or re-litigation of adjudicator's decisions and applications for injunctions or summary judgment, claims ought not to be commenced in the TCC without the parties having complied with the steps which are set out in the TCC Protocol.[2] The purpose of the protocol is, by an early exchange of information, to allow the parties to a claim to understand what the other side's point of view and case is so that a settlement can be reached or, if no settlement is reached, to allow for the parties to have a sufficient knowledge of the other side's case that the case can be properly managed.

22.05 Therefore there is now a whole stage of litigation which takes place pre-action and is controlled by the courts retrospectively.

Aims of the protocol regime

22.06 The aims of the protocols are now achievable because cases are no longer tried blind and there is, as a matter of course, a great deal more information available to the parties than used to be the case. This must be understood in the historical context, which sets the scene for understanding the aims of the pre-action protocols and how they are looked at by the courts, particularly if they are not complied with.

22.07 In the early 1980s, neither side had any idea what the other side's witnesses of fact were going to say. Allied with this was a culture whereby the lawyers for each side saw it as their professional duty to ensure that the other side did not know exactly what its real case was. A claimant (or plaintiff) was keen to ensure that the defendant had to face as wide a set of allegations as possible and the defendant was keen to see to it that, as far as possible, the claimant's case was 'pinned down'. Pleadings became an art, often designed to obscure or at least evade. The combination was devastating. Cases were started with no real knowledge of the risks attending the litigation because neither side was sure that it knew the extent of the other side's case.

22.08 This state of affairs was exacerbated, in ordinary litigation, outside the TCC (then known as the Official Referees' Court), by the procedure whereby all the preparatory stages of the case were dealt with by Queen's Bench Masters and a trial was served up to the judge to try. Whilst this did not affect TCC cases, it did affect the culture of civil trials, which seemed to regard it as the exception rather than the rule that trial judges would take an interest in their cases prior to the first day of the trial itself.

[2] TCC Protocol, para 1.2.

The end result was that litigation was as slow and laborious as it was possible to imagine. The evidence of the parties was got out by a process of 'Examination in Chief' which was prolonged, leisurely, and criminally expensive. The other party then had to cross-examine the witness with little or no notice of what that witness was going to say. Counsel had to do so having taken a contemporaneous note usually in unreadable longhand. This meant that there would be a less than thorough cross-examination as the amount of time for preparation to assimilate and deal with the evidence which was adduced was often minimal, given the almost complete absence of notice given as to what the case was really about. **22.09**

It was amazing that there was not an outcry caused by this state of affairs. It might have been accepted because of the deference then afforded to the legal process. **22.10**

Management of the case was almost impossible in ordinary construction cases since the issues to be decided were either deliberately obscured or were so as a matter of routine. **22.11**

Case management is a difficult art and, although the courts are keen to promote it, unless it is done efficiently and with insight, it can be very expensive when not done properly. **22.12**

Since the 1980s, the TCC has been a pioneer in case management beginning with the introduction of a requirement that witness statements be exchanged prior to trial.[3] Once it is known what witnesses are going to say, then proper case management becomes possible. The TCC's approach to case management is seen as a success, with costs incurred in litigation adjudged in Jackson LJ's Interim Report (the Jackson Report) as being proportionate to the importance of the issues (which includes any non-financial issues such as reputational value) before the Court.[4] **22.13**

The protocol and proceedings are best examined in the light of a worked example; so this chapter will take as a template, for illustrative purposes, a case where a building owner, in this case a university, has a defective structure in the form of a student's accommodation with external tiles which fall off into the road below. We will assume that the university had a contract with a contractor which built the building and an architect who designed it as well as a direct warranty with the tiling sub-contractor warranting the design of the external tiling. In this example the university has taken the advice of an architect who specializes in looking at defective structures and she has said that the tiles have fallen off the exterior of the building because the contractor has used glue which is not contract compliant. In this example we shall assume that the external tiling was completed five and a half years ago, the building was completed six years ago, and a document called a 'Final Certificate' was issued one year after completion. **22.14**

[3] The first case where this was ordered was *Victoria University of Manchester v Hugh Wilson and Lewis Womersley and Pochin Ltd* (1984) 2 Const LR 43; in that case, in addition to witness statements being exchanged, the experts met to discuss what they agreed and disagreed about probably for the first time in major construction litigation.

[4] The Jackson Report, Vol 1, ch 34, para 3.2.

The letter of claim

22.15 The TCC Protocol requires the claimant or its solicitor to send to all potential defendants a 'letter of claim'. The letter of claim is required to have contained within it the following information:[5]

(i) The name and address of the potential claimant and defendant.

(ii) A clear summary of the facts on which each claim is made.

(iii) The basis of each of the claims. The contractual terms relied upon as well as statutory provisions must be given. In addition all facts and matters which go to create a legal duty owed by the defendant to the claimant must be set out. So, for example, if the claim is made in negligence, the letter ought to set out how the duty of care arises.

(iv) The nature of the relief claimed. The protocol covers specifically three areas: (a) if damages are claimed, then there must be a breakdown showing how they are quantified; (b) if sums are claimed under a contract then there must be a statement or breakdown showing how they have been calculated; and (c) if an extension of time is claimed the period claimed. Other forms of relief ought to be specifically set out; if, for example, a declaration is sought then it must be set out clearly; if an injunction is sought, then the terms of the injunction sought ought to be precisely set out.

(v) Often a claim has been made and rejected in correspondence. If that has happened, then the letter ought to set out why the rejection of the claim is wrong. In our example, it may be that the contractor relies upon the final certificate as absolving it from liability. If that has been stated in correspondence, then the letter should set out why the final certificate does not afford the contractor a defence to the claim.

(vi) If an expert has been instructed and the claimant will wish to rely upon that expert's evidence in any subsequent proceedings which may be issued, then: (a) the expert should be identified; and (b) the issue to which that evidence will go should be stated.

22.16 This information gives the prospective defendant much more information than it would normally have in litigation. There are two principal areas where this is so. The first relates to pleadings and answers to cases; the second relates to experts.

22.17 First pleadings: the reasoning as to why a defence or answer to a claim is ill founded is not normally given until the pleadings reach the stage of a reply. Even then, the reply can often be opaque.

22.18 If one takes the example above, the letter of claim should set out exactly why the final certificate is ineffective. That might be for any number of reasons. Two reasons suffice to give an example. One reason might be because the final certificate clause in the contract does not give the certificate the effect of discharging the contractor from liability for defective work. Another reason might be because the certificate relied on by the contractor was not a final certificate at all, on its face. In pleadings a claimant in its reply might get away with

[5] TCC Protocol, para 3.

saying something as opaque as 'It is denied that the certificate of [date] is an effective discharge from this liability.' The claimant might even get away with saying nothing at all, on the basis that a reply joins issue with the defence anyway unless the contrary is stated. Neither the opaque sentence nor the simple leaving matters uncommented upon will tell the defendant why its final certificate point is bad.

However,with the protocol the claimant is required to deal with the point clearly and explain why the final certificate does not have the effect which the defendant says it has. **22.19**

So far as experts are concerned, it was never the case that a party had to identify the other side who was their expert. That meant that each party instructed an expert and each expert worked alone and separate from the other side's expert. That state of affairs had endless scope for confusion and irrelevant inquiry. Now, before an action is started, each side ought to know: (a) what expert the claimant has instructed; and (b) to what issue their evidence goes. **22.20**

Following receipt of the letter of claim the defendant should acknowledge receipt within fourteen days. If this does not happen, then the claimant can commence proceedings and the protocol becomes redundant. **22.21**

Objections taken to the claim

If there has been an acknowledgement, then the defendant may take an objection to proceedings on one of three grounds. If it wishes to do so, then it must do so within twenty-eight days after the letter of claim.[6] The three grounds for objection are: (i) the court lacks jurisdiction; (ii) the matter should be referred to arbitration; and (iii) the defendant is the 'wrong' party. **22.22**

Lack of jurisdiction is a rare objection; the court will usually have jurisdiction. The right to go to arbitration is the most frequent objection to court proceedings; albeit that it is not an objection to jurisdiction in fact. In substance it is since the court is required to stay proceedings on an application which is made by a party to an agreement which is governed by an arbitration clause. The 'wrong' party may be technical and cured by a correct reference (for example, a limited company is written to as opposed to a partnership). **22.23**

If these preliminary objections are taken, then a formal letter of response is not needed.[7] The objection either stops the prospective claimant in its tracks or the claimant must consider issuing proceedings to get the objections out of the way.[8] **22.24**

[6] TCC Protocol, para 4.2.1.

[7] TCC Protocol, para 4.2.2.

[8] The TCC Protocol is silent as to this. However this logically follows from the fact that, once the objection is taken, the TCC Protocol stops, unless the objection is then abandoned—TCC Protocol, para 4.2.3. That means that either the objection is taken to be good, in which case the matter goes no further, or the objection is challenged, in which case the only manner by which such challenge is to be resolved by the commencement of proceedings.

22.25 If there are no objections of one or more of the three kinds referred to, then the prospective defendant has between a minimum of twenty-eight days and a maximum of four months to send a letter of response.[9]

22.26 The times for this are usually agreed. The heavier the action and the more facts which are involved and the less prepared the prospective defendant, the longer the time needed.

Contents of the letter of response

22.27 Just as the contents of the letter of claim are prescribed, so are the contents of the letter of response. The letter of response must state:[10]

(i) What facts in the letter of claim which are agreed and what facts are not and the basis of the disagreement.
(ii) What claims are accepted and what claims are rejected and if rejected the grounds of objection.
(iii) If a claim is accepted in whole or in part whether the amounts claimed by way of quantum are accepted or rejected and if rejected the basis of rejection. If the terms of an injunction similar to that which is claimed are accepted in principle, then the terms of any undertaking which the defendant is prepared to give to the claimant ought to be identified. Similarly if the terms of a declaration which is sought by the claimant are not wholly accepted but different terms are accepted, then what declaratory relief is agreed as appropriate ought to be given.
(iv) If contributory negligence is alleged against the claimant, then a summary of the facts relied upon in support of the claim should be set out.
(v) If a counterclaim is going to be made, then the full nature of it should be set out in the same way as it would be in a letter of claim.
(vi) If experts have been instructed, then their names should be given and the issues to which their evidence relates should be given too.

22.28 This is designed to give the claimant all the information it needs before deciding whether to proceed with a claim.

Front loading of costs

22.29 The criticism of this regime is that it 'front loads' costs. Parties used to be able to commence litigation with relatively little preparation and expert and legal input. This regime ensures that a significant level of costs is incurred pre-action. There is no real way round this and no going back to the time when the costs of an action could be relatively modest at its start. The early exchange of information at the commencement of proceedings comes at this price. Where the pre-action process results in settlement and proceedings are therefore not issued, it would appear that the costs incurred are not recoverable at all.[11]

[9] TCC Protocol, para 4.3.1.
[10] Ibid.
[11] See *McGlinn v Waltham Contractors Ltd (No 3)* [2007] EWHC 149 (TCC), 111 Con LR 1, Coulson J.

Not surprisingly, there have been calls for reform of the pre-action process. In its submissions during the consultation process relating to the Jackson Report, the Technology and Construction Bar Association (TECBAR)[12] has suggested that the formal pre-action protocol process should be abolished and replaced with a more informal process which is limited to a summary letter of claim, a summary letter of response and a meeting.[13] Summaries are likely to be more cost effective, the current practice of long and exhaustive letters gives rise to a huge expense in themselves, thereby defeating their purpose. Jackson LJ himself has sought opinions on the possibility of having the 'pre-action' process take place after issue of the claim form: the claim form would be issued and then the action could be stayed for as long as necessary for carrying out the protocol process.[14] This would have the consequence of the supervision of the court in the 'pre-action' process. This may be beneficial given the very large costs which are expended at this pre-action stage. **22.30**

The counterclaim

If a counterclaim is made, then the claimant has a similar time to respond to that, in a letter of defence, as the prospective defendant has to respond to the claim.[15] **22.31**

The meeting of the parties (mediation)

Following this exchange of letters the parties should normally meet. The objective of the meeting is the same as the exchange of correspondence. The meeting, if it takes place, is to be without prejudice.[16] But the court can be told (if litigation follows) that a meeting did not take place. If the meeting took place the court can be told, who attended or refused to attend and any agreements which have been made.[17] **22.32**

The specific objectives of the meeting are:[18] **22.33**

(i) To identify the root cause of any disagreement.

(ii) To consider whether and if so how the dispute can be resolved without recourse to litigation.

(iii) If litigation is inevitable, to see what steps can be taken to ensure that the litigation is carried out in accordance with the overriding objective of the CPR to deal with the case justly and proportionately.

The meetings cannot be the subject of prescriptive statements of how they are to be conducted and the TCC Protocol does not make any. However, the protocol does give guidance about the identity of those who are expected to attend and what the parties should discuss. **22.34**

---

[12] See <http://www.tecbar.org/>.

[13] *Response of The Technology and Construction Bar Association to Lord Justice Jackson's Preliminary Review of Civil Litigation Costs* (the TECBAR response), para 14.

[14] The Jackson Report, Vol 1, ch 34, paras 4.1 to 4.6. This is the position reached where a party issues proceedings in the light of limitation concerns: see the TCC Protocol, para 6 and the section in this chapter entitled 'What to do if limitation is about to expire.'

[15] See the TCC Protocol, para 4.4.

[16] Ibid, para 5.7.

[17] Ibid, para 5.6.

[18] Ibid, para 5.2.

22.35 The representatives will include:[19]

(i) Either the party himself or herself, if an individual, or representative of the party who has authority to settle the matter or recommend settlement.
(ii) A legal representative of each party, if there are any instructed.
(iii) A representative of the insurers, if insurers are involved.
(iv) If a claim is made on behalf of another party (for example if a claim is made by a main contractor passing on a sub-contractor's claims) then that other party and its legal representative.

22.36 The TCC protocol and the Pre-Action Practice Direction require that each party considers with the other whether some form of alternative dispute resolution procedure (ADR) should be used rather than litigation.[20] The most common forms of ADR are mediation and 'early neutral evaluation' and sometimes a combination of both.

22.37 Although the TCC Protocol expressly recognises that no party can or should be forced to mediate or enter into any form of alternative dispute resolution,[21] it makes quite clear that parties may be required to explain the precise steps they took (or refused to take) to settle the dispute by a pre-action meeting.[22]

22.38 This leads into the question of how much information the Court may be entitled to obtain about the content of the ADR process and the weight to be given to any refusal by a party to mediate. An extreme case was *Dunnett v Railtrack Plc*,[23] in which a successful defendant who had refused to mediate was refused its costs by the Court of Appeal on the grounds that such was unreasonable conduct which could be sanctioned under CPR r 44.2(5). In later cases, the position has become more nuanced. The courts have considered, where a party refuses to mediate at all, whether it was reasonable to do so given, for example, the nature of the dispute, the merits of the case, the extent to which other forms of ADR were attempted, whether the costs of ADR would be disproportionately high, whether resort to ADR would have itself caused prejudice by causing delay and whether a mediation had a reasonable prospect of success.[24]

22.39 The court will consider the parties' conduct in respect of and (possibly) during the mediation (if any) in some detail. In *Hickman v Blake Lapthorn*,[25] a defendant's refusal to mediate was held not to be unreasonable when the merits of the case had been reviewed by experienced solicitors and counsel and their view was that the claimant's lowest offer figure (which was three times higher than the defendant's assessment) was too high. The absence of a Part 36 offer will be relevant: in *Hickman*, the claimant had suggested that he might accept

[19] Ibid, para 5.3.
[20] Ibid, para 5.4. Pre-Action Practice Direction, para 4.4(3).
[21] TCC Protocol, para 5.4.
[22] Ibid, para 5.6.
[23] [2002] EWCA Civ 303, [2002] 1 WLR 2434.
[24] *Halsey v Milton Keynes General NHS Trust* [2004] EWCA Civ 576, [2004] 1 WLR 3002; on the last criterion see also *Hurst v Leeming* [2002] EWHC 1051 (Ch), [2003] 1 Lloyd's Rep 379.
[25] [2006] EWHC 12 (QB) at [26]–[30] *per* Jack J.

a settlement of £150,000 and later recovered £130,000. The Court indicated that it could not be right that to avoid being vulnerable on costs a defendant should always be prepared to mediate and then pay more than the claim is worth simply to save costs; otherwise claimants could impose undue pressure to settle at a higher figure than the claim merits.[26] The Courts will further consider how the parties act during and after the mediation. In *Earl of Malmesbury v Strutt & Parker*,[27] the court considered that both parties had behaved unreasonably and that, where the failure to mediate was due to the attitudes taken on either side, it is not open to one party to claim that the failure should be taken into account in the order as to costs.[28]

If the court is prepared to consider in detail the run-up to and content of any mediation, will it require or allow evidence from any mediator? Applicable mediation rules may seek to prevent this but not necessarily successfully. In *Farm Assist Ltd (In Liquidation) v The Secretary of State for the Environment, Food and Rural Affairs (No 2)*,[29] the parties had settled a dispute in mediation. The claimant then brought proceedings, seeking to set aside the settlement agreement on the basis that it had been entered into under economic duress at the mediation. Ramsey J refused an application by the mediator to set aside a witness summons on her. (Neither party objected to the mediator being called to give evidence.) The circumstances were highly unusual but the court's reasoning that it was in the interests of justice for evidence to be given by the mediator might be said to apply with force to disputes concerning costs, which may well be more substantial than sums recovered in the main proceedings.[30] **22.40**

Where the parties have engaged in mediation, they may seek the costs of that mediation. In order to do so, they would need first to show that those costs form part of 'the costs of or incidental to the litigation' in order to activate the court's jurisdiction to award costs under the Supreme Court Act 1981, s 51. Where the mediation is a stand-alone process conducted years before the proceedings began, the costs may not fall within the scope of recoverable costs.[31] Where the mediation is considered to have been part of pre-action conduct under the TCC Protocol, the requirement is likely to be made out and a party's unreasonable abandonment of mediation may lead to an adverse order for the **22.41**

[26] *Hickman v Blake Lapthorn* [2006] EWHC 12 (QB) at [18] *per* Jack J.

[27] [2008] EWHC 424 (QB), 118 Con LR 68.

[28] *Hickman v Blake Lapthorn* [2006] EWHC 12 (QB) at [69] *per* Jack J. The English courts used the test of 'reasonableness', as opposed to 'good faith' when considering the parties' actions (contrast the 'good faith' test proposed by the Supreme Court of West Australia in *Capolingua v Phylum Pty Ltd* (1991) 5 WAR 137). However, as can be seen from the decision in *Earl of Malmesbury v Strutt & Parker*, the Court may consider whether the parties did indeed mediate or seek to mediate in good faith. But it can only do so consistently with the confidentiality implicit in the mediation process. Things said and done in a mediation (with few exceptions) cannot be referred to in the proceedings. See *Farm Assist Ltd (In Liquidation) v The Secretary of State for the Environment, Food and Rural Affairs (No 2)* [2009] EWHC 1002 (TCC), (2009) 125 Con LR 154.

[29] [2009] EWHC 1002 (TCC).

[30] In *Earl of Malmesbury v Strutt & Parker*, the claimant recovered just over £900,000 exclusive of interest. The total costs were just under £6m.

[31] See *Lobster Group Ltd v Heidelberg Graphic Equipment Ltd* [2008] EWHC 413 (TCC) at [15]–[16] *per* Coulson J.

wasted costs.[32] The parties of course may have agreed under the mediation agreement that each party will bear its own costs of the mediation; if so, the court would not normally award those costs.[33]

What the parties should try to agree

22.42 If the matter is to go to court because ADR is not agreed or is unsuccessful in disposing of the claims then the parties should provide the following information and try to agree:[34]

(i) Whether expert evidence is necessary and if so how best to minimize expense. This includes consideration of the instruction of a joint expert and who that expert should be.[35]
(ii) The scope and extent of disclosure with a view to saving costs.[36]
(iii) How the litigation can be conducted thereby saving costs and avoiding delay.[37]

22.43 The Pre-Action Practice Direction now provides that, during the pre-action process, the parties should consider the use of a single joint expert or an 'agreed expert' (ie the parties agree the identity of the expert but only one party instructs that expert and pays his or her costs). The Pre-Action Practice Direction envisages that, where the parties do not agree to the appointment of a single joint expert, the party seeking expert evidence should give the other party a list of one or more experts in the proposed field and that within fourteen days of receipt, the other party should indicate to which experts it objects: the first party 'should' then instruct another expert from the list.[38] It is as yet unclear to what extent this is likely to impact upon practice in the TCC. It would seem odd if a party could exclude another's proposed expert simply by objecting to his or her appointment. It may instead be that persistence with an expert to whom objection has been taken would need to be justified when permission is sought to call that person's evidence.

22.44 As to disclosure, there is of course the possibility, where disclosure is refused at the pre-action stage, of an application to the court for an order for pre-action disclosure under CPR r 31.16. Such orders have been described as unusual.[39] The Pre-Action Practice Direction, however, specifically gives the failure without good reason to make disclosure of documents requested to be disclosed as an example of what may lead to a finding of non-compliance with pre-action protocols.[40] It is first necessary to demonstrate that the jurisdictional threshold criteria in CPR r 31.16(3) are established and then to persuade the court to

32 *Roundstone Nurseries Ltd v Stephenson Holdings Ltd* [2009] EWHC 1431 (TCC).

33 *Lobster Group Ltd v Heidelberg Graphic Equipment Ltd* [2008] EWHC 413 (TCC) at [15]–[20] *per* Coulson J.

34 TCC Protocol, para 5.5.

35 Ibid, para 5.5(i); Pre-Action Practice Direction, para 9.4 and Annex C.

36 TCC Protocol, para 5.5(ii). The Pre-Action Practice Direction provides (at para 9.2) that any documents disclosed must not be used (without the disclosing party's agreement in writing) for any purpose other than resolving the matter.

37 TCC Protocol, para 5.5(iii).

38 Pre-Action Practice Direction, para 9.4 and Annex C, paras 4 to 5.

39 *Anglo Irish Bank Corp Plc v West LB AG* [2009] EWHC 207 (Comm).

40 Pre-Action Practice Direction, para 4.4(4).

exercise its discretion to grant the order.[41] Since the threshold criteria include demonstrating that the respondent's duty by way of standard disclosure would extend to the documents or classes of documents of which the applicant seeks disclosure, it is necessary to have regard to the case intimated during the pre-action correspondences. Where for example an allegation of fraud is not made in the letter of claim or any draft particulars of claim circulated between the parties, it shall be more difficult to demonstrate that documents relevant to such an issue would be the subject of standard disclosure in due course.[42] By contrast, where an issue has clearly been identified by the claimant, an order may be made if it would assist the claimant to plead a more focussed case.[43]

It is open to question whether pre-action disclosure is less likely to be granted in construction cases. In *Birse Construction Ltd v HLC Engenharia E Gestão De Projectos SA*,[44] Jackson J (then the judge in charge of the TCC) noted that the projects which give rise to litigation in the TCC typically generate extensive documentation and that disclosure is a labour-intensive exercise and major head of costs: disclosure before the proper time is therefore not something which should be lightly ordered.[45] Holding that regard was to be had to the parties' conduct under the TCC Protocol and that this protocol demanded and usually obtained a high level of co-operation between opposing parties, he stated that he would not expect an order for pre-action disclosure to be appropriate in most cases in the TCC. Similarly, in a case in the Chancery Division, but in respect of which the underlying dispute was a construction dispute, Briggs J also had regard to the fact that the proposed disclosure exercise would be onerous and considered that in any event the disclosure would not result in any significant saving of costs or time in respect of the proposed litigation.[46] **22.45**

However, the test remains as set out in CPR r 31.16. As in other courts, the TCC shall have regard to the extent of pre-action cooperation in practice. Thus, whether (or to what extent) an application for pre-action disclosure may succeed will depend in part on whether the respondent to the application could and should have made its case clear in the pre-action conduct to date.[47] In *Resthaven Properties Ltd v Kier Regional Ltd*,[48] Ramsey J made an order for pre-action disclosure although the respondent had already provided documentation of the category sought, since the circumstances indicated that it did so informally and hence may have failed to ensure that all the relevant documentation was provided. By a targeted **22.46**

---

[41] See *Black v Sumitomo Corp* [2001] EWCA Civ 1819, [2002] 1 WLR 1562, [2003] 3 All ER 643.

[42] See *Cheshire Building Society v Dunlop Haywards (DHL) Ltd* [2007] EWHC 403 (QB).

[43] *Total E&P Soudan SA v Edmonds* [2007] EWCA Civ 50 at [27] *per* Tuckey LJ.

[44] [2006] EWHC 1258 (TCC).

[45] Ibid, at [25]–[29].

[46] *Hands v Morrison Construction Services Ltd* [2006] EWHC 2018 (Ch).

[47] In *Briggs & Forrester Electrical Ltd v Southfield School for Girls* [2005] EWHC 1734 (TCC), [2005] BLR 468, HHJ Coulson QC (as he then was) refused an application in respect of some categories of documentation sought, but granted it in relation to quantum documentation, commenting at [55] that, following their letter of claim, the respondents should have produced a schedule or schedules identifying the full quantum claimed under each head of claim, and demonstrating clearly how each headline figure had been made up. The quantum in the letter of claim was simply a collection of figures, many of which were estimates, without any further schedule or explanatory documentation.

[48] [2009] EWHC 542 (TCC).

application for disclosure, it may therefore be possible for the applicant to show that the request is not unfairly onerous.

22.47 Finally, where a party enters into a funding arrangement as defined by CPR r 43.2(1)(k) (most typically a form of Conditional Fee Agreement or CFA), notice should be given to the other parties as soon as possible.[49] Failure to do so risks forfeiting an entitlement to any uplift in the event of success in the proceedings.[50]

22.48 The Pre-Action Practice Direction states that, in giving directions for the management of claims and when making costs orders, the Court will have regard to compliance with the Practice Direction and any relevant protocol and may ask for an explanation for any non-compliance.[51] What matters above all is whether there has been substantial compliance with the relevant principles and requirements.[52] The Court will take into account the proportionality of the steps taken compared to the size and importance of the matter and its urgency.[53]

What to do if the limitation period is about to expire

22.49 There is no doubt that the pre-action protocol regime has caused cases to settle before they get caught up in the legal process. The way cases are begun is now different. Pre-action correspondence is less aggressive and unhelpful than it was, as lawyers come to appreciate and use a more constructive approach to sort out claims.

22.50 However, the pre-action steps are time consuming. The time they take to go through may mean that a claim is in danger of becoming statute barred. In that event the claimant can issue proceedings without going through the protocol steps and then apply to the court for directions.[54] The claimant should seek to agree this with the other parties.[55] The TCC judge will normally stay the action in order to allow for the protocol steps to be taken.[56]

22.51 A related question arises as to what the claimant should do when the pre-action process is not complete but the investigation of the matter is still some way from identifying the nature of the claim and the proper defendant(s) to it.

22.52 The claimant has four months within which to serve a claim form.[57] There may be a conflict between the need to avoid the claim becoming statute barred and the countervailing need to identify with sufficient particularity the claim to be made. In such circumstances,

---

49 Pre-Action Practice Direction, para 9.3.
50 CPR r 44.3B(1)(c).
51 Pre-Action Practice Direction, paras 4.1 and 4.2.
52 Ibid, para 4.3(1).
53 Ibid, para 4.3 (2) and 4.3(3).
54 TCC Protocol, para 6.
55 Pre-Action Practice Direction, para 9.6.
56 See TCC Guide, para 2.3.2. It is not unknown, where a limitation issue is evident from the claim form, for the court staff to contact the claimant's representatives to prompt them to apply for directions.
57 CPR r 7.5(1).

the claimant may wish to issue the claim form but not to serve it within the four month period and hence applies to extend that period.[58]

The Court of Appeal has indicated that, where the reason for the delay in service is incompetence on the part of the claimant's legal representatives, that is a powerful reason for refusing to grant an extension of time.[59] Nor is the need to particularize the claim alone sufficient grounds. The Court of Appeal has held that the claimant's desire to see the proposed defendant's response to its letter of claim or delay in the receipt of evidence to be used for the purposes of drafting pleadings were no justification for the grant of an extension.[60] In *Hoddinott v Persimmon Homes (Wessex) Ltd*,[61] the claimant claimed damages for trespass and damage to land and/or for nuisance and/or negligence. It sought an extension on the ground that, although it could serve the claim form in time, an extension would enable its claim to be fully particularized and for any resulting litigation to be conducted more economically and expediently. The Court of Appeal held that such was not a good reason for granting an extension. In those cases, the Court of Appeal indicated that the fact that the defendant might, by the grant of an extension, be deprived of the benefit of a statutory time bar, was an important factor. **22.53**

By contrast, in a more recent case in the TCC, *Imperial Cancer Research Fund v Ove Arup & Partners Ltd*,[62] Ramsey J upheld the grant of an extension of time for service of the claim form in circumstances where, although the allegedly defective work had been carried out years previously, there was further damage much more recently and the consequential investigations were at the stage at the time of issue of the claim form that the claimant could not properly make allegations of professional negligence against the defendants as opposed to other members of the construction team. Further, it appeared to be unclear whether the statutory limitation period would come into play to bar the claim during the period of the proposed extension. **22.54**

It may be more difficult in TCC cases to identify whether a statutory time bar will apply to all claims in the proceedings. For that reason, and because the investigation of defects may not (without any delay on the part of the claimant or of its advisors) be resolved at the time of issue of the claim form, there may be more scope for the TCC exercising its discretion in favour of the grant of an extension than in respect of other claims. However, the test applicable remains the same in all cases, and claimants must in any event bear in mind that under paragraph 6 of the TCC Protocol, they should apply for directions on the issue of the claim form. In such cases, instead of granting an extension of time for service of the claim form, the Court may direct that the proceedings be stayed following such service, in order for the pre-action process to be completed and the claim to be properly particularized in due course. A failure to do so can be costly. In *Charles Church Developments Ltd v Stent* **22.55**

[58] CPR r 7.6.

[59] *Hashtroodi v Hancock* [2004] EWCA Civ 652, [2004] 1 WLR 3206.

[60] *Collier v Williams (Marshall v Maggs; Leeson v Marsden, Glass v Surrendran)* [2006] EWCA Civ 20, [2007] 1 All ER 991, [2006] 1 WLR 1945.

[61] [2007] EWCA Civ 1203, [2008] 1 WLR 806.

[62] [2009] EWHC 1453 (TCC).

Foundations Ltd,[63] the claimant had failed to comply with the pre-action protocol and over a year passed before a stay was imposed to enable ADR. Ramsey J ordered that the claimant pay 50 percent of the defendant's costs from the service of the claim form to the commencement of the stay and (whatever the outcome of the litigation) to bear 50 percent of its own costs up to that date.[64]

Pre-action Protocols: the Professional Negligence Protocol and its significance

22.56 Although the TCC tries cases involving professional negligence, usually against architects, surveyors, engineers, and lawyers concerned with construction projects, the Professional Negligence Protocol exists as a separate protocol for professional negligence claims.

22.57 This expressly excludes construction professionals.[65] However the Professional Negligence Protocol contains a statement of matters which ought to be included within the letter of claim, which is relevant to claims against professionals and ought helpfully to be included within the letter of claim which is sent to them under the TCC Protocol.

22.58 The professional negligence letter of claim should include:[66]

(a) The identity of any other parties involved in the dispute or a related dispute.
(b) A clear chronological summary (including key dates) of the facts on which the claim is based enclosing key documents being identified, copied, and enclosed.
(c) The allegations against the professional. What has he done wrong? What did he fail to do?
(d) An explanation of how the error which is alleged caused the loss.
(e) An estimate of the financial loss suffered and how it is calculated, supporting documents being copied, and enclosed. If details cannot be supplied, then there should be an explanation as to why and a statement of when they will be supplied.
(f) Confirmation whether an expert has been appointed, providing the identity of the expert and discipline together with the date upon which he was appointed.
(g) A request that the letter be forwarded to insurers, if any.

22.59 The particular matters which are included within the Professional Negligence Protocol's letter of claim and are not included within the TCC Protocol for a letter of claim and ought helpfully to be included in one involving an allegation of professional negligence against a constructional professional are at (b) and (c) above. The matters at (d) to (g), if the action is in professional negligence, should also be considered for inclusion within the letter.

22.60 Paragraph (b) provides for a chronology. This is normally essential for the understanding of any case and there seems to be no reason why one should not be provided in TCC cases as well.

[63] [2007] EWHC 855 (TCC).

[64] This sort of order is unlikely to be treated as unusual: Jackson LJ referred to the judgment with apparent approval in the Jackson Report: see Vol 2, ch 43, para 3.11.

[65] Professional Negligence Protocol, A1.

[66] Professional Negligence Protocol, B2.2.

Paragraph (c) arises from the fact that an allegation of what the professional has done wrong is essential in any professional negligence action. The twin questions are: (i) what is it that the professional did or ought not to have done? and (ii) what is it that the professional did not do which he ought to have done? No professional negligence action can proceed without one or both of these twin questions being answered. **22.61**

The case as to what the professional did, or failed to do, must be backed up by an expert's report in all cases save those which involve allegations against lawyers, where expert evidence is normally inadmissible.[67] The court itself usually decides whether the lawyer is or is not negligent without reference to an expert. Any appointed expert must be qualified in the discipline in which the professional concerned practiced.[68] Therefore, for example, an expert surveyor will not be able to give expert evidence against an architect. **22.62**

Going to court is a last resort since it is an expensive process and not always predictable. If, however, ADR does not cause an action to settle, then the matter must go to arbitration or litigation. Arbitration is dealt with separately in Chapter 21. **22.63**

The effect of non-compliance with the protocols

The Pre-Action Practice Direction also provides that a court may order the following sanctions in the event of non-compliance with relevant protocols and the Pre-Action Practice Direction itself: **22.64**

(a) a stay of proceedings until steps which ought to have been taken have been taken (a sanction more obviously geared to address failings on the part of the claimant than of the defendant);

(b) the party at fault pays the costs, or part of the costs, of the other party or parties, including on the indemnity basis;

(c) depriving an otherwise successful claimant of some or all of the interest claimed or awarding interest at a lower rate by reason of that party's non-compliance;

(d) where the party at fault is a defendant, and an order for the payment of a sum of money is subsequently made in favour of the claimant, an order that the defendant pay interest on all or part of that sum at a higher rate, not exceeding 10 percent above base rate, than would otherwise have been awarded.

It can be seen therefore that the court's primary sanction against any party which does not observe the protocols relate to costs.[69] The court is unlikely to await until the end of trial to impose that sanction: in *Charles Church Developments Ltd v Stent Foundations Ltd*,[70] Ramsey J stated that the courts should generally deal with the cost consequences of failure to comply with a pre-action protocol at an early stage.[71]

---

[67] See Oliver J in *Midland Bank Trust Co Ltd v Hett Stubbs & Kemp* [1979] 1 Ch 384 at 402 referred to below.

[68] *Investors in Industry v South Bedfordshire District Council* [1986] 1 All ER 787.

[69] TCC Protocol, para 1.4.

[70] [2007] EWHC 855 (TCC).

[71] Ibid, at [38]. Jackson LJ referred to this statement with approval in the Jackson Report: see Vol 2, ch 43, para 3.11.

The Specialist Court: The TCC

The judges

22.65 Construction litigation exploded in quantity during the 1970s and 1980s. In the late 1960s construction litigation was tried by one of three Official Referees to whom cases were referred by High Court judges. Official Referees were judges of circuit status who had no original jurisdiction of their own. They were a referral service to whom cases of technical complexity were referred or cases where the primary matters in issue related to quantum. The referee was then required to report to the High Court judge who then gave judgment. This practice became obsolete when cases, having been transferred to Official Referees, remained with them to be tried to judgment by them.

22.66 In 1980, it became possible for writs (claim forms) to be issued directly before the Official Referees. This came about because the practice had grown up whereby writs (now claim forms) were issued in the Queen's Bench Division of the High Court followed almost immediately by the issue of a summons to seek to transfer the action to the Official Referee.

22.67 The explosion in the quantity of construction related litigation meant that the allocated judges were not able to cope with the load and so more were appointed. By 1997 their number had grown from three to seven, and by 2005 there were eight full-time circuit judges assigned to sit as TCC judges nationwide (five in London and one each in Liverpool, Birmingham, and Manchester).

22.68 It became a curious anomaly that Official Referees came to try the longest and most difficult cases. The cases were difficult in terms of technical content and also difficult in terms of the legal problems which they threw up. The development of the law of tort was one matter of continuing legal difficulty, but the contractual disputes involving more than one party meant that the litigation being tried by the Official Referees was of greater difficulty than nearly all other cases tried in the Queen's Bench Division.

22.69 In addition to complexity was weight. Construction cases usually dwarfed, in terms of amounts in issue, the normal business of the Queen's Bench Division and indeed most litigation conducted in the Commercial Court.

22.70 The Official Referees were, from 1980, in charge of their cases from inception to completion. This came about through the accident of history in that cases were, as stated above, originally referred to them and it was their job to take the case to its conclusion.

22.71 This care of cases from beginning to end meant that each case, having been assigned to its judge, was the responsibility of that judge. In the Queen's Bench Division cases at their interim stages are dealt with by Masters and, except in special circumstances, the trial judge has nothing to do with the case until the trial is called upon. Official Referees, being in charge of their cases throughout were able to, and did, apply case management and indeed they were the pioneers of case management. Nearly all case management techniques now in use in the courts were developed by the Official Referees.

The Technology and Construction Court (TCC) was created out of the Official Referees' courts and to this day they conduct the business of construction or construction related litigation in this country. **22.72**

The TCC can best be considered by reference to London and outside London. **22.73**

The TCC in London

The judges of the court sit principally in London, where Ramsey J heads the TCC nationwide.[72] There are two other High Court judges who hear TCC business regularly (Coulson J and Akenhead J), and TCC cases are also heard by other designated High Court judges identified in the TCC Guide.[73] Edwards-Stuart J. was appointed on 31 July 2009 and is expected to hear TCC business. These judges sit in addition to the circuit judges in London.[74] The current arrangements stem from the reform of the TCC announced by Lord Woolf CJ on 7 June 2005,[75] following which Jackson J was assigned full time to the TCC (whereas previously he was assigned to sit for only six months of the year). Each case started in London or transferred there will be classified by Ramsey J as fit for trial by either a High Court Judge (HCJ) or a Senior Circuit Judge (SCJ). The most complex and heavy cases will be classified as HCJ and will be managed or tried by him or another HCJ. The other cases will be classified as SCJ and will be allocated to a SCJ by the judge in charge of the TCC. **22.74**

This is a major breakthrough. The legal profession and users of the court had long complained that the TCC's position was anomalous with the heaviest cases escaping the attention of HCJs. **22.75**

No doubt the list of judges nominated to hear HCJ cases will change, but the basic point is that if the HCJ in charge of the TCC does not have space to try an HCJ case, then another HCJ will be allocated to it. **22.76**

The TCC outside London

Other circuit judges of the TCC sit virtually full time in Birmingham, Liverpool, and Manchester and other regional trial centres have part time judges who are authorized to sit as TCC judges. **22.77**

Any solicitor is well advised to issue proceedings in one of the trial centres where the TCC judges are full time or virtually full time. It does no good to a case to have it assigned to a judge whose priority is sitting as a judge in the Crown Court.[76] **22.78**

[72] On 1 September 2007 Ramsey J succeeded Jackson J (since elevated to the Court of Appeal), who had succeeded Forbes J in October 2004 as the judge in charge of the TCC. Dyson J was the judge in charge before him. Both Dyson and Forbes JJ sat as TCC judges for a significant period of the year, although not full time. Dyson J, now Dyson LJ, sits in the Court of Appeal and regularly sits on construction appeals.

[73] Currently Burton, Elias, Field, Ouseley, Simon, Christopher Clarke, and Teare JJ.

[74] Currently HHJ Wilcox and HHJ Toulmin CMG QC.

[75] See 'Statement by the Lord Chief Justice of 7 June 2005' (2005) 100 Con LR 209.

[76] However the trial centres in Manchester and Birmingham have judges who sit in the Mercantile List, and Mercantile Judges have experience in construction cases.

22.79 The provisions for classification of cases to HCJ and SCJ status do not apply to those cases which are commenced in trial centres outside London. Therefore if it is thought that the case ought to be tried by an HCJ, solicitors ought to issue proceedings in London. There is a provision in Lord Woolf's statement that cases begun outside London are not overseen by the judge in charge of the TCC, but if it is thought by the judge sitting there that the case should be managed by an HCJ then it can be referred to Ramsey J for classification. If, therefore, solicitors want a case to be tried by an HCJ, then, at the moment, the claim form should be issued in London.

22.80 Lord Woolf's statement also provides that, if a claim is begun in or transferred to London, the parties may make brief representations by letter as to the appropriate classification.

Small claims, fast track claims, and multi track claims

22.81 Civil court business is divided up into three main tracks. Small claims and fast track claims are the least complex and are of lowest value. This book is not intended to deal with the procedure of small and fast track claims. What follows deals with claims which are assigned to the multi track and are of a value of over £50,000.

How to Frame a Claim: The Essentials of a Claim

Introduction

22.82 A construction claim brought in court will have its own qualities which are unique to it. But the essentials are the same for all of them, namely, that the claim must set out in as concise a form as possible exactly why the claimant is entitled to the relief sought and how the quantum of the claim is calculated.

The elements of pleading: the claim

22.83 Each element of pleading ought to be set out in sequentially numbered paragraphs. It is a matter of judgment as to how short the paragraphs are. The best policy is usually to keep each paragraph short with one concept, term of the contract, or allegation in it. This allows for each paragraph to be pleaded to with precision in the defence with each relevant allegation being either admitted or denied or otherwise commented upon. Whatever the nature of the claim, the particulars of claim (or the claim form itself) should state whether the claimant has complied with Section IV of the Pre-Action Protocol.[77] Further, the pleader should be careful not to be lengthy, pleading material facts rather than recite narrative. The Jackson Report raises the possibility that the court in the TCC may refuse costs in relation to over-lengthy pleadings.[78]

[77] Pre-Action Practice Direction, para 9.7. This refers also to Section III but that applies only to cases not subject to any Pre-Action Protocol. Section IV deals with pre-action steps.

[78] The Jackson Report, Vol 1, ch 34, paras 5.8 to 5.9. The TECBAR response agrees: see para 27.

The elements needed for claims for damages for breach of contract or claims for sums due under the contract are normally as follows: **22.84**

(a) *A paragraph or section identifying the parties.* It ought to identify who are the claimant and the defendant (or defendants) and all other relevant parties. The architect under a building contract, the engineer under an engineering contract, and if relevant the project manager or clerk of works. This paragraph may be put in at the beginning of the pleading; sometimes it is best placed after the contract which is the subject matter of the claim is identified.
(b) *A paragraph which identifies and defines technical terms.* This ought to be relatively short and in many commonplace contractual disputes it will be unnecessary. If, however, more than half a dozen terms need to be defined, it is best to place that in an appendix.
(c) *The contract sued upon should be pleaded in summary.* If it is a written contract, the date of the contract, between which parties it is, what the consideration was for the contract, usually the contract sum, and what the contract was for—whether it is a building contract or a contract for professional services for example.
(d) *The architect, project manager, surveyor, or engineer* under the contract should be set out, if not identified separately in the paragraph which identified the parties.
(e) *The express terms of the contract which are relied upon* to found the relief claimed should be set out. This is sometimes done in extenso, but it is far preferable that only the essential terms are set out, and thought is given in relation to those as to whether the term should be set out in full. A summary of the term is not normally helpful and it is usually better to plead the express term in full, but a lengthy quote from standard terms and conditions is not normally welcome or useful. The terms of the contract which are essential to the understanding of the contract and the relief claimed are required. It is tempting for the pleader to adopt the line 'if in doubt set it out'; the better course is to ask 'do I need this?' or 'why does this term appear in the pleading?'
(f) *The express terms should be followed by any implied terms* which are relied upon. These should be succinctly and precisely formulated. If the term is implied by statute, then the statutory provision should be referred to. If the implication is for the normal business efficacy reasons, then it is not normally necessary to say so.

The above paragraphs will be common to most claims, be they damages for breach of contract or sums due thereunder. In the latter case implied terms are not normally a necessary ingredient since the express terms of the contract will govern the timing and amount of payment. **22.85**

In the case of a claim for damages for breach of contract, following the above sections there should then be a section or paragraph which identifies which of the terms of the contract are breached. The first part of the paragraph should be short and merely make the identification. Below that identification, probably under a heading entitled 'Particulars' should come in numbered and sequential sub-paragraphs the respects in which the contract was breached. If there are different terms breached by different acts or omissions, then there **22.86** should be short sub-headings referring to different terms of the contract with the breaches identified under each.

22.87 The section which identifies the terms of the contract breached, which may be lengthy and have numerous sub-paragraphs, should be followed by the quantification of the loss.

Quantum and linkage

22.88 The quantum of the claim may be the most difficult to set out. If the claim is for defects to a contractor's work each defect may have a different remedial work cost. Should this be the case, then it is better to set out the quantum in a schedule with the remedial work cost set alongside the defect.

22.89 The link between breach and damage should be clearly stated. Without such clear statement the likelihood is that the claim will fail.

22.90 If the claim is for sums due under the contract, then the term of the contract giving rise to the sum claimed should be shown in a separate section followed by a section which clearly explains why and how the sum claimed arises. In cases of valuation, again a schedule is most likely the most convenient way of setting the claim out. The link between the term of the contract relied upon and the amount claimed should be clearly shown. The claims can often be divided into two; that is: (a) claims for the valuation of work done, and (b) claims for loss and expense. The two separate claims must be treated separately.

22.91 If interest is to be claimed, it should be pleaded, and the basis of the claim (whether under contract or statute) should be set out.[79]

Elements of pleading: the defence

22.92 As with the pre-action correspondence, the defence should set out, clearly and fully, exactly why the claim is not accepted, or if it is partially accepted, exactly what is accepted and what is not. That does not mean that there needs to be a long narrative, but each paragraph of the claim should be dealt with.

22.93 This is easier to do with precision if each paragraph of the claim is short.

22.94 The following questions need to be asked by reference to each paragraph of the particulars of claim: (a) Is it accepted and therefore can it be admitted? (b) To the extent that it is admitted, is the paragraph incomplete? If so can a fuller picture be given? (c) If the paragraph is not accepted as accurate, why is this so? Is it because the events alleged never took place or is it because the events alleged did take place but in a different context and to a different purpose? (d) Can a positive case be made explaining why the particular allegation in this paragraph is not accepted?

[79] CPR r 16.4.

It is sometimes necessary for a counter-narrative to be given to that which is set out in the particulars of claim. This should be as brief as possible and the pleader should aim to be no longer than about two pages in length. It is, of course, impossible to be prescriptive but a short telling of the story is more likely to be absorbed than a lengthy one. **22.95**

The important thing for the pleader of the defence to remember is that s/he is not just seeking to get his or her client's story across, s/he is also making sure that the judge knows exactly what the issues in the case are. What is it that s/he has to try? Is it a factual dispute? Is it a dispute on valuation? Where is the hinge point? This is necessary so that the judge and both sides know exactly what evidence is needed to try the case. Is it factual or expert? If, as is usually the case, it is both what is the most important divergence? This is especially important when it comes to the consideration of witness statements and experts' reports. **22.96**

Claims Consultants and the Dangers of Defective Formulation of Claims

It has long been the case that contractors who are in trouble on certain projects engage claims consultants to advance their claims. Some of the better ones are not only well aware of the need for claims to be properly formulated, but they also have a record of putting forward well-presented claims. **22.97**

Unfortunately, a significant minority do not and a great deal of their client's money is wasted because of the failure to observe the guidelines set out above together with a very prolix narrative or narratives of what is alleged to have happened under the contract. The key to seeing whether the claims are being properly presented is whether the key elements of 'duty', 'breach', and 'damage' are clearly set out with the necessary linkage between the three. Long narratives and extensive documentary appendices are the hallmark of such unhelpful claims. When they are transferred, either with a small modification or with none at all, to a court, the claim is likely to get into trouble and cause a great deal of wasted costs to be expended by the client who has the claim put forward and by the side which is trying to address and defend the claim. **22.98**

Schedules: 'Scott Schedules'

Many claims are tried by reference to schedules which set out on a point by point basis the claims and counterclaims of each party. They are often referred to in the TCC as 'Scott Schedules', named after the judge who first put them into operation. They involve a series of columns, side by side with headings which set out what is identified. **22.99**

Two types of schedule are illustrated here. One for defects and the other for loss and expense or an extension of time. These are of a classic nature, but there will have to be adaptations for each different case to take account of the peculiarities of each case. **22.100**

22.101 The defects Scott Schedule should have the following column headings or headings to similar effect:

(a) Column 1: Item number.
(b) Column 2: Description of defect.
(c) Column 3: Contractual term which is breached.
(d) Column 4: Statement of whether this is a defect of design, poor materials, or poor workmanship.
(e) Column 5: Particulars of remedial work which is required to put the pleaded defect right.
(f) Column 6: Particulars of the cost of the remedial work.

The loss and expense and extension of time schedule should have the following headings or headings to similar effect:

(g) Column 1: Item number.
(h) Column 2: Description of delaying or disrupting event. If it is an architect's instruction it should be specifically identified. If it is a late instruction or the late provision of information the column should have the following sub-headings:
    (i) the date the information was required and brief reasons for it;
    (ii) the date the information was requested;
    (iii) the date the information was supplied.
(i) Column 3: Description of what effect the delaying event had on the progress of the works. If it is delay, then the estimated amount of delay to the works should be given; if it is disruption then the nature of the disruption should be given.

22.102 Scott Schedules are now set out in landscape format to illustrate the above. They have been filled in by reference to our worked example. The particulars which should go into the schedules will vary according to what is really in issue and what is genuinely a source of misunderstanding or what is needed to be known by the opposing party. The pleaded facts in the sample schedules will need adapting to the particular situation should be regarded as illustrative only.

22.103 Once the schedule is served then the defendant should plead to it, setting out its case by reference to each item number.

Defects Schedule

Item No	Description of Defect	Term of Contract and Specification Breached	Remedial Work Done/ Required	Cost of Remedial Work	Defendant's Comments on Breach of Contract	Defendant's Comments on Remedial Work and Cost of Remedial Work	Judge's Comments
1.	Tiles to western wall of the Art building are loose and falling off.	Clause 5 of the Specification requiring the back of the tiles to have adhesive properly applied to the full back of the tiles. This was not done causing them to come loose.	Hack off all external tiles to the western wall and re-apply in accordance with the contract.	X square metres @ £Y per spare metre = £Z	Adhesive was applied properly, save in the case of two tiles only. The tiles are falling off because the design did not allow for movable joints.	Remedial work should be £(Y–A) per square metre. The Defendant is only responsible for the replacement of two tiles.	

Delays Schedule

Item No	Delaying Event	**Effect of the Event**	Total Delay	Description of Disruption	Defendant's Comments	Judge's Comments
1.	AI No 55 changing the foundations from trench fill to piles.	Foundations were due to commence on 1 January 2006. They could not begin until 1 March 2006 because of the needs to mobilize a piling rig and gang. The piling work took five additional weeks more than the trench fill foundations would have taken.	Nine weeks' delay in commencement and five weeks in execution of the work.	The site was idle for nine weeks and following trades were slowed in the five weeks which followed.	(1) The contractor was not planning to commence work until 1 March 2006 in any event. (2) No following trades were delayed since the sequence of work executed was in accordance with Programme 1A dated 2 April 2006.	

Case Management

22.104 Once the defence is served, the claimant should get the case listed before the judge for directions as to how the case is to be prepared and timetabled for trial. This can also be done before the defence is served. This may be done if the pre-action letters set out clearly what the issues are. It may also be done if the claimant wishes the defendant to plead to specific schedules. There is no requirement that the matter is to be taken to the judge after service of the defence, but this will usually be the best time because the issues on the pleadings will be crystallized.

Issues for consideration at a case management conference

22.105 The following issues will then normally be for consideration:

(a) Are there any other pleadings to be exchanged? Does a Scott Schedule have to be served and if one has already been served should it be pleaded to by the defendant and if so how long should the defendant be allowed to do so?
(b) What are the issues in the case? Are they factual only? Are they expert only? If there are issues of fact, then there will have to be an exchange of witness statements between the parties so that each knows who is being called as a witness and what each witness is going to say. The court will also consider whether there should be provision for responsive statements, although it will be alert to whether those are really needed.[80] If they are expert issues, then the court and the parties will have to consider whether this is a case for a single joint expert or whether this is a case for each party having their own experts.
(c) Are the issues ones of law only? If so is there any need to give further directions other than to get the question of law tried. If the case does come down to a question of law, or the answer to a question or questions of law will unlock the dispute, then it is likely that a preliminary issue will be appropriate. Before such an issue is ordered to be tried the parties should formulate the issue as precisely as possible.
(d) Is this a case where the action should be tried in sections? Part 1 before Part 2 etc; or is it a case for a preliminary issue or preliminary issues of fact, expert opinion or law?

22.106 The objective of the first meeting before the judge is to try to get directions down to a trial date, be it of a preliminary issue or of the whole of the action.[81]

---

[80] The Jackson Report raises the possibility that responsive witness statements should not be allowed, unless some genuinely new issue arises which the witness needs to deal with: see Vol 1, ch 34, para 5.7. It further suggests that the TCC may refuse costs in relation to prolix or unnecessary witness statements: ibid, paras 5.6 and 5.9.

[81] One feature of litigation in other courts, the List of Issues, has not found favour in the TCC. That is unlikely to change: in the Jackson Report, Jackson LJ commented that in four years at the TCC he did not recall any case in which lists of issues submitted by the parties were of any great assistance: see Vol 1, ch 34, para 5.5. The problem with lists of issues is that they are frequently prepared as a compendium of the pleadings. Whilst that may make for an exhaustive list of what the judge has to try, it is not an exercise in distillation.

Directions

A classic form of directions given is: **22.107**

(i) The defendants shall plead to the claimant's Schedule A by [date];

(ii) Disclosure shall be by list; lists to be served by the parties upon each other by [date];

(iii) Each party has leave to call witnesses of fact conditional upon the substance of that evidence being provided to the other side in the form of a witness statement by [date]. The witness statements shall stand as their evidence in chief;

(iv) Each party has leave to call expert evidence in discipline x and discipline y conditional upon the service on the other side of the substance of that evidence in the form of a report by [date];

(v) Experts of like discipline to meet without prejudice with a view to narrowing issues by the week beginning [date];

(vi) Experts to produce a schedule or report in convenient form setting out the points of agreement and disagreement by [date];

(vii) Trial date shall be [date]; Estimated length [x days];

(viii) There will be a trial of the following preliminary issue [set out precisely] on date x, estimated length k days.

Nearly all directions given by the TCC judges are a variation of the directions given above. **22.108** Disclosure is currently made on the basis of standard disclosure by which each party is required to disclose the documents on which he relies; and the documents which adversely affect his own case, adversely affect another party's case or support another party's case.[82] This may change. The Jackson Report suggests that disclosure in multi track civil litigation may cause the most problems[83] and recommends various options for change, including the introduction of 'issues based' disclosure (currently the subject of trial in the Commercial Court).[84] Although Jackson LJ's comments were not specifically directed at the TCC, all TCC claims are allocated to the multi track under CPR r 60.6(1) and if change comes, it will presumably be introduced to such cases too.

It is now usual for the parties' experts to meet prior to the writing and exchange of their **22.109** reports. Some judges prefer the meetings to take place after the reports have been exchanged because the experts' issues are better crystallized in them. However, the better view is probably that the meetings and production of the schedule of agreement and disagreement should take place prior to the reports being written or finalized. This is because experts, like any other human being, find it easier to abandon a previously fixed view if they have not

[82] CPR r 31.6.

[83] The Jackson Report, Vol 2, ch 41, para 3.2.

[84] Ibid, s 6. The TECBAR response supports the status quo: see paras 28 to 30. By contrast, the Bar Council's submission (entitled *Review of Civil Litigation Costs: The Bar Council's Response* ('the Bar Council Response') suggests that the Bar generally supports the proposals to introduce 'issues based' disclosure, to introduce more rigorous case management and for the allocation of a single judge to any substantial case at the outset: see para 121. The latter two mirror what is already done in the TCC.

gone into writing supporting it. The expert concerned is less likely to feel vulnerable to cross-examination about his or her change of views.[85]

Experts

What is expert evidence?

22.110 Expert evidence is opinion evidence relating to a matter in issue requiring the opinion of someone of a skill or calling on a matter upon which the court cannot normally inform itself. Put another way, expert evidence goes to a matter upon which a layman would require instruction in the essentials of the necessary field of expertise to make a properly informed decision.[86] Expert evidence is admissible pursuant to s 3 of the Civil Evidence Act 1972.

22.111 The nature and admissibility of expert evidence was helpfully defined by Mance J in *The Ardent*[87] *quoting from the Australian criminal case, R v Bonython*[88] where King CJ said:

> Before admitting the opinion of a witness into evidence as expert testimony, the Judge must consider and decide two questions. The first is whether the subject matter of the opinion falls within the class of subjects upon which expert testimony is permissible. This first question may be divided into two parts: (a) whether the subject matter of the opinion is such that a person without instruction or experience in the area of knowledge or human experience would be able to form a sound judgment on the matter without the assistance of witnesses possessing special knowledge or experience in the area and (b) whether the subject matter of the opinion forms part of a body of knowledge or experience which is sufficiently organized or recognized to be accepted as a reliable body of knowledge or experience, a special acquaintance with which of the witness would render his opinion of assistance to the Court. The second question is whether the witness has acquired by study or experience sufficient knowledge of the subject to render his opinion of value in resolving the issue before the Court.

22.112 The most useful and cited judicial statement relating to expert evidence is that of Oliver J in *Midland Bank Trust Co Ltd v Hett Stubbs & Kemp*,[89] where he said:

> Clearly if there is some practice in a particular profession, some accepted standard of conduct which is laid down by a professional institute or sanctioned by common usage, evidence of that can and ought to be received.

Protocol for the instruction of experts to give evidence in civil claims

22.113 The Civil Justice Council in June 2005 issued a 'Protocol for Instruction of Experts to give Evidence in Civil Claims' ('Experts' Protocol') to be effective from 5 September 2005. The Protocol governs both the experts and those who instruct them. It defines the duties of parties instructing experts and of the experts themselves.

---

[85] The TCC Guide recommends their meeting at least once before the exchange of reports. See para 13.5.4 of the TCC Guide.
[86] *Re M and R (Minors)* [1994] 4 All ER 239 at 253 *per* Butler-Sloss LJ.
[87] [1997] 2 Lloyd's Rep 547.
[88] [1984] SASR 45 at 46.
[89] [1979] 1 Ch 384 at 402.

Independence of experts

Experts' opinions should be independent regardless of the pressures of litigation. A useful test of independence is whether the expert would express the same opinion if given the same instructions by the opposing party.[90] **22.114**

The expert ought ordinarily to be independent of the parties. The independence of the expert is a matter which goes to weight rather than admissibility. A judge may reject expert evidence on the basis that it does not come from an independent source. But that is no reason why the evidence should be ruled out before it is given and why it should not be received at all by the court[91] and weighed with the rest of the evidence. **22.115**

The admissibility of this evidence is subject to one important caveat. If the expert shows no sign of understanding his duty to the court (as set out in *The Ikarian Reefer*[92]), does not understand that he must comply with Lord Wilberforce's dictum in *Whitehouse v Jordan*[93] where he said that the expert evidence should be and should be seen to be the independent product of the expert 'uninfluenced in form or content by the exigencies of litigation' and does not understand his responsibilities so that the litigation can progress in accordance with the overriding objective or the witness is otherwise discredited, then the court can (and it is submitted should) refuse to admit the evidence.[94] **22.116**

It is nevertheless bad practice (save in the exceptional cases referred to later in this paragraph) for any party to call an expert who is not independent and seen to be independent. Even if a judge is persuaded to receive the evidence, that does not mean that he will accept it. Evidence which is from an expert who is perceived as not wholly independent of the party on whose behalf he is called is likely to have difficulty in being accepted, although such acceptance is not impossible. At the start of any case, and before any significant costs are expended, it is relatively easy in most cases to ensure that any proposed expert is truly independent. Problems may arise when the pool of experts available is small, as is the case in areas of specialist engineering. In such cases it may be nearly impossible to engage an expert without a connection with the party on whose behalf evidence is being given. Should it be necessary to call as an expert someone who is 'in house' and therefore not independent, then the expert concerned should be given training so that the expert has a full awareness if his or her duties.[95] **22.117**

---

[90] See Experts' Protocol, s 4 and particularly paras 4.3 and 4.4.

[91] *Factortame & Ors, R (on the application of) v Secretary of State for Transport* [2002] EWCA Civ 932, [2003] QB 381, [2002] 3 Costs LR 467, [2002] 3 WLR 1104, [2002] 4 All ER 97, [2003] BLR 1 and particularly paras 67–70. See also *Field & Anor v Leeds City Council* [1999] EWCA Civ 3013, (2000) 32 HLR 618, [2000] 1 EGLR 54.

[92] [1993] 2 Lloyd's Rep 68 at 81, see extract below.

[93] [1981] 1 WLR 246 at 256–7.

[94] See for example *Stevens v Gullis and Pile* [2000] 1 All ER 527, 73 Con LR 42, [1999] BLR 394, CA. The expert's evidence was disallowed for this reason, and he was not allowed to give evidence of fact because he was 'so discredited that it would be pointless for his evidence to be included on the hearing of the claim . . .' An even more striking example (in the field of copyright relating to architectural design) is *Pearce v Ove Arup Partnership Ltd (No 2)* [2001] All ER (D) 32 (Nov).

[95] *Factortame & Ors, R (on the application of) v Secretary of State for Transport* [2002] EWCA Civ 932, [2003] QB 381, [2002] 3 Costs LR 467, [2002] 3 WLR 1104, [2002] 4 All ER 97, [2003] BLR 1 *per* Lord Woolf MR para 19.

Nature of the evidence

22.118 Opinion evidence, not evidence of fact, is the essence of what is the subject of an expert's report. Therefore the expert will rely upon either his observation (of, say, a defect to a structure or a photograph of a defect to a structure) or what he is told.

22.119 In rare cases experts can also give factual evidence, but good practice would dictate that the expert's factual evidence is given in a separate witness statement. If the evidence of the expert is of what s/he has seen of the defective structure, the evidence given will still be treated as expert evidence, even though strictly speaking the evidence of what is seen by the expert is of fact.

22.120 Usually, in construction cases, expert evidence will involve the opinion of an architect, engineer, project manager, surveyor, or accountant. In such matters, the court will not have the expertise itself to make a finding. An engineer may give an opinion on the mechanism of failure of a structure. S/he will do so on the basis of what s/he has seen of the structure and been told.

22.121 Opinion evidence setting out what the expert him- or herself would have done in a given situation is not admissible.[96]

22.122 A skill or calling relating to an area of expertise is required. Therefore opinions which relate to everyday life or matters of observation by laymen are not the subject of expert evidence, since they do not fall within the definition of expert evidence. But questions relating to expertise acquired through study or experience will be admissible expert evidence.

22.123 Study as well as experience is an acceptable qualification for the giving of expert evidence. However if it comes with no relevant experience, the weight given to the opinion may be less than it would be if the expert had experience. In most cases there is no problem in that expert architects who are called as witnesses will have extensive experience. Parties will not usually instruct an inexperienced architect.

22.124 The problems relating to experience are normally felt when the expert concerned has not been in full-time practice for some time. This frequently happens with the most experienced experts whose time is often taken up giving evidence rather than practising. Partners in firms of architects and engineers will normally have day-to-day architectural experience. But if they are senior, they will often retire from the partnership and take on a consultancy role. In that event the career drift may be to a period of exclusively giving expert evidence, with the result that, before long, the expert concerned does not have the up-to-date experience of practice for the giving of the best evidence.[97]

22.125 A matter which the court cannot normally inform itself is a significant touchstone. Therefore, as has been stated, the court will not normally receive expert evidence relating

[96] See Oliver J in *Midland Bank Trust Co Ltd v Hett Stubbs & Kemp* [1979] 1 Ch 384 at 402.

[97] The Technology and Construction Solicitors Association Expert Witness Protocol (now superseded by the Expert Protocol) recommended in para 10 that the Association encourages experts 'who maintain professional practices'.

to legal expertise.[98] The judge will be taken to know the law and the standard of care to be expected of a lawyer.

Control of expert evidence

The court has control over the amount of expert evidence which is to be given.[99] The court can exclude relevant evidence if it so chooses.[100] It can also order the appointment of a single joint expert rather than permit the usual procedure whereby each side will have an expert.[101] That control continues throughout the proceedings: if, for example, a party wishes to change its expert and that involves opening up the court's order granting permission to call expert evidence, the court may grant permission only on conditions as to costs or on the condition that the party disclose the report of the previous expert.[102] **22.126**

The TCC judges will consider whether or not the case requires expert evidence and whether the issues require expert evidence.[103] It frequently happens that parties assume that an expert's report is needed when, in fact, it is not because the real question in the case is one of fact. **22.127**

Allied to this is the scope of the expert's evidence. Usually the judges will give directions that the parties have leave to call, for example, architectural expert evidence. This may suffice. But the better practice is to define the issues to which the evidence relates. That way the ambit of the reports is more limited and focused.[104] **22.128**

---

[98] See Oliver J in *Midland Bank Trust v Hett Stubbs & Kemp*, n 96 above, at 402:

> I must say that I doubt the value, or even the admissibility, of this sort of evidence, which seems to be becoming customary in cases of this type. The extent of the legal duty in any given situation must, I think, be a question of law for the Court. Clearly if there is some practice in a particular profession, some accepted standard of conduct which is laid down by a professional institute or sanctioned by common usage, evidence of that can and ought to be received. But evidence which really amounts to no more than an expression of opinion by a particular practitioner of what he thinks that he would have done had he been placed, hypothetically and without the benefit of hindsight, in the position of the Defendants, is of little assistance to the Court; whilst evidence of the witness' view of what, as a matter of law, the solicitor's duty was in the particular circumstances of the case is, I should have thought, inadmissible, for that is the very question which it is the Court's function to decide.

[99] CPR r 35.1, see also TCC Guide, s 13.

[100] See CPR r 32.1(2): 'The Court may use its power under this rule to exclude evidence which would otherwise be admissible.' CPR r 31.1(1) provides: 'The Court may control the evidence by giving directions as to (a) the issues on which it requires evidence; (b) the nature of the evidence which it requires to decide those issues; and (c) the way in which the evidence is to be placed before the court.' And see the judgment of the Court of Appeal in *Grobbelaar v News Group Newspaper Ltd* (Unreported, 9 July 1999) at p 13 per Potter LJ '. . . it can no longer be argued that the civil court has no jurisdiction to exclude relevant and admissible evidence in a civil case.'

[101] See CPR r 35.7.

[102] *Beck v Ministry of Defence* [2003] EWCA Civ 1043. This includes earlier drafts: *Hajigeorgiou v Vasiliou* [2005] EWCA Civ 236, [2005] 1 WLR 2195 at [30] *per* Dyson LJ. Where the court's permission did not identify a named expert, the change of experts may not involve opening up the court's order and there is no basis for the imposition of any conditions by the court: ibid at [20]–[21] *per* Dyson LJ.

[103] See TCC Guide, para 6.7. See also para 6 of the Experts' Protocol.

[104] See Thorpe LJ in *Vernon v Bosley* [1998] 1 FLR 297 at 300:

> It is unfortunate that the trial was hijacked by the expert witnesses as it was. To pursue the metaphor, expert witnesses are armed with the court's readiness to receive the expert evidence which it needs in order to reach a fully informed decision, whatever the nature of the topic may be. But their evidence ceases to be useful, and it may become counter-productive, when it is not marshalled by reference

22.129 The remarks of Dyson J in *Pozzolanic Lytag Ltd v Bryan Hobson Associates*[105] are an apposite warning, given before the advent of the CPR, but no less useful today:

> I strongly suspect, however, that in many cases insufficient thought is given by parties (and in particular their legal representatives) first to the question whether an expert is really necessary at all, and secondly to what issues the evidence of the expert should be directed . . . In my view the only issue to which expert evidence could properly have been directed was whether there is a common practice in the engineering profession as to what engineers, who are engaged as project managers, do in relation to the insurance obligations of contractors. That would have been a short point, which would have resulted in short reports. Instead of this, the experts prepared quite elaborate reports dealing with a number of other issues, which were inappropriate, and which no doubt added very considerably to the costs of this litigation. Thus [one expert] reviewed some of the correspondence in the case. [Another expert] produced a report which runs to 44 pages (excluding annexures), much of which is taken up with a recitation of events and extracts from the correspondence. The experts plainly went beyond what the [judge] had authorized. In view of the imminent implementation of the Woolf reforms, it is now opportune for everyone to take a hard look at the whole question of expert evidence. It seems to me that all have a role to play in this: case management judges, legal representatives and the experts themselves. Prolix experts' reports directed to issues with which they should not be concerned merely add to the expense of litigation. Everything possible should be done to discourage this. In appropriate cases, this will include making special orders for costs.

Single joint expert: generally

22.130 The CPR provides that the court can order that evidence can be taken from a single joint expert.[106] The Court of Appeal in *Daniels v Walker*[107] held that the procedure raises no human rights issues. As in that case, in construction cases, the order of such expert reports is frequent in smaller claims, on the grounds of proportionality, and is often used in relation to matters of quantum and technical matters. In cases of size the use of a single joint expert is rare when it comes to key questions of liability (such as whether a professional was negligent), but often the judge managing the case will encourage the parties to use a single joint expert on matters such as quantum.

22.131 Even in larger cases consideration of a single joint expert is encouraged in the early stages of controversy and also when matters are not expected to be in dispute (quantum, for example).[108] In *Quarmby Electrical Ltd v John Trant*,[109] Jackson J gave guidance as to the use of a single joint expert in smaller value construction cases as follows:

> I fully accept that in the larger construction cases the device of a single joint expert is generally reserved for subordinate issues or relatively uncontroversial matters. However, in the

---

to the issues in the particular case and kept within the limits so fixed. Identifying those limits is the primary responsibility of counsel and, if necessary, of the judge. When they are exceeded, or not made sufficiently clear, then the hearing is likely to be lengthened and the judge's task becomes more onerous than it already is. . .

105 [1999] BLR 267 at 275.

106 See CPR 35, PD 35 and CPR rr 35.7 and 35.8 and PD 35, para 5.

107 [2000] 1 WLR 1382.

108 Experts' Protocol, para 17.3.

109 [2005] EWHC 608 (TCC).

> smaller cases . . . if expert assistance is required, it is difficult to see any alternative to the use of a single joint expert in respect of the technical issues. If adversarial experts had been instructed to prepare reports and then give oral evidence in the present case, I do not see how there could have been a trial at all. The respective experts' fees and the trial costs would have become prohibitive. In lower value cases such as this one, I commend the use of single joint experts. The judge, of course, remains the decider of the case. He is not bound by everything which the single joint expert may say. However, the judge is able to perform his functions within more sensible costs parameters.
>
> The Civil Procedure Rules enable both parties to put written questions to a single expert: see Rule 35.6. This facility was used in the present case. Part 35 of the Civil Procedure Rules and the accompanying practice direction are silent on the matter of a single joint expert being called to give oral evidence. The commentary at paragraph 35.7.1 of the current edition of the White Book states:
>
>> If a single joint expert is called to give oral evidence at trial, it is submitted, although the rule and the practice direction do not make this clear, that both parties will have the opportunity to cross-examine him/her, but with a degree of restraint, given that the expert has been instructed by the parties.
>
> It must be a matter for the discretion of the judge whether oral examination of a single joint expert is appropriate. In a case where the single joint expert is dealing with major issues, such oral examination might be appropriate and proportionate. In such a case it is the practice of other TCC judges to whom I have spoken, and indeed of myself, for the judge to call the expert, and then for both sides to cross-examine. However, where the report of the single joint expert comes down strongly on the side of one party, it may be appropriate to allow only the other party to cross-examine.
>
> Before leaving the topic of single joint experts I wish to make four further comments:
>
> (1) The choice of single joint expert is important. He should be someone in whom both parties have confidence.
> (2) If the case is one in which it might become appropriate for the single joint expert to give oral evidence and be cross-examined, it is desirable to alert the expert to this possibility when he is invited to accept instructions.
> (3) Experience shows that quite often the instruction of a single joint expert leads to settlement of the whole litigation.
> (4) The procedure for dealing with single joint experts should, so far as possible, be addressed at case management hearings in advance of trial. Also provision should be made for securing payment of the fees of single joint experts before they undertake work.

The above guidance may be taken as a definitive statement of the approach to single joint **22.132** experts in construction cases today. As set out earlier above, the Pre-Action Practice Direction now indicates that the party seeking to appoint an expert (who may be an 'agreed' expert) should send a list of experts in the field of expertise, to which the other party may object.

Choice of experts

The use of experts is one of the most important aspects of construction litigation. First, the **22.133** expert must be chosen. This is the source of the most anxious consideration because the expert must be not only good in his or her chosen field but must also be good forensically. It is idle to pretend that forensic skill does not matter, but it is an overrated virtue. The most important attribute of any expert, aside from his professional integrity, is his or her mastery of the subject. The forensic 'game', which cross-examination has caused the giving of evidence to become, matters less than the consistency of the view the expert holds with his or her professional experience.

22.134 But the forensic skill does matter to this extent. Many experts find that they freeze in the witness box. They can be flattered by a skilful questioner, who rewards their favourable answer with praise and damns their unfavourable answer with a forensic buffeting, into making admissions which a pause for reflection would cause them not to make. There is no perfect balance when identifying the right expert for the case.

22.135 Difficulties with experts and their evidence has led to strict protocols as to how they are instructed and how they are to go about their business.

Duties of experts to court and client

22.136 The most striking requirement is that the expert is a servant not of his client but of the court.[110] The primary duty is to the court, and this confers with it immunities as well as responsibilities, including immunity from suit. That immunity, in addition to the normal witness immunity, extends to the giving of a report and actions in experts' meetings.[111] In one first instance case, *Phillips v Symes*, it has been held that the immunity does not relate to wasted costs orders being made against experts.[112] It is submitted that the decision is wrong and is inconsistent with *Stanton v Callaghan*,[113] which affirmed witness immunity. If, as *Stanton* decides, an expert receives witness immunity in relation to experts' meetings, it is inappropriate to tamper with that immunity on the basis that the claim is for wasted costs rather than damages.

---

[110] CPR r 35.3 and Experts' Protocol, para 4.1.

[111] See *Stanton v Callaghan* [2000] 1 QB 75. See particularly Chadwick LJ at 101–2:

> It is of importance to the administration of justice, and to those members of the public who seek access to justice, that trials should take no longer than is necessary to do justice in the particular case, and that, to that end, time in court should not be taken up with a consideration of matters which are not truly in issue. It is in that context that experts are encouraged to identify, in advance of the trial, those parts of their evidence on which they are, and those on which they are not, in agreement. Provision for a joint statement, reflecting agreement after a meeting of experts has taken place, is made by Ord 38, r 38. In my view, the public interest in facilitating full and frank discussion between experts before trial does require that each should be free to make proper concessions without fear that any departure from advice previously given to the party who has retained him will be seen as evidence of negligence. That, as it seems to me, is an area in which public policy justifies immunity. The immunity is needed in order to avoid the tension between a desire to assist the court and fear of the consequences of a departure from previous advice.

See also Otton LJ at 108: 'It is clear to me that in order to enable the court to arrive at the fair and just result in the way set out by Lord Morris of Borth-y-Gest in *Rondel v Worsley* [1969] 1 AC 191, 253 expert witnesses must be given full opportunity to uphold their duty to the court, and achieve it in the way set out by Cresswell J and Lord Woolf MR, in an atmosphere free from threats of suit from disappointed clients.' See also *Meadow v General Medical Council* [2006] EWHC 146 (Admin), [2006] 1 WLR 1452 applied in *Marc Tufano v Vincenti* [2006] EWHC 1496 (QB).

[112] *Phillips v Symes* [2004] EWHC 2330 (Ch) Peter Smith J at [95] said: 'It seems to me that the administration of justice, especially, in spite of the clearly defined duties now enshrined in CPR 35 and PD 35, it would be quite wrong of the Court to remove from itself the power to make a costs order in appropriate [sic] against an Expert who, by his evidence, causes significant expense to be incurred, and does so in flagrant disregard of his duties to the Court'.

[113] [2000] 1 QB 1.

It is a source of some regret to many and dismay to the courts that, in spite of the many strictures by judges over the years, a good proportion of experts see themselves as mouthpieces of their client and people whose obligation is to argue their client's case. This is the case, even today in spite of the fact that the duty is owed to the court and not the client. It is still a frequent criticism of experts that they become advocates for their party's cause rather than givers of objective evidence. **22.137**

The authoritative guidance as to the duties of experts is set out in *National Justice Cia Naviera SA v Prudential Assurance Co Ltd, The Ikarian Reefer*[114] where Cresswell J said: **22.138**

> *The duties and responsibilities of expert witnesses*
>
> The duties and responsibilities of expert witnesses in civil cases include the following:
>
> (1) Expert evidence presented to the Court should be, and should be seen to be, the independent product of the expert uninfluenced as to form or content by the exigencies of litigation (*Whitehouse v Jordan* [1981] 1 WLR 246 at p 256, per Lord Wilberforce).
> (2) An expert witness should provide independent assistance to the Court by way of objective unbiased opinion in relation to matters within his expertise (see *Polivitte Ltd v Commercial Union Assurance Co plc* [1987] 1 Lloyd's Rep 379, 386 *per* Garland J and *Re J* [1990] FCR 193 *per* Cazalet J). An expert witness in the High Court should never assume the role of an advocate.
> (3) An expert witness should state the facts or assumption upon which his opinion is based. He should not omit to consider material facts which could detract from his concluded opinion (*Re J* above).
> (4) An expert witness should make it clear when a particular question or issue falls outside his expertise.
> (5) If an expert's opinion is not properly researched because he considers that insufficient data is available, then this must be stated with an indication that the opinion is no more than a provisional one (*Re J* above). In cases where an expert witness who has prepared a report could not assert that the report contained the truth, the whole truth and nothing but the truth without some qualification, that qualification should be stated in the Report (*Derby & Co Ltd and Others v Weldon and Others* (1990) *The Times*, 9 November, *per* Staughton LJ).
> (6) If, after exchange of reports, an expert witness changes his view on a material matter having read the other side's expert's report or for any other reason, such change of view should be communicated (through legal representatives) to the other side without delay and when appropriate to the court.
> (7) Where expert evidence refers to photographs, plans, calculations, analyses, measurements, survey reports or other similar documents, these must be provided to the opposite party at the same time as the exchange of reports.

That statement was approved by the Court of Appeal[115] and has been substantially reproduced in the Practice Direction of the CPR relating to experts.[116] **22.139**

---

[114] [1993] 2 Lloyd's Rep 68 at 81.
[115] [1995] 1 Lloyd's Rep 455 at 496.
[116] See CPR PD 35.

Duties of experts: discussions between experts

22.140 The court has power to order discussions between experts, which can also be done by agreement.[117]

22.141 The purpose of the discussions is to narrow the issues if agreement cannot be reached. It is also to identify those issues where they agree and where they disagree and why, giving reasons and identifying action which may be taken to resolve any disagreement.[118] It is essential that this process is not truncated and it is clear that there is truly no agreement before a statement of disagreement is stated. Those instructing experts cannot tell the experts not to agree anything or indeed to defer agreement.[119]

Instruction of experts: the party-appointed expert

22.142 The instruction of an expert who is not objective is likely to backfire when he or she undergoes a proper forensic examination. So it is not in the interest of any party to instruct such an expert.

22.143 Argument and lack of objectivity usually is caught out in the forensic process and damning judicial comment is a stain on the expert's reputation which is impossible to eradicate. It may be thought that the reason for experts continuing to argue their client's cause rather than give impartial evidence is because it is profitable to do so. That view is too cynical. The most likely explanation is that the expert finds himself getting caught up so much in 'the cause' that the litigation becomes, that all objectivity ceases at some step along the way.

22.144 Instructions to an expert should require objectivity as a condition of the employment. The instructions are disclosable to the other side in the litigation.[120] It is hard to see what more the courts can do to ensure that experts are objective. Their report must refer to the guidelines of the *Ikarian Reefer* which are set out together with the relevant rules of court and state that they have understood them and will apply them in the giving of their evidence.

22.145 In summary: (a) the instructions to the expert are disclosable to the court, and (b) the expert must set out in his report that he understands the court guidelines governing expert evidence and has applied them in the drafting of his report.

22.146 The Experts' Protocol now gives specific guidance as to what should be contained in the instructions to experts.[121] In summary these requirements are:

(a) Basic information such as names, addresses, telephone numbers and the identity of all the parties.

(b) The nature and extent of expertise called for.

(c) The purpose of requesting the report; a description of the matters to be investigated.

(d) Pleadings, relevant disclosure, and witness statements.

117 CPR r 35.12 and Experts' Protocol, para 18.1.
118 Experts' Protocol, para 18.3.
119 See ibid, para 18.7 and see also the duties of independence.
120 CPR PD 35, para 4.
121 Experts' Protocol s 8.

(e) A statement of where the proceedings are, if they have been issued and whether they are contemplated.

(f) An outline programme for the completion and delivery of each stage of the expert's work.

(g) When proceedings have started the claim number, the track to which the claim has been allocated (in cases with which this book is concerned this will be almost always multi track in the TCC), the name of the court, the dates of hearings, and case management conferences or pre-trial reviews.

Instructions to the single joint expert

This is, of course, a cooperative effort.[122] **22.147**

The instructions should be agreed, but absent agreement, each party may give instructions. But differences in instructions ought to be as narrow as possible. Separate instructions should be copied to the other party or parties. However, the parties ought to agree: (a) what documents he or she must have, and (b) what assumptions he or she must make.[123] **22.148**

Fees must be dealt with openly and the liability of the parties jointly and severally ought to be made clear. If an order has been made limiting the amount of fees, then that should be stated.[124] **22.149**

Disclosure of instructions to experts

Instructions to experts have to be disclosed to the other party to the litigation.[125] The instructions should be in writing. **22.150**

Duties of experts on receipt of instructions: the single joint expert

All parties should be told of any material steps taken by the expert and all correspondence should be copied to all parties.[126] The report should be sent to all parties simultaneously.[127] **22.151**

In addition to the duty to the court, an expert must maintain independence, impartiality and transparency to all the parties.[128] **22.152**

All meetings or conferences should be joint unless agreed in writing otherwise.[129] **22.153**

The expert can seek instructions from the court.[130] **22.154**

---

122 In accordance with para 13.4.6 of the TCC Guide.
123 Experts' Protocol, paras 17.7 and 17.8.
124 Experts' Protocol, para 17.9.
125 CPR PD 35, para 4.
126 Experts' Protocol, para 17.11.
127 Ibid, para 17.15.
128 Ibid, para 17.12.
129 Ibid, para 17.13.
130 Ibid, para 17.14.

Duties of experts on receipt of instructions: the party-instructed expert

22.155 If the instructions received are not clear the expert should request clarification. They should not act unless the instructions are clear and they can indicate that this is the position if necessary.[131]

22.156 Experts should confirm without delay whether they accept the instructions. They should also inform those instructing them without delay if:

(a) Instructions are not acceptable because they are unclear, or ask for work outside their expertise or impose unrealistic deadlines.
(b) They consider the instructions are insufficient to complete their work.
(c) They become aware they may not be able to fulfil any of the terms of their employment.
(d) Their instructions or their work have placed them potentially in conflict with their duties as an expert.[132]
(e) They are satisfied that they cannot comply with any orders which have been made.

Rights of experts

22.157 Experts have the right to ask the court for directions, although such a request should normally be discussed with those instructing them first.[133]

22.158 The request for directions can include the situation where the expert has not been provided with the information which they require.

Factual witnesses as experts

22.159 It frequently happens in construction litigation that a witness of fact has a considerable level of expertise in one or other subject area (architect or engineer). Those witnesses are usually neither independent nor aware of the duties of experts to the court. It has been the practice in the TCC to receive that evidence as part of the narrative. Statements which are reasonably related to the facts within that witness's knowledge and relevant comments based upon that witness's experience will normally be received.[134]

---

[131] Ibid, para 8.2.

[132] In *Toth v Jarman* [2006] EWCA Civ 1028, [2006] 4 All ER 1276, the Court of Appeal stated that an expert should produce his CV when he provides his report, and that CV should give details of any employment or activity which raises a possible conflict of interest. Potential conflicts should be drawn to the attention of the court as early as possible and it was for the court, not the expert or instructing party, to decide whether there was a conflict or not.

[133] Ibid, s 11 and CPR r 35.14.

[134] *Multiplex Construction Ltd v Cleveland Bridge UK Ltd and another* [2008] EWHC 2220 (TCC). See the cases cited therein by Jackson J, *Lusty Architects v Finsbury Securities* (1991) 58 BLR 66, CA; and *DN v Greenwich LBC* [2004] EWCA Civ 1659, [2005] 1 FCR 112.

A View

22.160 At an early stage the parties should consider whether the trial judge would benefit from a view of the site. Often it is strictly unnecessary because an understanding of the site can be gained from photographs and plans and maps. However it is surprising how often the parties and the judges find that a view which at first seemed to be of marginal help was in fact of great assistance. This may be because there is nothing quite like 'boots on the ground' and sight of what is actually in issue as opposed to mere description.

22.161 The rules of views which are normally laid down are that what the judge sees is evidence, but what he hears from any of the parties, their representatives or experts, is not evidence. Whether or not the lawyers should accompany the judge on the view will be a matter of judgment, having regard to the size of the visiting party set against the size of the site and questions of cost. The advantage of having the lawyers there is that they see what the judge sees when s/he sees it and that is invaluable when the trial is under way and there has to be reference to what is on the site or what bits of machinery were around and where they were.

Disclosure

22.162 Disclosure is a key stage of the litigation process. It used to be that every piece of paper, however tangentially relevant, had to be disclosed and indeed was disclosed. Lists became long and unwieldy and the task of disclosure, then known as discovery, became a cottage industry in itself.

22.163 This almost indiscriminate disclosure was then followed on by equally indiscriminate copying of every document which would then end up in a trial bundle of overwhelming size which few were ever capable of reading, let alone absorbing and deploying.

22.164 Disclosure is now subject to a requirement of proportionality. A reasonable and proportionate search for relevant documents must be made, rather than an exhaustive search.

22.165 Once disclosed the documents must be listed. The provisions as to listing are set out in CPR r 31.10(3) which provides that the documents must be set out in a convenient order and manner and as concisely as possible. In smaller cases they will need to be numbered consecutively.

22.166 The important aspect of any listing, be it by file or by document, is that the documents must be readily identifiable upon an inspection or a re-inspection. This is subject to the caveat that sometimes this process becomes too burdensome and some compromise must be reached between a strict interpretation of the rules and common sense and cooperation.

22.167 In construction cases the documents will often be easily broken down into categories:

(a) Site meeting minutes of the main contract and applicable sub-contracts;
(b) Correspondence common to the parties;
(c) Personal files of the main people concerned with the project;

(d) Internal memoranda relevant to the project of the key personnel;
(e) Internal memoranda of the design team;
(f) Correspondence between main contractor and sub-contractors;
(g) Correspondence between the employer and its professionals;
(h) Drawings, specifications, and Bills of Quantity;
(i) Drafts of the abovementioned internal memoranda;
(j) Diaries of all key personnel involved in the project;
(k) Valuation and measurement documents.

22.168 These documents form easily definable categories which should be capable of being listed and made available for inspection in a methodical way. In the past complaints would be made that listing was by file, but this is not an acceptable complaint if the file falls into one of the categories identified above.[135] If listing by category becomes disproportionate, then directions from the court can be sought if there is no agreement as to what kind of list is acceptable. The position will rarely cause dispute in practice if both parties have to disclose a similar amount of documentation. However, if one party has little or few documents and the other has many, then the party with few documents may be tempted to press for a list which gives a greater breakdown of the documents disclosed than proportionality might permit. However, it must be remembered that the list ought to be so presented that, following inspection, any document can be identified for the purposes of making a request for a copy. So if a file is disclosed, it is good practice to paginate the documents. Thus a request can be made for, for example, 'File 56 page 301' after inspection.

Notice to Admit Facts

22.169 A party can serve a notice on another party seeking the admission of facts. This must be no later than twenty-one days before trial.[136] This is under-used. Getting all facts, so far as possible, agreed before trial saves enormous sums in costs. Strictly speaking, facts which are in issue can be identified from the pleadings, but after an action has proceeded for a while the non-admission of certain facts may be thought, by even the most unyielding opponent, unwise. So in any action, whether such a notice should be served and what is contained within it ought to be thought of carefully. The danger with such notices is a scattergun approach with all facts bound up in a notice. When that happens the notice will become ineffective.

Preservation and Sampling of Evidence

22.170 The court has power to order the preservation of property[137] and the taking of samples[138] including the taking of samples on someone else's property.[139]

135 CPR PD 31, para 3.2.
136 CPR r 32.18; for discretion as to costs see CPR r 44.3.
137 CPR r 25.1(1)(c)(i).
138 CPR r 25.1(1)(c)(iii).
139 CPR r 25.1(1)(d).

Normally questions of preservation and sampling of properties are agreed. This may be because the power the court has is so clear and the need for the property and samples are obvious. This is particularly the case where the trial judge will be the person hearing any opposed application. The poor impression which resistance conveys is the best disincentive to obstruction. Some tests may be onerous, however. A load test on a roof to check that it can take a specified weight may be difficult to have ordered since there may be a risk of destruction. Non-destructive testing has little potential for dispute. However, tests which may involve damage to property will be more difficult to order. It is submitted that if there is a real risk of significant damage to property, then unless that damage can be readily compensated for by money a test will not be ordered. However, the ordering of tests is a matter for the discretion of the judge. **22.171**

Summary Judgment

The court may grant summary judgment without a trial.[140] It is often a difficult decision for the practitioner as to whether to seek summary judgment or a preliminary issue. The problem with summary judgment is that courts of the less robust kind will permit the matter to go to trial on the grounds that the full facts might justify the position of either the claimant or the defendant. Other courts might, with the same set of facts, give summary judgment. The safer course, when the questions at issue are substantially of law, is to seek the trial of a preliminary issue. The costs of an abortive summary judgment application are often prohibitive. Summary judgment is now comparatively rare in construction cases (aside from adjudication enforcement which is in its own category). The rise of adjudication has assisted in giving the cash flow squeezed sub-contractor a quicker remedy. **22.172**

Offers to Settle; Payments into Court and Costs

The costs of litigation are ruinous. By the time any case (save those of the highest value) has proceeded even a short distance the costs will come to match or exceed the amounts in issue. The result of this has been many judicial dicta despairing of the lack of proportion of parties going to litigation.[141] **22.173**

Part 36 Offer by a Claimant

Therefore the courts encourage settlements. One startling mechanism is the 'Part 36 Offer' by a claimant. A claimant can offer to settle its claim for a sum less than claimed. Provided the offer is made in proper form, then should the claimant succeed in recovering more than the offer, the claimant will be entitled to recover interest on any sum recovered up to 10 percent above base.[142] The claimant will also be entitled to indemnity costs. **22.174**

---

140 CPR Pt 24.

141 See, for example, Ward LJ in *Carver v BAA Plc* [2008] EWCA Civ 412 at para 35 of the judgment saying 'To have incurred about £80,000 in costs to contest a claim under £5,000 fills one with despair'.

142 CPR r 36.21.

22.175 If proceedings have not begun and the matter is in the pre-action protocol stage then a prospective defendant can make an offer to settle and if he or she does so the offer will be taken into account when making any order as to costs.

Part 36 Offer

22.176 The most common technique for defendants in litigation, once begun, to reduce costs exposure is to make an offer of settlement pursuant to CPR Part 36. That procedure superseded the procedure for a payment into court after April 2007. If then the claimant does not accept the offer within twenty-one days and following a trial it recovers significantly less than the amount offered (in the sense that it is worse off than if it had accepted the offer) then the defendant will be entitled to recover all its costs of defending the action from the time the offer was made. Under the old procedure money had to be paid into court. As stated above that has now been abolished and defendants must make offers to settle. However, there has been a substantive change in how these offers are looked at. It used to be the position that if the claimant recovered a penny more than the offer, then it would recover all of its costs from the 'unsuccessful' defendant, notwithstanding that the claimant continued its fight simply to recover a penny more than what it was offered. That is no longer the position. The court will look at matters in the round and the decision of the Court of Appeal in *Carver v BAA Plc* [2008] EWCA Civ 412 is instructive. In that case the claimant recovered £51 more than what had been offered. Although this was a recovery of more than what was offered, that sum was more than off-set by the irrecoverable costs of pursuing the action and the claimant had to pay the costs incurred since that offer was made. In relation to an earlier offer which was £400 less than recovered, the first instance judge made no order for costs and had regard to the general conduct of the parties. This was upheld by the Court of Appeal[143] applying CPR 36.14.

Defendants' offer otherwise than a Part 36 Offer

22.177 If the case does not involve a money claim, then the offer will be taken into account. If the offer puts the claimant in a better or identical position than obtained following a fought trial, then the claimant which rejected the offer will normally have to pay the costs of the action from the date it was made.[144] The rationale behind this is that the trial was futile because the claimant could have obtained what he or she wanted without recourse to legal process.

22.178 The CPR allows the court to take into account offers to settle which are not made in accordance with the CPR[145] but the CPR give no guidance as to how that discretion is to be exercised. The judicial statements in *Carver* suggest that a broad and non-technical approach will be taken, and the primary question will relate to how reasonable the offer was in substance and how reasonable it was for the claimant to reject it and continue.

---

[143] This decision was applied by Jackson J in *Multiplex Construction (UK) Ltd v Cleveland Bridge and another* [2008] EWHC 2280. In that case an issues based costs order was not made but a proportionate based order was made. That case is a classic example of construction litigation fighting to a standstill with victory pyrrhic and costs ruinous and disproportionate.

[144] See *Calderbank v Calderbank* [1976] Fam 93.

[145] CPR r 36.1(2).

If an offer is withdrawn after the twenty-one-day period, when it has lapsed anyway, the court will still exercise its discretion and order the party which did not 'beat' the offer to pay the costs of the action.[146] **22.179**

Costs: do costs follow the event and the amounts of costs recoverable? When can they be recovered?

The normal rule in litigation is that costs 'follow the event' or rather the unsuccessful party pays the costs of the successful party.[147] That normal rule can be departed from and the court has a discretion as to whether or not to do so. The circumstances of such departure include: (a) the parties' conduct, (b) whether success has only been as to part, and (c) whether any offer has been made under Part 36 of the CPR. **22.180**

Previously, unless the circumstances were exceptional, if a claimant recovered a significant sum of money, then it recovered all its court-assessed costs. Still the test can be expressed as simply in many cases. A party can say 'this is a litigation, and the simple test of who is the winner is to ask: Who has to pay/who has to write out the cheque?', relying on *Day v Day* [2006] EWCA Civ 415, paras 16–17, and *Tamares (Vincent Square) Ltd v Fairpoint (Vincent Square) Ltd* [2007] EWCA Civ 1309, paras 15 and 31. **22.181**

The current law has been usefully summarized in *HLB Kidsons (a firm) v Lloyd's Underwriters subscribing to Lloyd's Policy No 621/PKID00101 and others* [2007] EWHC 2699 (Comm), *per* Gloster J, at paras 10 and 11: **22.182**

> [10] . . . The court's discretion as to costs is a wide one. The aim always is to 'make an order that reflects the overall justice of the case' (*Travellers' Casualty v Sun Life* [2006] EWHC 2885 (Comm) at paragraph 11 per Clarke J. As [Counsel] submitted, the general rule remains that costs should follow the event, ie that 'the unsuccessful party will be ordered to pay the costs of the successful party': CPR 44.3(2). In *Kastor Navigation v Axa Global Risks* [2004] EWCA Civ 277, [2005] 2 All ER (Comm) 720, [2004] 2 Lloyd's Rep 119, the Court of Appeal affirmed the general rule and noted that the question of who is the 'successful party' for the purposes of the general rule must be determined by reference to the litigation as a whole; see para 143, per Rix LJ. The court may, of course, depart from the general rule, but it remains appropriate to give 'real weight' to the overall success of the winning party: *Scholes Windows v Magnet (No 2)* [2000] ECDR 266 at 268. As Longmore LJ said in *Barnes v Time Talk* [2003] EWCA Civ 402, [2003] BLR 331 at para 28, it is important to identify at the outset who is the 'successful party'. Only then is the court likely to approach costs from the right perspective. The question of who is the successful party 'is a matter for the exercise of common sense': *BCCI v Ali (No 4)* 149 NLJ 1222, per Lightman J. Success, for the purposes of the CPR, is 'not a technical term but a result in real life' (*BCCI v Ali (No 4)* (supra)). The matter must be looked at 'in a realistic . . . and . . . commercially sensible way': *Fulham Leisure Holdings v Nicholson Graham & Jones* [2006] EWHC 2428 (Ch) at para 3 per Mann J.
>
> [11] There is no automatic rule requiring reduction of a successful party's costs if he loses on one or more issues. In any litigation, especially complex litigation such as the present case, any winning party is likely to fail on one or more issues in the case. As Simon Brown LJ said in *Budgen v Andrew Gardner Partnership* [2002] EWCA Civ 1125 at para 35: 'the court can

[146] *Trustees of Stokes Pension Fund v Western Power Distribution* [2005] EWCA Civ 854, [2005] 1 WLR 3595 at [41] and [42] per Dyson LJ.

[147] CPR r 44.3(2)(a).

properly have regard to the fact that in almost every case even the winner is likely to fail on some issues'. Likewise in *Travellers' Casualty* (supra), Clarke J said at para 12:

> If the successful Claimant has lost out on a number of issues it may be inappropriate to make separate orders for costs in respect of issues upon which he has failed, unless the points were unreasonably taken. It is a fortunate litigant who wins on every point.

22.183 The courts can make issue based orders reflecting the relative success or failure of the parties on the issues raised, in particular, where it is difficult to identify the 'winner' of the litigation.[148] The discretion is in the judge and that discretion is wide. But it is often difficult to ascribe particular costs to particular issues, particularly when there are degrees of overlap. Further, it has been stated in the TCC (following Simon Brown LJ in *Budgen* cited above) that issue based costs should not be confused with the fact that a party invariably wins some points but not others. Coulson J suggested in *Murphy & Sons Ltd v Johnston Precast Ltd (Formerly Johnston Pipes Ltd) (No 2 Costs)*[149] that, where for example, a defendant took ten points and was successful on seven of them, such that the claim against him failed, the claimant should only be entitled to an issue-based costs order if the three points on which the defendant was unsuccessful made a material difference to the length of the trial, and therefore the costs incurred on both sides. In the event of any difficulty, therefore, the court can order that the costs judge sort out the quantum of costs and make a simple issue based order. However, it is submitted that such a course is often disproportionate and the better course is for the judge to make an order that a successful party only recovers a percentage of its costs, such a reduction to reflect the lack of success on some issues and all other relevant factors. This view has support in *Dyson Technology Ltd v Ben Strutt* [2007] EWHC 17856 (Chan), where Patten J said:

> There was much to be said for the application of the general rule that costs should follow the event, and for keeping to the simple formula of orders for a stated proportion of the costs or a stated amount of costs in cases where recognition of a limited degree of success by one or the other party was called for.

It is therefore unwise for a party to take a point or an issue which is likely to take up a significant proportion of the court's time merely in the hope that the costs will 'come out in the wash' of the overall costs of the litigation. Equally it is prudent for a defending party to keep issue based costs records, if possible, if it sees that a significant proportion of the costs is likely to be spent on an issue which it will win.

Costs in the event of a settlement

22.184 Occasionally parties will start a case in the courts and then attempt a compromise. In the course of a compromise they will often leave out the question of costs and ask the court to determine the question. The courts are then presented with the tension between seeking to encourage compromise and trying to avoid making an agreement for parties who have not in fact agreed a proper settlement. Sometimes the court is driven to find that there has not

[148] *McGlinn v Waltham Contractors Ltd (No 4)* [2007] EWHC 698 (TCC), 112 Con LR 148.
[149] [2008] EWHC 3104 (TCC) at [10]–[11].

been a complete agreement. Usually it will seek to ensure that finality is reached and give effect to settlement.

Thus in *Brawley v Marczynski* [2002] EWCA Civ 756, [2003] 1 WLR 813; Longmore LJ, at paragraph 21, endorsed two principles for this situation (set out in a judicial review case):

> (iv) At each end of the spectrum there will be cases where it is obvious which side would have won had the substantive issues been fought to a conclusion. In between, the position will, in differing degrees be less clear. How far the court will be prepared to look into the previously unresolved substantive issues will depend on the circumstances of the particular case, not least the amount of costs at stake and the conduct of the parties.
>
> (v) In the absence of a good reason to make any other order the fall back is to make no order as to costs.

This dictum was applied by the Court of Appeal in *Bray (t/a the Building Company) v Bishop & Anor* [2009] EWCA Civ 768.[150]

Standard basis costs

The usual order for costs is that they be paid on the 'Standard Basis'.[151] On a standard basis the court will only order costs which are reasonably incurred and proportionate to the matters in issue and will resolve any doubt about reasonableness and proportionality in favour of the paying party. When the costs are to be paid on an indemnity basis any doubt as to whether the costs are reasonably incurred will be resolved in favour of the receiving party, but there is no requirement of proportionality.[152] There is no exhaustive guidance as to when the court will make a costs order on an indemnity basis.[153] The circumstances in which such an order is to be made are wide. The authorities suggest that the award on an indemnity basis is for cases where the court wishes to indicate its disapproval of the conduct of the paying party.[154] Relevant factors include the advancement of a speculative case or the aggressive pursuit of unjustified claims of dishonesty.[155] However, the grant of indemnity costs is not the only way in which unreasonable conduct may be sanctioned. Where, for example, a successful claimant has been dishonest in relation to an issue, the courts may instead disallow a part of its costs or order that he pay the other party's costs attributable to **22.185**

[150] See also *BCT Software Solutions v C Brewer & Sons* [2003] EWCA Civ 939; and *Venture Finance Plc v Mead & Ors* [2005] EWCA Civ 325 cited therein; and *Promar International Ltd v Clarke* [2006] EWCA Civ 332.

[151] CPR r 44.4.

[152] In *Fourie v Le Roux* [2007] UKHL 1, [2007] 1 WLR 320, Lord Scott suggested obiter in the House of Lords that the distinction between standard and indemnity costs was not, according to the language of the relevant rules, very great. Lord Scott stated (at [39]) that the only relevant difference appeared to be that, where the indemnity basis had been ordered, the onus lies on the payer to show any unreasonableness: he considered further that the criterion of proportionality seemed to add very little to the reasonableness criterion. There are differing accounts as to the extent to which the result is dissimilar in practice.

[153] See *Munkenbeck and Marshall v McAlpine* 44 Con LR 30 where the Court of Appeal expressly declined to give guidance.

[154] See *Reid Minty v Gordon Taylor* [2002] 1 WLR 2800; and *Brawley v Marczynski* [2003] 1 WLR 813 at [13].

[155] See *Three Rivers District Council v Bank of Credit and Commerce International SA* [2006] EWHC 816 (Comm), which contains a general review of the authorities.

proving that dishonesty.[156] The Jackson Report suggests that the TCC may refuse costs in relation to prolix or unnecessary pleadings or witness statements.[157] The normal order is an order on a standard basis; therefore there must be something in the case to take it out of the ordinary.[158] Merely losing a case does not take it out of the ordinary. At least half of all litigants lose their cases. However, pursuing a case which is doomed to fail is another matter and abandoning the position and compromising the claim in the run-up to trial may be too little too late. In two TCC cases, *Wates Construction Ltd v HGP Greentree Allchurch Evans Ltd*,[159] and *EQ Projects Ltd v Alavi (t/a Merc London)*,[160] claims were abandoned at or close to trial. The court awarded indemnity costs from the date when it was held that the losing party should have realised its case was bound to fail.

22.186 The CPR provide that, unless the court considers it unjust to do so, indemnity costs will be ordered from the date of expiry of the period of the Part 36 Offer, if a claimant's Part 36 Offer is not accepted and the claimant achieves a result in terms at least as advantageous as that offer: this applies also to a claimant's Part 36 Offer which suggests settlement as a proportion of the claim.[161] This will apply as well if the Part 36 Offer is made pre-action.[162] Similar rules apply where the defendant's Part 36 Payment is not accepted and the claimant fails to obtain a judgment more advantageous than that offer.[163] In such circumstances (by contrast with the position in respect of indemnity costs generally), the award of indemnity costs does not necessarily imply any criticism of the paying party's conduct.[164] When the claimant's Part 36 Offer is not accepted the court can also order interest on those costs and also on the principal sum in issue at a rate up to 10 percent above base rate.[165] At that stage, the court shall consider all the circumstances of the case, including the conduct of the losing party and there is therefore scope for the award of interest on a punitive and not merely compensatory basis.[166]

22.187 There is a potential lacuna, being the situation where a party accepts the other's Part 36 out of time but before judgment. CPR r 36.10 states that an offer can be accepted out of time but is silent as to the basis on which costs should be ordered. In *Fitzpatrick Contractors Ltd v Tyco Fire and Integrated Solutions (UK) Ltd*,[167] Coulson J considered such a situation, in

---

[156] *Bank of Tokyo-Mitsubishi UFJ Ltd v Baskan Gida Sanayi Ve Pazarlama AS* [2009] EWHC 1696 (Ch) at [15]–[19] *per* Briggs J.

[157] The Jackson Report, Vol 1, ch 34, paras 5.6 to 5.9.

[158] See *Excelsior Commercial and Industrial Holdings v Salisbury Hammer Aspden & Johnson* [2002] EWCA Civ 879.

[159] [2005] EWHC 2174 (TCC), (2005) 105 Con LR 47.

[160] [2006] EWHC 29 (TCC), [2006] BLR 30.

[161] CPR r 36.14(1)(b), (2) and (3). See also *Read v Edmed* [2004] EWHC 3274 (QB), *The Times*, 13 December 2004, a decision under the previously worded CPR r 36.21 but one which the TCC accepted applied to the new regime under CPR 36.14: see *Fitzpatrick Contractors Ltd v Tyco Fire and Integrated Solutions (UK) Ltd* [2009] EWHC 274 (TCC), 123 Con LR 69, [2009] BLR 144.

[162] *Huck v Robson* [2002] EWCA Civ 398, [2003] 1 WLR 1340.

[163] CPR r 36.14(1)(a).

[164] *McPhilemy v Times Newspapers No 2* [2001] EWCA Civ 933, [2002] 1 WLR 934 at [9] *per* Chadwick LJ.

[165] CPR r 36.14(2) and (3).

[166] CPR r 36.14(4); *Hammond Suddards (a firm) v Jebb* [2006] EWCA Civ 777.

[167] [2009] EWHC 274 (TCC), 123 Con LR 69, [2009] BLR 144.

which the Part 36 Offer had been accepted almost a year after the offer was made and only shortly before trial. He held that there was no presumption that indemnity costs would be ordered and, on the contrary, indicated that, unless the losing party's conduct brings CPR r 44.3 into play, costs would be ordered on the standard basis.

Where the court has ordered a split trial, should costs be dealt with at the end of the first substantive hearing or only at the end of the proceedings? The Court of Appeal has indicated that, in the case of a split trial on liability and quantum, it would generally be safer to defer consideration of costs until the final result is known.[168] Disputes in the TCC may, however, involve multiple defendants, each having played a different role on a building project and, where the issues in relation to such roles differ, a separate hearing may be ordered to address them, following which a costs order may be made without awaiting the further hearing.[169] **22.188**

Whatever the basis on which costs are ordered, the court may order interest to be paid on costs.[170] The purpose of awarding interest is to enable the court to ensure that the receiving party is compensated properly for the real cost to it of having conducted the litigation successfully.[171] As least in recent years, the custom appears in commercial cases to award interest at 1 percent above base rate.[172] **22.189**

Costs can also be pre-emptively capped.[173] The CPR provides for this[174] but it has not been a facility used much in construction cases. However where the justice of the case requires it, there can be costs capping. If, for example, there is a heavy over-estimate of costs in the allocation questionnaire then that might well trigger off an application. In libel cases the use of conditional fee arrangements has triggered off applications for costs capping. However in construction cases there is often a disproportionate amount of costs expenditure (including in one case in the Court of Appeal costs on each side in a two day case on a point of law argued fully below were £150,000) and there is no reason why protective steps should not be taken to ensure that the costs of an action are properly proportionate and are seen to be so. The Jackson Report suggests that the court might cap the costs of each stage of the litigation process. So when giving directions for disclosure, service of witness statements, service of expert reports, the court would attach a price tag to each activity.[175] **22.190**

---

168 *Weill v Mean Fidler* [2003] EWCA Civ 1058.

169 *McGlinn v Waltham Contractors Ltd* [2007] EWHC 294 (TCC), [2007] BLR 188.

170 CPR r 44.3(6)(g).

171 *Media Ltd v Highbury Leisure Publishing Ltd* [2005] EWHC 283 (Ch) at [11]–[12] per Laddie J; see also *McPhilemy v Times Newspapers No 2* [2001] EWCA Civ 933, [2002] 1 WLR 934 at [23] *per* Chadwick LJ.

172 See, for example, *ABCI (formerly Arab Business Consortium International Finance and Investment Co) v Banque Franco-Tunisienne* [2003] EWCA Civ 205, [2003] Lloyd's Rep 146; *HLB Kidsons (a firm) v Lloyd's Underwriters subscribing to Lloyd's Policy No 621/PKID00101* [2007] EWHC 2699 (Comm), [2007] All ER (D) 341 (Nov); *Fitzpatrick Contractors Ltd v Tyco Fire and Integrated Solutions (UK) Ltd* [2009] EWHC 274 (TCC), 123 Con LR 69, [2009] BLR 144.

173 *King v Telegraph Group Ltd* [2005] 1 WLR 2282.

174 CPR r 44.18.

175 The Jackson Report, Vol 2, ch 48, para 3.21. Although Jackson LJ suggested that this approach may possibly be welcomed by litigants in the TCC (ibid, para 3.23), the TECBAR response has been negative (see para 8).

Judgments and Appeals and Reasons

22.191 Courts must give reasons for their decisions[176] so that the parties know why the winner has won and the loser has lost. That does not mean that the reasons have to be elaborate or exhaustive, but they must be set out in sufficient detail to show the principles upon which the judge has acted and the reasons which have led him or her to the decision made.[177] Whilst a judgment ought to be comprehensible on a first reading, the adequacy of the reasons is not judged by that test. It must be tested in the context of the knowledge and understanding of the evidence and submissions at the trial.[178] If a judgment does not give proper reasons (as, for example, as to why particular witnesses of fact are to be disbelieved) then the matter will be remitted for a re-trial.[179]

22.192 With regard to expert evidence, it is not sufficient to simply say that one expert's evidence is preferred over the other. There should be an explanation as to why one expert's evidence is accepted and the other's is not. If the judge does not give sufficient reasons, then there should be an application to the trial judge for reasons to be given, and the judge can then consider whether additional reasons can be given.

22.193 If an application for leave is made to the Court of Appeal, then the court may remit it to the trial judge with an invitation to provide additional reasons for the decision. However the Court of Appeal will not allow an appeal for inadequate reasoning if it is possible to divine the reasoning by an examination of the evidence and submissions, including the submissions of the parties on the appeal. Therefore it is not wise to seek to appeal for lack of reasons unless it truly is the case that the reasoning is lacking and cannot be found from that examination.[180] In that event, even though the judge will not have done his or her duty, the appeal will be dismissed.

22.194 The need for reasons has always been part of the common law,[181] but it has become acute and brought into focus with the incorporation of Art 6 of the European Convention of Human Rights. Nevertheless, the requirement for reasons should not be taken too far. As Longmore LJ said in *Aerospace Publishing Ltd & Anor v Thames Water Utilities Ltd* [2007] EWCA Civ 3:

> In a complex factual case such as the present it will often be comparatively easy for an appellant to allege that a judgment is imperfectly or inadequately reasoned on one aspect or

---

[176] *Flannery v Halifax Estate Agencies Ltd* [2000] 1 WLR 377.

[177] See *Battista v Bassano* [2007] EWCA Civ 370, [2007] All ER (D) 111 (Feb), a building case.

[178] *Harris v CDMR Purfleet Ltd* [2009] EWCA Civ 1645. In that case leave to appeal had been granted on the basis that the reasons given by the judge appeared to be inadequate, see [2008] EWCA Civ 758. On the hearing of the full appeal the reasons were held to be adequate in the light of the court's knowledge and understanding of the evidence and submissions at trial, *English v Emery Reimbold & Strick Ltd* [2002] EWCA Civ 605, [2002] 1 WLR 2409 applied.

[179] For a stark example of this in action, see *Re F (Children)* [2006] EWCA Civ 792.

[180] *English v Emery Reimbold & Strick Ltd* [2002] 1 WLR 2409.

[181] *R v Knightsbridge Crown Court ex parte International Sporting Club (London) Ltd* [1982] QB 304; *R v Harrow Crown Court ex parte Dave* [1994] 1 WLR 98; *Flannery v Halifax Estate Agencies Ltd* [2000] 1 WLR 377.

> another and even to persuade this court, on an unopposed permission application, that that is arguably so. Appellants must, however, be aware that there is no obligation on a judge to give a particular response to every submission made (judgments in this country are quite long enough already) and that, unless it becomes apparent in the course of the appeal that a serious injustice has been done, appeals on the ground of inadequacy of reasons in complex factual disputes are likely to fail. One shudders to think what the costs of this appeal have been and it is important to emphasize that the fact that one division of this court gave permission to appeal is very far indeed from being any guarantee of success.

Reference should also be made to *West v Wilkinson & Anor* [2008] EWCA Civ 1005, a road traffic case where no reasons for the apportionment of liability were given. The Court of Appeal did not interfere with the apportionment even though no reasons were given because there was material before the judge to permit him to make the apportionment he did. Lack of reasons may give rise to leave to appeal being granted, but not necessarily being allowed. In that case the court said that leave to appeal was rightly granted because of the lack of reasons but it dismissed the appeal on the grounds stated. **22.195**

Permission is needed from either the judge or the Court of Appeal and that will normally only be granted if there is a good arguable case that the judge was wrong. Appeals from TCC judges are hard to mount on questions of fact because of the high margin of appreciation given to the fact finder,[182] given the complexity of the cases and the specialist nature of the tribunal. Lord Bingham MR in *Virgin Management Ltd v De Morgan Group*[183] said: **22.196**

> The reason why rights of appeal against the factual decisions of official referees have been subject to special conditions and restrictions is obvious. Official referees were originally appointed to conduct protracted and painstaking enquiries into complex and detailed factual issues. The investigation of multifarious claims and cross claims arising out of construction contracts is one example. The taking of accounts is another. The assessment of damages following an established breach of contract is yet another. We agree with Waller LJ when he said, giving the judgment of the Court in *Moody v Ellis* (1983) 26 BLR 39 at page 46, with reference to the old rule which only allowed factual appeals relating to fraud and breach of professional duty,
>
> > In our judgment the policy behind the rule is clear. In the ordinary way the official referee's decision on fact is to be final. He had to go into very complicated and detailed factual questions which take a great deal of time. As a matter of policy, when an official referee, with his great experience in such matters, makes a decision, it should ordinarily be final. . . .
>
> Whether a ground will be regarded as having a reasonable prospect of success will of course depend on what the point is. A prospective appellant will have difficulty in showing that the test is satisfied if he is seeking to challenge:
>
> (1) an official referee's findings of primary fact based on his evaluation of oral evidence, not because of any peculiar immunity enjoyed by official referees but because of the weight invariably given to the trial judge's factual conclusions unless and until they are shown to be wrong;

[182] See *Yorkshire Water Services Ltd v Taylor Woodrow Construction Northern Ltd* [2005] EWCA Civ 894 at [29] and cases cited therein.

[183] (1994) 68 BLR 32 at 33–35.

(2) the fine detail of an official referee's factual investigation: while the official referee's findings of this nature are no longer final, the Court of Appeal will not readily enter upon an enquiry of this kind;
(3) findings of fact falling within an official referee's area of specialized expertise, particularly if the official referee has had the advantage of inspecting the site or the subject matter of the dispute in question.

We do not suggest that these are no-go areas, only that the burden on a prospective appellant in these areas will be hard to discharge. . .

Leave to appeal on questions of fact should not, however, be granted simply because an appeal on legal grounds is proceeding anyway. Applications for leave should always be made and considered with reference to specific issues of fact. They should not be granted unless it appears that the challenge, if successful, would be likely to affect the official referee's overall decision . . .[184]

22.197 It is therefore not enough to point out error; the error must be material to the outcome and be a true error and a finding which the TCC judge was not entitled to make, rather than a weighing of the evidence leading to a result that another judge would not have arrived at. Furthermore the courts will not look too closely at a judgment. Even if there are small errors in the judge's detailed reasons, that will not be sufficient to overturn the judgement when the thrust of the judge's findings are clear.[185] The law in relation to what might be described as 'primary' facts and 'secondary' facts on appeals from the TCC is different from other courts in that there is not the distinction which is normally made between the two so far as TCC judgments are concerned. In other courts the Court of Appeal has a relatively free rein in dealing with inferences because the Court of Appeal is in as good a position as the trial judge to make the evaluation. However this is not the case in fact heavy TCC cases where the difference between the two is hard to evaluate.[186]

[184] For the importance of this point see particularly *Thomson v Christie's* [2005] EWCA Civ 555 at [141] where May LJ (with whom Smith and Jonathan Parker LJJ agreed) said: 'But, even accepting that individual points such as these are amenable to judicial appellate evaluation whatever the expert opinion, no appellate court should cherry pick a few such points so as to disagree with a composite first instance decision which, in the nature of a jig-saw, depended on the interlocking of a very large number of individual pieces, each the subject of oral expert evidence which the appellate court has not heard.' The relevant fact must be evaluated for its importance in the jigsaw.

[185] See, for example, *McLoughlin v Jones and others* [2006] EWCA Civ 1167 where a detailed attack failed.

[186] *Yorkshire Water Services Ltd v Taylor Woodrow Construction Northern Ltd* [2005] EWCA Civ 894 at [30] where May LJ (returning to the jigsaw theme) said:

> The technical complexity of the present case is well beyond the practical judicial experience of most judges. To talk in this appeal about primary facts and inferences, the credibility of oral evidence and other matters normally relevant to an appeal against relatively uncomplicated findings of fact has something of an air of unreality about it. The well-known principles are of course relevant and to be applied. But no appellate court would solemnly set about deciding in this case which of the judge's findings were findings of primary fact and which inferences. A broader approach would of necessity be needed. Most of the evidence was technical and complex. Most of the witnesses, including those giving factual evidence, were experts. Credibility was scarcely in issue in judging who was telling the truth. But credibility was centrally in issue in the sense of judging which strands of evidence, opinion and argument were persuasive. It was here that the judge had the unique advantage both of hearing the evidence, opinion and argument expressed, and of being able, through becoming steeped in the subject, to put together in 557 paragraphs of dense technical prose the interlocking pieces of a highly complicated jigsaw. Understanding the jargon is an effort in itself.

Potential Costs Reform: The Jackson Report

Specific areas of possible reform in respect of TCC business suggested by the Jackson Report have been identified above. The report also raises broader issues in relation to litigation costs. The most fundamental relate to the possibility of departing from the '100 percent principle', under which a claimant's award of damages would be reduced by irrecoverable but reasonably incurred costs. Jackson LJ notes that 100 percent damages recovery is largely taken for granted in England and Wales, but is a rarity elsewhere.[187] As shown above, the ability to penalize specified misconduct already exists under the CPR but the wholesale departure from the '100 percent principle' would be a dramatic change. Another potential reform suggested is to the principle of 'loser pays', under which the costs are generally borne by the unsuccessful party. The Jackson Report suggests that liability for costs might be based not on success but on the conduct and reasonableness of the parties, in particular, in relation to whether reasonable attempts have been made to settle the proceedings or, alternatively, on the basis that the claimant has no general liability for defendant's costs unless and until a defendant had made a Part 36 Offer, but from that point the current entitlements and costs sanctions would apply.[188] Finally, there is the prospect of expanding the fixed costs regime to cover multi-track cases (including TCC cases) involving small or medium size enterprises (SMEs) and 'small business disputes'.[189] That could mean introducing fixed costs to almost all TCC litigation for under the definition of SMEs proposed it would appear that over 98 percent of all UK companies would fall within that category.[190] This may be a somewhat surprising change, given that the Jackson Report states that TCC litigation is already generally conducted in a proportionate manner.[191] What is certain is that the drive to keep costs proportionate to the issues at stake is going to be of continuing importance. **22.198**

[187] The Jackson Report, Vol 2, ch 46, para 2.2.

[188] Ibid, Vol 2, ch 46, s 6.

[189] Ibid, Vol 1, ch 23, para 3.8; ch 29, para 1.3 and s 3.

[190] Ibid, Vol 1, ch 29, paras 3.1 to 3.3.

[191] Ibid, Vol 1, ch 34, para 3.2. The Bar Council Response is unfavourable, citing the broad definition of SMEs and the uncertainty as to defining a 'small business dispute': see paras 62 to 67.

23

ADJUDICATION

Introduction

Unsurprisingly, the courts have had to deal with a great many cases where parties have sought to enforce decisions of adjudicators. This started with a large number of decisions at first instance where there was a great deal of discussion of the basis upon which decisions are enforced or not enforced. **23.01**

There have been only a few decisions of the Court of Appeal. However, as the general approach of the courts has become clearer, the flow of cases has reduced. Not all the first instance decisions can be dealt with in this chapter. Many turn on their own facts and, it is true to say, are not necessarily rightly decided. Different judges have placed differing emphasis on the legislation and its effect. It is convenient to look at the purpose of the legislation and then to examine the principal decisions of the courts and the practical implications of those decisions. **23.02**

The relevant statutory provisions are ss 104 to 117 of the Housing Grants, Construction and Generation Act 1996 ('the 1996 Act') and SI 1998/649, being the Scheme for Construction Contracts (England and Wales) Regulations 1998 ('the Scheme'). These are set out in Appendices 1–3. The passing of the Local Democracy, Economic Development and Construction Act 2009 in November 2009 has meant that some of the provisions of the 1996 Act may be subject to change. Appendix 1 contains the unamended 1996 Act, which is still unamended at the time of going to press and may remain so. Should the Secretary of State bring into force the provisions of the Local Democracy, Economic Development and Construction Act 2009, then the amended provisions can be seen in Appendix 3. **23.03**

The 1996 legislation arose from the report of Sir Michael Latham, who in July 1993, was asked to undertake a review of procurement and contractual arrangements in the construction **23.04** industry in the United Kingdom. One of the more important of the recommendations was that there should be a speedier and cheaper means of resolving construction disputes. This was long overdue given the lengthy and expensive path which had to be trod in all construction disputes. Dyson J (as he then was) in *Macob Civil Engineering Ltd v Morrison Construction Ltd*[1] at paragraph 14 said:

> The intention of Parliament in enacting the Act was plain. It was to introduce a speedy mechanism for settling disputes in construction contracts on a provisional interim basis, and requiring the decisions of adjudicators to be enforced pending the final determination of disputes by arbitration, litigation or agreement: see s 108(3) of the Act and paragraph 23(2) of Part 1 of the Scheme. The timetable for adjudications is very tight (see s 108 of the Act). Many would say unreasonably tight, and likely to result in injustice. Parliament must be taken to have been aware of this. So far as procedure is concerned, the adjudicator is given a fairly free hand. It is true (but hardly surprising) that he is required to act impartially (s 108(2)(e) of the Act and paragraph 12(a) of Part 1 of the Scheme). He is, however, permitted to take the initiative in ascertaining the facts and the law (s 108(2)(f) of the Act and paragraph 13 of Part 1 of the Scheme). He may, therefore, conduct an entirely inquisitorial process, or he may, as in the present case, invite representations from the parties. It is clear that Parliament intended that the adjudication should be conducted in a manner which those familiar with the grinding detail of the traditional approach to the resolution of construction disputes apparently find difficult to accept. But Parliament has not abolished arbitration and litigation of construction disputes. It has merely introduced an intervening provisional stage in the dispute resolution process. Crucially, it has made it clear that decisions of adjudicators are binding and are to be complied with until the dispute is finally resolved.

23.05 That statement has since been cited with approval by the Court of Appeal.[2] It follows that the courts should not be too astute to find fault with the adjudication process, which is necessarily rough and ready. The other side of that coin is that adjudicators' decisions are not finally declarative of the rights of the parties. They are only an interim measure pending the finality of an arbitrator's award or a judgment of the court.

23.06 The policy of the 1996 Act is 'pay first, argue later'.[3] One of the chief mischiefs the Act was designed to deal with was the gross inequality of financial strength between the various parties in construction contracts, from developer to main contractor to sub-contractor, sub-sub-contractor, and so forth. The law did not permit a financially weak sub-contractor simply to stop work because it had not been paid sufficiently. Therefore, if a financially strong main contractor starved the sub-contractor of funds by making deductions on account of unjustified claims, there was little the sub-contractor could do about it. The reality was that summary judgment was rarely available as a remedy and litigation and arbitration was a slow and expensive process. So the sub-contractor, or to use the term 'financially weaker party', would be starved of cash whilst the works were proceeding and have to pay substantial sums to its lawyers in order to try and get restitution at the end of the job.

---

[1] (1999) 64 Con LR 1, [1999] BLR 93.

[2] *Tally Wiejl (UK) Ltd v Pegram Shopfitters Ltd* [2003] EWCA Civ 1750.

[3] *RJT Consulting Engineers v DM Engineering (Northern Ireland) Ltd* [2002] EWCA Civ 270, [2002] 1 WLR 2344; *Tally Wiejl (UK) Ltd v Pegram Shopfitters Ltd*, above, para [9].

It was hardly surprising that many contractors and sub-contractors would be forced either into insolvency or into a settlement which was worth to them far less than their true rights. **23.07**

Therefore, a contractor could obtain a job in free market competition but founder financially because the legal and arbitral system gave it no remedy which it could afford in the time in which it could properly survive. **23.08**

It was clear that something had to be done in order to redress the balance between the paying parties and the paid parties. The solution was to create a fresh legal framework for the resolution of construction disputes on an interim basis. Thus the emergence of 'pay first, argue later', which was a radical change from the old position of 'complete the work first and, if you are financially strong enough, litigate to receive the price later'. This new framework was adapted from the adjudication procedure which was written into the JCT Standard Forms of Sub-Contract,[4] which was designed to have similar results. The procedures in that standard form were used, but infrequently. One reason was that it originally related to sub-contracts. The JCT main contracts contained adjudication provisions by 1988, when a supplement was introduced into the JCT Standard Form with Contractor's Design, but frequently that provision was deleted and in reality few adjudications took place. The breakthrough was to have legislation which required adjudication in all forms of building contract, with few exceptions, and thereby make the process available to all who enter into building contracts. **23.09**

Adjudication in Practice

In order to use the adjudication provisions for the 1996 Act, being a statutory procedure, it is vital to ensure that the provisions of the 1996 Act apply to each specific set of facts. To that extent, there are a number of hurdles which must be overcome before any attempt at adjudication can be made and it is important for the practitioner to understand the various elements of the 1996 Act (and the underlying contract) and to apply them to the specific circumstances of the contractual situation. **23.10**

Is there a 'construction contract'?

The 1996 Act does not affect every contract; it only applies to construction contracts. The first question to consider, therefore, is whether there is a construction contract. **23.11**

The primary requirement is that there is an enforceable contract under English law and the cases on certainty of terms and intention to create legal relations are all applicable as much to determine whether there is a contract under the 1996 Act as they are under English contract law generally. **23.12**

For example, depending on the particular circumstances, a letter of intent requesting work to be undertaken in return for a reasonable sum may not constitute an enforceable contract **23.13**

[4] See for example the JCT 1980 Edition of the Standard Form for Nominated Sub-contractors NSC 4.a, clauses 21–4. However the NFBTE/Standard Form of Sub-contract which was then known as the 'Green Form' was the first to contain an adjudication provision.

if there is no obligation on the other party to undertake that work and, in the absence of payment, the party performing the work will only be able to commence a claim for payment on a quantum meruit basis through the courts, rather than through adjudication.

23.14 Similarly, where the parties have yet to agree all material terms of the contract or where there was not an objective intent on the part of both parties to enter into a contract, there may only be a series of orders for part of the work, and it will be a question of construction whether such orders in themselves can amount to enforceable construction contracts for those elements of the work.

23.15 What must be borne in mind is that every letter of intent and every contested contract situation must be looked at in its own particular circumstances; general principles can only be applied to such specific facts and wording. As will be seen later, such disputes as to the existence or otherwise of a construction contract are relatively frequent and provide fertile ground for jurisdictional challenges to an adjudication.

What are construction operations?

23.16 Section 104 of the 1996 Act defines a construction contract, somewhat tautologically, as an agreement for the carrying out of construction operations, arranging for others to carry out construction operations (whether as sub-contractor or otherwise), and providing labour for the carrying out of construction operations. These operations include design work, surveying work, advice on building or engineering, interior or exterior decoration, or laying out landscapes.[5]

23.17 Therefore, the provision of advice relating to a construction contract is subject to the 1996 Act and the adjudication provisions therein.

23.18 Having established that there is an agreement or contract in place that can come within the 1996 Act, one, therefore, has to consider whether this agreement or contract relates to the carrying out of 'construction operations' which itself is defined in s 105 of the 1996 Act.

23.19 Construction operations are defined in s 105(1), the definitions are lengthy and are set out in full in the section.[6] The definitions in s 105 are designed to include all construction and engineering operations which take place on a site where the structure is supposed to be built. They are also designed to include such things as the cleaning of buildings, the clearance and landscaping of the site, and painting internally or externally.

23.20 The section is based on s 567 of the Income and Corporation Taxes Act 1988 and is relatively self-explanatory and logical, as well as being wide ranging in its application. However, there are a number of points which need to be borne in mind before commencing or defending adjudication proceedings.

23.21 Of particular importance is the fact that s 105(1)(a) to (c), although relating to construction, repair, and demolition activities generally, have a pre-condition that such activities must relate to buildings or structures 'forming or to form part of the land'.

---

[5] See s 104(1) and (2).

[6] See App 1 below.

This is an important limitation on the definition of construction operations and requires careful thought. For example, HH Judge Thornton QC in *Palmers Ltd v ABB Power Construction Ltd*[7] found that the fabrication of plant on site, prior to it being moved to its final position, can constitute construction operations within s 105(1)(b), given the fact that such plant was to form part of the land once operational. Conversely, in *Gibson Lea Retail Interiors Ltd v Makro Self Service Wholesalers Ltd*[8] it was held that shopfitting items were not fixtures in the particular circumstances of that case (where the display units were fitted to the floor merely to ensure their stability), albeit shopfitting cannot be excluded from potential construction operations. The question of whether shopfitting is a construction operation all depends on the degree of fixity of the items to the land/existing fixtures at the shop. **23.22**

By analogy, therefore, the erection of portable site cabins, which are only placed on the ground, would not fall within construction operations as they could not be said to form part of the land. On the other hand, the erection of a tower crane at a site would be a construction operation under sub-section (b) because, although the crane is not to form part of the works, for the period of the works it is to form part of the land, having sufficient connection with the land that it cannot easily be moved. **23.23**

Sub-section (c) extends the definition to include mechanical and electrical works and perhaps extends the concept further than one of pure construction, to include the installation of such equipment within buildings. **23.24**

The definition in sub-section (c) is more limited than (a) or (b), but has to be read in the context of those sub-sections. Dyson J in *Nottingham Community Housing Association Ltd v Powerminster Ltd*[9] rejected the argument that maintenance of heating systems is not covered by the definition of construction operations on the basis that sub-section (c) did not include a reference to maintenance. Sub-section (a) clearly refers to 'maintenance' and is unaffected by the definition of mechanical and electrical work in sub-section (c). **23.25**

Sub-section (d) also extends the definition of construction operations to cleaning of buildings, but only whilst 'carried out in the course of their construction, alteration, repair, extension or restoration'. It is, therefore, important to determine whether a particular set of facts involving the cleaning of a building constitute maintenance, which would fall within the definition of 'construction operations' under sub-section (a), or whether such works are a one-off activity carried out as part, say, of the final clean before handover, which would fall within sub-section (d) or whether the cleaning works are simply a one off or irregular activity and fall outside the 1996 Act. **23.26**

Sub-section (e) appears to be a sweep-up provision, including a number of disparate activities, including site clearance and landscaping, albeit to fall within the ambit of the 1996 Act all of those operations must form an integral part of or be preparatory to the operations described in sub-sections (a) to (d). However, even if such operations stand alone, they **23.27**

[7] [1999] BLR 426.
[8] [2001] BLR 407.
[9] [2000] BLR 309.

may still fall within the definitions of construction operations in other sub-sections—for example, landscaping alone could be said to fall within sub-section (b) regardless of its connection with other activities.

What are not construction operations?

23.28 Section 105 goes further and specifically lists a number of operations which are not construction operations and which, therefore, do not fall within the 1996 Act. These exclusions include drilling for oil (as opposed to drilling for piles), extraction of minerals, off-site manufacture and delivery, and 'except under a contract which also provides for their installation' the making, installation, and repair of artistic works.

23.29 It is not always obvious as to why things are included or excluded within the definitions. Why drilling for oil is excluded and drilling pursuant to a piling contract is included is not easy to follow as a matter of principle. This is because the exclusions arise as a result of lobbying from specific industries which were able to show that there was already satisfactory provision for payment within their industry.[10]

23.30 In addition, and where particular care has to be taken, there is a specific exception relating to the provision of plant or machinery and any associated steelwork on a site where the primary activity is nuclear processing, power generation, or water/effluent treatment. Operations on a site where the principal activity is the production, processing, or bulk storage of chemicals, pharmaceuticals, or oils are similarly excluded.

23.31 Hence, the key question for any potential claimant is 'what is the principal activity on the site'. This is not always straightforward, as illustrated by the case of *Mitsui Babcock Energy Services Ltd v Foster Wheeler Energia OY*[11] which related to the construction of two boiler plants adjacent to an oil refinery. The boiler plants were to be operated by a company which supplied energy to the oil refinery but the boilers themselves were not directly involved in the process of producing oil at the refinery. It was argued that, because the installation of the boilers was on a site leased to a separate company, the primary activity of that separate site did not fall within the exclusion of s 105(2)(c)(ii). However, Lord Hardie found that, on the specific facts of the case, the installation of the boiler plant was to further the primary activity of processing chemicals and oil within the overall chemical complex. As such, the exclusion within sub-section (c) applied and the installations of the boilers were not 'construction operations' within the 1996 Act.

23.32 Conversely, in *ABB Zantingh Ltd v Zedal Building Services Ltd*[12] the dispute concerned the wiring for standby generators to a site where printing works were to be erected. The standby generators only operated if the permanent power was cut off. The court found, perhaps unsurprisingly, that the 'primary activity' of the site was the printing works, and not power generation; as such the exclusion in s 105(2)(c) was not relevant and the 1996 Act applied.

---

[10] Ibid, para 29.
[11] 2001 SLT 1158.
[12] [2001] BLR 66.

The court also found that the construction site was not to be limited to an area surrounded by a fence containing the generators but constituted the whole area occupied by the head contractor.

There has been conflicting case law on the question of whether or not a contract for a construction operation (as defined in s 105(1) of the 1996 Act) should be covered by the 1996 Act in circumstances where it forms part of an overall project or scheme which includes operations which s 105(2) expressly states are not construction operations. **23.33**

In the *Palmers v ABB Power Construction Ltd*[13] case, s 105(2) of the 1996 Act was construed narrowly so that a sub-contract for scaffolding which was required as part of the construction of an item of plant was held to be covered by the 1996 Act, even though the main contract for the construction of the plant itself was excluded from the 1996 Act by s 105(2)(c) because the primary activity on the site was to be power generation. By contrast, in the case of *ABB Power Construction Ltd v Norwest Holst Engineering Ltd*,[14] s 105(2) of the 1996 Act was given a broad construction meaning that all of the construction operations necessary to achieve the aims and purposes of the owner or main contractor would be exempt so that a sub-contract providing for the provision of scaffolding or paint systems to prevent corrosion of the plant would not be covered by the 1996 Act. **23.34**

Ramsey J considered the difference between the approaches adopted in these cases in *North Midland Construction plc v AE&E Lentjes UK Ltd*[15] and concluded that the narrow construction of s105(2) is to be preferred. As a result, it is now clear that a contract for a construction operation (as defined in s 105(1) of the 1996 Act) is covered by the 1996 Act even in circumstances where it forms part of an overall project or scheme which includes operations which s 105(2) expressly states are not construction operations. **23.35**

Sometimes the decisions relating to the demarcation between what is or is not a construction operation are narrow. Each case must be looked at on its own facts to see whether the activity contracted for falls within s 105(1) or is excluded by s 105(2) of the Act. **23.36**

An example of this is *ABB Power Construction Ltd v Norwest Holst Engineering Ltd*,[16] which involved the off-site fabrication of insulation to clad pipework, which clearly fell within s 105(1) because the provision of insulation was an integral part of the construction of pipework, but because the contract in question only provided for its fabrication, and not its installation, the works were excluded from the provisions of the 1996 Act by way of s 105(2)(d). **23.37**

In the overwhelming majority of cases, there will be no difficulty in describing the work contracted for as a construction operation within the meaning of the 1996 Act. **23.38**

[13] [1999] BLR 426.
[14] (2001) 17 Const LJ 246.
[15] [2009] EWHC 1371 (TCC).
[16] (2001) 17 Const LJ 246.

Definition of a construction contract

23.39 As set out above, the 1996 Act does not affect every contract.

23.40 If the contract provides for the carrying out of construction operations, as defined in s 105, the contract will be deemed to be a construction contract and will potentially be covered by the 1996 Act.

23.41 Construction contracts apply whether the work in relation to construction operations is to be carried out by one of the parties or whether one of the parties is merely to arrange for others, by way of sub-contract or otherwise, to undertake such construction operations.

23.42 A contractor does not need to take any responsibility for design, management, or the quality of materials for a contract to be deemed to be a construction contract. A contract for the provision of labour for carrying out such construction operations is deemed to be a construction contract (but a contract of employment is not). This can include the supply of labour by an employment agency, even where the employment agency may not know in advance the type of operations to be undertaken by those individuals it provides.

23.43 Similarly, the majority of professional appointments, as long as they relate to construction operations, are also deemed to be construction contracts. For example, a contract for the design or management of the construction of a bridge would inevitably be deemed to be a construction contract under s 105(1)(a), whereas a contract for the design or management of the construction of a floating barge may not, depending on the circumstances, as the floating barge would not form part of the land and may not be integral to other construction operations.

23.44 Administration of a construction contract does fall within s 105,[17] whereas the provision of claim support or expert evidence in relation to a construction dispute is outside the definition.[18]

23.45 Similarly, a contract for a structural survey to a building, as part of the design of remedial or renovation works, would be treated as a construction contract but a contract for a valuation survey of the same building, perhaps as part of the initial feasibility study, would not be a construction contract.

23.46 As with the definition of 'construction operations', s 104 also provides for a number of exclusions from the definition of a 'construction contract'.

23.47 Sub-section (3) specifically provides that contracts of employment do not constitute construction contracts.

23.48 Sub-section (4) enables the Secretary of State to set out agreements which are not construction contracts for the purpose of the 1996 Act. To date, exclusions have been ordered

---

[17] See *Gillies Ramsay Diamond v PJW Enterprise Ltd* [2003] BLR 48.

[18] See *Fence Gate Ltd v James R Knowles Ltd* (2001) 84 Con LR 206.

under the Construction Contracts (England and Wales) Exclusion Order 1998 (made on 6 March 1998) in relation to seven types or categories of construction contract:[19]

(i) agreements under ss 38 and 278 of the Highways Act 1980;

(ii) planning agreements under ss 106, 106A, and 299A of the Town and Country Planning Act 1990;

(iii) agreements under s 104 of the Water Industry Act 1991;

(iv) externally financed development agreements within the meaning of s 1 of the National Health Service (Private Finance) Act 1997;

(v) the top tier of Private Finance Initiative contracts;

(vi) certain finance agreements, which are defined as including a contract of insurance and any contract where the principal obligation includes the lending of money or the provision of bond finance agreements; and

(vii) development agreements.[20]

The fact that only the top tier of Private Finance Initiative contracts are excluded from the 1996 Act has led to some problems given the way that such projects are usually structured. In particular, the building contract in PFI projects frequently includes provisions which are designed to prevent the contractor taking any steps to enforce its claims against its special purpose vehicle employer until the employer's claims against the relevant authority have been resolved. This is because the employer relies exclusively on the relevant authority to provide it with the funds to meet the claims of the building contractor. However, in the case of *Midland Expressway Ltd v (1) Carillion Construction Ltd (2) Alfred McAlpine Construction Ltd (3) Balfour Beatty Group Ltd (4) AMEC Capital Projects Limited (5) John E Price (No 2)*,[21] it was held that such provisions are unenforceable as they seek to prevent the contractor from referring a dispute to adjudication at any time contrary to s 108 of the 1996 Act. This means that the contractor can pursue its claims against the special purpose vehicle employer by adjudication in circumstances where, in the absence of contractual provision to the contrary, the employer has no equivalent right to commence adjudication proceedings against the relevant authority. As a result, the contractor can obtain an adjudicator's decision within twenty-eight days ordering the employer to make a payment to the contractor, while the employer can only obtain the funds to make the payment through the agreement of the relevant authority or by the much slower routes of litigation or arbitration. **23.49**

Section 104(3) permits the Secretary of State to amend or repeal the provisions of s 104(1) to (3) of the 1996 Act as to the agreements which are construction contracts. So far no order has been made by the Secretary of State. **23.50**

[19] The prospective s 106A of the 1996 Act, as provided for by the Local Democracy, Economic Development and Construction Act 2009 will have the same effect but is expected to provide for the expansion of the types of PFI contracts which fall outside the ambit of adjudication.

[20] See the case of *Captiva Estates Ltd v Rybarn Ltd (in administration)* [2006] BLR 66 for an application of the wide definition of development agreements.

[21] [2005] EWHC 2963 (TCC).

23.51 Sir Michael Latham was asked to review the operation of the 1996 Act and it is possible that there may be changes in the definitions set out in s 104(1) to (3) and a change to the exclusions made by SI 1998/648 but it is doubtful that such orders will be made.

Construction operations and other operations

23.52 Section 104(5) specifically provides that only construction operations within an agreement are covered by the 1996 Act.

23.53 As such, care has to be taken in deciding whether a dispute relates to a construction operation within that contract or not. For example, a dispute relating to the supply and installation of lighting to a building would constitute a construction operation, whereas a dispute under the same contract relating to the subsequent provision of spare lights or light fittings to take account of future wear and tear could be said to fall within the exclusion in s 105(2)(d)(iii) as there is no requirement for the installation of such fittings. As such, any dispute for the provision of spares could not be adjudicated, whereas a dispute relating to the original supply and installation of the lights under the contract would fall within the 1996 Act.

Time and place

23.54 Section 104(6) specifically provides that the contract must have been entered into after 1 May 1998, the commencement date of the 1996 Act.

23.55 In addition, to fall within the 1996 Act, the construction contract must relate to construction operations carried out in England, Wales or Scotland (or Northern Ireland under the Construction Contracts (Northern Ireland) Order 1997). Therefore, contracts entered into for work to be undertaken outside these countries do not fall within the 1996 Act.

23.56 In *Staveley Industries plc v Odebrecht Oil & Gas Services Ltd*,[22] the contract provided for the supply and installation of telecoms equipment in living quarter modules for an oil rig which was to be towed to the Gulf of Mexico on completion. It was held that the modules did not 'form part of the land' at the place of construction (Teeside, England) as they were moveable and as such, they could not constitute 'construction operations' in England.

Residential occupiers

23.57 A wide exclusion, which applies to all construction contracts, excluding them from the adjudication provisions, is that relating to residential occupiers under s 106. A construction contract with a residential occupier is defined as:

> 106(2) . . . a construction contract which principally relates to operations on a dwelling which one of the parties to the contract occupies or intends to occupy, as his residence.
>
> In this subsection 'dwelling' means a dwelling-house or a flat; and for this purpose 'dwelling house' does not include a building containing a flat; and 'flat' means separate and self-contained premises constructed or adapted for use for residential purposes and forming part of a building from some other part of which the premises are divided horizontally.

[22] (2001) WL 332113.

As such, a contract between an individual and a contractor to build a house that he or she will occupy will not fall within the 1996 Act, whereas a contract between a developer and a contractor or a contractor and a sub-contractor for a number of dwellings will not be excluded—as there is no intention to occupy by the parties. **23.58**

Similarly, a company cannot be a residential occupier, as determined in *Absolute Rentals Ltd v Gencor Enterprises Ltd*[23] *and Edenbooth Ltd v Cre8 Developments Ltd.*[24] This does sometimes have unforeseen consequences where a company is used as a tax vehicle for construction work to a dwelling, on behalf of an individual, as the adjudication provisions will still apply. **23.59**

In addition, the construction contract must principally relate to operations on a dwelling. In the case of *Samuel Thomas Construction Ltd v Bick & (J&B) Developments*,[25] the parties agreed to enter into a contract for the conversion of two barns into dwellings. One of the contracting parties was to occupy one of the barns. It was estimated that 65 percent of the contract sum related to the work on that barn. **23.60**

The fact that the barn was not a 'dwelling' at the start of the works was not relevant. However, the second barn, although to be converted to a dwelling, was not to be occupied by one of the contracting parties. As such, the contract overall fell within the provisions of the 1996 Act as the contract related to work on both barns and could not be said to relate principally to operations on a dwelling for a residential occupier. **23.61**

There is little justification for the exclusion of construction contracts with residential occupiers from the scope of the 1996 Act. Small contracts with homeowners are just as in need of temporary and speedy dispute resolution as larger commercial contracts. Whilst it might be said that the hegemony of the paying party is less of a problem than in commercial contracts, such a statement cannot be justified. Small builders can be starved of cash with just as unsatisfactory consequences as the larger ones. **23.62**

Contract in writing

It was a cardinal provision of the 1996 Act that the construction agreement shall be in writing. There is provision for prospective repeal of this provision in the Local Democracy, Economic Development and Construction Act 2009. The law will be looked at on the basis that the prospective repeal is not yet in force because the repeal will only come into force pursuant to an order of the Secretary of State. The fact of the extensive jurisprudence which follows may be said to be a good policy reason for the requirement's eventual repeal. But, as stated above, the repeal is prospective only and may not come into force. **23.63**

The definition of what contracts shall be treated as being made in writing is set out in s 107 of the 1996 Act. That provides that there is an agreement in writing if the agreement is made in writing or by exchange of communications in writing, evidenced in writing or **23.64**

23 (2001) 17 Const LJ 322.
24 [2008] EWHC 2963 (TCC).
25 HHJ Overend 28 January 2000, [2000] CILL 1637, Lawtel Document No AC7201298.

made orally by reference to written terms.[26] Thus an oral agreement that a standard form of building contract shall apply to the construction works is, for the purposes of the 1996 Act, an agreement in writing. Furthermore, if the parties exchange written submissions in any proceedings (including adjudication proceedings) and one party alleges an oral agreement and that agreement is not denied, then the lack of denial is to be taken as an agreement in writing.[27] However, this does not apply where the exchange of submissions is in an adjudication where that adjudication is the first time that the issue was raised.[28]

23.65 The 1996 Act states that an agreement is evidenced in writing if the oral agreement is recorded by one of the parties, or by a third party, with the authority of the parties to the agreement.[29]

23.66 A narrow interpretation of the words 'in writing' was taken by the Court of Appeal in *RJT Consulting Engineers Ltd v DM Engineering (Northern Ireland) Ltd*[30] where the majority view was that the whole of the agreement had to be evidenced in writing. Ward and Robert Walker LJJ found that all the terms of the construction contract had to be evidenced in writing for a contract to fall within s 107. Auld LJ was of the minority view that only the material terms of the agreement must be recorded in writing and, therefore, unrelated issues did not need to be recorded, but this approach was not shared by the majority.

23.67 *RJT* was followed by *Carillion Construction Ltd v Devonport Royal Dockyard Ltd (No 1)*[31] where it was held that a written contract which is amended by oral agreement is not evidenced in writing for the purposes of the 1996 Act.[32]

23.68 Following the *Carillion* case, the Court of Appeal considered the matter further in *Pegram Shopfitters Ltd v Tally Wiejl (UK) Ltd*[33] which related to a battle of the forms argument as to the appropriate terms that applied to the construction contract between the parties, on the agreed basis that the construction contract had been formed.

23.69 The claimant argued that the contract was based on the JCT Standard Form of Prime Cost Contract 1998 while the defendant contended that the contract was based upon a standard form that they had put forward.

23.70 May LJ considered that a more realistic possibility was that there was no contract at all and, as such, the 1996 Act could not apply. May LJ referred to the first instance decision of Judge Thornton QC and his characterization of the situation as a construction contract whose terms cannot be readily ascertained as suggestive of a real possibility that there was no written construction contract at all.

[26] See s 107.
[27] See s 107(5).
[28] *Grovedeck Ltd v Capital Demolition Ltd* [2000] BLR 181.
[29] See s 107(4).
[30] [2002] BLR 217.
[31] [2003] BLR 79.
[32] In the case of *Treasure & Son Ltd v Martin Dawes* [2008] EWHC 2181 (TCC), the court concluded that an oral variation to a written contract does not give rise to a valid jurisdictional challenge in a contractual as opposed to statutory adjudication.
[33] [2004] BLR 65.

May LJ rejected the contention that the identification of the precise terms of the construction contract was a 'matter of detail which did not impinge the existence of the contract' on the basis that this was 'palm tree contractual analysis'. **23.71**

On the basis of the current position under the *RJT* and *Tally Weijl* decisions, a large number of construction contracts are excluded from the scope of the 1996 Act. This is due to the fact that, in practice, it is common for parties in the construction industry to agree contracts which are partly written and partly oral. In particular, it is common for sub-contracts not to be entirely evidenced in writing. Unfortunately, given that sub-contractors are often the type of small businesses which the 1996 Act was intended to protect from poor payment practices, this means that a large number of the construction contracts which the Act was intended to cover are excluded from the scope of the 1996 Act. It is this clear policy objective which underlies the prospective repeal of the requirement that the clause be in writing by the Local Democracy, Economic Development and Construction Act 2009.

The Right to Adjudication: How It Arises

Any party to a construction contract which is not a construction contract excluded from the ambit of the 1996 Act has the statutory right to refer a dispute arising under the contract for adjudication under a procedure which complies with the 1996 Act.[34] **23.72**

The right of reference of disputes under construction contracts has been provided for by s 108(1) to (5) of the 1996 Act. It does so in four stages: **23.73**

(i) by s 108(2) it provides that each contract shall: (a) enable a party to give notice of his intention to refer a dispute to adjudication; (b) provide a timetable with the object of securing the appointment of the adjudicator and referral of the dispute to him within seven days; (c) require the adjudicator to reach a decision within twenty-eight days (or forty-two days with the consent of the claimant) or longer with the consent of the parties; (d) require the adjudicator to act impartially; and (e) enable the adjudicator to take the initiative in ascertaining the facts and the law;

(ii) by s 108(3) it requires that each contract shall provide that the decision of the adjudicator is binding until final determination by litigation or arbitration;

(iii) by s 108(4) it requires each contract to exempt the adjudicator from liability for all acts or omissions save those in bad faith; and

(iv) by s 108(5), if the contract does not make provision for and comply with the above requirements then the adjudication provisions of the scheme shall apply.

The statute does not imply the provisions of the scheme directly into all contracts provided that the contract expressly makes the provisions required under s 108(2) to (4) of the 1996 Act. Therefore, the parties are at liberty to make their own adjudication provisions. But the contractual provisions will not apply if they do not satisfy s 108. Section 108(5) is clear that **23.74**

34 See s 108(1).

unless all the provisions of s 108(2) to (4) are contained in the contract, then the scheme applies. The provision could have been to the effect that to the extent that the contract did not contain those provisions, then they will be implied into the contract. However, the statute ensures instead that the scheme applies wholesale unless the provisions set out in s 108 are expressly contained in the contract.

23.75 Judge Toulmin CMG QC in *John Mowlem & Company plc v Hydra Tight Ltd*[35] stated:

> I have considered whether, if some parts of the sub-contract comply with the Act, they can be retained and the Act can be used in substitution for or to fill in those parts of the sub-contract which are contrary to the Act. But the words of the Act are clear. Either a party complies in its own terms and conditions with the requirements of Sections 108(1) to (4) or the provisions of the Scheme apply.

23.76 In addition to a close analysis of ss 104–107 of the 1996 Act in the light of the specific facts of the project, to determine whether the parties are able to get over the initial threshold for adjudication, another key element to consider is, therefore, the adjudication provisions, if any, in the contract itself.

23.77 As discussed above, in the absence of any express adjudication provisions, the adjudication procedures in Part 1 of the scheme take effect as implied terms of the contract under s 114(4) of the 1996 Act. However, the parties are free to provide for their own adjudication procedure in the contract or rely upon the statutory provisions. Where the parties have sought to set out their own specific provisions—either by referring to an existing set of rules or by incorporating their own bespoke provisions—it is important to check that such adjudication provisions comply with the requirements of s 108 in order to be clear whether the provisions set out in the contract or the provisions of the scheme apply.

23.78 If there is any requirement for mediation or discussions between chief executives before adjudication can take place, then such adjudication rules do not comply with the Act, which is to allow adjudication to take place at any time. Clauses that seek to defer when a dispute can be said to have arisen are also non-compliant and will result in the application of the scheme in place of such provisions.[36]

23.79 Provisions in a contractual procedure which are more extensive than the 1996 Act requirements do not necessarily make such provisions non-compliant. For example:

(1) although the 1996 Act only envisages a single dispute being referred to adjudication, procedures which provide for multiple issues (albeit, defined as a single dispute) rather than a single dispute being referred have been held to be compliant with the 1996 Act;[37]

(2) a provision restricting the selection of an adjudicator from a named list in a contract has been deemed to be compliant;[38]

[35] 17 Const LJ 358.

[36] See ibid where the Notice of Dissatisfaction procedure under the ICE conditions was agreed to be non-compliant with the 1996 Act.

[37] *Barr Ltd v Law Mining Ltd* 80 Con LR 135.

[38] *John Mowlem & Co plc v Hydra Tight Ltd* 17 Const LJ 358.

(3) provisions in the TeCSA rules edition 2.0 enabling an adjudicator to rule on his or her own jurisdiction were held to be compliant with the 1996 Act in *Nordot Engineering Services Ltd v Siemens plc*[39] *and Farebrother Building Services Ltd v Frogmore Investments Ltd*;[40] and

(4) in *Bridgeway Construction Ltd v Tolent Construction Ltd*[41] *(Tolent)* it was held that a contractual provision requiring the claimant to be responsible for all costs and expenses of the adjudication did not conflict with the provisions of the 1996 Act. There is provision for prospective outlawing of these types of contractual provision in the Local Democracy, Economic Development and Construction Act 2009. The 2009 Act amendments may be brought into force by order of the Secretary of State. It is submitted that *Tolent* was wrongly decided since the practical effect of such clauses is to either discourage or emasculate the right to adjudicate which must be contrary to the purpose of the statute.

Having determined what adjudication provisions apply, either the scheme or the contractual provisions, one has to consider the available procedure for any adjudication before moving forward to consider potential proceedings, as it is still possible to fall outside the contractual jurisdiction of any appointed adjudicator by failing to follow such provisions. There can often be material differences between the scheme and not only bespoke provisions, but also the institutional rules. The impact of such differences will be referred to in the discussion of the substantive adjudication proceedings. **23.80**

Notice of Intention to Refer a Dispute to Adjudication ('adjudication notice')

Introduction

Once it is established that the 1996 Act applies to the contract in question, it is important to consider how and when a claim can be made in adjudication proceedings. Adjudication proceedings are commenced by one party serving on the other party an adjudication notice. However, there are a number of hurdles—both procedural and substantive—which must be overcome before an adjudication notice can be served. **23.81**

When should a notice be served?

The statute permits reference 'at any time'. There is, therefore, no time when a dispute cannot be referred to an adjudicator. No restriction is stipulated for when the dispute can be referred. The term 'at any time' means just that.[42] Therefore, a dispute, may arise before the works commence or long afterwards. Of course, if the dispute relates to defects which appear towards the end of the limitation period, it may be unwise to commence an adjudication without concurrent court proceedings since time may expire otherwise. **23.82**

[39] TCC, 14 April 2000.

[40] (2001) WL 1251778.

[41] (2000) WL 1027055.

[42] See *Connex South Eastern Ltd v MJ Building Services* [2005] 2 All ER 870, [2005] EWCA Civ 193.

23.83 It follows that an adjudication cannot be stayed or otherwise impeded because of the legal concept of 'abuse of process'.[43] Abuse of process is not a concept which is known to the 1996 Act. Given that the process is designed to be of a short duration, there can be no question of such legal process concepts as dismissal for want of prosecution.

23.84 It has been suggested by the Court of Appeal[44] (obiter) that there may be circumstances when a party loses the right to adjudication by waiver or estoppel. It is submitted that this is incorrect and would run contrary to the policy of the statute. The terms of the 1996 Act are so framed that they cannot be contracted out of. It would be strange if what cannot be deliberately contracted out of could be otherwise lost. This point has not yet been argued or subject to judicial decision.

23.85 If the express terms of the contract do not allow the parties to the contract to refer a dispute to adjudication at any time, then the provisions of the scheme apply.

23.86 The scheme provides that any party to a construction contract may give written notice of his intention to refer any dispute arising under the contract to adjudication and such a notice may be given at any time in accordance with the 1996 Act.

23.87 Section 115 of the 1996 Act provides that the parties are free to agree on the manner of the service of the adjudication notice and so it is important to take care to ensure that the adjudication notice is served in accordance with the terms of the contract.

23.88 To the extent that the contract does not set out provisions for service of the adjudication notice, s 115(3) of the 1996 Act provides that: 'A notice or other document may be served on a person by any effective means.'

23.89 Section 115(4) of the 1996 Act provides that:

> If a notice or other document is addressed, pre-paid and delivered by post:
> (a) to the addressee's last known principal residence or, if he is or has been carrying on a trade, profession or business, his last known principal business address, or
> (b) where the addressee is a body corporate, to the body's registered or principal office,
> it shall be treated as effectively served.

23.90 It is important that the adjudication notice is properly served either in accordance with the terms of the contract (if the terms make provision for the service of the adjudication notice) or in accordance with the terms of the 1996 Act.[45]

Limitations on timing: the crystallization of a dispute

23.91 As stated above, the 1996 Act provides that a party to a construction contract may refer a 'dispute' to adjudication at any time. This means that a dispute between the parties to the contract must have arisen before the adjudication notice can be served.

[43] Ibid, para 40.

[44] Ibid, *per* Dyson LJ at para 39.

[45] See the cases of *Primus Build Ltd v (1) Pompey Centre Ltd (2) Slidesilver Ltd* [2009] EWHC 1487 (TCC); and *Nageh v Richard Giddings & another* [2006] EWHC 3240 (TCC) on the question of the proper service of adjudication notices.

As to the meaning of 'dispute' there is a wealth of case law, first originating in relation to arbitration and more recently in relation to adjudication under the 1996 Act. The provisions for arbitration in the Arbitration Act 1996 and its predecessors and provisions in the 1996 Act for adjudication are materially different. **23.92**

Under the Arbitration Act, if the parties have agreed that arbitration is to determine any disputes and, in default, one party commences court proceedings, the responding party can seek an automatic stay of any such court proceedings commenced in breach of the arbitration provisions. Conversely, adjudication under the 1996 Act is a right—it is not an obligation. There is, therefore, no ability to stay any court proceedings brought in advance or even at the same time as adjudication proceedings. **23.93**

In addition, the nature of an arbitration decision is usually that it is to be final and binding on the parties, subject to appeal in limited circumstances. Conversely, an adjudication is only to have what can be described as 'temporary finality' pending any subsequent court or arbitration proceedings where the dispute is considered afresh—there is no appeal, as such, of the adjudication decision. **23.94**

These are material differences between arbitration and adjudication and need to be borne in mind when reading the case law on the meaning of the word 'dispute' and the apparent conflicting decisions at first instance. **23.95**

The leading authority on the meaning of 'dispute' in the context of the Arbitration Act 1996 was *Halki Shipping Corp v Sopex Oils Ltd*[46] where the majority of the Court of Appeal approved the approach adopted in *Hayter v Nelson Home Insurance*[47] where Saville J reasoned that the words 'dispute' and 'differences' are not confined to cases where it cannot immediately be determined whether one party is right or wrong. Saville J used the analogy of two men arguing over the result of the University Boat Race in a particular year and stated that the fact that one person could easily and immediately demonstrate the correct result in that particular year did not mean that a dispute did not in fact exist between the two men. In short, any form of disagreement—or failure to admit the other side's claim—constitutes a dispute and gives rise to an ability to stay any court proceedings in favour of arbitration proceedings under the Arbitration Act 1996, thereby preventing the claiming party from seeking summary judgment under CPR Part 24. **23.96**

It is this apparent ability to frustrate and avoid a quick court decision by requiring the arbitration provisions to be followed—simply by 'masterly inactivity' *per* Templeman LJ in *Ellerine Bros Pty Ltd v Klinger*[48] which has given rise to concern when applied to the materially different, some would say, provisions for adjudication under the 1996 Act. **23.97**

A body of case law has grown up in relation to adjudication which demonstrates potentially two conflicting schools of judicial opinion reflecting policy considerations in interpreting the proper meaning of the word 'dispute' under the 1996 Act.[49] **23.98**

---

46 [1998] 1 WLR 726.

47 [1990] 2 Lloyd's Rep 265.

48 [1982] 2 All ER 737.

49 See P Sheridan and D Helps, 'Construction Act Review' (2003) 19 CLJ 6, 319.

23.99 There has been the so called 'narrow' interpretation of dispute, as applied in *Fastrack Contractors Ltd v Morrison Construction Ltd*,[50] which appeared designed to take on board the risk of 'ambush' by the claimant.[51] On the other hand, another line of thought viewed such an approach as being inconsistent with the *Halki* decision and detrimental to the underlying objectives of the 1996 Act requiring the provision of a swift, effective, and straightforward remedy appropriate to payment disputes and disputes during the currency of a project.

23.100 In short, the differences of approach could be formulated as follows:

> For there to be a dispute for the purpose of exercising the statutory right to adjudication it must be clear that a point has emerged from the process of discussion or negotiation that has ended and that there is something which needs to be decided. (*Sindall Ltd v Solland*);[52]
>
> or
>
> . . . there is a dispute once money is claimed unless and until the defendants admit the sum is due and payable. (*Halki Shipping Corp v Sopex Oils Ltd*).[53]

23.101 However, it is submitted that any conflict in the decisions has been brought to an end by the Court of Appeal decision in *Amec Civil Engineering Ltd v Secretary of State for Transport*.[54]

23.102 This case concerned a challenge under the Arbitration Act 1996 and at first instance Jackson J had given a review of the authorities on the word 'dispute' under the 1996 Act. The judge set out seven propositions, which distil the law correctly:[55]

> 1. The word 'dispute' which occurs in many arbitration clauses and also in section 108 of the [1996 Act] should be given its normal meaning. It does not have some special or unusual meaning conferred upon it by lawyers.
> 2. Despite the simple meaning of the word 'dispute', there has been much litigation over the years as to whether or not disputes existed in particular situations. This litigation has not generated any hard-edged legal rules as to what is or is not a dispute. However, the accumulating judicial decisions have produced helpful guidance.
> 3. The mere fact that one party ('the claimant') notifies the other party ('the respondent') of a claim does not automatically and immediately give rise to a dispute. It is clear, both as a matter of language and from judicial decisions, that a dispute does not arise unless and until it emerges that the claim is not admitted.
> 4. The circumstances from which it may emerge that a claim is not admitted are Protean. For example, there may be an express rejection of the claim. There may be discussions between the parties from which objectively it is to be inferred that the claim is not admitted. The respondent may prevaricate, thus giving rise to the inference that he does

[50] [2000] BLR 168.

[51] See the decision of HH Judge Seymour QC in *Edmund Nuttall Ltd v RG Carter Ltd* [2002] EWHC 400 (TCC). Not followed by Akenhead J in *Balfour Beatty Engineering Services (HY) Ltd v Shepherd Construction Ltd* [2009] EWHC 2218 (TCC).

[52] (2001) TCLR 30.

[53] [1998] 1 WLR 726.

[54] [2005] BLR 227.

[55] The propositions were endorsed by the Court of Appeal in *Collins (Contractors) Ltd v Baltic Quay Management (1994) Ltd* [2004] EWCA Civ 1757, [2005] BLR 62 at [63] *per* Clarke LJ and in the appeal against Jackson J's decision at [2005] EWCA Civ 291 at [31].

not admit the claim. The respondent may simply remain silent for a period of time, thus giving rise to the same inference.

5. The period of time for which a respondent may remain silent before a dispute is to be inferred depends heavily upon the facts of the case and the contractual structure. Where the gist of the claim is well known and it is obviously controversial, a very short period of silence may suffice to give rise to this inference. Where the claim is notified to some agent of the respondent who has a legal duty to consider the claim independently and then give a considered response, a longer period of time may be required before it can be inferred that mere silence gives rise to a dispute.
6. If the claimant imposes upon the respondent a deadline for responding to the claim, that deadline does not have the automatic effect of curtailing what would otherwise be a reasonable time for responding. On the other hand, a stated deadline and the reasons for its imposition may be relevant factors when the court comes to consider what is a reasonable time for responding.
7. If the claim as presented by the claimant is so nebulous and ill-defined that the respondent cannot sensibly respond to it, neither silence by the respondent nor even an express non-admission is likely to give rise to a dispute for the purposes of arbitration or adjudication.

May LJ was broadly content to accept the judge's propositions as correctly stating the law, albeit with his own further observations. **23.103**

May LJ referred to the decision of Clarke LJ (now Lord Clarke of Stone-cum-Ebony) in *Collins (Contractors) Ltd v Baltic Quay Management (1994) Ltd*[56] and Clarke LJ's commentary on Jackson J's propositions. Whilst stating that 'it all depends on the circumstances of a particular case', Clarke LJ took the opportunity to confirm that the mere making of a claim does not amount to a dispute and that a dispute will only be held to exist once it can be reasonably claimed as being not admitted. This is the key factor to determine. **23.104**

In particular, Clarke LJ noted that Jackson J had not endorsed the suggestion that a dispute could not arise until negotiations or discussions had been concluded. Clarke LJ felt that such negotiations and discussions were more likely to be consistent with the existence of a dispute which had yet to be resolved rather than with an absence of a dispute. In particular, he implied that a robust approach should be adopted in determining whether a dispute has arisen when he stated: 'It also appears to me that the Court is likely to be willing readily to infer that a claim is not admitted and that a dispute exists so that it can be referred to arbitration or adjudication'. **23.105**

May LJ's analysis of the word 'dispute' dealt in large part with an arbitration dispute in the context of the ICE conditions although he also reiterated that the existence of a dispute involves affording a party a reasonable time to respond to a claim and what may constitute a reasonable time depends on the facts of the case and the relevant contractual structure. **23.106**

Rix LJ agreed with May LJ albeit his reasoning differed to an extent. Rix LJ pointed to the fact that adjudication is seen as an *additional* provisional layer of dispute resolution pending final litigation or arbitration and that, as such, there was a legitimate concern to ensure that the point at which this additional complexity had been properly reached should not be **23.107**

[56] [2004] EWCA Civ 1757, [2005] BLR 62.

too readily anticipated. The suggestion from Rix LJ's judgment is that he would be willing to favour a broad meaning to the word 'dispute' in the arbitration context, but a narrower meaning in the context of adjudication.

23.108 It is essential that a dispute has crystallized before the adjudication notice is served because the courts will refuse to enforce any decision of an adjudicator which is based on an adjudication notice which is issued before a dispute has arisen in respect of the issues set out in the adjudication notice.

23.109 In practical terms, the timing of the issue of the adjudication notice will be entirely dependent upon the claimant being confident that a dispute has crystallized which can be referred to adjudication.

23.110 As stated above, the Court of Appeal has now given very clear guidance on the principles which should be applied by the courts when deciding whether or not a dispute had arisen prior to the issue of an adjudication notice.[57]

23.111 However, while the principles that the courts will apply to deciding whether or not a dispute has arisen are now relatively clear, what is less certain is how those principles will be applied to the facts of each case and what steps parties can take to ensure that a dispute does crystallize before an adjudication notice is issued.

23.112 A discussion of each of the key principles to emerge from the Court of Appeal decisions and how parties can apply those principles to ensure, as far as possible, that a dispute has crystallized before an adjudication notice is issued is set out below.

Key principles

The word 'dispute' is to be given its normal meaning

23.113 The Court of Appeal has made it clear that the word 'dispute' should be given its normal meaning by the courts.

23.114 However, while there is no legal definition of the word 'dispute' the previous court decisions have produced careful guidance as to the circumstances in which the courts are likely to consider that a dispute has arisen and these should be carefully considered.

Notification of a claim is not enough

23.115 In order for a dispute to arise, a claim must be submitted by one party to the contract to another party to the contract for their consideration. If no claim is submitted then no dispute can arise.

23.116 The practical implications of this principle are demonstrated by the case of *Maxi Construction Management Ltd v Mortons Rolls Ltd*.[58] In this case, the claimant contended that it was entitled to an interim payment in respect of its application for payment number 10. However, it was held that under the terms of the contract, the claimant's application was a

[57] *Amec Civil Engineering Ltd v Secretary of State for Transport* [2005] BLR 227; and *Collins (Contractors) Ltd v Baltic Quay Management (1994) Ltd* [2004] EWCA Civ 1757, [2005] BLR 63, [2004] 2 All ER 982.

[58] [2001] CILL 1784.

request for the employer's agent to value the work and not a claim for payment. As a result, the court concluded that no dispute had arisen as no claim had been made and not admitted.

It is, therefore, extremely important for parties to consider the terms of the contract and ensure that a formal claim for, say, payment or an extension of time, is made prior to any assertion being made that a dispute has arisen. **23.117**

The mere fact that one party ('the claimant') notifies the other party ('the respondent') of a claim does not automatically and immediately give rise to a dispute. A dispute does not arise unless and until it emerges that the claim is not admitted by the respondent. **23.118**

This means that it is not open to a claimant to simply issue a letter claiming, for example, an extension of time and to then immediately issue an adjudication notice. This is because a dispute will not arise until it is clear whether or not the claim for an extension of time is not admitted by the respondent. **23.119**

The reasons that the Court of Appeal has adopted this approach are clear: **23.120**

(i) the Court of Appeal wants to avoid claims being referred to adjudication in circumstances where a respondent may be prepared to admit the claim but has not had a proper opportunity to consider it. This is to avoid costs being incurred which could otherwise have been avoided; and

(ii) the Court of Appeal wants to ensure that a respondent is given a reasonable time to consider the claim before it is referred to adjudication. This is to minimize the risk that a respondent will be 'ambushed' and so not be given a reasonable opportunity to defend itself.

Rix LJ made his concerns in this respect clear in his judgment in the *Amec* case[59] when he said that: **23.121**

> Unlike the arbitration context, adjudication is likely to occur at an early stage . . . but there is the different concern that parties may be plunged into an expensive contest, the timing provisions of which are tightly drawn, before they, and particularly the respondent, are ready for it. In this context there has been an understandable concern that the respondent should have a reasonable time in which to respond to any claim.

If the respondent expressly rejects the claim by stating, for example, that the sum claimed is not due or that the contractor is not entitled to an extension of time, then a dispute will have crystallized. **23.122**

However, the practical difficulty with adopting a narrow definition of the word 'dispute' is that it provides potential respondents with an opportunity to attempt to delay the crystallization of a dispute by requesting further information and by refusing to respond to the claim. **23.123**

It is, therefore, relatively rare for an adjudication notice to be issued in response to an express rejection of a claim. **23.124**

[59] [2005] BLR 227.

23.125 However, the Court of Appeal has given some guidance on the circumstances in which it will be prepared to find that a dispute has arisen despite the fact that a respondent has not expressly rejected the claim.

Discussions between the parties may infer that the claim is not admitted

23.126 There may be discussions between the parties from which objectively it is to be inferred that the claim is not admitted. It is a little difficult to give an example of the type of discussions which may have taken place which give rise to a dispute and each case will depend upon its facts. However, if, for example, the employer and the contract administrator make it clear in discussions that they do not consider that certain works constitute a variation to the scope of works included in a building contract, the court may be prepared to infer that a claim by the contractor that those works are a variation is not admitted even where there has been no express rejection of the claim.

23.127 As stated above, Clarke LJ made it clear in the case of *Collins (Contractors) Ltd v Baltic Quay Management (1994) Ltd*,[60] that, contrary to some earlier cases, it is not correct that a dispute may not arise until negotiations or discussions have been concluded and that, in fact, negotiations and discussions are likely to be evidence that a dispute, albeit as yet unresolved, exists. Clarke LJ's view in this respect was endorsed by May LJ in the case of *Amec Civil Engineering Ltd v Secretary of State for Transport*.[61]

23.128 The practical problem for potential claimants is that it will be difficult to predict whether or not the courts will be prepared to infer that a claim has been rejected on the basis of discussions that have taken place. In addition, there may be difficulties in proving the nature and content of the discussions that took place in the absence of an agreed written record of the discussions.

Prevarication by the respondent may infer that he does not admit the claim

23.129 The respondent may prevaricate giving rise to the inference that he does not admit the claim. A good example of the type of situation envisaged arose in the case of *Amec Civil Engineering Ltd v Secretary of State for Transport*.[62]

23.130 In this case, the Highways Agency sent a formal letter of claim to Amec on 6 December 2002 asking Amec to confirm by close of business on 10 December 2002 whether or not it accepted that it was responsible for some defective work. On 10 December 2002, Amec responded by stating that it was not in any position to comment on liability.

23.131 The Court of Appeal was prepared to hold that although the letter of 10 December 2002 did not amount to a denial of liability, a dispute had arisen because Amec did not admit the claim.

23.132 The question of whether or not prevarication gives rise to an inference that a respondent does not admit a claim will always depend upon the specific facts and circumstances of each individual case. For example, in this case, in reaching its decision, the Court of Appeal took

---

60 [2004] EWCA Civ 1757, [2005] BLR 63, [2004] 2 All ER 982.
61 [2005] BLR 227.
62 Ibid.

into account the extensive correspondence and discussions that had taken place, the fact that it was highly unlikely that Amec would admit liability for the claim and that the limitation period for bringing the claim was about to expire.

This case gives a clear indication that respondents will not be allowed to postpone the crystallization of a dispute by constantly stating that they are not in a position to confirm whether or not they admit or deny the claim. It appears from the *Amec* case that where a sizeable claim has been notified to the other party which alleges negligence or a material breach of contract, it is more likely that a dispute will arise sooner than in a complex claim for, say, an extension of time. **23.133**

Silence by the respondent may infer that the claim is not admitted

The respondent may simply remain silent for a period of time giving rise to an inference that he does not admit the claim. The Court of Appeal has indicated that the period of time for which a respondent may remain silent before a dispute is to be inferred depends heavily upon the facts of the case and the contractual structure. **23.134**

Where the basis of the claim is well known and it is obviously controversial, a very short period of silence may suffice to give rise to this inference. For example, in the *Amec* case[63] the Court of Appeal considered that it was inconceivable that Amec would admit liability for the defects for which the Highways Agency was alleging that it was responsible and so a relatively short period of silence was considered to be enough to infer that the claim was not admitted. **23.135**

In addition, if a contract administrator has certified that a payment is due to a contractor and no withholding notice has been issued, a very short period of silence following a demand for payment after the final date for payment is likely to give rise to an inference of a dispute. **23.136**

However, where the claim is notified to an architect or engineer who has a legal duty to consider the claim independently and then give a considered response, a longer period of time may be required before it can be inferred that mere silence gives rise to a dispute. For example, if a contractor claims an extension of time and the contract does not specify the period of time within which the contract administrator has to decide whether or not to grant an extension of time, then a dispute will not arise until the contract administrator has had a reasonable opportunity to consider the claim and to make a decision. **23.137**

The clearest example of the way in which this has been applied in practice occurred in the case of *R Durtnell & Sons Ltd v Kaduna Ltd*.[64] In this case Judge Seymour QC refused to enforce the decision of an adjudicator to award the claimant an extension of time because the claimant had issued the adjudication notice before the expiry of the twelve-week period during which the architect had to make his decision on the claimant's application for an extension of time under the terms of the contract. **23.138**

---

63 Ibid.
64 [2003] BLR 225.

23.139 Judge Seymour QC concluded that it could not be inferred that the respondent did not admit the claimant's claim in circumstances where the contractual period for responding to the claim had not expired. If, however, the architect had failed to make a decision by the end of the twelve-week contractual period, then it is highly likely that the court would have concluded that the claim had not been admitted and that a dispute had arisen.

23.140 Again, practical difficulties can arise for claimants because the twelve-week period for responding to the claim may be stated in the contract to run from the time when the architect has received sufficient particulars of the claim and not from the date when the claim is made. This can, of course, lead to uncertainties as to when sufficient particulars have been provided and when the timescale begins to run.

23.141 In addition, claims are often made in circumstances where there is no contractual provision setting out a timetable within which the other party to the contract or their agent must respond. In those circumstances, it will be a question of fact in each case whether or not the potential respondent had been given a reasonable opportunity to respond to the claim prior to the issue of the adjudication notice.

23.142 The difficulty for the claimant is that this means that there will always be some uncertainty as to the decision which the court will make on the particular facts. It is, therefore, usually in the interests of a respondent to delay responding to a claim. The respondent may then seek to persuade the adjudicator that he does not have jurisdiction and should resign as no dispute has arisen. If the adjudicator proceeds to make a decision then the respondent may seek to resist enforcement of the decision of the adjudicator on the grounds that a dispute had not arisen as it had not had a reasonable opportunity to consider the claim prior to the commencement of the adjudication.

23.143 The courts have taken a reasonably robust approach in relation to such cases and there are only a handful of cases in which the courts have been prepared to conclude that a respondent was not given a reasonable opportunity to consider a claim before the adjudication notice was issued.

23.144 Clarke LJ supported the adoption of such a robust approach when he stated in the *Amec* case[65] that: 'It also appears to me that the Court is likely to be willing readily to infer that a claim is not admitted and that a dispute exists so that it can be referred to arbitration or adjudication.'

The imposition of a deadline for responding does not automatically curtail a reasonable time for responding

23.145 The courts have stated that if the claimant imposes upon the respondent a deadline for responding to the claim, that deadline does not have the automatic effect of curtailing what would otherwise be a reasonable time for responding. However, a stated deadline and the reasons for its imposition may be relevant factors when the court comes to consider what is a reasonable time for responding.

[65] [2005] BLR 227.

An example of this in practice is evident from the decision of Judge Humphrey LLoyd QC in the case of *Sindall Ltd v Solland*.[66] In that case, Judge LLoyd QC held that a dispute had not arisen where a large amount of documentation was submitted to a contract administrator in support of a claim for an extension of time with only a seven-day ultimatum being given for a response. The contract administrator responded by saying that it was considering the new information but the adjudication notice was, nevertheless, issued on the expiry of the seven-day period. **23.146**

Judge LLoyd QC held that no dispute had arisen in relation to the extension of time claim because: 'A person in the position of the Contract Administrator must be given sufficient time to make up its mind before one can fairly draw the inference that the absence of a useful reply means that there is a dispute.' **23.147**

It is, therefore, important to ensure that any deadline that is imposed is reasonable in all of the circumstances of the case and can be justified to the court if necessary. **23.148**

Nebulous and ill-defined claims will not give rise to a dispute even if met with an express non-admission

One tactic that may be adopted is to refer to claims in correspondence prior to the issue of an adjudication notice in such wide terms that it would be difficult to argue that the detailed claim had not been included within the terms of the claim asserted in correspondence. **23.149**

However, care has to be taken in relation to this tactic as the Court of Appeal has stated in the cases of *Collins (Contractors) Ltd v Baltic Quay Management (1994) Ltd*[67] and *Amec Civil Engineering Ltd v Secretary of State for Transport*[68] that if the claim as presented by the claimant is so nebulous and ill-defined that the respondent cannot sensibly respond to it, neither silence by the respondent nor even an express non-admission is likely to give rise to a dispute for the purposes of arbitration or adjudication. **23.150**

Again, the application of this principle will depend on the facts of each case and this gives rise to further uncertainty for claimants when preparing for adjudications. **23.151**

How can a claimant crystallize a dispute in the light of the key principles?

What is key is that although a claim may be made, a dispute cannot arise until the respondent has had the opportunity of either denying or not admitting the claim. **23.152**

The question which arises is what approach should be adopted by a potential claimant in order to ensure that a dispute has crystallized prior to serving an adjudication notice to avoid subsequent jurisdictional disputes and difficulties with enforcement. **23.153**

The whole purpose of adjudication is to provide a quick and relatively cheap method of resolving disputes. This purpose is undermined if costs are wasted on court proceedings challenging jurisdiction and if the lack of certainty as to when a dispute arises can be used by respondents to seek to delay adjudication. **23.154**

---

66 (2001) 3 TCLR 30.

67 ([2004] 2 All ER 982, [2004] EWCA Civ 1757, [2005] BLR 63.

68 [2005] BLR 227.

23.155 It is important that a claim is made to the other party. The way in which that claim is made will depend upon the terms of the contract and the nature of the claim.

23.156 If, for example, the claim is for an extension of time, then the contract may require the contractor to make that claim to the architect or engineer in a particular form. Where a clear and supported claim has been put forward to an architect or engineer, a subsequent rejection of the claim by the architect or engineer will immediately give rise to a dispute.

23.157 It may be that the claim which is made to the architect or engineer will not be as detailed as the referral notice and will not attach all of the evidence and information that would usually be attached to a referral notice given the familiarity of the architect/engineer with the project.

23.158 However, if it is intended to refer the dispute to adjudication if the claim is rejected, it is advisable to ensure that the same claim is referred to adjudication as is made to the architect/engineer in order to avoid the difficulties which arose in the *Nuttall* case.[69] Note that case was not followed in *Balfour Beatty Engineering Services (HY) Ltd v Shepherd Construction Ltd*.[70]

23.159 In this respect, the courts will understand that the claimant will need to include a fuller explanation of the claim and provide the adjudicator with all of the relevant documentary evidence when the dispute is referred to adjudication. In *William Verry (Glazing Systems) Ltd v Furlong Homes Ltd*[71] Judge Coulson QC (now Coulson J) stated that it was acceptable for a party to expand its explanation of its claim to enable an adjudicator who is unfamiliar with the project to understand it.

23.160 However, it is important to ensure that the scope for jurisdictional challenges, on the basis that the claim that is submitted to the architect/engineer is different from the claim that is made in adjudication, is kept to an absolute minimum. To this extent, it will almost inevitably be more difficult to assert that the nature of the claim has changed if the claim to the architect/engineer is fully and coherently explained and any documentary evidence to which the architect/engineer does not have access is provided. The nature and extent of the explanation will depend upon the nature of the claim. If the claim is for payment based on progress then the explanation and supporting evidence is likely to be quite limited. Conversely, if the claim is for a substantial extension of time based on a complicated factual, legal, and programming analysis, then the explanation is likely to be fuller and the extent of the supporting documentation is likely to be larger.

23.161 Where a claim does not have to be made to the architect/engineer, then a letter of claim should be submitted to the other party to the contract. The letter of claim should:

(i) set out as clearly and coherently as possible the legal and factual basis of the claim; and

---

[69] *Edmund Nuttall Ltd v RG Carter Ltd* [2002] EWHC 400 (TCC).

[70] [2009] EWHC 2218 (TCC).

[71] [2005] EWHC 138.

(ii) ensure that the claim is fully described and is consistent with any witness statements, expert reports, or other evidence that will be submitted to the adjudicator. This is to limit the scope for the respondent to argue that the claim submitted to the adjudicator is different from the claim that has been set out in correspondence; and

(iii) set a reasonable deadline for responding to that claim.

It will never be possible to be completely certain that the time imposed for the response is going to be considered to be reasonable by a court. However, a number of steps can be taken to ensure that the claimant has the best possible prospects of convincing a court that a reasonable time has been offered. For example, if legal advisors and/or expert advisors have been instructed, they should be asked to advise on the likely length of time that they would need to review the material and to respond to it. If, for example, an extension of time claim is being made, then any expert programmer who has been instructed to advise on the claim can be asked how long they would usually expect to have to respond to the material which has been submitted to the other side. **23.162**

The letter should set out the basis upon which the reasonable time has been calculated. For example, if the general basis of the claim has been set out in previous correspondence or discussions and this is being used to justify imposing a relatively short deadline for a response on the respondent, then the letter should specifically refer to the correspondence or discussions. **23.163**

It is common for the respondent to adopt a position in which it claims to be unable to admit or not admit a claim on the basis of a lack of evidence or a lack of understanding of the claim as presented. It is in these circumstances that questions of jurisdiction and the ability to commence adjudication proceedings are likely to be most difficult and where the particular facts, timing of claims, and contents of claims will be most relevant. The question of the length of time provided to the respondent to admit or not admit the claim will revolve completely around the form and content of the claim submitted in the first instance and the context in which that claim was submitted—had there been numerous or detailed discussions at site meetings or in previous correspondence to flag up, at least, the outline nature of the dispute? **23.164**

The greater the clarity of the claim and the more detailed and specific the supporting evidence, the more likely it is that the courts will be prepared to accept a shorter period of time for the claim to be considered prior to there being a deemed non-acceptance—either through silence or prevarication by the respondent. **23.165**

This obliges the claimant to ensure that, as far as possible, all claims submitted are logically and coherently put together and contain all the necessary elements to enable the respondent to understand how and why the claim is being made. **23.166**

With the various types of claim, however, there will also be varying levels of complexity which, in themselves, will necessarily impact upon the length of time it takes to admit or not admit the claim. For example, where a debt claim arises under a contract where there has been a refusal to pay by the employer and an absence of a withholding notice, the employer should not need more than a very few days within which to respond positively or negatively to the claim before adjudication proceedings can be commenced. **23.167**

23.168 At the other end of the spectrum, a complex variations and extension of time claim, involving lengthy critical path analysis and numerous programmes for the various time slices of the project based upon expert programming and engineering/architectural evidence, necessarily will involve the respondent in analysing a greater bulk of detailed factual and expert evidence almost certainly with the help of some external expert assistance. As such, a period of some weeks can be required to enable the respondent to understand the issues and to admit or not admit the claim before a dispute could be said to have arisen and adjudication proceedings can be commenced.

23.169 If it appears likely that a dispute will arise as to whether or not the claim referred to adjudication is the same as the claim submitted to the respondent, a claimant may wish to consider serving a draft referral notice together with all of the documents which will be sent to the adjudicator.

23.170 If the referral notice and the documents in support of the referral notice are subsequently sent to the adjudicator in the same form, then it will be very difficult for the respondent to argue that the claim which is referred to adjudication is in any way different from the claim that has previously been made in correspondence.

23.171 If the respondent does respond to the claim and, as a result, any changes have to be made to the claim set out in the letter of claim or draft referral notice, then the claimant will need to consider whether or not a revised version of the letter of claim or the referral notice should be forwarded to the respondent for its consideration prior to the adjudication notice being issued. This will be necessary if, as a result of the response, the claimant needs to materially change the claim set out in correspondence. This might occur if, for example, the claimant maintains its claim for an extension of time for the period set out in correspondence but relies on different events to support that claim from those set out in the correspondence.

23.172 The potential disadvantage of sending a detailed letter of claim or draft referral notice is that the claimant will lose the benefit of the respondent having to respond to the claim within the short period of time allowed in adjudication proceedings. This potential disadvantage should, however, be weighed against the following considerations:

23.173 (i) the claimant is more likely to succeed if it produces a thorough and detailed claim which is supported by detailed contemporaneous documents. As Max Abrahamson stated in *Engineering Law and the ICE Contracts*, 4th edn: 'A party to a dispute . . . will learn three lessons (often too late): the importance of records, the importance of records and the importance of records';

(ii) the courts are far more likely to enforce the adjudicator's decision if it is clear that the respondent has been given a proper opportunity to consider the claim prior to the issue of the adjudication notice; there is little point in a claimant incurring the costs of an adjudication if it will only find that those costs are wasted due to a refusal by the courts to enforce the decision; and

(iii) if the respondent's advisors consider that the respondent has no reasonable prospect of convincing a court not to enforce the adjudicator's decision, then the claimant may well find that the respondent complies with the decision without the need for the claimant to incur the costs and delay associated with enforcing the decision.

The scope of the dispute

Although the Act states that a 'dispute' rather than 'disputes' may be referred to adjudication, the courts have adopted a wide definition of the word 'dispute' in this context. **23.174**

Judge Thornton QC stated in the case of *Fastrack Contractors Ltd v Morrison Construction Ltd*[72] that: 'In other words, the "dispute" is whatever claims, heads of claim, issues, contentions or causes of action that are then in dispute which the Referring Party has chosen to crystallise into an adjudication reference.' **23.175**

This means that if, for example, there is a dispute as to what sum is due at the final account stage, then all of the issues in dispute relating to the level of the final account, including the contractor's entitlement to an extension of time, the extent to which the work has been varied and the value of the works, can be referred to the adjudicator for a decision at the same time. **23.176**

This issue was also considered in the case of *Balfour Kilpatrick Ltd v Glauser International SA*.[73] In this case it was argued that the adjudication notice was defective because it sought to refer more than one dispute to adjudication whereas the TeCSA Rules (version 1.3) which applied in this case permitted only one dispute to be referred. **23.177**

Judge Gilliland QC adopted a similar approach to Judge Thornton QC in *Fastrack* and decided that as r 11 of the TeCSA Rules referred to the 'matters identified in the Notice' the Rules did allow a range of issues to be referred to adjudication as one dispute. **23.178**

The scope of the dispute will depend upon what has been claimed by one party and rejected by the other party. If a contractor has, for example, made a claim for an extension of time and that claim has been rejected then a dispute will arise as to that claim. However, the fact that there is a dispute about the extension of time to which the contractor is entitled does not mean that the contractor can necessarily ask the adjudicator to determine the level of the prolongation costs to which the contractor is entitled in respect of the extension of time. A dispute would only arise in respect of the prolongation costs if a claim has been made to the employer or its agent in respect of those costs and rejected by the employer or its agent. **23.179**

The importance of the adjudication notice

The adjudication notice is the document which defines the scope of the dispute which the adjudicator has jurisdiction to decide. **23.180**

It is, therefore, extremely important that the adjudication notice is carefully drafted to properly identify the dispute which the claimant requires the adjudicator to decide and that the dispute identified in the adjudication notice is the dispute which has crystallized. **23.181**

The potential problems that are faced in drafting an adjudication notice are amply demonstrated by reference to a simple example. If a notice of adjudication requires the adjudicator to decide 'whether or not £1m is due to the contractor in accordance with the Construction **23.182**

[72] [2000] BLR 168.
[73] TCC, 27 July 2000.

Contract' then the adjudicator may only answer that question in two ways. The adjudicator may either conclude that £1m is due to the contractor or he may decide that £1m is not due to the contractor. It is not, however, open to the adjudicator to decide that the sum of £750,000 is due to the contractor rather than the sum of £1m.

23.183 In those circumstances, it would be better for the contractor to ask the adjudicator for an order that the sum of £1m 'or such other sum as the Adjudicator finds to be due' be paid to the contractor.

23.184 This seemingly legalistic and artificial distinction was recognized by Judge Thornton QC in the case of *Fastrack Contractors Ltd v Morrison Construction Ltd*[74] when he stated that: 'A claim can often be made without its quantification having been finalised or even attempted. The subsequent dispute can then be in the form: "What sum is due?" as opposed to "is £x due . . .".'

23.185 In addition, Judge Thornton QC recognized in that case that it is open to a claimant to split up its claims as it wishes. In that particular case, the claimant chose to bring an adjudication in respect of sums that it considered were due in respect of its interim application no 12. It then subsequently began a separate adjudication in relation to sums that it considered were due under interim application no 13 and for damages in respect of what it claimed to be a repudiatory breach of contract. Judge Thornton QC held that it is open to a claimant to split up related claims in this way.

23.186 An indication of the importance of carefully considering the scope of the dispute which is referred to adjudication was actually given by Judge Thornton QC in this case when he said that:

> It is doubtful that Fastrack could have recovered anything if it had confined its notice to the sum due under interim application no. 13 since the subcontract had, by then come to an end given Fastrack's prior acceptance of Morrison's repudiation. Thus, the interim payment procedures had been replaced by an overall entitlement to damages of the balance of the sums already due plus such loss of profit as had occurred by virtue of the repudiation.

23.187 This means that Fastrack would not have succeeded if it had limited the scope of the dispute to the sums which it claimed to be due in accordance with its interim application no 13, rather than widening the scope of the dispute to include its claim for general damages in respect of the repudiation.

23.188 In the case of *KNS Industrial Services (Birmingham) Ltd v Sindall Ltd*,[75] Judge Humphrey LLoyd QC stated that:

> A party to a dispute who identifies the dispute in simple or general terms has to accept that any ground that exists which might justify the action complained of is comprehended within the dispute for which adjudication is sought. It takes the risk that its bluff may be called in an unexpected manner. The further documents which come into existence following the notice of adjudication (such as the referral) . . . do not cut down or, indeed enlarge, the dispute (unless they contain an agreement so to do). . .

---

[74] [2000] BLR 168.

[75] (2001) 17 Const LJ 170.

> The adjudicator is appointed to decide the dispute which is the subject of the notice and that notice determines his jurisdiction. The adjudicator's jurisdiction does not therefore derive from the further documents . . . although those documents are likely to help the adjudicator to find out what needs to be decided in order to arrive at a conclusion on the dispute.

Two principles emerge from this judgment: (i) if the dispute is widely defined, then it is likely to leave it open to the respondent to raise any reasonable arguments in defence to the claim; and (ii) it is the adjudication notice which defines the scope of the dispute which the adjudicator is appointed to decide and any errors in the identification of that dispute cannot be corrected in the referral notice. **23.189**

Widely defined dispute

A good example of the danger of drafting an adjudication notice in very wide terms is provided in the case of *William Verry (Glazing Systems) Ltd v Furlong Homes Ltd.*[76] **23.190**

In that case, Furlong requested in its adjudication notice that the adjudicator decide: (i) that the extension of time granted by Furlong to 2 February 2004 is correct; and (ii) the final account figure, including Furlong's entitlement to liquidated damages. **23.191**

Prior to the adjudication, Verry had made applications for an extension of time up until 24 June 2004 but had only been awarded an extension of time to 2 February 2004. In its response to the referral notice, Verry claimed an extension of time to 27 July 2004. The adjudicator decided that Verry was entitled to an extension of time to 27 July 2004. **23.192**

Furlong argued in the subsequent enforcement proceedings that the adjudicator did not have jurisdiction to consider Verry's claim for an extension of time to 27 July 2004 as no dispute had arisen in relation to that claim because it had only previously applied for an extension of time to 24 June 2004. **23.193**

The judge held that the claim for an extension of time to 27 July 2004 was not a new claim as it was merely an enhancement of the claim for an extension of time to 24 June 2004 with slightly longer periods of delay ascribed to the events which had previously been identified in the earlier claim. **23.194**

In any event, Judge Coulson QC went on to consider whether or not the scope of the dispute referred to the adjudicator entitled him to consider the overall extension of time to which Verry was entitled or whether the scope of the dispute was limited to considering whether or not the extension of time that had been previously granted to Verry was correct on the basis of the information that had been produced by Verry up to the date of the adjudication notice. **23.195**

In this respect, Judge Coulson held that the scope of the dispute referred to the adjudicator in the adjudication notice did entitle him to consider Verry's overall entitlement to an extension of time because: **23.196**

(i) the adjudication notice asked the adjudicator to decide that 'the extension of time granted by Furlong to the 2nd February 2004 is correct'. The use of the word 'is' made

[76] [2005] EWHC 138.

it clear that there was no attempt to limit the dispute to earlier claims or information previously provided by Verry and considered by Furlong; and

(ii) the adjudication notice requested that the adjudicator decide Furlong's entitlement to liquidated damages and the adjudicator could not make such a decision without deciding Verry's entitlement to an extension of time.

23.197 It is important to note that the courts are unlikely to be sympathetic to a claimant who defines the dispute in very narrow terms in an attempt to limit the defences which can be put forward by the respondent. In this case Judge Coulson made it clear that his conclusion was influenced by the fact that Verry were responding to the claim and were not themselves the claimant. The judge held that it should be open to Verry to:

> defend themselves as best they could against the allegation that their only entitlement to an extension of time was down to 2 February 2004 and that liquidated damages should be deducted for the period of delay thereafter. In my judgment, they were not to be taken as having agreed that they could only defend themselves in the adjudication with half a shield, able to rely on some matters of fact but being artificially prevented from referring to others.

Adjudication notice defines the scope of the dispute

23.198 In construing the adjudication notice the courts have made it clear that they will interpret adjudication notices like any other commercial document and that it must be: '. . . construed against the underlying factual background from which it springs and which will be known to both parties.'[77]

23.199 The importance of carefully drafting an adjudication notice was demonstrated in the case of *David McLean Housing Contractors Ltd v Swansea Housing Association Ltd*.[78]

23.200 In that case, the claimant's adjudication notice did not seek a decision that a particular sum should be paid but required the adjudicator to decide the amount of money to be included in the next payment certificate. When the next payment certificate was issued it was, therefore, open to the defendant to issue a notice of withholding in respect of that certificate.

23.201 Exactly the same issue arose in *FW Cook Ltd v Shimizu (UK) Ltd*[79] in which an adjudication notice was interpreted to require the adjudicator to decide the way in which sums in issue were to be treated in a final account rather than requiring an immediate payment of those sums.

Drafting the adjudication notice

23.202 The information which needs to be included in any adjudication notice is set out in the scheme or the contractual adjudication provisions.

23.203 For example, the scheme requires the adjudication notice to set out:

(i) the nature and brief description of the dispute and of the parties involved;
(ii) details of where and when the dispute has arisen;

[77] *Fastrack Contractors Ltd v Morrison Construction Ltd* [2000] BLR 168.
[78] [2002] BLR 125.
[79] [2000] BLR 199.

(iii) the nature of the redress sought; and
(iv) the names and addresses of the parties to the construction contract.

The importance of this is indicated by the case of *Jerome Engineering Ltd v Lloyd Morris Electrical Ltd*.[80] In that case, the claimant served an adjudication notice which did not include any express request for relief and simply made reference to a failure to make payment in accordance with the terms of the contract. The adjudicator published a decision requiring the respondent to make a payment to the claimant. The respondent refused to comply with the decision on the grounds that the adjudicator did not have jurisdiction to order any payment to be made given that the adjudication notice did not seek any relief and, in particular, an award of payment of money. **23.204**

The rules that governed this adjudication did not require the adjudication notice to include details of the relief sought. The details of the relief sought were only required to be set out in the referral notice. The adjudication notice was required to set out details of the dispute but the court chose to draw a distinction between the words 'dispute' and 'relief' so that it could not be said that the relief sought in the referral notice in any way changed the nature of the dispute that was set out in the adjudication notice. **23.205**

However, the position would have been different if the adjudication had been governed by the scheme as this requires details of the relief claimed to be set out in the notice of adjudication. **23.206**

It is essential that a great deal of care is taken when drafting the adjudication notice, as it defines the dispute which the claimant requires the adjudicator to decide, and there will be no opportunities to amend the adjudication notice either during the course of the adjudication or prior to the enforcement of the adjudicator's decision. **23.207**

It is important that the adjudication notice properly identifies the dispute for the adjudicator and that the dispute that is identified for the adjudicator to decide is the dispute that has crystallized between the parties. **23.208**

In addition to the legal issues about whether or not a dispute has crystallized, a number of practical issues should also be taken into account when drafting the adjudication notice and deciding on the scope of the dispute which the adjudicator should be required to decide. **23.209**

As stated above, it is open to the claimant to define the dispute for the adjudicator as widely or as narrowly as the party chooses. If, for example, there is a dispute as to the final account, there are no legal reasons why all of the issues relating to that final account should be referred to the adjudicator at the same time. There may, in fact, be a number of very good reasons for not asking the adjudicator to decide all of the issues at once. **23.210**

[80] [2002] CILL 1827, [2000] EWHC 152 (TCC). Applied by Akenhead J in *Workspace Management Ltd v YJL London Ltd* [2009] EWHC 2017 (TCC) where he said: 'The starting point for any consideration of jurisdiction is the Notice of Adjudication: see *Jerome Engineering Limited v Lloyd Morris Electrical Limited* [2002] CILL 1827'.

23.211 A number of judges have been critical of parties referring all of the issues included in a final account dispute to adjudication at the same time. For example, in the case of *William Verry (Glazing Systems) Ltd v Furlong Homes Ltd*,[81] Judge Coulson stated that:

> It was, to use the vernacular, a 'kitchen sink' final account adjudication. Whilst such adjudications are not expressly prohibited by the Housing Grants, Construction and Regeneration Act 1996 as it presently stands, there is little doubt that composite and complex disputes such as this cannot easily be accommodated within the summary procedure of adjudication. A Claimant should think very carefully before using the adjudication process to try and obtain some sort of perceived tactical advantage in final account negotiations and, in so doing, squeezing a wide-ranging final account dispute into a procedure for which it is fundamentally unsuited.

23.212 In practical terms, a claimant should consider whether or not it will be possible for the adjudicator to reach a decision on a large number of complex issues within the short adjudication period and, if not, consider whether or not the issues can be broken down and referred to adjudication separately.

23.213 For example, in a final account dispute, it may be possible to refer disputes about the extension of time claimed and disputes about whether or not certain works constitute variations to adjudication separately. There are a number of advantages in taking this approach:

(i) a court is more likely to be sympathetic to the claimant in enforcement proceedings if the court can see that the claimant has not sought a tactical advantage by referring all of the disputes to adjudication at once;
(ii) different disputes can be referred to different adjudicators with different expertise. For example, the parties may wish to agree that a programming expert is appointed as the adjudicator to deal with the extension of time claim and that a quantity surveyor is appointed to deal with any quantum issues; and
(iii) a claimant is unlikely to succeed if the adjudicator does not have time to properly consider and understand the claim.

23.214 Ultimately, a decision about what dispute to refer to the adjudicator will depend upon what the claimant is hoping to achieve and what the claimant considers an adjudicator will be able to understand and make a decision on in a relatively short period of time.

23.215 A party has to weigh up at the outset the pros and cons of making one large complex claim, perhaps involving detailed architectural and engineering variations under the contract, as well as the impact of such variations on the completion date and the provision of detailed valuation evidence regarding such variations and loss and expense matters. Such a claim could involve a variety of expert evidence in the fields of architecture, engineering, programming, and quantum, as well as a wide range of factual issues over the course of the project. The time for such a claim to be considered and for a response to be provided is likely to be a number of weeks if not months. Conversely, however, had the claim been limited in the first instance to whether variations had been instructed or not, the response time, being limited to such a discrete issue, is going to be far shorter and the claimant will

[81] [2005] EWHC 138.

have the ability to refer at least an element of the dispute to adjudication proceedings at a far earlier stage, whilst setting out the subsequent issues on a rolling basis either at the same time as any adjudication proceedings or afterwards, once a decision has been provided.

Usually, claimants will want to ensure that the adjudicator is asked to order that a payment is made to the claimant (rather than order, for example, that a certificate be issued against which a withholding notice can subsequently be issued). They will also want to ensure that the adjudicator is asked to decide what sum is due in relation to the payment and not simply asked whether or not a certain sum is due. This is in order to avoid any arguments that the adjudicator can only decide whether or not that specific sum is due and cannot order that a different sum is due. **23.216**

Remedies which can be sought

The adjudication notice should include details of the remedies sought if this is required by the rules that apply to the adjudication or if the rules are silent as to whether the remedies sought should be specified. **23.217**

The rules applicable set out the powers that are available to the adjudicator. **23.218**

Paragraph 20 of the scheme provides that the adjudicator: **23.219**

(i) shall decide the matters in dispute;

(ii) may open up, revise, and review any decision taken or any certificate given by any person referred to in the contract unless the contract states that the decision or certificate is final and conclusive; and

(iii) may decide that any of the parties to the dispute is liable to make a payment under the contract and when that payment is due and the final date for payment.

An adjudicator may make a declaration as to the parties' legal rights, order a party to carry out an act (such as issuing a certificate) or order a party to make a payment to another party. **23.220**

An adjudicator cannot grant an injunction (as this is a power reserved to the courts). However, an adjudicator does have the power to make a decision including an order for specific performance.[82] **23.221**

Nominating and Appointing the Adjudicator

The contract may state the individual who will act as the adjudicator or the parties may agree who will be appointed to act as the adjudicator. One of the key criticisms of adjudication has been that the standard of adjudicators is not always as high and as consistent as it should be. Parties should, therefore, seriously consider the merits of agreeing the identity of the adjudicator or, at the very least, agreeing the qualifications and appropriate expertise that the adjudicator should have. **23.222**

[82] See *Multiplex Construction (UK) Ltd v Mott MacDonald Ltd* [2007] EWHC 20 (TCC).

23.223 Alternatively, the contract may name a body to nominate the adjudicator or the parties may agree on a body to appoint the adjudicator. Any express agreement between the parties as to the identity of the adjudicator or the identity of a body to appoint the adjudicator prevails over the scheme.

23.224 However, if no agreement is reached and the construction contract does not provide for how the adjudicator should be nominated, then the scheme applies and the claimant may apply to any adjudicator nominating body to nominate an adjudicator.

23.225 Each body has a slightly different procedure which has to be followed in order for it to nominate an adjudicator, and so it is important for the claimant to ensure that it is fully familiar with the requirements and complies with them.

23.226 If an application has to be made to a nominating body, there is no reason why either party cannot make representations to the body as to the expertise required. For example, it may be important that a programming expert, a quantity surveyor, or a lawyer is appointed, depending on the nature of the dispute.

23.227 The adjudicator will be nominated by the appointed body and then advise the parties that the nomination is accepted.

23.228 The rules usually provide that the dispute is to be referred to the adjudicator within seven days of the adjudication notice being issued. In the case of *Hart Investments Ltd v Fidler & Another*,[83] it was held that a failure to comply with this timescale even by one day rendered the adjudicator's decision to be a nullity. However, the court also made it clear in the case of *Cubitt Building & Interiors Ltd v Fleetglade Ltd*[84] that the courts will interpret the rules and the scheme in a sensible and commercial way. In that case, it was held that an adjudicator's decision was enforceable where through no fault of the claimant, the adjudicator was not appointed by the nominating body until late on the seventh day following the service of the adjudication notice and so the referral notice was not served until the eighth day after the service of the adjudication notice. In the case of *William Verry Ltd v North West London Communal Mikvah*,[85] Judge Thornton QC held that the 1996 Act did not prevent the parties from agreeing rules which allow for the time for the referral to be issued later than seven days after the adjudication notice. In this case, it was held that it was acceptable for the claimant to issue the referral eight days after the adjudication notice pursuant to a direction of the adjudicator as the adjudication rules allowed the adjudicator to adjust the timetable for the adjudication.

23.229 In order to avoid any challenge to the enforceability of a decision of an adjudicator on the grounds that the referral notice has been served late, the claimant should ensure, if at all possible, that the adjudicator and/or the other party has agreed to late service prior to the service of the referral notice.

---

[83] [2007] BLR 30.
[84] [2006] EWHC 3413 (TCC).
[85] [2004] BLR 308.

Responding To an Adjudication Notice

Adjudications take place over a very short period of time and once the adjudication notice has been served there is limited time for the respondent to consider what the response should be and what form it should take. The first question which needs to be considered is whether the jurisdiction of the adjudicator should be challenged. If the answer is negative, then it is necessary to proceed to prepare a substantive response as soon as possible. **23.230**

If a decision is taken to challenge the jurisdiction of the adjudicator, then this needs to be addressed as early as possible in order to avoid any allegations of waiver and to ensure that the right to raise a challenge at a later stage is protected. In addition, it is crucial to decide what form the challenge should take, bearing in mind the different pros and cons of each method. **23.231**

Challenging Jurisdiction

No power to decide own jurisdiction

To start with, it is necessary to recognize that an adjudicator does not have the jurisdiction to decide his own jurisdiction and any such decision is not binding,[86] albeit some rules do give the adjudicator this power.[87] If the court decides that there was no jurisdiction to make a decision in the first place, the decision of the adjudicator is a nullity.[88] **23.232**

By contrast, under s 30 of the Arbitration Act 1996, unless otherwise agreed, an arbitrator has the power to rule on his own jurisdiction in relation to whether there is a valid arbitration agreement, whether the tribunal is properly constituted and what matters have been submitted to arbitration in accordance with the arbitration agreement. That ruling can only be challenged if there is an arbitral appeal procedure but will otherwise be binding subject to appeal on a point of law under s 69 of the Arbitration Act or a challenge to jurisdiction under s 67.[89] **23.233**

In practice, a party will often set out its grounds for challenging jurisdiction to an adjudicator and will invite an adjudicator to consider those arguments. If the adjudicator decides that he has no jurisdiction, then he will resign. **23.234**

However, a better course of action may be for a party to reserve its position in relation to jurisdiction and to then decide whether to take part in the adjudication. That party will then raise the question of jurisdiction when the other party seeks to enforce the adjudicator's decision in court through the summary judgment procedure. The decision as to how to proceed will depend on the nature of any argument as to jurisdiction, as well as the costs which may be incurred and the overall time taken for the adjudication. In addition, the **23.235**

---

[86] *Homer Burgess Ltd v Chirex (Annan) Ltd* [2000] BLR 124.

[87] For example, the TeCSA rules.

[88] *The Project Consultancy Group v The Trustees of the Gray Trust* [1999] BLR 377.

[89] *Azov Shipping Co v Baltic Shipping Co Ltd (No 1)* [1999] 1 All ER 476.

decision as to how to proceed may depend on whether the decision sought from the adjudicator is a declaration or an order for the payment of a sum of money.

23.236 A decision ordering payment will be enforced through court proceedings, which is when it would be challenged. However, a declaratory decision will usually be a step in a further claim for payment or serve as a defence. It will, therefore, not be enforced in the same way and in order to challenge such a decision it will be necessary to apply for a declaration from the court that the decision is wrong or that the adjudicator had no jurisdiction to make it.

Determining jurisdiction

23.237 In *Fastrack Contractors Ltd v Morrison Construction Ltd*[90] Judge Thornton QC identified the options available to a party who wishes to challenge the adjudicator's jurisdiction. He stated as follows:

> If a party challenges the entire jurisdiction of the adjudicator, as Morrison does, it has four options. Firstly, it can agree to widen the jurisdiction of the adjudicator so as to refer the dispute as to the adjudicator's jurisdiction to the same adjudicator. If the Claimant agrees to that course, and the appointed adjudicator accepts the reference to him of this second dispute, the jurisdiction of the adjudicator could then be resolved as part of the reference. The challenging party could, secondly, refer the dispute as to jurisdiction to a second adjudicator. This would not put a halt to the first adjudication, if that already led to an appointment, since the adjudicator has a statutory duty, unless both parties agree otherwise, to decide the reference in a very short timescale. The challenging party could, thirdly, seek a declaration from the court that the proposed adjudication lacked jurisdiction. This option is of little utility unless the adjudicator has yet to be appointed or the parties agree to put the adjudication into abeyance pending the relatively speedy determination of the jurisdiction by the court. The Technology and Construction Court can, for example, resolve questions of that kind within days of them being referred to it. Fourthly, the challenging party could reserve its position, participate in the adjudication and then challenge any attempt to enforce the adjudicator's decision on jurisdictional grounds.

Agreeing to give the adjudicator power to determine his own jurisdiction

23.238 On the above basis, the first option is to simply agree that the adjudicator will also have the power to decide on his own jurisdiction, in the same way that an arbitrator does under the Arbitration Act 1996. The benefit of such a course of action is that it is relatively straightforward and does not result in any material additional costs. A party choosing to proceed on this basis will have to notify the claimant as soon as possible (to avoid allegations of waiver) that while it does not accept that the adjudicator has jurisdiction to decide the dispute, it is willing to be bound by the same adjudicator's decision as to his own jurisdiction.

23.239 The adjudication will then proceed as usual, save that the first question which the adjudicator will have to address is the question of his own jurisdiction. Arguments as to the adjudicator's jurisdiction will then be made either as part of the overall submissions by the parties or the parties could agree that submissions should be limited, in the first instance, to the issue of jurisdiction until the adjudicator has decided on his own jurisdiction. The benefit

[90] [2000] BLR 168 at para 31.

of the latter course of action is that it saves the costs of full arguments if the ultimate decision is that there is no jurisdiction and the adjudication does not go ahead.

The difficulty with this course of action is that in many cases a challenge to jurisdiction is part of an overall attempt to avoid a dispute being determined through adjudication or by a specific adjudicator. There is also the risk that, on some level, an adjudicator will feel inclined to find that he/she has jurisdiction so that the dispute can continue to be resolved. However, this option is likely to be the quickest and most cost-effective way of resolving jurisdictional issues. **23.240**

Referring the question of jurisdiction to another adjudicator

The second option identified by Judge Thornton QC is to have the dispute as to jurisdiction referred to a second adjudicator. Whilst, in principle, this is a course of action that would be available to the parties, it is unlikely to be an attractive option given the short timescales within which the Technology and Construction Court can now deal with jurisdictional challenges. **23.241**

Referring the question of jurisdiction to the court

The third option identified by Judge Thornton QC is to refer any question as to jurisdiction to the Technology and Construction Court. The benefit of this course of action is that the issue of jurisdiction is determined by the High Court and will be binding as a court judgment. In the first two options outlined above, there is a risk that a party will seek to challenge the decision on jurisdiction, thereby leading to it being re-heard by a court, but at a later stage with more costs being incurred. **23.242**

As pointed out by Judge Thornton QC, the Technology and Construction Court is familiar with these types of disputes and it is possible to obtain a decision from a judge within a relatively short period of time. By way of example, in the case of *Palmers Ltd v ABB Power Construction Ltd*[91] the court rapidly heard and gave a declaration on the jurisdiction question. However, in the case of *The Dorchester Hotel Ltd v Vivid Interiors Ltd*,[92] the court indicated that while it has power to grant declarations in relation to jurisdiction and natural justice during the course of an adjudication, it will exercise those powers sparingly and only in the clearest cases. **23.243**

In addition, currently the only fees payable are the usual court fees and there is no need to pay an hourly rate as is the case with adjudicators. Court proceedings may, therefore, be less costly than a second adjudication. However, this does mean that the parties no longer control the proceedings in the same way as in an adjudication. The timetable will depend on the court's timetable and it will also be necessary to use barristers or solicitor-advocates, which may increase the costs. Importantly, court decisions become a matter of public record and parties will lose the limited degree of confidentiality which may be afforded in an adjudication. **23.244**

[91] [1999] BLR 426.
[92] [2009] BLR 135.

23.245 Like the previous options, going to court will not prevent the main adjudication from going ahead. However, it is possible to agree that the adjudication should be suspended until a decision is given.

Reserving the position

23.246 The final option identified by Judge Thornton QC is for a party to simply reserve its position, participate in the adjudication and then, depending on the outcome of the adjudication, challenge any attempt to enforce the adjudicator's decision on the basis that the adjudicator had no jurisdiction. In such cases it is crucial to make it absolutely clear that there is no agreement as to jurisdiction and that the right to challenge the adjudicator's jurisdiction is reserved. The benefit of this course of action is that if the challenging party believes that it has a strong defence to the substantive claim, it is able to rely on that defence in the hope that it will be successful in the adjudication. It will, therefore, only be required to challenge the adjudicator's jurisdiction in the event that it is unsuccessful in defending the substantive adjudication. If this option is chosen, it is important for the party that wishes to challenge the adjudicator's jurisdiction to make and maintain its reservation of position in clear terms to avoid an argument that it has in some way waived its right to challenge the adjudicator's jurisdiction.[93]

23.247 The disadvantage of this course of action is that a party will incur the full costs and delays of going through the adjudication, all of which may be wasted if it is ultimately found that the adjudicator has no jurisdiction.

Additional option

23.248 It should be noted that there is an additional possibility which was not identified by Judge Thornton QC. It is, of course, also open to a party to refuse to take part in the adjudication, either by simply stating that it does not accept the adjudicator's jurisdiction or without providing any reasons. Although the adjudicator will then go ahead and make a decision, it will still be possible to challenge the adjudicator's decision when the claimant seeks to enforce it through summary judgment proceedings in court. However, a party should not decide on this course of action unless it is extremely confident of its ability to demonstrate that the adjudicator had no jurisdiction.

23.249 Unless the challenge succeeds, the challenging party will be faced with an adjudication decision following proceedings where it had not made any representations on its behalf. It is, therefore, much more likely that the decision will not be in its favour. However, the benefit of such a course of action is that no expenses or delays are incurred in relation to the main adjudication proceedings.

Fees

23.250 In the light of the case of *Mr Christopher Michael Linnett v Halliwells LLP*,[94] it is clear that a party who challenges the jurisdiction of an adjudicator will be jointly and severally liable

[93] See *OSC Building Services Ltd v Interior Dimensions Contracts Ltd* [2009] EWHC 248 (TCC).
[94] [2009] 1 CLC 157.

for the adjudicator's fees unless the challenge is valid and the party takes no part in the adjudication.

Waiver

As outlined above, it is necessary to ensure that the right to challenge is not waived by any steps taken before or during the adjudication. Regardless of the precise course of action taken by a party seeking to challenge an adjudicator's jurisdiction, such a party must make it absolutely clear in writing that it does not accept that the adjudicator has jurisdiction to hear the dispute. It is also important for such a party to ensure that it does not take any steps which could be regarded as a waiver of its objection to the adjudicator's jurisdiction. Such a waiver can take place through a party's conduct and its correspondence and in some cases parties have discovered that it is no longer open to them to challenge the adjudicator's jurisdiction, since the court had regarded the earlier steps as a waiver of such a right. **23.251**

This was the position in the Scottish case of *Watson Building Services Ltd v Harrison*,[95] where the Outer House, Court of Session, considered a petition based, inter alia, on the challenger's jurisdiction. Lady Paton accepted an argument that the petitioners chose to agree that the question of jurisdiction should be decided by the same adjudicator who heard the main dispute and it was, therefore, not open to them to later argue that he had no jurisdiction. This was due to the fact that the response referred to the possibility of the adjudicator determining that he had jurisdiction to hear this matter, in which case it was accepted that the adjudicator could and should proceed to determine the main dispute. **23.252**

Similarly, where a party disputing jurisdiction sends written submissions in relation to the question of jurisdiction and invites an adjudicator to decide on this issue, it will be bound by the adjudicator's decision (unless it expressly reserves its position),[96] albeit on an interim basis pending final determination by the court. In another case, the service of a counter notice and further correspondence was held by the court to be a waiver of the right to challenge the adjudicator's jurisdiction.[97] In *R Durtnell & Sons Ltd v Kaduna Ltd*,[98] it was held that a jurisdictional objection is not waived unless the party has full knowledge of the availability of the argument. However, an invitation to an adjudicator to correct part of his decision will be regarded as a waiver of any jurisdictional objections.[99] **23.253**

The adjudicator's response to a challenge

The adjudicator, when confronted with a challenge to his jurisdiction, faces the unenviable task of deciding to what extent he can consider such a challenge. While it is clear that, unless given such a power, the adjudicator cannot decide on his/her own jurisdiction, the courts have accepted that a practical approach must be taken to avoid such challenges **23.254**

[95] 2001 SLT 846.

[96] *Whiteways Contractors (Sussex) Ltd v Impresa Castelli Construction UK Ltd* 75 Con LR 92, [2000] CILL 1664, Judge Bowsher QC.

[97] *Cowlin Construction Ltd v CFW Architects* [2003] BLR 241.

[98] [2003] BLR 225.

[99] *Shimizu Europe Ltd v Automajor Ltd* [2002] BLR 113.

paralysing the adjudication procedure. This approach has been reflected in the guidelines issued to adjudicators by various industry bodies such as the RICS.

23.255 It has, therefore, been held that if an adjudicator finds that there is an arguable case that he has jurisdiction then he should continue until a court order determines otherwise.[100] Accordingly, when faced with a challenge to jurisdiction, an adjudicator will consider the arguments on their face. If he is confident that the challenge has no basis, then he should inform the parties of his decision and continue with the adjudication. If he finds that he has no jurisdiction, or that he is unable to decide on the basis of available facts, then the adjudicator must resign.

23.256 In *Farebrother Building Services Ltd v Frogmore Investments Ltd*,[101] it was determined that the adjudicator can inquire into his jurisdiction and decide whether or not to continue, but his decision on a jurisdictional objection is not binding on the parties, unless the rules governing the adjudication provide otherwise.

23.257 A robust approach was taken by Lady Paton in *Watson Building Services Ltd v Harrison*. She held that the adjudicator, although not a lawyer, can and should consider the contract terms, form a view about their meaning and import, and make decisions accordingly. She also stated she had difficulty with a submission that the adjudicator could not determine a dispute about what is the contract. In her view, Parliament intended that adjudicators could and should determine the meaning and import of the terms of the construction contract under which the dispute arises.

23.258 In practice, unless the jurisdictional objection is indisputable and undermines the whole proceedings, most adjudicators will continue with the proceedings and deal with the objection in their decision. It will often be the case that adjudicators will not only consider the arguments themselves but will also contact the parties and possibly hold a short hearing or telephone conference in order to deal with the jurisdictional issues. However, the taking of such a step is likely to impact on the adjudication timetable and the adjudicator's ability to deal with the substantive dispute.

The Referral

23.259 After the notice of adjudication, the referral notice is the most important document that will be prepared by the claimant.

23.260 The referral notice is often stated to be the equivalent of the particulars of claim which are served in court or arbitration proceedings. However, this is a misleading way to refer to a referral notice. The particulars of claim in court or arbitration proceedings effectively set out the skeleton of the claimant's case in relation to the facts and the law and will be expanded upon by further written and oral submissions during the course of the proceedings. However, it is quite possible that the referral notice will be the only submission which is provided to the adjudicator on behalf of the claimant.

[100] *ABB Power Construction Ltd v Norwest Holst Engineering Ltd* 77 Con LR 20.

[101] [2001] CILL 1762, Lawtel Document No AC0300703.

It is, therefore, extremely important that the document fully sets out the detail of the legal and factual basis of the claimant's case. The case should be set out succinctly and logically and should include details of all of the legal and factual arguments upon which the claimant seeks to rely. **23.261**

The adjudicator only has a very short period of time to ensure that he fully understands the claimant's case and to ask any questions that he has upon the case. If the adjudicator misunderstands the case then it is quite possible that the claimant will not realize this and will not have an opportunity to make further submissions before the decision is made. It is, therefore, important that the document is drafted in a clear and persuasive style and is cross-referenced to all of the relevant documents. The referral notice should, for example: **23.262**

(i) refer the adjudicator to the relevant parts of the contract and explain how those parts of the contract impact upon the claim;
(ii) refer the adjudicator to relevant legal authorities and explain how these are relevant to the claim;
(iii) refer the adjudicator to all of the relevant correspondence, meeting minutes, notes, photographs, and other contemporaneous documents and explain how they support the claimant's case; and
(iv) refer the adjudicator to any relevant witness statements and/or expert reports and explain how they support the claimant's case.

If the claimant is aware of the nature of the likely defence that will be put forward then the claimant may wish to explain why that defence is incorrect. **23.263**

Overall, it is important to remember that the claimant will only succeed if the adjudicator considers that it is more likely than not that its case is correct. **23.264**

When preparing the referral notice it is important to consider all of the elements of the case which must be proved and to ensure that the referral notice and the supporting documentation covers all of those elements satisfactorily. **23.265**

In a contractual claim, it will be important to prove the parties entered into the contract, the terms of that contract, that a breach of contract has occurred, and the loss which the claimant has suffered as a result. **23.266**

In an extension of time claim, it will be necessary to prove that the parties entered into a contract, the terms of that contract, that a relevant event has occurred, the delay that was caused by the relevant event, and the effect of the delay caused by that relevant event. **23.267**

In relation to variations, it will be necessary to prove that the parties entered into a contract, the terms of the contract, the scope of works originally included in the contract, the revised scope of works, and the failure to certify costs in respect of the variations. **23.268**

How each of the elements is proved will depend upon the facts of each case. Where there is a written contract which has been signed by both parties, the production of a copy of that contract will be evidence that the parties entered into the contract and what the terms of that contract are. If the contract is evidenced in writing then the written evidence of the **23.269** contract will need to be produced and there may, in addition, need to be a witness statement explaining how the written documents evidence that a contract was entered into between the parties and the terms of the contract.

23.270 The breach of the contract may be proved by a mixture of factual and expert evidence. In a claim against an architect for defective design, the production of the drawings and other relevant documents will be evidence of the architect's design. Expert evidence will be needed to prove that the design was not completed with reasonable skill and care and the impact of the defective design. The expert will need to explain, for example, whether or not the defective design will result in the failure of the structure or whether the impact will be less serious.

23.271 Invoices may be produced to show the losses suffered as a result of the defective design including the cost of employing another architect to properly complete the design, the cost of a contractor removing any abortive work, and any costs associated with any delay caused to the completion of the project. Again, expert evidence may be required to prove the extent of any delay that the defective design caused to the project and the costs associated with that delay.

23.272 In an extension of time claim, contemporaneous documentary evidence (such as site reports, correspondence and photographs), and witness evidence will be used to prove that a relevant event occurred, when it occurred and the delay that was caused by the relevant event. This is likely to be supported by expert programming evidence.

23.273 It is important that the referral notice complies with any of the relevant rules governing the adjudication.

Witness statements and expert reports

23.274 One of the issues which inevitably has to be considered prior to beginning an adjudication is whether or not witness statements and experts reports should be submitted to the adjudicator with the referral notice.

23.275 The problem with submitting witness statements and experts' reports is that they will be prepared in advance of any formal disclosure exercise of the type which is carried out in court or arbitration proceedings. This means that the witnesses and the experts will prepare the statements and reports without having had the benefit of seeing all of the documents in the possession of the other party and without having had the benefit of seeing all of the arguments raised by the other party.

23.276 The problem with this is that any witnesses must sign a statement of truth stating that they believe the facts stated in their witness statement are true and any expert must sign a statement confirming that the opinions expressed in the expert's report represent his true and complete professional opinion. In court or arbitration proceedings witness statements and experts' reports are signed when the nature of each party's case has been set out in detailed statements of case and each party has disclosed all of the documents which are relevant to the dispute.

23.277 Clearly, in adjudication it is possible that the witness statements and experts' reports will be produced before the respondent has set out the nature of its case in any detail and there is a risk that documents will be disclosed by the respondent which reminds the witness that its recollection is not correct or causes the expert to change his opinion. The witness statements and experts' reports may then be used by the respondent to undermine the evidence of the witness and the expert both in the adjudication itself and possibly in any subsequent court or arbitration proceedings.

It is often, therefore, considered preferable to avoid relying on witness statements or experts' reports in adjudication. However, in order to succeed in an adjudication, it is necessary for a party to be able to prove all of the elements of its case. If the adjudication is, for example, in relation to an allegation of professional negligence, then it is difficult to see how an adjudicator could possibly conclude that a claimant has proved its case unless it produces an expert's report supporting the claim that the respondent has not acted with reasonable skill and care. **23.278**

In addition, in claims which depend upon complicated factual issues and expert opinion, it is almost inevitable that an adjudicator will attach more weight to signed witness statements and experts' reports than he will to submissions prepared by a party or its representative. **23.279**

Confidentiality

The question of whether or not documentation produced in the course of an adjudication and the adjudicator's decision is confidential depends entirely on the contractual provisions and on any agreement that is made between the parties. The 1996 Act does not provide that such documents are confidential. **23.280**

For example, paragraph 18 of the scheme provides that: **23.281**

> The adjudicator and any party to the dispute shall not disclose to any other person any information or document provided to him in connection with the adjudication which the party supplying it has indicated is to be treated as confidential, except to the extent that it is necessary for the purposes of, or in connection with, the adjudication.

It is, therefore, necessary for the party disclosing the information to indicate that it requires the information to be treated as confidential. **23.282**

By contrast, the TeCSA Rules (version 2.0) state that: **23.283**

> The Adjudication and all matters arising in the course thereof are and will be kept confidential by the Parties except insofar as necessary to implement or enforce any decision of the Adjudicator or as may be required for the purpose of any subsequent proceedings.

There is, therefore, no need to specifically indicate that a document should be kept confidential if the adjudication is governed by the TeCSA Rules. Nevertheless, the confidentiality provision does not prevent either party from disclosing information or documentation produced for the purpose of the adjudication in any enforcement proceedings. It is not clear in what circumstances it would be argued that documents are required for the purpose of any subsequent proceedings but it is arguable that documentation such as the referral notice, witness statements, and experts' report are discloseable in related court proceedings in any event, particularly if they undermine or support the case of the parties. A witness **23.284** statement or expert's report produced in an adjudication which is inconsistent with a claim or defence advanced in subsequent court proceedings would almost certainly be disclosed as supporting or undermining the case of one or both of the parties.

Response Document

23.285 If the jurisdiction is not challenged, then a substantive response must be prepared within the timetable allowed by the adjudicator or the relevant rules governing the adjudication. There are many ways to present such a response and thought should be given as to how to present the arguments in the most effective way.

23.286 When a dispute comes to be determined, it is often difficult to identify which facts each party must prove and whether a simple denial will suffice. It can often be tempting to simply state a position and call on the other side to disprove it. However, care should be taken so that each party is aware of what it is required to prove, and how such proof needs to be provided if that party is to succeed.

Burden of proof

23.287 It is a general principle of English law that a party who asserts a fact must prove that fact.[102] This rule applies to adjudication proceedings in the same way as in arbitration or court proceedings.

23.288 In any adjudication, the claimant will, therefore, be arguing that the adjudicator should reach a particular conclusion which the claimant is putting forward. Such conclusion will be based on certain facts such as, for example, the existence of a defect. It is these facts which constitute the assertions which that party must prove.

23.289 This means that the burden of proof, as it is referred to, is generally on the claimant to demonstrate that the facts are as they are presenting them and that they are, therefore, entitled to whatever remedy or relief they are claiming in the adjudication once the correct legal principles are applied to these facts.

23.290 It rests on the claimant, therefore, to put forward adequate evidence to persuade the adjudicator, within the time of the restricted timetable of adjudication, that its claim is correct. Once the claimant has put forward its evidence, as proof of its assertions, the respondent will need to decide whether to pursue a positive or a negative case, as discussed below.

Standard of proof

23.291 The standard of proof which the party seeking to assert a fact needs to meet is to persuade the adjudicator 'on the balance of probabilities' that the facts as stated are correct. This has been stated to mean that the evidence must be such that the judge/adjudicator can say it is

[102] *Joseph Constantine Steamship Line Ltd v Imperial Smelting Corp Ltd* [1942] AC 154 *per* Lord Maugham: 'The burden of proof in any particular case depends on the circumstances in which the claim arises. In general the rule which applies is *ei qui affirmat non ie qui negat incumvit proeatio* [proof rests on he who affirms not he who denies]. It is an ancient rule founded on the considerations of good sense and should not be departed from without strong reason.'

'more probable than not'[103] or 'more likely than not'[104] that the case put forward is the correct one.

If the probabilities are equal then the burden has not been discharged and the claimant will not have proved its case.[105] On the same basis, if, for example, it can be shown that there are three possible causes for a building defect (one only of which was caused by the defendant) but it cannot be established which one actually caused the problem, the burden will not be discharged and the case will fail. **23.292**

In practice, this means that the respondent only needs to place enough doubt in the adjudicator's mind about the claimant's case that the adjudicator considers that there is an equal chance that the factual case presented by the claimant is correct or not. **23.293**

Positive/negative case

In order for the respondent to persuade the adjudicator that the issue should not be decided in the claimant's favour, the respondent can adopt one or both of two different approaches: **23.294**

(i) it can put forward an alternative version of events that is more likely to have occurred than the version put forward by the claimant (a 'positive case')—if the adjudicator prefers this alternative version of events, then he or she cannot logically reach the conclusion that the claimant has proved its case; or

(ii) it can put forward reasons why the version of events put forward by the claimant is unlikely to have occurred (a 'negative case')—if this negative case is successful, then the adjudicator will not be able to conclude that the claimant's case is 'more likely than not' and, again, the claimant will not have proved its case.

The different methods of employing a negative and positive case can be seen in the example of an extension of time claim by a contractor against an employer. **23.295**

In an extension of time claim, the contractor would be arguing that he was directly prevented from completing the works by the date for completion by events of a type for which the contract allows him to be granted an extension of time. The contractor might put forward a programme showing the critical path on the project and how the events in question affected that critical path. **23.296**

In any defence to this claim, the employer might raise the following arguments: **23.297**

(i) a negative case criticizing the programme put forward by the claimant—for example the logic links—to show that this could not be correct; and/or

(ii) a positive case putting forward an alternative critical path with supporting evidence.

Given the practicalities of the adjudication process it is relatively easy for the respondent to put forward a persuasive negative case. This is because: **23.298**

103 *Miller v Ministry of Pensions* [1947] 2 All ER 372.

104 *Secretary of State for the Home Department v Rehman* [2003] 1 AC 153.

105 *Miller v Ministry of Pensions*, n 103 above.

(i) it is difficult for the claimant to put forward significant amounts of further information to resist any arguments which the respondent raises without risking straying outside of the 'dispute' that has been referred; and

(ii) even if the claimant is able to put in a reply, the timetable for the adjudication would mean that such reply would need to be prepared very quickly.

23.299 An effective way to cast doubt in the adjudicator's mind concerning the claimant's case is to concentrate on drawing out any factual discrepancies and flaws in the evidence put forward by the claimant.

23.300 Depending on the type of matter or dispute being referred, the claimant's evidence could amount to a substantial quantity of documentation. Any contemporaneous records or documentation is unlikely to have been produced with adjudication in mind. Considering the practical difficulties in keeping contemporaneous documentary records of the project as it progresses, it is likely that this documentation will be incomplete in some respects and also contain minor discrepancies. Any such issues can be usefully drawn out in any response.

23.301 At the same time, it should be appreciated that an adjudicator will be more impressed by a positive case, supported by evidence, than by a negative case which merely questions the claim but does not put forward an alternative version. Ideally, therefore, while the claim should be questioned as far as possible, it is often advantageous, where possible, to put forward a positive case as well as a negative case.

Witness statements/expert evidence

23.302 Whether witness statements or expert evidence should be used in any defence will depend upon the circumstances of the dispute and whether it is likely that any adjudication decision will be re-considered by the courts or in arbitration.

23.303 The advantage of producing witness statements and expert evidence is that it provides credible evidence on its face, especially if it comes from an independent expert. Adjudicators recognize that parties' representatives will argue their own case and tend to place more reliance on external experts who are seen as more independent and who have a reputation to protect. In addition, this evidence is unlikely to be challenged in cross-examination, as there is unlikely to be a hearing held by the adjudicator. This means that there is less scope to explore any inconsistencies or defects in the evidence, in the same way as in court or arbitration proceedings.

23.304 As discussed above, the claimant needs to prove its case on the balance of probabilities, or, in other words, that it is 'more likely than not' that its case is true. If the respondent can put forward conflicting evidence to create sufficient doubt in the adjudicator's mind so that he does not feel the claim has been proven, then on a strict legal interpretation the adjudicator should refuse the claim.

23.305 As the adjudicator will be deciding in most cases the dispute on paper without a hearing, if both sides' evidence looks credible at first glance, it will be hard for him or her to establish which party has the strongest case in the event of a complete conflict of evidence. As mentioned above, the absence of the ability to cross-examine witnesses means that it is more difficult to decide which party's evidence to prefer. However, this proposition should be treated with caution, as it is open to the adjudicator to question a party and its witnesses. While this is uncommon, this does come within the adjudicator's powers to investigate the facts under the 1996 Act and some of the institutional rules allow for this possibility. In any event, there may be disadvantages to providing such evidence if the dispute continues to arbitration or litigation proceedings.

Due to the restricted timescales involved in adjudication, it is likely that any evidence submitted in a defence will need to be prepared within a short period of time. Clearly, the scope for preparation of this evidence will be much reduced from that which would be available in arbitration or litigation proceedings. This means that it often cannot be prepared with the same precision and attention to detail as would be the case in court proceedings. It is for that reason that it is preferable to begin preparations for an adjudication once it is clear a dispute will not be resolved, rather than wait until an adjudication is commenced. **23.306**

In addition, as such statements are not based on a mutual disclosure of documents (as would be the case in court proceedings), they are likely to be less complete and to miss some of the relevant facts. In court proceedings witness statements and experts' reports are only produced after full statements of case are filed and disclosure provided, which ensures they are more accurate and address the relevant issues. **23.307**

If witnesses or experts sign up to evidence as part of adjudication proceedings, and then the evidence needs to be further refined in the future due to further investigation or consideration of the issues, this is likely to be picked up by the other side in any subsequent proceedings. This is likely to undermine the credibility of the evidence and the weight which it will hold in the mind of the judge or arbitrator. **23.308**

Arguments not referred to in previous correspondence

There has been some dispute among commentators on adjudication on the issue of whether a respondent could raise issues in its adjudication defence which it had not previously referred to in correspondence before the adjudication. **23.309**

The issue seems to have been generally settled by the decision in *William Verry (Glazing Systems) Ltd v Furlong Homes Ltd.*[106] The position is that the respondent is able to raise any points it wishes to in order to defend itself against the claim which is being made. The adjudicator's decision will be a nullity if he or she fails to take into account any such defence.[107] However, the defendant will not be able to rely on any cross claims in circumstances where it has failed to serve a valid withholding notice as required by s 111 of the 1996 Act and the dispute which the claimant refers to adjudication is whether or not the defendant is entitled to withhold the sum claimed rather than whether or not the cross **23.310**

[106] [2005] EWHC 138 (TCC). See also *KNS Industries Services (Birmingham) Ltd v Sindall Ltd* (2001) 17 Const LJ 170; and *Construction Group Centre Ltd v Highland Council* [2002] BLR 476.

[107] See *Quartzelec Ltd v Honeywell Control Systems Ltd* [2009] BLR 328.

claims in themselves are meritorious.[108] It should be noted that the courts will not take a technical view in relation to withholding notices.[109]

23.311 The key as to what arguments can be raised by the respondent will, therefore, be related to the scope of the adjudication as defined by the adjudication notice, rather than being restricted to the arguments which the respondent has raised previously.

Practical considerations

23.312 In addition to the substantive considerations raised above, it is also worth considering what form the response should take, to ensure that any arguments are presented in the most persuasive way.

23.313 To start with, as the adjudicator will be unfamiliar with the dispute, it is worth including a brief introduction to the contract and the relevant works, setting out the most relevant events and dates. A chronology could be very useful and could also be included. When corresponding with the other side it can be assumed that the facts are known. However, such assumptions should not be made when writing a response.

23.314 Similarly, any technical issues should be explained by starting with general basic principles and then developing the argument and adding detail. If an expert report is provided, then reference can be made to its conclusions.

23.315 Where there are legal issues in dispute, it will be helpful to set out the legal arguments in full and refer to any relevant case law.

23.316 In general, the response should be user-friendly and easy to read. It should, therefore, be paginated and separated by tabs with a clear contents page. The text should identify, by page number, the location by page number of any documents referred to.

Reply

23.317 There is no provision within the Construction Act, the scheme, or the other published adjudication rules providing for the submission of a reply to the response by the claimant. However, adjudication is intended to be a flexible process, as well as quick, and in reality it will be up to the adjudicator as to whether he or she will allow any further submissions.

23.318 It is common for parties to put forward further submissions if they believe it to be necessary. Unfortunately, this will often lead to yet another counter submission. Adjudicators are cautious in prohibiting such further submissions since they do not want to be accused of failing to observe the requirements of natural justice. In most cases a common sense approach will mean that such additional submissions are allowed but at some point the adjudicator will state that any further submissions will not be considered. Bearing in mind the tight timetable and the duty to give a decision within twenty-eight days, it is clear why

---

[108] *Letchworth Roofing Co v Stirling Building Co* [2009] EWHC 1119 (TCC).

[109] *Thomas Vale Construction v Brookside Syston Ltd* [2006] EWHC 3637 (TCC); and *Windglass Windows Ltd v Capital Skyline Construction Ltd & Anor* [2009] EWHC 2022 (TCC).

at some point the submissions must stop so that the adjudicator has sufficient time to consider the case and make his decision.

Hearing

Adjudicators generally have wide-ranging powers as to the conduct of the adjudication and can, therefore, meet with the parties and hold a hearing if they wish to do so. **23.319**

If an adjudicator does decide to hold a hearing then the parties will wish to consider a number of practical issues including who should attend the hearing on behalf of the party. While a barrister or a solicitor may be able to present submissions in a clear, concise and effective manner, it can often be more convincing if the parties or the employees of the parties directly involved in the dispute can attend the hearing and put their case to the adjudicator in person, particularly in respect of the factual issues which they are best placed to know. **23.320**

The adjudicator has power to issue any directions that he/she feels to be appropriate including allowing the parties to cross-examine each other as to the evidence that has been put forward and to make submissions. **23.321**

It is important to ensure that parties prepare thoroughly for any adjudication hearing. If the adjudicator has not specified the purpose of the hearing and how the hearing should be conducted, then those questions should be raised so that thorough preparation can take place. An adjudicator may use such a hearing to assess the credibility of witnesses and to test and examine the legal and factual submissions that have been put forward. It is, therefore, important that whoever attends the hearing on behalf of each party is fully familiar with the legal and factual issues and is able, as far as possible, to answer any questions raised by the adjudicator. **23.322**

The parties should also ensure that they are fully familiar with the case which is being made against them so that they are in a position to ask questions of the other party which may expose any weaknesses in the other party's case. **23.323**

While the adjudicator may ask questions to test the evidence and the legal submissions of each party, the adjudicator should be careful to ensure that s/he does not begin to introduce new arguments which have not been made by the parties without, at least, giving the parties an opportunity to properly consider those arguments and to respond to them. If adjudicators do introduce new arguments and the parties are not given a proper opportunity to consider and respond to them then there is a chance that any adjudication decision will not be enforceable due to a breach of natural justice. (See paragraph 23.424 et seq for a discussion of the principles that the courts have applied to the question of natural justice in the context of adjudication). **23.324**

Enforcement of an Adjudicator's Decision

In certain cases, obtaining an adjudicator's decision is sufficient to resolve the disputed issue between the parties. This will be the case if the decision only has a declaratory effect or if it is a negative decision in the sense that it finds that no liability to make a **23.325** payment arises. If that is the case, no further steps are required although the decision may be referred to as a defence in any related proceedings. As a matter of principle, under s 108 of the 1996 Act a decision is binding on the parties once it is made, until finally determined by court or arbitration proceedings, and no other steps are required.

23.326 However, in the majority of cases the decision will order one party to pay a sum of money to the other party. In practice, a losing party will sometimes refuse to make payment, especially if it has argued that the adjudicator did not have jurisdiction to make the decision or that the decision was wrong. In such cases it will be necessary to take an additional step and enforce the decision. Unlike a judgment by the court, or an arbitrator's decision,[110] an adjudicator's decision derives its power from the contract between the parties and requires a claim for breach of contract to be enforced.

23.327 The most common way, and often the best way, of enforcing an adjudicator's decision is through an application for summary judgment. Less common is enforcement through an injunction or by way of a statutory demand. These are discussed below in more detail.

Enforcement prior to the 1996 Act

23.328 A useful starting point is the courts' approach to enforcement of adjudication decisions prior to the 1996 Act. Although, as will be discussed below, most adjudications now fall within the 1996 Act, there are still a large number of construction contracts which do fall outside the 1996 Act. The courts do not appear to have given this distinction any significance, but the repeated reliance on Parliament's intentions when the courts consider enforcement could give rise to an argument that non-1996 Act adjudications should be enforced differently, albeit this is unlikely to succeed in reality. This is addressed further below.

23.329 Adjudication was in existence before the 1996 Act came into force, as a pure contractual obligation arising out of an express agreement between the parties. The enforcement of the adjudicator's decision was based on the contractual nature of the obligation to adjudicate and the duty the parties have to do whatever is necessary to give effect to their contractual obligations.[111] However, in one judgment, an adjudicator's decision was contrasted with an arbitrator's award and described as having 'an ephemeral and subordinate character'.[112] Nonetheless, the coming into force of the 1996 Act and the way in which it operates has not changed the basic principle that adjudication arises from a contractual relationship[113] and that an adjudicator's decision relies on the contractual relationship to be enforced.

---

[110] Which can be enforced like a court judgment under s 66 of the Arbitration Act 1996.

[111] *Bremer Vulkan Schiffbau und Maschinenfabrik v South India Shipping Corp Ltd* [1981] AC 909 in relation to arbitration. See also the general implied duty to cooperate as set out in para 13-011 of *Chitty on Contracts*, 29th edn.

[112] *A Cameron v John Mowlem and Co plc* (1990) 25 Con LR 11.

[113] A detailed consideration of the nature of such decisions was carried out by Mark McGaw in 1992, prior to the Construction Act ('Adjudicators, Experts and Keeping Out of Court' (1992) 8 Const LJ 332). McGaw relied on two cases concerning the status of an adjudicator's decision, *Archital Luxfer Ltd v Horton Construction Ltd;* and *A Cameron Ltd v John Mowlem and Co* to argue that summary judgment was an available route to enforcement in appropriate cases and was proven to be right.

Enforcement following the 1996 Act

Although the basis for the adjudication process is contractual, it is impossible to ignore its statutory roots and the aims behind the 1996 Act. The Latham Report recommended adjudication as a quick and cheap method of resolving disputes trying to avoid the problem of costly and lengthy legal battles. As a result, the 1996 Act provided for a reference to adjudication to be made at any time. This right is implied as a term of the contract by s 108 of the 1996 Act. **23.330**

Significantly, however, no guidance is provided as to how such adjudication decisions would be enforced. The only guidance is in the scheme's attempt to use the same mechanism as s 42 of the Arbitration Act 1996. **23.331**

It was, therefore, left to the courts to set out how adjudication decisions could be enforced. The question of enforcement was clarified following the decision of Dyson J (as he then was) in the case of *Macob Civil Engineering Ltd v Morrison Construction Ltd*.[114] Although the actual decision concerned enforcement of a peremptory order under s 42 of the Arbitration Act (as modified by the scheme), it was held that the court would enforce adjudication decisions by granting summary judgment[115] as provided for under CPR Part 24. It is clear from the judgment that the objectives of the 1996 Act were regarded as paramount and that the courts would accordingly do their utmost to ensure that decisions are enforced.[116] **23.332**

The basis for enforcement was also considered in the case of *Bovis Lend Lease Ltd v Triangle Development Ltd*.[117] In that case it was disputed whether an adjudicator's decision ordering payment was enforceable or whether it had been superseded by contractual provisions allowing a payment to be withheld following determination. Judge Thornton QC held that the payment could be withheld, on the basis that the contractual obligation to comply with adjudicator's decisions could be overridden by another contractual term. However, this seems to have been rejected in the *Levolux AT Ltd v Ferson Contractors Ltd*[118] Court of Appeal case. **23.333**

It has been suggested that adjudicator's decisions may be subject to judicial review. It is submitted that this is an incorrect view. An adjudicator's decision is something which can be either enforced or not. It is unnecessary and inappropriate for a dissatisfied party to go to court for a quashing order relating to the decision. If a party is dissatisfied with the decision, then his or her only remedy is to avoid enforcement. So far there has been no reported **23.334**

---

114 (1999) 64 Con LR 1.

115 Note, however, the different approach of HH Judge Humphrey LLoyd QC in *Glencot Development and Design Co Ltd v Ben Barratt & Son (Contractors) Ltd* 80 Con LR 14, finding an adjudicator's decision to be merely an expression of contractual liabilities which does not create an independent cause of action.

116 As before, enforcement through an injunction or a statutory demand is unlikely to be effective or commonly used. To some extent, there is also now an obligation to cooperate where the scheme for Construction Contracts (at Pt I, para 14) applies, which could be used to support enforcement also on the basis of the obligation to cooperate.

117 [2003] BLR 31.

118 [2002] BLR 341.

decision of the High Court where a decision of an adjudicator has been subjected to judicial review.

23.335 The correct legal procedure is for the beneficiary of the decision to apply to the High Court for summary judgment pursuant to CPR Part 24 for the decision to be enforced. Usually the decisions relate to payment of money and such enforcement cases are heard by the Technology and Construction Court. The decisions relate to the ascertainment of contractual obligations and, therefore, private law rather than public law is the route to be preferred.

Enforcing parts of an adjudicator's decision and multiple adjudications

23.336 It is common to find several disputes being referred to an adjudicator in a single adjudication. As a result, the adjudicator's decision will encompass several issues, some of which may be accepted by all the parties and some of which one of the parties may seek to challenge. This raises the question of whether a decision can be enforced partially and to what extent parties are required to comply with a decision which is only challenged in part.[119]

23.337 In *Homer Burgess Ltd v Chirex (Annan) Ltd*,[120] part of the adjudicator's decision was held by the courts to have been made without jurisdiction. However, the case suggested that this did not affect the part of the decision which was made within the adjudicator's jurisdiction. In *Shimizu Europe Ltd v Automajor Ltd*,[121] a party made part payment of the sum ordered by the adjudicator but resisted payment of the outstanding sum on the basis of a jurisdictional objection. Judge Seymour QC rejected the jurisdictional challenge but went on to state that it was not open to a party to simultaneously approbate and reprobate the adjudicator's decision. A party, therefore, has to accept or contest the whole of the decision. However, he also accepted that in certain circumstances (for example, a decision on how much money was due and what extension of the time should be granted) it may be possible to accept one part of a decision but to challenge the other.

23.338 The same issue arose again before Judge Seymour QC in *R Durtnell & Sons Ltd v Kaduna Ltd*,[122] where he confirmed that the doctrine of approbation and reprobation is not applicable to decisions of adjudicators.

23.339 In the more recent cases of *Cantillon Ltd v Urvasco Ltd*[123] *and Bovis Lend Lease Ltd v Trustees of the London Clinic Ltd*,[124] the court gave a clear indication that, where a decision is given in relation to a number of issues, the fact that a decision on one of the issues is unenforceable will not render the decision in relation to the other issues unenforceable.

119 See *Balfour Beatty Construction Ltd v Serco Ltd* [2004] EWHC 3336 (TCC), Jackson J; and *JPA Design and Build Ltd v Sentosa (UK) Ltd and others* [2009] EWHC 2312 (TCC)

120 [2000] BLR 124.

121 [2002] BLR 113.

122 [2003] BLR 225.

123 [2008] BLR 250.

124 [2009] EWHC 64 (TCC).

In *Hart (t/a D W Hart & Son) v Smith & Anor* [2009] EWHC 2223 (TCC) HH Judge Toulmin QC had to consider the approach of the courts in enforcement proceedings to multiple adjudications and in paras 38 and 39 referred to two previous decisions on the subject which accurately reflect the current position: **23.340**

> In *Interserve Industrial Services Ltd v Cleveland Bridge (UK) Ltd* [2006] EWHC 741 (TCC) ('Interserve'), at Paragraph 43 of his judgment, Jackson J set out the following principles relating to set off in multiple adjudications with which I respectfully agree.
>
>> Where the parties to a construction contract engage in successive adjudications, each focussed upon the party's current rights and remedies, in my view the correct approach is as follows: at the end of each adjudication, absence special circumstances, the losing party must comply with the adjudicator's decision. He cannot withhold payment on the grounds of his anticipated recovery in a future adjudication based on different issues. I reach this conclusion both on the express terms of the Act and also from the line of authority referred to in this judgment.

In his recent judgment in *H S Works v Enterprise Managed Services* [2009] EWHC 729 (TCC), (2009) 124 Con LR 69, Akenhead J set out in paragraph 40 the questions to be asked and answered:

> 1) Are both decisions valid?
> 2) If yes, are both decisions enforceable?
> 3) If yes, the court should give effect to both decisions but in a way which does not favour one party or the other.
> 4) How each decision is to be enforced is a matter for the Court exercising its duty to act fairly.

Summary judgment

Once the decision has been made, the adjudicator no longer has to take any further steps and it will then be for the party seeking enforcement to decide on the appropriate procedure for enforcement. In most cases, the rules chosen by the parties will reflect the requirement of s 108(3) of the 1996 Act stating that the adjudicator's decision is binding on the parties but not final. Paragraph 24 of the scheme provides that the decision, if expressed in peremptory terms, may be enforced under provisions similar to s 42 of the Arbitration Act 1996. However, this route is now unlikely to be followed. **23.341**

Any uncertainty as to the proper way of enforcement was dealt with in the case of *Macob Civil Engineering Ltd v Morrison Construction Ltd*,[125] when Dyson J (as he then was) confirmed that a decision of an adjudicator is enforced summarily regardless of any procedural irregularity, error or breach of natural justice (Dyson J's ruling as to the effect of the breach of natural justice has not been followed subsequently). Since then, enforcement of an adjudicator's decision has been by commencing proceedings in the Technology and Construction Court, followed by an immediate application under Part 24 of the CPR for summary judgment. **23.342**

[125] (1999) 64 Con LR 1.

23.343 It is, therefore, necessary to consider how summary judgment can be applied for under CPR Part 24 and what are the procedural requirements for such an application. To start with, CPR Part 24.2 sets out the test for granting summary judgment as follows:

> The court may give summary judgment against a claimant or defendant on the whole of a claim or on a particular issue if:
>
> (a) it considers that:
> (i) the claimant has no real prospect of succeeding on the claim or issue; or
> (ii) the defendant has no real prospect of successfully defending the claim or issue; and
> (b) there is no other compelling reason why the case or issue should be disposed of at trial.

23.344 It is, therefore, necessary to establish that the defendant has no real prospect of successfully defending a failure to abide by an adjudicator's decision and that there is no other reason why the case should go for trial. Once this is established, the court will give judgment in favour of the claimant which can then be enforced in the same way as any other court judgment.

23.345 In order to prove that there is no real prospect of successfully defending the claim, it will be necessary to show that the defendant does not have a defence that is better than merely arguable.[126] It should be noted that the hearing of a summary judgment application is not supposed to be a brief rehearsal of the trial itself. The court will usually only consider written evidence, in the form of witness statements and documents, and will decide on the basis of such evidence whether the defence has sufficient merit to allow the case to proceed to trial (see Lord Woolf MR in *Swain v Hillman*[127]). In order to resist enforcement of an adjudicator's decision, the defendant must, therefore, show that his chances of success are realistic rather than merely fanciful.

23.346 The inclusion of the word 'real' in the test for summary judgment means that it is necessary to show a case that is better than merely arguable (*International Finance Corp v Utexafrica*[128] and *ED&F Man Liquid Products Ltd v Patel*[129]). It is not necessary to show that the case will probably succeed at trial. While the hearing of the application is not a summary trial and will not include cross-examination of witnesses and a consideration of the probative value of the evidence, the court does not have to simply accept whatever a party says in his statement of case or witness statements.

23.347 The other test which needs to be satisfied is that there is no other compelling reason for a trial. This test is unlikely to be relevant or applied in applications for the enforcement of an adjudicator's decision.

23.348 As to the burden of proof, it was established in *ED&F Man Liquid Products Ltd v Patel* that the overall burden of proof rests on the claimant to establish that there are grounds to believe that the defendant has no real prospect of success and that there is no other compelling reason for trial. It is for that reason that it is necessary to state that the application is

[126] *International Finance Corp v Utexafrica Sprl* [2001] LTL 16 May.
[127] [2001] 1 All ER 91.
[128] [2003] CLC 1361.
[129] [2003] EWCA Civ 472.

made because the claimant believes that on the evidence the defendant has no real prospect of successfully defending the claim. The claimant will satisfy his burden of proof by adducing credible evidence in support of his application. It will then be for the defendant to produce other evidence which will establish some real prospect of success or some other reason for a trial.

An application for enforcement could also be supported by any express contractual terms which require payment. This was used in the case of *Levolux AT Ltd v Ferson Contractors Ltd*,[130] where Mantell LJ held that the terms of a contract must be construed so as to give effect to the adjudicator's decision. This was in a situation where the parties had expressly agreed to be bound by the adjudicator's decision. **23.349**

Although an application for summary judgment will usually ask the court to give judgment on the whole of the claim, in certain circumstances it may also be advisable to request an interim payment at the same time. This helps to address any concerns as to whether summary judgment could be obtained and, in such a case, if it can be shown that an amount of money would have to be paid in any event, then the court may order an interim payment even if it is not willing to order summary judgment (see for example *Glencot Development & Design Co Ltd v Ben Barrett & Son (Contractors) Ltd*[131]). Similarly, in certain cases it may be possible to ask for summary judgment only in relation to one issue rather than the whole claim. This may be useful, for example, where the adjudicator's decision has been challenged in respect of some of the disputes but not all. **23.350**

Procedure for applications under CPR Part 24

An application for summary judgment can only be issued once proceedings have been commenced. Accordingly, the first step is to commence proceedings by way of a Part 7 Claim Form and Particulars of Claim in the Technology and Construction Court. Such a claim is relatively straightforward and needs to set out the following elements: **23.351**

(i) the existence of a contract which falls within the 1996 Act or which contains a contractual adjudication provision;
(ii) details of the dispute which has arisen;
(iii) the circumstances of the adjudication notice and referral to adjudication and the fact that the adjudication took place;
(iv) the adjudicator's decision in favour of the claimant; and
(v) the defendant's failure to comply with the decision of the adjudicator.

The claim will also include a claim for interest in respect of the period from when the amount was payable until payment. **23.352**

Once the claim form and particulars of claim containing the above elements have been filed with the court, they will then have to be served on the defendant. The provisions regarding service are set out in the CPR Part 6. It is necessary to ensure that service is effected correctly in order to avoid any disputes as to whether the claim documents have been served. **23.353**

[130] [2003] BLR 118.
[131] [2001] BLR 207.

23.354 Together with the claim form and particulars of claim, the claimant will also need to serve what is known as a 'Response Pack', which contains some guidance for the defendant as to how to respond to such a claim and, importantly, an 'Acknowledgement of Service Form'. Under CPR Parts 10 and 15, the defendant then has a choice of either acknowledging service of the claim within fourteen days of service (in which case it will have twenty-eight days from the date of service to file its defence) or filing its defence within the required period of fourteen days from service. In most cases, defendants will choose to file an acknowledgement of service rather than a defence, in order to obtain more time in which to file the defence.

23.355 However, the Technology and Construction Court Guide requires claimants to apply, at the time the claim form and particulars of claim are filed, to the court without notice for an order:

(i) abridging time for the filing of the acknowledgment of service;

(ii) giving permission to the claimant to apply for summary judgment before the acknowledgment of service is filed; and

(iii) setting a date for the hearing of the summary judgment application and giving directions as to the steps to be taken in advance of that hearing.

23.356 The Technology and Construction Court Guide appends the standard directions which are usually given by the court in response to such an application.

23.357 However, in the absence of such directions, the claimant must wait until the acknowledgement of service has been filed, in which the defendant confirms that it intends to dispute the claim, before applying for summary judgment. Otherwise, an application for summary judgment can only be made once the defence has been served.[132]

23.358 An application for summary judgment is made by an application notice in accordance with CPR Part 23, supported by a witness statement setting out the evidence which is relevant for the application. A draft of the order sought should also be filed. Filing the application also requires the appropriate court fee to be paid.

23.359 If the application is made before a defence has been filed, there is no need to file a defence until the hearing of the application for summary judgment. Once the date has been fixed for the hearing of the summary judgment application, it is necessary to give the defendant at least fourteen days' notice of the date fixed for the hearing and of any issues which it is proposed the court will decide at the hearing.[133]

23.360 Practice form N244 can be used for the purpose of the application notice, if it complies with the formal requirements in paragraph 2.1 of the Practice Direction supplementing CPR Part 23. In addition, it must state that it is an application for summary judgment made under CPR Part 24, state what order the applicant is seeking and, briefly, why he is

---

[132] CPR r 24.4(l).

[133] CPR r 24.4(3).

seeking that order, identify the written evidence on which the applicant relies and draw the attention of the defendant to CPR Part 24.5(1).

In most cases an application for summary judgment will be heard by a TCC judge and if the application is intended to be made before a judge not of the TCC, the application notice should state so. **23.361**

The evidence in support of the application should be in the form of a witness statement complying with the requirements of CPR Part 32. In addition, it must also identify precisely any point of law or provision in a document on which the applicant relies, state that it is made because the applicant believes that on the evidence the defendant has no real prospect of succeeding on the claim and that the claimant knows of no other reason why disposal of the claim should await trial.[134] **23.362**

The witness statement must be served together with the application. If the defendant then wishes to rely on any written evidence at the hearing, such evidence will also be in the form of a witness statement and must be filed and served on the other parties at least seven days before the date of the hearing.[135] The party seeking enforcement of the adjudicator's decision then has the option of putting in further witness statements in reply, which must be filed and served at least three days before the date of the hearing.[136] In practice, the applying party will arrange for a paginated bundle containing the claim form, particulars of claim, the application notice, and all of the written evidence served by the parties to be filed with the court prior to the hearing. **23.363**

In addition, it is necessary to consider the question of costs. The general rule is that costs follow the event and, therefore, the successful party will expect its costs to be paid by the other party. Part 44 of the CPR deals with the general rules on costs but in most summary judgment applications costs will be awarded on the basis of summary assessment under paragraph 48.16 of the Practice Direction to CPR Part 48. **23.364**

This requires any party seeking its costs to file and serve a schedule of its costs at least twenty-four hours before the hearing.[137] The schedule must include details of the fee earner's grades and hourly rates, as well as any disbursements and counsel's fees. Failure to file such a schedule means a party may not recover its costs, even if it wins the application. **23.365**

Following the hearing, the court will decide whether to grant summary judgment, in which case a court order will be made, giving judgment and putting an end to the proceedings. If the application fails, the court will then make directions as to the further conduct of the trial and, in particular, the date by which the defence has to be filed if it has not been filed already. **23.366**

In addition, under CPR Part 24.6, the court may also attach conditions to any order it makes. This could be useful when the court is minded to allow a defendant to defend a **23.367**

[134] Para 2(3) of the Practice Direction supplementing Pt 24.
[135] CPR r 24.5(1).
[136] CPR r 24.5(2).
[137] Practice Direction to Pt 44, para 13(4).

claim at trial rather than give summary judgment, but has concerns as to the defendant's financial position and ability to make a payment in the main claim and any costs orders. In such circumstances, the court may order that the defendant is required to deposit a certain amount of money with the court before the dispute can proceed to trial, failing which summary judgment will be granted.

Proceedings under CPR Part 8

23.368 The usual way to commence proceedings under the CPR is through a claim form issued under Part 7. However, a claim form can also be issued under CPR Part 8 where the claim is unlikely to involve a substantial dispute of fact. In many cases this can be suitable for enforcement of an adjudicator's decision because all that the court will be looking at is whether the adjudicator had the jurisdiction to make his decision and the court, importantly, is not concerned with the underlying facts or the correctness of the adjudicator's decision. If Part 8 proceedings are used, then the same procedure outlined above for an application for summary judgment is to be followed, save that a defendant does not have the option of filing a defence.

Statutory demand

23.369 In addition, an adjudicator's decision ordering payment could be enforced by a statutory demand followed by winding up proceedings. In practice, this is of little help since a genuine cross claim would result in the demand being set aside.[138] Similarly, while an application could be made for an injunction to enforce a decision,[139] this is rarely used as an enforcement method.[140]

23.370 Under the Insolvency Act 1986, it is possible for a creditor to apply for a company to be wound-up on the basis that it is unable to pay its debts (s 122(1)(f)). Section 123 of the Insolvency Act sets out the basis on which it can be determined whether or not a company can pay its debts, since this is not simply done by reference to its accounts. On this basis, it is sufficient to demonstrate that a written demand, in a prescribed form, has been served on a company for a sum exceeding £750 and that no payment has been made within the following three-week period. The written demand is known as a statutory demand and it can only be used in relation to an amount which is payable under a contract or has been determined in a judgment. It cannot be used in circumstances where it is not possible to identify the specific sum owed. Since an adjudicator will often specify the precise amount which is to be paid, it is possible to issue a statutory demand in order to enforce an adjudicator's decision.

---

[138] r 6.5(4) of the Insolvency Rules 1986.

[139] The power to grant such an injunction was recognized by the House of Lords in *Channel Tunnel Group Ltd v Balfour Beatty Construction Ltd* [1993] 1 All ER 664. In practice a court is unlikely to grant such an injunction where payment is ordered.

[140] This option was subsequently rejected by Dyson J in *Macob Civil Engineering Ltd v Morrison Construction Ltd* [1999] BLR 93, but a conditional injunction was granted by Bowsher J in *Drake & Scull Engineering Ltd v McLaughlin & Harvey Plc* (1993) 60 BLR 102.

The benefit of taking this route is that issuing a statutory demand is relatively simple and does not result in any substantial costs. Coupled with the threat of winding up, issuing a statutory demand is usually an effective way of recovering undisputed debts. However, if the statutory demand does not result in payment, subsequent winding up proceedings can be costly and are unlikely to be effective if a company is insolvent. This is because, as a general principle, on a company's insolvency secured creditors will be paid first with unsecured creditors receiving a proportionate share of any remaining funds, regardless of which creditor actually applied for the winding up and incurred the costs of such an application. **23.371**

In addition, there are various grounds on which a debtor can apply for the statutory demand to be set aside or to restrain the creditor from applying for a winding up of the company. The grounds for setting aside a statutory demand are set out in r 6.5(4) of the Insolvency Rules 1986 and they include the following: **23.372**

(i) if the debtor has a counterclaim, set off, or cross demand equalling or exceeding the debt;
(ii) if the debt is disputed on grounds which appear to the court to be substantial;
(iii) if it appears that the creditor holds security in respect of the debt; or
(iv) if the court is satisfied on other grounds that the demand ought to be set aside.

The case of *George Parke v Fenton Gretton Partnership*[141] concerned a statutory demand served in an attempt to enforce payment of an adjudicator's decision. The debtor argued that it had a valid cross claim exceeding the amount of the adjudicator's decision and argued that he had commenced proceedings in the Technology and Construction Court to recover the alleged overpayment. Judge Boggis QC recognized that adjudicators' decisions had to be enforced summarily without looking at the merits, but then proceeded to find that there was a valid cross claim, confirmed by the existing proceedings in the Technology and Construction Court and that, therefore, the statutory demand should be set aside. **23.373**

In the case of Oakley v Air Clear Environmental Ltd[142] enforcement of an adjudicator's decision was sought through a statutory demand. On the facts of this case it was held that the adjudicator had not been validly appointed and that his decision was a nullity. Accordingly, the statutory demand was not based on a valid debt and was set aside. **23.374**

However, it will not always be easy to demonstrate a genuine cross claim which will result in the setting aside of a statutory demand. In *Guardi Shoes Ltd v Datum Contracts*,[143] a statutory demand was served in order to enforce an adjudicator's decision, followed by the issuing of a winding up petition. The debtor then sought an injunction on the basis that the statutory demand and petition were an abuse of process because there was a valid cross claim. In this case, the injunction was refused on the basis that the cross claim was presented too late and it was not clear whether there was in fact a valid cross claim. **23.375**

---

[141] [2001] CILL 1712.
[142] TCC, 4 October 2001.
[143] [2003] CILL 1934.

23.376 Similarly, in *Jamil Mohammed v Dr Michael Bowles*,[144] a statutory demand was served in order to enforce an adjudicator's decision. Enforcement was resisted on the basis that the adjudicator had no jurisdiction since the contract was with a residential occupier. It was held that the form of contract did contain an adjudication clause and that it was not for the bankruptcy court to look behind the adjudicator's decision. The court expressed a view that if the debtor believed that this was a real issue, then the proper course would have been to apply to the court for a declaration.

Does an adjudicator's decision create a debt or an independent cause of action?

23.377 A question which is yet to be clarified is the precise status of an adjudicator's decision. The 1996 Act does not provide any guidance as to what is the status of an adjudicator's decision. It is merely stated that such decisions are binding until finally determined.

23.378 In *VHE Construction plc v RBSTB Trust Co Ltd*,[145] there were two sequential adjudications. The sum determined in the first adjudication was substantially reduced in the second adjudication and it was necessary to consider the effect of the second adjudication and, specifically, whether the amount in the first adjudication was reduced or whether it remained valid but that the difference between the first and second adjudication decision was to be immediately repaid. The court preferred the second option, on the basis that the second adjudicator did not have the power to revise the first adjudicator's decision.

23.379 A different view was expressed by Judge LLoyd QC in *David McLean Housing Contractors Ltd v Swansea Housing Association Ltd*,[146] who stated that the cause of action in proceedings for enforcement is not the breach of any obligation to comply with the decision but the breach of the obligation under the contract which was the subject of the dispute referred to adjudication.

23.380 The judge re-affirmed his view in the case of *Glencot Development & Design Co Ltd v Ben Barrett & Son (Contractors) Ltd*,[147] where Judge LLoyd QC held that the cause of action is the right or the obligation in dispute. He did not agree with the approach taken in *VHE Construction*, holding that enforcement was in respect of a right under the contract and that the adjudicator's decision was not a cause of action by itself.

23.381 The same issue was briefly considered in *Parsons Plastics (Research and Development) Ltd v Purac Ltd*,[148] where the Court of Appeal concluded that it is acceptable for a party to set off against the adjudicator's decision any claims it had against the other party which had not been determined by the adjudicator. However, it should be noted that this was not a statutory adjudication and that in cases where the 1996 Act applies, the effect of s 111 will be that such set off will not be possible unless a valid withholding notice has been served. In addition, the contract in that case provided that a set off or an abatement was available against an adjudicator's decision.

[144] TCC, 11 March 2003.
[145] [2000] BLR 187.
[146] [2002] BLR 125.
[147] [2001] BLR 207.
[148] [2002] BLR 334.

Finally, the Court of Appeal also considered the status of an adjudicator's decision in the case of *Levolux AT Ltd v Ferson Contractors Ltd.*[149] The defendants in this case sought to rely on a term of the contract that provided for no further payment to be made to defend a refusal to make a payment in accordance with an adjudicator's decision. It was held by Mantell LJ that the terms of the contract should be construed so as to give effect to the adjudicator's decision and held that such clauses must be read so they do not apply to amounts due by reason of the adjudicator's decision. **23.382**

Non-1996 Act adjudications

As mentioned above, it is worth noting that the courts have enforced adjudication decisions which do not fall within the 1996 Act in the same way as statutory adjudications. Contractual adjudication arises through an express agreement by the parties. In contrast, statutory adjudication arises through the implication of an adjudication agreement by law (unless a statutory compliant procedure has already been agreed by the parties). While the source of the obligation may be different, it is clear that in both cases the obligation is contractual and it is difficult to see why these contractual obligations should not be enforced in the same way when applying the test in CPR Part 24. **23.383**

This was recognized in the case of *Nordot Engineering Services Ltd v Siemens plc*[150] in which the judge considered an agreement to adjudicate which did not come within the 1996 Act. The judge made it clear that he treated such an agreement as equivalent to a statutory agreement to adjudicate and considered that it should be enforced in the same way. This decision was approved by the Court of Appeal in *Thomas Fredrics (Construction) Ltd v Keith Wilson*,[151] who regarded the 'ad hoc submission' to adjudication as reflecting s 108(3) of the 1996 Act. **23.384**

The same approach was taken by Judge Seymour QC in the case of *Galliford Try Construction Ltd v Michael Heal Associates Ltd.*[152] In that case, the judge found that an agreement to submit disputes to adjudication arose as a result of the exchange of correspondence between the parties' solicitors. The judge then went on to consider the effect in law of such an agreement, finding that the parties were agreeing to the process which was to be provided under the 1996 Act.[153] **23.385**

Significantly, this approach suggests that an agreement to submit disputes to adjudication is likely to be interpreted by the courts as an implied agreement to submit disputes to adjudication, and have them enforced, as if they were governed by the 1996 Act.[154] However, in *Galliford Try v Michael Heal Associates Ltd*,[155] the parties did not refer to the 1996 Act **23.386**

[149] [2003] BLR 118.

[150] [2001] CILL 1778.

[151] [2003] EWCA Civ 1494.

[152] [2003] EWHC 2886.

[153] Note, however, Judge Seymour's suggestion at [29] that terms implied by law may be different from terms implied to express the intentions of the parties.

[154] Judge Seymour QC took the same view in *RSL (South West) Ltd v Stansell Ltd* [2003] EWHC 1390 (TCC) at [31] and [37], when considering the application of natural justice principles to an adjudication, noting the pure contractual nature of adjudication.

[155] [2003] EWHC 2886.

in their correspondence, referring only to the Construction Industry Council Model Adjudication Procedure.

23.387 The same principle was subsequently applied in *Westminster Building Co Ltd v Beckingham*.[156] In that case, Judge Thornton QC considered a contract in relation to a dwelling house which did not fall under the 1996 Act.[157] It was found by the judge that the agreement between the parties incorporated the terms of the IFC 84 Form, including an adjudication provision. On that basis, the judge found that the adjudicator's decision was valid and awarded summary judgment in favour of the claimant.[158]

23.388 However, in *Parsons Plastics (Research and Development) Ltd v Purac Ltd*[159] the fact that it was not a statutory adjudication was found to mean that the 1996 Act provisions on withholding did not apply and there could be a set off.

23.389 It is clear from these decisions that while the courts do recognize the difference between contractual and statutory adjudication, they regard the enforcement of such decisions as identical[160] in practical terms. The courts have also gone further, finding that a contractual agreement to adjudicate a dispute can be interpreted as having been intended by the parties to have the same effect as statutory adjudication, even if no reference is made to the 1996 Act.

23.390 Putting the 1996 Act aside, the courts' approach must be considered in the context of CPR Part 24. It is provided that summary judgment can be applied for in circumstances where the defendant has no real prospect of success and there is no other reason why a trial should be ordered.[161] It is therefore clear that when an adjudicator's decision comes before the court, whether statutory or contractual, all that the court is concerned with is whether the defendant has a real prospect of defending such a claim. Strictly speaking, Parliament's intentions in passing the 1996 Act are irrelevant to the court's assessment of the defendant's case. In circumstances where the adjudicator's decision is valid and no challenge can be made to his jurisdiction, a court will enforce such a decision.

23.391 At the same time it must also be recognized that the courts' approach remains heavily influenced by the aims of the 1996 Act. It is clear that the courts still treat adjudication under the 1996 Act as a special case, trying to promote what are perceived to be Parliament's aims in passing the legislation[162] and in doing so often ignoring manifest errors. It can, therefore,

---

[156] [2004] BLR 163.

[157] Contracts with residential occupiers are excluded under s 106(2) of the 1996 Act.

[158] Interestingly, the judge also considered whether the adjudication clause was unfair within the meaning of the Unfair Terms in Consumer Contracts Regulations 1999. Having regard to previous decisions and the provisions of the Regs, the judge concluded that the adjudication clause was not unfair and therefore binding.

[159] [2002] BLR 334.

[160] See Judge Seymour's approach in *Bryen & Langley Ltd v Martin Rodney Boston* [2004] EWHC 2450 (TCC), where contractual adjudication was considered in the light of cases decided in statutory adjudications.

[161] CPR r 24.2.

[162] Note that the same approach has not been taken in other cases of terms implied by law, eg terms as to fitness and quality under the Sale of Goods Act 1979.

be questioned whether it is justified to enforce all adjudication decisions in the same way as under those contracts which Parliament saw fit to include in the 1996 Act.

Bearing in mind Judge Seymour QC's approach in *Galliford Try Construction Ltd v Michael Heal Associates Ltd*,[163] it is still worth noting that a term usually will not be implied if it is not necessary, even if it would be reasonable to do so.[164] Finding an implied term that contractual adjudications are to be enforced in the same way as statutory adjudications cannot be justified simply because it is a reasonable and practical approach. **23.392**

However, implied terms can also come into effect through market practice in specific industries.[165] A better argument could, therefore, be made to the effect that when parties in the construction industry agree to refer disputes to adjudication, they expect such adjudications to be treated in the same way as adjudications under the 1996 Act. The popularity of adjudication supports the view that parties are happy to accept the courts' approach, recognizing that while it may not be perfect, it does provide a quick and cheap way of resolving disputes. **23.393**

Nonetheless, it is worth noting that the courts' approach to enforcement has come under some criticism and it is possible that parties to a contract which does not come within the 1996 Act (for example, for the construction of power generation plants) may seek to have the benefit of adjudication, while at the same time limiting the courts' power of enforcement. **23.394**

One way of doing so would be to agree that decisions should not be enforced in cases of manifest error, as is commonly done with expert determination, as this may prevent decisions such as *Bouygues UK Ltd v Dahl-Jensen UK Ltd*.[166] Other limitations could relate to when an adjudication could be brought (for example, not at any time but within certain time limits), pre-conditions to bringing an adjudication (for example, a requirement to attempt mediation or deposit funds) or limiting adjudication only to payment disputes (as originally intended). **23.395**

As long as such agreements are not governed by the 1996 Act, the courts will have to accept such contractual terms as binding. However, such a development is undesirable in potentially creating two separate bodies of law governing how adjudication decisions are dealt with and a collection of 'bespoke' adjudication agreements, seeking to have the benefits of the 1996 Act approach but not its perceived disadvantages where this may be inappropriate (for example, professional negligence claims). **23.396**

In addition, cases demonstrate that in practice adjudication often appears as the contractual method of resolving disputes in residential contracts. This is a result of contractors seeking to use a process with which they are familiar and which they feel often favours **23.397**

[163] [2003] EWHC 2866.

[164] However, see the modern approach on implied terms set out in Ch 1 following *Attorney General of Belize & Ors v Belize Telecom Ltd & Anor* [2009] UKPC 11. They are a matter of construction. But necessity rather than reasonableness will usually be the deciding factor unless as a matter of construction it must be held otherwise.

[165] *Chitty on Contracts*, para 13-018.

[166] [2000] BLR 522.

their position. While the courts' approach should not cause any difficulties in situations where both parties are commercial entities with experience of the construction industry, it will be more difficult to justify the same robust approach being applied to contracts with residential occupiers. Such occupiers have no experience of the construction industry and are rarely in a position to defend themselves effectively in an adjudication.

23.398 However, while the courts did conclude that an adjudication clause in a contract with a residential occupier was unenforceable under the Unfair Terms in Consumer Contracts Regulations 1999 in the case of *Picardi v Cuniberti*,[167] they have more often reached the opposite conclusion.[168]

Enforcement outside England and Wales

23.399 Where enforcement is sought under a contract which comes within the 1996 Act and the parties are based in England and Wales, then the preferable route is through summary judgment as outlined above. However, effective enforcement could become complicated if it is sought against a party which is not based in the United Kingdom.

23.400 The first question will be whether the foreign party takes part in the adjudication. If it does not, then it is still open to the adjudicator to make a decision. However, even if the party did take part in the adjudication, this does not guarantee that the foreign party will not seek to avoid its obligations, especially if it has no assets within the United Kingdom.

23.401 In such a case, the question is whether it would be possible to enforce an adjudicator's decision outside England and Wales. Since adjudication is based on contract, it should be possible to enforce the decision abroad through an action for breach of contract, subject to such an action being allowed under the relevant private international law legal principles.

23.402 A better course of action is likely to be to seek and obtain summary judgment in England and Wales. Once the judgment is obtained it can be enforced like any judgment. Although this will be a question for the applicable law at the place of enforcement, most countries tend to recognize and enforce valid decisions of the courts of England and Wales. Within the EU such matters are governed by EU Regulation 44/2001 which aims to simplify and expedite the process under which Member States recognize and enforce judgments obtained in other Member States.

The Slip Rule

23.403 In some cases, a party will seek to avoid enforcement if it discovers an error in the decision. Similarly, a party may seek to enforce a decision but discover it contains an error. Consideration, therefore, also needs to be given to cases where the adjudicator has made a clerical or arithmetical error. In arbitration this is dealt with under s 57 of the Arbitration

[167] [2003] BLR 487.

[168] See *Lovell Projects Ltd v Legg & Carver* [2003] BLR 452; *Westminster Building Co Ltd v Beckingham* [2004] BLR 163; *Bryen Langley Ltd v Martin Rodney Boston* [2005] EWCA Civ 973; and *Steve Domsalla (t/a Domsalla Building Services) v Kenneth Dyason* [2007] BLR 348.

Act 1996, where provision is made for correction by the Tribunal. Although some errors are inadvertent, not all of them will come within the ambit of the slip rule and where the error is of a fundamental nature it will impact on whether the decision can be enforced.

In *Joinery Plus Ltd v Laing Ltd*,[169] the same adjudicator considered two adjudications between the same parties but relating to different sub-contracts. The adjudicator referred to the first decision as a template and in his second adjudication decision made wrong references throughout, including references to the wrong contractual terms. It was held by the court that the adjudicator failed to decide a dispute on the basis of the contract and that his error went to the basis of his jurisdiction. As a result his decision could not be enforced. **23.404**

The 1996 Act does not make any provision for correcting an error in the decision after it has been reached. However this may be subject to change if section 108(3A) of the 1996 Act is brought into force.[170] The law is considered on the basis that the Secretary of State has not brought the section into force–as indeed it had not been brought into force at the time of going to press. It was suggested in *Outwing Construction Ltd v H Randell & Son Ltd*[171] that once a decision has been given the adjudicator becomes functus officio and has no power to make any further changes. However, it has subsequently been held that an adjudicator has the power to correct slips which do not alter the reasoning or intention of his decision, notwithstanding the absence of a statutory provision. This was decided in *Bloor Construction (UK) Ltd v Bowmer & Kirkland (London) Ltd*.[172] Judge Toulmin QC considered the circumstances under which an adjudicator may alter his own decision, to correct a clerical mistake or error arising from an accidental slip or omission. **23.405**

He relied on the powers given to arbitrators under the Arbitration Act 1996 and held that, in the absence of any specific agreement to the contrary, a term should be implied into the contract that the adjudicator may, on his own initiative or on the application of a party, correct an error arising from an accidental error or omission. **23.406**

Although the *Bloor* decision has been the subject of some criticism, it subsequently received further support in the case of *Edmund Nuttall Ltd v Sevenoaks District Council*[173] where Dyson J held that it was arguable that the adjudicator had the power to correct his error. The same question came up again in the case of *CIB Properties Ltd v Birse Construction Ltd*[174] also before Judge Toulmin QC. He followed his earlier decision and concluded that in relation to a slip or alleged slip the relevant questions were: (1) is the adjudicator prepared to acknowledge that he has made a mistake and correct it? (2) is the mistake a genuine slip which failed to give effect to his first thoughts? He held that if the answer to both questions is yes then, subject to the time within which the correction is made and questions of prejudice, the court can, if the justice of the case requires, give effect to the amendment to **23.407**

169 [2003] BLR 184.

170 s 108(3A) may be inserted into the 1996 Act pursuant to the Local Democracy, Economic Development and Construction Act 2009 should the Secretary of State bring that provision into force.

171 [1999] BLR 156.

172 [2000] BLR 314.

173 [2000] WL 1544581.

174 [2005] BLR 173.

rectify the slip. If the adjudicator does not acknowledge the mistake then the adjudicator's decision will be enforced as it stands and no amendment will be made.

23.408 In the case of *YCMS Ltd (t/a Young Construction Management Services) v Mr and Mrs Grabiner*,[175] while the court made it clear that an adjudicator can correct an arithmetical error in a calculation in accordance with the slip rule, it is not open to him or her to produce an entirely different calculation having reconsidered his or her conclusion. In addition, the court indicated that the correction of an error within 48 hours of the error being pointed out would usually be considered to be reasonable.

23.409 In conclusion, it appears that an adjudicator may correct a clerical mistake or error in his decision arising from an accidental slip or omission. The correction can be made by the adjudicator of his own volition or as a result of an application by one of the parties. It also appears that the correction must be made within a reasonable time, although it is less clear what would be considered a reasonable time and this is likely to depend on the circumstances of each case. Finally, the correction can only be made if it does not prejudice any of the parties and it requires the adjudicator to acknowledge and correct the error himself.

Resisting Enforcement

23.410 The objections to enforcement which have been taken are:

(i) the construction contract with which the adjudication was concerned pre-dated the coming into force of the 1996 Act and, therefore, the 'adjudication' was a nullity;

(ii) the adjudicator was not properly appointed, making the decision a nullity;

(iii) the contract pursuant to which the decision was made did not concern a construction contract and, therefore, fell outside the scheme;

(iv) the adjudicator's decision did not address the issue referred;

(v) the decision was made contrary to natural justice;

(vi) there was no dispute between the parties at the time of the service of the adjudication notice and consequently there was no proper reference to adjudication; the adjudicator's decision was issued out of time; and the insolvency of the party seeking to enforce the decision.

23.411 As mentioned above, the courts take a very robust approach to the enforcement of adjudicators' decisions in order to give effect to the policy of the 1996 Act and it is very rare for challenges to be successful. In addition, it is becoming common for the courts to order the defendant to pay the claimant's costs on the indemnity basis where the court considers that an unmeritorious defence to enforcement proceedings has been advanced.[176]

---

[175] [2009] BLR 211.

[176] See, for example, *Gray & Sons Builders (Bedford) Ltd v Essential Box Co Ltd* [2006] EWHC 2520 (TCC); and *Harris Calnan Construction Co Ltd v Ridgewood (Kensington) Ltd* [2007] EWHC 2738 (TCC).

The construction contract pre-dates the 1996 Act

Section 104(6) of the 1996 Act provides that the 1996 Act applies to contracts entered into after the commencement of Part II. The objection that the contract was entered into prior to the relevant date (1 May 1998) was taken frequently shortly after the coming into force of the 1996 Act and the operation of the scheme. However, the 1996 Act has been in force for so long now that this is, for all practical purposes, a potentially very rare objection. The principles applicable to contract formation are relevant for the purposes of considering this question (see Chapter 1). However it must be noted that work can commence before a contract is entered into and the contract entered into will then have retrospective effect.[177] Therefore, the touchtone is not the date of commencement of work on site. If work is done pursuant to a letter of intent and then a contract is executed, then the later date will be applicable for the purposes of seeing if the 1996 Act applies. **23.412**

Problems may occur if a contract is said only to come into force when certain conditions have been met. There may then be an academic debate as to whether the construction contract was entered into on a certain date and became operative at a later date or whether a deed was signed on one date with the provision that the contract would not be entered into until a later date.[178] These debates may be more of form than substance, the key issue is whether or not the contract was entered into after 1 May 1998 and that is a question of fact. This is only likely to be a live issue with respect to claims for defective work for contracts under seal where the limitation period is twelve years. All other types of dispute are now almost certain to relate to post May 1998 contracts. **23.413**

No proper appointment

The lack of proper appointment is a valid basis for resisting enforcement because, in those circumstances, the person who has acted as an adjudicator has not in fact been an adjudicator within the meaning of the 1996 Act. As stated above, pursuant to the scheme, for there to be a valid appointment there must be served on the other party by the claimant an adjudication notice which is written notice of an intention to refer any dispute arising under the contract to adjudication. The notice must set out the nature of and a brief description of the dispute and the parties involved, details of where and when the dispute arose, the nature of redress sought, and the names and addresses of the parties to the contract. **23.414**

If an adjudicator has not been agreed (by being named in the contract or otherwise) then an appropriate nominating body must select the adjudicator. The nominating body can be named in the contract, failing such provision any body (which is not a party to the dispute and is not an individual person) which holds itself out publicly as a body which will select an adjudicator can be asked by the claimant to nominate an adjudicator.[179] Once an adjudicator has been selected, then the claimant must refer the dispute in writing to the adjudicator. This reference is known as the referral notice. **23.415**

---

[177] *Trollope & Colls Ltd v Atomic Power Construction Ltd* [1963] 1 WLR 1035.

[178] cf *The Atlas Ceiling and Partition Co Ltd v Crowngate Estates (Cheltenham) Ltd* Judge Thornton QC [2000] CILL 1639.

[179] See paras 2, 5, and 6 of the scheme reproduced in App 2 below.

23.416 The above steps, if taken correctly and in sequence, will ensure that the adjudicator is validly appointed and thus competent to produce a valid decision. A failure to carry out the steps properly will ensure that the adjudicator is not validly appointed and thus any decision made will not be enforceable.[180]

23.417 The procedure set out in the scheme requires that the adjudication notice must pre-date the application to the nominating body for an adjudicator.[181] When the scheme is operative, the application to the nominating body prior to the service of the adjudication notice will make the proceedings a nullity.[182]

23.418 There is no reason why parties should not follow the correct chronological sequence, so challenges on these grounds will be rare.

Not a construction contract

23.419 The types of contract which fall within the 1996 Act are defined in s 105 of the 1996 Act. The 1996 Act is designed to catch all construction contracts, which are contracts for the carrying out of 'construction operations'. These are extensively defined and encompass nearly all construction and engineering contracts, save for those with residential occupiers of dwellings. The classes of contract which come within the scheme and those which do not are contained in ss 104 to 107 of the 1996 Act.

23.420 However, it is important to note that this will only be a defence to enforcement in statutory as opposed to contractual adjudications.

Not addressing the issue referred

23.421 As has already been seen, the courts will enforce decisions which are within jurisdiction even if plainly wrong.[183] But decisions which fall outside the jurisdiction of the adjudicator will not be enforced because any such decision would be a nullity.[184]

23.422 The adjudicator's decision stems from the adjudication notice and the referral of that dispute to him or her. An adjudicator cannot derive his/her jurisdiction from anything other than the reference to him/her as defined in the adjudication notice. Therefore, an adjudicator who resolves a dispute which is not the one which is referred has produced a decision which is a nullity.[185]

---

[180] In relation to sequence see for example *IDE Contracting v RG Carter Cambridge Ltd* [2004] BLR 172, [2004] EWHC 36 (TCC) where Judge Havery QC decided that approaching the nominating body prior to service of the notice of adjudication deprived the adjudicator of jurisdiction. This has been followed by Christopher Clarke J in *Vision Homes Ltd v Lancsville Construction Ltd* [2009] EWHC 2042 (TCC).

[181] See paras 1 and 2 of the scheme.

[182] *Ide Contracting Ltd v RG Carter Cambridge Ltd* [2004] BLR 172 (TCC). Cf *Palmac Contracting Ltd v Park Lane Estates Ltd* [2005] EWHC B1 (TCC).

[183] However, see *Steve Domsalla (t/a Domsalla Building Services v Kenneth Dyason* [2007] BLR 348 in which it was held that the court can refuse to enforce an adjudicator's decision on the basis of an error made within his or her jurisdiction where it is a contractual as opposed to statutory adjudication.

[184] *Bouygues (UK) Ltd v Dahl-Jensen (UK) Ltd* [2001] 1 All ER (Comm) 1041, [2000] BLR 522, CA.

[185] See *Ballast plc v The Burrell Company* [2001] BLR 529; *C&B Scene Concept Design v Isobars* (2002) 82 Con LR 154 at paras 29–30.

A common pitfall is an adjudication notice which is narrowly drawn which causes the respondent to bring up matters which are not expressly referred to in the adjudication notice. For example, a contractor may complain about non-payment on the ground of under-valuation and the employer may as an alternative defence state that it is entitled to contra-charge liquidated damages. The question may arise as to whether the claim to recover liquidated damages falls within the adjudication notice and, therefore, the dispute referred. Equally a contractor may require a determination of what a final account valuation is but not put in its adjudication notice a complaint that it is disputed that the final account sum has not been paid. If only a valuation is sought, then the adjudicator will be outside his or her jurisdiction if she/he seeks to order payment of that sum.[186] If a decision is outside jurisdiction then it is not binding.[187] **23.423**

Natural Justice and the Human Rights Act 1998

Natural justice

No statute will be construed as allowing for decisions to be made affecting the rights of any party if it is made contrary to natural justice. But it is impossible to say that the statutory scheme involves or permits a breach of natural justice. Inherent within it is the risk that the adjudicator will not have the time to come to a decision which finally decides the rights of the parties, particularly when the claimant commits an 'ambush' on the respondent by putting before it a mass of documentary material which is hard or nearly impossible to deal with in the time allowed. **23.424**

But the mischief which the 1996 Act seeks to deal with must not be forgotten. In any time-limited process it is always possible for one party to put a strain on it through the presentation of a mass of indigestible material. But without the time limitation the statutory scheme will not work. So it is submitted that objections to the enforcement of an adjudicator's decision on the basis that the respondent did not have sufficient time to respond to the 'ambush' of the claimant should be dismissed. The scheme provides for time limitations which are mandatory and the time limitations cannot of themselves make the scheme inoperable. It is not, therefore, contrary to natural justice if the respondent considers that it did not have sufficient time to respond to the reference. **23.425**

The conduct of the adjudicator is sometimes cited as a reason why a decision should not be enforced as being contrary to natural justice. Natural justice is an elastic concept which must be looked at in the context of what is being done. It applies to adjudication, but it does not apply to, for example, engineer's decisions made pursuant to clause 66 of the ICE conditions.[188] But the applicability of natural justice is contextual. There is a requirement for a speedy decision by a lay tribunal. The parties cannot expect the perfection or near **23.426**

[186] *FW Cook Ltd v Shimizu* [2000] BLR 199 (TCC).

[187] *Thomas-Frederic's (Construction) Ltd v Wilson* (2003) 91 Con LR 161.

[188] See *AMEC v Secretary of State for Transport* [2005] EWCA Civ 291 at [46] *per* May LJ with whom Hooper LJ agreed (at [87]); note Rix LJ dubitante on this point at [83]–[84].

perfection which attends an arbitration. The courts will not apply too strict a test to the conduct of an adjudicator. Obvious unfairness will vitiate a decision.

23.427 The question of natural justice has been considered in two principal contexts. First, that it is easy for a party to assert a breach of natural justice and potentially disrupt the speedy enforcement of decisions because the inquiry into whether there has been a breach will be prolonged.[189] Secondly, there is a line which, if crossed, will mean that natural justice has not been properly observed and, therefore, voids the decision. The effect of the authorities and the 1996 Act,[190] and the scheme may be summarized as follows:

(1) minimum standards of fairness are expected of adjudicators. The adjudicator must have a term within his or her contract that he or she will act impartially;

(2) a basic standard is that a party must not be judge in his or her own cause;

(3) another is that each party must have an opportunity to put its case;

(4) an adjudicator is not tied to court procedure;

(5) fairness will be judged in each case against the test of whether the proceedings can be objectively justified as constituting fair justice between the parties or, put another way, would a fair minded person describe the proceedings as being tainted with bias; and

(6) a fair procedure does not require the adjudicator to ask for the comments of the parties of all matters in every case, fairness must have regard to the volume of material to be considered against the deadline faced by the adjudicator.

23.428 It is simply not possible for the courts to define all of the circumstances in which they would consider a breach of natural justice to be sufficiently serious to warrant refusing to enforce the decision of an adjudicator.[191]

23.429 However, a degree of certainty can be gained by a review of the previous decisions that the courts have made. For example, the courts will be prepared to refuse to enforce a decision of an adjudicator on the grounds that there has been a breach of natural justice where:

(1) the adjudicator has made his or herdecision on the basis of an argument or analysis of his or her own on which he or she has not given the parties an opportunity to comment[192]

---

[189] See *Macob Civil Engineering Ltd v Morrison Construction Ltd* [1999] BLR 93; and *Carillion Construction Ltd v Devonport Royal Dockyard Ltd* [2005] BLR 310.

[190] See particularly s 108(2)(e) which states that the contract must require the adjudicator to act impartially.

[191] See *Balfour Beatty Construction Ltd v London Borough of Lambeth* [2002] EWHC 597; *RSL (South West) Ltd v Stansell Ltd* [2003] EWHC 1390 (TCC); *Amec Capital Projects Ltd v Whitefriars City Estates Ltd* [2005] BLR 1; *Kier Regional Ltd (t/a Wallis) v City & General (Holborn) Ltd* [2006] EWHC 848 (TCC); and *Cantillon Ltd v Urvasco Ltd* [2008] EWHC 282 (TCC); *Multiplex Constructions (UK) Ltd v West India Quay Development Company (Eastern) Ltd* [2006] EWHC 1569 (TCC); *Primus Build Ltd v Pompey Centre Ltd & Anor* [2009] EWHC 1487 (TCC).

[192] See *Balfour Beatty Construction Ltd v London Borough of Lambeth* [2002] BLR 288; *Ardmore Construction ltd v Taylor Woodrow Construction Ltd* [2006] CILL 2309. However, it should be noted that in *Carillion Construction Ltd v Devonport Royal Dockyard Ltd* [2005] BLR 310 at p 35, the Court of Appeal approved Jackson J's statement that:

> It is often not practicable for an adjudicator to put to the parties his provisional conclusions for comment. Very often these provisional conclusions will represent some intermediate position, for

(2) the adjudicator has held telephone conversations with one of the parties in the absence of the other party and failed properly to disclose the nature of those telephone conversations to the party that did not take part in them;[193]

(3) the adjudicator has obtained assistance from an expert but failed to give the parties an opportunity to comment on that expert's final report;[194]

(4) the adjudicator has not given one party the opportunity to comment on information s/he has obtained from the other party or third parties;[195]

(5) the adjudicator has decided disputes between other parties which are related to the dispute in question and arise out of the same project;[196]

(6) the adjudicator has acted as a mediator in the dispute between the parties prior to making his or her decision;[197]

(7) one party has not been given a reasonable opportunity to consider information served by the other party;[198]

(8) the adjudicator criticises the failure of a witness to give oral evidence without requiring a party to arrange for that witness to give oral evidence;[199]

(9) the adjudicator has failed to take into account a defence because he or she wrongly believed that the defence had been determined by an earlier adjudicator's decision;[200] and

(10) the adjudicator failed to consider submissions on the grounds that s/he wrongly concluded that s/he did not have the power to extend the time for the provision of those submissions.[201]

It has been suggested, at first instance, that some disputes, particularly those which arise at the end of a project, may be too complex to permit a fair adjudication process.[202] It is submitted that this is an incorrect view. The 1996 Act proceeded on the basis that some disputes would be complex, fact heavy, and of a nature that it would be impossible to get a perfect result within the time limits set. It would be quite wrong to stigmatize the adjudication decisions relating to such disputes as being unfair in those circumstances. The statute **23.430**

which neither party was contending. It will only be in an exceptional case such as *Balfour Beatty v Lambeth London Borough Council* that an adjudicator's failure to put his provisional conclusions to the parties will constitute such a serious breach of the rules of natural justice that the court will decline to enforce his decision.

193 *Discain Project Services Ltd v Opecprime Development Ltd* [2001] BLR 285 (TCC). See also *Costain Ltd v Strathclyde Builders Ltd* 2004 SLT 102; (2005) 100 Con LR 41 in which an adjudicator's decision was not enforced because the adjudicator failed to disclose advice which he had received from a lawyer.

194 *RSL (South West) Ltd v Stansell Ltd* [2003] EWHC 1390 (TCC).

195 *Woods Hardwick Ltd v Chiltern Air-Conditioning Ltd* [2001] BLR 23.

196 *Pring & St Hill Ltd v CJ Hafner* [2002] EWHC 1775 (TCC); (2004) 20 Const LJ 402.

197 *Glencot Development & Design Ltd v Ben Barrett & Son Contractors Ltd* [2001] BLR 207 (TCC).

198 *London & Amsterdam Properties v Waterman Partnership* [2004] BLR 179; and *McAlpine PPS Pipeline Systems Joint Space Venture v Transco Plc* [2004] EWHC 2030 (TCC); [2004] BLR 352.

199 *A&S Enterprises Ltd v Kema Holdings Ltd* [2004] EWHC 3365 (TCC), [2005] BLR 76.

200 *Quietfield Ltd v Vascroft Construction Ltd* [2007] BLR 67.

201 *CJP Builders Ltd v William Verry Ltd* [2008] BLR 545.

202 *London and Amsterdam Properties v Waterman Partnership* [2004] BLR 179 at para 146, Judge Wilcox in the TCC.

requires accelerated procedures and permits any dispute to be referred, and not just those which are relatively simple. It would be contrary to principle for the courts to refuse to enforce an adjudication decision on the grounds that the dispute was so complex that the accelerated process provided for by adjudication was 'unfair'.

Human Rights Act 1998

23.431 Section 6(1) of the Human Rights Act 1998 provides that it is unlawful for a public authority to act in a way which is incompatible with a Convention right.

23.432 Article 6 of the Convention provides that:

> In the determination of his civil rights and obligations or of any criminal charge against him, everyone is entitled to a fair and public hearing within a reasonable time by an independent and impartial tribunal established by law. Judgment shall be pronounced publicly but the press and public may be excluded from all or part of the trial in the interest of morals, public order or national security in a democratic society, where the interests of juveniles or the protection of the private life of the parties so require, or to the extent strictly necessary in the opinion of the court in special circumstances where publicity would prejudice the interests of justice.

23.433 The twin questions, therefore, arise as to whether the adjudicator is a public authority for the purposes of the Convention and whether the 1996 Act is contrary to Art 6.

23.434 There have been three first instance decisions which all held that Article 6 was inapplicable to the adjudication process.[203] All have held that the Human Rights Act and the Convention do not apply to adjudication. It is interesting that those decisions have not gone to appeal and the point has not been taken in any other case since.

23.435 There are respectable arguments to the effect that the Convention does apply and the adjudicator is a public authority. The principal reason is that the adjudicator has the right to determine the civil rights of a party (albeit temporarily) and the fact that these rights are temporary or subject to review by the court or arbitrator does not affect matters. However, the provisions in relation to fairness, discussed above, mean that, in the circumstances, a fair hearing is likely to be obtained by the parties. That leaves the potential objection to enforcement that the hearing did not take place in public.

23.436 If, however, an adjudication takes place in private and the parties do not object, the courts are likely to infer that the parties have waived their Convention rights. Therefore, the question of the applicability of Article 6 is likely to remain of academic interest only and the courts will look at fairness in the same way as it looks at natural justice and say whether there has been a fair hearing applying the principles which have grown up since the *Discain* case.

No dispute at the time of the service of the adjudication notice

23.437 It is axiomatic that if there is no dispute at the time of the adjudication notice, then there cannot be jurisdiction in the adjudicator because there is no dispute to resolve.

---

[203] *Elanay Contracts Ltd v The Vestry* [2001] BLR 33; *Austin Hall Building Ltd v Buckland Securities Ltd* [2001] BLR 27; *RG Carter v Edmund Nuttall Ltd* [2002] BLR 359 at para 30. But see *Balfour Beatty Engineering Services (HY) Ltd v Shepherd Construction Ltd* [2009] EWHC 2218 (TCC).

23.438 The law defining a dispute is now well settled. It is often artificial, for a party which is denying liability to a party seeking payment, to say that there was no dispute at the time of the reference to an adjudicator. The courts will not be astute to accept artificial arguments which have the sole effect of delaying the coming into existence of a dispute.

23.439 Nevertheless if there is no dispute, there is no jurisdiction. Accordingly, claimants should be careful to ensure that they have a dispute in existence before the adjudication notice is served. A short time lost at the beginning of a process, to ensure that there are no jurisdiction arguments at the end of it, is time well spent. If the supposedly paying party can show that it is arguable for the purposes of CPR Part 24 that there was no dispute at the time of the adjudication notice, then the courts will not be able to enforce the decision summarily.

Adjudicator's decision issued out of time

23.440 Section 108(2)(c) of the Act requires construction contracts governed by the Act to require the adjudicator to reach a decision '. . . *within 28 days of referral or such longer period as is agreed by the parties after the dispute has been referred*.'

23.441 There have inevitably been a number of cases in which the adjudicator has failed to provide his or her decision within the timetable set out in the Act. As a result, parties have sought to resist the enforcement of those decisions on the grounds that the adjudicator lacked jurisdiction to make a decision once the time limit for making it had expired.

23.442 Unfortunately, this is an area in which there was initially a lack of consistency between the decisions of the courts.[204]

23.443 An appeal court was given the opportunity to consider this issue for the first time in the Scottish case of *Ritchie Brothers (PWC) Ltd v David Philp (Commercials) Ltd*[205]and concluded that an adjudicator's jurisdiction ceases on the expiry of the time limit for the adjudicator to reach his decision.

23.444 This decision has since been followed by the English courts in the cases of *Hart Investments Ltd v Fidler*,[206] *Epping Electrical Co Ltd v Briggs and Forrester (Plumbing Services) Ltd*,[207] and *Mott MacDonald Ltd v London & Regional Properties Ltd*[208] in which it has been determined that the adjudicator's decision will be valid provided that it is made within the correct time limit and communicated to the parties within a day or two after the time limit.

Insolvency

23.445 The question arises as to whether an adjudicator's decision ought to be enforced when the receiving party either is insolvent or is about to become insolvent. Were a decision to be

[204] See *St Andrews Bay Development Ltd v HBG Management Ltd* 2003 SLT 740; 2003 SCLR 526; *Bloor Construction (United Kingdom) London Ltd v Bowmer & Kirkland (London) Ltd* [2000] BLR 314; *Barnes & Elliot Ltd v Taylor Woodrow Holdings Ltd* [2004] BLR 111; and *Simons Construction Ltd v Aardvark Developments Ltd* [2004] BLR 117.

[205] [2005] BLR 384.

[206] [2007] BLR 303.

[207] [2007] BLR 126.

[208] [2007] EWHC 1055 (TCC).

enforced in those circumstances, the paying party will have to prove in a liquidation and may receive only a dividend or a nil return if it establishes that, in fact, no money was payable in the end.

23.446 It must be remembered that the purpose of the 1996 Act was to ensure that financial risks transferred from the party holding the cash to the party seeking it. Furthermore, the smaller sub-contractors seeking payment are likely to be less financially secure than their main contractor counterparts.

23.447 The Court of Appeal has said that r 4.90 of the Insolvency Rules 1986, which provides for the set off of mutual credits and debits, makes a compelling case for refusing summary judgment on a claim arising out of an adjudication and causing all claims and cross claims to be resolved in the liquidation.[209]

23.448 However, there has been no decided case where this suggestion has been taken up. Instead the courts have granted a stay of execution in cases where there is credible evidence of insolvency[210] or an inability of the receiving party to repay the sums ordered to be paid by an adjudicator. This follows from an application of RSC Order 47 which permits the grant of a stay of execution of a judgment if the court is satisfied that the applicant for a judgment would be unable to repay the money.

23.449 However this is not immutable. If, for example, the financial position of the claimant is the same as when it signed the contract, the grant of a stay would not be just.[211]

23.450 The principles which are to be applied in an application for a stay of execution of a summary judgment of an adjudicator's decision have been summarized as follows:[212]

(a) Adjudication (whether pursuant to the 1996 Act or the consequential amendments to the standard forms of building and engineering contracts) is designed to be a quick and inexpensive method of arriving at a temporary result in a construction dispute;

(b) In consequence, adjudicators' decisions are intended to be enforced summarily and the claimant (being the successful party in the adjudication) should not generally be kept out of its money.

(c) In an application to stay the execution of summary judgment arising out of an Adjudicator's decision, the Court must exercise its discretion under Order 47 with considerations (a) and (b) firmly in mind;

(d) The probable inability of the claimant to repay the judgment sum (awarded by the Adjudicator and enforced by way of summary judgment) at the end of the substantive trial, or arbitration hearing, may constitute special circumstances within the meaning of Ord 47 r 1(1)(a) rendering it appropriate to grant a stay;

---

209 See *Bouygues v Dahl-Jensen* [2000] BLR 522 para 35 *per* Chadwick LJ.

210 For example *Rainford House Ltd v Cadogan Ltd* [2001] BLR 417.

211 *Herschell Engineering Ltd v Breen Property Ltd* (QBD, 28 July 2000).

212 *Wimbledon Construction Co 2000 Ltd v Derek Vago* [2005] EWHC 1086 (TCC); *JPA Design and Build Ltd v Sentosa (UK) Ltd* [2009] EWHC 2312 (TCC). Note that this decision has been applied in *Balfour Beatty Engineering Services (HY) Ltd v Shepherd Construction Ltd* [2009] EWHC 2218 (TCC); and *SG South Ltd v King's Head Cirencester LLP and another* [2009] EWHC 2645 (TCC)

(e) If the enforcing party is in insolvent liquidation, or there is no dispute on the evidence that it is insolvent, then a stay of execution will usually be granted; and

(f) Even if the evidence of the financial position of the enforcing party suggested that it is probable that it would be unable to repay the judgment sum when it fell due, that would not usually justify the grant of a stay if:
   (i) its financial position is the same or similar to its financial position at the time that the relevant contract was made; or
   (ii) its financial position is due, either wholly, or in significant part, to the defendant's failure to pay those sums which were awarded by the adjudicator.

The following qualifications have been made in relation to sub-paragraph (e) in the case of *Mead General Building Ltd (subject to a company voluntary arrangement) v Dartmoor Properties Ltd*:[213] **23.451**

(i) if a company is in insolvent liquidation then judgment should not be entered at all due to the operation of the Insolvency Rules; and

(ii) a judgment can be entered where the claimant is in administrative receivership but a stay can be granted where the receivership indicates that the claimant would not be able to repay any sums paid out in accordance with the judgment in due course.

In addition, it was held in that case that while the fact that a claimant is subject to a company voluntary arrangement ('CVA') is a relevant matter to be taken into account when considering whether or not to grant a stay of execution, it is not of itself enough to justify the granting of a stay. In this case, a stay was not granted because the claimant's financial problems had been caused by the defendant's failure to pay the sums that the adjudicator had found to be due, the CVA was due to the failure to pay those sums, the CVA supervisor considered the claimant to be a viable ongoing concern and the claimant was currently trading successfully. **23.452**

The above summary is a correct statement of first instance authority. The points are, it is submitted, correct for appellate purposes. **23.453**

There must be cogent evidence adduced to show that the party seeking payment is, in fact, unable to repay the money. **23.454**

Costs Provisions

The original purpose of adjudication, being a cheap remedy, meant that no provision was made within the statute or the scheme for the payment of costs. The result is that unless there is a contrary agreement, each side pays its own costs of the adjudication. **23.455**

However, there is no provision in either the 1996 Act or the scheme which prevents the parties from making contractual provision for the payment of costs. Some forms expressly give the adjudicator the power to award costs. Others provide that the claimant must pay **23.456**

[213] [2009] BLR 225.

the costs of both sides. The effectiveness of this clause has been upheld at first instance.[214] There is provision for prospective outlawing of these types of contractual provision (*Tolent* clauses) in the Local Democracy, Economic Development and Construction Act 2009. It is submitted however that *Tolent* was wrongly decided since the practical effect of such clauses is to discourage the free exercise of the right to adjudicate, which must be contrary to the whole purpose of the statute.

Interaction with Litigation and Arbitration

23.457 It is clear that the courts will stay court proceedings if they have been commenced in breach of an agreement to refer a dispute to adjudication before the commencement of court proceedings.[215]

23.458 Under s 108(3) of the 1996 Act, an adjudicator's decision is to be binding until the dispute is finally determined by legal proceedings, arbitration, or by an agreement. However, the 1996 Act does not provide any guidance as to how any subsequent proceedings are dealt with. Nonetheless, it seems to be accepted that a court or an arbitrator is not required to review the decision of the adjudicator in the same way that an appeal court would do. The requirement is to decide the dispute afresh.

23.459 The first issue for a challenging party to decide is what remedy it will be seeking. If the claim has failed completely, then the remedy will often be the same one sought in the adjudication, subject to any additional information or updating. However, in cases where the adjudication was successful on some of the issues then the claimant will need to decide whether to make the original claim in full or to make a claim only in relation to the parts which were unsuccessful in the adjudication.

23.460 In that respect, Judge LLoyd QC's approach in *Glencot Development v Ben Barratt*,[216] *David McLean Housing Ltd v Swansea Housing Association Ltd*,[217] and in *Balfour Beatty v Lambeth LBC*[218] suggests that enforcement is based on the underlying rights which exist under the contract. On this basis, proceedings for enforcement of an adjudicator's decision should also be treated as proceedings for payment under the contract itself, with the result that no new proceedings are necessary. However, it is not clear what would happen if, for example, there was a valid arbitration clause, as this may be the appropriate dispute resolution method for an underlying dispute. For that reason, it may be that Judge Hicks QC's approach in *VHE Construction* that a claim for enforcement is based simply on a breach of the obligation to comply with an adjudicators' decision is to be preferred.

23.461 In addition, it is also worth considering the requirements of the Pre-Action Protocol for Construction and Engineering Disputes. In most cases the parties would have complied with the requirements in the adjudication procedure and this is confirmed in paragraph 1.2

---

214 *Bridgeway Construction Ltd v Tolent Construction Ltd* (TCC, 11 April 2000).
215 *DGT Steel & Cladding Ltd v Speedwell Roofing and Interiors Ltd* [2007] BLR 371.
216 TCC, 2 and 13 February 2001.
217 [2002] BLR 125.
218 [2002] BLR 288.

of the Protocol which provides that compliance is not required where the proposed proceedings relate to the same or substantially the same issues as have been the subject of a recent adjudication under the 1996 Act. Although this appears to be limited to statutory adjudications, the better view would be that the same provisions should also be applied to contractual adjudications which do not fall within the 1996 Act.

Another question which needs to be considered is the application of the general rules as to costs under CPR Part 44. In the absence of an agreement, an adjudicator does not have the power to order a party's costs to be paid following adjudication. To date, attempts to recover the costs of any adjudication as general damages have been unsuccessful. However, it has been suggested in *Griffin v Midas Homes Ltd*[219] that the court or arbitrator could have the power to order the payment of such costs since the fact that they were incurred gives rise to a separate dispute. In any event, the reality is that it is possible to argue that any of the costs incurred in adjudication would have had to be incurred in any event in any arbitration or litigation proceedings and, accordingly, such costs should be recoverable. **23.462**

Limitation

Section 5 of the Limitation Act 1980 provides that in the case of a contract which is not signed as a deed the time limit for commencing legal proceedings is six years from the date on which the cause of action arose. **23.463**

If the contract is signed as a deed then the time limit for commencing proceedings is twelve years from the date of the cause of action accruing in accordance with s 8 of the Limitation Act 1980. However, the Limitation Act 1980 applies only to litigation and arbitration proceedings and has not been extended to apply to adjudication proceedings. **23.464**

The question arose in the recent case of *Connex South Eastern Ltd v MJ Building Services Group plc*[220] as to whether or not adjudication proceedings could be commenced after the expiry of the limitation period in the Limitation Act 1980. **23.465**

The question arises because s 108 of the 1996 Act provides that a dispute can be referred to adjudication at 'any time'. There is no limitation expressly imposed on the right to refer a dispute to adjudication at any time and, therefore, there does not appear to be any reason why a dispute cannot be referred to adjudication after the expiry of the limitation period set out under the Limitation Act 1980. **23.466**

Dyson LJ said in relation to this question that: **23.467**

> ... there is nothing to prevent a party from referring a dispute to adjudication at any time, even after the expiry of the relevant limitation period. Similarly, there is nothing to stop a party from issuing court proceedings after the expiry of the relevant limitation period. Just as a party who takes that course in court proceedings runs the risk that, if the limitation defence is pleaded, the claim will fail (and indeed may be struck out) so a party who takes that course

[219] (2002) 18 Const LJ 67.
[220] [2005] BLR 201.

in an adjudication runs the risk that, if the limitation defence is taken, the adjudicator will make an award in favour of the Respondent.

23.468 It appears, therefore, from this judgment that although the Limitation Act 1980 cannot be relied upon to prevent a party from beginning adjudication proceedings, an adjudicator can and should take into account the limitation period for commencing proceedings set out in the Limitation Act 1980 and that a claim should fail if it is commenced after the expiry of the relevant period.

23.469 The question arises as to how long a losing party to an adjudication has got to reclaim monies paid to the successful party in the adjudication. It has been held that the unsuccessful party has six years from the date of payment of the adjudicator's award.[221]

[221] *Jim Ennis Construction Ltd v Premier Asphalt Ltd* [2009] EWHC 1906 (TCC). That case was decided on the basis of an implied term but leaves open the question of whether and if so what limitation period would apply in Restitution. It is submitted that if the implied term analysis is correct, Restitution would not be an appropriate remedy and the question does not arise.

APPENDICES

APPENDIX 1

Housing Grants, Construction and Regeneration Act 1996

Part II
Construction Contracts

Introductory provisions

104 Construction contracts

(1) In this Part a 'construction contract' means an agreement with a person for any of the following—
   (a) the carrying out of construction operations;
   (b) arranging for the carrying out of construction operations by others, whether under subcontract to him or otherwise;
   (c) providing his own labour, or the labour of others, for the carrying out of construction operations.

(2) References in this Part to a construction contract include an agreement—
   (a) to do architectural, design, or surveying work, or
   (b) to provide advice on building, engineering, interior or exterior decoration or on the laying-out of landscape,

in relation to construction operations.

(3) References in this Part to a construction contract do not include a contract of employment (within the meaning of the Employment Rights Act 1996).

(4) The Secretary of State may by order add to, amend or repeal any of the provisions of subsection (1), (2) or (3) as to the agreements which are construction contracts for the purposes of this Part or are to be taken or not to be taken as included in references to such contracts.
No such order shall be made unless a draft of it has been laid before and approved by a resolution of each House of Parliament.

(5) Where an agreement relates to construction operations and other matters, this Part applies to it only so far as it relates to construction operations.

An agreement relates to construction operations so far as it makes provision of any kind within subsection (1) or (2).

(6) This Part applies only to construction contracts which—
   (a) are entered into after the commencement of this Part, and
   (b) relate to the carrying out of construction operations in England, Wales or Scotland.

(7) This Part applies whether or not the law of England and Wales or Scotland is otherwise the applicable law in relation to the contract.

105 Meaning of 'construction operations'

(1) In this Part 'construction operations' means, subject as follows, operations of any of the following descriptions—
   (a) construction, alteration, repair, maintenance, extension, demolition or dismantling of buildings, or structures forming, or to form, part of the land (whether permanent or not);
   (b) construction, alteration, repair, maintenance, extension, demolition or dismantling of any works forming, or to form, part of the land, including (without prejudice to the foregoing) walls, roadworks, power-lines, [electronic communications apparatus], aircraft runways, docks and harbours, railways, inland waterways, pipe-lines, reservoirs, water-mains, wells, sewers, industrial plant and installations for purposes of land drainage, coast protection or defence;

(c) installation in any building or structure of fittings forming part of the land, including (without prejudice to the foregoing) systems of heating, lighting, air-conditioning, ventilation, power supply, drainage, sanitation, water supply or fire protection, or security or communications systems;

(d) external or internal cleaning of buildings and structures, so far as carried out in the course of their construction, alteration, repair, extension or restoration;

(e) operations which form an integral part of, or are preparatory to, or are for rendering complete, such operations as are previously described in this subsection, including site clearance, earth-moving, excavation, tunnelling and boring, laying of foundations, erection, maintenance or dismantling of scaffolding, site restoration, landscaping and the provision of roadways and other access works;

(f) painting or decorating the internal or external surfaces of any building or structure.

(2) The following operations are not construction operations within the meaning of this Part—

(a) drilling for, or extraction of, oil or natural gas;

(b) extraction (whether by underground or surface working) of minerals; tunnelling or boring, or construction of underground works, for this purpose;

(c) assembly, installation or demolition of plant or machinery, or erection or demolition of steel-work for the purposes of supporting or providing access to plant or machinery, on a site where the primary activity is—

(i) nuclear processing, power generation, or water or effluent treatment, or

(ii) the production, transmission, processing or bulk storage (other than warehousing) of chemicals, pharmaceuticals, oil, gas, steel or food and drink;

(d) manufacture or delivery to site of—

(i) building or engineering components or equipment,

(ii) materials, plant or machinery, or

(iii) components for systems of heating, lighting, air-conditioning, ventilation, power supply, drainage, sanitation, water supply or fire protection, or for security or communications systems,

except under a contract which also provides for their installation;

(e) the making, installation and repair of artistic works, being sculptures, murals and other works which are wholly artistic in nature.

(3) The Secretary of State may by order add to, amend or repeal any of the provisions of subsection (1) or (2) as to the operations and work to be treated as construction operations for the purposes of this Part.

(4) No such order shall be made unless a draft of it has been laid before and approved by a resolution of each House of Parliament.

106 Provisions not applicable to contract with residential occupier

(1) This Part does not apply—

(a) to a construction contract with a residential occupier (see below), or

(b) to any other description of construction contract excluded from the operation of this Part by order of the Secretary of State.

(2) A construction contract with a residential occupier means a construction contract which principally relates to operations on a dwelling which one of the parties to the contract occupies, or intends to occupy, as his residence.

In this subsection 'dwelling' means a dwelling-house or a flat; and for this purpose—

'dwelling-house' does not include a building containing a flat; and

'flat' means separate and self-contained premises constructed or adapted for use for residential purposes and forming part of a building from some other part of which the premises are divided horizontally.

(3) The Secretary of State may by order amend subsection (2).

(4) No order under this section shall be made unless a draft of it has been laid before and approved by a resolution of each House of Parliament.

107 Provisions applicable only to agreements in writing

(1) The provisions of this Part apply only where the construction contract is in writing, and any other agreement between the parties as to any matter is effective for the purposes of this Part only if in writing.
The expressions 'agreement', 'agree' and 'agreed' shall be construed accordingly.

(2) There is an agreement in writing—
  (a) if the agreement is made in writing (whether or not it is signed by the parties),
  (b) if the agreement is made by exchange of communications in writing, or
  (c) if the agreement is evidenced in writing.

(3) Where parties agree otherwise than in writing by reference to terms which are in writing, they make an agreement in writing.

(4) An agreement is evidenced in writing if an agreement made otherwise than in writing is recorded by one of the parties, or by a third party, with the authority of the parties to the agreement.

(5) An exchange of written submissions in adjudication proceedings, or in arbitral or legal proceedings in which the existence of an agreement otherwise than in writing is alleged by one party against another party and not denied by the other party in his response constitutes as between those parties an agreement in writing to the effect alleged.

(6) References in this Part to anything being written or in writing include its being recorded by any means.

Adjudication

108 Right to refer disputes to adjudication

(1) A party to a construction contract has the right to refer a dispute arising under the contract for adjudication under a procedure complying with this section.
For this purpose 'dispute' includes any difference.

(2) The contract shall—
  (a) enable a party to give notice at any time of his intention to refer a dispute to adjudication;
  (b) provide a timetable with the object of securing the appointment of the adjudicator and referral of the dispute to him within 7 days of such notice;
  (c) require the adjudicator to reach a decision within 28 days of referral or such longer period as is agreed by the parties after the dispute has been referred;
  (d) allow the adjudicator to extend the period of 28 days by up to 14 days, with the consent of the party by whom the dispute was referred;
  (e) impose a duty on the adjudicator to act impartially; and
  (f) enable the adjudicator to take the initiative in ascertaining the facts and the law.

(3) The contract shall provide that the decision of the adjudicator is binding until the dispute is finally determined by legal proceedings, by arbitration (if the contract provides for arbitration or the parties otherwise agree to arbitration) or by agreement.
The parties may agree to accept the decision of the adjudicator as finally determining the dispute.

(4) The contract shall also provide that the adjudicator is not liable for anything done or omitted in the discharge or purported discharge of his functions as adjudicator unless the act or omission is in bad faith, and that any employee or agent of the adjudicator is similarly protected from liability.

(5) If the contract does not comply with the requirements of subsections (1) to (4), the adjudication provisions of the Scheme for Construction Contracts apply.

(6) For England and Wales, the Scheme may apply the provisions of the Arbitration Act 1996 with such adaptations and modifications as appear to the Minister making the scheme to be appropriate.
For Scotland, the Scheme may include provision conferring powers on courts in relation to adjudication and provision relating to the enforcement of the adjudicator's decision.

Payment

109 Entitlement to stage payments

(1) A party to a construction contract is entitled to payment by instalments, stage payments or other periodic payments for any work under the contract unless—
   (a) it is specified in the contract that the duration of the work is to be less than 45 days, or
   (b) it is agreed between the parties that the duration of the work is estimated to be less than 45 days.
(2) The parties are free to agree the amounts of the payments and the intervals at which, or circumstances in which, they become due.
(3) In the absence of such agreement, the relevant provisions of the Scheme for Construction Contracts apply.
(4) References in the following sections to a payment under the contract include a payment by virtue of this section.

110 Dates for payment

(1) Every construction contract shall—
   (a) provide an adequate mechanism for determining what payments become due under the contract, and when, and
   (b) provide for a final date for payment in relation to any sum which becomes due.

   The parties are free to agree how long the period is to be between the date on which a sum becomes due and the final date for payment.
(2) Every construction contract shall provide for the giving of notice by a party not later than five days after the date on which a payment becomes due from him under the contract, or would have become due if—
   (a) the other party had carried out his obligations under the contract, and
   (b) no set-off or abatement was permitted by reference to any sum claimed to be due under one or more other contracts,

   specifying the amount (if any) of the payment made or proposed to be made, and the basis on which that amount was calculated.
(3) If or to the extent that a contract does not contain such provision as is mentioned in subsection (1) or (2), the relevant provisions of the Scheme for Construction Contracts apply.

111 Notice of intention to withhold payment

(1) A party to a construction contract may not withhold payment after the final date for payment of a sum due under the contract unless he has given an effective notice of intention to withhold payment.

   The notice mentioned in section 110(2) may suffice as a notice of intention to withhold payment if it complies with the requirements of this section.
(2) To be effective such a notice must specify—
   (a) the amount proposed to be withheld and the ground for withholding payment, or
   (b) if there is more than one ground, each ground and the amount attributable to it, and must be given not later than the prescribed period before the final date for payment.
(3) The parties are free to agree what that prescribed period is to be.

   In the absence of such agreement, the period shall be that provided by the Scheme for Construction Contracts.
(4) Where an effective notice of intention to withhold payment is given, but on the matter being referred to adjudication it is decided that the whole or part of the amount should be paid, the decision shall be construed as requiring payment not later than—
   (a) seven days from the date of the decision, or
   (b) the date which apart from the notice would have been the final date for payment, whichever is the later.

112 Right to suspend performance for non-payment

(1) Where a sum due under a construction contract is not paid in full by the final date for payment and no effective notice to withhold payment has been given, the person to whom the sum is due has the right (without prejudice to any other right or remedy) to suspend performance of his obligations under the contract to the party by whom payment ought to have been made ('the party in default').

(2) The right may not be exercised without first giving to the party in default at least seven days' notice of intention to suspend performance, stating the ground or grounds on which it is intended to suspend performance.

(3) The right to suspend performance ceases when the party in default makes payment in full of the amount due.

(4) Any period during which performance is suspended in pursuance of the right conferred by this section shall be disregarded in computing for the purposes of any contractual time limit the time taken, by the party exercising the right or by a third party, to complete any work directly or indirectly affected by the exercise of the right.

Where the contractual time limit is set by reference to a date rather than a period, the date shall be adjusted accordingly.

113 Prohibition of conditional payment provisions

(1) A provision making payment under a construction contract conditional on the payer receiving payment from a third person is ineffective, unless that third person, or any other person payment by whom is under the contract (directly or indirectly) a condition of payment by that third person, is insolvent.

(2) For the purposes of this section a company becomes insolvent—

[(a) when it enters administration within the meaning of Schedule B1 to the Insolvency Act 1986,]

(b) on the appointment of an administrative receiver or a receiver or manager of its property under Chapter I of Part III of that Act, or the appointment of a receiver under Chapter II of that Part,

(c) on the passing of a resolution for voluntary winding-up without a declaration of solvency under section 89 of that Act, or

(d) on the making of a winding-up order under Part IV or V of that Act.

(3) For the purposes of this section a partnership becomes insolvent—

(a) on the making of a winding-up order against it under any provision of the Insolvency Act 1986 as applied by an order under section 420 of that Act, or

(b) when sequestration is awarded on the estate of the partnership under section 12 of the Bankruptcy (Scotland) Act 1985 or the partnership grants a trust deed for its creditors.

(4) For the purposes of this section an individual becomes insolvent—

(a) on the making of a bankruptcy order against him under Part IX of the Insolvency Act 1986, or

(b) on the sequestration of his estate under the Bankruptcy (Scotland) Act 1985 or when he grants a trust deed for his creditors.

(5) A company, partnership or individual shall also be treated as insolvent on the occurrence of any event corresponding to those specified in subsection (2), (3) or (4) under the law of Northern Ireland or of a country outside the United Kingdom.

(6) Where a provision is rendered ineffective by subsection (1), the parties are free to agree other terms for payment.

In the absence of such agreement, the relevant provisions of the Scheme for Construction Contracts apply.

Supplementary provisions

114 The Scheme for Construction Contracts

(1) The Minister shall by regulations make a scheme ('the Scheme for Construction Contracts') containing provision about the matters referred to in the preceding provisions of this Part.
(2) Before making any regulations under this section the Minister shall consult such persons as he thinks fit.
(3) In this section 'the Minister' means—
    (a) for England and Wales, the Secretary of State, and
    (b) for Scotland, the [Secretary of State].
(4) Where any provisions of the Scheme for Construction Contracts apply by virtue of this Part in default of contractual provision agreed by the parties, they have effect as implied terms of the contract concerned.
(5) Regulations under this section shall not be made unless a draft of them has been approved by resolution of each House of Parliament.

115 Service of notices, &c

(1) The parties are free to agree on the manner of service of any notice or other document required or authorised to be served in pursuance of the construction contract or for any of the purposes of this Part.
(2) If or to the extent that there is no such agreement the following provisions apply.
(3) A notice or other document may be served on a person by any effective means.
(4) If a notice or other document is addressed, pre-paid and delivered by post—
    (a) to the addressee's last known principal residence or, if he is or has been carrying on a trade, profession or business, his last known principal business address, or
    (b) where the addressee is a body corporate, to the body's registered or principal office, it shall be treated as effectively served.
(5) This section does not apply to the service of documents for the purposes of legal proceedings, for which provision is made by rules of court.
(6) References in this Part to a notice or other document include any form of communication in writing and references to service shall be construed accordingly.

116 Reckoning periods of time

(1) For the purposes of this Part periods of time shall be reckoned as follows.
(2) Where an act is required to be done within a specified period after or from a specified date, the period begins immediately after that date.
(3) Where the period would include Christmas Day, Good Friday or a day which under the Banking and Financial Dealings Act 1971 is a bank holiday in England and Wales or, as the case may be, in Scotland, that day shall be excluded.

117 Crown application

(1) This Part applies to a construction contract entered into by or on behalf of the Crown otherwise than by or on behalf of Her Majesty in her private capacity.
(2) This Part applies to a construction contract entered into on behalf of the Duchy of Cornwall notwithstanding any Crown interest.
(3) Where a construction contract is entered into by or on behalf of Her Majesty in right of the Duchy of Lancaster, Her Majesty shall be represented, for the purposes of any adjudication or other proceedings arising out of the contract by virtue of this Part, by the Chancellor of the Duchy or such person as he may appoint.
(4) Where a construction contract is entered into on behalf of the Duchy of Cornwall, the Duke of Cornwall or the possessor for the time being of the Duchy shall be represented, for the purposes of any adjudication or other proceedings arising out of the contract by virtue of this Part, by such person as he may appoint.

APPENDIX 2

Scheme for Construction Contracts (England and Wales) Regulations 1998

(SI 1998/649)

CONSTRUCTION, ENGLAND AND WALES

(Made 6th March 1998; Coming into force 1st May 1998)

The Secretary of State, in exercise of the powers conferred on him by sections 108(6), 114 and 146(1) and (2) of the Housing Grants, Construction and Regeneration Act 1996, and of all other powers enabling him in that behalf, having consulted such persons as he thinks fit, and draft Regulations having been approved by both Houses of Parliament, hereby makes the following Regulations:

Citation, commencement, extent and interpretation

1 (1) These Regulations may be cited as the Scheme for Construction Contracts (England and Wales) Regulations 1998 and shall come into force at the end of the period of 8 weeks beginning with the day on which they are made (the 'commencement date').

(2) These Regulations shall extend only to England and Wales.

(3) In these Regulations, 'the Act' means the Housing Grants, Construction and Regeneration Act 1996.

The Scheme for Construction Contracts

2 Where a construction contract does not comply with the requirements of section 108(1) to (4) of the Act, the adjudication provisions in Part I of the Schedule to these Regulations shall apply.

3 Where—

(a) the parties to a construction contract are unable to reach agreement for the purposes mentioned respectively in sections 109, 111 and 113 of the Act, or

(b) a construction contract does not make provision as required by section 110 of the Act,

the relevant provisions in Part II of the Schedule to these Regulations shall apply.

4 The provisions in the Schedule to these Regulations shall be the Scheme for Construction Contracts for the purposes of section 114 of the Act.

SCHEDULE
THE SCHEME FOR CONSTRUCTION CONTRACTS

PART I
ADJUDICATION

Notice of Intention to seek Adjudication

1 (1) Any party to a construction contract (the 'referring party') may give written notice (the 'notice of adjudication') of his intention to refer any dispute arising under the contract, to adjudication.

(2) The notice of adjudication shall be given to every other party to the contract.

(3) The notice of adjudication shall set out briefly—

(a) the nature and a brief description of the dispute and of the parties involved,

(b) details of where and when the dispute has arisen,

(c) the nature of the redress which is sought, and

(d) the names and addresses of the parties to the contract (including, where appropriate, the addresses which the parties have specified for the giving of notices).

2 (1) Following the giving of a notice of adjudication and subject to any agreement between the parties to the dispute as to who shall act as adjudicator—

(a) the referring party shall request the person (if any) specified in the contract to act as adjudicator, or

(b) if no person is named in the contract or the person named has already indicated that he is unwilling or unable to act, and the contract provides for a specified nominating body to select a person, the referring party shall request the nominating body named in the contract to select a person to act as adjudicator, or

(c) where neither paragraph (a) nor (b) above applies, or where the person referred to in (a) has already indicated that he is unwilling or unable to act and (b) does not apply, the referring party shall request an adjudicator nominating body to select a person to act as adjudicator.

(2) A person requested to act as adjudicator in accordance with the provisions of paragraph (1) shall indicate whether or not he is willing to act within two days of receiving the request.

(3) In this paragraph, and in paragraphs 5 and 6 below, an 'adjudicator nominating body' shall mean a body (not being a natural person and not being a party to the dispute) which holds itself out publicly as a body which will select an adjudicator when requested to do so by a referring party.

3 The request referred to in paragraphs 2, 5 and 6 shall be accompanied by a copy of the notice of adjudication.

4 Any person requested or selected to act as adjudicator in accordance with paragraphs 2, 5 or 6 shall be a natural person acting in his personal capacity. A person requested or selected to act as an adjudicator shall not be an employee of any of the parties to the dispute and shall declare any interest, financial or otherwise, in any matter relating to the dispute.

5 (1) The nominating body referred to in paragraphs 2(1)(b) and 6(1)(b) or the adjudicator nominating body referred to in paragraphs 2(1)(c), 5(2)(b) and 6(1)(c) must communicate the selection of an adjudicator to the referring party within five days of receiving a request to do so.

(2) Where the nominating body or the adjudicator nominating body fails to comply with paragraph (1), the referring party may—

(a) agree with the other party to the dispute to request a specified person to act as adjudicator, or

(b) request any other adjudicator nominating body to select a person to act as adjudicator.

(3) The person requested to act as adjudicator in accordance with the provisions of paragraphs (1) or (2) shall indicate whether or not he is willing to act within two days of receiving the request.

6 (1) Where an adjudicator who is named in the contract indicates to the parties that he is unable or unwilling to act, or where he fails to respond in accordance with paragraph 2(2), the referring party may—

(a) request another person (if any) specified in the contract to act as adjudicator, or

(b) request the nominating body (if any) referred to in the contract to select a person to act as adjudicator, or

(c) request any other adjudicator nominating body to select a person to act as adjudicator.

(2) The person requested to act in accordance with the provisions of paragraph (1) shall indicate whether or not he is willing to act within two days of receiving the request.

7 (1) Where an adjudicator has been selected in accordance with paragraphs 2, 5 or 6, the referring party shall, not later than seven days from the date of the notice of adjudication, refer the dispute in writing (the 'referral notice') to the adjudicator.

(2) A referral notice shall be accompanied by copies of, or relevant extracts from, the construction contract and such other documents as the referring party intends to rely upon.

(3) The referring party shall, at the same time as he sends to the adjudicator the documents referred to in paragraphs (1) and (2), send copies of those documents to every other party to the dispute.

8 (1) The adjudicator may, with the consent of all the parties to those disputes, adjudicate at the same time on more than one dispute under the same contract.

(2) The adjudicator may, with the consent of all the parties to those disputes, adjudicate at the same time on related disputes under different contracts, whether or not one or more of those parties is a party to those disputes.

(3) All the parties in paragraphs (1) and (2) respectively may agree to extend the period within which the adjudicator may reach a decision in relation to all or any of these disputes.

(4) Where an adjudicator ceases to act because a dispute is to be adjudicated on by another person in terms of this paragraph, that adjudicator's fees and expenses shall be determined in accordance with paragraph 25.

9 (1) An adjudicator may resign at any time on giving notice in writing to the parties to the dispute.

(2) An adjudicator must resign where the dispute is the same or substantially the same as one which has previously been referred to adjudication, and a decision has been taken in that adjudication.

(3) Where an adjudicator ceases to act under paragraph 9(1)—

(a) the referring party may serve a fresh notice under paragraph 1 and shall request an adjudicator to act in accordance with paragraphs 2 to 7; and

(b) if requested by the new adjudicator and insofar as it is reasonably practicable, the parties shall supply him with copies of all documents which they had made available to the previous adjudicator.

(4) Where an adjudicator resigns in the circumstances referred to in paragraph (2), or where a dispute varies significantly from the dispute referred to him in the referral notice and for that reason he is not competent to decide it, the adjudicator shall be entitled to the payment of such reasonable amount as he may determine by way of fees and expenses reasonably incurred by him. The parties shall be jointly and severally liable for any sum which remains outstanding following the making of any determination on how the payment shall be apportioned.

10 Where any party to the dispute objects to the appointment of a particular person as adjudicator, that objection shall not invalidate the adjudicator's appointment nor any decision he may reach in accordance with paragraph 20.

11 (1) The parties to a dispute may at any time agree to revoke the appointment of the adjudicator. The adjudicator shall be entitled to the payment of such reasonable amount as he may determine by way of fees and expenses incurred by him. The parties shall be jointly and severally liable for any sum which remains outstanding following the making of any determination on how the payment shall be apportioned.

(2) Where the revocation of the appointment of the adjudicator is due to the default or misconduct of the adjudicator, the parties shall not be liable to pay the adjudicator's fees and expenses.

Powers of the adjudicator

12 The adjudicator shall—

(a) act impartially in carrying out his duties and shall do so in accordance with any relevant terms of the contract and shall reach his decision in accordance with the applicable law in relation to the contract; and

(b) avoid incurring unnecessary expense.

13 The adjudicator may take the initiative in ascertaining the facts and the law necessary to determine the dispute, and shall decide on the procedure to be followed in the adjudication. In particular he may—

(a) request any party to the contract to supply him with such documents as he may reasonably require including, if he so directs, any written statement from any party to the contract supporting or supplementing the referral notice and any other documents given under paragraph 7(2),

(b) decide the language or languages to be used in the adjudication and whether a translation of any document is to be provided and if so by whom,

(c) meet and question any of the parties to the contract and their representatives,

(d) subject to obtaining any necessary consent from a third party or parties, make such site visits and inspections as he considers appropriate, whether accompanied by the parties or not,
(e) subject to obtaining any necessary consent from a third party or parties, carry out any tests or experiments,
(f) obtain and consider such representations and submissions as he requires, and, provided he has notified the parties of his intention, appoint experts, assessors or legal advisers,
(g) give directions as to the timetable for the adjudication, any deadlines, or limits as to the length of written documents or oral representations to be complied with, and
(h) issue other directions relating to the conduct of the adjudication.

14 The parties shall comply with any request or direction of the adjudicator in relation to the adjudication.

15 If, without showing sufficient cause, a party fails to comply with any request, direction or timetable of the adjudicator made in accordance with his powers, fails to produce any document or written statement requested by the adjudicator, or in any other way fails to comply with a requirement under these provisions relating to the adjudication, the adjudicator may—

(a) continue the adjudication in the absence of that party or of the document or written statement requested,
(b) draw such inferences from that failure to comply as circumstances may, in the adjudicator's opinion, be justified, and
(c) make a decision on the basis of the information before him attaching such weight as he thinks fit to any evidence submitted to him outside any period he may have requested or directed.

16 (1) Subject to any agreement between the parties to the contrary, and to the terms of paragraph (2) below, any party to the dispute may be assisted by, or represented by, such advisers or representatives (whether legally qualified or not) as he considers appropriate.

(2) Where the adjudicator is considering oral evidence or representations, a party to the dispute may not be represented by more than one person, unless the adjudicator gives directions to the contrary.

17 The adjudicator shall consider any relevant information submitted to him by any of the parties to the dispute and shall make available to them any information to be taken into account in reaching his decision.

18 The adjudicator and any party to the dispute shall not disclose to any other person any information or document provided to him in connection with the adjudication which the party supplying it has indicated is to be treated as confidential, except to the extent that it is necessary for the purposes of, or in connection with, the adjudication.

19 (1) The adjudicator shall reach his decision not later than—

(a) twenty eight days after the date of the referral notice mentioned in paragraph 7(1), or
(b) forty two days after the date of the referral notice if the referring party so consents, or
(c) such period exceeding twenty eight days after the referral notice as the parties to the dispute may, after the giving of that notice, agree.

(2) Where the adjudicator fails, for any reason, to reach his decision in accordance with paragraph (1)

(a) any of the parties to the dispute may serve a fresh notice under paragraph 1 and shall request an adjudicator to act in accordance with paragraphs 2 to 7; and
(b) if requested by the new adjudicator and insofar as it is reasonably practicable, the parties shall supply him with copies of all documents which they had made available to the previous adjudicator.

(3) As soon as possible after he has reached a decision, the adjudicator shall deliver a copy of that decision to each of the parties to the contract.

Adjudicator's decision

20 The adjudicator shall decide the matters in dispute. He may take into account any other matters which the parties to the dispute agree should be within the scope of the adjudication or which are matters under the contract which he considers are necessarily connected with the dispute. In particular, he may—

(a) open up, revise and review any decision taken or any certificate given by any person referred to in the contract unless the contract states that the decision or certificate is final and conclusive,

(b) decide that any of the parties to the dispute is liable to make a payment under the contract (whether in sterling or some other currency) and, subject to section 111(4) of the Act, when that payment is due and the final date for payment,

(c) having regard to any term of the contract relating to the payment of interest decide the circumstances in which, and the rates at which, and the periods for which simple or compound rates of interest shall be paid.

21 In the absence of any directions by the adjudicator relating to the time for performance of his decision, the parties shall be required to comply with any decision of the adjudicator immediately on delivery of the decision to the parties in accordance with this paragraph.

22 If requested by one of the parties to the dispute, the adjudicator shall provide reasons for his decision.

Effects of the decision

23 (1) In his decision, the adjudicator may, if he thinks fit, order any of the parties to comply peremptorily with his decision or any part of it.

(2) The decision of the adjudicator shall be binding on the parties, and they shall comply with it until the dispute is finally determined by legal proceedings, by arbitration (if the contract provides for arbitration or the parties otherwise agree to arbitration) or by agreement between the parties.

24 Section 42 of the Arbitration Act 1996 shall apply to this Scheme subject to the following modifications—

(a) in subsection (2) for the word 'tribunal' wherever it appears there shall be substituted the word 'adjudicator',

(b) in subparagraph (b) of subsection (2) for the words 'arbitral proceedings' there shall be substituted the word 'adjudication',

(c) subparagraph (c) of subsection (2) shall be deleted, and

(d) subsection (3) shall be deleted.

25 The adjudicator shall be entitled to the payment of such reasonable amount as he may determine by way of fees and expenses reasonably incurred by him. The parties shall be jointly and severally liable for any sum which remains outstanding following the making of any determination on how the payment shall be apportioned.

26 The adjudicator shall not be liable for anything done or omitted in the discharge or purported discharge of his functions as adjudicator unless the act or omission is in bad faith, and any employee or agent of the adjudicator shall be similarly protected from liability.

Part II
Payment

Entitlement to and amount of stage payments

1 Where the parties to a relevant construction contract fail to agree—

(a) the amount of any instalment or stage or periodic payment for any work under the contract, or

(b) the intervals at which, or circumstances in which, such payments become due under that contract, or

(c) both of the matters mentioned in sub-paragraphs (a) and (b) above,

the relevant provisions of paragraphs 2 to 4 below shall apply.

2 (1) The amount of any payment by way of instalments or stage or periodic payments in respect of a relevant period shall be the difference between the amount determined in accordance with sub-paragraph (2) and the amount determined in accordance with sub-paragraph (3).

(2) The aggregate of the following amounts—

(a) an amount equal to the value of any work performed in accordance with the relevant construction contract during the period from the commencement of the contract to the end of the relevant period (excluding any amount calculated in accordance with subparagraph (b)),

(b) where the contract provides for payment for materials, an amount equal to the value of any materials manufactured on site or brought onto site for the purposes of the works during the period from the commencement of the contract to the end of the relevant period, and

(c) any other amount or sum which the contract specifies shall be payable during or in respect of the period from the commencement of the contract to the end of the relevant period.

(3) The aggregate of any sums which have been paid or are due for payment by way of instalments, stage or periodic payments during the period from the commencement of the contract to the end of the relevant period.

(4) An amount calculated in accordance with this paragraph shall not exceed the difference between—

(a) the contract price, and

(b) the aggregate of the instalments or stage or periodic payments which have become due.

Dates for payment

3 Where the parties to a construction contract fail to provide an adequate mechanism for determining either what payments become due under the contract, or when they become due for payment, or both, the relevant provisions of paragraphs 4 to 7 shall apply.

4 Any payment of a kind mentioned in paragraph 2 above shall become due on whichever of the following dates occurs later—

(a) the expiry of 7 days following the relevant period mentioned in paragraph 2(1) above, or

(b) the making of a claim by the payee.

5 The final payment payable under a relevant construction contract, namely the payment of an amount equal to the difference (if any) between—

(a) the contract price, and

(b) the aggregate of any instalment or stage or periodic payments which have become due under the contract,

shall become due on the expiry of—

(a) 30 days following completion of the work, or

(b) the making of a claim by the payee,

whichever is the later.

6 Payment of the contract price under a construction contract (not being a relevant construction contract) shall become due on

(a) the expiry of 30 days following the completion of the work, or

(b) the making of a claim by the payee,

whichever is the later.

7 Any other payment under a construction contract shall become due

(a) on the expiry of 7 days following the completion of the work to which the payment relates, or

(b) the making of a claim by the payee, whichever is the later.

Final date for payment

8 (1) Where the parties to a construction contract fail to provide a final date for payment in relation to any sum which becomes due under a construction contract, the provisions of this paragraph shall apply.

(2) The final date for the making of any payment of a kind mentioned in paragraphs 2, 5, 6 or 7, shall be 17 days from the date that payment becomes due.

Notice specifying amount of payment

9 A party to a construction contract shall, not later than 5 days after the date on which any payment—

(a) becomes due from him, or

(b) would have become due, if—

(i) the other party had carried out his obligations under the contract, and

(ii) no set-off or abatement was permitted by reference to any sum claimed to be due under one or more other contracts,

give notice to the other party to the contract specifying the amount (if any) of the payment he has made or proposes to make, specifying to what the payment relates and the basis on which that amount is calculated.

Notice of intention to withhold payment 10

Any notice of intention to withhold payment mentioned in section 111 of the Act shall be given not later than the prescribed period, which is to say not later than 7 days before the final date for payment determined either in accordance with the construction contract, or where no such provision is made in the contract, in accordance with paragraph 8 above.

Prohibition of conditional payment provisions

11 Where a provision making payment under a construction contract conditional on the payer receiving payment from a third person is ineffective as mentioned in section 113 of the Act, and the parties have not agreed other terms for payment, the relevant provisions of—

(a) paragraphs 2, 4, 5, 7, 8, 9 and 10 shall apply in the case of a relevant construction contract, and

(b) paragraphs 6, 7, 8, 9 and 10 shall apply in the case of any other construction contract.

Interpretation

12 In this Part of the Scheme for Construction Contracts—

'claim by the payee' means a written notice given by the party carrying out work under a construction contract to the other party specifying the amount of any payment or payments which he considers to be due and the basis on which it is, or they are calculated;

'contract price' means the entire sum payable under the construction contract in respect of the work;

'relevant construction contract' means any construction contract other than one—

(a) which specifies that the duration of the work is to be less than 45 days, or

(b) in respect of which the parties agree that the duration of the work is estimated to be less than 45 days;

'relevant period' means a period which is specified in, or is calculated by reference to the construction contract or where no such period is so specified or is so calculable, a period of 28 days;

'value of work' means an amount determined in accordance with the construction contract under which the work is performed or where the contract contains no such provision, the cost of any work performed in accordance with that contract together with an amount equal to any overhead or profit included in the contract price;

'work' means any of the work or services mentioned in section 104 of the Act.

Explanatory Note

(This note is not part of the Order)

Part II of the Housing Grants, Construction and Regeneration Act 1996 makes provision in relation to construction contracts. Section 114 empowers the Secretary of State to make the Scheme for Construction Contracts. Where a construction contract does not comply with the requirements of sections 108 to 111 (adjudication of disputes and payment provisions), and section 113 (prohibition of conditional payment provisions), the relevant provisions of the Scheme for Construction Contracts have effect.

The Scheme which is contained in the Schedule to these Regulations is in two parts. Part I provides for the selection and appointment of an adjudicator, gives powers to the adjudicator to gather and consider information, and makes provisions in respect of his decisions. Part II makes provision with respect to payments under a construction contract where either the contract fails to make provision or the parties fail to agree—

(a) the method for calculating the amount of any instalment, stage or periodic payment,
(b) the due date and the final date for payments to be made, and
(c) prescribes the period within which a notice of intention to withhold payment must be given.

APPENDIX 3

Housing Grants, Construction and Regeneration Act 1996 with prospective amendments and repeals enacted by the Local Democracy, Economic Development and Construction Act 2009

[These amendments and repeals are to come into force on a date to be prescribed by the relevant Secretary of State.]

PART II
CONSTRUCTION CONTRACTS

Introductory provisions

104 Construction contracts

(1) In this Part a 'construction contract' means an agreement with a person for any of the following—
  (a) the carrying out of construction operations;
  (b) arranging for the carrying out of construction operations by others, whether under subcontract to him or otherwise;
  (c) providing his own labour, or the labour of others, for the carrying out of construction operations.

(2) References in this Part to a construction contract include an agreement—
  (a) to do architectural, design, or surveying work, or
  (b) to provide advice on building, engineering, interior or exterior decoration or on the laying-out of landscape,

in relation to construction operations.

(3) References in this Part to a construction contract do not include a contract of employment (within the meaning of the Employment Rights Act 1996).

(4) The Secretary of State may by order add to, amend or repeal any of the provisions of subsection (1), (2) or (3) as to the agreements which are construction contracts for the purposes of this Part or are to be taken or not to be taken as included in references to such contracts.
No such order shall be made unless a draft of it has been laid before and approved by a resolution of each House of Parliament.

(5) Where an agreement relates to construction operations and other matters, this Part applies to it only so far as it relates to construction operations.
An agreement relates to construction operations so far as it makes provision of any kind within subsection (1) or (2).

(6) This Part applies only to construction contracts which—
  (a) are entered into after the commencement of this Part, and
  (b) relate to the carrying out of construction operations in England, Wales or Scotland.

(7) This Part applies whether or not the law of England and Wales or Scotland is otherwise the applicable law in relation to the contract.

105 Meaning of 'construction operations'

(1) In this Part 'construction operations' means, subject as follows, operations of any of the following descriptions—

(a) construction, alteration, repair, maintenance, extension, demolition or dismantling of buildings, or structures forming, or to form, part of the land (whether permanent or not);

(b) construction, alteration, repair, maintenance, extension, demolition or dismantling of any works forming, or to form, part of the land, including (without prejudice to the foregoing) walls, roadworks, power-lines, [electronic communications apparatus], aircraft runways, docks and harbours, railways, inland waterways, pipe-lines, reservoirs, water-mains, wells, sewers, industrial plant and installations for purposes of land drainage, coast protection or defence;

(c) installation in any building or structure of fittings forming part of the land, including (without prejudice to the foregoing) systems of heating, lighting, air-conditioning, ventilation, power supply, drainage, sanitation, water supply or fire protection, or security or communications systems;

(d) external or internal cleaning of buildings and structures, so far as carried out in the course of their construction, alteration, repair, extension or restoration;

(e) operations which form an integral part of, or are preparatory to, or are for rendering complete, such operations as are previously described in this subsection, including site clearance, earth-moving, excavation, tunnelling and boring, laying of foundations, erection, maintenance or dismantling of scaffolding, site restoration, landscaping and the provision of roadways and other access works;

(f) painting or decorating the internal or external surfaces of any building or structure.

(2) The following operations are not construction operations within the meaning of this Part—

(a) drilling for, or extraction of, oil or natural gas;

(b) extraction (whether by underground or surface working) of minerals; tunnelling or boring, or construction of underground works, for this purpose;

(c) assembly, installation or demolition of plant or machinery, or erection or demolition of steel-work for the purposes of supporting or providing access to plant or machinery, on a site where the primary activity is—

(i) nuclear processing, power generation, or water or effluent treatment, or

(ii) the production, transmission, processing or bulk storage (other than warehousing) of chemicals, pharmaceuticals, oil, gas, steel or food and drink;

(d) manufacture or delivery to site of—

(i) building or engineering components or equipment,

(ii) materials, plant or machinery, or

(iii) components for systems of heating, lighting, air-conditioning, ventilation, power supply, drainage, sanitation, water supply or fire protection, or for security or communications systems,

except under a contract which also provides for their installation;

(e) the making, installation and repair of artistic works, being sculptures, murals and other works which are wholly artistic in nature.

(3) The Secretary of State may by order add to, amend or repeal any of the provisions of subsection (1) or (2) as to the operations and work to be treated as construction operations for the purposes of this Part.

(4) No such order shall be made unless a draft of it has been laid before and approved by a resolution of each House of Parliament.

106 Provisions not applicable to contract with residential occupier

(1) This Part does not apply—

(a) to a construction contract with a residential occupier (see below), ~~or~~

~~(b) to any other description of construction contract excluded from the operation of this Part by order of the Secretary of State.~~[1]

(2) A construction contract with a residential occupier means a construction contract which principally relates to operations on a dwelling which one of the parties to the contract occupies, or intends to occupy, as his residence.

In this subsection 'dwelling' means a dwelling-house or a flat; and for this purpose—

'dwelling-house' does not include a building containing a flat; and

'flat' means separate and self-contained premises constructed or adapted for use for residential purposes and forming part of a building from some other part of which the premises are divided horizontally.

(3) The Secretary of State may by order amend subsection (2).

(4) No order under this section shall be made unless a draft of it has been laid before and approved by a resolution of each House of Parliament.

[106A Power to disapply provisions of this Part:

(1) The Secretary of State may by order provide that any or all of the provisions of this Part, so far as extending to England and Wales, shall not apply to any description of construction contract relating to the carrying out of construction operations (not being operations in Wales) which is specified in the order.

(2) The Welsh Ministers may by order provide that any or all of the provisions of this Part, so far as extending to England and Wales, shall not apply to any description of construction contract relating to the carrying out of construction operations in Wales which is specified in the order.

(3) The Scottish Ministers may by order provide that any or all of the provisions of this Part, so far as extending to Scotland, shall not apply to any description of construction contract which is specified in the order.

(4) An order under this section shall not be made unless a draft of it has been laid before and approved by resolution of—

(a) in the case of an order under subsection (1), each House of Parliament;

(b) in the case of an order under subsection (2), the National Assembly for Wales;

(c) in the case of an order under subsection (3), the Scottish Parliament][2].

107 Provisions applicable only to agreements in writing

~~(1) The provisions of this Part apply only where the construction contract is in writing, and any other agreement between the parties as to any matter is effective for the purposes of this Part only if in writing.~~

~~The expressions 'agreement', 'agree' and 'agreed' shall be construed accordingly.~~

~~(2) There is an agreement in writing—~~

~~(a) if the agreement is made in writing (whether or not it is signed by the parties),~~

~~(b) if the agreement is made by exchange of communications in writing, or~~

~~(c) if the agreement is evidenced in writing.~~

~~(3) Where parties agree otherwise than in writing by reference to terms which are in writing, they make an agreement in writing.~~

~~(4) An agreement is evidenced in writing if an agreement made otherwise than in writing is recorded by one of the parties, or by a third party, with the authority of the parties to the agreement.~~

[1] Repealed by the Local Democracy, Economic Development and Construction Act 2009.

[2] Inserted by the Local Democracy, Economic Development and Construction Act 2009.

~~(5) An exchange of written submissions in adjudication proceedings, or in arbitral or legal proceedings in which the existence of an agreement otherwise than in writing is alleged by one party against another party and not denied by the other party in his response constitutes as between those parties an agreement in writing to the effect alleged.~~

~~(6) References in this Part to anything being written or in writing include its being recorded by any means.~~[3]

Adjudication

108 Right to refer disputes to adjudication

(1) A party to a construction contract has the right to refer a dispute arising under the contract for adjudication under a procedure complying with this section.

For this purpose 'dispute' includes any difference.

(2) The contract shall [include provision in writing so as to][4]—

 (a) enable a party to give notice at any time of his intention to refer a dispute to adjudication;
 (b) provide a timetable with the object of securing the appointment of the adjudicator and referral of the dispute to him within 7 days of such notice;
 (c) require the adjudicator to reach a decision within 28 days of referral or such longer period as is agreed by the parties after the dispute has been referred;
 (d) allow the adjudicator to extend the period of 28 days by up to 14 days, with the consent of the party by whom the dispute was referred;
 (e) impose a duty on the adjudicator to act impartially; and
 (f) enable the adjudicator to take the initiative in ascertaining the facts and the law.

(3) The contract shall provide [in writing][5] that the decision of the adjudicator is binding until the dispute is finally determined by legal proceedings, by arbitration (if the contract provides for arbitration or the parties otherwise agree to arbitration) or by agreement.

The parties may agree to accept the decision of the adjudicator as finally determining the dispute.

[(3A) The contract shall include provision in writing permitting the adjudicator to correct his decision so as to remove a clerical or typographical error arising by accident or Omission][6].

(4) The contract shall also provide [in writing][7] that the adjudicator is not liable for anything done or omitted in the discharge or purported discharge of his functions as adjudicator unless the act or omission is in bad faith, and that any employee or agent of the adjudicator is similarly protected from liability.

(5) If the contract does not comply with the requirements of subsections (1) to (4), the adjudication provisions of the Scheme for Construction Contracts apply.

(6) For England and Wales, the Scheme may apply the provisions of the Arbitration Act 1996 with such adaptations and modifications as appear to the Minister making the scheme to be appropriate.
For Scotland, the Scheme may include provision conferring powers on courts in relation to adjudication and provision relating to the enforcement of the adjudicator's decision.

[108A Adjudication costs: effectiveness of provision

(1) This section applies in relation to any contractual provision made between the parties to a construction contract which concerns the allocation as between those parties of costs relating to the adjudication of a dispute arising under the construction contract.

[3] Repealed by the Local Democracy, Economic Development and Construction Act 2009.
[4] Inserted by the Local Democracy, Economic Development and Construction Act 2009.
[5] Inserted by the Local Democracy, Economic Development and Construction Act 2009.
[6] Inserted by the Local Democracy, Economic Development and Construction Act 2009.
[7] Inserted by the Local Democracy, Economic Development and Construction Act 2009.

(2) The contractual provision referred to in subsection (1) is ineffective unless—
  (a) it is made in writing, is contained in the construction contract and confers power on the adjudicator to allocate his fees and expenses as between the parties, or
  (b) it is made in writing after the giving of notice of intention to refer the dispute to adjudication.][8]

Payment

109 Entitlement to stage payments

(1) A party to a construction contract is entitled to payment by instalments, stage payments or other periodic payments for any work under the contract unless—
  (a) it is specified in the contract that the duration of the work is to be less than 45 days, or
  (b) it is agreed between the parties that the duration of the work is estimated to be less than 45 days.
(2) The parties are free to agree the amounts of the payments and the intervals at which, or circumstances in which, they become due.
(3) In the absence of such agreement, the relevant provisions of the Scheme for Construction Contracts apply.
(4) References in the following sections to a payment ~~under~~[9] [provided for by][10] the contract include a payment by virtue of this section.

110 Dates for payment

(1) Every construction contract shall—
  (a) provide an adequate mechanism for determining what payments become due under the contract, and when, and
  (b) provide for a final date for payment in relation to any sum which becomes due.
The parties are free to agree how long the period is to be between the date on which a sum becomes due and the final date for payment.
[(1A) The requirement in subsection (1)(a) to provide an adequate mechanism for determining what payments become due under the contract, or when, is not satisfied where a construction contract makes payment conditional on—
  (a) the performance of obligations under another contract, or
  (b) a decision by any person as to whether obligations under another contract have been performed.
(1B) In subsection (1A)(a) and (b) the references to obligations do not include obligations to make payments (but see section 113).
(1C) Subsection (1A) does not apply where—
  (a) the construction contract is an agreement between the parties for the carrying out of construction operations by another person, whether under sub-contract or otherwise, and
  (b) the obligations referred to in that subsection are obligations on that other person to carry out those operations.
(1D) The requirement in subsection (1)(a) to provide an adequate mechanism for determining when payments become due under the contract is not satisfied where a construction contract provides for the date on which a payment becomes due to be determined by reference to the giving to the person to whom the payment is due of a notice which relates to what payments are due under the contract.][11]

[8] Inserted by the Local Democracy, Economic Development and Construction Act 2009.
[9] Repealed by the Local Democracy, Economic Development and Construction Act 2009.
[10] Inserted by the Local Democracy, Economic Development and Construction Act 2009.
[11] Inserted by the Local Democracy, Economic Development and Construction Act 2009.

~~(2) Every construction contract shall provide for the giving of notice by a party not later than five days after the date on which a payment becomes due from him under the contract, or would have become due if—~~

~~(a) the other party had carried out his obligations under the contract, and~~

~~(b) no set-off or abatement was permitted by reference to any sum claimed to be due under one or more other contracts, specifying the amount (if any) of the payment made or proposed to be made, and the basis on which that amount was calculated.~~[12]

(3) If or to the extent that a contract does not contain such provision as is mentioned in subsection (1) ~~or (2)~~,[13] the relevant provisions of the Scheme for Construction Contracts apply.

[110A Payment notices: contractual requirements

(1) A construction contract shall, in relation to every payment provided for by the contract—

(a) require the payer or a specified person to give a notice complying with subsection (2) to the payee not later than five days after the payment due date, or

(b) require the payee to give a notice complying with subsection (3) to the payer or a specified person not later than five days after the payment due date.

(2) A notice complies with this subsection if it specifies—

(a) in a case where the notice is given by the payer—

(i) the sum that the payer considers to be or to have been due at the payment due date in respect of the payment, and

(ii) the basis on which that sum is calculated;

(b) in a case where the notice is given by a specified person—

(i) the sum that the payer or the specified person considers to be or to have been due at the payment due date in respect of the payment, and

(ii) the basis on which that sum is calculated.

(3) A notice complies with this subsection if it specifies—

(a) the sum that the payee considers to be or to have been due at the payment due date in respect of the payment, and

(b) the basis on which that sum is calculated.

(4) For the purposes of this section, it is immaterial that the sum referred to in subsection (2)(a) or (b) or (3)(a) may be zero.

(5) If or to the extent that a contract does not comply with subsection (1), the relevant provisions of the Scheme for Construction Contracts apply.

(6) In this and the following sections, in relation to any payment provided for by a construction contract—

"payee" means the person to whom the payment is due;

"payer" means the person from whom the payment is due;

"payment due date" means the date provided for by the contract as the date on which the payment is due;

"specified person" means a person specified in or determined in accordance with the provisions of the contract.

110B Payment notices: payee's notice in default of payer's notice

(1) This section applies in a case where, in relation to any payment provided for by a construction contract—

(a) the contract requires the payer or a specified person to give the payee a notice complying with section 110A(2) not later than five days after the payment due date, but

(b) notice is not given as so required.

---

[12] Repealed by the Local Democracy, Economic Development and Construction Act 2009.

[13] Repealed by the Local Democracy, Economic Development and Construction Act 2009.

(2) Subject to subsection (4), the payee may give to the payer a notice complying with section 110A(3) at any time after the date on which the notice referred to in subsection (1)(a) was required by the contract to be given.

(3) Where pursuant to subsection (2) the payee gives a notice complying with section 110A(3), the final date for payment of the sum specified in the notice shall for all purposes be regarded as postponed by the same number of days as the number of days after the date referred to in subsection (2) that the notice was given.

(4) If—

(a) the contract permits or requires the payee, before the date on which the notice referred to in subsection (1)(a) is required by the contract to be given, to notify the payer or a specified person of—

(i) the sum that the payee considers will become due on the payment due date in respect of the payment, and

(ii) the basis on which that sum is calculated, and

(b) the payee gives such notification in accordance with the contract, that notification is to be regarded as a notice complying with section 110A(3) given pursuant to subsection (2) (and the payee may not give another such notice pursuant to that subsection).][14]

111 Notice of intention to withhold payment

~~(1) A party to a construction contract may not withhold payment after the final date for payment of a sum due under the contract unless he has given an effective notice of intention to withhold payment.~~

~~The notice mentioned in section 110(2) may suffice as a notice of intention to withhold payment if it complies with the requirements of this section.~~

~~(2) To be effective such a notice must specify—~~

~~(a) the amount proposed to be withheld and the ground for withholding payment, or~~

~~(b) if there is more than one ground, each ground and the amount attributable to it, and must be given not later than the prescribed period before the final date for payment.~~

~~(3) The parties are free to agree what that prescribed period is to be.~~

~~In the absence of such agreement, the period shall be that provided by the Scheme for Construction Contracts.~~

~~(4) Where an effective notice of intention to withhold payment is given, but on the matter being referred to adjudication it is decided that the whole or part of the amount should be paid, the decision shall be construed as requiring payment not later than—~~

~~(a) seven days from the date of the decision, or~~

~~(b) the date which apart from the notice would have been the final date for payment, whichever is the later.~~[15]

[111 Requirement to pay notified sum

(1) Subject as follows, where a payment is provided for by a construction contract, the payer must pay the notified sum (to the extent not already paid) on or before the final date for payment.

(2) For the purposes of this section, the "notified sum" in relation to any payment provided for by a construction contract means—

(a) in a case where a notice complying with section 110A(2) has been given pursuant to and in accordance with a requirement of the contract, the amount specified in that notice;

(b) in a case where a notice complying with section 110A(3) has been given pursuant to and in accordance with a requirement of the contract, the amount specified in that notice;

(c) in a case where a notice complying with section 110A(3) has been given pursuant to and in accordance with section 110B(2), the amount specified in that notice.

---

[14] Inserted by the Local Democracy, Economic Development and Construction Act 2009.

[15] Repealed by the Local Democracy, Economic Development and Construction Act 2009.

(3) The payer or a specified person may in accordance with this section give to the payee a notice of the payer's intention to pay less than the notified sum.

(4) A notice under subsection (3) must specify—

(a) the sum that the payer considers to be due on the date the notice is served,

And

(b) the basis on which that sum is calculated. It is immaterial for the purposes of this subsection that the sum referred to in paragraph (a) or (b) may be zero.

(5) A notice under subsection (3)—

(a) must be given not later than the prescribed period before the final date for payment, and

(b) in a case referred to in subsection (2)(b) or (c), may not be given before the notice by reference to which the notified sum is determined.

(6) Where a notice is given under subsection (3), subsection (1) applies only in respect of the sum specified pursuant to subsection (4)(a).

(7) In subsection (5), "prescribed period" means—

(a) such period as the parties may agree, or

(b) in the absence of such agreement, the period provided by the Scheme for Construction Contracts.

(8) Subsection (9) applies where in respect of a payment—

(a) a notice complying with section 110A(2) has been given pursuant to and in accordance with a requirement of the contract (and no notice under subsection (3) is given), or

(b) a notice under subsection (3) is given in accordance with this section, but on the matter being referred to adjudication the adjudicator decides that more than the sum specified in the notice should be paid.

(9) In a case where this subsection applies, the decision of the adjudicator referred to in subsection (8) shall be construed as requiring payment of the additional amount not later than—

(a) seven days from the date of the decision, or

(b) the date which apart from the notice would have been the final date for

payment,

whichever is the later.

(10) Subsection (1) does not apply in relation to a payment provided for by a construction contract where—

(a) the contract provides that, if the payee becomes insolvent the payer need not pay any sum due in respect of the payment, and

(b) the payee has become insolvent after the prescribed period referred to in subsection (5)(a).

(11) Subsections (2) to (5) of section 113 apply for the purposes of subsection (10) of this section as they apply for the purposes of that section.][16]

112 Right to suspend performance for non-payment

(1) ~~Where a sum due under a construction contract is not paid in full by the final date for payment and no effective notice to withhold payment has been given~~[17]. [Where the requirement in section 111(1) applies in relation to any sum but is not complied with,][18] the person to whom the sum is due has the right (without prejudice to any other right or remedy) to suspend performance of [any or all of][19] his obligations under the contract to the party by whom payment ought to have been made ('the party in default').

---

16 Inserted by the Local Democracy, Economic Development and Construction Act 2009.

17 Repealed by the Local Democracy, Economic Development and Construction Act 2009.

18 Inserted by the Local Democracy, Economic Development and Construction Act 2009.

19 Inserted by the Local Democracy, Economic Development and Construction Act 2009.

(2) The right may not be exercised without first giving to the party in default at least seven days' notice of intention to suspend performance, stating the ground or grounds on which it is intended to suspend performance.

(3) The right to suspend performance ceases when the party in default makes payment in full of ~~the amount due~~[20] [the sum referred to in subsection (1)][21].

[(3A) Where the right conferred by this section is exercised, the party in default shall be liable to pay to the party exercising the right a reasonable amount in respect of costs and expenses reasonably incurred by that party as a result of the exercise of the right.][22]

(4) Any period during which performance is suspended in pursuance of, [or in consequence of the exercise of][23] the right conferred by this section shall be disregarded in computing for the purposes of any contractual time limit the time taken, by the party exercising the right or by a third party, to complete any work directly or indirectly affected by the exercise of the right.

Where the contractual time limit is set by reference to a date rather than a period, the date shall be adjusted accordingly.

113 Prohibition of conditional payment provisions

(1) A provision making payment under a construction contract conditional on the payer receiving payment from a third person is ineffective, unless that third person, or any other person payment by whom is under the contract (directly or indirectly) a condition of payment by that third person, is insolvent.

(2) For the purposes of this section a company becomes insolvent—
- (a) when it enters administration within the meaning of Schedule B1 to the Insolvency Act 1986,
- (b) on the appointment of an administrative receiver or a receiver or manager of its property under Chapter I of Part III of that Act, or the appointment of a receiver under Chapter II of that Part,
- (c) on the passing of a resolution for voluntary winding-up without a declaration of solvency under section 89 of that Act, or
- (d) on the making of a winding-up order under Part IV or V of that Act.

(3) For the purposes of this section a partnership becomes insolvent—
- (a) on the making of a winding-up order against it under any provision of the Insolvency Act 1986 as applied by an order under section 420 of that Act, or
- (b) when sequestration is awarded on the estate of the partnership under section 12 of the Bankruptcy (Scotland) Act 1985 or the partnership grants a trust deed for its creditors.

(4) For the purposes of this section an individual becomes insolvent—
- (a) on the making of a bankruptcy order against him under Part IX of the Insolvency Act 1986, or
- (b) on the sequestration of his estate under the Bankruptcy (Scotland) Act 1985 or when he grants a trust deed for his creditors.

(5) A company, partnership or individual shall also be treated as insolvent on the occurrence of any event corresponding to those specified in subsection (2), (3) or (4) under the law of Northern Ireland or of a country outside the United Kingdom.

(6) Where a provision is rendered ineffective by subsection (1), the parties are free to agree other terms for payment.

In the absence of such agreement, the relevant provisions of the Scheme for Construction Contracts apply.

---

[20] Repealed by the Local Democracy, Economic Development and Construction Act 2009.
[21] Inserted by the Local Democracy, Economic Development and Construction Act 2009.
[22] Inserted by the Local Democracy, Economic Development and Construction Act 2009.
[23] Inserted by the Local Democracy, Economic Development and Construction Act 2009.

Supplementary provisions

114 The Scheme for Construction Contracts

(1) The Minister shall by regulations make a scheme ('the Scheme for Construction Contracts') containing provision about the matters referred to in the preceding provisions of this Part.
(2) Before making any regulations under this section the Minister shall consult such persons as he thinks fit.
(3) In this section 'the Minister' means—
  (a) for England and Wales, the Secretary of State, and
  (b) for Scotland, the [Secretary of State].
(4) Where any provisions of the Scheme for Construction Contracts apply by virtue of this Part in default of contractual provision agreed by the parties, they have effect as implied terms of the contract concerned.
(5) Regulations under this section shall not be made unless a draft of them has been approved by resolution of each House of Parliament.

115 Service of notices, &c

(1) The parties are free to agree on the manner of service of any notice or other document required or authorised to be served in pursuance of the construction contract or for any of the purposes of this Part.
(2) If or to the extent that there is no such agreement the following provisions apply.
(3) A notice or other document may be served on a person by any effective means.
(4) If a notice or other document is addressed, pre-paid and delivered by post—
  (a) to the addressee's last known principal residence or, if he is or has been carrying on a trade, profession or business, his last known principal business address, or
  (b) where the addressee is a body corporate, to the body's registered or principal office, it shall be treated as effectively served.
(5) This section does not apply to the service of documents for the purposes of legal proceedings, for which provision is made by rules of court.
(6) References in this Part to a notice or other document include any form of communication in writing and references to service shall be construed accordingly.

116 Reckoning periods of time

(1) For the purposes of this Part periods of time shall be reckoned as follows.
(2) Where an act is required to be done within a specified period after or from a specified date, the period begins immediately after that date.
(3) Where the period would include Christmas Day, Good Friday or a day which under the Banking and Financial Dealings Act 1971 is a bank holiday in England and Wales or, as the case may be, in Scotland, that day shall be excluded.

117 Crown application

(1) This Part applies to a construction contract entered into by or on behalf of the Crown otherwise than by or on behalf of Her Majesty in her private capacity.
(2) This Part applies to a construction contract entered into on behalf of the Duchy of Cornwall notwithstanding any Crown interest.
(3) Where a construction contract is entered into by or on behalf of Her Majesty in right of the Duchy of Lancaster, Her Majesty shall be represented, for the purposes of any adjudication or other proceedings arising out of the contract by virtue of this Part, by the Chancellor of the Duchy or such person as he may appoint.
(4) Where a construction contract is entered into on behalf of the Duchy of Cornwall, the Duke of Cornwall or the possessor for the time being of the Duchy shall be represented, for the purposes of any adjudication or other proceedings arising out of the contract by virtue of this Part, by such person as he may appoint.

APPENDIX 4

Protocol for the Instruction of Experts to give Evidence in Civil Claims (Annex to the PD 35—Experts and Assessors from the Civil Procedure Rules)

1. Introduction

Expert witnesses perform a vital role in civil litigation. It is essential that both those who instruct experts and experts themselves are given clear guidance as to what they are expected to do in civil proceedings. The purpose of this Protocol is to provide such guidance. It has been drafted by the Civil Justice Council and reflects the rules and practice directions current [in June 2005], replacing the Code of Guidance on Expert Evidence. The authors of the Protocol wish to acknowledge the valuable assistance they obtained by drawing on earlier documents produced by the Academy of Experts and the Expert Witness Institute, as well as suggestions made by the Clinical Dispute Forum. The Protocol has been approved by the Master of the Rolls.

2. Aims of Protocol

2.1 This Protocol offers guidance to experts and to those instructing them in the interpretation of and compliance with Part 35 of the Civil Procedure Rules (CPR 35) and its associated Practice Direction (PD 35) and to further the objectives of the Civil Procedure Rules in general. It is intended to assist in the interpretation of those provisions in the interests of good practice but it does not replace them. It sets out standards for the use of experts and the conduct of experts and those who instruct them. The existence of this Protocol does not remove the need for experts and those who instruct them to be familiar with CPR 35 and PD 35.

2.2 Experts and those who instruct them should also bear in mind para 1.4 of the Practice Direction on Protocols which contains the following objectives, namely to:

- (a) encourage the exchange of early and full information about the expert issues involved in a prospective legal claim;
- (b) enable the parties to avoid or reduce the scope of litigation by agreeing the whole or part of an expert issue before commencement of proceedings; and
- (c) support the efficient management of proceedings where litigation cannot be avoided.

3. Application

3.1 This Protocol applies to any steps taken for the purpose of civil proceedings by experts or those who instruct them on or after 5th September 2005.

3.2 It applies to all experts who are, or who may be, governed by CPR Part 35 and to those who instruct them. Experts are governed by Part 35 if they are or have been instructed to give or prepare evidence for the purpose of civil proceedings in a court in England and Wales (CPR 35.2).

3.3 Experts, and those instructing them, should be aware that some cases may be 'specialist proceedings' (CPR 49) where there are modifications to the Civil Procedure Rules. Proceedings may also be governed by other Protocols. Further, some courts have published their own Guides which supplement the Civil Procedure Rules for proceedings in those courts. They contain provisions affecting expert evidence. Expert witnesses and those instructing them should be familiar with them when they are relevant.

3.4 Courts may take into account any failure to comply with this Protocol when making orders in relation to costs, interest, time limits, the stay of proceedings and whether to order a party to pay a sum of money into court.

Limitation

3.5 If, as a result of complying with any part of this Protocol, claims would or might be time barred under any provision in the Limitation Act 1980, or any other legislation that imposes a time limit for the bringing an action, claimants may commence proceedings without complying with this Protocol. In such circumstances, claimants who commence proceedings without complying with all, or any part, of this Protocol must apply, giving notice to all other parties, to the court for directions as to the timetable and form of procedure to be adopted, at the same time as they request the court to issue proceedings. The court may consider whether to order a stay of the whole or part of the proceedings pending compliance with this Protocol and may make orders in relation to costs.

4. Duties of experts

4.1 Experts always owe a duty to exercise reasonable skill and care to those instructing them, and to comply with any relevant professional code of ethics. However when they are instructed to give or prepare evidence for the purpose of civil proceedings in England and Wales they have an overriding duty to help the court on matters within their expertise (CPR 35.3). This duty overrides any obligation to the person instructing or paying them. Experts must not serve the exclusive interest of those who retain them.

4.2 Experts should be aware of the overriding objective that courts deal with cases justly. This includes dealing with cases proportionately, expeditiously and fairly (CPR 1.1). Experts are under an obligation to assist the court so as to enable them to deal with cases in accordance with the overriding objective. However the overriding objective does not impose on experts any duty to act as mediators between the parties or require them to trespass on the role of the court in deciding facts.

4.3 Experts should provide opinions which are independent, regardless of the pressures of litigation. In this context, a useful test of 'independence' is that the expert would express the same opinion if given the same instructions by an opposing party. Experts should not take it upon themselves to promote the point of view of the party instructing them or engage in the role of advocates.

4.4 Experts should confine their opinions to matters which are material to the disputes between the parties and provide opinions only in relation to matters which lie within their expertise. Experts should indicate without delay where particular questions or issues fall outside their expertise.

4.5 Experts should take into account all material facts before them at the time that they give their opinion. Their reports should set out those facts and any literature or any other material on which they have relied in forming their opinions. They should indicate if an opinion is provisional, or qualified, or where they consider that further information is required or if, for any other reason, they are not satisfied that an opinion can be expressed finally and without qualification.

4.6 Experts should inform those instructing them without delay of any change in their opinions on any material matter and the reason for it.

4.7 Experts should be aware that any failure by them to comply with the Civil Procedure Rules or court orders or any excessive delay for which they are responsible may result in the parties who instructed them being penalised in costs and even, in extreme cases, being debarred from placing the experts' evidence before the court. In Phillips v Symes Peter Smith J held that courts may also make orders for costs (under section 51 of the Supreme Court Act 1981) directly against expert witnesses who by their evidence cause significant expense to be incurred, and do so in flagrant and reckless disregard of their duties to the Court.

5. Conduct of Experts instructed only to advise

5.1 Part 35 only applies where experts are instructed to give opinions which are relied on for the purposes of court proceedings. Advice which the parties do not intend to adduce in litigation is likely to be confidential; the Protocol does not apply in these circumstances.

5.2 The same applies where, after the commencement of proceedings, experts are instructed only to advise (e.g. to comment upon a single joint expert's report) and not to give or prepare evidence for use in the proceedings.

5.3 However this Protocol does apply if experts who were formerly instructed only to advise are later instructed to give or prepare evidence for the purpose of civil proceedings.

6. The Need for Experts

6.1 Those intending to instruct experts to give or prepare evidence for the purpose of civil proceedings should consider whether expert evidence is appropriate, taking account of the principles set out in CPR Parts 1 and 35, and in particular whether:

- (a) it is relevant to a matter which is in dispute between the parties.
- (b) it is reasonably required to resolve the proceedings (CPR 35.1);
- (c) the expert has expertise relevant to the issue on which an opinion is sought;
- (d) the expert has the experience, expertise and training appropriate to the value, complexity and importance of the case; and whether
- (e) these objects can be achieved by the appointment of a single joint expert (see section 17 below).

6.2 Although the court's permission is not generally required to instruct an expert, the court's permission is required before experts can be called to give evidence or their evidence can be put in (CPR 35.4).

7. The appointment of experts

7.1 Before experts are formally instructed or the court's permission to appoint named experts is sought, the following should be established:

- (a) that they have the appropriate expertise and experience;
- (b) that they are familiar with the general duties of an expert;
- (c) that they can produce a report, deal with questions and have discussions with other experts within a reasonable time and at a cost proportionate to the matters in issue;
- (d) a description of the work required;
- (e) whether they are available to attend the trial, if attendance is required; and
- (f) there is no potential conflict of interest.

7.2 Terms of appointment should be agreed at the outset and should normally include:

- (a) the capacity in which the expert is to be appointed (e.g. party appointed expert, single joint expert or expert advisor);
- (b) the services required of the expert (e.g. provision of expert's report, answering questions in writing, attendance at meetings and attendance at court);
- (c) time for delivery of the report;
- (d) the basis of the expert's charges (either daily or hourly rates and an estimate of the time likely to be required, or a total fee for the services);
- (e) travelling expenses and disbursements;
- (f) cancellation charges;
- (g) any fees for attending court;
- (h) time for making the payment; and
- (i) whether fees are to be paid by a third party.
- (j) if a party is publicly funded, whether or not the expert's charges will be subject to assessment by a costs officer.

7.3 As to the appointment of single joint experts, see section 17 below.

7.4 When necessary, arrangements should be made for dealing with questions to experts and discussions between experts, including any directions given by the court, and provision should be made for the cost of this work.

7.5 Experts should be informed regularly about deadlines for all matters concerning them. Those instructing experts should promptly send them copies of all court orders and directions which may affect the preparation of their reports or any other matters concerning their obligations.

Conditional and Contingency Fees

7.6 Payments contingent upon the nature of the expert evidence given in legal proceedings, or upon the outcome of a case, must not be offered or accepted. To do so would contravene experts' overriding duty to the court and compromise their duty of independence.

7.7 Agreement to delay payment of experts' fees until after the conclusion of cases is permissible as long as the amount of the fee does not depend on the outcome of the case.

8. Instructions

8.1 Those instructing experts should ensure that they give clear instructions, including the following:

- (a) basic information, such as names, addresses, telephone numbers, dates of birth and dates of incidents;
- (b) the nature and extent of the expertise which is called for;
- (c) the purpose of requesting the advice or report, a description of the matter(s) to be investigated, the principal known issues and the identity of all parties;
- (d) the statement(s) of case (if any), those documents which form part of standard disclosure and witness statements which are relevant to the advice or report;
- (e) where proceedings have not been started, whether proceedings are being contemplated and, if so, whether the expert is asked only for advice;
- (f) an outline programme, consistent with good case management and the expert's availability, for the completion and delivery of each stage of the expert's work; and
- (g) where proceedings have been started, the dates of any hearings (including any Case Management Conferences and/or Pre-Trial Reviews), the name of the court, the claim number and the track to which the claim has been allocated.

8.2 Experts who do not receive clear instructions should request clarification and may indicate that they are not prepared to act unless and until such clear instructions are received.

8.3 As to the instruction of single joint experts, see section 17 below.

9. Experts' Acceptance of Instructions

9.1 Experts should confirm without delay whether or not they accept instructions. They should also inform those instructing them (whether on initial instruction or at any later stage) without delay if:

- (a) instructions are not acceptable because, for example, they require work that falls outside their expertise, impose unrealistic deadlines, or are insufficiently clear;
- (b) they consider that instructions are or have become insufficient to complete the work;
- (c) they become aware that they may not be able to fulfil any of the terms of appointment;
- (d) the instructions and/or work have, for any reason, placed them in conflict with their duties as an expert; or
- (e) they are not satisfied that they can comply with any orders that have been made.

9.2 Experts must neither express an opinion outside the scope of their field of expertise, nor accept any instructions to do so.

10. Withdrawal

10.1 Where experts' instructions remain incompatible with their duties, whether through incompleteness, a conflict between their duty to the court and their instructions, or for any other substantial and significant reason, they may consider withdrawing from the case. However, experts should not withdraw without first discussing the position fully with those who instruct them and considering carefully whether it would be more appropriate to make a written request for directions from the court. If experts do withdraw, they must give formal written notice to those instructing them.

11. Experts' Right to ask Court for Directions

11.1 Experts may request directions from the court to assist them in carrying out their functions as experts. Experts should normally discuss such matters with those who instruct them before making any such request. Unless the court otherwise orders, any proposed request for directions should be copied to the party instructing the expert at least seven days before filing any request to the court, and to all other parties at least four days before filing it. (CPR 35.14).

11.2 Requests to the court for directions should be made by letter, containing.

- (a) the title of the claim;
- (b) the claim number of the case;
- (c) the name of the expert;
- (d) full details of why directions are sought; and
- (e) copies of any relevant documentation.

12. Power of the Court to Direct a Party to Provide Information

12.1 If experts consider that those instructing them have not provided information which they require, they may, after discussion with those instructing them and giving notice, write to the court to seek directions (CPR 35.14).

12.2 Experts and those who instruct them should also be aware of CPR 35.9. This provides that where one party has access to information which is not readily available to the other party, the court may direct the party who has access to the information to prepare, file and copy to the other party a document recording the information. If experts require such information which has not been disclosed, they should discuss the position with those instructing them without delay, so that a request for the information can be made, and, if not forthcoming, an application can be made to the court. Unless a document appears to be essential, experts should assess the cost and time involved in the production of a document and whether its provision would be proportionate in the context of the case.

13 Contents of Experts' Reports

13.1 The content and extent of experts' reports should be governed by the scope of their instructions and general obligations, the contents of CPR 35 and PD 35 and their overriding duty to the court.

13.2 In preparing reports, experts should maintain professional objectivity and impartiality at all times.

13.3 PD 35, para 2 provides that experts' reports should be addressed to the court and gives detailed directions about the form and content of such reports. All experts and those who instruct them should ensure that they are familiar with these requirements.

13.4 Model forms of Experts' Reports are available from bodies such as the Academy of Experts or the Expert Witness Institute.

13.5 Experts' reports must contain statements that they

- (i) understand their duty to the court and have complied and will continue to comply with it; and
- (ii) are aware of the requirements of Part 35 and Practice Direction 35, this protocol and the Practice Direction on pre-action conduct.

Experts' reports must also be verified by a statement of truth. The form of the statement of truth is as follows—

> 'I confirm that I have made clear which facts and matters referred to in this report are within my own knowledge and which are not. Those that are within my own knowledge I confirm to be true. The opinions I have expressed represent my true and complete professional opinions on the matters to which they refer.'

This wording is mandatory and must not be modified.

Qualifications

13.6 The details of experts' qualifications to be given in reports should be commensurate with the nature and complexity of the case. It may be sufficient merely to state academic and professional qualifications. However, where highly specialised expertise is called for, experts should include the detail of particular training and/or experience that qualifies them to provide that highly specialised evidence.

Tests

13.7 Where tests of a scientific or technical nature have been carried out, experts should state:

(a) the methodology used; and
(b) by whom the tests were undertaken and under whose supervision, summarising their respective qualifications and experience.

Reliance on the work of others

13.8 Where experts rely in their reports on literature or other material and cite the opinions of others without having verified them, they must give details of those opinions relied on. It is likely to assist the court if the qualifications of the originator(s) are also stated.

Facts

13.9 When addressing questions of fact and opinion, experts should keep the two separate and discrete.

13.10 Experts must state those facts (whether assumed or otherwise) upon which their opinions are based. They must distinguish clearly between those facts which experts know to be true and those facts which they assume.

13.11 Where there are material facts in dispute experts should express separate opinions on each hypothesis put forward. They should not express a view in favour of one or other disputed version of the facts unless, as a result of particular expertise and experience, they consider one set of facts as being improbable or less probable, in which case they may express that view, and should give reasons for holding it.

Range of opinion

13.12 If the mandatory summary of the range of opinion is based on published sources, experts should explain those sources and, where appropriate, state the qualifications of the originator(s) of the opinions from which they differ, particularly if such opinions represent a well-established school of thought.

13.13 Where there is no available source for the range of opinion, experts may need to express opinions on what they believe to be the range which other experts would arrive at if asked. In those circumstances, experts should make it clear that the range that they summarise is based on their own judgement and explain the basis of that judgement.

Conclusions

13.14 A summary of conclusions is mandatory. The summary should be at the end of the report after all the reasoning. There may be cases, however, where the benefit to the court is heightened by placing a short summary at the beginning of the report whilst giving the full conclusions at the end. For example, it can assist with the comprehension of the analysis and with the absorption of the detailed facts if the court is told at the outset of the direction in which the report's logic will flow in cases involving highly complex matters which fall outside the general knowledge of the court.

Basis of report: material instructions

13.15 The mandatory statement of the substance of all material instructions should not be incomplete or otherwise tend to mislead. The imperative is transparency. The term 'instructions' includes all material which solicitors place in front of experts in order to gain advice. The omission from the statement of 'off-the-record' oral instructions is not permitted. Courts may allow cross-examination about the instructions if there are reasonable grounds to consider that the statement may be inaccurate or incomplete.

14. After receipt of experts' reports

14.1 Following the receipt of experts' reports, those instructing them should advise the experts as soon as reasonably practicable whether, and if so when, the report will be disclosed to other parties; and, if so disclosed, the date of actual disclosure.

14.2 If experts' reports are to be relied upon, and if experts are to give oral evidence, those instructing them should give the experts the opportunity to consider and comment upon other reports within their area of expertise and which deal with relevant issues at the earliest opportunity.

14.3 Those instructing experts should keep experts informed of the progress of cases, including amendments to statements of case relevant to experts' opinion.

14.4 If those instructing experts become aware of material changes in circumstances or that relevant information within their control was not previously provided to experts, they should without delay instruct experts to review, and if necessary, update the contents of their reports.

15. Amendment of reports

15.1 It may become necessary for experts to amend their reports:

(a) as a result of an exchange of questions and answers;
(b) following agreements reached at meetings between experts; or
(c) where further evidence or documentation is disclosed.

15.2 Experts should not be asked to, and should not, amend, expand or alter any parts of reports in a manner which distorts their true opinion, but may be invited to amend or expand reports to ensure accuracy, internal consistency, completeness and relevance to the issues and clarity. Although experts should generally follow the recommendations of solicitors with regard to the form of reports, they should form their own independent views as to the opinions and contents expressed in their reports and exclude any suggestions which do not accord with their views.

15.3 Where experts change their opinion following a meeting of experts, a simple signed and dated addendum or memorandum to that effect is generally sufficient. In some cases, however, the benefit to the court of having an amended report may justify the cost of making the amendment.

15.4 Where experts significantly alter their opinion, as a result of new evidence or because evidence on which they relied has become unreliable, or for any other reason, they should amend their reports to reflect that fact. Amended reports should include reasons for amendments. In such circumstances those instructing experts should inform other parties as soon as possible of any change of opinion.

15.5 When experts intend to amend their reports, they should inform those instructing them without delay and give reasons. They should provide the amended version (or an addendum or memorandum) clearly marked as such as quickly as possible.

16. Written Questions to Experts

16.1 The procedure for putting written questions to experts (CPR 35.6) is intended to facilitate the clarification of opinions and issues after experts' reports have been served. Experts have a duty to provide answers to questions properly put. Where they fail to do so, the court may impose sanctions against the party instructing the expert, and, if, there is continued non-compliance, debar a party from relying on the report. Experts should copy their answers to those instructing them.

16.2 Experts' answers to questions automatically become part of their reports. They are covered by the statement of truth and form part of the expert evidence.

16.3 Where experts believe that questions put are not properly directed to the clarification of the report, or are disproportionate, or have been asked out of time, they should discuss the questions with those instructing them and, if appropriate, those asking the questions. Attempts should be made to resolve such problems without the need for an application to the court for directions.

Written requests for directions in relation to questions

16.4 If those instructing experts do not apply to the court in respect of questions, but experts still believe that questions are improper or out of time, experts may file written requests with the court for directions to assist in carrying out their functions as experts (CPR 35.14). See Section 11 above.

17. Single Joint Experts

17.1 CPR 35 and PD35 deal extensively with the instruction and use of joint experts by the parties and the powers of the court to order their use (see CPR 35.7 and 35.8, PD 35, para 5).

17.2 The Civil Procedure Rules encourage the use of joint experts. Wherever possible a joint report should be obtained. Consideration should therefore be given by all parties to the appointment of single joint experts in all cases where a court might direct such an appointment. Single joint experts are the norm in cases allocated to the small claims track and the fast track.

17.3 Where, in the early stages of a dispute, examinations, investigations, tests, site inspections, experiments, preparation of photographs, plans or other similar preliminary expert tasks are necessary, consideration should be given to the instruction of a single joint expert, especially where such matters are not, at that stage, expected to be contentious as between the parties. The objective of such an appointment should be to agree or to narrow issues.

17.4 Experts who have previously advised a party (whether in the same case or otherwise) should only be proposed as single joint experts if other parties are given all relevant information about the previous involvement.

17.5 The appointment of a single joint expert does not prevent parties from instructing their own experts to advise (but the costs of such expert advisers may not be recoverable in the case).

Joint instructions

17.6 The parties should try to agree joint instructions to single joint experts, but, in default of agreement, each party may give instructions. In particular, all parties should try to agree what documents should be included with instructions and what assumptions single joint experts should make.

17.7 Where the parties fail to agree joint instructions, they should try to agree where the areas of disagreement lie and their instructions should make this clear. If separate instructions are given, they should be copied at the same time to the other instructing parties.

17.8 Where experts are instructed by two or more parties, the terms of appointment should, unless the court has directed otherwise, or the parties have agreed otherwise, include:

- (a) a statement that all the instructing parties are jointly and severally liable to pay the experts' fees and, accordingly, that experts' invoices should be sent simultaneously to all instructing parties or their solicitors (as appropriate); and
- (b) a statement as to whether any order has been made limiting the amount of experts' fees and expenses (CPR 35.8(4)(a)).

17.9 Where instructions have not been received by the expert from one or more of the instructing parties the expert should give notice (normally at least 7 days) of a deadline to all instructing parties for the receipt by the expert of such instructions. Unless the instructions are received within the deadline the expert may begin work. In the event that instructions are received after the deadline but before the signing off of the report the expert should consider whether it is practicable to comply with those instructions without adversely affecting the timetable set for delivery of the report and in such a manner as to comply with the proportionality principle. An expert who decides to issue a report without taking into account instructions received after the deadline should inform the parties who may apply to the court for directions. In either event the report must show clearly that the expert did not receive instructions within the deadline, or, as the case may be, at all.

Conduct of the single joint expert

17.10 Single joint experts should keep all instructing parties informed of any material steps that they may be taking by, for example, copying all correspondence to those instructing them.

17.11 Single joint experts are Part 35 experts and so have an overriding duty to the court. They are the parties' appointed experts and therefore owe an equal duty to all parties. They should maintain independence, impartiality and transparency at all times.

17.12 Single joint experts should not attend any meeting or conference which is not a joint one, unless all the parties have agreed in writing or the court has directed that such a meeting may be held and who is to pay the experts' fees for the meeting.

17.13 Single joint experts may request directions from the court—see Section 11 above.

17.14 Single joint experts should serve their reports simultaneously on all instructing parties. They should provide a single report even though they may have received instructions which contain areas of conflicting fact or allegation. If conflicting instructions lead to different opinions (for example, because the instructions require experts to make different assumptions of fact), reports may need to contain more than one set of opinions on any issue. It is for the court to determine the facts.

Cross-examination

17.15 Single joint experts do not normally give oral evidence at trial but if they do, all parties may cross-examine them. In general written questions (CPR 35.6) should be put to single joint experts before requests are made for them to attend court for the purpose of cross-examination.

18. Discussions between Experts

18.1 The court has powers to direct discussions between experts for the purposes set out in the Rules (CPR 35.12). Parties may also agree that discussions take place between their experts.

18.2 Where single joint experts have been instructed but parties have, with the permission of the court, instructed their own additional Part 35 experts, there may, if the court so orders or the parties agree, be discussions between the single joint experts and the additional Part 35 experts. Such discussions should be confined to those matters within the remit of the additional Part 35 experts or as ordered by the court.

18.3 The purpose of discussions between experts should be, wherever possible, to:

(a) identify and discuss the expert issues in the proceedings;
(b) reach agreed opinions on those issues, and, if that is not possible, to narrow the issues in the case;
(c) identify those issues on which they agree and disagree and summarise their reasons for disagreement on any issue; and
(d) identify what action, if any, may be taken to resolve any of the outstanding issues between the parties.

Arrangements for discussions between experts

18.4 Arrangements for discussions between experts should be proportionate to the value of cases. In small claims and fast-track cases there should not normally be meetings between experts. Where discussion is justified in such cases, telephone discussion or an exchange of letters should, in the interests of proportionality, usually suffice. In multi-track cases, discussion may be face to face, but the practicalities or the proportionality principle may require discussions to be by telephone or video conference.

18.5 The parties, their lawyers and experts should co-operate to produce the agenda for any discussion between experts, although primary responsibility for preparation of the agenda should normally lie with the parties' solicitors.

18.6 The agenda should indicate what matters have been agreed and summarise concisely those which are in issue. It is often helpful for it to include questions to be answered by the experts. If agreement cannot be reached promptly or a party is unrepresented, the court may give directions for the drawing up of the agenda. The agenda should be circulated to experts and those instructing them to allow sufficient time for the experts to prepare for the discussion.

18.7 Those instructing experts must not instruct experts to avoid reaching agreement (or to defer doing so) on any matter within the experts' competence. Experts are not permitted to accept such instructions.

18.8 The parties' lawyers may only be present at discussions between experts if all the parties agree or the court so orders. If lawyers do attend, they should not normally intervene except to answer questions put to them by the experts or to advise about the law.

18.9 The content of discussions between experts should not be referred to at trial unless the parties agree (CPR 35.12(4)). It is good practice for any such agreement to be in writing.

18.10 At the conclusion of any discussion between experts, a statement should be prepared setting out:

(a) a list of issues that have been agreed, including, in each instance, the basis of agreement;
(b) a list of issues that have not been agreed, including, in each instance, the basis of disagreement;
(c) a list of any further issues that have arisen that were not included in the original agenda for discussion;
(d) a record of further action, if any, to be taken or recommended, including as appropriate the holding of further discussions between experts.

18.11 The statement should be agreed and signed by all the parties to the discussion as soon as may be practicable.

18.12 Agreements between experts during discussions do not bind the parties unless the parties expressly agree to be bound by the agreement (CPR 35.12(5)). However, in view of the overriding objective, parties should give careful consideration before refusing to be bound by such an agreement and be able to explain their refusal should it become relevant to the issue of costs.

19. Attendance of Experts at Court

19.1 Experts instructed in cases have an obligation to attend court if called upon to do so and accordingly should ensure that those instructing them are always aware of their dates to be avoided and take all reasonable steps to be available.

19.2 Those instructing experts should:

(a) ascertain the availability of experts before trial dates are fixed;
(b) keep experts updated with timetables (including the dates and times experts are to attend) and the location of the court;
(c) give consideration, where appropriate, to experts giving evidence via a video-link.
(d) inform experts immediately if trial dates are vacated.

19.3 Experts should normally attend court without the need for the service of witness summonses, but on occasion they may be served to require attendance (CPR 34). The use of witness summonses does not affect the contractual or other obligations of the parties to pay experts' fees.

APPENDIX 5

Technology and Construction Court Guide

SECOND EDITION
(REVISED WITH EFFECT FROM 1ST OCTOBER 2007)

INDEX

APPENDICES

Section 1. Introduction

1.1 Purpose of Guide

1.1.1 This new edition of the Technology and Construction Court ('TCC') Guide is intended to provide straightforward, practical guidance to the conduct of litigation in the TCC. Whilst it is intended to be comprehensive, it naturally concentrates on the most important aspects of such litigation. It therefore cannot cover all the procedural points that may arise. It does, however, describe the main elements of the practice that is likely to be followed in most TCC cases.

1.1.2 The Guide reflects the flexible framework within which litigation in the TCC is habitually conducted. The requirements set out in the Guide are designed to ensure effective management of proceedings in the TCC. It must always be remembered that, if parties fail to comply with these requirements, the court may impose sanctions including orders for costs.

1.1.3 In respect of those procedural areas for which specific provision is not made in this Guide, the parties, together with their advisors, will be expected to act reasonably and in accordance with both the spirit of the Guide and the overriding objective at **CPR Rule 1.1**

1.1.4 It is not the function of the Guide to summarise the Civil Procedure Rules ('the CPR'), and it should not be regarded as a substitute for the CPR. The parties and their advisors are expected to familiarise themselves with the CPR and, in particular, to understand the importance of the 'overriding objective' of the CPR. The TCC endeavours to ensure that all its cases are dealt with justly and with proper proportionality. This includes ensuring that the parties are on an equal footing; taking all practicable steps to save expenditure; dealing with the dispute in ways which are proportionate to the size of the claim and cross-claim and the importance of the case to the parties; and managing the case throughout in a way that takes proper account of its complexity and the different financial positions of the parties. The court will also endeavour to ensure expedition, and to allot to each case an appropriate share of the court's resources.

1.1.5 This new edition of the TCC Guide has been prepared in consultation with the judges of the TCC in London, Cardiff, Birmingham, Manchester and Leeds, and with the advice and support of TECBAR, TeCSA, the Society for Computers and Law and the TCC Users' Committees in London, Cardiff, Birmingham, Manchester, Liverpool and Leeds. The TCC Guide is published with the approval of the Head of Civil Justice and the deputy Head of Civil Justice.

1.2 The CPR

1.2.1 Proceedings in the TCC are governed by the CPR and the supplementary Practice Directions. **CPR Part 60** and its associated **Practice Direction** deal specifically with the practice and procedure of the TCC.

1.2.2 Other parts of the CPR that frequently arise in TCC cases include **Part 8** (Alternative Procedure for Claims); **Parts 12 and 13** (Default Judgment and Setting Aside); **Part 17** (Amendments); **Part 20** (Counterclaims and Other Additional Claims); **Part 24** (Summary Judgment); **Part 25** (Interim Remedies and Security for Costs); **Part 26** (Case Management); **Part 32** (Evidence); **Part 35** (Experts and Assessors) and **Part 62** (Arbitration Claims).

1.3 The TCC

1.3.1 *What are TCC Claims?* **CPR Rules 60.1** (2) and (3) provide that a TCC claim is a claim which (i) involves technically complex issues or questions (or for which trial by a TCC judge is desirable) and (ii) has been issued in or transferred into the TCC specialist list. Paragraph 2.1 of the TCC Practice Direction identifies the following as examples of the types of claim which it may be appropriate to bring as TCC claims—

a. building or other construction disputes, including claims for the enforcement of the decisions of adjudicators under the Housing Grants, Construction and Regeneration Act 1996;
b. engineering disputes;
c. claims by and against engineers, architects, surveyors, accountants and other specialised advisors relating to the services they provide;
d. claims by and against local authorities relating to their statutory duties concerning the development of land or the construction of buildings;
e. claims relating to the design, supply and installation of computers, computer software and related network systems;
f. claims relating to the quality of goods sold or hired, and work done, materials supplied or services rendered;
g. claims between landlord and tenant for breach of a repairing covenant;

h. claims between neighbours, owners and occupiers of land in trespass, nuisance, etc.
i. claims relating to the environment (for example, pollution cases);
j. claims arising out of fires;
k. claims involving taking of accounts where these are complicated; and
l. challenges to decisions of arbitrators in construction and engineering disputes including applications for permission to appeal and appeals.

It should be noted that this list is not exhaustive and other types of claim may well be appropriate for the TCC.

1.3.2 *The Court.* Both the High Court and the county courts deal with TCC business. Circuit judges and recorders only have jurisdiction to manage and try TCC cases if they have been nominated by the Lord Chancellor pursuant to section 68 (1) (a) of the Supreme Court Act 1981. It should be noted that those circuit judges who have been nominated pursuant to section 68 (1) of the Supreme Court Act 1981 fall into two categories: 'full time' TCC judges and 'part time' TCC judges. 'Full time' TCC judges spend most of their time dealing with TCC business, although they will do other work when there is no TCC business requiring their immediate attention. 'Part time' TCC judges are circuit judges who are only available to sit in the TCC for part of their time. They have substantial responsibilities outside the TCC.

In respect of a court centre where there is no full time TCC judge, the term 'principal TCC judge' is used in this Guide to denote the circuit judge who has principal responsibility for TCC work.

The phrase 'Technology and Construction Court' or 'TCC' or 'the court' is used in this Guide to denote any court which deals with TCC claims. All of the courts which deal with TCC claims form a composite group of courts. When those courts are dealing with TCC business, **CPR Part 60, its accompanying Practice Direction and this Guide** govern the procedures of those courts. The High Court judge in charge of the TCC ('the Judge in Charge'), although based principally in London, has overall responsibility for the judicial supervision of TCC business in those courts.

1.3.3 *The TCC in London.* The principal centre for TCC work is the High Court in London at St Dunstan's House, 133–137 Fetter Lane, London, EC4A 1HD. The Judge in Charge of the TCC sits principally at St Dunstan's House together with five full time TCC judges. Subject to paragraph 3.7.1 below, any communication or enquiry concerning a TCC case, which is proceeding at St Dunstan's House, should be directed to the clerk of the judge who is assigned to that case. The various contact details for the judges' clerks are set out in **Appendix D**.

The TCC judges who are based at St Dunstan's House will, when appropriate, sit at court centres outside London.

TCC county court cases in London are brought in (or transferred to) the Central London Civil Justice Centre, 13–14 Park Crescent, London W1N 4HT.

1.3.4 *District Registries.* TCC claims can be brought in the High Court outside London in any District Registry, although the Practice Direction states that it is preferable that, wherever possible, such claims should be issued in one of the following District Registries: Birmingham, Bristol, Cardiff, Chester, Exeter, Leeds, Liverpool, Newcastle, Nottingham and Salford (Manchester). There are full-time TCC Judges in Birmingham, Liverpool and Salford (Manchester). Contact details are again set out in **Appendix D**. There are part time TCC judges and/or recorders nominated to deal with TCC business available at most court centres throughout England and Wales.

In a number of regions a 'TCC liaison district judge' has been appointed. It is the function of the TCC liaison district judge:

a. To keep other district judges in that region well informed about the role and remit of the TCC (in order that appropriate cases may be transferred to the TCC at an early, rather than late, stage).
b. To deal with any queries from colleagues concerning the TCC or cases which might merit transfer to the TCC.

c. To deal with any subsidiary matter which a TCC judge directs should be determined by a district judge pursuant to rule 60.1 (5) (b) (ii).
d. To deal with urgent applications in TCC cases pursuant to paragraph 7.2 of the Practice Direction (i.e. no TCC judge is available and the matter is of a kind that falls within the district judge's jurisdiction).

1.3.5 *County Courts outside London.* TCC claims may also be brought in those county courts which are specified in the **Part 60 Practice Direction**. The specified county courts are: Birmingham, Bristol, Cardiff, Chester, Exeter, Leeds, Liverpool, Newcastle, Nottingham and Salford (Manchester). Contact details are again set out in **Appendix D**.

Where TCC proceedings are brought in a county court, statements of case and applications should be headed:

'In the ... County Court

Technology and Construction Court'

This heading is important because in TCC cases (subject to the limited exceptions mentioned in paragraph 1.3.4 above) district judges do not have jurisdiction to hear applications or make orders.

1.3.6 *The division between High Court and county court TCC cases.* As a general rule TCC claims for more than £50,000 are brought in the High Court, whilst claims for lower sums are brought in the county court. However, this is not a rigid dividing line. The monetary threshold for High Court TCC claims tends to be higher in London than in the regions. Regard must also be had to the complexity of the case and all other circumstances. Arbitration claims and claims to enforce or challenge adjudicators' awards are generally (but not invariably) brought in the High Court. The scale of fees differs in the High Court and the county court. This is a factor which should be borne in mind in borderline cases.

1.4 The TCC Users' Committees

1.4.1 The continuing ability of the TCC to meet the changing needs of all those involved in TCC litigation depends in large part upon a close working relationship between the TCC and its users.

1.4.2 *London.* The Judge in Charge chairs two meetings a year of the London TCC Users' Committee. The judge's clerk acts as secretary to the Committee and takes the minutes of meetings. That Committee is made up of representatives of the London TCC judges, the barristers and solicitors who regularly use the Court, the professional bodies, such as architects, engineers and arbitrators, whose members are affected by the decisions of the Court, and representatives of both employers and contractors' groups.

1.4.3 *Outside London.* There are similar meetings of TCC Users' Committees in Birmingham, Salford (Manchester), Liverpool, Cardiff and Leeds. Each Users' Committee is chaired by the full time TCC judge or the principal TCC judge in that location.

1.4.4 The TCC regards these channels of communication as extremely important and all those who are concerned with the work of the Court are encouraged to make full use of these meetings. Any suggestions or other correspondence raising matters for consideration by the Users' Committee should, in the first instance, be addressed to the clerk to the Judge in Charge at St. Dunstan's House or to the clerk to the appropriate TCC judge outside London.

1.5 Specialist Associations

1.5.1 There are a number of associations of legal representatives which are represented on the Users' Committees and which also liaise closely with the Court. These contacts ensure that the Court remains responsive to the opinions and requirements of the professional users of the Court.

1.5.2 The relevant professional organisations are the TCC Bar Association ('TECBAR') and the TCC Solicitors Association ('TeCSA'). Details of the relevant contacts at these organisations are set out on their respective websites, namely <http://www.tecbar.org/> and <http://www.tecsa.org.uk/>.

Section 2. Pre-Action Protocol

2.1 Introduction

2.1.1 There is a Pre-Action Protocol for Construction and Engineering Disputes. Where the dispute involves a claim against architects, engineers or quantity surveyors, this Protocol prevails over the Professional Negligence Pre-Action Protocol: see paragraph 1.1 of the Protocol for Construction and Engineering Disputes and paragraph A.1 of the Professional Negligence Pre-Action Protocol. The current version of the Construction and Engineering Pre-Action Protocol ('the Protocol') is set out in volume 1 of the White Book at section C.

2.1.2 The purpose of the Protocol is to encourage the frank and early exchange of information about the prospective claim and any defence to it; to enable parties to avoid litigation by agreeing a settlement of the claim before the commencement of proceedings; and to support the efficient management of proceedings where litigation cannot be avoided.

2.1.3 *Proportionality.* The overriding objective (CPR rule 1.1) applies to the pre-action period. The Protocol must not be used as a tactical device to secure advantage for one party or to generate unnecessary costs. In lower value TCC claims (such as those likely to proceed in the county court), the letter of claim and the response should be simple and the costs of both sides should be kept to a modest level. In all cases the costs incurred at the Protocol stage should be proportionate to the complexity of the case and the amount of money which is at stake. The Protocol does not impose a requirement on the parties to marshal and disclose all the supporting details and evidence that may ultimately be required if the case proceeds to litigation.

2.2 To Which Claims Does The Protocol Apply?

2.2.1 The court will expect all parties to have complied in substance with the provisions of the Protocol in all construction and engineering disputes. The only exceptions to this are identified in paragraph 2.3 below.

2.2.2 The court regards the Protocol as setting out normal and reasonable pre-action conduct. Accordingly, whilst the Protocol is not mandatory for a number of the claims noted by way of example in **paragraph 1.3.1** above, such as computer cases or dilapidations claims, the court would, in the absence of a specific reason to the contrary, expect the Protocol generally to be followed in such cases prior to the commencement of proceedings in the TCC.

2.3 What Are The Exceptions?

2.3.1 A claimant does not have to comply with the Protocol if his claim:

a. is to enforce the decision of an adjudicator;
b. includes a claim for interim injunctive relief;
c. will be the subject of a claim for summary judgment pursuant to **Part 24** of the CPR; or
d. relates to the same or substantially the same issues as have been the subject of a recent adjudication or some other formal alternative dispute resolution procedure.

2.3.2 In addition, a claimant need not comply with any part of the Protocol if, by so doing, his claim may become time-barred under the Limitation Act 1980. In those circumstances, a claimant should commence proceedings without complying with the Protocol and must, at the same time, apply for specific directions as to the timetable and form of procedure to be adopted. The court may order a stay of those proceedings pending completion of the steps set out in the Protocol.

2.4 What Are The Essential Ingredients Of The Protocol?

2.4.1 *The Letter of Claim.* The letter of claim must comply with Section 3 of the Protocol. Amongst other things, it must contain a clear summary of the facts on which each claim is based; the basis on which each claim is made; and details of the relief claimed, including a breakdown showing how any damages have been quantified. The claimant must also provide the names of experts already instructed and on whom he intends to rely.

2.4.2 *The Defendant's Response.* The defendant has 14 days to acknowledge the letter of claim and 28 days (from receipt of the letter of claim) either to take any jurisdiction objection or to respond in substance to the letter of claim. Paragraph 4.3.1 of the Protocol enables the parties to agree an extension of the 28 day period up to a maximum of 3 months. In any case of substance it is quite usual for an extension of time to be agreed for the defendant's response. The letter of response must comply with section 4 of the Protocol. Amongst other things, it must state which claims are accepted, which claims are rejected and on what basis. It must set out any counterclaim to be advanced by the defendant. The defendant should also provide the names of experts who have been instructed and on whom he intends to rely. If the defendant fails either to acknowledge or to respond to the letter of claim in time, the claimant is entitled to commence proceedings.

2.4.3 *Pre-action Meeting.* The Construction and Engineering Protocol is the only Protocol under the CPR that generally requires the parties to meet, without prejudice, at least once, in order to identify the main issues and the root causes of their disagreement on those issues. The purpose of the meeting is to see whether, and if so how, those issues might be resolved without recourse to litigation or, if litigation is unavoidable, what steps should be taken to ensure that it is conducted in accordance with the over-riding objective. At or as a result of the meeting, the parties should consider whether some form of alternative dispute resolution ('ADR') would be more suitable than litigation and if so, they should endeavour to agree which form of ADR to adopt. Although the meeting is 'without prejudice', any party who attended the meeting is at liberty to disclose to the Court at a later stage that the meeting took place; who attended and who refused to attend, together with the grounds for their refusal; and any agreements concluded between the parties.

2.5 What Happens To The Material Generated By The Protocol?

2.5.1 The letter of claim, the defendant's response, and the information relating to attendance (or otherwise) at the meeting are not confidential or 'without prejudice' and can therefore be referred to by the parties in any subsequent litigation. The detail of any discussion at the meeting(s) and/or any note of the meeting cannot be referred to the court unless all parties agree.

2.5.2 Normally the parties should include in the bundle for the first case management conference: (a) the letter of claim, (b) the response, and (c) any agreed note of the pre-action meeting: see **Section 5** below. The documents attached to or enclosed with the letter and the response should not be included in the bundle.

2.6 What If One Party Has Not Complied With The Protocol?

2.6.1 There can often be a complaint that one or other party has not complied with the Protocol. The court will consider any such complaints once proceedings have been commenced. If the court finds that the claimant has not complied with one part of the Protocol, then the court may stay the proceedings until the steps set out in the Protocol have been taken.

2.6.2 **Paragraph 2.3 of the Practice Direction in respect of Protocols** (section C of volume 1 of the White Book) makes plain that the court may make adverse costs orders against a party who has failed to comply with the Protocol. The court will exercise any sanctions available with the object of placing the innocent party in no worse a position than he would have been if the Protocol had been complied with.

2.6.3 The court is unlikely to be concerned with minor infringements of the Protocol or to engage in lengthy debates as to the precise quality of the information provided by one party to the other during the Protocol stages. The court will principally be concerned to ensure that, as a result of the Protocol stage, each party to any subsequent litigation has a clear understanding of the nature of the case that it has to meet at the commencement of those proceedings.

2.7 Costs of compliance with the Protocol

2.7.1 If compliance with the Protocol results in settlement, the costs incurred will not be recoverable from the paying party, unless this is specifically agreed.

2.7.2 If compliance with the Protocol does not result in settlement, then the costs of the exercise cannot be recovered as costs, unless:

- those costs fall within the principles stated by Sir Robert Megarry V-C in *Re Gibson's Settlement Trusts* [1981] Ch 179; or
- the steps taken in compliance with the Protocol can properly be attributable to the conduct of the action.

Section 3. Commencement and Transfer

3.1 Claim Forms

3.1.1 All proceedings must be started using a claim form under **CPR Part** 7 or **CPR Part 8**. All claims allocated to the TCC are assigned to the Multi-Track: see **CPR Rule 60.6** (**1**).

3.2 Part 7 Claims

3.2.1 The **Part** 7 claim form must be marked 'Technology and Construction Court' in the appropriate place on the form.

3.2.2. Particulars of Claim may be served with the claim form, but this is not a mandatory requirement. If the Particulars of Claim are not contained in or served with the claim form, they must be served within **14 days** after service of the claim form.

3.2.3 A claim form must be verified by a statement of truth, and this includes any amendment to a claim form, unless the court otherwise orders.

3.3 Part 8 Claims

3.3.1 The **Part 8** claim form must be marked 'Technology and Construction Court' in the appropriate place on the form.

3.3.2 A **Part 8** claim form will normally be used where there is no substantial dispute of fact, such as the situation where the dispute turns on the construction of the contract or the interpretation of statute. For example, claims challenging the jurisdiction of an adjudicator or the validity of his decision are sometimes brought under **Part 8**. In those cases the relevant primary facts are often not in dispute. **Part 8** claims will generally be disposed of on written evidence and oral submissions.

3.3.3 It is important that, where a claimant uses the **Part 8** procedure, his claim form states that **Part 8** applies and that the claimant wishes the claim to proceed under **Part 8**.

3.3.4 A statement of truth is again required on a **Part 8** claim form.

3.4 Service

3.4.1 Claim forms issued in the TCC at St Dunstan's House in London are to be served by the claimant, not by the Registry. In some other court centres claim forms are served by the court, unless the claimant specifically requests otherwise.

3.4.2 The different methods of service are set out in **CPR Part 6** and the accompanying Practice Direction.

3.4.3 Applications for an extension of time in which to serve a claim form are governed by **CPR Rule** 7.6. The evidence required on an application for an extension of time is set out in **paragraph 8.2 of Practice Direction A supplementing CPR Part** 7.

3.4.4 When the claimant has served the claim form, he must file a certificate of service: **Rule 6.14** (2). This is necessary if, for instance, the claimant wishes to obtain judgment in default (**CPR Part 12**).

3.4.5 Applications for permission to serve a claim form out of the jurisdiction are subject to **Rules 6.19–6.31** inclusive.

3.5 Acknowledgment of Service

3.5.1 A defendant must file an acknowledgment of service in response to both **Part** 7 and **Part 8** claims. Save in the special circumstances that arise when the claim form has been served out of the jurisdiction, the period for filing an acknowledgment of service is **14 days** after service of the claim form.

3.6 Transfer

3.6.1 Proceedings may be transferred from any Division of the High Court or from any specialist list to the TCC pursuant to **CPR rule 30.5**. The order made by the transferring court should be expressed as being subject to the approval of a TCC judge. The decision whether to accept such a transfer must be made by a TCC judge: see **rule 30.5 (3)**. Many of these applications are uncontested, and may conveniently be dealt with on paper. Transfers from the TCC to other Divisions of the High Court or other specialist lists are also governed by **CPR rule 30.5**. In London there are quite often transfers between the Commercial Court and the TCC, in order to ensure that cases are dealt with by the most appropriate judge. Outside London there are quite often transfers between the TCC and the mercantile courts.

3.6.2 A TCC claim may be transferred from the High Court to one of the county courts noted above, and from any county court to the High Court, if the criteria stated in **CPR Rule 30.3** are satisfied. In ordinary circumstances, proceedings will be transferred from the TCC in the High Court to the TCC in an appropriate county court if the amount of the claim does not exceed £50,000.

3.6.3 Where no TCC judge is available to deal with a TCC claim which has been issued in a district registry or one of the county courts noted above, the claim may be transferred to another district registry or county court or to the High Court TCC in London (depending upon which court is appropriate).

3.7 Assignment

3.7.1 Where a claim has been issued at or transferred to the TCC at St Dunstan's House in London, the Judge in Charge of the TCC ('the Judge in Charge') shall with the assistance of court staff classify the case either 'HCJ' or 'SCJ'.

i. If the case is classified 'HCJ', it shall be managed and tried either by the Judge in Charge or by another High Court judge, who will be identified after consultation between the Judge in Charge and the Vice-President of the Queen's Bench Division. The clerical administration of 'HCJ' cases will be carried out by the Case Administration Unit ('CAU') of the TCC at St Dunstan's House. The CAU will also deal with the listing of all applications and trials in such cases.
ii. If the case is classified 'SCJ', it shall be managed and tried by one of the senior circuit judges, who is a full time TCC judge in London. Cases in the latter category will either (a) be assigned by the Judge in Charge to a specific senior circuit judge or (b) be assigned to a senior circuit judge by operation of the rota. The assigned judge will have primary responsibility for the management of that case.
iii. Although continuity of judge is regarded as important, it will sometimes be necessary for there to be a change of assigned judge. If no judge is available during the period fixed for trial, then the case may be tried by one of the deputy judges or recorders who has been nominated by the Lord Chancellor under section 68 (1) (a) of the Supreme Court Act 1981.

3.7.2 When classifying a case 'HCJ' or 'SCJ', the Judge in Charge will take into account the following matters, as well as all the circumstances of the case:

1. The size and complexity of the case.
2. The nature and importance of any points of law arising.
3. The amount of money which is at stake.
4. Whether the case is one of public importance.
5. Whether the case has an international element or involves overseas parties.
6. The limited number of High Court judges and the needs of other court users, both civil and criminal.

Most TCC cases in London will be classified 'SCJ'. The Judge in Charge may change the classification of any case from 'HCJ' to 'SCJ' or from 'SCJ' to 'HCJ', if it becomes appropriate to do so. There will be a band of cases near the borderline between 'HCJ' and 'SCJ', where the classification will be liable to change depending upon the settlement rate of other cases and the availability of judges.

3.7.3 When proceedings are commenced in, or transferred to, the TCC at St Dunstan's House in London, any party to those proceedings may write to the court setting out matters relevant to classification. Any such letter should be clear and concise and should be copied to all other parties. A defendant who wishes to send such a letter should do so as soon as he becomes aware of the proceedings. Any party who believes that a case has been wrongly classified (whether 'HCJ' or 'SCJ') should write to the court promptly setting out his grounds for that belief. All letters referred to in this paragraph are referred to the judge in charge of the TCC or (in his absence) to the other TCC High Court judge for consideration.

3.7.4 When a case has been assigned to a named senior circuit judge in the TCC at St Dunstan's House, all communications to the court about the case (save for communications in respect of fees) shall be made to that judge's clerk. When a TCC case has been assigned to a named High Court judge, all such communications shall be made to the CAU or to the judge's clerk as appropriate. A Protocol on HMCS website sets out the procedure which will be followed when the CAU accepts a 'provisional booking' for a High Court judge. All communications in respect of fees, however, should be sent to the Registry. All statements of case and applications should be marked with the name of the assigned judge.

3.7.5 There are full time TCC judges at Birmingham, Liverpool and Salford (Manchester). There are principal TCC judges at other court centres outside London. TCC cases at these court centres are assigned to judges either (a) by direction of the full time or principal TCC judge or (b) by operation of a rota. It will not generally be appropriate for the Judge in Charge (who is based in London) to consider TCC cases which are commenced in, or transferred to, court centres outside London. Nevertheless, if any TCC case brought in a court centre outside London appears to require management and trial by a High Court judge, then the full time or principal TCC judge at that court centre should refer the case to the Judge in Charge for a decision as to its future management and trial.

3.7.6 When a TCC case has been assigned to a named circuit judge at a court centre other than St Dunstan's House, all communications to the court about the case (save for communications in respect of fees) shall be made to that judge's clerk. All communications in respect of fees should be sent to the relevant registry. All statements of case and applications should be marked with the name of the assigned judge.

Section 4. Access to the Court

4.1 General Approach

4.1.1 There may be a number of stages during the case management phase when the parties will make applications to the court for particular orders: see **Section 6** below. There will also be the need for the court to give or vary directions, so as to enable the case to progress to trial.

4.1.2 The court is acutely aware of the costs that may be incurred when both parties prepare for an oral hearing in respect of such interlocutory matters and is always prepared to consider alternative, and less expensive, ways in which the parties may seek the court's assistance.

4.1.3 There are certain stages in the case management phase when it will generally be better for the parties to appear before the assigned judge. Those are identified at **Section 4.2** below. But there are other stages, and/or particular applications which a party may wish to make, which could conveniently be dealt with by way of a telephone hearing (**Section 4.3** below) or by way of a paper application (**Section 4.4** below).

4.2 Hearings in Court

4.2.1 *First Case Management Conference.* The court will normally require the parties to attend an oral hearing for the purposes of the first Case Management Conference. This is because there may be matters which the judge would wish to raise with the parties arising out of the answers to the case management information sheets and the parties' proposed directions: see section 5.4 below. Even in circumstances where the directions and the case management timetable may be capable of being agreed by the parties and the court, the assigned judge may still wish to consider a range of case management matters face-to-face with the parties, including the possibility of ADR. See **paragraphs** 7.2.3, 7.3.2, **8.1.3**, **11.1**, **13.3**, **13.4** and **16.3.2** below.

4.2.2 Whilst the previous paragraph sets out the ideal position, it is recognised that in low value cases the benefits of personal attendance might be outweighed by the costs involved. This is particularly so at court centres outside London, where the parties may have to travel substantial distances to court. Ultimately, the question whether personal attendance should be dispensed with at any particular case management conference must be decided by the judge, after considering any representations made and the circumstances of that particular case.

4.2.3 *Pre-trial Review.* It will normally be helpful for the parties to attend before the judge on a Pre-trial Review ('PTR'). It is always preferable for Counsel or other advocates who will be appearing at the trial to attend the PTR. Again, even if the parties can agree beforehand any outstanding directions and the detailed requirements for the management of the trial, it is still of assistance for the judge to raise matters of detailed trial management with the parties at an oral hearing. In appropriate cases, e.g. where the amount in issue is disproportionate to the costs of a full trial, the judge may wish to consider with the parties whether there are other ways in which the dispute might be resolved.

4.2.4 Whether or not other interlocutory applications require to be determined at an oral hearing will depend on the nature and effect of the application being made. Disputed applications for interim payments, summary judgment and security for costs will almost always require an oral hearing. Likewise, the resolution of a contested application to enforce an adjudicator's decision will normally be heard orally. At the other end of the scale, applications for extensions of time for the service of pleadings or to comply with other orders of the court can almost always be dealt with by way of a telephone hearing or in writing.

4.3 Telephone Hearings

4.3.1 Depending on the nature of the application and the extent of any dispute between the parties, the Court is content to deal with many case management matters and other interlocutory applications by way of a telephone conference.

4.3.2 Whilst it is not possible to lay down mandatory rules as to what applications should be dealt with in this way (rather than by way of an oral hearing in court), it may be helpful to identify certain situations which commonly arise and which can conveniently be dealt with by way of a telephone conference.

a. If the location of the court is inconvenient for one or more of the parties, or the value of the claim is low, then the CMC and the PTR could, in the alternative to the procedure set out in **Section 4.2** above, take place by way of a telephone conference. The judge's permission for such a procedure would have to be sought in advance.
b. If the parties are broadly agreed on the orders to be made by the court, but they are in dispute in respect of one or two particular matters, then a telephone hearing is a convenient way in which those outstanding matters can be dealt with by the parties and the assigned judge.
c. Similarly, specific arguments about costs, once a substantive application has been disposed of, or arguments consequential on a particular judgment or order having been handed down, may also conveniently be dealt with by way of telephone hearing. Other applications which, depending on their size and importance, may conveniently be dealt with by way of a tele-

phone hearing include limited applications in respect of disclosure and specific applications as to the scope and content of factual or expert evidence exchanged by the parties.

4.3.3 Telephone hearings are not generally suitable for matters which are likely to last for more than an hour, although the judge may be prepared, in an appropriate case, to list a longer application for a telephone hearing.

4.3.4 *Practical matters.* Telephone hearings can be listed at any time between 8.30 a.m. and 5.30 p.m., subject to the convenience of the parties and the availability of the judge. Any party, who wishes to have an application dealt with by telephone, should make such request by letter or e-mail to the judge's clerk, sending copies to all other parties. Except in cases of urgency, the judge will allow a period of three days for the other parties to comment upon that request before deciding whether to deal with the application by telephone.

4.3.5 If permission is given for a telephone hearing, the court will normally indicate which party is to make all the necessary arrangements. In most cases, it will be the applicant. The procedure to be followed in setting up and holding a telephone hearing is that set out in **section 6 of the Practice Direction supplementing CPR Part 23**. The party making arrangements for the telephone hearing must ensure that all parties and the judge have a bundle for that hearing with identical pagination.

It is vital that the judge has all the necessary papers, in good time before the telephone conference, in order that it can be conducted efficiently and effectively.

4.4 Paper Applications

4.4.1 **CPR rule 23.8 and section 11 of the accompanying Practice Direction** enable certain applications to be dealt with in writing. Parties in a TCC case are encouraged to deal with applications in writing, whenever practicable. Applications for both abridgments of time and extensions of time can generally be dealt with in writing, as well as all other variations to existing directions which are wholly or largely agreed. Disputes over particular aspects of disclosure and evidence may also be capable of being resolved in this way.

4.4.2 If a party wishes to make an application to the court, it should ask itself the question: 'Can this application be conveniently dealt with in writing?' If it can, then the party should issue the application and make its (short) written submissions both in support of its application and why it should be dealt with on paper. The application, any supporting evidence and the written submissions should be provided to all parties, as well as the court. These must include a draft of the precise order sought.

4.4.3 The party against whom the application is made, and any other interested party, should respond within **3 days** dealing both with the substantive application and the request for it to be dealt with in writing.

4.4.4 The court can then decide whether or not to deal with the application in writing. If the parties are agreed that the court should deal with it in writing, it will be rare for the court to take a different view. If the parties disagree as to whether or not the application should be dealt with in writing, the court can decide that issue and, if it decides to deal with it in writing can go on to resolve the substantive point on the basis of the parties' written submissions.

4.4.5 Further guidance in respect of paper applications is set out in **Section 6.7** below.

4.4.6 It is important for the parties to ensure that all documents provided to the court are also provided to all the other parties, so as to ensure that both the court and the parties are working on the basis of the same documentation. The pagination of any bundle which is provided to the court and the parties must be identical.

4.5 E-mail Communications

4.5.1 The general rules relating to communication and filing of documents by e-mail are set out in **CPR Part 5, Practice Direction B.**

4.5.2 The judges' clerks all have e-mail addresses identified in **Appendix D**. They welcome communication from the parties electronically. By agreement with the judge's clerk, it may also be possible to provide documents to the Court in this way. However, it should be noted that HM Court Service has imposed a blanket restriction of 2MB on the length of any e-mail, including its attachments. This equates to approximately 40 pages of normal typescript.

4.5.3 Depending on the particular circumstances of an individual trial, the assigned judge may ask for an e-mail contact address for each of the parties and may send e-mail communications to that address. In addition, the judge may provide a direct contact e-mail address so that the parties can communicate directly with him out of court hours. In such circumstances, the judge and the parties should agree the times at which the respective e-mail addresses can be used.

4.5.4 Every e-mail communication to and from the judge must be simultaneously copied to all the other parties.

4.6 Video Conferencing

4.6.1 In appropriate cases, particularly where there are important matters in dispute and the parties' representatives are a long distance from one another and/or the court, the hearing may be conducted by way of a Video Conference ('VC'). Prior arrangements will be necessary for any such hearing.

4.6.2 In London, a VC can be arranged through the VC suite at the Royal Courts of Justice. However, this facility is popular and will need to be booked well in advance of the hearing. Alternatively, there are a number of other VC suites in the Strand/Fleet Street area which would be suitable. Details of these facilities are available from the judges' clerks.

4.6.3 Outside London, a VC can be arranged at the following TCC courts with the requisite facilities: Birmingham, Bristol, Cardiff, Central London, Chester, Exeter, Leeds, Liverpool, Newcastle-upon-Tyne, Nottingham and Salford (Manchester).

4.7 Contacting the court out of hours

4.7.1 Occasionally it is necessary to contact a TCC judge out of hours. For example, it may be necessary to apply for an injunction to prevent the commencement of building works which will damage adjoining property; or for an order to preserve evidence. A case may have settled and it may be necessary to inform the judge, before he/she spends an evening or a weekend reading the papers.

4.7.2 *At St Dunstan's House.* RCJ Security has been provided with the telephone numbers and other contact information of all the TCC judges based at St Dunstan's House and of the court manager. If contact is required with a judge out of hours, the initial approach should be to RCJ Security on 0207–947–6000. Security will then contact the judge and/or the court manager and pass on the message or other information. If direct contact with the judge or court manager is required, RCJ Security must be provided with an appropriate contact number. This number will then be passed to the judge and/or the court manager, who will decide whether it is appropriate for him or her to speak directly to the contacting party.

4.7.3 *At other court centres.* At the Central London Civil Justice Centre and at all court centres outside London there is a court officer who deals with out of hours applications.

4.8 Lodging documents

4.8.1 In general documents should be lodged in hard copy and not sent by email or fax. This causes unnecessary duplication as well as additional work for hard-pressed court staff. Only if matters are urgent may documents be sent by either email or fax, with a hard copy sent by way of confirmation and marked as such. In certain cases, the court may ask for documents to be submitted in electronic form by email or otherwise, where that is appropriate.

Section 5. Case Management and the First CMC

5.1 General

5.1.1 The general approach of the TCC to case management is to give directions at the outset and then throughout the proceedings to serve the overriding objective of dealing with cases justly. The judge to whom the case has been assigned has wide case management powers, which will be exercised to ensure that:

- the real issues are identified early on and remain the focus of the ongoing proceedings;
- a realistic timetable is ordered which will allow for the fair and prompt resolution of the action;
- costs are properly controlled and reflect the value of the issues to the parties and their respective financial positions.

5.1.2 In order to assist the judge in the exercise of his case management functions, the parties will be expected to co-operate with one another at all times. See **CPR rule 1.3**. Costs sanctions may be applied, if the judge concludes that one party is not reasonably co-operating with the other parties.

5.1.3 A hearing at which the judge gives general procedural directions is a case management conference ('CMC'). CMCs are relatively informal and business-like occasions. Counsel are not robed. Representatives sit when addressing the judge.

5.1.4 The following procedures apply in order to facilitate effective case management:

- Upon commencement of a case in the TCC, it is allocated automatically to the multi-track. The provisions of **CPR Part 29** apply to all TCC cases.
- The TCC encourages a structured exchange of proposals and submissions for CMCs in advance of the hearing, so as to enable the parties to respond on an informed basis to proposals made.
- The judges of the TCC operate pro-active case management. In order to avoid the parties being taken by surprise by any judicial initiative, the judge will consider giving prior notification of specific or unusual case management proposals to be raised at a case management conference.

5.1.5 The TCC's aim is to ensure that the trial of each case takes place before the judge who has managed the case since the first CMC. Whilst continuity of judge is not always possible, because of the need to double- or triple-book judges, or because cases can sometimes overrun their estimated length through no fault of the parties, this remains an aspiration of case management within the TCC.

5.2 The Fixing of the First CMC

5.2.1 Where a claim has been started in the TCC, or where it has been transferred into the TCC, **paragraph 8.1 of the Part 60 Practice Direction** requires the court to fix the first CMC within **14 days** of the earliest of

- the filing by the defendant of an acknowledgement of service or
- the filing by the defendant of the defence or
- the date of the order transferring the case to the TCC. If some defendants but not others are served with proceedings, the claimant's solicitors should so inform the court and liaise about the fixing of the first CMC.

5.2.2 This means that the first CMC takes place relatively early, sometimes before the defendant has filed a defence. However, if, as will usually be the case, the parties have complied with the protocol (**Section 2** above) they will have a good idea of each other's respective positions, and an effective CMC can take place. If, on the other hand, there has been a failure to comply with the protocol, or there are other reasons why the issues are not clearly defined at the outset, then it may be important for the judge to be involved at an early stage.

5.2.3 Despite the foregoing considerations, it is sometimes apparent to the parties that it will be more cost effective to postpone the first CMC until after service of the defence or the defences. If any of the parties wishes to delay the first CMC until then, they can write to the judge's clerk explaining why a delayed CMC is appropriate. If such a request is agreed by the other party or parties, it is likely that the judge will grant the request.

5.3 The Case Management Information Sheet and Other Documents

5.3.1 All parties are expected to complete a detailed response to the case management information sheet sent out by the Registry when the case is commenced/transferred. A copy of a blank case management information sheet is attached as **Appendix A**. It is important that all parts of the form are completed, particularly those sections (e.g. concerned with estimated costs) that enable the judge to give directions in accordance with the overriding objective.

5.3.2 The Registry will also send out a blank standard directions form to each party. A copy is attached at **Appendix B**. This sets out the usual directions made on the first CMC. The parties should fill them in, indicating the directions and timetable sought. The parties should return both the questionnaire and the directions form to the court, so that the areas (if any) of potential debate at the CMC can be identified. The parties are encouraged to exchange proposals for directions and the timetable sought, with a view to agreeing the same before the CMC for consideration by the court.

5.3.3 If the case is large or complex, it is helpful for the advocates to prepare a Note to be provided to the judge the day before the CMC which can address the issues in the case, the suggested directions, and the principal areas of dispute between the parties. If such a Note is provided, it is unnecessary for the claimant also to prepare a Case Summary as well.

5.3.4 In smaller cases, a Case Summary for the CMC, explaining briefly the likely issues, can be helpful. Such Case Summaries should be non-contentious and should (if this is possible without incurring disproportionate cost) be agreed between the parties in advance of the hearing.

5.4 Checklist of Matters likely to be considered at the first CMC

5.4.1 The following checklist identifies the matters which the judge is likely to want to consider at the first CMC, although it is not exhaustive:

- The need for, and content of, any further pleadings. This is dealt with in **Section 5.5** below.
- The outcome of the Protocol process, and the possible further need for ADR. ADR is dealt with in **Section** 7 below.
- The desirability of dealing with particular disputes by way of a Preliminary Issue hearing. This is dealt with in **Section 8** below.
- Whether the trial should be in stages (e.g. stage 1 liability and causation, stage 2 quantum). In very heavy cases this may be necessary in order to make the trial manageable. In more modest cases, where the quantum evidence will be extensive, a staged trial may be in the interest of all parties.
- The appropriate orders in respect of the disclosure of documents. This is dealt with in **Section 11** below.
- The appropriate orders as to the exchange of written witness statements. This is dealt with in **Section 12** below. It should be noted that, although it is normal for evidence-in-chief to be given by way of the written statements in the TCC, the judge may direct that evidence about particular disputes (such as what was said at an important meeting) should be given orally without reference to such statements.
- Whether it is appropriate for the parties to rely on expert evidence and, if so, what disciplines of experts should give evidence, and on what issues. This may be coupled with an order relating to the carrying out of inspections, the obtaining of samples, the conducting of experiments, or the performance of calculations. Considerations relating to expert evidence are dealt with in **Section 13** below. The parties must be aware that, in accordance with the overriding objective,

the judge will only give the parties permission to rely on expert evidence if it is both necessary and appropriate, and, even then, will wish to ensure that the scope of any such evidence is limited as far as possible.

- In certain cases the possibility of making a costs cap order. See **Section 16.3** below.
- The appropriate timetable for the taking of the various interlocutory steps noted above, and the fixing of dates for both the PTR and the trial itself (subject to **paragraph 5.4.2** below). The parties will therefore need to provide the judge with an estimate for the length of the trial, assuming all issues remain in dispute. Unless there is good reason not to, the trial date will generally be fixed at the first CMC (although this may be more difficult at court centres with only one TCC judge). Therefore, to the extent that there are any relevant concerns as to availability of either witnesses or legal representatives, they need to be brought to the attention of the court on that occasion. The length of time fixed for the trial will depend on the parties' estimates, and also the judge's own view.

If the parties' estimate of trial length subsequently changes, they should inform the clerk of the assigned judge immediately.

5.4.2 The fixing of the trial date at the CMC is usually as a provisional fixture. Therefore no trial fee is payable at this stage. The court should at the same time specify a date upon which the fixture will cease to be 'provisional' and, therefore, the trial fee will become payable. This should ordinarily be two months before the trial date. It should be noted that:

- if the trial fee is not paid within 14 days of the due date, then the whole claim will be struck out: see **CPR rule 3.7 (1) (a) and (4)**;
- if the court is notified at least 14 days before the trial date that the case is settled or discontinued, then the trial fee, which has been paid, shall be refunded: see **fee 2.2 in Schedule 1 to the Civil Proceedings Fees Order 2004**. For all other purposes other than payment of the trial fee, the provisional date fixed at the CMC shall be regarded as a firm date.

5.4.3 Essentially, the judge's aim at the first CMC is to set down a detailed timetable which, in the majority of cases, will ensure that the parties need not return to court until the PTR.

5.5 Further Pleadings

5.5.1 *Defence.* If no defence has been served prior to the first CMC, then (except in cases where judgment in default is appropriate) the court will usually make an order for service of the defence within a specified period. The defendant must plead its positive case. Bare denials and non-admissions are, save in exceptional circumstances, unacceptable.

5.5.2 *Further Information.* If the defendant wants to request further information of the Particulars of Claim, the request should, if possible, be formulated prior to the first CMC, so that it can be considered on that occasion. All requests for further information should be kept within reasonable limits, and concentrate on the important parts of the case.

5.5.3 *Reply.* A reply to the defence is not always necessary. However, where the defendant has raised a positive defence on a particular issue, it may be appropriate for the claimant to set out in a reply how it answers such a defence. If the defendant makes a counterclaim, the claimant's defence to counterclaim and its reply (if any) should be in the same document.

5.5.4 *Part 20 Claims.* The defendant should, at the first CMC, indicate (so far as possible) any Part 20 claims that it is proposing to make, whether against the claimant or any other party. Part 20 claims are required to be pleaded in the same detail as the original claim. They are a very common feature of TCC cases, because the widespread use of sub-contractors in the UK construction industry often makes it necessary to pass claims down a contractual chain. Defendants are encouraged to start any necessary Part 20 proceedings as soon as possible. It is undesirable for applications to join Part 20 defendants, to be made late in the proceedings.

5.6 Scott Schedules

5.6.1 It can sometimes be appropriate for elements of the claim, or any Part 20 claim, to be set out by way of a Scott Schedule. For example, claims involving a final account or numerous alleged defects or items of disrepair, may be best formulated in this way, which then allows for a detailed response from the defendant. Sometimes, even where all the damage has been caused by one event, such as a fire, it can be helpful for the individual items of loss and damage to be set out in a Scott Schedule. The secret of an effective Scott Schedule lies in the information that is to be provided. This is defined by the column headings. The judge may give directions for the relevant column headings for any Schedule ordered by the court. It is important that the defendant's responses to any such Schedule are as detailed as possible. Each party's entries on a Scott Schedule should be supported by a statement of truth.

5.6.2 Nevertheless, before any order is made or agreement is reached for the preparation of a Scott Schedule, both the parties and the court should consider whether this course (a) will genuinely lead to a saving of cost and time or (b) will lead to a wastage of costs and effort (because the Scott Schedule will simply duplicating earlier schedules, pleadings or expert reports). A Scott Schedule should only be ordered by the court, or agreed by the parties, in those cases where it is appropriate and proportionate.

5.6.3 When a Scott Schedule is ordered by the court or agreed by the parties, the format must always be specified. The parties must co-operate in the physical task of preparation. Electronic transfer between the parties of their respective entries in the columns will enable a clear and user-friendly Scott Schedule to be prepared, for the benefit of all involved in the trial.

5.7 Agreement Between the Parties

5.7.1 Many, perhaps most, of the required directions at the first CMC may be agreed by the parties. If so, the judge will endeavour to make orders in the terms which have been agreed, unless he considers that the agreed terms fail to take into account important features of the case as a whole, or the principles of the CPR. The agreed terms will always, at the very least, form the starting-point of the judge's consideration of the orders to be made at the CMC. If the agreed terms are submitted to the judge 3 days in advance of the hearing date, it may be possible to avoid the need for a hearing altogether.

5.7.2 The approach outlined in **paragraph 5.7.1** above is equally applicable to all other occasions when the parties come before the court with a draft order that is wholly or partly agreed.

5.8 Drawing Up of Orders

5.8.1 Unless the court itself draws up the order, it will direct one party (usually the claimant or applicant) to do so within a specified time. That party must draw up the order and lodge it with the court for approval. Once approved, the order will be stamped by the court and returned to that party for service upon all other parties.

5.9 Further CMC

5.9.1 In an appropriate case, the judge will fix a review CMC, to take place part way through the timetable that has been set down, in order to allow the court to review progress, and to allow the parties to raise any matters arising out of the steps that have been taken up to that point. This will not, however, be ordered automatically.

5.9.2 Each party will be required to give notice in writing to the other parties and the court of any order which it will be seeking at the review CMC, two days in advance of the hearing.

5.10 The Permanent Case Management Bundle

5.10.1 In conjunction with the judge's clerk, the claimant's solicitor is responsible for ensuring that, for the first CMC and at all times thereafter, there is a permanent bundle of copy documents available to the judge, which contains:

- any relevant documents resulting from the Pre-Action Protocol;
- the claim form and all statements of case;
- all orders;
- all completed case management information sheets.

5.10.2 The permanent case management bundle can then be supplemented by the specific documents relevant to any particular application that may be made. Whether these supplementary documents should (a) become a permanent addition to the case management bundle or (b) be set on one side, will depend upon their nature. The permanent case management bundle may remain at court and be marked up by the judge; alternatively, the judge may direct that the permanent case management bundle be maintained at the offices of the claimant's solicitors and provided to the court when required.

Section 6. Applications after the First CMC

6.1 Relevant parts of the CPR

6.1.1 The basic rules relating to all applications that any party may wish to make are set out in **CPR Part 23** and its accompanying **Practice Directions.**

6.1.2 **Part** 7 of the **Practice Direction** accompanying **CPR Part 60** is also of particular relevance.

6.2 Application Notice

6.2.1 As a general rule, any party to proceedings in the TCC wishing to make an application of any sort must file an application notice (**rule 23.3**) and serve that application notice on all relevant parties as soon as practicable after it has been filed (**rule 23.4**). Application notices should be served by the parties, unless (as happens in some court centres outside London) service is undertaken by the court.

6.2.2 The application notice must set out in clear terms what order is sought and, more briefly, the reasons for seeking that order: see **rule 23.6.**

6.2.3 The application notice must be served at least **3 days** before the hearing at which the Court deals with the application: **rule** 23.7 (1). Such a short notice period is only appropriate for the most straight-forward type of application.

6.2.4 Most applications, in particular applications for summary judgment under **CPR Part 24** or to strike out a statement of case under **CPR rule 3.4**, will necessitate a much longer notice period than 3 **days.** In such cases, it is imperative that the applicant obtain a suitable date and time for the hearing of the application from the assigned judge's clerk before the application notice is issued. The applicant must then serve his application notice and evidence in support sufficiently far ahead of the date fixed for the hearing of the application for there to be time to enable the respondent to serve evidence in response. Save in exceptional circumstances, there should be a minimum period of **10 working days** between the service of the notice (and supporting evidence) and the hearing date. If any party considers that there is insufficient time before the hearing of the application or if the time estimate for the application itself is too short, that party must notify the Judge's clerk and the hearing may then be refixed by agreement.

6.2.5 When considering the application notice, the judge may give directions in writing as to the dates for the provision or exchange of evidence and any written submissions or skeleton arguments for the hearing.

6.3 Evidence in Support

6.3.1 The application notice when it is served must be accompanied by all evidence in support: **rule** 23.7 (2).

6.3.2 Unless the CPR expressly requires otherwise, evidence will be given by way of witness statements. Such statements must be verified by a statement of truth signed by the maker of the statement: **rule 22.1.**

6.4 Evidence in opposition and Evidence in reply

6.4.1 Likewise, any evidence in opposition to the application should, unless the rules expressly provide otherwise, be given by way of witness statement verified by a statement of truth.

6.4.2 It is important to ensure that the evidence in opposition to the application is served in good time before the hearing so as to enable:

- the court to read and note up the evidence;
- the applicant to put in any further evidence in reply that may be considered necessary. Such evidence should be served at least **5 working days** before the hearing.

6.4.3 Any evidence in reply should be served not less than **3 working days** before the hearing. Again, if there are disputes as to the time taken or to be taken for the preparation of evidence prior to a hearing, or any other matters in respect of a suitable timetable for that hearing, the court will consider the written positions of both parties and decide such disputes on paper. It will not normally be necessary for either a separate application to be issued or a hearing to be held for such a purpose.

6.4.4 If the hearing of an application has to be adjourned because of delays by one or other of the parties in serving evidence, the court is likely to order that party to pay the costs straight away, and to make a summary assessment of those costs.

6.5 Application Bundle

6.5.1 The bundle for the hearing of anything other than the most simple and straightforward application should consist of:

- the permanent case management bundle (see **Section 5.8** above);
- the witness statements provided in support of the application, together with any exhibits;
- the witness statements provided in opposition to the application together with exhibits;
- any witness statements in reply, together with exhibits.

6.5.2 The permanent case management bundle will either be with the court or with the claimant's solicitors, depending on the order made at the first CMC: see **paragraph 5.9** above. If it is with the claimant's solicitors, it should be provided to the court not less than **2 working days** before the hearing. In any event, a paginated bundle (see **paragraph 6.5.4** below) containing any material specific to the application should also be provided to the court not less than **2 working days** before the hearing, unless otherwise directed by the judge. A failure to comply with this deadline may result in the adjournment of the hearing, and the costs thrown away being paid by the defaulting party.

6.5.3 In all but the simplest applications, the court will expect the parties to provide skeleton arguments and copies of any authorities to be relied on. The form and content of the skeleton argument is principally a matter for the author, although the judge will expect it to identify the issues that arise on the application, the important parts of the evidence relied on, and the applicable legal principles. For detailed guidance as to the form, content and length of skeleton arguments, please see paragraph 7.11.12 of the Queen's Bench Guide; Appendix 3 of the Chancery Guide; and Appendix 9 of the Commercial Court Guide.

6.5.4 For an application that is estimated to last ½ day or less, the skeleton should be provided no later than **1 pm on the last working day before the hearing**. It should be accompanied by photocopies of the authorities relied on.

6.5.5 For an application that is estimated to last more than ½ day, the skeleton should be provided no later than **4 pm one clear working day before the hearing**. It should be accompanied by photocopies of the authorities relied on.

6.5.6 The time limits at **paragraphs 6.5.4** and **6.5.5** above will be regarded as the latest times by which such skeletons should be provided to the court. Save in exceptional circumstances, no extension to these periods will be permitted.

6.5.7 *Pagination.* It is generally necessary for there to be a paginated bundle for the hearing. Where the parties have produced skeleton arguments, these should be cross-referred to the bundle page numbers.

6.6 Hearings

6.6.1 Arbitration applications may be heard in private: see **CPR rule 62.10**. All other applications will be heard in public in accordance with **CPR rule 39.2**, save where otherwise ordered.

6.6.2 Provided that the application bundle and the skeletons have been lodged in accordance with the time limits set out above, the parties can assume that the court will have a good understanding of the points in issue. However, the court will expect to be taken to particular documents relied on by the parties and will also expect to be addressed on any important legal principles that arise.

6.6.3 It is important that the parties ensure that every application is dealt with in the estimated time period. Since many applications are dealt with on Fridays, it causes major disruption if application hearings are not disposed of within the estimated period. If the parties take too long in making their submissions, the application may be adjourned, part heard, and the Court may impose appropriate costs sanctions.

6.6.4 At the conclusion of the hearing, unless the court itself draws up the order, it will direct the applicant's solicitor to do so within a specified period.

6.7 Paper Applications

6.7.1 Contested applications are usually best disposed of at an oral hearing (either in court or by telephone). However, as noted in **Section 4** above, some applications may be suitable for determination on paper. The procedure for dealing with paper applications is outlined in **paragraph 4.4** above.

6.7.2 In addition, certain simple applications (particularly in lower value cases) arising out of the management of the proceedings may be capable of being dealt with by correspondence without the need for any formal application or order of the court. This is particularly true of applications to vary procedural orders, which variations are wholly or largely agreed, or proposals to vary the estimated length of the trial. In such cases, the applicant should write to the other parties indicating the nature of its application and to seek their agreement to it. If, however, it emerges that there is an issue to be resolved by the court, then a formal application must be issued and dealt with in the normal manner.

6.8 Consent Orders

6.8.1 Consent Orders may be submitted to the Court in draft for approval and initialling without the need for attendance.

6.8.2 Two copies of the draft order should be lodged, at least one of which should be signed. The copies should be undated.

6.8.3 As noted above, whilst the parties can agree between themselves the orders to be made either at the Case Management Conference or the Pre-Trial Review, it is normally necessary for the Court to consider the case with the parties (either at an oral hearing or by way of a telephone conference) on those occasions in any event.

6.8.4 Generally, when giving directions, the court will endeavour to identify the date by which the relevant step must be taken, and will not simply provide a period during which that task should be performed. The parties should therefore ensure that any proposed consent order also identifies particular dates, rather then periods, by which the relevant steps must be taken.

6.9 Costs

6.9.1 Costs are dealt with generally at **Section 16** below.

6.9.2 The costs of any application which took a day or less to be heard and disposed of will be dealt with summarily, unless there is a good reason for the court not to exercise its powers as to the summary assessment of costs.

6.9.3 Accordingly, it is necessary for parties to provide to the court and to one another their draft statements of costs no later than **24 hours** before the start of the application hearing. Any costs which are incurred after these draft statements have been prepared, but which have not been allowed for (e.g. because the hearing has exceeded its anticipated length), can be mentioned at the hearing.

Section 7. ADR

7.1 General

7.1.1 The court will provide encouragement to the parties to use alternative dispute resolution ('ADR') and will, whenever appropriate, facilitate the use of such a procedure. In this Guide, ADR is taken to mean any process through which the parties attempt to resolve their dispute, which is voluntary. In most cases, ADR takes the form of mediation conducted by a neutral mediator. Alternative forms of ADR include formal inter-party negotiations or (occasionally) early neutral evaluations. In an early neutral evaluation either a judge or some other neutral person receives a concise presentation from each party and then states his own evaluation of the case.

7.1.2 Although the TCC is an appropriate forum for the resolution of all IT and construction/ engineering disputes, the use of ADR can lead to a significant saving of costs and may result in a settlement which is satisfactory to all parties.

7.1.3 Legal representatives in all TCC cases should ensure that their clients are fully aware of the benefits of ADR and that the use of ADR has been carefully considered prior to the first CMC.

7.2 Timing

7.2.1 ADR may be appropriate before the proceedings have begun or at any subsequent stage.

7.2.2 The TCC Pre-Action Protocol (**Section 2** above) itself provides for a type of ADR, because it requires there to be at least one face-to-face meeting between the parties before the commencement of proceedings. At this meeting, there should be sufficient time to discuss and resolve the dispute. As a result of this procedure having taken place, the court will not necessarily grant a stay of proceedings upon demand and it will always need to be satisfied that an adjournment is actually necessary to enable ADR to take place.

7.2.3 However, at the first CMC, the court will want to be addressed on the parties' views as to the likely efficacy of ADR, the appropriate timing of ADR, and the advantages and disadvantages of a short stay of proceedings to allow ADR to take place. Having considered the representations of the parties, the court may order a short stay to facilitate ADR at that stage. Alternatively, the court may simply encourage the parties to seek ADR and allow for it to occur within the timetable for the resolution of the proceedings set down by the court.

7.2.4 At any stage after the first CMC and prior to the commencement of the trial, the court, will, either on its own initiative or if requested to do so by one or both of the parties, consider afresh the likely efficacy of ADR and whether or not a short stay of the proceedings should be granted, in order to facilitate ADR.

7.3 Procedure

7.3.1 In an appropriate case, the court may indicate the type of ADR that it considers suitable, but the decision in this regard must be made by the parties. In most cases, the appropriate ADR procedure will be mediation.

7.3.2 If at any stage in the proceedings the court considers it appropriate, an ADR order in the terms of **Appendix E** may be made. If such an order is made at the first CMC, the court may go on to give directions for the conduct of the action up to trial (in the event that the ADR fails). Such directions may include provision for a review CMC.

7.3.3 The court will not ordinarily recommend any individual or body to act as mediator or to perform any other ADR procedure. In the event that the parties fail to agree the identity of a mediator or other neutral person pursuant to an order in the terms of **Appendix E**, the court may select such a person from the lists provided by the parties. To facilitate this process, the court would also need to be furnished with the C.V's of each of the individuals on the lists.

7.3.4 Information as to the types of ADR procedures available and the individuals able to undertake such procedures is available from TeCSA, TECBAR, the Civil Mediation Council, and from some TCC court centres outside London.

7.4 Non-Cooperation

7.4.1 *Generally.* At the end of the trial, there may be costs arguments on the basis that one or more parties unreasonably refused to take part in ADR. The court will determine such issues having regard to all the circumstances of the particular case. In ***Halsey v Milton Keynes General NHS Trust*** **[2004] EWCA Civ 576; [2004] 1 WLR 3002**, the Court of Appeal identified six factors that may be relevant to any such consideration:

a. the nature of the dispute;
b. the merits of the case;
c. the extent to which other settlement methods have been attempted;
d. whether the costs of the ADR would be disproportionately high;
e. whether any delay in setting up and attending the ADR would have been prejudicial;
f. whether the ADR had a reasonable prospect of success.

7.4.2 *If an ADR Order Has Been Made.* The court will expect each party to co-operate fully with any ADR which takes place following an order of the court. If any other party considers that there has not been proper co-operation in relation to arrangements for the mediation, the complaint will be considered by the court and cost orders and/or other sanctions may be ordered against the defaulting party in consequence. However, nothing in this paragraph should be understood as modifying the rights of all parties to a mediation to keep confidential all that is said or done in the course of that mediation.

7.5 Early Neutral Evaluation

7.5.1 An early neutral evaluation ('ENE') may be carried out by any appropriately qualified person, whose opinion is likely to be respected by the parties. In an appropriate case, and with the consent of all parties, a TCC judge may provide an early neutral evaluation either in respect of the full case or of particular issues arising within it. Such an ENE will not, save with the agreement of the parties, be binding on the parties.

7.5.2 If the parties would like an ENE to be carried out by the court, then they can seek an appropriate order from the assigned judge either at the first CMC or at any time prior to the commencement of the trial.

7.5.3 The assigned judge may choose to do the ENE himself. In such instance, the judge will take no further part in the proceedings once he has produced the ENE, unless the parties expressly agree otherwise. Alternatively, the assigned judge will select another available TCC judge to undertake the ENE.

7.5.4 The judge undertaking the ENE will give appropriate directions for the preparation and conduct of the ENE. This may include a stay of the substantive proceedings whilst the ENE is carried out. The ENE may be carried out entirely on paper. Alternatively, there may be an oral hearing (either with or without evidence). The parties should agree whether the entire ENE procedure is to be without prejudice, or whether it can be referred to at any subsequent trial or hearing.

Section 8. Preliminary Issues

8.1 General

8.1.1 The hearing of Preliminary Issues ('PI'), at which the court considers and delivers a binding judgment on particular issues in advance of the main trial, can be an extremely cost-effective and efficient way of narrowing the issues between the parties and, in certain cases, of resolving disputes altogether.

8.1.2 Some cases listed in the TCC lend themselves particularly well to this procedure. A PI hearing can address particular points which may be decisive of the whole proceedings; even if that is not the position, it is often possible for a PI hearing to cut down significantly on the scope (and therefore the costs) of the main trial.

8.1.3 At the first CMC the court will expect to be addressed on whether or not there are matters which should be taken by way of Preliminary Issues in advance of the main trial. Subject to **paragraph 8.5** below, it is not generally appropriate for the court to make an order for the trial of preliminary issues until after the defence has been served. After the first CMC, and at any time during the litigation, any party is at liberty to raise with any other party the possibility of a PI hearing and the court will consider any application for the hearing of such Preliminary Issues. In many cases, although not invariably, a PI order will be made with the support of all parties.

8.1.4 Whilst, for obvious reasons, it is not possible to set out hard and fast rules for what is and what is not suitable for a PI hearing, the criteria set out in **Section 8.2** below should assist the parties in deciding whether or not some or all of the disputes between them will be suitable for a PI hearing.

8.1.5 *Drawbacks of preliminary issues in inappropriate cases.* If preliminary issues are ordered inappropriately, they can have adverse effect. Evidence may be duplicated. The same witnesses may give evidence before different judges, in the event that there is a switch of assigned judge. Findings may be made at the PI hearing, which are affected by evidence called at the main hearing. The prospect of a PI hearing may delay the commencement of ADR or settlement negotiations. Also two trials are more expensive than one. For all these reasons, any proposal for preliminary issues needs to be examined carefully, so that the benefits and drawbacks can be evaluated. Also the court should give due weight to the views of the parties when deciding whether a PI hearing would be beneficial.

8.1.6 *Staged trials.* The breaking down of a long trial into stages should be differentiated from the trial of preliminary issues. Sometimes it is sensible for liability (including causation) to be tried before quantum of damages. Occasionally the subject matter of the litigation is so extensive that for reasons of case management the trial needs to be broken down into separate stages.

8.2 Guidelines

8.2.1 *The Significance of the Preliminary Issues.* The court would expect that any issue proposed as a suitable PI would, if decided in a particular way, be capable of:

- resolving the whole proceedings or a significant element of the proceedings; or
- significantly reducing the scope, and therefore the costs, of the main trial; or
- significantly improving the possibility of a settlement of the whole proceedings.

8.2.2 *Oral Evidence.* The court would ordinarily expect that, if issues are to be dealt with by way of a PI hearing, there would be either no or relatively limited oral evidence. If extensive oral evidence was required on any proposed PI, then it may not be suitable for a PI hearing. Although it is difficult to give specific guidance on this point, it is generally considered that a PI hearing in a smaller case should not take more than about 2 days, and in a larger and more complex case, should not take more than about 4 days.

8.3 Common Types of Preliminary Issue

The following are commonly resolved by way of a PI hearing:

a. Disputes as to whether or not there was a binding contract between the parties.
b. Disputes as to what documents make up or are incorporated within the contract between the parties and disputes as to the contents or relevance of any conversations relied on as having contractual status or effect.
c. Disputes as to the proper construction of the contract documents or the effect of an exclusion or similar clause.
d. Disputes as to the correct application of a statute or binding authority to a situation where there is little or no factual dispute.
e. Disputes as to the existence and/or scope of a statutory duty.
f. Disputes as to the existence and/or scope of a duty of care at common law in circumstances where there is no or little dispute about the relevant facts.

8.4 Other Possible Preliminary Issues

The following can sometimes be resolved by way of a preliminary issue hearing, although a decision as to whether or not to have such a hearing will always depend on the facts of the individual case:

8.4.1 *A Limitation Defence.* It is often tempting to have limitation issues resolved in advance of the main trial. This can be a good idea because, if a complex claim is statute-barred, a decision to that effect will lead to a significant saving of costs. However, there is also a risk that extensive evidence relevant to the limitation defence (relating to matters such as when the damage occurred or whether or not there has been deliberate concealment) may also be relevant to the liability issues within the main trial. In such a case, a preliminary issue hearing may lead to a) extensive duplication of evidence and therefore costs and b) give rise to difficulty if the main trial is heard by a different judge.

8.4.2 *Causation and 'No Loss' Points.* Causation and 'No Loss' points may be suitable for a PI hearing, but again their suitability will diminish if it is necessary for the court to resolve numerous factual disputes as part of the proposed PI hearing. The most appropriate disputes of this type for a PI hearing are those where the defendant contends that, even accepting all the facts alleged by the claimant, the claim must fail by reason of causation or the absence of recoverable loss.

8.4.3 *One-Off Issues.* Issues which do not fall into any obvious category, like economic duress, or misrepresentation, may be suitable for resolution by way of a PI hearing, particularly if the whole case can be shown to turn on them.

8.5 Use of PI as an adjunct to ADR

8.5.1 Sometimes parties wish to resolve their dispute by ADR, but there is one major issue which is a sticking point in any negotiation or mediation. The parties may wish to obtain the court's decision on that single issue, in the expectation that after that they can resolve their differences without further litigation.

8.5.2 In such a situation the parties may wish to bring proceedings under **CPR Part 8**, in order to obtain the court's decision on that issue. Such proceedings can be rapidly progressed. Alternatively, if the issue is not suitable for **Part 8** proceedings, the parties may bring proceedings under **Part 7** and then seek determination of the critical question as a preliminary issue. At the first CMC the position can be explained and the judge can be asked to order early trial of the proposed preliminary issue, possibly without the need for a defence or any further pleadings.

8.6 Precise Wording of PI

8.6.1 If a party wishes to seek a PI hearing, either at the first CMC or thereafter, that party must circulate a precise draft of the proposed preliminary issues to the other parties and to the court well in advance of the relevant hearing.

8.6.2 If the court orders a PI hearing, it is likely to make such an order only by reference to specific and formulated issues, in order to avoid later debate as to the precise scope of the issues that have been ordered. Of course, the parties are at liberty to propose amendments to the issues before the PI hearing itself, but if such later amendments are not agreed by all parties, they are unlikely to be ordered.

8.7 Appeals

8.7.1 When considering whether or not to order a PI hearing, the court will take into account the effect of any possible appeal against the PI judgment, and the concomitant delay caused.

8.7.2 At the time of ordering preliminary issues, both the parties and the court should specifically consider whether, in the event of an appeal against the PI judgment, it is desirable that the trial of the main action should (a) precede or (b) follow such appeal. It should be noted, however, that the first instance court has no power to control the timetable for an appeal. A first instance court's power to extend time under **CPR rule 52.4 (2) (a)** for filing an appellant's notice is effectively limited to 14 days (see **paragraph 5.19 of the Practice direction supplementing Part 52**). The question whether an appeal should be (a) expedited or (b) stayed is entirely a matter for the Court of Appeal. Nevertheless, the Court of Appeal will take notice of any 'indication' given by the lower court in this regard.

Section 9. Adjudication Business

9.1 Introduction

9.1.1 The TCC is ordinarily the court in which the enforcement of an adjudicator's decision and any other business connected with adjudication is undertaken. Adjudicators' decisions predominantly arise out of adjudications which are governed by the mandatory provisions of the Housing Grants, Construction and Regeneration Act 1996 ('HGCRA'). These provisions apply automatically to any construction contract as defined in the legislation. Some Adjudicators' decisions arise out of standard form contracts which contain adjudication provisions, and others arise from ad-hoc agreements to adjudicate. The TCC enforcement procedure is the same for all three kinds of adjudication.

9.1.2 In addition to enforcement applications, declaratory relief is sometimes sought in the TCC at the outset of an adjudication in respect of matters such as the jurisdiction of the adjudicator or the validity of the adjudication. This kind of application is dealt with in **Paragraph 9.4** below.

9.1.3 The HGCRA provides for a mandatory 28-day period within which the entire adjudication process must be completed, unless a) the referring party agrees to an additional 14 days, or b) both parties agree to a longer period. In consequence, the TCC has moulded a rapid procedure for enforcing an adjudication decision that has not been honoured. Other adjudication proceedings are ordinarily subject to similar rapidity.

9.2 Procedure In Enforcement Proceedings

9.2.1 Unlike arbitration business, there is neither a practice direction nor a claim form concerned with adjudication business. The enforcement proceedings normally seek a monetary judgment so that **CPR Part 7** proceedings are usually appropriate. However, if the enforcement proceedings are known to raise a question which is unlikely to involve a substantial dispute of fact and no monetary judgment is sought, **CPR Part 8** proceedings may be used instead.

9.2.2 The TCC has fashioned a procedure whereby enforcement applications are dealt with promptly. The details of this procedure are set out below.

9.2.3 The claim form should identify the construction contract, the jurisdiction of the adjudicator, the procedural rules under which the adjudication was conducted, the adjudicator's decision, the relief sought and the grounds for seeking that relief.

9.2.4 The claim form should be accompanied by an application notice that sets out the procedural directions that are sought. Commonly, the claimant's application will seek an abridgement of time for the various procedural steps, and summary judgment under **CPR Part 24**. The claim form and the application should be accompanied by a witness statement or statements setting out the evidence relied on in support of both the adjudication enforcement claim and the associated procedural application. This evidence should ordinarily include a copy of the adjudicator's decision.

9.2.5 The claim form, application notice and accompanying documents should be lodged in the appropriate registry or court centre clearly marked as being a 'paper without notice adjudication enforcement claim and application for the urgent attention of a TCC judge'. The parties will be informed that the enforcement proceedings will be assigned to a named judge. That judge will then manage the proceedings up to and including any hearing. He will ordinarily provide his directions made in connection with the procedural application within **3 working days** of the receipt of the application notice at the courts.

9.2.6 The procedural application is dealt with by a TCC judge on paper, without notice. The paper application and the consequent directions should deal with:

a. the abridged period of time in which the defendant is to file an acknowledgement of service;
b. the time for service by the defendant of any witness statement in opposition to the relief being sought;
c. an early return date for the hearing of the summary judgment application and a note of the time required or allowed for that hearing; and

d. identification of the judgment, order or other relief being sought at the hearing of the adjudication claim.

The order made at this stage will always give the defendant liberty to apply.

9.2.7 A direction providing for a date by which the claim form, supporting evidence and court order providing for the hearing are to be served on the defendant should ordinarily also be given when the judge deals with the paper procedural application.

9.2.8 The directions will ordinarily provide for an enforcement hearing within **28 days** of the directions being made and for the defendant to be given at least **14 days** from the date of service for the serving of any evidence in opposition to the adjudication application. In more straightforward cases, the abridged periods may be less.

9.2.9 Draft standard directions of the kind commonly made by the court on a procedural application by the claimant in an action to enforce the decision of an adjudicator are attached as **Appendix F**.

9.2.10 The claimant should, with the application, provide an estimate of the time needed for the hearing of the application. This estimate will be taken into account by the judge when fixing the date and length of the hearing. The parties should, if possible jointly, communicate any revised time estimate to the court promptly and the judge to whom the case has been allocated will consider whether to refix the hearing date or alter the time period that has been allocated for the hearing.

9.2.11 If the parties cannot agree on the date or time fixed for the hearing, a paper application must be made to the judge to whom the hearing has been allocated for directions.

9.3 The Enforcement Hearing

9.3.1 Where there is any dispute to be resolved at the hearing, the judge should be provided with copies of the relevant sections of the HGCRA, the adjudication procedural rules under which the adjudication was conducted, the adjudicator's decision and copies of any adjudication provisions in the contract underlying the adjudication.

9.3.2 Subject to any more specific directions given by the court, the parties should lodge, **by 4.00 p.m. one clear working day before the hearing**, a bundle containing the documents that will be required at the hearing and copies of any authorities which are to be relied on. The parties should also file and serve short skeleton arguments, summarising their respective contentions as to why the adjudicator's decision is or is not enforceable or as to any other relief being sought. For a hearing that is expected to last half a day or less, the skeletons should be provided **no later than 1 p.m. on the last working day before the hearing.** For a hearing that is estimated to last more than half a day, the skeletons should be provided **no later than 4 p.m. one clear working day before the hearing.**

9.3.3 The parties should be ready to address the court on the limited grounds on which a defendant may resist an application seeking to enforce an adjudicator's decision or on which a court may provide any other relief to any party in relation to an adjudication or an adjudicator's decision.

9.4 Other Proceedings Arising Out Of Adjudication

9.4.1 As noted above, the TCC will also hear any applications for declaratory relief arising out of the commencement of a disputed adjudication. Commonly, these will concern:

a. Disputes over the jurisdiction of an adjudicator. It can sometimes be appropriate to seek a declaration as to jurisdiction at the outset of an adjudication, rather than both parties incurring considerable costs in the adjudication itself, only for the jurisdiction point to emerge again at the enforcement hearing.
b. Disputes over whether there is a written contract between the parties or, in appropriate cases, whether there is a construction contract within the meaning of the Act.
c. Disputes over the permissible scope of the adjudication, and, in particular, whether the matters which the claimant seeks to raise in the adjudication are the subject of a pre-existing dispute between the parties.

9.4.2 Any such application will be immediately assigned to a named judge. In such circumstances, given the probable urgency of the application, the judge will usually require the parties to attend a CMC **within 2 working days** of the assignment of the case to him, and he will then give the necessary directions to ensure the speedy resolution of the dispute.

9.4.3 It sometimes happens that one party to an adjudication commences enforcement proceedings, whilst the other commences proceedings under Part 8, in order to challenge the validity of the adjudicator's award. This duplication of effort is unnecessary and it involves the parties in extra costs, especially if the two actions are commenced at different court centres. Accordingly there should be sensible discussions between the parties or their lawyers, in order to agree the appropriate venue and also to agree who shall be claimant and who defendant. All the issues raised by each party can and should be raised in a single action.

Section 10. Arbitration

10.1 Arbitration Claims in the TCC

10.1.1 'Arbitration claims' are any application to the court under the Arbitration Act 1996 and any other claim concerned with an arbitration that is referred to in **CPR 62.2(1)**. Common examples of arbitration claims are challenges to an award on grounds of jurisdiction under section 67, challenges to an award for serious irregularity under section 68 or appeals on points of law under section 69 of the Arbitration Act 1996. Arbitration claims may be started in the TCC, as is provided for in **paragraph 2.3 of the Practice Direction—Arbitration** which supplements **CPR Part 62**.

10.1.2 In practice, arbitration claims arising out of or connected with a construction or engineering arbitration (or any other arbitration where the subject matter involved one or more of the categories of work set out in **paragraph 1.3.1 above**) should be started in the TCC. The only arbitration claims that must be started in the Commercial Court are those (increasingly rare) claims to which the old law (i.e. the pre-1996 Act provisions) apply: see **CPR rule 62.12**.

10.1.3 The TCC follows the practice and procedure for arbitration claims established by **CPR Part 62** and (broadly) the practice of the Commercial Court as summarised by **Section O of the Admiralty and Commercial Court Guide**. In the absence of any specific directions given by the court, the automatic directions set out in **section 6 of the Practice Direction supplementing CPR Part 62** govern the procedures to be followed in any arbitration claim from the date of service up to the substantive hearing.

10.2 Leave to appeal

10.2.1 Where a party is seeking to appeal a question of law arising out of an award pursuant to section 69 of the Arbitration Act 1996 and the parties have not in their underlying contract agreed that such an appeal may be brought, the party seeking to appeal must apply for leave to appeal pursuant to sections 69(2), 69(3) and 69(4) of that Act. That application must be included in the arbitration claim form as explained in **paragraph 12 of the Practice Direction**.

10.2.2 In conformity with the practice of the Commercial Court, the TCC will normally consider any application for permission to appeal on paper after the defendant has had an appropriate opportunity to answer in writing the application being raised.

10.2.3 The claimant must include within the claim form an application for permission to appeal. No separate application notice is required.

10.2.4 The claim form and supporting documents must be served on the defendant. The judge will not consider the application for permission to appeal until (a) a certificate of service has been filed at the appropriate TCC registry or court centre and (b) a further **28 days** have elapsed, so as to enable the defendant to file written evidence in opposition. Save in exceptional circumstances, the only material admissible on an application for permission to appeal is (a) the award itself and any documents annexed to the award and (b) evidence relevant to the issue whether any identified question of law is of general public importance.

10.2.5 If necessary, the judge dealing with the application will direct an oral hearing with a date for the hearing. That hearing will, ordinarily, consist of brief submissions by each party. The judge dealing with the application will announce his decision in writing or, if a hearing has been directed, at the conclusion of the hearing with brief reasons if the application is refused.

10.2.6 Where the permission has been allowed in part and refused in part:

a. Only those questions for which permission has been granted may be raised at the hearing of the appeal.
b. Brief reasons will be given for refusing permission in respect of the other questions.

10.2.7 If the application is granted, the judge will fix the date for the appeal, and direct whether the same judge or a different judge shall hear the appeal.

10.3 Appeals where leave to appeal is not required

10.3.1 Parties to a construction contract should check whether they have agreed in the underlying contract that an appeal may be brought without leave, since some construction and engineering standard forms of contract so provide. If that is the case, the appeal may be set down for a substantive hearing without leave being sought. The arbitration claim form should set out the clause or provision which it is contended provides for such agreement and the claim form should be marked 'Arbitration Appeal—Leave not required'.

10.3.2 Where leave is not required, the claimant should identify each question of law that it is contended arises out of the award and which it seeks to raise in an appeal under section 69. If the defendant does not accept that the questions thus identified are questions of law or maintains that they do not arise out of the award or that the appeal on those questions may not be brought for any other reason, then the defendant should notify the claimant and the court of its contentions and apply for a directions hearing before the judge nominated to hear the appeal on a date prior to the date fixed for the hearing of the appeal. Unless the judge hearing the appeal otherwise directs, the appeal will be confined to the questions of law identified in the arbitration claim form.

10.3.3 In an appropriate case, the judge may direct that the question of law to be raised and decided on the appeal should be reworded, so as to identify more accurately the real legal issue between the parties.

10.4 The hearing of the appeal

10.4.1 Parties should ensure that the court is provided only with material that is relevant and admissible to the point of law. Again, save in exceptional circumstances, this will be limited to the award and any documents annexed to the award: see *Hok Sport Ltd v Aintree Racecourse Ltd* [2003] BLR 155 at 160. However, the court should also receive any document referred to in the award, which the court needs to read in order to determine a question of law arising out of the award: see *Kershaw Mechanical Services Ltd v Kendrick Construction Ltd* [2006] EWHC (TCC).

10.4.2 On receiving notice of permission being granted, or on issuing an arbitration claim form in a case where leave to appeal is not required, the parties should notify the court of their joint estimate or differing estimates of the time needed for the hearing of the appeal.

10.4.3 The hearing of the appeal is to be in open court unless an application (with notice) has previously been made that the hearing should be wholly or in part held in private and the court has directed that this course should be followed.

10.5 Section 68 applications—Serious Irregularity

10.5.1 In some arbitration claims arising out of construction and engineering arbitrations, a party will seek to appeal a question of law and, at the same time, seek to challenge the award under section 68 of the Arbitration Act 1996 on the grounds of serious irregularity. This raises questions of procedure, since material may be admissible in a section 68 application which is inadmissible on an application or appeal under section 69. Similarly, it may not be appropriate for all applications to be heard together. A decision is needed as to the order in which the applications should be heard, whether there should be one or more separate hearings to deal with them and whether or not the same judge should deal with all applications. Where a party intends to raise applications under both sections of the Arbitration Act 1996, they should be issued in the same arbitration claim form or in separate claim forms issued together. The court should be informed that separate applications are intended and asked for directions as to how to proceed.

10.5.2 The court will give directions as to how the section 68 and section 69 applications will be dealt with before hearing or determining any application. These directions will normally be given in writing but, where necessary or if such is applied for by a party, the court will hold a directions hearing at which directions will be given. The directions will be given following the service of any documentation by the defendant in answer to all applications raised by the claimant.

10.6 Successive awards and successive applications

10.6.1 Some construction and engineering arbitrations give rise to two or more separate awards issued at different times. Where arbitration applications arise under more than one of these awards, any second or subsequent application, whether arising from the same or a different award, should be referred to the same judge who has heard previous applications. Where more than one judge has heard previous applications, the court should be asked to direct to which judge any subsequent application is to be referred.

10.7 Other applications and Enforcement

10.7.1 All other arbitration claims, and any other matter arising in an appeal or an application concerning alleged serious irregularity, will be dealt with by the TCC in the same manner as is provided for in **CPR Part 62, Practice Direction? Arbitration** and **Section O of The Admiralty and Commercial Courts Guide.**

10.7.2 All applications for permission to enforce arbitration awards are governed by **Section III of Part 62 (rules 62.17–62.19).**

10.7.3 An application for permission to enforce an award in the same manner as a judgment or order of the court may be made in an arbitration claim form without notice and must be supported by written evidence in accordance with **rule 62.18(6)**. Two copies of the draft order must accompany the application, and the form of the order sought must correspond to the terms of the award.

10.7.4 An order made without notice giving permission to enforce the award:

(a) must give the defendant 14 days after service of the order (or longer, if the order is to be served outside the jurisdiction) to apply to set it aside;
(b) must state that it may not be enforced until after the expiry of the 14 days (or any longer period specified) or until any application to set aside the order has been finally disposed of: rule **62.18(9) and (10).**

10.7.5 On considering an application to enforce without notice, the judge may direct that, instead, the arbitration claim form must be served on specified parties, with the result that the application will then continue as an arbitration claim in accordance with the procedure set out in **Section I of Part 62:** see **rule 62.18(1)–(3).**

Section 11. Disclosure

11.1 Standard Disclosure

11.1.1 The appropriate disclosure and inspection orders to be made will normally be considered and made at the first case management conference. This is governed by **CPR Part 31** and the Practice Direction supplementing it. This procedure provides for standard disclosure, being disclosure and inspection in accordance with **CPR Part 31** of:

a. the documents upon which a party relies;
b. the documents which adversely affect his or another party's case or support another party's case; and
c. the documents which a party is required to disclose by any relevant practice direction.

11.2 Limiting disclosure and the cost of disclosure

11.2.1 In many cases being conducted in the TCC, standard disclosure will not be appropriate. This may for any one or more of the following reasons:

a. The amount of documentation may be considerable, given the complexity of the dispute and the underlying contract or contracts, and the process of giving standard disclosure may consequently be disproportionate to the issues and sums in dispute.
b. The parties may have many of the documents in common from their previous dealings so that disclosure is not necessary or desirable.
c. The parties may have provided informal disclosure and inspection of the majority of these documents, for example when complying with the pre-action Protocol.
d. The cost of providing standard disclosure may be disproportionate.

In such cases, the parties should seek to agree upon a more limited form of disclosure or to dispense with formal disclosure altogether. Such an agreement could limit disclosure to specified categories of documents or to such documents as may be specifically applied for.

11.2.2 Where disclosure is to be provided, the parties should consider whether it is necessary for lists of documents to be prepared or whether special arrangements should be agreed as to the form of listing and identifying disclosable documents, the method, timing and location of inspection and the manner of copying or providing copies of documents. Where documents are scattered over several locations, or are located overseas or are in a foreign language, special arrangements will also need to be considered. Thought should also be given to providing disclosure in stages or to reducing the scope of disclosure by providing the relevant material in other forms.

11.2.3 Electronic data and documents give rise to particular problems as to searching, preserving, listing, inspecting and other aspects of discovery and inspection. These problems should be considered and, if necessary made the subject of special directions. Furthermore, in appropriate cases, disclosure, inspection and the provision of copies of hard copies may be undertaken using information technology. Attention is drawn to the relevant provisions in **CPR Part 31**, to the Admiralty and Commercial Courts Guide concerned with Electronic Disclosure, and to the TeCSA IT Protocol which provide guidance in relation to these matters. In appropriate cases the TCC will order that the provisions concerning electronic disclosure contained in section E of the Admiralty and Commercial Courts Guide shall apply.

11.2.4 All these matters should be agreed between the parties. If it is necessary to raise any of these matters with the court they should be raised, if possible, at the first CMC. If points arise on disclosure after the first CMC, they may well be capable of being dealt with by the court on paper.

11.3 Service using information technology

11.3.1 The parties should consult with each other before the first CMC with a view to arranging the service and (where required) filing of pleadings, schedules, witness statements, experts' reports, disclosure lists and other documents in computer readable form as well as in hard copy. The parties should also consider whether to maintain a common running index, so that every document which has been exchanged between the parties has a unique reference number. Any agreement reached on these matters should be recorded and made the subject of an order for directions. Where agreement is not possible, the parties should raise these matters for decision at a CMC.

Section 12. Witness Statements and Factual Evidence for Use at Trial

12.1 Witness statements

12.1.1 Witness statements should be prepared generally in accordance with **CPR Part 22.1** (documents verified by a statement of truth) and **CPR Part 32** (provisions governing the evidence of witnesses) and their practice directions, particularly **paragraphs 17 to 22 of the Practice Direction supplementing CPR Part 32.**

12.1.2 Unless otherwise directed by the court, witness statements should *not* have annexed to them copies of other documents and should *not* reproduce or paraphrase at length passages from other documents. The only exception arises where a specific document needs to be annexed to the statement in order to make that statement reasonably intelligible.

12.1.3 When preparing witness statements, attention should be paid to the following matters:

a. Even when prepared by a legal representative or other professional, the witness statement should be, so far as practicable, in the witness's own words.
b. The witness statement should indicate which matters are within the witness's own knowledge and which are matters of information and belief. Where the witness is stating matters of hearsay or of either information or belief, the source of that evidence should also be stated.
c. The witness statement must include a statement by the witness that he believes the facts stated to be true.
d. A witness statement should be no longer than necessary and should not be argumentative.

12.2 Other matters concerned with witness statements

12.2.1 *Foreign language.* If a witness is not sufficiently fluent in English to give his evidence in English, the witness statement should be in his or her own language and an authenticated translation provided. Where the witness has a broken command of English, the statement may be drafted by others so as to express the witness's evidence as accurately as possible. In that situation, however, the witness statement should indicate that this process of interpolation has occurred and also should explain the extent of the witness's command of English and how and to what parts of the witness statement the process of interpolation has occurred.

12.2.2 *Reluctant witness.* Sometimes a witness is unwilling or not permitted or is unavailable to provide a witness statement before the trial. The party seeking to adduce this evidence should comply with the provisions of **CPR rule 32.9** concerned with the provision of witness summaries.

12.2.3 *Hearsay.* Parties should keep in mind the need to give appropriate notice of their intention to rely on hearsay evidence or the contents of documents without serving a witness statement from their maker or from the originator of the evidence contained in those documents. The appropriate procedure is contained in **CPR rules 33.1–33.5**.

12.3 Cross-referencing

12.3.1 Where a substantial number of documents will be adduced in evidence or contained in the trial bundles, it is of considerable assistance to the court and to all concerned if the relevant page references are annotated in the margins of the copy witness statements. It is accepted that this is a time-consuming exercise, the need for which will be considered at the PTR, and it will only be ordered where it is both appropriate and proportionate to do so. See further **paragraphs 14.5.1** and **15.2.3** below.

12.4 Video link

12.4.1 If any witness (whose witness statement has been served and who is required to give oral evidence) is located outside England and Wales or would find a journey to court inconvenient or impracticable, his evidence might be given via a video link. Thought should be given before the PTR to the question whether this course would be appropriate and proportionate. Such evidence is regularly received by the TCC and facilities for its reception, whether in appropriate court premises or at a convenient venue outside the court building, are now readily available.

12.4.2 Any application for a video link direction and any question relating to the manner in which such evidence is to be given should be dealt with at the PTR. Attention is drawn to the Video-conferencing Protocol set out at Annex 3 to the Practice Direction—Witness Evidence. The procedure described in Annex 3 is followed by the TCC.

Section 13. Expert Evidence

13.1 Nature of expert evidence

13.1.1 Expert evidence is evidence as to matters of a technical or scientific nature and will generally include the opinions of the expert. The quality and reliability of expert evidence will depend upon (a) the experience and the technical or scientific qualifications of the expert and (b) the accuracy of the factual material that is used by the expert for his assessment. Expert evidence is dealt with in detail in **CPR Part 35** ('Experts and Assessors') and in the **Practice Direction supplementing Part 35**. Particular attention should be paid to all these provisions, given the detailed reliance on expert evidence in most TCC actions. Particular attention should also be paid to the 'Protocol for the instruction of experts to give evidence in civil claims' published by the Civil Justice Council in June 2005. This protocol has been approved by the Master of the Rolls.

13.1.2 The provisions in **CPR Part 35** are concerned with the terms upon which the court may receive expert evidence. These provisions are principally applicable to independently instructed expert witnesses. In cases where a party is a professional or a professional has played a significant part in the subject matter of the action, opinion evidence will almost inevitably be included in the witness statements. Any points arising from such evidence (if they cannot be resolved by agreement) can be dealt with by the judge on an application or at the PTR.

13.2 Control of expert evidence

13.2.1 Expert evidence is frequently needed and used in TCC cases. Experts are often appointed at an early stage. Most types of case heard in the TCC involve more than one expertise and some, even when the dispute is concerned with relatively small sums, involve several different experts. Such disputes include those concerned with building failures and defects, delay and disruption, dilapidations, subsidence caused by tree roots and the supply of software systems. However, given the cost of preparing such evidence, the parties and the court must, from the earliest pre-action phase of a dispute until the conclusion of the trial, seek to make effective and proportionate use of experts. The scope of any expert evidence must be limited to what is necessary for the requirements of the particular case.

13.2.2 At the first CMC, or thereafter, the court may be asked to determine whether the cost of instructing experts is proportionate to the amount at issue in the proceedings, and the importance of the case to the parties. In dealing with any issues of proportionality, the court should be provided with estimates of the experts' costs.

13.2.3 The parties should also be aware that the court has the power to limit the amount of the expert's fees that a party may recover pursuant to CPR 35.4 (4).

13.3 Prior to and at the first CMC

13.3.1 There is an unresolved tension arising from the need for parties to instruct and rely on expert opinions from an early pre-action stage and the need for the court to seek, wherever possible, to reduce the cost of expert evidence by dispensing with it altogether or by encouraging the appointment of jointly instructed experts. This tension arises because the court can only consider directing joint appointments or limiting expert evidence long after a party may have incurred the cost of obtaining expert evidence and have already relied on it. Parties should be aware of this tension. So far as possible, the parties should avoid incurring the costs of expert evidence on uncontroversial matters or matters of the kind referred to in paragraph 13.4.3 below, before the first CMC has been held.

13.3.2 In cases where it is not appropriate for the court to order a single joint expert, it is imperative that, wherever possible, the parties' experts co-operate fully with one another. This is particularly important where tests, surveys, investigations, sample gathering or other technical methods of obtaining primary factual evidence are needed. It is often critical to ensure that any laboratory testing or experiments are carried out by the experts together, pursuant to an agreed procedure. Alternatively, the respective experts may agree that a particular firm or laboratory shall carry out specified tests or analyses on behalf of all parties.

13.3.3 Parties should, where possible, disclose initial or preliminary reports to opposing parties prior to any pre-action protocol meeting, if only on a without prejudice basis. Such early disclosure will assist in early settlement or mediation discussions and in helping the parties to define and confine the issues in dispute with a corresponding saving in costs.

13.3.4 Before and at the first CMC and at each subsequent pre-trial stage of the action, the parties should give careful thought to the following matters:

a. The number disciplines and identity of the expert witnesses they are considering instructing as their own experts or as single joint experts.
b. The precise issues which each expert is to address in his/her reports, to discuss without prejudice with opposing parties' experts and give evidence about at the trial.
c. The timing of any meeting, agreed statement or report.
d. Any appropriate or necessary tests, inspections, sampling or investigations that could be undertaken jointly or in collaboration with other experts. Any such measures should be preceded by a meeting of relevant experts at which an appropriate testing or other protocol is devised. This would cover (i) all matters connected with the process in question and its recording and (ii) the sharing and agreement of any resulting data or evidence. Any common method of analysis, investigation or reporting where it is appropriate or proportionate that such should be adopted by all relevant experts. An example of this would be an agreement as to the method to be used to analyse the cause and extent of any relevant period of delay in a construction project, where such is in issue in the case. The availability and length of time that experts will realistically require to complete the tasks assigned to them.

13.3.5 In so far as the matters set out in the previous paragraph cannot be agreed, the court will give appropriate directions. In giving permission for the reception of any expert evidence, the court will ordinarily order the exchange of such evidence, with a definition of the expert's area of expertise and a clear description of the issues about which that expert is permitted to give evidence. It is preferable that, at the first CMC or as soon as possible thereafter, the parties should provide the court with the name(s) of their expert(s).

13.4 Single joint experts

13.4.1 An order may be made, at the first CMC or thereafter, that a single joint expert should address particular issues between the parties. Such an order would be made pursuant to **CPR Parts 35.7 and 35.8**.

13.4.2 Single joint experts are not usually appropriate for the principal liability disputes in a large case, or in a case where considerable sums have been spent on an expert in the pre-action stage. They are generally inappropriate where the issue involves questions of risk assessment or professional competence.

13.4.3 On the other hand, single joint experts can often be appropriate:

a. in low value cases, where technical evidence is required but the cost of adversarial expert evidence may be prohibitive;
b. where the topic with which the single joint expert's report deals is a separate and self-contained part of the case, such as the valuation of particular heads of claim;
c. where there is a subsidiary issue, which requires particular expertise of a relatively uncontroversial nature to resolve;
d. where testing or analysis is required, and this can conveniently be done by one laboratory or firm on behalf of all parties.

13.4.4 Where a single joint expert is to be appointed or is to be directed by the court, the parties should attempt to devise a protocol covering all relevant aspects of the appointment (save for those matters specifically provided for by **CPR rules 35.6, 35.7 and 35.8**).

13.4.5 The matters to be considered should include: any ceiling on fees and disbursements that are to be charged and payable by the parties; how, when and by whom fees will be paid to the expert on an interim basis pending any costs order in the proceedings; how the expert's fees will be secured; how the terms of reference are to be agreed; what is to happen if terms of reference cannot be agreed; how and to whom the jointly appointed expert may address further enquiries and from whom he should seek further information and documents; the timetable for preparing any report or for undertaking any other preparatory step; the possible effect on such timetable of any supplementary or further instructions. Where these matters cannot be agreed, an application to the court, which may often be capable of being dealt with as a paper application, will be necessary.

13.4.6 The usual procedure for a single joint expert will involve:

a. The preparation of the expert's instructions. These instructions should clearly identify those issues or matters where the parties are in conflict, whether on the facts or on matters of opinion. If the parties can agree joint instructions, then a single set of instructions should be delivered to the expert. However, rule 35.8 expressly permits separate instructions and these are necessary where joint instructions cannot be agreed
b. The preparation of the agreed bundle, which is to be provided to the expert. This bundle must include CPR Part 35, the Practice Direction supplementing Part 35 and the section 13 of the TCC Guide.
c. The preparation and production of the expert's report.
d. The provision to the expert of any written questions from the parties, which the expert must answer in writing.

13.4.7 In most cases the single joint expert's report, supplemented by any written answers to questions from the parties, will be sufficient for the purposes of the trial. Sometimes, however, it is necessary for a single joint expert to be called to give oral evidence. In those circumstances, the usual practice is for the judge to call the expert and then allow each party the opportunity to cross-examine. Such cross-examination should be conducted with appropriate restraint, since the witness has been instructed by the parties. Where the expert's report is strongly in favour of one party's position, it may be appropriate to allow only the other party to cross-examine.

13.5 Meetings of experts

13.5.1 The desirability of holding without prejudice meetings between experts at all stages of the pre-trial preparation should be kept in mind. The desired outcome of such meetings is to produce a document whose contents are agreed and which defines common positions or each expert's differing position. The purpose of such meetings includes the following:

a. to define a party's technical case and to inform opposing parties of the details of that case;
b. to clear up confusion and to remedy any lack of information or understanding of a party's technical case in the minds of opposing experts;
c. to identify the issues about which any expert is to give evidence;
d. to narrow differences and to reach agreement on as many 'expert' issues as possible; and
e. to assist in providing an agenda for the trial and for cross examination of expert witnesses, and to limit the scope and length of the trial as much as possible.

13.5.2 In many cases it will be helpful for the parties' respective legal advisors to provide assistance as to the agenda and topics to be discussed at an experts' meeting. However, (save in exceptional circumstances and with the permission of the judge) the legal advisors must not attend the meeting. They must not attempt to dictate what the experts say at the meeting.

13.5.3 Experts' meetings can sometimes usefully take place at the site of the dispute. Thought is needed as to who is to make the necessary arrangements for access, particularly where the site is occupied or in the control of a non-party. Expert meetings are often more productive, if (a) the expert of one party (usually the claimant) is appointed as chairman and (b) the experts exchange in advance agendas listing the topics each wishes to raise and identifying any relevant material which they intend to introduce or rely on during the meeting.

13.5.4 It is generally sensible for the experts to meet at least once before they exchange their reports.

13.6 Experts' Joint Statements

13.6.1 Following the experts' meetings, and pursuant to **CPR 35.12 (3)**, the judge will almost always require the experts to produce a signed statement setting out the issues which have been agreed, and those issues which have not been agreed, together with a short summary of the reasons for their disagreement. In any TCC case in which expert evidence has an important role to play, this statement is a critical document and it must be as clear as possible.

13.6.2 It should be noted that, even where experts have been unable to agree very much, it is of considerable importance that the statement sets out their disagreements and the reasons for them. Such disagreements as formulated in the joint statement are likely to form an important element of the agenda for the trial of the action.

13.6.3 Whilst the parties' legal advisors may assist in identifying issues which the statement should address, those legal advisors must not be involved in either negotiating or drafting the experts' joint statement.

13.7 Experts' Reports

13.7.1 It is the duty of an expert to help the court on matters within his expertise. This duty overrides any duty to his client: **CPR rule 35.3**. Each expert's report must be independent and unbiased.

13.7.2 The parties must identify the issues with which each expert should deal in his report. Thereafter, it is for the expert to draft and decide upon the detailed contents and format of his report, so as to conform with **section 2 of the Part 35 Practice Direction**. It is appropriate, however, for the party instructing an expert to indicate that the report (a) should be as short as is reasonably possible; (b) should not set out copious extracts from other documents; (c) should identify the source of any opinion or data relied upon; and (d) should not annex or exhibit more than is reasonably necessary to support the opinions expressed in the report.

Section 14. The Pre-Trial Review

14.1 Timing and Attendance

14.1.1 The Pre-Trial Review ('PTR') will usually be fixed for a date that is 4–6 weeks in advance of the commencement of the trial itself. It is vital that the advocates, who are going to conduct the trial, should attend the PTR and every effort should be made to achieve this. It is usually appropriate for the PTR to be conducted by way of an oral hearing or, at the very least, a telephone conference.

14.2 Documents

14.2.1 The parties must complete the PTR Questionnaire (a copy of which is at **Appendix C** attached) and return it in good time to the court. In addition, the judge may order the parties to provide other documents for the particular purposes of the PTR.

14.2.2 In an appropriate case, the advocates for each party should prepare a Note for the PTR, which addresses:

- any outstanding directions or interlocutory steps still to be taken;
- the issues for determination at the trial;
- the most efficient way in which those issues might be dealt with at the trial, including all questions of timetabling of witnesses.

These Notes should be provided to the court **by 4 p.m. one clear day before the PTR**.

14.2.3 The parties should also ensure that, for the PTR, the court has an up-to-date permanent case management bundle, together with a bundle of the evidence (factual and expert) that has been exchanged. This Bundle should also be made available to the court **by 4 p.m. one clear day before the PTR**.

14.3 Outstanding Directions

14.3.1 It can sometimes be the case that there are still outstanding interlocutory steps to be taken at the time of the PTR. That will usually mean that one, or more, of the parties has not complied with an earlier direction of the court. In that event, the court is likely to require prompt compliance, and may make costs orders to reflect the delays.

14.3.2 Sometimes a party will wish to make an application to be heard at the same time as the PTR. Such a practice is unsatisfactory, because it uses up time allocated for the PTR, and it gives rise to potential uncertainty close to the trial date. It is always better for a party, if it possibly can, to make all necessary applications well in advance of the PTR. If that is not practicable, the court should be asked to allocate additional time for the PTR, in order to accommodate specific applications. If additional time is not available, such applications will not generally be entertained.

14.4 Issues

14.4.1 The parties should, if possible, provide the judge at the PTR with an agreed list of the Issues for the forthcoming trial (including, where appropriate, a separate list of technical issues to be covered by the experts).

14.4.2 If the parties are unable to agree the precise formulation of the issues, they should provide to the court their respective contentions as to what the issues are, and why.

14.4.3 In order to determine the best way to deal with the trial, it is necessary for the issues to be identified. If the precise formulation of the issues is a matter of dispute between the parties, the judge will note the parties' respective contentions, but is unlikely to give a ruling on this matter at the PTR.

14.5 Timetabling and Trial Logistics

14.5.1 Much of the PTR will be devoted to a consideration of the appropriate timetable for the trial, and other logistical matters. These will commonly include:

- Directions in respect of oral and written openings.
- Sequence of oral evidence; for example, whether all the factual evidence should be called before the expert evidence.
- Timetabling of oral evidence. (To facilitate this exercise, the advocates should tell the judge which witnesses need to be cross-examined and which evidence can be agreed.)
- Whether any form of time limits should be imposed. (Since the purpose of time limits is to ensure that that the costs incurred and the resources devoted to the trial are proportionate, this is for the benefit of the parties. The judge will endeavour to secure agreement to any time limits imposed.)
- Directions in respect of the trial bundle: when it should be agreed and lodged; the contents and structure of the bundle; avoidance of duplication; whether witness statements and/or expert reports should be annotated with cross references to page numbers in the main bundle (see **paragraph 12.3** above); and similar matters.
- Whether there should be a core bundle; if so how it should be prepared and what it should contain. (The court will order a core bundle in any case where (a) there is substantial documentation and (b) having regard to the issues it is appropriate and proportionate to put the parties to cost of preparing a core bundle).
- Rules governing any email communication during trial between the parties and the court.
- Any directions relating to the use of simultaneous transcription at trial (this subject to agreement between the parties).

14.5.2 The topics identified in paragraph 14.5.1 are discussed in greater detail in section 15 below.

Section 15. The Trial

15.1 Arrangements prior to the trial—witnesses

15.1.1 Prior to the trial the parties' legal representatives should seek to agree on the following matters, in so far as they have not been resolved at the PTR: the order in which witnesses are to be called to give evidence; which witnesses are not required for cross examination and whose evidence in consequence may be adduced entirely from their witness statements; the timetable for the trial and the length of time each advocate is to be allowed for a brief opening speech. When planning the timetable, it should be noted that trials normally take place on Mondays to Thursdays, since Fridays are reserved for applications.

15.1.2 The witnesses should be notified in advance of the trial as to: (a) when each is required to attend court and (b) the approximate period of time for which he or she will be required to attend.

15.1.3 It is the parties' responsibility to ensure that their respective witnesses are ready to attend court at the appropriate time. It is never satisfactory for witnesses to be interposed, out of their proper place. It would require exceptional circumstances for the trial to be adjourned for any period of time because of the unavailability of a witness.

15.2 Opening notes, trial bundle and oral openings

15.2.1 *Opening notes.* Unless the court has ordered otherwise, each party's advocate should provide an opening note, which outlines that party's case in relation to each of the issues identified at the PTR. Each opening note should indicate which documents (giving their page numbers in the trial bundle) that party considers that the judge should pre-read. The claimant's opening note should include a neutral summary of the background facts, as well as a chronology and cast list. The other parties' opening notes should be shorter and should assume familiarity with the factual background. In general terms, all opening notes should be of modest length and proportionate to the size and complexity of the case. Subject to any specific directions at the PTR, the claimant's opening note should be served two working days before the start of the trial; the other parties opening notes should be served by 1 p.m on the last working day before the trial.

15.2.2 *Trial bundles.* Subject to any specific directions at the PTR, the trial bundles should be delivered to court at least three working days before the hearing. It is helpful for the party delivering the trial bundles to liaise in advance with the judge's clerk, in order to discuss practical arrangements, particularly when a large number of bundles are to be delivered. The parties should provide for the court an agreed index of all trial bundles. There should also be an index at the front of each bundle. This should be a helpful guide to the contents of that bundle. (An interminable list, itemising every letter or sheet of paper is not a helpful guide. Nor are bland descriptions, such as 'exhibit 'JT3', of much help to the bundle user.) The spines of bundles should be clearly labelled.

15.2.3 As a general rule the trial bundles should be clearly divided between statements of case, orders, contracts, witness statements, expert reports and correspondence/ minutes of meetings. The correspondence/minutes of meetings should be in a separate bundle or bundles and in chronological order. Documents should only be included, if they are relevant to the issues in the case or helpful as background material. Documents should not be duplicated. Exhibits to witness statements should generally be omitted, since the documents to which the witnesses are referring will be found elsewhere in the bundles. The bundles of contract documents and correspondence/minutes of meetings should be paginated, so that every page has a discrete number. The other bundles could be dealt with in one of two ways:

- The statements of case, witness statements and expert reports could be placed in bundles and continuously paginated.
- Alternatively, the statements of case, witness statements and expert reports could be placed behind tabbed divider cards, and then the internal numbering of each such document can be used at trial. If the latter course is adopted, it is vital that the internal page numbering of each expert report continues sequentially through the appendices to that report.

The ultimate objective is to create trial bundles, which are user friendly and in which any page can be identified with clarity and brevity (e.g. 'bundle G page 273' or 'defence page 3' or 'Dr Smith page 12'). The core bundle, if there is one (as to which see paragraph 14.5.1 above), will be a separate bundle with its own pagination

15.2.4 *Opening speeches.* Subject to any directions made at the PTR, each party will be permitted to make an opening speech. These speeches should be prepared and presented on the basis that the judge will have pre-read the opening notes and the documents identified by the parties for pre-reading. The claimant's advocate may wish to highlight the main features of the claimant's case and/or to deal with matters raised in the other parties' opening notes. The other parties' advocates will then make shorter opening speeches, emphasising the main features of their own cases and/or responding to matters raised in the claimant's opening speech.

15.2.5 It is not usually necessary or desirable to embark upon legal argument during opening speeches. It is, however, helpful to foreshadow those legal arguments which (a) explain the relevance of particular parts of the evidence or (b) will assist the judge in following a party's case that is to be presented during the trial.

15.2.6 *Narrowing of issues.* Experience shows that often that the issues between the parties progressively narrow as the trial advances. Sometimes this process begins during the course of opening speeches. Weaker contentions may be abandoned and responses to those contentions may become irrelevant. The advocates will co-operate in focussing their submissions and the evidence on the true issues between the parties, as those issues are thrown into sharper relief by the adversarial process.

15.3 Simultaneous transcription

15.3.1 Many trials in the TCC, including the great majority of the longer trials, are conducted with simultaneous transcripts of the evidence being provided. There are a number of transcribing systems available. It is now common for a system to be used involving simultaneous transcription onto screens situated in court. However, systems involving the production of the transcript in hard or electronic form at the end of the day or even after a longer period of time are also used. The parties must make the necessary arrangements with one of the companies who provide this service. The court can provide a list, on request, of all companies who offer such a service.

15.3.2 In long trials or those which involve any significant amount of detailed or technical evidence, simultaneous transcripts are helpful. Furthermore, they enable all but the shortest trials to be conducted so as to reduce the overall length of the trial appreciably, since the judge does not have to note the evidence or submissions in longhand as the trial proceeds. Finally, a simultaneous transcript makes the task of summarising a case in closing submissions and preparing the judgment somewhat easier. It reduces both the risk of error or omission and the amount of time needed to prepare a reserved judgment.

15.3.3 If possible, the parties should have agreed at or before the PTR whether a simultaneous transcript is to be employed. It is usual for parties to agree to share the cost of a simultaneous transcript as an interim measure pending the assessment or agreement of costs, when this cost is assessable and payable as part of the costs in the case. Sometimes, a party cannot or will not agree to an interim cost sharing arrangement. If so, it is permissible for one party to bear the cost, but the court cannot be provided with a transcript unless all parties have equal access to the transcript. Unlike transcripts for use during an appeal, there is no available means of obtaining from public funds the cost of a transcript for use at the trial.

15.4 Time limits

15.4.1 Generally trials in the TCC are conducted under some form of time limit arrangement. Several variants of time limit arrangements are available, but the TCC has developed the practice of imposing flexible guidelines in the form of directions as to the sharing of the time allotted for the trial. These are not mandatory but an advocate should ordinarily be expected to comply with them.

15.4.2 The practice is, in the usual case, for the court to fix, or for the parties to agree, at the PTR or before trial an overall length of time for the trial and overall lengths of time within that period for the evidence and submissions. The part of those overall lengths of time that will be allocated to each party must then be agreed or directed.

15.4.3 The amount of time to be allotted to each party will not usually be the same. The guide is that each party should have as much time as is reasonably needed for it to present its case and to test and cross examine any opposing case, but no longer.

15.4.4 Before the trial, the parties should agree a running order of the witnesses and the approximate length of time required for each witness. A trial timetable should be provided to the court when the trial starts and, in long trials, regularly updated.

15.4.5 The practice of imposing a strict guillotine on the examination or cross examination of witnesses, is not normally appropriate. Flexibility is encouraged, but the agreed or directed time limits should not ordinarily be exceeded without good reason. It is unfair on a party, if that party's advocate has confined cross-examination to the agreed time limits, but an opposing party then greatly exceeds the corresponding time limits that it has been allocated.

15.4.6 An alternative form of time limit, which is sometimes agreed between the parties and approved by the court, is the 'chess clock arrangement'. The available time is divided equally between the parties, to be used by the parties as they see fit. Thus each side has X hours. One representative on each side operates the chess clock. The judge has discretion 'to stop the clock' in exceptional circumstances. A chess clock arrangement is only practicable in a two-party case.

15.5 Oral evidence

15.5.1 Evidence in chief is ordinarily adduced by the witness confirming on oath the truth and accuracy of the previously served witness statement or statements. A limited number of supplementary oral questions will usually be allowed (a) to give the witness an opportunity to become familiar with the procedure and (b) to cover points omitted by mistake from the witness statement or which have arisen subsequent to its preparation.

15.5.2 In some cases, particularly those involving allegations of dishonest, disreputable or culpable conduct or where significant disputes of fact are not documented or evidenced in writing, it is desirable that the core elements of a witness's evidence-in-chief are given orally. The giving of such evidence orally will often assist the court in assessing the credibility or reliability of a witness.

15.5.3 If any party wishes such evidence to be given orally, a direction should be sought either at the PTR or during the openings to that effect. Where evidence in chief is given orally, the rules relating to the use of witness statements in cross-examination and to the adducing of the statement in evidence at any subsequent stage of the trial remain in force and may be relied on by any party.

15.5.4 It is usual for all evidence of fact from all parties to be adduced before expert evidence and for the experts to give evidence in groups with all experts in a particular discipline giving their evidence in sequence. Usually, but not invariably, the order of witnesses will be such that the claimant's witnesses give their evidence first, followed by all the witnesses for each of the other parties in turn. If a party wishes a different order of witnesses to that normally followed, the agreement of the parties or a direction from the judge must be obtained in advance.

15.5.5 In a multi-party case, attention should be given (when the timetable is being discussed) to the order of cross-examination and to the extent to which particular topics will be covered by particular cross-examiners. Where these matters cannot be agreed, the order of cross-examination will (subject to any direction of the judge) follow the order in which the parties are set out in the pleadings. The judge will seek to limit cross examination on a topic which has been covered in detail by a preceding cross examination.

15.5.6 The coaching of witnesses or the suggestion of answers that may be given, before that witness starts to give evidence, is not permitted. Any prior discussion between the lawyers and the witness about the giving of evidence should be confined to factual information about the evidence-giving process. In short, witness familiarisation is permissible, but witness coaching is not. The boundary between witness familiarisation and witness coaching is discussed by the Court of Appeal in *R v Momodou* [2005] EWCA Crim 177 at [61]–[62]. Once a witness has started giving evidence, he cannot discuss the case or his evidence either with the lawyers or with anyone else until he has finally left the witness box. Occasionally a dispensation is needed (for example, an expert may need to participate in an experts' meeting about some new development). In those circumstances the necessary dispensation will either be agreed between the advocates or ordered by the judge.

15.6 Submissions during the trial

15.6.1 Submissions and legal argument should be kept to a minimum during the course of the trial. Where these are necessary, (a) they should, where possible, take place when a witness is not giving evidence and (b) the judge should be given forewarning of the need for submissions or legal argument. Where possible, the judge will fix a time for these submissions outside the agreed timetable for the evidence.

15.7 Closing submissions

15.7.1 The appropriate form of closing submissions can be determined during the course of the trial. Those submissions may take the form of (a) oral closing speeches or (b) written submission alone or (c) written submissions supplemented by oral closing speeches. In shorter or lower value cases, oral closing speeches immediately after the evidence may be the most cost effective way to proceed. Alternatively, if the evidence finishes in the late afternoon, a direction for written closing submissions to be delivered by specified (early) dates may avoid the cost of a further day's court hearing. In longer and heavier cases the judge may (in consultation with the advocates) set a timetable for the delivery of sequential written submissions (alternatively, an exchange of written submissions) followed by an oral hearing. In giving directions for oral and/or written closing submissions, the judge will have regard to the circumstances of the case and the overriding objective.

15.7.2 It is helpful if, in advance of preparing closing submissions, the parties can agree on the principal topics or issues that are to be covered. It is also helpful for the written and oral submissions of each party to be structured so as to cover those topics in the same order.

15.7.3 It is both customary and helpful for the judge to be provided with a photocopy of each authority and statutory provision that is to be cited in closing submissions.

15.8 Views

15.8.1 It is sometimes necessary or desirable for the judge to be taken to view the subject-matter of the case. In normal circumstances, such a view is best arranged to take place immediately after the openings and before the evidence is called. However, if the subject matter of the case is going to be covered up or altered prior to the trial, the view must be arranged earlier. In that event, it becomes particularly important to avoid a change of judge. Accordingly, the court staff will note on the trial diary the fact that the assigned judge has attended a view. In all subsequent communications between the parties and court concerning trial date, the need to avoid a change of judge must be borne firmly in mind.

15.8.2 The matters viewed by the judge form part of the evidence that is received and may be relied on in deciding the case. However, nothing said during the view to (or in the earshot of) the judge, has any evidential status, unless there has been an agreement or order to that effect.

15.8.3 The parties should agree the arrangements for the view and then make those arrangements themselves. The judge will ordinarily travel to the view unaccompanied and, save in exceptional circumstances when the cost will be shared by all parties, will not require any travelling costs to be met by the parties.

15.9 Judgments

15.9.1 Depending on the length and complexity of the trial, the judge may (a) give judgment orally immediately after closing speeches; (b) give judgment orally on the following day or soon afterwards; or (c) deliver a reserved judgment in writing at a later date.

15.9.2 *Where judgment is reserved.* The judge will normally indicate at the conclusion of the trial what arrangements will be followed in relation to (a) the making available of any draft reserved judgment and (b) the handing down of the reserved judgment in open court. If a judgment is reserved, it will be handed down as soon as possible. Save in exceptional circumstances, any reserved judgment will be handed down within 3 months of the conclusion of the trial. Any enquiries as to the progress of a reserved judgment should be addressed in the first instance to the judge's clerk, with notice of that enquiry being given to other parties. If concerns remain following the judge's response to the parties, further enquiries or communication should be addressed to the judge in charge of the TCC.

15.9.3 If the judge decides to release a draft judgment in advance of the formal hand down, this draft judgment will be confidential to the parties and their legal advisers. Solicitors and counsel on each side should send to the judge a note (if possible, agreed) of any clerical errors or slips which they note in the judgment. However, this is not to be taken as an opportunity to re-argue the issues in the case.

15.10 Disposal of judge's bundle after conclusion of the case

15.10.1 The judge will have made notes and annotations on the bundle during the course of the trial. Accordingly, the normal practice is that the entire contents of the judge's bundle are disposed of as confidential waste. The empty ring files can be recovered by arrangement with the judge's clerk.

15.10.2 If any party wishes to retrieve from the judge's bundle any particular items of value which it has supplied (e.g. plans or photographs), a request for these items should be made to the judge's clerk promptly at the conclusion of the case. If the judge has not made annotations on those particular items, they will be released to the requesting party.

Section 16. Costs

16.1 General

16.1.1 All disputes as to costs will be resolved in accordance with **CPR Part 44**, and in particular **CPR rule 44.3**.

16.1.2 The judge's usual approach will be to determine which party can be properly described as 'the successful party', and then to investigate whether there are any good reasons why that party should be deprived of some or all of their costs.

16.1.3 It should be noted that, in view of the complex nature of TCC cases, a consideration of the outcome on particular issues or areas of dispute can sometimes be an appropriate starting point for any decision on costs.

16.2 Summary Assessment of Costs

16.2.1 Interlocutory hearings that last one day or less will usually be the subject of a summary assessment of costs in accordance with **CPR 44.7** and **section 13 of the Costs Practice Direction**. The parties must ensure that their statements of costs, on which the summary assessment will be based, are provided to each other party, and the Court, no later than **24 hours** before the hearing in question: see **paragraph 6.9.3** above.

16.2.2 The Supreme Court Costs Office Guide to the Summary Assessment of Costs sets out clear advice and guidance as to the principles to be followed in any summary assessment. Generally summary assessment proceeds on the standard basis. In making an assessment on the standard basis, the court will only allow a reasonable amount in respect of costs reasonably incurred and any doubts must be resolved in favour of the paying party.

16.2.3 In arguments about the hourly rates claimed, the judge will have regard to the principles set out by the Court of Appeal in ***Wraith v Sheffield Forgemasters Ltd*** [1998] 1 WLR 132: i.e. the judge will consider whether the successful party acted reasonably in employing the solicitors who had been instructed and whether the costs they charged were reasonable compared with the broad average of charges made by similar firms practising in the same area.

16.2.4 In addition, when considering hourly rates, the judge in the TCC may have regard to the guideline rates published from time to time by TecSA.

16.2.5 The court will also consider whether unnecessary work was done or an unnecessary amount of time was spent on the work.

16.2.6 It may be that, because of pressures of time, and/or the nature and extent of the disputes about the level of costs incurred, the court is unable to carry out a satisfactory summary assessment of the costs. In those circumstances, the court will direct that costs be assessed on the standard (or indemnity) basis and will order an amount to be paid on account of costs under **CPR Rule 44.3 (8)**.

16.3 Costs Cap Orders

16.3.1 In exercising case management powers, the judge may make costs cap orders which, in normal circumstances, will be prospective only. He should only do so, however, where there is a real and substantial risk that, without such an order:

a. costs will be disproportionately or unreasonably incurred and
b. such costs cannot be controlled by conventional case management and a detailed assessment of costs after a trial, and
c. it is just to make such an order.

See **CPR rule 3.1** and the notes to that rule in the White Book headed 'Prospective costs cap orders'.

16.3.2 The possibility of a costs cap order should be considered at the first CMC. The later such an order is sought, the more difficult it may be to impose an effective costs cap.

16.4 Costs: Miscellaneous

16.4.1 The court may at any stage order any party to file and serve on the other parties an estimate of costs: see **CPR rule 3.1 (2) (ll)** and **section 6 of the Costs Practice Direction**. The case management information sheet for the first CMC requires such costs information. This information allows the court properly to exercise its case management functions. In appropriate cases (and where it is proportionate to do so) the judge will exercise his power under **paragraph 3 of the Costs Practice Direction** to direct the parties to file estimates of costs prepared in such a way as to demonstrate the likely effects of giving or not giving or not giving a particular case management direction.

16.4.2 Pursuant to **CPR Rule 44.2** and **Section 7 of the Costs Practice Direction**, solicitors have a duty to tell their clients within 7 days if an order for costs was made against the clients and they were not present at the hearing, explaining how the order came to be made. They must also give the same information to anyone else who has instructed them to act on the case or who is liable to pay their fees.

Section 17. Enforcement

17.1 General

17.1.1 The TCC is concerned with the enforcement of judgments and orders given by the TCC and with the enforcement of adjudicators' decisions and arbitrators' awards. Adjudication and arbitration enforcement have been dealt with in, respectively, sections 9 and 10 above.

17.2 High Court

17.2.1 *London.* A party wishing to make use of any provision of the **CPR** concerned with the enforcement of judgments and orders made in the TCC in London can use the TCC Registry in London or any other convenient TCC District Registry listed in **Appendix A**.

17.2.2 *Outside London.* Where the judgment or order in respect of which enforcement is sought was made by a judge of the TCC out of London, the party seeking enforcement should use the Registry of the court in which the judgment or order was made.

17.2.3 Where orders are required or sought to support enforcement of a TCC judgment or order, a judge of the TCC is the appropriate judge for that purpose. If available, the judge who gave the relevant judgment or made the relevant order is the appropriate judge to whom all applications should be addressed.

17.3 County Court

17.3.1 A TCC county court judgment (like any other county court judgment):

a. if for less than £600, must be enforced in the county court;
b. if for between £600 and £4999, can be enforced in either the county court or the High Court, at the option of the judgment creditor;
c. if for £5,000 or more, must be enforced in the High Court.

17.3.2 If a judgment creditor in a TCC county court wishes to transfer any enforcement proceedings to any other county court (whether a TCC county court or not), he must make a written request to do so pursuant to **section 2 of the Practice Direction supplementing Part 70**. Alternatively, at the end of the trial the successful party may make an oral application to the trial judge to transfer the proceedings to some other specified county court for the purposes of enforcement.

17.4 Enforcement on paper

17.4.1 Where the application or order is unopposed or does not involve any substantial dispute, the necessary order should be sought by way of a paper application.

Section 18. The TCC Judge as Arbitrator

18.1 General

18.1.1 Section 93(1) of the Arbitration Act 1996 provides that a judge of the TCC (previously an Official Referee) may 'if in all the circumstances he thinks fit, accept appointment as a sole arbitrator or as an umpire by or by virtue of an arbitration agreement.' Judges of the TCC may accept appointments as arbitrators or umpires pursuant to these statutory provisions. A judge of the TCC may also accept an appointment as a member of a three-member panel of arbitrators pursuant to these provisions.

18.1.2 A TCC judge cannot accept such an appointment unless the Lord Chief Justice 'has informed him that, having regard to the state of (TCC) business, he can be made available': see section 93(3) of the 1996 Act.

18.1.3 Application should be made in the first instance to the judge whose acceptance of the appointment is sought. If the judge is willing to accept the appointment, he will make application on behalf of the appointing party or parties, through the judge in charge of the TCC, to the Lord Chief Justice for his necessary approval. He will inform the party or parties applying for his appointment once the consent or refusal of consent has been obtained.

18.1.4 Subject to the workload of the court, such requests for judge arbitrators will generally be accepted. Particular advantages have been noted by both TECBAR and TeCSA in the appointment of a TCC judge to act as arbitrator where the dispute centres on the proper interpretation of a clause or clauses within one of the standard forms of building and engineering contracts.

18.2 Arbitration Management and Fees

18.2.1 Following the appointment of the judge arbitrator, the rules governing the arbitration will be decided upon, or directed, at the First Preliminary Meeting, when other appropriate directions will be given. The judge arbitrator will manage the reference to arbitration in a similar way to a TCC case.

18.2.2 The judge sitting as an arbitrator will sit in a TCC court room (suitably rearranged) unless the parties and the judge arbitrator agree to some other arrangement.

18.2.3 Fees are payable to the Court Service for the judge arbitrator's services and for any accommodation provided. The appropriate fee for the judge arbitrator, being a daily rate, is published in the Fees Order and should be paid through the TCC Registry.

18.3 Appeals

18.3.1 **Section 2 of Schedule 2 to the Arbitration Act 1996** provides that any appeal from a judge arbitrator is to be heard, in the first instance, by the Court of Appeal.

APPENDIX A

Case Management Information Sheet

This Appendix is the same as Appendix A to the Part 60 Practice Direction <http://www.justice.gov.uk/civil/procrules_fin/contents/practice_directions/pd_part60.htm>. In electronic versions of the TCC Guide it is not possible to copy this Appendix. Practitioners should refer either to the Part 60 Practice Direction or to published hard copies of this Guide.

APPENDIX B

Case Management Directions Form

Action no HT-............
Delete or amend the following directions, as appropriate to the circumstances of the case.

1. Trial date................. For the purposes of payment of the trial fee, but for no other purposes, this date is provisional. This date will cease to be provisional and the trial fee will become payable on... [usually be 2 months before the trial date].
2. Estimated length of trial..................
3. Directions, if appropriate, (a) for the trial of any preliminary issues or (b) for the trial to be divided into stages ...
4. This action is to be [consolidated] [managed and tried with] action no... The lead action shall be... All directions given in the lead action shall apply to both actions, unless otherwise stated.
5. Further statements of case shall be filed and served as follows:
   - Defence and any counterclaim by 4 p.m. on...
   - Reply (if any) and defence to counterclaim (if any) by 4 p.m. on...
6. Permission to make the following amendments...
7. Disclosure of documents by 5 p.m. on... [Standard disclosure dispensed with/ limited/ varied as follows...]. Specific directions in respect of electronic disclosure...
8. There shall be a Scott Schedule in respect of defects/ items of damage/ other...
   - The column headings shall be as follows...
   - Claimant/ defendant to serve Scott Schedule by 5 p.m. on...
   - Defendant/ claimant to respond to Scott Schedule by 5 p.m. on...
9. Signed statements of witnesses of fact to be served by 5 p.m. on... [Supplementary statements of witnesses of fact to be served by 5 p.m. on...]
10. [Supplementary statements of witnesses of fact to be served by 5 p.m. on ...]
11. The parties have permission to call the following expert witnesses in respect of the following issues:
    - ...
    - ...
    - ...

12. In respect of any expert evidence permitted under paragraph 10:
    - Directions for carrying out inspections/ taking samples/ conducting experiments/ performance of calculations shall be...
    - Experts in like fields to hold discussions in accordance with rule 35.12 by...
    - Experts' statements rule 35.12 (3) to be prepared and filed by 5 p.m. on...
    - Experts' reports to be served by 5 p.m. on...
13. A single joint expert shall be appointed by the parties to report on the following issue(s).... The following directions shall govern the appointment of the single joint expert:
    - ...
    - ...
14. The following documents shall be provided to the court electronically or in computer readable form, as well as in hard copy...
15. A review case management conference shall be held on... at... a.m./ p.m. Time allowed...
16. The pre-trial review shall be held on... at... a.m./p.m. Time allowed...
17. The above dates and time limits may be extended by agreement between the parties. Nevertheless:
    - The dates and time limits specified in paragraphs... may not be extended by more than... days without the permission of the court.
    - The dates specified in paragraph 1 (trial) and paragraph 15 (pre-trial review) cannot be varied without the permission of the court.
18. Liberty to restore.
19. Costs in the case.
20. Claimant's solicitors to draw up this order by... [Delete if order is to be drawn up by the court.]

APPENDIX C

Pre-Trial Review Questionnaire

This Appendix is the same as Appendix C to the Part 60 Practice Direction <http://www.justice.gov.uk/civil/procrules_fin/contents/practice_directions/pd_part60.htm#IDA3JGWB>. In electronic versions of the TCC Guide it is not possible to copy this Appendix. Practitioners should refer either to the Part 60 Practice Direction or to published hard copies of this Guide.

APPENDIX D

Contact Details for Technology and Construction Court

...[1]

APPENDIX E

Draft ADR Order

1. By [ ] the parties shall exchange lists of three neutral individuals who have indicated their availability to conduct a mediation/ENE in this case prior to [ ].
2. By [ ] the parties shall agree an individual from the exchanged lists to conduct the mediation/ENE by [ ]. If the parties are unable to agree on the neutral individual, they will apply to the Court in writing by [ ] and the Court will choose one of the listed individuals to conduct the mediation/ENE.

[1] This Appendix provides contact details of the various judges of the TCC and is not reproduced here since they change so frequently. The current details are to be found in Paragraph 2C-151 onwards of volume 2 of the White Book.

3. There will be a stay of the proceedings until [ ] to allow the mediation/ENE to take place. On or before that date, the Court shall be informed as to whether or not the case has been finally settled. If it has not been finally settled, the parties will:
   a. comply with all outstanding directions made by the Court;
   b. attend for a review CMC on [ ].

APPENDIX F

Draft Directions in Adjudication Enforcement Proceedings

Upon reading the Claim Form, Particulars of Claim, the Claimant's without notice application dated the day of 200 and the evidence in support thereof

IT IS ORDERED THAT:

1. The Claimant's solicitor shall [as soon as practicable after receipt of this Order/ by 4pm on day of] serve upon the Defendant
   a. The Claim Form and Response Pack
   b. This Order
   c. The Claimant's Application Pursuant to Part 24 and the Claimant's evidence in support.
2. The time for the Defendant to file its acknowledgement of service is abridged to [ ] days.
3. The Claimant hereby has permission to issue an application pursuant to CPR Part 24 without an acknowledgement of service or Defence having been filed.
4. The Part 24 application will be heard on the day of at am/pm at. Estimated Length of Hearing hours]
5. Any further evidence in relation to the Part 24 Application shall be served and filed
   a. By the Defendant, [14 days after the service of the documents in Paragraph 1 above/ at least 5 working days before the date fixed for the hearing of the Application] [on day the day of]
   b. By the Claimant, in response to that of the Defendant, [at least 3 working days before the date fixed for the hearing of the Application] [on day the day of 200] and in either case no later than 4.00 pm upon that day.
6. The Claimant's solicitor shall file a paginated bundle comprising
   a. The witness statements provided in support of the application, together with any exhibits;
   b. The witness statements provided in opposition to the application together with exhibits;
   c. Any witness statements in reply, together with exhibits;
   d. Photocopies of relevant authorities. This bundle is to be provided no later than [2 working days before the hearing of the Application] [on day of].
7. The parties shall file and serve skeleton arguments by no later than [4.00 pm one clear working day before the hearing/ 1pm the last working day before the hearing]* [on day the day of]
8. The costs of and incidental to these directions are reserved to the Part 24 hearing. Permission to apply in respect of such costs in the absence of such hearing.
9. The parties have permission to apply to the court on 48 hours written notice to the other to seek to set aside or vary these directions.

* Depending whether the hearing is estimated to last in excess of 1/2 day or not.

INDEX